# Interdisciplinary Applied Mathematics

## Volume 8

*Editors*
**J.E. Marsden**   **L. Sirovich**
**S. Wiggins**

*Geophysics and Planetary Science*

*Mathematical Biology*
**L. Glass,** J.D. Murray

*Mechanics and Materials*
**S.S. Antman,** R.V. Kohn

*Systems and Control*
**S.S. Sastry,** P.S. Krishnaprasad

Problems in engineering, computational science, and the physical and biological sciences are using increasingly sophisticated mathematical techniques. Thus, the bridge between the mathematical sciences and other disciplines is heavily traveled. The correspondingly increased dialog between the disciplines has led to the establishment of the series: *Interdisciplinary Applied Mathematics*

The purpose of this series is to meet the current and future needs for the interaction between various science and technology areas on the one hand and mathematics on the other. This is done, firstly, by encouraging the ways that mathematics may be applied in traditional areas, as well as point towards new and innovative areas of applications; secondly, by encouraging other scientific disciplines to engage in a dialog with mathematicians outlining their problems to both access new methods as well as to suggest innovative developments within mathematics itself.

The series will consist of monographs and high level texts from researchers working on the interplay between mathematics and other fields of science and technology.

# Interdisciplinary Applied Mathematics

James Keener    James Sneyd

# Mathematical Physiology

With 360 Illustrations

 Springer

James Keener
Department of Mathematics
University of Utah
Salt Lake City, UT 84112
USA
keener@math.utah.edu

James Sneyd
Institute of Information and Mathematic Sciences
Massey University, Albany Campus
North Shore Mail Centre
Auckland, New Zealand
j.sneyd@massey.ac.nz

*Editors*

J.E. Marsden
Control and Dynamical Systems
Mail Code 107-81
California Institute of Technology
Pasadena, CA 91125, USA

L. Sirovich
Division of Applied Mathematics
Brown University
Providence, RI 02912, USA

S. Wiggins
Control and Dynamical Systems
Mail Code 107-81
California Institute of Technology
Pasadena, CA 91125, USA

Cover illustration: "Musculature of the Human Male" by Andreas Vesalius.

Mathematics Subject Classification (1991): 92Cxx

Library of Congress Cataloging-in-Publication Data
Keener, James P.
    Mathematical physiology / James Keener, James Sneyd.
        p. cm.—(Interdisciplinary applied mathematics; v. 8)
    Includes bibliographical references and index.
    ISBN 0-387-98381-3 (alk. paper)
    1. Physiology—Mathematics. I. Sneyd, James. II. Title. III. Series.
    QP33.6.M36K44 1998
    571'. 01'51—DC21          98-14499

Printed on acid-free paper.

Printed in the United States of America.     (AU/MVY)

9 8 7 6 5

ISBN 0-387-98381-3

*springeronline.com*

# Preface

It can be argued that of all the biological sciences, physiology is the one in which mathematics has played the greatest role. From the work of Helmholtz and Frank in the last century through to that of Hodgkin, Huxley, and many others in this century, physiologists have repeatedly used mathematical methods and models to help their understanding of physiological processes. It might thus be expected that a close connection between applied mathematics and physiology would have developed naturally, but unfortunately, until recently, such has not been the case.

There are always barriers to communication between disciplines. Despite the quantitative nature of their subject, many physiologists seek only verbal descriptions, naming and learning the functions of an incredibly complicated array of components; often the complexity of the problem appears to preclude a mathematical description. Others want to become physicians, and so have little time for mathematics other than to learn about drug dosages, office accounting practices, and malpractice liability. Still others choose to study physiology precisely because thereby they hope not to study more mathematics, and that in itself is a significant benefit. On the other hand, many applied mathematicians are concerned with theoretical results, proving theorems and such, and prefer not to pay attention to real data or the applications of their results. Others hesitate to jump into a new discipline, with all its required background reading and its own history of modeling that must be learned.

But times are changing, and it is rapidly becoming apparent that applied mathematics and physiology have a great deal to offer one another. It is our view that teaching physiology without a mathematical description of the underlying dynamical processes is like teaching planetary motion to physicists without mentioning or using Kepler's laws; you can observe that there is a full moon every 28 days, but without Kepler's laws you cannot determine when the next total lunar or solar eclipse will be nor when

Halley's comet will return. Your head will be full of interesting and important facts, but it is difficult to organize those facts unless they are given a quantitative description. Similarly, if applied mathematicians were to ignore physiology, they would be losing the opportunity to study an extremely rich and interesting field of science.

To explain the goals of this book, it is most convenient to begin by emphasizing what this book is not; it is not a physiology book, and neither is it a mathematics book. Any reader who is seriously interested in learning physiology would be well advised to consult an introductory physiology book such as Guyton and Hall (1996) or Berne and Levy (1993), as, indeed, we ourselves have done many times. We give only a brief background for each physiological problem we discuss, certainly not enough to satisfy a real physiologist. Neither is this a book for learning mathematics. Of course, a great deal of mathematics is used throughout, but any reader who is not already familiar with the basic techniques would again be well advised to learn the material elsewhere.

Instead, this book describes work that lies on the border between mathematics and physiology; it describes ways in which mathematics may be used to give insight into physiological questions, and how physiological questions can, in turn, lead to new mathematical problems. In this sense, it is truly an interdisciplinary text, which, we hope, will be appreciated by physiologists interested in theoretical approaches to their subject as well as by mathematicians interested in learning new areas of application.

It is also an introductory survey of what a host of other people have done in employing mathematical models to describe physiological processes. It is necessarily brief, incomplete, and outdated (even before it was written), but we hope it will serve as an introduction to, and overview of, some of the most important contributions to the field. Perhaps some of the references will provide a starting point for more in-depth investigations.

Unfortunately, because of the nature of the respective disciplines, applied mathematicians who know little physiology will have an easier time with this material than will physiologists with little mathematical training. A complete understanding of all of the mathematics in this book will require a solid undergraduate training in mathematics, a fact for which we make no apology. We have made no attempt whatever to water down the models so that a lower level of mathematics could be used, but have instead used whatever mathematics the physiology demands. It would be misleading to imply that physiological modeling uses only trivial mathematics, or vice versa; the essential richness of the field results from the incorporation of complexities from both disciplines.

At the least, one needs a solid understanding of differential equations, including phase plane analysis and stability theory. To follow everything will also require an understanding of basic bifurcation theory, linear transform theory (Fourier and Laplace transforms), linear systems theory, complex variable techniques (the residue theorem), and some understanding of partial differential equations and their numerical simulation. However, for those whose mathematical background does not include all of these topics, we have included references that should help to fill the gap. We also make extensive use of asymptotic methods and perturbation theory, but include explanatory material to help the novice understand the calculations.

This book can be used in several ways. It could be used to teach a full-year course in mathematical physiology, and we have used this material in that way. The book includes enough exercises to keep even the most diligent student busy. It could also be used as a supplement to other applied mathematics, bioengineering, or physiology courses. The models and exercises given here can add considerable interest and challenge to an otherwise traditional course.

The book is divided into two parts, the first dealing with the fundamental principles of cell physiology, and the second with the physiology of systems. After an introduction to basic biochemistry and enzyme reactions, we move on to a discussion of various aspects of cell physiology, including the problem of volume control, the membrane potential, ionic flow through channels, and excitability. Chapter 5 is devoted to calcium dynamics, emphasizing the two important ways that calcium is released from stores, while cells that exhibit electrical bursting are the subject of Chapter 6. This book is not intentionally organized around mathematical techniques, but it is a happy coincidence that there is no use of partial differential equations throughout these beginning chapters.

Spatial aspects, such as synaptic transmission, gap junctions, the linear cable equation, nonlinear wave propagation in neurons, and calcium waves, are the subject of the next few chapters, and it is here that the reader first meets partial differential equations. The most mathematical sections of the book arise in the discussion of signaling in two- and three-dimensional media—readers who are less mathematically inclined may wish to skip over these sections. This section on basic physiological mechanisms ends with a discussion of the biochemistry of RNA and DNA and the biochemical regulation of cell function.

The second part of the book gives an overview of organ physiology, mostly from the human body, beginning with an introduction to electrocardiology, followed by the physiology of the circulatory system, blood, muscle, hormones, and the kidneys. Finally, we examine the digestive system, the visual system, ending with the inner ear.

While this may seem to be an enormous amount of material (and it is!), there are many physiological topics that are not discussed here. For example, there is almost no discussion of the immune system and the immune response, and so the work of Perelson, Goldstein, Wofsy, Kirschner, and others of their persuasion is absent. Another glaring omission is the wonderful work of Michael Reed and his collaborators on axonal transport; this work is discussed in detail by Edelstein-Keshet (1988). The study of the central nervous system, including fascinating topics like nervous control, learning, cognition, and memory, is touched upon only very lightly, and the field of pharmacokinetics and compartmental modeling, including the work of John Jacquez, Elliot Landaw, and others, appears not at all. Neither does the wound-healing work of Maini, Sherratt, Murray, and others, or the tumor modeling of Chaplain and his colleagues. The list could continue indefinitely. Please accept our apologies if your favorite topic (or life's work) was omitted; the reason is exhaustion, not lack of interest.

As well as noticing the omission of a number of important areas of mathematical physiology, the reader may also notice that our view of what "mathematical" means appears to be somewhat narrow as well. For example, we include very little discussion of statistical methods, stochastic models, or discrete equations, but concentrate almost wholly on continuous, deterministic approaches. We emphasize that this is not from any inherent belief in the superiority of continuous differential equations. It results rather from the unpleasant fact that choices had to be made, and when push came to shove, we chose to include work with which we were most familiar. Again, apologies are offered.

Finally, with a project of this size there is credit to be given and blame to be cast; credit to the many people, like the pioneers in the field whose work we freely borrowed, and many reviewers and coworkers (Andrew LeBeau, Matthew Wilkins, Richard Bertram, Lee Segel, Bruce Knight, John Tyson, Eric Cytrunbaum, Eric Marland, Tim Lewis, J.G.T. Sneyd, Craig Marshall) who have given invaluable advice. Particular thanks are also due to the University of Canterbury, New Zealand, where a significant portion of this book was written. Of course, as authors we accept all the blame for not getting it right, or not doing it better.

University of Utah                                                                                                     James Keener
University of Michigan                                                                                             James Sneyd

# Acknowledgments

With a project of this size it is impossible to give adequate acknowledgment to everyone who contributed: My family, whose patience with me is herculean; my students, who had to tolerate my rantings, ravings, and frequent mistakes; my colleagues, from whom I learned so much and often failed to give adequate attribution. Certainly the most profound contribution to this project was from the Creator who made it all possible in the first place. I don't know how He did it, but it was a truly astounding achievement. To all involved, thanks.

University of Utah                                                    James Keener

Between the three of them, Jim Murray, Charlie Peskin and Dan Tranchina have taught me almost everything I know about mathematical physiology. This book could not have been written without them, and I thank them particularly for their, albeit unaware, contributions. Neither could this book have been written without many years of support from my parents and my wife, to whom I owe the greatest of debts.

University of Michigan                                                James Sneyd

# Contents

# PART I

# CELLULAR PHYSIOLOGY

# CHAPTER 1

# Biochemical Reactions

Cells can do lots of wonderful things. Individually they can move, contract, excrete, reproduce, signal or respond to signals, and carry out the energy transactions necessary for this activity. Collectively they perform all of the numerous functions of any living organism necessary to sustain life. Yet all of what cells do can be described in terms of a few basic natural laws. The fascination with cells is that although the rules of behavior are relatively simple, they are applied to an enormously complex network of interacting chemicals and substrates. The effort of many lifetimes has been consumed in unraveling just a few of these reaction schemes, and there are many more mysteries yet to be uncovered.

## 1.1 The Law of Mass Action

The fundamental "law" of a chemical reaction is the law of mass action. This "law" describes the rate at which chemicals, whether large macromolecules or simple ions, collide and interact to form different chemical combinations. Suppose that two chemicals, say A and B, react upon collision with each other to form product C,

$$A + B \xrightarrow{k} C. \tag{1.1}$$

The rate of this reaction is the rate of accumulation of product, $\frac{d[C]}{dt}$. This rate is the product of the number of collisions per unit time between the two reactants and the probability that a collision is sufficiently energetic to overcome the free energy of activation of the reaction. The number of collisions per unit time is taken to be proportional to the product of the concentrations of A and B with a factor of proportionality that depends on the geometrical shapes and sizes of the reactant molecules and on the temperature of the mixture. Combining these factors we have

$$\frac{d[C]}{dt} = k[A][B].$$                                                                          (1.2)

The identification of (1.2) with the reaction (1.1) is called the *law of mass action*, and the constant $k$ is called the *rate constant* for the reaction. However, the law of mass action is not a law in the sense that it is inviolable, but rather it is a useful model, much like Ohm's law or Newton's law of cooling. As a model, there are situations where it is not valid. For example, at high concentrations, doubling the concentration of one reactant need not double the overall reaction rate, and at extremely low concentrations, it may not be appropriate to represent concentration as a continuous variable.

   While it is typical to denote reactions as proceeding in only one direction, with most biochemical reactions, reverse reactions also take place, so that the reaction scheme for A, B, and C should have been written as

$$A + B \underset{k_-}{\overset{k_+}{\rightleftarrows}} C$$                                                        (1.3)

with $k_+$ and $k_-$ denoting the forward and reverse rate constants of reaction, respectively. If the reverse reaction is slow compared to the forward reaction, it is often ignored, and only the primary direction is displayed. Since the quantity A is consumed by the forward reaction and produced by the reverse reaction, the rate of change of [A] for this bidirectional reaction is

$$\frac{d[A]}{dt} = k_-[C] - k_+[A][B].$$                                                           (1.4)

At equilibrium, concentrations are not changing, so that $[C]_{eq} = \frac{k_+}{k_-}[A]_{eq}[B]_{eq}$. If there are no other reactions involving A and C, then $[A] + [C] = A_0$ is constant, and

$$[C] = A_0 \frac{[B]}{K_{eq} + [B]}.$$                                                               (1.5)

The number $K_{eq} = k_-/k_+$ is called the *equilibrium constant*, and it relates to the relative preference for the chemicals to be in the combined state C compared to the disassociated state. The equilibrium constant has units of concentration. If $K_{eq}$ is small, then there is a high affinity between A and B. Notice from (1.5) that when $[B] = K_{eq}$, half of A is in the bound state.

   Unfortunately, the law of mass action cannot be used in all situations because not all chemical reaction mechanisms are known with sufficient detail. In fact, a vast number of chemical reactions cannot be described by mass action kinetics. Those reactions that follow mass action kinetics are called *elementary reactions* because presumably, they proceed directly from collision of the reactants. Reactions that do not follow mass action kinetics usually proceed by a complex mechanism consisting of two or more elementary reaction steps. It is often the case with biochemical reactions that the elementary reaction schemes are not known or are very complicated to write down.

## 1.2 Enzyme Kinetics

To see where some of the more complicated reaction schemes come from, we consider a reaction that is catalyzed by an enzyme. Enzymes are catalysts (generally proteins) that help convert other molecules called *substrates* into products, but they themselves are not changed by the reaction. Their most important features are catalytic power, specificity, and regulation. Enzymes accelerate the conversion of substrate into product by lowering the free energy of activation of the reaction. For example, enzymes may aid in overcoming charge repulsions and allowing reacting molecules to come into contact for the formation of new chemical bonds. Or, if the reaction requires breaking of an existing bond, the enzyme may exert a stress on a substrate molecule, rendering a particular bond more easily broken. Enzymes are particularly efficient at speeding up biological reactions, giving increases in speed of up to 10 million times or more. They are also highly specific, usually catalyzing the reaction of only one particular substrate or closely related substrates. Finally, they are typically regulated by an enormously complicated set of positive and negative feedback systems, thus allowing precise control over the rate of reaction. A detailed presentation of enzyme kinetics, including many different kinds of models, can be found in Dixon and Webb (1979). Here, we present some of the simplest models.

One of the first things to realize about enzyme reactions is that they do not follow the law of mass action directly. For as the concentration of substrate (S) is increased, the rate of the reaction increases only to a certain extent, reaching a maximal reaction velocity at high substrate concentrations. This is in contrast to the law of mass action, which, when applied directly to the reaction of S with the enzyme E predicts that the reaction velocity increases linearly as [S] increases.

A model to explain the deviation from the law of mass action was first proposed by Michaelis and Menten (1913). In their reaction scheme, the enzyme E converts the substrate S into the product P through a two-step process. First E combines with S to form a complex C which then breaks down into the product P releasing E in the process. The reaction scheme is represented schematically by

$$S + E \underset{k_{-1}}{\overset{k_1}{\rightleftharpoons}} C \xrightarrow{k_2} P + E.$$

It is important to note that, although this appears to imply that P cannot combine with E to form the complex, this is not the case. In fact, nearly all enzymes increase the speed of the reaction in both directions. Typically, however, reaction rates are measured under conditions where P is continually removed, which effectively prevents the reverse reaction from occurring. Thus, to determine the kinetic parameters from experimental data it suffices to assume that no reverse reaction occurs. Nevertheless, one must keep in mind that this is not the case *in vivo*, and thus the expressions we derive for reaction velocities can be applied only with great care to the physiology of intact cells. That being said, we ignore these complexities in the remainder of this book.

There are two similar ways to analyze this equation; the equilibrium approximation, and the quasi-steady-state approximation. Because these methods give similar results it is easy to confuse these two approaches, so it is worthwhile to understand their differences.

We begin by defining $s = [S], c = [C], e = [E]$, and $p = [P]$. The law of mass action applied to this reaction mechanism yields four differential equations for the rates of change of $s, c, e$, and $p$ as

$$\frac{ds}{dt} = k_{-1}c - k_1 se, \tag{1.6}$$

$$\frac{de}{dt} = (k_{-1} + k_2)c - k_1 se, \tag{1.7}$$

$$\frac{dc}{dt} = k_1 se - (k_2 + k_{-1})c, \tag{1.8}$$

$$\frac{dp}{dt} = k_2 c. \tag{1.9}$$

Notice that $p$ can be found by direct integration, and there is a conserved quantity since $\frac{de}{dt} + \frac{dc}{dt} = 0$, so that $e + c = e_0$, where $e_0$ is the total amount of available enzyme.

## 1.2.1   The Equilibrium Approximation

In their original analysis, Michaelis and Menten assumed that the substrate is in instantaneous equilibrium with the complex, and thus

$$k_1 se = k_{-1}c. \tag{1.10}$$

Since $e + c = e_0$, we then find that

$$c = \frac{e_0 s}{K_s + s}, \tag{1.11}$$

where $K_s = k_{-1}/k_1$. Hence, the velocity, $V$, of the reaction, i.e., the rate at which the product is formed, is given by

$$V = \frac{dp}{dt} = k_2 c = \frac{k_2 e_0 s}{K_s + s} = \frac{V_{max} s}{K_s + s}, \tag{1.12}$$

where $V_{max} = k_2 e_0$ is the maximum reaction velocity, attained when all the enzyme is complexed with the substrate.

At small substrate concentrations, the reaction rate is linear, at a rate proportional to the amount of available enzyme $e_0$. At large concentrations, however, the reaction rate saturates to $V_{max}$, so that the maximum rate of the reaction is limited by the amount of enzyme present and the dissociation rate constant $k_2$. For this reason, the dissociation reaction $C \xrightarrow{k_2} P + E$ is said to be *rate limiting* for this reaction. At $s = K_s$, the reaction rate is half that of the maximum.

It is important to note that (1.10) cannot be exactly correct at all times; if it were, then according to (1.6) substrate would not be used up, and product would not be formed. This points out the fact that (1.10) is an approximation. It also illustrates the need for a systematic way to make approximate statements, so that one has an idea of the magnitude and nature of the errors introduced in making such an approximation.

## 1.2.2 The Quasi-Steady-State Approximation

An alternative analysis of an enzymatic reaction was proposed by Briggs and Haldane (1925), and their analysis is now the basis for most present-day descriptions of enzyme reactions. Briggs and Haldane assumed that the rates of formation and breakdown of the complex were essentially equal at all times (except perhaps at the beginning of the reaction, as the complex is "filling up"). Thus, $dc/dt$ should be approximately zero. With this approximation, it is relatively simple to determine the velocity of the reaction.

To give this approximation a systematic mathematical basis, it is useful to introduce dimensionless variables

$$\sigma = \frac{s}{s_0}, \quad x = \frac{c}{e_0}, \quad \tau = k_1 e_0 t, \quad \kappa = \frac{k_{-1} + k_2}{k_1 s_0}, \quad \epsilon = \frac{e_0}{s_0}, \quad \alpha = \frac{k_{-1}}{k_1 s_0}, \tag{1.13}$$

in terms of which we obtain the system of two differential equations

$$\frac{d\sigma}{d\tau} = -\sigma + x(\sigma + \alpha), \tag{1.14}$$

$$\epsilon \frac{dx}{d\tau} = \sigma - x(\sigma + \kappa). \tag{1.15}$$

There are usually a number of ways that a system of differential equations can be nondimensionalized. This nonuniqueness is often a source of great confusion, as it is often not obvious which choice of dimensionless variables and parameters is "best." In Section 1.4 we discuss this difficult problem briefly.

The remarkable effectiveness of enzymes as catalysts of biochemical reactions is reflected by their small concentrations needed compared to the concentrations of the substrates. For this model, this means that $\epsilon$ is small, typically in the range of $10^{-2}$ to $10^{-7}$. Therefore, the reaction (1.15) is fast, equilibrates rapidly and remains in near-equilibrium even as the variable $\sigma$ changes. Thus, we take the *quasi-steady-state approximation* $\epsilon \frac{dx}{d\tau} = 0$. Notice that this is *not* the same as taking $\frac{dx}{d\tau} = 0$. However, because of the different scaling of $x$ and $c$, it is equivalent to taking $\frac{dc}{dt} = 0$ as suggested in the introductory paragraph. The quasi-steady-state approximation means that the variable $x$ is changing while restricted to some manifold described by setting the right-hand side of (1.15) to zero. This assumption is valid, provided that $\epsilon$ is small and $\frac{dx}{d\tau}$ is of order 1.

It follows from the quasi-steady-state approximation that

$$x = \frac{\sigma}{\sigma + \kappa}, \tag{1.16}$$

$$\frac{d\sigma}{d\tau} = -\frac{q\sigma}{\sigma + \kappa}, \tag{1.17}$$

where $q = \kappa - \alpha = \frac{k_2}{k_1 s_0}$. Equation (1.17) describes the rate of uptake of the substrate and is called a *Michaelis–Menten law*. In terms of the original variables, this law is

$$V = \frac{dp}{dt} = -\frac{ds}{dt} = \frac{k_2 e_0 s}{s + K_m} = \frac{V_{\max} s}{s + K_m}, \tag{1.18}$$

where $K_m = \frac{k_{-1} + k_2}{k_1}$. In quasi-steady state, the concentration of the complex satisfies

$$c = \frac{e_0 s}{s + K_m}. \tag{1.19}$$

Note the similarity between (1.12) and (1.18), the only difference being that the equilibrium approximation uses $K_s$, while the quasi-steady-state approximation uses $K_m$. Despite this similarity of form, it is important to keep in mind that the two results are based on different approximations.

As with the law of mass action, the Michaelis–Menten law (1.18) is not universally applicable but is a useful approximation. It may be applicable even if $\epsilon = e_0/s_0$ is not small (see, for example, Exercise 6), and in model building it is often invoked without regard to the underlying assumptions.

While the individual rate constants are difficult to measure experimentally, the ratio $K_m$ is relatively easy to measure because of the simple observation that (1.18) can be written in the form

$$\frac{1}{V} = \frac{1}{V_{\max}} + \frac{K_m}{V_{\max}} \frac{1}{s}. \tag{1.20}$$

In other words, $1/V$ is a linear function of $1/s$. Plots of this double reciprocal curve are called *Lineweaver–Burk plots*, and from such (experimentally determined) plots, $V_{\max}$ and $K_m$ can be found.

Although a Lineweaver–Burk plot makes it easy to determine $V_{\max}$ and $K_m$ from reaction rate measurements, it is not a simple matter to determine the reaction rate as a function of substrate concentration during the course of a single experiment. Substrate concentrations usually cannot be measured with sufficient accuracy or time resolution to permit the calculation of a reliable derivative. In practice, since it is more easily measured, the initial reaction rate is determined for a range of different initial substrate concentrations.

An alternative method to determine $K_m$ and $V_{\max}$ from experimental data is the direct linear plot (Eisenthal and Cornish-Bowden, 1974; Cornish-Bowden and Eisenthal, 1974). First we write (1.18) in the form

$$V_{\max} = V + \frac{V}{s} K_m, \tag{1.21}$$

and then treat $V_{max}$ and $K_m$ as variables for each experimental measurement of $V$ and $s$. (Recall that typically only the initial substrate concentration and initial velocity are used.) Then a plot of the straight line of $V_{max}$ against $K_m$ can be made. Repeating this for a number of different initial substrate concentrations and velocities gives a family of straight lines, which, in an ideal world free from experimental error, intersect at the single point $V_{max}$ and $K_m$ for that reaction. Of course, in reality, experimental error precludes an exact intersection, but $V_{max}$ and $K_m$ can be estimated from the median of the pairwise intersections.

## 1.2.3 Enzyme Inhibition

An enzyme inhibitor is a substance that inhibits the catalytic action of the enzyme. Enzyme inhibition is a common feature of enzyme reactions, and is an important means by which the activity of enzymes is controlled. Inhibitors come in many different types. For example, *irreversible inhibitors*, or *catalytic poisons*, decrease the activity of the enzyme to zero. This is the method of action of cyanide and many nerve gases. For this discussion, we restrict our attention to *competitive* inhibitors and *allosteric* inhibitors.

To understand the distinction between competitive and allosteric inhibition, it is useful to keep in mind that an enzyme molecule is usually a large protein, considerably larger than the substrate molecule whose reaction is catalyzed. Embedded in the large enzyme protein are one or more *active sites*, to which the substrate can bind to form the complex. In general, an enzyme catalyzes a single reaction or substrates with similar structures. This is believed to be a steric property of the enzyme that results from the three-dimensional shape of the enzyme allowing it to fit in a "lock-and-key" fashion with a corresponding substrate molecule.

If another molecule has a shape similar enough to that of the substrate molecule, it may also bind to the active site, preventing the binding of a substrate molecule, thus inhibiting the reaction. Because the inhibitor competes with the substrate molecule for the active site, it is called a competitive inhibitor.

However, because the enzyme molecule is large, it often has other binding sites, distinct from the active site, the binding of which affects the activity of the enzyme at the active site. These binding sites are called *allosteric* sites (from the Greek for "another solid") to emphasize that they are structurally different from the catalytic active sites. They are also called *regulatory sites* to emphasize that the catalytic activity of the protein is regulated by binding at this site. The ligand (any molecule that binds to a specific site on a protein, from Latin *ligare*, to bind) that binds at the allosteric site is called an *effector* or *modifier*, which, if it increases the activity of the enzyme, is called an allosteric activator, while if it decreases the activity of the substrate, it is called an allosteric inhibitor. The allosteric effect is presumed to arise because of a conformational change of the enzyme, that is, a change in the folding of the polypeptide chain, called an *allosteric transition*.

## Competitive Inhibition

In the simplest example of a competitive inhibitor, the reaction is stopped when the inhibitor is bound to the active site of the enzyme. Thus,

$$S + E \underset{k_{-1}}{\overset{k_1}{\rightleftharpoons}} C_1 \overset{k_2}{\longrightarrow} E + P,$$

$$E + I \underset{k_{-3}}{\overset{k_3}{\rightleftharpoons}} C_2.$$

From the law of mass action we get

$$\frac{ds}{dt} = -k_1 se + k_{-1} c_1, \tag{1.22}$$

$$\frac{di}{dt} = -k_3 ie + k_{-3} c_2, \tag{1.23}$$

$$\frac{dc_1}{dt} = k_1 se - (k_{-1} + k_2) c_1, \tag{1.24}$$

$$\frac{dc_2}{dt} = k_3 ie - k_{-3} c_2. \tag{1.25}$$

As before, it is not necessary to write an equation for the accumulation of the product. Furthermore, we know that $e + c_1 + c_2 = e_0$. To be systematic, the next step is to introduce dimensionless variables, and identify those reactions that are rapid and equilibrate rapidly to their quasi-steady states. However, from our previous experience (or from a calculation on a piece of scratch paper), we know, assuming the enzyme-to-substrate ratios are small, that the fast equations are those for $c_1$ and $c_2$. Hence, the quasi-steady states are found by (formally) setting $dc_1/dt = dc_2/dt = 0$ and solving for $c_1$ and $c_2$. Recall that this does not mean that $c_1$ and $c_2$ are unchanging, rather that they are changing in quasi-steady-state fashion, keeping the right-hand sides of these equations nearly zero. This gives

$$c_1 = \frac{K_i e_0 s}{K_m i + K_i s + K_m K_i}, \tag{1.26}$$

$$c_2 = \frac{K_m e_0 i}{K_m i + K_i s + K_m K_i}, \tag{1.27}$$

where $K_m = \frac{k_{-1} + k_2}{k_1}, K_i = k_{-3}/k_3$. Thus the velocity of the reaction is

$$V = k_2 c_1 = \frac{k_2 e_0 s K_i}{K_m i + K_i s + K_m K_i} = \frac{V_{max} s}{s + K_m (1 + i/K_i)}. \tag{1.28}$$

Notice that the effect of the inhibitor is to increase the effective equilibrium constant of the enzyme by the factor $1 + i/K_i$, from $K_m$ to $K_m(1 + i/K_i)$, thus decreasing the velocity of reaction, while leaving the maximum velocity unchanged.

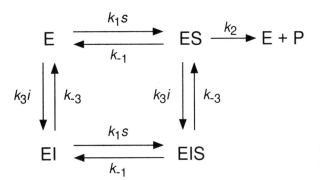

**Figure 1.1** Diagram of the possible states of an enzyme with one allosteric and one catalytic binding site.

### Allosteric Inhibitors

If the inhibitor can bind at an allosteric site, we have the possibility that the enzyme could bind both the inhibitor and the substrate simultaneously. In this case, there are four possible binding states for the enzyme, and transitions between them, as demonstrated graphically in Fig. 1.1.

The simplest analysis of this reaction scheme is the equilibrium analysis. (The more complicated quasi-steady-state analysis is posed as Exercise 2.) We define $K_s = k_{-1}/k_1$, $K_i = k_{-3}/k_3$, and let $x, y$, and $z$ denote, respectively, the concentrations of ES, EI and EIS. Then, it follows from the law of mass action that in steady state,

$$(e_0 - x - y - z)s - K_s x = 0, \tag{1.29}$$

$$(e_0 - x - y - z)i - K_i y = 0, \tag{1.30}$$

$$ys - K_s z = 0, \tag{1.31}$$

$$xi - K_i z = 0, \tag{1.32}$$

where $e_0 = e + x + y + z$ is the total amount of enzyme. Notice that this is a linear system of equations for $x, y$, and $z$. Although there are four equations, one is a linear combination of the other three (the system is of rank three), so that we can determine $x, y$, and $z$ as functions of $i$ and $s$, finding

$$x = \frac{e_0 K_i}{K_i + i} \frac{s}{K_s + s}. \tag{1.33}$$

It follows that the reaction rate, $V = k_2 x$, is given by

$$V = \frac{V_{\text{max}}}{1 + i/K_i} \frac{s}{K_s + s}, \tag{1.34}$$

where $V_{\text{max}} = k_2 e_0$. Thus, in contrast to the competitive inhibitor, the allosteric inhibitor decreases the maximum velocity of the reaction, while leaving $K_s$ unchanged. (Of course, the situation is more complicated if the quasi-steady-state approximation is used, and no such simple conclusion follows.)

## 1.2.4  Cooperativity

For many enzymes, the reaction velocity is not a simple hyperbolic curve, as predicted
by the Michaelis–Menten model, but often has a sigmoidal character. This can result from cooperative effects, in which the enzyme can bind more than one substrate
molecule but the binding of one substrate molecule affects the binding of subsequent
ones.

Originally, much of the theoretical work on cooperative behavior was stimulated
by the properties of hemoglobin, and this is often the context in which cooperativity is
discussed. A detailed discussion of hemoglobin and oxygen binding is given in Chapter
16, while here cooperativity is discussed in more general terms.

Suppose that an enzyme can bind two substrate molecules, so it can exist in one of
three states, namely as a free molecule E, as a complex with one occupied center $C_1$,
and as a complex with two occupied centers $C_2$. The reaction mechanism is represented
by

$$S + E \underset{k_{-1}}{\overset{k_1}{\rightleftarrows}} C_1 \overset{k_2}{\rightarrow} E + P, \tag{1.35}$$

$$S + C_1 \underset{k_{-3}}{\overset{k_3}{\rightleftarrows}} C_2 \overset{k_4}{\rightarrow} C_1 + P. \tag{1.36}$$

Using the law of mass action, one can write down the rate equations for the 5
concentrations [S], [E], [$C_1$], [$C_2$], and [P]. However, because the amount of product [P]
can be determined by quadrature, and because the total amount of enzyme molecule
is conserved, we only need three equations for the three quantities [S], [$C_1$], and [$C_2$].
These are

$$\frac{ds}{dt} = -k_1 se + k_{-1}c_1 - k_3 sc_1 + k_{-3}c_2, \tag{1.37}$$

$$\frac{dc_1}{dt} = k_1 se - (k_{-1} + k_2)c_1 - k_3 sc_1 + (k_4 + k_{-3})c_2, \tag{1.38}$$

$$\frac{dc_2}{dt} = k_3 sc_1 - (k_4 + k_{-3})c_2, \tag{1.39}$$

where $s = [S], c_1 = [C_1], c_2 = [C_2]$, and $e + c_1 + c_2 = e_0$.

Proceeding as before, we invoke the quasi-steady-state assumption that $dc_1/dt = dc_2/dt = 0$, and solve for $c_1$ and $c_2$ to get

$$c_1 = \frac{K_2 e_0 s}{K_1 K_2 + K_2 s + s^2}, \tag{1.40}$$

$$c_2 = \frac{e_0 s^2}{K_1 K_2 + K_2 s + s^2}, \tag{1.41}$$

where $K_1 = \frac{k_{-1}+k_2}{k_1}$ and $K_2 = \frac{k_4+k_{-3}}{k_3}$. The reaction velocity is thus given by

$$V = k_2 c_1 + k_4 c_2 = \frac{(k_2 K_2 + k_4 s)e_0 s}{K_1 K_2 + K_2 s + s^2}. \tag{1.42}$$

It is instructive to examine two extreme cases. First, if the active sites act independently and identically, then $k_1 = 2k_3 = 2k_+$, $2k_{-1} = k_{-3} = 2k_-$ and $2k_2 = k_4$, where $k_+$ and $k_-$ are the forward and backward reaction rates for the individual binding sites. The factors of 2 occur because two identical binding sites are involved in the reaction, doubling the amount of the reactant. In this case,

$$V = \frac{2k_2e_0(K+s)s}{K^2 + 2Ks + s^2} = 2\frac{k_2e_0s}{K+s}, \tag{1.43}$$

where $K = \frac{k_-+k_2}{k_+}$ is the equilibrium constant for the individual binding site. As expected, the rate of reaction is exactly twice that for the individual binding site.

In the opposite extreme, suppose that the binding of the first substrate molecule is slow, but that with one site bound, binding of the second is fast (this is large positive cooperativity). This can be modeled by letting $k_3 \to \infty$ and $k_1 \to 0$, while keeping $k_1k_3$ constant, in which case $K_2 \to 0$ and $K_1 \to \infty$ while $K_1K_2$ is constant. In this limit, the velocity of the reaction is

$$V = \frac{k_4e_0s^2}{K_m^2 + s^2} = \frac{V_{\max}s^2}{K_m^2 + s^2}, \tag{1.44}$$

where $K_m^2 = K_1K_2$, and $V_{\max} = k_4e_0$.

In general, if $n$ substrate molecules can bind to the enzyme, there are $n$ equilibrium constants, $K_1$ through $K_n$. In the limit as $K_n \to 0$ and $K_1 \to \infty$ while keeping $K_1K_n$ fixed, the rate of reaction is

$$V = \frac{V_{\max}s^n}{K_m^n + s^n}, \tag{1.45}$$

where $K_m^n = \Pi_{i=1}^n K_i$. This rate equation is known as the *Hill equation*. Typically, the Hill equation is used for reactions whose detailed intermediate steps are not known but for which cooperative behavior is suspected. The exponent $n$ and the parameters $V_{\max}$ and $K_m$ are usually determined from experimental data. Observe that

$$n \ln s = n \ln K_m + \ln \left( \frac{V}{V_{\max} - V} \right), \tag{1.46}$$

so that a plot of $\ln(\frac{V}{V_{\max}-V})$ against $\ln s$ (called a *Hill plot*) should be a straight line of slope $n$. Although the exponent $n$ suggests an $n$-step process (with $n$ binding sites), in practice it is not unusual for the best fit for $n$ to be noninteger.

An enzyme can also exhibit negative cooperativity, in which the binding of the first substrate molecule *decreases* the rate of subsequent binding. This can be modeled by decreasing $k_3$. In Fig. 1.2 we plot the reaction velocity against the substrate concentration for the cases of independent binding sites (no cooperativity), extreme positive cooperativity (the Hill equation), and negative cooperativity. From this figure it can be seen that with positive cooperativity, the reaction velocity is a sigmoidal function of the substrate concentration, while negative cooperativity primarily decreases the velocity.

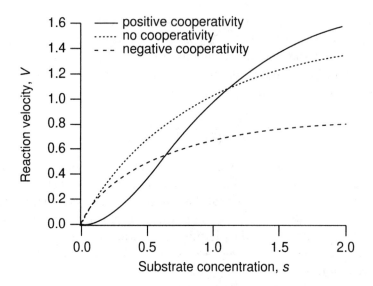

**Figure 1.2**  Reaction velocity plotted against substrate concentration, for three different cases. Positive cooperativity, $K_1 = 1000$, $K_2 = 0.001$; independent binding sites, $K_1 = 0.5$, $K_2 = 2$; and negative cooperativity, $K_1 = 0.5$, $K_2 = 100$. The other parameters were chosen as $e_0 = 1$, $k_2 = 1$, $k_4 = 2$. Concentration and time units are arbitrary.

### The Monod–Wyman–Changeux Model

Cooperative effects occur when the binding of one substrate molecule alters the rate of binding of subsequent ones. However, the above models give no explanation of how such alterations in the binding rate occur. The earliest mechanistic model proposed to account for cooperative effects in terms of the enzyme's conformation was that of Monod, Wyman, and Changeux (1965). Their model is based on the following assumptions about the structure and behavior of enzymes.

1. Cooperative proteins are composed of several identical reacting units, called *protomers*, that occupy equivalent positions within the protein.
2. Each protomer contains one binding site for each ligand.
3. The binding sites within each protein are equivalent.
4. If the binding of a ligand to one protomer induces a conformational change in that protomer, an identical conformational change is induced in all protomers.
5. The protein has two conformational states, usually denoted by R and T, which differ in their ability to bind ligands.

To illustrate how these assumptions can be quantified, we consider a protein with only two binding sites. Thus, the protein can exist in one of six states: $R_i, i = 0, 1, 2$, or $T_i, i = 0, 1, 2$, where the subscript $i$ is the number of bound ligands. For convenience, we also assume that $R_1$ cannot convert directly to $T_1$, or vice versa, and similarly for

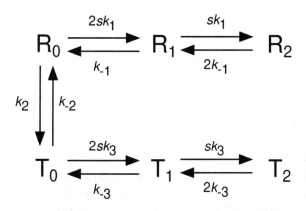

**Figure 1.3** Diagram of the states of the protein, and the possible transitions, in a six-state Monod–Wyman–Changeux model.

$R_2$ and $T_2$. The general case is left for Exercise 3. The states of the protein and the allowable transitions are illustrated in Fig. 1.3.

We now assume that all the reactions are in equilibrium. We let a lowercase letter denote a concentration, and thus $r_i$ and $t_i$ denote the concentrations of chemical species $R_i$ and $T_i$ respectively. Also, as before, we let $s$ denote the concentration of the substrate. Then, the fraction $Y$ of occupied sites (also called the *saturation function*) is

$$Y = \frac{r_1 + 2r_2 + t_1 + 2t_2}{2(r_0 + r_1 + r_2 + t_0 + t_1 + t_2)}. \tag{1.47}$$

Furthermore, with $K_i = k_{-i}/k_i$, for $i = 1, 2, 3$, we find that

$$r_1 = 2sK_1^{-1}r_0, \qquad r_2 = s^2K_1^{-2}r_0, \tag{1.48}$$

$$t_1 = 2sK_3^{-1}t_0, \qquad t_2 = s^2K_3^{-2}t_0. \tag{1.49}$$

Substituting these into (1.47) gives

$$Y = \frac{sK_1^{-1}(1 + sK_1^{-1}) + K_2^{-1}[sK_3^{-1}(1 + sK_3^{-1})]}{(1 + sK_1^{-1})^2 + K_2^{-1}(1 + sK_3^{-1})^2}, \tag{1.50}$$

where we have used that $r_0/t_0 = K_2$. More generally, if there are $n$ binding sites, then

$$Y = \frac{sK_1^{-1}(1 + sK_1^{-1})^{n-1} + K_2^{-1}[sK_3^{-1}(1 + sK_3^{-1})^{n-1}]}{(1 + sK_1^{-1})^n + K_2^{-1}(1 + sK_3^{-1})^n}. \tag{1.51}$$

In general, $Y$ is a sigmoidal function of $s$.

Some special cases are of interest. For example, if $K_3 = \infty$, so that the substrate cannot bind directly to the $T$ conformation, then

$$Y = \frac{sK_1^{-1}(1 + sK_1^{-1})}{(1 + sK_1^{-1})^2 + K_2^{-1}}, \tag{1.52}$$

or if $K_2 = \infty$, so that only the $R$ conformation exists, then

$$Y = \frac{s}{s + K_1}, \tag{1.53}$$

which is the Michaelis–Menten equation.

There are many other models of enzyme cooperativity, and the interested reader is referred to Dixon and Webb (1979) for a comprehensive discussion and comparison of other models in the literature.

# 1.3   Glycolysis and Glycolytic Oscillations

Metabolism is the process of extracting useful energy from chemical bonds. A metabolic pathway is the sequence of enzymatic reactions that take place in order to transfer chemical energy from one form to another. The common carrier of energy in the cell is the chemical *adenosine triphosphate* (ATP). ATP is formed by the addition of an inorganic phosphate group ($HPO_4^{2-}$) to *adenosine diphosphate* (ADP), or by the addition of two inorganic phosphate groups to *adenosine monophosphate* (AMP). The process of adding an inorganic phosphate group to a molecule is called *phosphorylation*. Since the three phosphate groups on ATP carry negative charges, considerable energy is required to overcome the natural repulsion of like-charged phosphates as additional groups are added to AMP. Thus, the hydrolysis (the cleavage of a bond by water) of ATP to ADP releases large amounts of energy.

Energy to perform chemical work is made available to the cell by the oxidation of glucose to carbon dioxide and water, with a net release of energy. Some of this energy is dissipated as heat, but fortunately, some of it is also stored in other chemical bonds. The overall chemical reaction for the oxidation of glucose can be written as

$$C_6H_{12}O_6 + 6O_2 \longrightarrow 6CO_2 + 6H_2O + \text{energy}, \tag{1.54}$$

but of course, this is not an elementary reaction. Instead, this reaction takes place in a series of enzymatic reactions, with three major reaction stages, *glycolysis*, the *Krebs cycle*, and the *electron transport* (or *cytochrome) system*.

Glycolysis involves 11 elementary reaction steps, each of which is an enzymatic reaction. Here we consider a simplified model of the initial steps. (To understand more of the labyrinthine complexity of glycolysis, interested readers are encouraged to consult a specialized book on biochemistry, such as Stryer, 1988.) The first three steps of glycolysis are (Fig. 1.4)

1. the phosphorylation of glucose to glucose 6-phosphate;
2. the isomerization of glucose 6-phosphate to fructose 6-phosphate; and
3. the phosphorylation of fructose 6-phosphate to fructose 1,6-bisphosphate.

This last reaction is catalyzed by the enzyme phosphofructokinase (PFK1).

PFK1 is an example of an allosteric enzyme as it is allosterically inhibited by ATP. Note that ATP is both a substrate of PFK1, binding at a catalytic site, and an allosteric inhibitor, binding at a regulatory site. The inhibition due to ATP is removed by AMP, and thus the activity of PFK1 increases as the ratio of ATP to AMP decreases. As PFK1 phosphorylates fructose 6-P, ATP is converted to ADP. ADP, in turn, is converted back

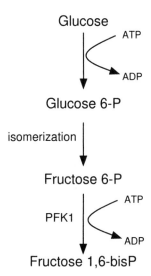

**Figure 1.4** The first three reactions in the glycolytic pathway.

to ATP and AMP by the reaction

$$2\text{ADP} \rightleftharpoons \text{ATP} + \text{AMP},$$

which is catalyzed by the enzyme adenylate kinase. Since there is normally little AMP in cells, the conversion of ADP to ATP and AMP serves to significantly decrease the ATP/AMP ratio, thus activating PFK1. This is an example of a positive feedback loop; the greater the activity of PFK1, the lower the ATP/AMP ratio, thus further increasing PFK1 activity.

It was discovered in 1980 that in some cell types, another important allosteric activator of PFK1 is fructose 2,6-bisphosphate (Stryer, 1988), which is formed from fructose 6-phosphate in a reaction catalyzed by phosphofructokinase 2 (PFK2), a different enzyme from phosphofructokinase (PFK1) (you were given fair warning about the labyrinthine nature of this process!). Of particular significance is that an abundance of fructose 6-phosphate leads to a corresponding abundance of fructose 2,6-bisphosphate, and thus a corresponding increase in the activity of PFK1. This is an example of a negative feedback loop, where an increase in the substrate concentration leads to a greater rate of substrate reaction and consumption. Clearly, PFK1 activity is controlled by an intricate system of reactions, the collective behavior of which is not obvious a priori.

Under certain conditions the rate of glycolysis is known to be oscillatory, or even chaotic (Nielsen et al., 1997). This biochemical oscillator has been known and studied experimentally for some time. For example, Hess and Boiteux (1973) devised a flow reactor containing yeast cells into which a controlled amount of substrate (either glucose or fructose) was continuously added. They measured the pH and fluorescence of the reactants, thereby monitoring the glycolytic activity, and they found ranges of continuous input under which glycolysis was periodic.

A mathematical model describing this oscillation was proposed by Sel'kov (1968) and later modified by Goldbeter and Lefever (1972). It is meant to capture only the positive feedback of ADP on PFK1 activity, and does not take into account the negative feedback process that was discovered more recently. (An interesting exercise would be to construct a more detailed model, including both positive and negative feedback processes, to see what difference this makes to the conclusions.) In the Sel'kov model, PFK1 is inactive in its unbound state but is activated by binding with several ADP molecules. Note that, for simplicity, the model does not take into account the conversion of ADP to AMP and ATP, but assumes that ADP activates PFK1 directly, since the overall effect is similar. In the active state, the enzyme catalyzes the production of ADP from ATP as fructose-6-P is phosphorylated. Sel'kov's reaction scheme for this process is as follows: PFK1 (denoted by E) is activated or deactivated by binding or unbinding with $\gamma$ molecules of ADP (denoted by $S_2$)

$$\gamma S_2 + E \underset{k_{-3}}{\overset{k_3}{\rightleftharpoons}} ES_2^\gamma,$$

and ATP (denoted $S_1$) can bind with the activated form of enzyme to produce a product molecule of ADP. In addition, there is assumed to be a steady supply rate of $S_1$, while product $S_2$ is irreversibly removed. Thus,

$$\overset{v_1}{\longrightarrow} S_1, \tag{1.55}$$

$$S_1 + ES_2^\gamma \underset{k_{-1}}{\overset{k_1}{\rightleftharpoons}} S_1 ES_2^\gamma \overset{k_2}{\longrightarrow} ES_2^\gamma + S_2, \tag{1.56}$$

$$S_2 \overset{v_2}{\longrightarrow}. \tag{1.57}$$

Applying the law of mass action to the Sel'kov kinetic scheme, we find five differential equations for the production of the five species $s_1 = [S_1], s_2 = [S_2], e = [E], x_1 = [ES_2^\gamma], x_2 = [S_1 ES_2^\gamma]$:

$$\frac{ds_1}{dt} = v_1 - k_1 s_1 x_1 + k_{-1} x_2, \tag{1.58}$$

$$\frac{ds_2}{dt} = k_2 x_2 - k_3 s_2^\gamma e + k_{-3} x_1 - v_2 s_2, \tag{1.59}$$

$$\frac{dx_1}{dt} = -k_1 s_1 x_1 + (k_{-1} + k_2) x_2 + k_3 s_2^\gamma e - k_{-3} x_1, \tag{1.60}$$

$$\frac{dx_2}{dt} = k_1 s_1 x_1 - (k_{-1} + k_2) x_2. \tag{1.61}$$

The fifth differential equation is not necessary, because the total available enzyme is conserved, $e + x_1 + x_2 = e_0$. Now we introduce dimensionless variables $\sigma_1 = \frac{k_1 s_1}{k_2 + k_{-1}}, \sigma_2 = (\frac{k_3}{k_{-3}})^{1/\gamma} s_2, u_1 = x_1/e_0, u_2 = x_2/e_0, t = \frac{k_2 + k_{-1}}{e_0 k_1 k_2} \tau$ and find

$$\frac{d\sigma_1}{d\tau} = v - \frac{k_2 + k_{-1}}{k_2} u_1 \sigma_1 + \frac{k_{-1}}{k_2} u_2, \tag{1.62}$$

$$\frac{d\sigma_2}{d\tau} = \alpha \left[ u_2 - \frac{k_{-3}}{k_2} \sigma_2^\gamma (1 - u_1 - u_2) + \frac{k_{-3}}{k_2} u_1 \right] - \eta \sigma_2, \tag{1.63}$$

$$\epsilon \frac{du_1}{d\tau} = u_2 - \sigma_1 u_1 + \frac{k_{-3}}{k_2 + k_{-1}} \left[ \sigma_2^\gamma (1 - u_1 - u_2) - u_1 \right], \tag{1.64}$$

$$\epsilon \frac{du_2}{d\tau} = \sigma_1 u_1 - u_2, \tag{1.65}$$

where $\epsilon = \frac{e_0 k_1 k_2}{(k_2 + k_{-1})^2}$, $v = \frac{v_1}{k_2 e_0}$, $\eta = \frac{v_2 (k_2 + k_{-1})}{k_1 k_2 e_0}$, $\alpha = \frac{k_2 + k_{-1}}{k_1} (\frac{k_3}{k_{-3}})^{1/\gamma}$. If we assume that $\epsilon$ is a small number, then both $u_1$ and $u_2$ are "fast" variables and can be set to their quasi-steady values,

$$u_1 = \frac{\sigma_2^\gamma}{\sigma_2^\gamma \sigma_1 + \sigma_2^\gamma + 1}, \tag{1.66}$$

$$u_2 = \frac{\sigma_1 \sigma_2^\gamma}{\sigma_2^\gamma \sigma_1 + \sigma_2^\gamma + 1} = f(\sigma_1, \sigma_2), \tag{1.67}$$

and with these quasi-steady values, the evolution of $\sigma_1$ and $\sigma_2$ is governed by

$$\frac{d\sigma_1}{d\tau} = v - f(\sigma_1, \sigma_2), \tag{1.68}$$

$$\frac{d\sigma_2}{d\tau} = \alpha f(\sigma_1, \sigma_2) - \eta \sigma_2. \tag{1.69}$$

The goal of the following analysis is to demonstrate that this system of equations has oscillatory solutions for some range of the supply rate $v$. First observe that because of saturation, the function $f(\sigma_1, \sigma_2)$ is bounded by 1. Thus, if $v > 1$, the solutions of the differential equations are not bounded. For this reason we consider only $0 < v < 1$. The nullclines of the flow are given by the equations

$$\sigma_1 = \frac{v}{1 - v} \frac{1 + \sigma_2^\gamma}{\sigma_2^\gamma} \qquad \left( \frac{d\sigma_1}{d\tau} = 0 \right), \tag{1.70}$$

$$\sigma_1 = \frac{1 + \sigma_2^\gamma}{\sigma_2^{\gamma - 1}(p - \sigma_2)} \qquad \left( \frac{d\sigma_2}{d\tau} = 0 \right), \tag{1.71}$$

where $p = \alpha / \eta$. These two nullclines are shown plotted as dotted and dashed curves respectively in Fig. 1.5.

The steady-state solution is unique and satisfies

$$\sigma_2 = pv, \tag{1.72}$$

$$\sigma_1 = \frac{v(1 + \sigma_2^\gamma)}{(1 - v)\sigma_2^\gamma}. \tag{1.73}$$

The stability of the steady solution is found by linearizing the differential equations about the steady-state solution and examining the eigenvalues of the linearized system. The linearized system has the form

$$\frac{d\tilde{\sigma}_1}{d\tau} = -f_1 \tilde{\sigma}_1 - f_2 \tilde{\sigma}_2, \tag{1.74}$$

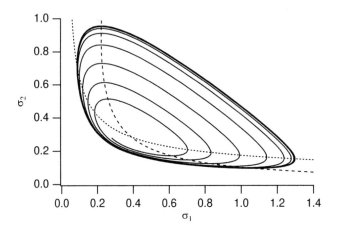

**Figure 1.5** Phase portrait of the Sel'kov glycolysis system with $v = 0.0285, \eta = 0.1, \alpha = 1.0$, and $\gamma = 2$. Dotted curve: $\frac{d\sigma_1}{d\tau} = 0$. Dashed curve: $\frac{d\sigma_2}{d\tau} = 0$.

$$\frac{d\tilde{\sigma}_2}{d\tau} = \alpha f_1 \tilde{\sigma}_1 + (\alpha f_2 - \eta)\tilde{\sigma}_2, \tag{1.75}$$

where $f_j = \frac{\partial f}{\partial \sigma_j}, j = 1, 2$, evaluated at the steady-state solution, and where $\tilde{\sigma}_i$ denotes the deviation from the steady-state value of $\sigma_i$. The characteristic equation for the eigenvalues $\lambda$ of the linear system (1.74)–(1.75) is

$$\lambda^2 - (\alpha f_2 - \eta - f_1)\lambda + f_1\eta = 0. \tag{1.76}$$

Since $f_1$ is always positive, the stability of the linear system is determined by the sign of $H = \alpha f_2 - \eta - f_1$, being stable if $H < 0$ and unstable if $H > 0$. Changes of stability, if they exist, occur at $H = 0$, and are Hopf bifurcations to periodic solutions with approximate frequency $\omega = \sqrt{f_1\eta}$.

The function $H(v)$ is given by

$$H(v) = \frac{(1 - v)}{(1 + y)}(\eta\gamma + (v - 1)y) - \eta, \tag{1.77}$$

$$y = (pv)^\gamma. \tag{1.78}$$

Clearly, $H(0) = \eta(\gamma - 1), H(1) = -\eta$, so for $\gamma > 1$, there must be at least one Hopf bifurcation point, below which the steady solution is unstable. Additional computations show that this Hopf bifurcation is supercritical, so that for $v$ slightly below the bifurcation point, there is a stable periodic orbit.

An example of this periodic orbit is shown in Fig. 1.5 with coefficients $v = 0.0285, \eta = 0.1, \alpha = 1.0$, and $\gamma = 2$. The evolution of $\sigma_1$ and $\sigma_2$ are shown plotted as functions of time in Fig. 1.6. This periodic orbit exists only in very small regions of parameter space, rapidly expanding until it contacts the $S_2 = 0$ axis, into which it collapses.

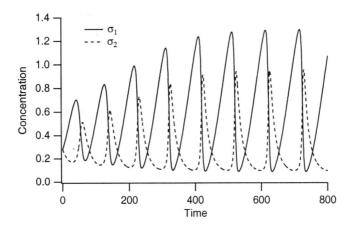

**Figure 1.6** Evolution of $\sigma_1$ and $\sigma_2$ for the Sel'kov glycolysis system toward a periodic solution. Parameters are the same as in Fig. 1.5.

While the Sel'kov model has certain features that are qualitatively correct, it fails to agree with the experimental results at a number of points. Hess and Boiteux (1973) report that for high and low substrate injection rates, there is a stable steady-state solution. There are two Hopf bifurcation points, one at the flow rate of 20 mM/hr and another at 160 mM/hr. The period of oscillation at the low flow rate is about 8 minutes and decreases as a function of flow rate to about 3 minutes at the upper Hopf bifurcation point. In contrast, the Sel'kov model has but one Hopf bifurcation point.

To reproduce these additional experimental features we consider a more detailed model of the reaction. In 1972, Goldbeter and Lefever proposed a model of Monod–Wyman–Changeux type that provided a more accurate description of the oscillations. More recently, by fitting a simpler model to experimental data on PFK1 kinetics in skeletal muscle, Smolen (1995) has shown that this level of complexity is not necessary; his model assumes that PFK1 consists of four independent, identical subunits, and reproduces the observed oscillations well. Despite this, we consider only the Goldbeter–Lefever model in detail, as it provides an excellent example of the use of Monod–Wyman–Changeux models.

In the Goldbeter–Lefever model of the phosphorylation of fructose-6-P, the enzyme PFK1 is assumed to be a dimer that exists in two states, an active state R and an inactive state T. The substrate, $S_1$, can bind to both forms, but the product, $S_2$, which is an activator, or positive effector, of the enzyme, binds only to the active form. The enzymatic forms of R carrying substrate decompose irreversibly to yield the product ADP. In addition, substrate is supplied to the system at a constant rate, while product is removed at a rate proportional to its concentration. The reaction scheme for this is as follows: let $T_j$ represent the inactive T form of the enzyme bound to $j$ molecules of substrate and let $R_{ij}$ represent the active form R of the enzyme bound to $i$ substrate molecules and $j$ product molecules. Then

$$R_{00} \underset{k_{-1}}{\overset{k_1}{\rightleftarrows}} T_0,$$

$$S_1 + R_{0j} \underset{k_{-2}}{\overset{2k_2}{\rightleftarrows}} R_{1j} \overset{k}{\longrightarrow} R_{0j} + S_2, \qquad S_1 + R_{1j} \underset{2k_{-2}}{\overset{k_2}{\rightleftarrows}} R_{2j} \overset{2k}{\longrightarrow} R_{1j} + S_2, \qquad j = 0, 1, 2,$$

$$S_2 + R_{00} \underset{k_{-2}}{\overset{2k_2}{\rightleftarrows}} R_{01}, \qquad S_2 + R_{01} \underset{2k_{-2}}{\overset{k_2}{\rightleftarrows}} R_{02},$$

$$S_1 + T_0 \underset{k_{-3}}{\overset{2k_3}{\rightleftarrows}} T_1, \qquad S_1 + T_1 \underset{2k_{-3}}{\overset{k_3}{\rightleftarrows}} T_2.$$

The possible receptor states are illustrated graphically in Fig. 1.7. In this system, the substrate $S_1$ holds the enzyme in the inactive state by binding with $T_0$ to produce $T_1$ and $T_2$, while product $S_2$ holds enzyme in the active state by binding with $R_{00}$ to produce $R_{01}$ and binding with $R_{01}$ to produce $R_{02}$. There is a factor two in the rates of reaction because a dimer with two available binding sites reacts like twice the same amount of monomer.

    The analysis of this reaction scheme is substantially more complicated than that of the Sel'kov scheme, although the idea is the same. We use the law of mass action to write differential equations for the fourteen chemical species. For example, the equation for

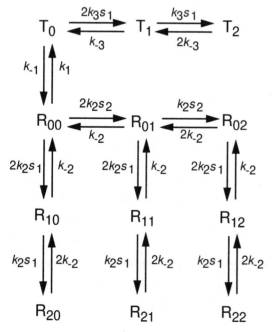

**Figure 1.7** Possible receptor states of the Goldbeter–Lefever model for glycolytic oscillations.

$s_1 = [S_1]$ is

$$\frac{ds_1}{dt} = v_1 - F, \qquad (1.79)$$

where

$$\begin{aligned} F = \ & k_{-2}(r_{10} + r_{11} + r_{12}) + 2k_{-2}(r_{20} + r_{21} + r_{22}) \\ & - 2k_2 s_1 (r_{00} + r_{01} + r_{02}) - k_2 s_1 (r_{10} + r_{11} + r_{12}) \\ & - 2k_3 s_1 t_0 - k_3 s_1 t_1 + k_{-3} t_1 + 2k_{-3} t_2, \end{aligned} \qquad (1.80)$$

and the equation for $r_{00} = [R_{00}]$ is

$$\frac{dr_{00}}{dt} = -(k_1 + 2k_2 s_1 + 2k_2 s_2)r_{00} + (k_{-2} + k)r_{10} + k_{-2} r_{01} + k_{-1} t_0. \qquad (1.81)$$

We then assume that all twelve of the intermediates are in quasi-steady state. This leads to a 12 by 12 linear system of equations, which, if we take the total amount of enzyme to be $e_0$, can be solved. We substitute this solution into the differential equations for $s_1$ and $s_2$ with the result that

$$\frac{ds_1}{dt} = v_1 - F(s_1, s_2), \qquad (1.82)$$

$$\frac{ds_2}{dt} = F(s_1, s_2) - v_2 s_2, \qquad (1.83)$$

where

$$F(s_1, s_2) = \left( \frac{2k_2 k_{-1} k e_0}{k + k_{-2}} \right) \left( \frac{s_1(1 + \frac{k_2}{k+k_{-2}} s_1)(k_2 s_2 + k_{-2})^2}{k_{-2}^2 k_1 (\frac{k_3}{k_{-3}} s_1 + 1)^2 + k_{-1}(1 + \frac{k_2}{k+k_{-2}} s_1)^2 (k_{-2} + k_2 s_2)^2} \right). \qquad (1.84)$$

Now we introduce dimensionless variables $\sigma_1 = \frac{k_2 s_1}{k_{-2}}, \sigma_2 = \frac{k_2 s_2}{k_{-2}}, t = \frac{\tau}{\tau_c}$ and parameters $\nu = \frac{k_2 v_1}{k_{-2} \tau_c}, \eta = \frac{v_2}{\tau_c}$, where $\tau_c = \frac{2k_2 k_{-1} k e_0}{k_1(k+k_{-2})}$, and arrive at the system (1.68)–(1.69), but with a different function $f(\sigma_1, \sigma_2)$, and with $\alpha = 1$. If, in addition, we assume that

1. the substrate does not bind to the T form ($k_3 = 0$, T is completely inactive),
2. $T_0$ is preferred over $R_{00}$ ($k_1 \gg k_{-1}$), and
3. if the substrate $S_1$ binds to the R form, then formation of product $S_2$ is preferred to dissociation ($k \gg k_{-2}$),

then we can simplify the equations substantially to obtain

$$f(\sigma_1, \sigma_2) = \sigma_1(1 + \sigma_2)^2. \qquad (1.85)$$

The nullclines for this system of equations are somewhat different from the Sel'kov system, being

$$\sigma_1 = \frac{\nu}{(1 + \sigma_2)^2} \qquad \left( \frac{d\sigma_1}{d\tau} = 0 \right), \qquad (1.86)$$

$$\sigma_1 = \frac{\eta \sigma_2}{(1 + \sigma_2)^2} \qquad \left( \frac{d\sigma_2}{d\tau} = 0 \right), \qquad (1.87)$$

and the unique steady-state solution is given by

$$\sigma_1 = \frac{v}{(1+\sigma_2)^2},$$ (1.88)

$$\sigma_2 = \frac{v}{\eta}.$$ (1.89)

The stability of the steady-state solution is again determined by the characteristic equation (1.76), and the sign of the real part of the eigenvalues is the same as the sign of

$$H = f_2 - f_1 - \eta = 2\sigma_1(1+\sigma_2)^2 - (1+\sigma_2) - \eta$$ (1.90)

evaluated at the steady state (1.86)–(1.87). Equation (1.90) can be written as the cubic polynomial

$$\frac{1}{\eta}y^3 - y + 2 = 0, \qquad y = 1 + \frac{v}{\eta}.$$ (1.91)

For $\eta$ sufficiently large, the polynomial (1.91) has two roots greater than 2, say, $y_1$ and $y_2$. Recall that $v$ is the nondimensional flow rate of substrate ATP. To make some correspondence with the experimental data, we assume that the flow rate $v$ is proportional to the experimental supply rate of glucose. This is not strictly correct, although ATP is produced at about the same rate that glucose is supplied. Accepting this caveat, we see that to match experimental data, we require

$$\frac{y_2 - 1}{y_1 - 1} = \frac{v_2}{v_1} = \frac{160}{20} = 8.$$ (1.92)

Requiring (1.91) to hold at $y_1$ and $y_2$ and requiring (1.92) to hold as well, we find numerical values

$$y_1 = 2.08, y_2 = 9.61, \eta = 116.7$$ (1.93)

corresponding to $v_1 = 126$ and $v_2 = 1005$.

At the Hopf bifurcation point, the period of oscillation is

$$T_i = \frac{2\pi}{\omega_i} = \frac{2\pi}{\sqrt{\eta}(1+\sigma_2)} = \frac{2\pi}{\sqrt{\eta}y_i}.$$ (1.94)

For the numbers (1.93), we obtain a ratio of periods $T_1/T_2 = 4.6$, which is acceptably close to the experimentally observed ratio $T_1/T_2 = 2.7$.

A typical phase portrait for the periodic solution that exists between the Hopf bifurcation points is shown in Fig. 1.8, and the concentrations of the two species are shown as functions of time in Fig. 1.9.

## 1.4   Appendix: Math Background

It is certain that some of the mathematical concepts and tools that we routinely invoke here are not familiar to all of our readers. In this first chapter alone, we have

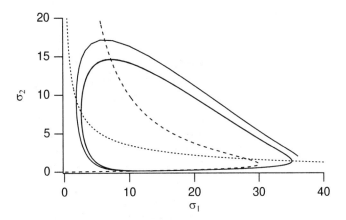

**Figure 1.8** Phase portrait of the Goldbeter–Lefever model with $\nu = 200$, $\eta = 120$. Dotted curve: $\frac{d\sigma_1}{d\tau} = 0$. Dashed curve: $\frac{d\sigma_2}{d\tau} = 0$.

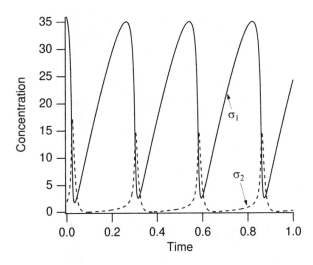

**Figure 1.9** Solution of the Goldbeter–Lefever model with $\nu = 200$, $\eta = 120$.

used nondimensionalization, phase-plane analysis, linear stability analysis, bifurcation theory, and asymptotic analysis, all the while assuming that these are familiar to the reader.

The purpose of this appendix is to give a brief guide to those techniques that are a basic part of the applied mathematician's toolbox but that may not be familiar to all our readers.

## 1.4.1 Basic Techniques

In any problem, there are a number of parameters that are dictated by the problem. However, it often happens that not all parameters are *independent*; that is, different variations in different parameters may lead to identical changes in the behavior of the model. Second, there may be parameters whose influence on a behavior is negligible and can be safely ignored for a given context.

The way to identify independent parameters and to determine their relative magnitudes is to nondimensionalize the problem. Unfortunately, having said that, we cannot describe a unique algorithm for nondimensionalization, because one does not exist; nondimensionalization is as much art as it is science.

There are, however, rules of thumb to apply. In any system of equations, there are a number of independent variables (time, space, etc.), dependent variables (concentrations, etc.) and parameters (rates of reaction, sizes of containers, etc.). Nondimensionalization begins by rescaling the independent and dependent variables by "typical" units, rendering them thereby dimensionless. One goal may be to ensure that the dimensionless variables remain of a fixed order of magnitude, not becoming too large or negligibly small. This usually requires some a priori knowledge about the solution, as it can be difficult to choose typical scales unless something is already known about typical solutions. Time and space scales can be vastly different depending on the context.

Once this selection of scales has been made, the governing equations are written in terms of the rescaled variables and dimensionless combinations of the remaining parameters are identified. The number of remaining free dimensionless parameters is usually less than the original number of physical parameters. The primary difficulty (at least to understand and apply the process) is that there is not necessarily a single way to scale and nondimensionalize the equations. Some scalings may highlight certain features of the solution, while other scalings may emphasize others. Nonetheless, nondimensionalization often provides a good starting point for the analysis of a model system.

An excellent discussion of scaling and nondimensionalization can be found in Lin and Segel (1988, Chapter 6). A great deal of more advanced work has also been done on this subject, particularly its application to the quasi-steady-state approximation, by Segel and his collaborators (Segel, 1988; Segel and Slemrod, 1989; Segel and Perelson, 1992; Segel and Goldbeter, 1994; Borghans et al., 1996; see also Frenzen and Maini, 1988).

Phase-plane analysis and linear stability analysis are standard fare in introductory courses on differential equations. A nice introduction to these topics for the biologically inclined can be found in Edelstein-Keshet (1988, Chapter 5) or Braun (1993, Chapter 4). A large number of books discuss the qualitative theory of differential equations, for example, Boyce and Diprima (1997), or at a more advanced level, Hale and Koçak (1991), or Hirsch and Smale (1974).

Bifurcation theory is a topic that is gradually finding its way into introductory literature. The most important terms to understand are those of *steady-state* bifurcations,

*Hopf* bifurcations, *homoclinic* bifurcations, and *saddle-node* bifurcations, all of which appear in this book. An excellent introduction to these concepts is found in Strogatz (1994, Chapters 3, 6, 7, 8). More advanced treatments include those in Guckenheimer and Holmes (1983), or Arnold (1983).

## 1.4.2 Asymptotic Analysis

Applied mathematicians love small parameters, because of the hope that the solution of a problem with a small parameter might be approximated by an *asymptotic representation*. A commonplace notation has emerged in which $\epsilon$ is often the small parameter. An asymptotic representation has a precise mathematical meaning. Suppose that $G(\epsilon)$ is claimed to be an asymptotic representation of $g(\epsilon)$, expressed as

$$g(\epsilon) = G(\epsilon) + O(\phi(\epsilon)). \tag{1.95}$$

The precise meaning of this statement is that there is a constant $A$ such that

$$\left| \frac{g(\epsilon) - G(\epsilon)}{\phi(\epsilon)} \right| \leq A \tag{1.96}$$

for all $\epsilon$ with $|\epsilon| \leq \epsilon_0$ and $\epsilon > 0$. The function $\phi(\epsilon)$ is called a *gauge function*, a typical example of which is a power of $\epsilon$.

### Perturbation Expansions

It is often the case that an asymptotic representation can be found as a development in powers of the small parameter $\epsilon$. Such representations are called *perturbation expansions*. Usually, a few terms of this power series representation suffice to give a good approximation to the solution. It should be kept in mind that under no circumstances does this power series development imply that a complete power series (with an infinite number of terms) exists or is convergent. Terminating the series at one or two terms is deliberate.

However, there are times when a full power series could be found and would be convergent in some nontrivial $\epsilon$ domain. Such problems are called *regular perturbation problems* because their solutions are regular, or analytic, in the parameter $\epsilon$.

There are numerous examples of regular perturbation problems, including all of those related to bifurcation theory. These problems are regular because their solutions can be developed in a convergent power series of some parameter.

There are, however, many problems with small parameters whose solutions are not regular, called *singular perturbation problems*. Singular perturbation problems are characterized by the fact that their dependence on the small parameter is not regular, but *singular*, and their convergence as a function of $\epsilon$ is not uniform.

Singular problems come in two basic varieties. Characteristic of the first type is a small region of width $\epsilon$ somewhere in the domain of interest (either space or time) in which the solution changes rapidly. For example, the solution of the boundary value problem

$$\epsilon u'' + u' + u = 0 \qquad (1.97)$$

subject to boundary conditions $u(0) = u(1) = 1$ is approximated by the asymptotic representation

$$u(x; \epsilon) = (1 - e)e^{-x/\epsilon} + e^{1-x} + O(\epsilon). \qquad (1.98)$$

Notice the nonuniform nature of this solution, as

$$e = \lim_{x \to 0^+} (\lim_{\epsilon \to 0^+} u(x; \epsilon)) \neq \lim_{\epsilon \to 0^+} (\lim_{x \to 0^+} u(x; \epsilon)) = 1. \qquad (1.99)$$

Here the term $e^{-x/\epsilon}$ is a *boundary layer correction*, as it is important only in a small region near the boundary at $x = 0$.

Other terms that are typical in singular perturbation problems are *interior layers* or *transition layers*, typified by expressions of the form $\tan(\frac{x-x_0}{\epsilon})$, and *corner layers*, locations where the derivative changes rapidly but the solution itself changes little. Transition layers are of great significance in the study of excitable systems (Chapter 4). While corner layers show up in this book, we do not study or use them in any detail.

Singular problems of this type can often be identified by the fact that the order of the system decreases if $\epsilon$ is set to zero. An example that we have already seen is the quasi-steady-state analysis used to simplify reaction schemes in which some reactions are significantly faster than others. Setting $\epsilon$ to zero in these examples reduces the order of the system of equations, signaling a possible problem. Indeed, solutions of these equations typically have *initial layers* near time $t = 0$. We take a closer look at this example below.

The second class of singular perturbation problems is that in which there are two scales in operation everywhere in the domain of interest. Problems of this type show up throughout this book. For example, action potential propagation in cardiac tissue is through a cellular medium whose detailed structure varies rapidly compared to the length scale of the action potential wave front. Physical properties of the cochlear membrane in the inner ear vary slowly compared to the wavelength of waves that propagate along it. For problems of this type, one must make explicit the dependence on multiple scales, and so solutions are often expressed as functions of two variables, say $x$ and $x/\epsilon$, which are treated as independent variables. Solution techniques that exploit the multiple-scale nature of the solution are called *multiscale methods* or *averaging methods*.

Detailed discussions of these asymptotic methods may be found in Murray (1984), Kevorkian and Cole (1996), and Holmes (1995).

### 1.4.3  Enzyme Kinetics and Singular Perturbation Theory

In most of the examples of enzyme kinetics discussed in this chapter, extensive use was made of the quasi-steady-state approximation (1.19), according to which the concentration of the complex remains constant during the course of the reaction. Although this assumption gives the right answers (which, some might argue, is justification enough), mathematicians have sought for ways to justify this approximation rigorously. Bowen et al. (1963) and Heineken et al. (1967) were the first to show that the quasi-steady-state approximation can be derived as the lowest-order term in an asymptotic expansion of the solution. This has since become one of the standard examples of the application of singular perturbation theory to biological systems, and it is discussed in detail by Rubinow (1973), Lin and Segel (1988), and Murray (1989), among others.

First, note that the quasi-steady-state assumption cannot be correct for all times if one starts the reaction from arbitrary initial concentrations. This is apparent with the Michaelis–Menten kinetics, for example, because the single first-order differential equation (1.18) describes the rate of conversion of substrate into product, but for this to be valid the quasi-steady-state approximation is assumed to hold. There must therefore be a brief period of time at the start of the reaction during which the quasi-steady-state equilibrium does not hold. During this initial time period the enzyme is "filling up" with substrate, until the concentration of complexed enzyme reaches the value given by the quasi-steady-state approximation. Since there is little enzyme compared to the total amount of substrate, the concentration of substrate remains essentially constant during this period.

For most biochemical reactions this transition to the quasi-steady-state happens so fast that it is not physiologically important, but for mathematical reasons, it is interesting to understand these kinetics for early times as well. To see how the reaction runs for early times from arbitrary initial conditions, we make a change of time scale, $\eta = \tau/\epsilon$. This change of variables expands the time scale on which we look at the reaction and allows us to study events that happen on a fast time scale. In the new time scale, the dimensionless reaction equations (1.14)–(1.15) become

$$\frac{d\sigma}{d\eta} = \epsilon(-\sigma + x(\sigma + \alpha)), \tag{1.100}$$

$$\frac{dx}{d\eta} = \sigma - x(\sigma + \kappa). \tag{1.101}$$

For small $\epsilon$, these equations are well approximated by the simplification

$$\frac{d\sigma}{d\eta} = 0, \tag{1.102}$$

$$\frac{dx}{d\eta} = \sigma - x(\sigma + \kappa). \tag{1.103}$$

Simply stated, this means that on this time scale the variable $\sigma$ does not change, so that $\sigma = 1$. Furthermore, we can solve for $x$ as

$$x = \frac{1}{1 + \kappa} + \left( x_0 - \frac{1}{1 + \kappa} \right) e^{-(1+\kappa)\eta}. \tag{1.104}$$

If $[E] = e_0$ at time $t = 0$, then $x_0 = 0$. In terms of the original time variable $\tau$, this solution is

$$x(\tau) = \frac{1}{1 + \kappa}(1 - e^{-(1+\kappa)\frac{\tau}{\epsilon}}),\qquad(1.105)$$

and it is valid only for times of order $\epsilon$. The exponential term is significant only when $\tau$ is small of order $\epsilon$. Thus, this simple analysis shows that if the reaction is started from arbitrary initial conditions, there is first a time span during which the enzyme products rapidly equilibrate, consuming little substrate, and after this initial "layer" the reaction proceeds according to Michaelis–Menten kinetics along the quasi-steady-state curve.

In this problem the analysis of the initial layer is relatively easy and not particularly revealing. However, this type of analysis will be of much greater importance later when we discuss the behavior of excitable systems.

## 1.5 EXERCISES

1. Consider an enzymatic reaction in which an enzyme can be activated or inactivated by the same chemical substance, as follows:

$$E + X \underset{k_{-1}}{\overset{k_1}{\rightleftharpoons}} E_1,\qquad(1.106)$$

$$E_1 + X \underset{k_{-2}}{\overset{k_2}{\rightleftharpoons}} E_2,\qquad(1.107)$$

$$E_1 + S \overset{k_3}{\longrightarrow} P + Q + E.\qquad(1.108)$$

   Suppose further that X is supplied at a constant rate and removed at a rate proportional to its concentration. Use quasi-steady-state analysis to find the nondimensional equation describing the degradation of X,

$$\frac{dx}{dt} = \gamma - x - \frac{\beta xy}{1 + x + y + \frac{\alpha}{\delta}x^2}.\qquad(1.109)$$

   Identify all the parameters and variables.

2. Using the quasi-steady-state approximation, show that the velocity of the reaction for an enzyme with an allosteric inhibitor (Section 1.2.3) is given by

$$V = \left(\frac{V_{max}K_3}{i + K_3}\right)\left(\frac{s(k_{-1} + k_3 i + k_1 s + k_{-3})}{k_1(s + K_1)^2 + (s + K_1)(k_3 i + k_{-3} + k_2) + k_2 k_{-3}/k_1}\right),\qquad(1.110)$$

   where $K_3 = k_{-3}/k_3$ and $K_1 = k_{-1}/k_1$.

3. (a) Derive the expression (1.51) for the fraction of occupied sites in a Monod–Wyman–Changeux model with $n$ binding sites.

   (b) The principle of detailed balance says that in order for a system of reactions to be in thermal equilibrium (as opposed to merely at a steady state), each individual reaction must be at equilibrium. Thus, for example, for the cycle of reactions

$$A \underset{k_{-1}}{\overset{k_1}{\rightleftharpoons}} B \underset{k_{-2}}{\overset{k_2}{\rightleftharpoons}} C \underset{k_{-3}}{\overset{k_3}{\rightleftharpoons}} A\qquad(1.111)$$

to be in thermal equilibrium, we must have $[A] = K_1[B]$, $[B] = K_2[C]$, and $[C] = K_3[A]$, where $K_i = k_{-i}/k_i$, $i = 1, 2, 3$. Hence, for this reaction cycle, detailed balance implies the relation $k_1 k_2 k_3 = k_{-1} k_{-2} k_{-3}$.

Modify the Monod–Wyman–Changeux model shown in Fig. 1.3 to include transitions between states $R_1$ and $T_1$, and between states $R_2$ and $T_2$. Use the principle of detailed balance to derive an expression for the equilibrium constant of each of these transitions. Do these transitions change the expression for $Y$, the fraction of occupied sites?

4. Suppose that a substrate can be broken down by two different enzymes with different kinetics. (This happens, for example, in the case of cAMP or cGMP, which can be hydrolyzed by two different forms of phosphodiesterase—see Chapter 22).

   (a) Write down the reaction scheme and differential equations, and nondimensionalize, to get a system of equations of the form

   $$\frac{d\sigma}{dt} = -\sigma(1 + \alpha) + x(\sigma + \mu_1) + \alpha y(\sigma + \mu_2), \tag{1.112}$$

   $$\epsilon \frac{dx}{dt} = \sigma(1 - x) - \lambda_1 x, \tag{1.113}$$

   $$\epsilon \frac{dy}{dt} = \alpha[\sigma(1 - y) - \lambda_2 y], \tag{1.114}$$

   where $x$ and $y$ are the nondimensionalized concentrations of the two complexes and $\alpha, \mu_1, \mu_2, \lambda_1$, and $\lambda_2$ are positive constants.

   (b) Apply the quasi-steady-state approximation to solve for $\sigma$, $x$, and $y$.

   (c) Rescale time and find the inner solution to lowest order in $\epsilon$. Construct a solution that is valid for all times.

5. For the system in the previous question, show that the solution can never leave the first quadrant $\sigma, x, y \geq 0$. By showing that $\sigma + \epsilon x + \epsilon y$ is always decreasing, show that the solution approaches the origin for large time.

6. For some enzyme reactions (for example, the hydrolysis of cAMP by phosphodiesterase in vertebrate retinal cones) the enzyme is present in large quantities, so that $e_0/s_0$ is not a small number. Show that when this is the case, the quasi-steady-state approximation may still be used, provided that $k_{-1}/k_1$ is much larger than $e_0$ or $s_0$, in which case the concentration of the complex is small compared to the total amount of substrate (Segel, 1988; Frenzen and Maini, 1988; Segel and Slemrod, 1989; Sneyd and Tranchina, 1989).

7. ATP is known to inhibit its own dephosphorylation. One possible way for this to occur is if ATP binds with the enzyme, holding it in an inactive state, via

$$S_1 + E \underset{k_{-5}}{\overset{k_5}{\rightleftharpoons}} S_1 E.$$

Add this reaction to the Sel'kov model for glycolysis and derive the equations governing glycolysis of the form (1.68)–(1.69). Explain from the model why this additional reaction is inhibitory.

8. The following reaction scheme is a simplified version of the Goldbeter–Lefever reaction scheme:

$$R_0 \underset{k_{-1}}{\overset{k_1}{\rightleftharpoons}} T_0,$$

$$S_1 + R_j \underset{k_{-2}}{\overset{k_2}{\rightleftharpoons}} C_j \overset{k}{\longrightarrow} R_j + S_2, \qquad j = 0, 1, 2,$$

$$S_2 + R_0 \underset{k_{-3}}{\overset{2k_3}{\rightleftharpoons}} R_1, \qquad S_2 + R_1 \underset{2k_{-3}}{\overset{k_3}{\rightleftharpoons}} R_2.$$

Show that, under appropriate assumptions about the ratios $k_1/k_{-1}$ and $\frac{k_{-2}+k_3}{k_2}$ the equations describing this reaction are of the form (1.68)–(1.69) with $f(\sigma_1, \sigma_2)$ given by (1.85).

9. Use the law of mass action and the quasi-steady-state assumption for the enzymatic reactions to derive a system of equations of the form (1.68)–(1.69) for the Goldbeter–Lefever model. Verify (1.84).

10. When much of the ATP is depleted in a cell, a considerable amount of cAMP is formed as a product of ATP degradation. This cAMP activates an enzyme phophorylase that splits glycogen, releasing glucose that is rapidly metabolized, replenishing the ATP supply. Devise a model for this control loop and determine conditions under which the production of ATP is oscillatory.

11. By looking for solutions to (1.14) and (1.15) of the form

$$\sigma = \sigma_0 + \epsilon\sigma_1 + \epsilon^2\sigma_2 + \cdots, \tag{1.115}$$

$$x = x_0 + \epsilon x_1 + \epsilon^2 x_2 + \cdots, \tag{1.116}$$

show that $\sigma_0$ and $x_0$ satisfy the quasi-steady-state approximation. Thus, the quasi-steady-state approximation is the lowest-order term in an asymptotic expansion for the solution. Typical initial conditions are $\sigma = 1$, $x = 0$. Does the lowest-order solution satisfy the initial conditions? Find $\sigma_1$ and $x_1$ and plot the solution to first order in $\epsilon$. The variables $\sigma$ and $x$ are called the *outer solution*, as they are valid for times outside some boundary layer around $\tau = 0$. Now rescale time by $\epsilon$, and use the same procedure to construct an asymptotic solution to (1.100) and (1.101), the so-called *inner solution*. Show that the inner solution satisfies the initial conditions. How can one construct a solution that satisfies the initial conditions and is valid for all times?

# CHAPTER 2

# Cellular Homeostasis

## 2.1 The Cell Membrane

The cell membrane provides a boundary separating the internal workings of the cell from its external environment. More importantly, it is selectively permeable, permitting the free passage of some materials and restricting the passage of others, thus regulating the passage of materials into and out of the cell. It consists of a double layer (a *bilayer*) of phospholipid molecules about 7.5 nm (=75 Å) thick (Fig. 2.1). The term *lipid* is used to specify a category of water-insoluble, energy rich macromolecules, typical of fats, waxes, and oils. Irregularly dispersed throughout the phospholipid bilayer are aggregates of globular proteins, which are apparently free to move within the layer, giving the membrane a fluid-like appearance. The membrane also contains water-filled pores with diameters of about 0.8 nm, as well as protein-lined pores, called *channels*, which allow passage of specific molecules. Both the intracellular and extracellular environments consist of, among many other things, a dilute aqueous solution of dissolved salts, primarily NaCl and KCl, which dissociate into $Na^+$, $K^+$, and $Cl^-$ ions. The cell membrane acts as a barrier to the free flow of these ions and maintains concentration differences of these ions. In addition, the cell membrane acts as a barrier to the flow of water.

Molecules can be transported across the cell membrane by passive or active processes. An active process is one that requires the expenditure of energy, while a passive process results solely from the inherent, random movement of molecules. There are three passive transport mechanisms to transport molecules through the cell membrane. *Osmosis* is the process by which water is transported through the cell membrane. Simple diffusion accounts for the passage of small molecules through pores and of lipid-

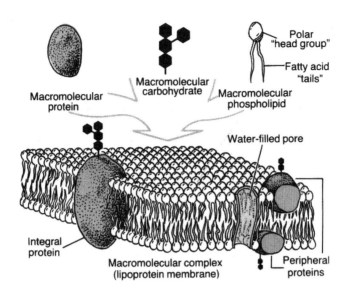

**Figure 2.1**   Schematic diagram of the cell membrane. (Davis et al., 1985, Fig. 3-1, p. 41.)

soluble molecules through the bilipid layer. For example, water, urea (a nitrogenous waste product of metabolism), and hydrated chloride ions diffuse through membrane pores. Oxygen and carbon dioxide diffuse through the membrane readily because they are soluble in lipids. Sodium and potassium ions pass through ion-specific channels, driven by diffusion and electrical forces. Some other mechanism must account for the transport of larger sugar molecules such as galactose, glucose, and sucrose, as they are too large to pass through membrane pores (Fig. 2.2). *Carrier-mediated diffusion* occurs when a molecule is bound to a carrier molecule that moves readily through the membrane. For example, the transport of glucose and amino acids across the cell membrane is believed to be by a carrier-mediated process.

Concentration differences are set up and maintained by active mechanisms that use energy to pump ions against their concentration gradient. One of the most important of these pumps is the $Na^+$–$K^+$ pump, which uses the energy stored in ATP molecules to pump $Na^+$ out of the cell and $K^+$ in. Another pump, the $Ca^{2+}$ ATPase, pumps $Ca^{2+}$ out of the cell or into the endoplasmic reticulum. There are also a variety of exchange pumps that use the energy inherent in the concentration gradient of one ion type to pump another ion type against its concentration gradient. For example, the $Na^+$–$Ca^{2+}$ exchanger removes $Ca^{2+}$ from the cell at the expense of $Na^+$ entry, and similarly for the $Na^+$–$H^+$ exchanger. Typical values for intracellular and extracellular ionic concentrations are given in Table 2.1.

Differences in ionic concentrations create a potential difference across the cell membrane that drives ionic currents. Water is also absorbed into the cell because of concentration differences of these ions and also because of other large molecules contained in the cell, whose presence provides an osmotic pressure for the absorption

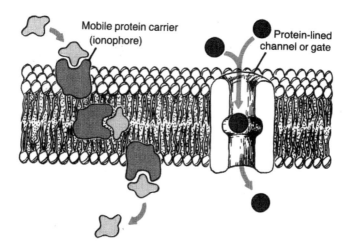

**Figure 2.2** Schematic diagram of the cell membrane containing a protein carrier and a protein-lined ionic channel. (Davis et al., 1985, Fig. 3-7, p. 45.)

**Table 2.1** Typical values for intracellular and extracellular ionic concentrations, from three different cell types. Concentrations are given in units of mM, and potentials are in units of mV. Extracellular concentrations for the squid giant axon are for seawater, while those for frog muscle and red blood cells are for plasma. Later in this chapter we discuss Nernst potentials and resting potentials. (Adapted from Mountcastle, 1974, Table 1-1.)

| | Squid Giant Axon | Frog Sartorius Muscle | Human Red Blood Cell |
|---|---|---|---|
| Intracellular concentrations | | | |
| $Na^+$ | 50 | 13 | 19 |
| $K^+$ | 397 | 138 | 136 |
| $Cl^-$ | 40 | 3 | 78 |
| $Mg^{2+}$ | 80 | 14 | 5.5 |
| Extracellular concentrations | | | |
| $Na^+$ | 437 | 110 | 155 |
| $K^+$ | 20 | 2.5 | 5 |
| $Cl^-$ | 556 | 90 | 112 |
| $Mg^{2+}$ | 53 | 1 | 2.2 |
| Nernst potentials | | | |
| $V_{Na}$ | +56 | +55 | +55 |
| $V_K$ | −77 | −101 | −86 |
| $V_{Cl}$ | −68 | −86 | −9 |
| Resting potentials | −65 | −99 | −6 to −10 |

of water. It is the balance of these forces that regulates both the cell volume and the
membrane potential.

## 2.2   Diffusion

To keep track of a chemical concentration or any other measurable entity, we must
track where it comes from and where it goes; that is, we must write a *conservation law*.
If $u$ is the amount of some chemical species, then the appropriate conservation law
takes the following form (in words):

**rate of change of $u$=local production of $u$ + accumulation of $u$ due to transport.**

If $\Omega$ is a region of space, then this conservation law can be written symbolically as

$$\frac{d}{dt} \int_{\Omega} u \, dV = \int_{\Omega} f \, dV - \int_{\partial\Omega} \mathbf{J} \cdot \mathbf{n} \, dA, \tag{2.1}$$

where $\partial\Omega$ is the boundary of the region $\Omega$, $\mathbf{n}$ is the outward unit normal to the boundary
of $\Omega$, $f$ represents the local production of $u$ per unit volume, and $\mathbf{J}$ is the flux of $u$.
According to the divergence theorem, if $\mathbf{J}$ is sufficiently smooth, then

$$\int_{\partial\Omega} \mathbf{J} \cdot \mathbf{n} \, dA = \int_{\Omega} \nabla \cdot \mathbf{J} \, dV, \tag{2.2}$$

so that if the volume in which $u$ is being measured is fixed but arbitrary, the integrals
can be dropped, with the result that

$$\frac{\partial u}{\partial t} = f - \nabla \cdot \mathbf{J}. \tag{2.3}$$

This, being a conservation law, is inviolable. However, there are many ways in which
the production term $f$ and the flux $\mathbf{J}$ can vary. Indeed, much of our study here is involved
in determining appropriate models for production and flux.

### 2.2.1   Fick's Law

The simplest description of the flux of a chemical species is

$$\mathbf{J} = -D\nabla u. \tag{2.4}$$

Equation (2.4) is called a *constitutive relationship*, and for chemical species it is called
*Fick's law*. The scalar $D$ is the *diffusion coefficient* and is characteristic of the solute
and the fluid in which it is dissolved. If $u$ represents the heat content of the volume,
(2.4) is called *Newton's law of cooling*. Fick's law is not really a law, but is a reasonable
approximation to reality if the concentration of the chemical species is not too high.
When Fick's law applies, the conservation equation becomes the reaction–diffusion
equation

$$\frac{\partial u}{\partial t} = \nabla \cdot (D\nabla u) + f, \tag{2.5}$$

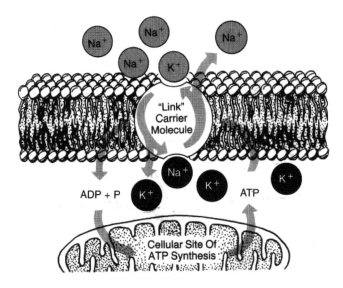

**Figure 2.3** Schematic diagram of the cell membrane containing a $Na^+ - K^+$ pump. (Davis, et al., 1985, Fig. 3-11, p. 49.)

or, if $D$ is a constant,

$$\frac{\partial u}{\partial t} = D\nabla^2 u + f. \tag{2.6}$$

## 2.2.2  Diffusion Coefficients

A quantitative understanding of diffusion was given by Einstein (1906) in his theory of Brownian motion. He showed that if a spherical solute molecule is large compared to the solvent molecule, then

$$D = \frac{kT}{6\pi\mu a}, \tag{2.7}$$

where $k$ is Boltzmann's constant, $T$ is the absolute temperature of the solution, $\mu$ is the coefficient of viscosity for the solute, and $a$ is the radius of the solute molecule. For nonspherical molecules, Einstein's formula generalizes to

$$D = \frac{kT}{f}, \tag{2.8}$$

where $f$ is the Stokes frictional coefficient of the particle and $f = 6\pi\mu a$ for a sphere. The molecular weight of a spherical molecule is

$$M = \frac{4}{3}\pi a^3 \rho, \tag{2.9}$$

where $\rho$ is the molecular density, so that in terms of molecular weight,

$$D = \frac{kT}{3\mu}\left(\frac{\rho}{6\pi^2 M}\right)^{1/3}. \tag{2.10}$$

The density of most large protein molecules is nearly constant (about $1.3 - 1.4$ g/cm$^3$), so that $DM^{1/3}$ is nearly the same for spherical molecules at a fixed temperature. The diffusion of small molecules, such as the respiratory gases, is different, being proportional to $M^{-1/2}$.

### 2.2.3   Diffusion Through a Membrane: Ohm's Law

We can use Fick's law to derive the chemical analogue of Ohm's law for a membrane of thickness $L$. Suppose that a membrane separates two large reservoirs of a dilute chemical, with concentration $c_l$ on the left (at $x = 0$), and concentration $c_r$ on the right (at $x = L$). According to the diffusion equation, in the membrane (assuming that the only gradients are transverse to the membrane)

$$\frac{\partial c}{\partial t} = D\frac{\partial^2 c}{\partial x^2}, \tag{2.11}$$

subject to boundary conditions $c(0,t) = c_l, c(L,t) = c_r$.

The full time-dependent solution can be found using separation of variables, but for our purposes here, the steady-state solution is sufficient. At steady state, $\frac{\partial c}{\partial t} = 0$, so that $\frac{\partial J}{\partial x} = -D\frac{\partial^2 c}{\partial x^2} = 0$, from which it follows that $J = -D\frac{\partial c}{\partial x} = $ constant, or that $c(x) = ax + b$, for some constants $a$ and $b$. Applying the boundary conditions, we find

$$c(x) = c_l + (c_r - c_l)\frac{x}{L}. \tag{2.12}$$

From Fick's law it follows that the flux of chemical is constant, independent of $x$, and is given by

$$J = \frac{D}{L}(c_l - c_r). \tag{2.13}$$

Note that a flux from left to right is counted as a positive flux. The ratio $L/D$ is the effective "resistance" (per unit area) of the membrane, and so $D/L$ is called the *conductance*, or *permeability*, per unit area.

## 2.3   Facilitated Diffusion

It is often the case that reactants in an enzymatic reaction (as in Chapter 1) are free to diffuse, so that one must keep track of the effects of both diffusion and reaction. Such problems, called *reaction–diffusion systems*, are of fundamental significance in physiology and are also important and difficult mathematically.

An important example in which both diffusion and reaction play a role is known as *facilitated diffusion*. Facilitated diffusion occurs when the flux of a chemical is amplified by a reaction that takes place in the diffusing medium. An example of facilitated diffusion occurs with the flux of oxygen in muscle fibers. In muscle fibers, oxygen is bound to myoglobin and is transported as oxymyoglobin, and this transport is greatly enhanced above the flow of oxygen in the absence of myoglobin.

This well-documented observation needs further explanation, because at first glance it seems counterintuitive. Myoglobin molecules are much larger (molecular weight M=16,890) than oxygen molecules (molecular weight M=32) and therefore have a much smaller diffusion coefficient ($D = 4.4 \times 10^{-7}$ and $D = 1.2 \times 10^{-5} \text{cm}^2/\text{s}$ for myoglobin and oxygen, respectively). The diffusion of oxymyoglobin would therefore seem to be much slower than the diffusion of free oxygen.

A simple model of this phenomenon is as follows. Suppose we have a slab reactor containing diffusing myoglobin. On the left (at $x = 0$) the oxygen concentration is held fixed at $s_0$, and on the right (at $x = L$) it is held at $s_L$, which is assumed to be less than $s_0$.

If $f$ is the rate of uptake of oxygen into oxymyoglobin, then equations governing the concentrations of $s = [O_2], e = [Mb], c = [MbO_2]$ are

$$\frac{\partial s}{\partial t} = D_s \frac{\partial^2 s}{\partial x^2} - f, \tag{2.14}$$

$$\frac{\partial e}{\partial t} = D_e \frac{\partial^2 e}{\partial x^2} - f, \tag{2.15}$$

$$\frac{\partial c}{\partial t} = D_c \frac{\partial^2 c}{\partial x^2} + f. \tag{2.16}$$

It is reasonable to take $D_e = D_c$, since myoglobin and oxymyoglobin are nearly identical in molecular weight and structure. Since myoglobin and oxymyoglobin remain inside the slab, it is also reasonable to specify the boundary conditions $\partial e/\partial x = \partial c/\partial x = 0$ at $x = 0$ and $x = L$. Because it reproduces the oxygen saturation curve (discussed in Chapter 16), we assume that the reaction of oxygen with myoglobin is governed by the elementary reaction

$$O_2 + Mb \underset{k_-}{\overset{k_+}{\rightleftharpoons}} MbO_2,$$

so that (from the law of mass action) $f = -k_- c + k_+ se$. The total amount of myoglobin is conserved by the reaction, so that at steady state $e + c = e_0$ and (2.15) is superfluous.

At steady state,

$$0 = s_t + c_t = D_s s_{xx} + D_c c_{xx}, \tag{2.17}$$

and thus there is a second conserved quantity, namely

$$D_s \frac{ds}{dx} + D_c \frac{dc}{dx} = -J, \tag{2.18}$$

which follows by integrating (2.17) once with respect to $x$. The constant $J$ (which is yet to be determined) is the sum of the flux of free oxygen and the flux of oxygen in the complex oxymyoglobin, and therefore represents the total flux of oxygen. Integrating (2.18) with respect to $x$ between $x = 0$ and $x = L$, we can express the total flux $J$ in terms of boundary values of the two concentrations as

$$J = \frac{D_s}{L}(s_0 - s_L) + \frac{D_c}{L}(c_0 - c_L), \tag{2.19}$$

although the values $c_0$ and $c_L$ are as yet unknown.

To further understand this system of equations, we introduce dimensionless variables, $\sigma = \frac{k_+}{k_-}s$, $u = c/e_0$, and $x = Ly$, in terms of which (2.14) and (2.16) become

$$\epsilon_1 \sigma_{yy} = \sigma(1 - u) - u = -\epsilon_2 u_{yy}, \tag{2.20}$$

where $\epsilon_1 = \frac{D_s}{e_0 k_+ L^2}$, $\epsilon_2 = \frac{D_c}{k_- L^2}$.

Reasonable numbers for the uptake of oxygen by myoglobin (Wittenberg, 1966) are $k_+ = 1.4 \times 10^{10} \text{cm}^3 \text{M}^{-1} \text{s}^{-1}$, $k_- = 11 \text{ s}^{-1}$, and $L = 0.022$ cm in a solution with $e_0 = 1.2 \times 10^{-5} \text{ M/cm}^3$. (These numbers are for an experimental setup in which the concentration of myoglobin was substantially higher than what naturally occurs in living tissue.) With these numbers we estimate that $\epsilon_1 = 1.5 \times 10^{-7}$, and $\epsilon_2 = 8.2 \times 10^{-5}$. Clearly, both of these numbers are small, suggesting that oxygen and myoglobin are at quasi-steady state throughout the medium, with

$$c = e_0 \frac{s}{K + s}, \tag{2.21}$$

where $K = k_-/k_+$. Now we substitute (2.21) into (2.19) to find the flux

$$J = \frac{D_s}{L}(s_0 - s_L) + \frac{D_c}{L} e_0 \left( \frac{s_0}{K + s_0} - \frac{s_L}{K + s_L} \right) \tag{2.22}$$

$$= \frac{D_s}{L}(s_0 - s_L) \left( 1 + \frac{D_c}{D_s} \frac{e_0 K}{(s_0 + K)(s_L + K)} \right) \tag{2.23}$$

$$= \frac{D_s}{L}(1 + \mu\rho)(s_0 - s_L), \tag{2.24}$$

where $\rho = \frac{D_c}{D_s} \frac{e_0}{K}$, $\mu = \frac{K^2}{(s_0 + K)(s_L + K)}$.

In terms of dimensionless variables the full solution is given by

$$\sigma(y) + \rho u(y) = y[\sigma(1) + \rho u(1)] + (1 - y)[\sigma(0) + \rho u(0)], \tag{2.25}$$

$$u(y) = \frac{\sigma(y)}{1 + \sigma(y)}. \tag{2.26}$$

Now we see how diffusion can be facilitated by an enzymatic reaction. In the absence of a diffusing carrier, $\rho = 0$ and the flux is purely Fickian, as in (2.13). However, in the presence of carrier, diffusion is enhanced by the factor $\mu\rho$. The maximum enhancement possible is at zero concentration, when $\mu = 1$. With the above numbers for myoglobin, this maximum enhancement is substantial, being $\rho = 560$. If the oxygen supply is sufficiently high on the left side (near $x = 0$), then oxygen is stored as oxymyoglobin. Moving to the right, as the total oxygen content drops, oxygen is released by the myoglobin. Thus, even though the bound oxygen diffuses slowly compared to free oxygen, the quantity of bound oxygen is high (provided that $e_0$ is large compared to the half saturation level $K$), so that lots of oxygen is transported. We can also understand that to take advantage of the myoglobin-bound oxygen, the concentration of oxygen must drop to sufficiently low levels so that myoglobin releases its stored oxygen.

To explain it another way, note from (2.22) that $J$ is the sum of two terms, the usual ohmic flux term and an additional term that depends on the diffusion coefficient of $MbO_2$. The total oxygen flux is the sum of the flux of free oxygen and the flux of oxygen bound to myoglobin. Clearly, if myoglobin is free to diffuse, the total flux is thereby increased.

In Fig. 2.4A are shown the dimensionless free oxygen concentration $\sigma$ and the dimensionless bound oxygen concentration $u$ plotted as functions of position. Notice that the free oxygen content falls at first, indicating higher free oxygen flux, and the bound oxygen decreases more rapidly at larger $y$. Perhaps easier to interpret is Fig. 2.4B, where the dimensionless flux of free oxygen and the dimensionless flux of bound oxygen are shown as functions of position. Here we can see that as the free oxygen concentration drops, the flux of free oxygen also drops, but the flux of bound oxygen increases. For large $y$, most of the flux is due to the bound oxygen. For these figures, $\rho = 10, \sigma(0) = 2.0, \sigma(1) = 0.1$.

One mathematical detail that was ignored in this discussion is the validity of the quasi-steady-state solution (2.21) as an approximation of (2.20). Usually, when one makes an approximation to boundary value problems in which the order of the system is reduced (as here where the order is four, and drops by two when $\epsilon_1$ and $\epsilon_2$ are ignored), there are difficulties with the solution at the boundary, because the boundary conditions cannot, in general, be met. Such problems, discussed briefly in Chapter 1

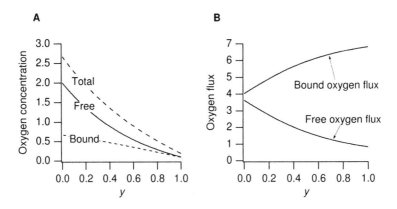

**Figure 2.4**  A: Free oxygen content $\sigma(y)$ and bound oxygen content $u(y)$ as a function of $y$. B: Free oxygen flux $-\sigma'(y)$ and bound oxygen flux $-\rho u'(y)$ plotted as a function of $y$.

in the context of enzyme kinetics, are called *singular perturbation problems*, because the behavior of the solutions as functions of the small parameters is not regular, but singular (certain derivatives become infinitely large as the parameters approach zero). In this problem, however, there are no boundary layers, and the quasi-steady-state solution is a uniformly valid approximation to the solution. This occurs because the boundary conditions on $c$ are of no-flux (Neumann) type, rather than of fixed (Dirichlet) type. That is, since the value of $c$ is not specified by the boundary conditions, $c$ is readily adjusted so that there are no boundary layers. Only a slight correction to the quasi-steady-state solution is needed to meet the no-flux boundary conditions, but this correction affects only the derivative, not the value, of $c$ in a small region near the boundaries.

## 2.3.1  Facilitated Diffusion in Muscle Respiration

Even at rest, muscle fibers consume oxygen. This is because ATP is constantly consumed to maintain a nonzero membrane potential across a muscle cell wall, and this consumption of energy requires the constant metabolizing of sugar, which consumes oxygen. Although sugar can be metabolized anaerobically, the waste product of this reaction is lactic acid, which is toxic to the cell. In humans, the oxygen consumption of live muscle tissue at rest is about $5 \times 10^{-8}$ mol/cm$^3$s, and the concentration of myoglobin is about $2.8 \times 10^{-7}$ mol/cm$^3$. Thus, when myoglobin is fully saturated, it contains only about a 5 s supply of oxygen. Further, the oxygen at the exterior of the muscle cell must penetrate to the center of the cell to prevent the oxygen concentration at the center falling to zero, a condition called *oxygen debt*.

To explain how myoglobin aids in providing oxygen to a muscle cell and helps to prevent oxygen debt, we examine a model of oxygen consumption that includes the effects of diffusion of oxygen and myoglobin. We suppose that a muscle fiber is a long circular cylinder (radius $a = 2.5 \times 10^{-3}$ cm) and that diffusion takes place only in the radial direction. We suppose that the oxygen concentration at the boundary of the fiber is a fixed constant and that the distribution of chemical species is radially symmetric. With these assumptions, the steady-state equations governing the diffusion of oxygen and oxymyoglobin are

$$D_s \frac{1}{r} \frac{d}{dr} \left( r \frac{ds}{dr} \right) - f - g = 0, \tag{2.27}$$

$$D_c \frac{1}{r} \frac{d}{dr} \left( r \frac{dc}{dr} \right) + f = 0, \tag{2.28}$$

where, as before, $s = [O_2]$, $c = [MbO_2]$, and $f = -k_- c + k_+ se$. The coordinate $r$ is in the radial direction. The new term in these equations is the constant $g$, corresponding to the constant consumption of oxygen. The boundary conditions are $s = s_a$, $dc/dr = 0$ at $r = a$, and $ds/dr = dc/dr = 0$ at $r = 0$. For muscle, $s_a$ is typically $3.5 \times 10^{-8}$ mol/cm$^3$ (corresponding to the partial pressure 20 mm Hg). Numerical values for the parameters in this model are difficult to obtain, but reasonable numbers are $D_s = 10^{-5} \text{ cm}^2/\text{s}, D_c = 5 \times 10^{-7} \text{ cm}^2/\text{s}, k_+ = 2.4 \times 10^{10} \text{ cm}^3/\text{mol} \cdot \text{s}$, and $k_- = 65/\text{s}$ (Wyman, 1966).

Introducing nondimensional variables $\sigma = \frac{k_+}{k_-}s, u = c/e_0$, and $r = ay$, we obtain the differential equations

$$\epsilon_1 \frac{1}{y}\frac{d}{dy}\left(y\frac{d\sigma}{dy}\right) - \gamma = \sigma(1-u) - u = -\epsilon_2 \frac{1}{y}\frac{d}{dy}\left(y\frac{du}{dy}\right), \tag{2.29}$$

where $\epsilon_1 = \frac{D_s}{e_0 k_+ a^2}, \epsilon_2 = \frac{D_c}{k_- a^2}, \gamma = g/k_-$. Using the parameters appropriate for muscle, we estimate that $\epsilon_1 = 2.3 \times 10^{-4}, \epsilon_2 = 1.2 \times 10^{-3}, \gamma = 3.3 \times 10^{-3}$. While these numbers are not as small as for the experimental slab described earlier, they are still small enough to warrant the approximation that the quasi-steady state (2.21) holds in the interior of the muscle fiber.

It also follows from (2.29) that

$$\epsilon_1 \frac{1}{y}\frac{d}{dy}\left(y\frac{d\sigma}{dy}\right) + \epsilon_2 \frac{1}{y}\frac{d}{dy}\left(y\frac{du}{dy}\right) = \gamma. \tag{2.30}$$

We integrate (2.30) twice with respect to $y$ to find

$$\epsilon_1 \sigma + \epsilon_2 u = A \ln y + B + \frac{\gamma}{4}y^2. \tag{2.31}$$

The constants $A$ and $B$ are determined by boundary conditions. Since we want the solution to be bounded at the origin, $A = 0$, and $B$ is related to the concentration at the origin.

Now suppose that there is just enough oxygen at the boundary to prevent oxygen debt. In this model, oxygen debt occurs if $\sigma$ falls to zero. Marginal oxygen debt occurs if $\sigma = u = 0$ at $y = 0$. For this boundary condition, we take $A = B = 0$. Then the concentration at the boundary must be at least as large as $\sigma_0$, where, using the quasi-steady state $\sigma(1 - u) = u$,

$$\sigma_0 + \rho \frac{\sigma_0}{\sigma_0 + 1} = \frac{\gamma}{4\epsilon_1}, \tag{2.32}$$

and where $\rho = \epsilon_2/\epsilon_1$. Otherwise, the center of the muscle is in oxygen debt. Note also that $\sigma_0$ is a decreasing function of $\rho$, indicating a reduced need for external oxygen because of facilitated diffusion.

A plot of this critical concentration $\sigma_0$ as a function of the scaled consumption $\frac{\gamma}{4\epsilon_1}$ is shown in Fig. 2.5. For this plot $\rho = 5$, which is a reasonable estimate for muscle. The dashed curve is the critical concentration when there is no facilitated diffusion ($\rho = 0$). The easy lesson from this plot is that facilitated diffusion decreases the likelihood of oxygen debt, since the external oxygen concentration necessary to prevent oxygen debt is smaller in the presence of myoglobin than without.

A similar lesson comes from Fig. 2.6, where the internal free oxygen content $\sigma$ is shown, plotted as a function of radius $y$. The solid curves show the internal free oxygen with facilitated diffusion, and the dashed curve is without. The smaller of the two solid curves and the dashed curve have exactly the critical external oxygen concentration, showing clearly that in the presence of myoglobin, oxygen debt is less likely at a

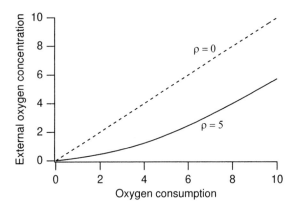

**Figure 2.5** Critical concentration $\sigma_0$ plotted as a function of oxygen consumption $\frac{\gamma}{4\epsilon_1}$. The dashed curve is the critical concentration with no facilitated diffusion.

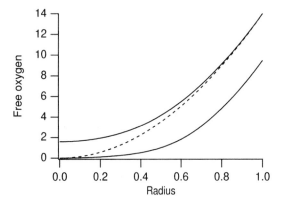

**Figure 2.6** Free oxygen $\sigma$ as a function of radius $y$. Solid curves show oxygen concentration in the presence of myoglobin ($\rho = 5$), the lower of the two having the critical external oxygen concentration. The dashed curve shows the oxygen concentration without facilitation at the critical external concentration level.

given external oxygen concentration. The larger of the two solid curves has the same external oxygen concentration as the dashed curve, showing again the contribution of facilitation toward preventing oxygen debt. For this figure, $\rho = 5$, $\gamma/\epsilon_1 = 14$.

## 2.4  Carrier-Mediated Transport

Some substances are insoluble in the cell membrane and yet pass through by a process called *carrier-mediated transport*. It is also called *facilitated diffusion* in many physiology books, although we prefer to reserve this expression for the process described in the previous section. Carrier-mediated transport is the means by which some sugars cross the cell membrane to provide an energy source for the cell. For example, glucose, the most important of the sugars, combines with a carrier protein at the outer boundary of the membrane, and by means of a conformational change is released from the inner boundary of the membrane.

There are three types of carrier-mediated transport. Carrier proteins that transport a single solute from one side of the membrane to the other are called *uniports*. Other proteins function as coupled transporters by which the simultaneous transport of two solute molecules is accomplished, either in the same direction (called a *symport*) or in the opposite direction (called an *antiport*).

## 2.4.1 Glucose Transport

Although the details are not certain, the transport of glucose across the lipid bilayer of the cell membrane is thought to occur when the carrier molecule alternately exposes the solute binding site first on one side and then on the other side of the membrane. It is considered highly unlikely that the carrier molecule actually diffuses back and forth through the membrane.

We can model the process of glucose transport as follows: We suppose that the population of enzymatic carrier proteins $C$ has two conformational states, $C_i$ and $C_e$, with its glucose binding site exposed on the cell interior (subscript $i$) or exterior (subscript $e$) of the membrane, respectively. The glucose substrate on the interior $S_i$ can bind with $C_i$ and the glucose substrate on the exterior can bind with enzyme $C_e$ to form the complex $P_i$ or $P_e$, respectively. Finally, a conformational change transforms $P_i$ into $P_e$ and vice versa. These statements are summarized in

$$S_i + C_i \underset{k_-}{\overset{k_+}{\rightleftarrows}} P_i \underset{k}{\overset{k}{\rightleftarrows}} P_e \underset{k_+}{\overset{k_-}{\rightleftarrows}} S_e + C_e, \tag{2.33}$$

$$C_i \underset{k}{\overset{k}{\rightleftarrows}} C_e. \tag{2.34}$$

We further suppose that the glucose is supplied at the constant rate $J$ on the exterior and taken away at the same rate from the interior.

Following mass action kinetics, the differential equations describing these kinetics are

$$\frac{ds_i}{dt} = k_- p_i - k_+ s_i c_i - J, \tag{2.35}$$

$$\frac{ds_e}{dt} = k_- p_e - k_+ s_e c_e + J, \tag{2.36}$$

$$\frac{dp_i}{dt} = k p_e - k p_i + k_+ s_i c_i - k_- p_i, \tag{2.37}$$

$$\frac{dp_e}{dt} = k p_i - k p_e + k_+ s_e c_e - k_- p_e, \tag{2.38}$$

$$\frac{dc_i}{dt} = k c_e - k c_i + k_- p_i - k_+ s_i c_i, \tag{2.39}$$

$$\frac{dc_e}{dt} = k c_i - k c_e + k_- p_e - k_+ s_e c_e. \tag{2.40}$$

where $s_i = [S_i]$, $p_i = [P_i]$, etc. There are two degeneracies in these equations; first, the total amount of receptor is conserved, and thus $p_i + p_e + c_i + c_e = c_0$, where $c_0$ is a constant, and second, the total amount of glucose is conserved, and thus $s_i + s_e + p_i + p_e = $ constant.

In steady state there are six equations with seven undetermined quantities, where the flow rate $J$ is considered unknown. However, notice that the steady-state versions of equations (2.36)–(2.39) and the conservation law $p_i + p_e + c_i + c_e = c_0$ constitute a linear system for the five unknowns $p_i, p_e, c_i, c_e$, and $J$. Thus, $J$ can be found as a function of $s_i$ and $s_e$ to be

$$J = \frac{1}{2} K_d K k_+ C_0 \frac{s_e - s_i}{(s_i + K + K_d)(s_e + K + K_d) - K_d^2}, \tag{2.41}$$

where $K = k_-/k_+$ and $K_d = k/k_+$. Since $k$ is the rate at which conformational change takes place, it acts like a diffusion coefficient in that it reflects the effect of random thermal activity at the molecular level.

In this model we have assumed that the two conformational states of the carrier molecule are equally likely, because otherwise energy would be required to run the exchanger, and that the affinity of the glucose binding site is unchanged by the conformational change. We have also assumed that it is impossible for glucose to cross the membrane by simple diffusion.

The nondimensional flux is

$$j = \frac{\sigma_e - \sigma_i}{(\sigma_i + 1 + \kappa)(\sigma_e + 1 + \kappa) - \kappa^2}, \tag{2.42}$$

where $\sigma_i = s_i/K, \sigma_e = s_e/K, \kappa = K_d/K$. A plot of this nondimensional flux is shown in Fig. 2.7, plotted as a function of extracellular glucose $\sigma_e$, with fixed intracellular glucose and fixed $\kappa$. We can see that the rate of transport is limited by saturation of the enzyme kinetics (this saturation is observed experimentally) and thermal conformational change is crucial to the transport process, as transport $J$ drops to zero if $K_d = 0$. The binding affinity of the carrier protein for glucose $(k_+)$, and hence the flux of glucose, is controlled by insulin.

Models for symport and antiport transporters follow in similar fashion. For a symport or antiport, the protein carrier has multiple binding sites, which can be exposed to the intracellular or extracellular space. A change of conformation exchanges the location of all of the participating binding sites, from inside to outside, or vice versa. An example of an antiport is the sodium–calcium exchanger in muscle cells and nerve cells, which exchanges one calcium ion for three sodium ions. Presumably, the sodium–calcium exchanger has three sodium binding sites and one calcium binding site, which are always on opposite sides of the membrane. An example of a symport is the sodium-driven glucose symport that transports glucose and sodium from the lumen of the gut to the intestinal epithelium. A similar process occurs in epithelial cells lining the proximal tubules in the kidney, to remove glucose and amino acids from the filtrate (discussed in Chapter 20). Five different amino acid cotransporters have been identified.

If there are $k$ binding sites that participate in the exchange, then there are $2^k$ possible combinations of bound and unbound sites. The key assumption that makes this model of transport work is that only the completely unbound or completely bound

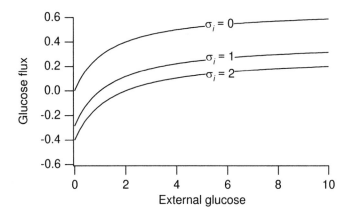

**Figure 2.7** Plot of the (nondimensional) flux of glucose as a function of extracellular glucose, for fixed intracellular glucose, with $\kappa = K_d/K = 0.5$.

carrier participates in a conformational change. Thus, there is a carrier molecule, say C, with two conformations, $C_i$ and $C_e$, and a fully bound complex P, also with two conformations, $P_i$ and $P_e$, and possible transformation between the two conformations,

$$C_i \underset{k_{-c}}{\overset{k_c}{\rightleftarrows}} C_e, \qquad P_i \underset{k_{-p}}{\overset{k_p}{\rightleftarrows}} P_e. \tag{2.43}$$

In addition, there are $2^k$ possible combinations of binding and unbinding in each of the two conformations. For example, with two substrates S and T, and one binding site for each, we have the complexes C, SC, CT, and SCT = P. The possible reactions are summarized in Fig. 2.8.

Unfortunately, the analysis of this fully general reaction scheme is quite complicated. However, it simplifies significantly if we assume that the intermediates can be

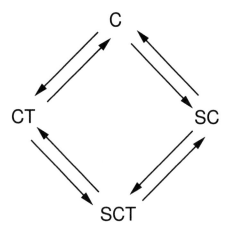

**Figure 2.8** States and possible transitions of a transporter with two substrates, S and T, and one binding site for each.

safely ignored and postulate the reaction scheme

$$mS + nT + C \underset{k_-}{\overset{k_+}{\rightleftharpoons}} P. \tag{2.44}$$

Now the result for a symport is strikingly similar to the uniport flux, with

$$J = \frac{1}{2} K_d K k_+ C_0 \frac{s_e^m t_e^n - s_i^m t_i^n}{(s_i^m t_i^n + K + K_d)(s_e^m t_e^n + K + K_d) - K_d^2}, \tag{2.45}$$

where the flux of $s$ is $mJ$ and the flux of $t$ is $nJ$. Here we have set $k_c = k_{-c} = k_p = k_{-p} = k$ and then $K = k_-/k_+$ and $K_d = k/k_+$.

For an antiport, the subscripts on one of the substances must be exchanged, to give

$$J = \frac{1}{2} K_d K k_+ C_0 \frac{s_e^m t_i^n - s_i^m t_e^n}{(s_i^m t_e^n + K + K_d)(s_e^m t_i^n + K + K_d) - K_d^2}. \tag{2.46}$$

Antiports and symports are passive pumps because no chemical energy is consumed by them, although they are often described as secondarily active pumps because they exploit a chemical gradient that requires energy to establish. Thus, for example, the sodium–calcium exchanger uses the sodium gradient to pump calcium against its gradient, although energy is required to establish the sodium gradient in the first place.

The efficiency of this type of exchanger is determined by the coefficients $m$ and $n$. Consider, for example, the sodium–calcium exchanger, taking S to be sodium and T to be calcium, with $m = 3$ and $n = 1$. For this exchanger, flux is positive (sodium flows inward and calcium flows outward) if

$$\frac{t_i}{t_e} > \left(\frac{s_i}{s_e}\right)^3. \tag{2.47}$$

Observe that the curve $y = x^3$ is smaller than the curve $y = x$ on the interval $0 < x < 1$. It follows that a sodium exchanger with $m = 3$ is better able to act as a pump than a (hypothetical) sodium exchanger with $m = 1$, because with $s_i/s_e < 1$ fixed, it is able to function at lower values of $t_i/t_e$.

While this answer is interesting, the analysis is flawed because of the assumption that the rates of conformational change $k_c$, $k_{-c}$, $k_p$, and $k_{-p}$ are identical. For ionic cotransporters this is not correct because with each exchange there is a net flux of charge. If there is a difference of potential across the membrane (which, as we will see, is typical), then the exchange of an ion is resisted by a potential increase, so the forward and backward conformational changes cannot take place at the same rates. We will revisit this problem in Chapter 3 after the problem of ion flow across a potential difference has been addressed.

## 2.5   Active Transport

The carrier-mediated transport described above is always down electrochemical gradients, and so is identified with diffusion. Any process that works against gradients requires the expenditure of energy. Perhaps the most important example of active (energy-consuming) transport is the sodium–potassium pump. This pump acts as an antiport, actively pumping sodium ions out of the cell against its steep electrochemical gradient and pumping potassium ions in. As we will see later in this chapter, this pump is used to regulate the cell volume and to maintain a membrane potential. Indeed, almost a third of the energy requirement of a typical animal cell is consumed in fueling this pump; in electrically active nerve cells, this figure approaches two-thirds of the cell's energy requirement.

This pumping activity uses energy by the dephosphorylation of ATP into ADP through the overall reaction scheme

$$\text{ATP} + 3\text{Na}_i^+ + 2\text{K}_e^+ \longrightarrow \text{ADP} + \text{P}_i + 3\text{Na}_e^+ + 2\text{K}_i^+, \tag{2.48}$$

with subscript $e$ or $i$ denoting extracellular or intracellular concentrations respectively. The details of the sodium–potassium pump (or $\text{Na}^+$–$\text{K}^+$ ATPase) are thought to be as follows. In its dephosphorylated state, sodium binding sites are exposed to the intracellular space. When sodium ions are bound, the carrier protein is phosphorylated by the hydrolysis of ATP (step 1; see Fig. 2.9). This induces a change of conformation, exposing the sodium binding sites to the extracellular space and reducing the binding affinity of these sites, thereby causing the release of the bound sodium. Simultaneously, the potassium binding sites are exposed to the extracellular medium, so that potassium is bound (step 2). When potassium is bound, the carrier is dephosphorylated, inducing the reverse conformational change and exposing the potassium binding site to the cytosol (step 3). Potassium is released to the cytosol when the binding affinity for potassium decreases (step 4). Some estimates for the affinities are shown in Table 2.2.

To illustrate how to turn this verbal description into a mathematical model, we consider a simplified case in which there is a single binding site for sodium and potassium, leading to a one-for-one exchange, rather than the three-for-two exchange that actually occurs. We denote the carrier molecule by C, and assume the reactions

$$\text{Na}_i^+ + \text{C} \underset{k_{-1}}{\overset{k_1}{\rightleftharpoons}} \text{NaC} \underset{\text{ATP}\nearrow\searrow\text{ADP}}{\longrightarrow} \text{NaCP} \underset{k_{-2}}{\overset{k_2}{\rightleftharpoons}} \text{Na}_e^+ + \text{CP}, \tag{2.49}$$

$$\text{CP} + \text{K}_e^+ \underset{k_{-3}}{\overset{k_3}{\rightleftharpoons}} \text{KCP} \underset{k_{-4}}{\overset{k_4}{\rightleftharpoons}} \text{P} + \text{KC}, \tag{2.50}$$

**Table 2.2** Some estimated equilibrium constants for the sodium–potassium pump.

| Ion | $K_d$ (mM) | |
|-----|-----|-----|
| Na$^+$ | 1.3 | Inside |
| K$^+$ | 12 | Inside |
| Na$^+$ | 32 | Outside |
| K$^+$ | 0.14 | Outside |

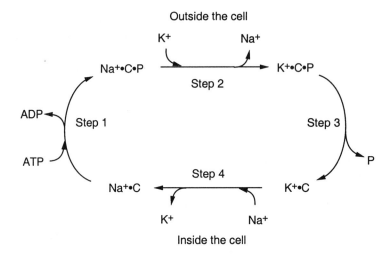

**Figure 2.9**  Schematic diagram of reactions for the sodium–potassium pump.

$$\text{KC} \underset{k_{-5}}{\overset{k_5}{\rightleftharpoons}} \text{K}_i^+ + \text{C}. \tag{2.51}$$

We apply the law of mass action to these kinetics, assume that intracellular sodium and extracellular potassium are supplied at the constant rate $J$ and that intracellular potassium and extracellular sodium are also removed at the constant rate $J$, and then find that in steady state the flow of ions through the pump is given by

$$J = C_0 \frac{[\text{Na}_i^+][\text{K}_e^+]K_1 K_2 - [\text{Na}_e^+][\text{K}_i^+]K_{-1}K_{-2}[\text{P}]}{([\text{K}_e^+]K_2 + [\text{K}_i^+]K_{-2})K_n + ([\text{Na}_i^+]K_1 + [\text{Na}_e^+]K_{-1})K_k}, \tag{2.52}$$

where $K_1 = k_1 k_2 k_p$, $K_{-1} = k_{-1} k_{-2} k_{-p}$, $K_2 = k_3 k_4 k_5$, $K_{-2} = k_{-3} k_{-4} k_{-5}$, $K_n = k_{-1} k_{-p} + k_2 k_{-1} + k_2 k_p$, and $K_k = k_{-3} k_{-4}[\text{P}] + k_{-3} k_5 + k_4 k_5$. The rate constants $k_p$ and $k_{-p}$ are the forward and backward rate constants for the hydrolysis of ATP. As before, the total concentration of carrier molecule is denoted by $C_0$.

Because ATP is much more energetic than ADP, we expect the reverse reaction rate $k_{-p}$ to be small compared to the forward reaction rate $k_p$. If we ignore the reverse reaction (take $K_{-1} = 0$), we find

$$J = C_0 K_1 K_2 \frac{[\text{Na}_i^+][\text{K}_e^+]}{([\text{K}_e^+]K_2 + [\text{K}_i^+]K_{-2})K_n + [\text{Na}_i^+]K_1 K_k}, \tag{2.53}$$

which is independent of the extracellular sodium concentration. As expected, this flux exhibits the features of an enzymatic reaction, being nearly linear at small concentrations of intracellular sodium and saturating at large concentrations. However, if we include the effects of the reverse reactions, we see that it should be possible to run the pump backward by maintaining sufficiently high levels of extracellular sodium and intracellular potassium, so that the energy stored in the electrochemical gradients can be extracted. Indeed, when this is the case experimentally, ATP is synthesized from ADP.

Other important pumps are $Ca^{2+}$ ATPases and transporters that keep the intracellular concentration of $Ca^{2+}$ low. Calcium is extremely important to the operation of cells (as will be discussed in Chapter 5). Internal free calcium is maintained at low concentrations ($10^{-7}$ M) compared to high concentrations of extracellular calcium ($10^{-3}$ M). The flow of $Ca^{2+}$ down its steep concentration gradient in response to extracellular signals is one means of transmitting signals rapidly across the plasma membrane. The $Ca^{2+}$ gradient is maintained in part by $Ca^{2+}$ pumps in the membrane that actively transport $Ca^{2+}$ out of the cell. One of these is an ATPase, while the other is a passive antiporter that is driven by the $Na^+$ electrochemical gradient. The best-understood $Ca^{2+}$ pump is an ATPase in the *sarcoplasmic reticulum* of muscle cells (Exercise 9). This $Ca^{2+}$ pump has been found to function in a way similar to the sodium–potassium pump. In fact, the carriers for these two are known from DNA sequencing to be homologous proteins.

# 2.6  The Membrane Potential

The principal function of the active transport processes described above is to regulate the intracellular ionic composition of the cell. For example, the operation of the $Na^+$–$K^+$ pump results in high intracellular $K^+$ concentrations and low intracellular $Na^+$ concentrations. As we will see, this is necessary for a cell's survival, as without such regulation, cells could not control their volume. However, before we consider models for cell volume regulation, we consider the effects of ionic separation. It is a consequence of the control of cell volume by ionic transport that the cell develops a potential difference across its membrane.

## 2.6.1  The Nernst Equilibrium Potential

One of the most important equations in electrophysiology is the Nernst equation, which describes how a difference in ionic concentration between two phases can result in a potential difference between the phases. We do not derive the Nernst equation from first principles, but give a nonrigorous derivation in Section 2.6.2. Derivations of the Nernst equation using the theory of chemical equilibrium thermodynamics can be found in standard physical chemistry textbooks (for example, Levine, 1978; Denbigh, 1981).

Suppose we have two reservoirs containing the same ion S, but at different concentrations, as shown schematically in Fig. 2.10. The reservoirs are separated by a semipermeable membrane. The solutions on each side of the membrane are assumed to be electrically neutral (at least initially), and thus each ion S is balanced by another ion, S', with opposite sign. For example, S could be $Na^+$, while S' could be $Cl^-$. Because we ultimately wish to apply the Nernst equation to cellular membranes, we call the left of the membrane the inside and the right the outside.

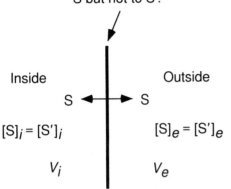

Cell membrane permeable to
S but not to S'.

Inside                    Outside

S ⟷ S

$[S]_i = [S']_i$          $[S]_e = [S']_e$

$V_i$                     $V_e$

**Figure 2.10** Schematic diagram of a membrane separating two solutions with different ionic concentrations.

If the membrane is permeable to S but not to S', the concentration difference across the membrane results in a flow of S from one side to another, say, from left to right. However, because S' cannot diffuse through the membrane, the diffusion of S causes a buildup of charge across the membrane. This charge imbalance, in turn, sets up an electric field that opposes the further diffusion of S through the membrane. Equilibrium is reached when the electric field exactly balances the diffusion of S. Note that at steady state there will be more S ions on one side than on the other, and thus neither side of the membrane is exactly electrically neutral. However, although the diffusion of S causes an electric potential to develop, it is important to realize that only a small amount of S moves across the membrane. To a good approximation the concentrations of S on either side of the membrane remain unchanged, the solutions on either side of the membrane remain electrically neutral, and the small excess charge accumulates near the interface.

At equilibrium the potential difference, $V_S$, across the membrane is given by the *Nernst potential*,

$$V_S = \frac{RT}{zF} \ln\left(\frac{[S]_e}{[S]_i}\right) = \frac{kT}{zq} \ln\left(\frac{[S]_e}{[S]_i}\right),$$   (2.54)

where subscripts $i$ and $e$ denote internal and external concentrations respectively. $R$ is the universal gas constant, $T$ is the absolute temperature, $F$ is Faraday's constant, $k$ is Boltzmann's constant, $q$ is the charge on a proton, and $z$ is the charge on the ion S. Values of these constants, and their units, are given in the Appendix. One particularly important relationship is

$$k = \frac{R}{N_A},$$   (2.55)

where $N_A$ is Avogadro's number. Because of this, the Nernst equation can be written in the two equivalent forms shown above. Throughout this book we follow the standard convention and define the potential difference across the cell membrane as

$$V = V_i - V_e, \tag{2.56}$$

i.e., the intracellular minus the extracellular potential. When $V = V_S$, there is no net current of S between the phases, as the diffusion of S is exactly balanced by the electric potential difference.

Typical concentrations (in this case, for squid axon) are 397, 50, and 40 mM for potassium, sodium, and chloride, respectively, in the intracellular space, and 20, 437, and 556 mM in the extracellular space. With these concentrations, the Nernst potentials for squid nerve axon are $V_{Na} = 56$ mV, $V_K = -77$ mV, $V_{Cl} = -68$ mV (using $RT/F = 25.8$ mV at $27°C$. See Table 2.1).

The Nernst equation is independent of how the ions move through the membrane and is dependent only on the concentration difference. In this sense, it is a "universal" law. Any equation that expresses the transmembrane current of S in terms of the membrane potential, no matter what its form, must have a reversal potential of $V_S$; i.e., the current must be zero at the Nernst potential $V = V_S$. However, although this is true when only a single ion species crosses the membrane, the situation is considerably more complicated when more than one type of ion can cross the membrane. In this case, the membrane potential that generates zero total current does not necessarily have no net current for each individual ion. For example, a current of S in one direction might be balanced by a current of S' in the same direction. Hence, when multiple ion types can diffuse through the membrane, the phases are not, in general, at equilibrium, even when there is no total current. Therefore, the arguments of chemical equilibrium used to derive the Nernst equation cannot be used, and there is no universal expression for the reversal potential in the multiple ion case. In this case, the reversal potential depends on the model used to describe the individual transmembrane ionic flows (see Chapter 3).

## 2.6.2 Electrodiffusion: The Goldman–Hodgkin–Katz Equations

In general, the flow of ions through the membrane is driven by concentration gradients and also by the electric field. The contribution to the flow from the electric field is given by *Planck's equation*

$$\mathbf{J} = -u\frac{z}{|z|}c\nabla\phi, \tag{2.57}$$

where $u$ is the *mobility* of the ion, defined as the velocity of the ion under a constant unit electric field; $z$ is the valence of the ion, so that $z/|z|$ is the sign of the force on the ion; $c$ is the concentration of S; and $\phi$ is the electrical potential, so that $-\nabla\phi$ is the electrical field.

There is a relationship, determined by Einstein, between the ionic mobility $u$ and Fick's diffusion constant:

$$D = \frac{uRT}{|z|F}. \tag{2.58}$$

When the effects of concentration gradients and electrical gradients are combined, we obtain the *Nernst–Planck equation*

$$\mathbf{J} = -D\left(\nabla c + \frac{zF}{RT}c\nabla\phi\right). \tag{2.59}$$

If the flow of ions and the electric field are transverse to the membrane, we can view (2.59) as the one-dimensional relation

$$J = -D\left(\frac{dc}{dx} + \frac{zF}{RT}c\frac{d\phi}{dx}\right). \tag{2.60}$$

## The Nernst equation

The Nernst equation can be derived from the Nernst–Planck electrodiffusion equation (2.60). When the flux $J$ is zero, we find

$$-D\left(\frac{dc}{dx} + \frac{zF}{RT}c\frac{d\phi}{dx}\right) = 0, \tag{2.61}$$

so that

$$\frac{1}{c}\frac{dc}{dx} + \frac{zF}{RT}\frac{d\phi}{dx} = 0. \tag{2.62}$$

Now suppose that the cell membrane extends from $x = 0$ (the inside) to $x = L$ (the outside), and let subscripts $i$ and $e$ denote internal and external quantities respectively. Then, integrating from $x = 0$ to $x = L$ we get

$$\ln(c)\big|_{c_i}^{c_e} = \frac{zF}{RT}(\phi_i - \phi_e), \tag{2.63}$$

and thus the potential difference across the membrane, $V = \phi_i - \phi_e$, is given by

$$V = \frac{RT}{zF}\ln\left(\frac{c_e}{c_i}\right), \tag{2.64}$$

which is the Nernst equation.

This derivation of the Nernst equation relies on the Nernst–Planck electrodiffusion equation, and so is not a derivation from first principles. The derivation from first principles can be given, but it is beyond the scope of this text. The interested reader is referred to Levine (1978) or Denbigh (1981) for the details.

## The constant field approximation

In general, the electric potential $\phi$ is determined by the local charge density, and so $J$ must be found by solving a coupled system of equations (this is discussed in detail in Chapter 3). However, a useful result is obtained by assuming that the electric field in the membrane is constant, and thus decoupled from the effects of charges moving through the membrane. Suppose we have two reservoirs separated by a semipermeable membrane of thickness $L$, such that the potential difference across the membrane is $V$. On the left of the membrane (the inside) $[S] = c_i$, and on the right (the outside)

[S] $= c_e$. If the electric field is constant through the membrane, we have $\partial \phi / \partial x = -V/L$, where $V = \phi(0) - \phi(L)$ is the membrane potential.

At steady state and with no production of ions, the flux must be constant. In this case, the Nernst–Planck equation (2.59) is an ordinary differential equation for the concentration $c$,

$$\frac{dc}{dx} - \frac{zFV}{RTL}c + \frac{J}{D} = 0, \tag{2.65}$$

whose solution is

$$\exp\left(\frac{-zVFx}{RTL}\right)c(x) = -\frac{JRTL}{DzVF}\left[\exp\left(\frac{-zVFx}{RTL}\right) - 1\right] + c_i, \tag{2.66}$$

where we have used the left boundary condition $c(0) = c_i$. To satisfy the boundary condition $c(L) = c_e$, it must be that

$$J = \frac{D}{L}\frac{zFV}{RT}\frac{c_i - c_e \exp\left(\frac{-zVF}{RT}\right)}{1 - \exp\left(\frac{-zVF}{RT}\right)}, \tag{2.67}$$

where $J$ is the flux density with units (typically) of moles per area per unit time. This flux density becomes an electrical current density (current per unit area) when multiplied by $zF$, the number of charges carried per mole, and thus

$$I_S = P_S \frac{z^2 F^2}{RT} V \frac{c_i - c_e \exp\left(\frac{-zFV}{RT}\right)}{1 - \exp\left(\frac{-zFV}{RT}\right)}, \tag{2.68}$$

where $P_S = D/L$ is the permeability of the membrane to S. This is the famous Goldman–Hodgkin–Katz (GHK) current equation. It plays an important role in models of cellular electrical activity.

This flow is zero if the diffusively driven flow and the electrically driven flow are in balance, which occurs, provided that $z \neq 0$, if

$$V = V_S = \frac{RT}{zF}\ln\left(\frac{c_e}{c_i}\right), \tag{2.69}$$

which is, as expected, the Nernst potential.

If there are several ions that are separated by the same membrane, then the flow of each of these is governed separately by its own current–voltage relationship. In general there is no potential at which these currents are all zero. However, the potential at which the net electrical current is zero is called the Goldman–Hodgkin–Katz potential. For a collection of ions all with valence $z = \pm 1$, we can calculate the GHK potential directly. For zero net electrical current, it must be that

$$0 = \sum_{z=1} P_j \frac{c_i^j - c_e^j \exp\left(\frac{-VF}{RT}\right)}{1 - \exp\left(\frac{-VF}{RT}\right)} + \sum_{z=-1} P_j \frac{c_i^j - c_e^j \exp\left(\frac{VF}{RT}\right)}{1 - \exp\left(\frac{VF}{RT}\right)}, \tag{2.70}$$

where $P_j = D_j/L$. This expression can be solved for $V$, to get

$$V = -\frac{RT}{F} \ln \left( \frac{\sum_{z=-1} P_j c_e^j + \sum_{z=1} P_j c_i^j}{\sum_{z=-1} P_j c_i^j + \sum_{z=1} P_j c_e^j} \right). \tag{2.71}$$

For example, if the membrane separates sodium ($Na^+, z = 1$), potassium ($K^+, z = 1$), and chloride ($Cl^-, z = -1$) ions, then the GHK potential is

$$V_r = -\frac{RT}{F} \ln \left( \frac{P_{Na}[Na^+]_i + P_K[K^+]_i + P_{Cl}[Cl^-]_e}{P_{Na}[Na^+]_e + P_K[K^+]_e + P_{Cl}[Cl^-]_i} \right). \tag{2.72}$$

It is important to emphasize that neither the GHK potential nor the GHK current equation are universal expressions like the Nernst equation. Both depend on the assumption of a constant electric field, and other models give different expressions for the transmembrane current and reversal potential. In Chapter 3 we present a detailed discussion of other models of ionic current and compare them to the GHK equations. However, the importance of the GHK equations is so great, and their use so widespread, that their separate presentation here is justified.

## 2.6.3 Electrical Circuit Model of the Cell Membrane

Since the cell membrane separates charge, it can be viewed as a capacitor. The capacitance of any insulator is defined as the ratio of the charge across the capacitor to the voltage potential necessary to hold that charge, and is denoted by

$$C_m = \frac{Q}{V}. \tag{2.73}$$

From standard electrostatics (Coulomb's law), one can derive the fact that for two parallel conducting plates separated by an insulator of thickness $d$, the capacitance is

$$C_m = \frac{k\epsilon_0}{d}, \tag{2.74}$$

where $k$ is the dielectric constant for the insulator and $\epsilon_0$ is the permittivity of free space. The capacitance of cell membrane is typically found to be 1.0 $\mu F/cm^2$. Using that $\epsilon_0 = (10^{-9}/(36\pi))F/m$, we calculate that the dielectric constant for cell membrane is about 8.5, compared to $k = 3$ for oil.

A simple electrical circuit model of the cell membrane is shown in Fig. 2.11. It is assumed that the membrane acts like a capacitor in parallel with a resistor (although not necessarily ohmic). Since the current is defined by $dQ/dt$, it follows from (2.73) that the capacitive current is $C_m dV/dt$, provided that $C_m$ is constant. Since there can be no net buildup of charge on either side of the membrane, the sum of the ionic and capacitive currents must be zero, and so

$$C_m \frac{dV}{dt} + I_{ion} = 0, \tag{2.75}$$

where $V = V_i - V_e$.

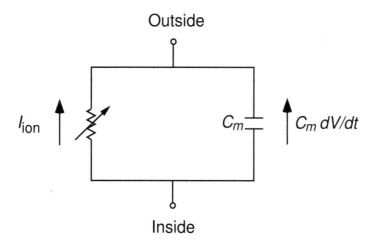

**Figure 2.11** Electrical circuit model of the cell membrane.

We will meet this equation many times in this book, as it is the basis for much of theoretical electrophysiology. A significant challenge is to determine the form of $I_{ion}$. We have already derived one possible choice, the GHK current equation (2.68), and others will be discussed in Chapter 3.

Another common model describes $I_{ion}$ as a linear function of the membrane potential. In Chapter 3 we will see how a linear $I$–$V$ curve can be derived from more realistic models; however, because it is used so widely, we present a brief, heuristic, derivation here. Consider the movement of an ion S across a membrane. We assume that the potential drop across the membrane has two components. First, the potential drop due to concentration differences is given by the Nernst equation

$$V_S = \frac{RT}{zF} \ln \left( \frac{[S]_e}{[S]_i} \right), \tag{2.76}$$

and, second, the potential drop due to an electrical current is $rI_S$ (if the channel is ohmic), where $r$ is the channel resistance and $I_S$ is the transmembrane current (positive outward) of S. Summing these two contributions we find

$$V = rI_S + V_S, \tag{2.77}$$

and solving for the current, we get the current–voltage relationship

$$I_S = g(V - V_S), \tag{2.78}$$

where $g = 1/r$ is the *membrane conductance*. The current $I_S$ and conductance $g$ are usually specified per unit area of membrane, being the product of the single channel conductance times the number of channels per unit area of membrane.

## 2.7  Osmosis

Suppose two chambers of water are separated by a rigid porous membrane. Because it is porous, water can flow between the two chambers. If the two chambers are topped by pistons, then water can be driven between the two chambers by applying different pressures to the two pistons. In general there is a linear relationship between the pressure difference and the flux of water through the membrane, given by

$$rQ = P_1 - P_2, \tag{2.79}$$

where $Q$ is the flux (volume per unit time) of water from chamber one to chamber two, $P_1$ and $P_2$ are the applied pressures for chambers one and two, respectively, and $r$ is the flow resistance of the membrane (not the same as the resistance to flow of ions). The expression (2.79) is actually a definition of the flow resistance $r$, and this linear relationship is analogous to Ohm's law relating current and voltage in a conductor, and therefore it is useful but not universally true.

Now suppose that a solute is added to chamber one, say. In standard chemistry texts, it is shown that the presence of a solute lowers the chemical potential of the solvent, thus lowering its effective pressure. If $\pi_s$ is the effective pressure due to the presence of the dissolved molecules, then the flow rate is modified by the solute via

$$rQ = P_1 - \pi_s - P_2, \tag{2.80}$$

where

$$\pi_s = kcT. \tag{2.81}$$

Here, $k$ is Boltzmann's constant, $c$ is the concentration of the solute (in molecules per unit volume), and $T$ is the absolute temperature. Note that if $c$ is expressed in the more usual units of moles per volume, the expression for the osmotic pressure then becomes

$$\pi_s = RcT, \tag{2.82}$$

where $R = kN_A$ is the universal gas constant, and $N_A$ is Avogadro's number. Using that $c = n/v$, where $v$ is the volume, we find that (2.81) becomes

$$\pi_s v = nkT, \tag{2.83}$$

which is the same as the ideal gas law. The quantity $\pi_s$ is known as the *osmotic pressure*, and the flux of water due to osmotic pressure is called *osmosis*. If $P_1 = P_2$, the effect of the osmotic pressure is to draw water into chamber one, causing an increase in its volume.

Osmotic pressure is determined by the number of particles per unit volume of fluid, and not the mass of the particles. The unit that expresses the concentration in terms of number of particles is called the *osmole*. One osmole is 1 gram molecular weight (a mole) of an undissociated solute. Thus, 180 grams of glucose (1 gram molecular weight) is 1 osmole of glucose, since glucose does not dissociate in water. On the other hand, 1 gram molecular weight of sodium chloride, 58.5 grams, is 2 osmoles, since it dissociates into 2 moles of osmotically active ions in water.

A solution with 1 osmole of solute dissolved in a kilogram of water is said to have osmolality of 1 osmole per kilogram. Since it is difficult to measure the amount of water in a solution, a more common unit of measure is osmolarity, which is the osmoles per liter of aqueous solution. In dilute conditions, such as in the human body, osmolarity and osmolality differ by less than one percent. At body temperature, $37° C$, a concentration of 1 osmole per liter of water has an osmotic pressure of 19,300 mm Hg.

Suppose two columns (of equal cross-section) of water are separated at the bottom by a rigid porous membrane. If $n$ molecules of sugar are dissolved in column one, what will be the height difference between the two columns after they achieve steady state? At steady state there is no flux between the two columns, so at the level of the membrane, $P_1 - \pi_s = P_2$. Now, $P_1$ and $P_2$ are related to the height of the column of water through $P = \rho g h$, where $\rho$ is the density of the fluid, $g$ is the gravitational constant, and $h$ is the height of the column. We suppose that the density of the two columns is the same, unaffected by the presence of the dissolved molecule, so we have

$$\rho g h_2 = \rho g h_1 - \frac{nkT}{h_1 A}, \tag{2.84}$$

where $A$ is the cross-sectional area of the columns. Since fluid is conserved, $h_1 + h_2 = 2h_0$, where $h_0$ is the height of the two columns of water before the sugar was added. From these, we find a single quadratic equation for $h_1$:

$$h_1^2 - h_0 h_1 - \frac{nkT}{2\rho g A} = 0. \tag{2.85}$$

The positive root of this equation is $h_1 = h_0/2 + \frac{1}{2}\sqrt{h_0^2 + \frac{2nkT}{\rho g A}}$ , so that

$$h_1 - h_2 = \sqrt{h_0^2 + \frac{2nkT}{\rho g A}} - h_0. \tag{2.86}$$

## 2.8 Control of Cell Volume

The principal function of the ionic pumps is to set up and maintain concentration differences across the cell membrane, concentration differences that are necessary for the cell to control its volume. In this section we justify these statements by means of a simple model in which the volume of the cell is regulated by the balance between ionic pumping and ionic flow down concentration gradients (Tosteson and Hoffman, 1960; Jakobsson, 1980; Hoppensteadt and Peskin, 1992).

Because the cell membrane is a thin lipid bilayer, it is incapable of withstanding any hydrostatic pressure differences. This is a potentially fatal weakness. If the intracellular concentrations of various ions and larger molecules become too large, osmotic forces cause the entry of water into the cell, causing it to swell and burst (this is what happens to many cells when their pumping machinery is disabled). Thus, for cells to survive, they must regulate their intracellular ionic composition precisely (Macknight, 1988; Reuss, 1988).

An even more difficult problem for some cells is to transport large quantities of water, ions, or other molecules while maintaining a steady volume. For example, $Na^+$-transporting epithelial cells, found (among other places) in the urinary bladder, the colon, and nephrons of the kidney, are designed to transport large quantities of $Na^+$ from the lumen of the gut or the nephron to the blood. Indeed, these cells can transport an amount of $Na^+$ equal to their entire intracellular contents in one minute. However, the rate of transport varies widely, depending on the concentration of $Na^+$ on the mucosal side. Thus, these cells must regulate their volume and ionic composition under a wide variety of conditions and transport rates (Schultz, 1981).

## 2.8.1  A Pump–Leak Model

We begin by modeling the active and passive transport of ionic species across the cell membrane. We have already derived two equations for ionic current as a function of membrane potential: the GHK current equation (2.68) and the linear relationship (2.78). For our present purposes it is convenient to use the linear expression for ionic currents. Active transport of $Na^+$ and $K^+$ is performed, by and large, by the $Na^+$–$K^+$ pump, which pumps three sodium ions out of the cell in exchange for the entry of two potassium ions.

Combining the expressions for active and passive ion transport, we find that the $Na^+$, $K^+$, and $Cl^-$ currents are given by

$$I_{Na} = g_{Na}\left[V - \frac{RT}{F}\ln\left(\frac{[Na^+]_e}{[Na^+]_i}\right)\right] + 3pq, \tag{2.87}$$

$$I_K = g_K\left[V - \frac{RT}{F}\ln\left(\frac{[K^+]_e}{[K^+]_i}\right)\right] - 2pq, \tag{2.88}$$

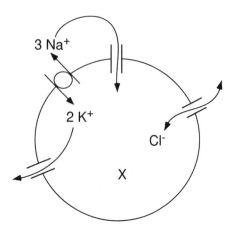

**Figure 2.12**  Schematic diagram of the pump–leak model.

$$I_{Cl} = g_{Cl} \left[ V + \frac{RT}{F} \ln \left( \frac{[Cl^-]_e}{[Cl^-]_i} \right) \right], \tag{2.89}$$

where $p$ is the rate at which the ion exchange pump works and $q$ is the charge of a single ion.

We can express these current–voltage equations as differential equations by noting that an outward ionic current of ion $A^{z+}$ affects the intracellular concentration of that ion through

$$I_A = -\frac{d}{dt}(zqw[A^{z+}]), \tag{2.90}$$

with $w$ denoting the cell volume. Thus we have

$$-\frac{d}{dt}(qw[Na^+]_i) = g_{Na} \left[ V - \frac{RT}{F} \ln \left( \frac{[Na^+]_e}{[Na^+]_i} \right) \right] + 3pq, \tag{2.91}$$

$$-\frac{d}{dt}(qw[K^+]_i) = g_K \left[ V - \frac{RT}{F} \ln \left( \frac{[K^+]_e}{[K^+]_i} \right) \right] - 2pq, \tag{2.92}$$

$$\frac{d}{dt}(qw[Cl^-]_i) = g_{Cl} \left[ V + \frac{RT}{F} \ln \left( \frac{[Cl^-]_e}{[Cl^-]_i} \right) \right]. \tag{2.93}$$

The total charge across the membrane is

$$C_m S V = qw([Na^+]_i + [K^+]_i - [Cl^-]_i) - qw_e([Na^+]_e + [K^+]_e - [Cl^-]_e) + z_x q X. \tag{2.94}$$

Here $w_e$ represents the extracellular volume for one cell, $C_m$ is the capacitance per unit area of the cell membrane, $X$ represents the number of large negatively charged molecules (with valence $z_x \leq -1$) that are trapped inside the cell, and $S$ is the surface area of the cell.

Finally, there is a flow of water across the membrane driven by osmotic pressure, given by

$$r\frac{dw}{dt} = RT \left( [Na^+]_i - [Na^+]_e + [K^+]_i - [K^+]_e + [Cl^-]_i - [Cl^-]_e + \frac{X}{w} \right). \tag{2.95}$$

Here we have assumed that the mechanical pressure difference across the membrane is zero, and we have also assumed that the elastic restoring force for the membrane is negligible.

Before we analyze this system of equations, it is valuable to make a few physical observations. First, both the extracellular and intracellular media are nearly in electroneutrality; that is, they have zero net charge. If this were not so, there would be extremely large electrostatic forces, which would quickly restore zero net charge. The only place where electroneutrality is violated is in a thin region near the membrane, and the amount of charge here is so small, relatively speaking, that it does not affect the overall assumption of electroneutrality. To see that this stored charge is quite small, consider a cylindrical piece of squid axon of typical radius 500 $\mu$m. With a capacitance of 1 $\mu$F/cm$^2$ and a typical membrane potential of 100 mV, the total charge is

$Q = C_m V = \pi \times 10^{-8}$ C/cm. In comparison, the charge associated with intracellular potassium ions at 400 mM is about $0.1\,\pi$ C/cm, showing a relative charge deflection of about $10^{-7}$. Another way to see that the relative charge deflection from electroneutrality must be quite small is to write (2.94) in dimensionless variables and then observe that the dimensionless capacitance $c = \frac{C_m RTS}{w[\mathrm{Cl}^-]_i Fq}$ is small, on the order of $10^{-8}$ for a spherical cell of radius 50 microns. A capacitance this small can be neglected with impunity in (2.94).

For this model, we assume that sodium, potassium, and chloride are in electroneutrality in the extracellular region. In view of the numbers for squid axon, this assumption is not quite correct, indicating that there must be other ions around to maintain electrical balance. In the intracellular region, sodium, potassium, and chloride are not even close to being in electrical balance, but here, electroneutrality is maintained by the large negatively charged proteins trapped within the cell's interior.

With the assumption of electroneutrality, (2.94) reduces to two separate equations:

$$[\mathrm{Na}^+]_e + [\mathrm{K}^+]_e - [\mathrm{Cl}^-]_e = 0, \tag{2.96}$$

$$[\mathrm{Na}^+]_i + [\mathrm{K}^+]_i - [\mathrm{Cl}^-]_i + z_x \frac{X}{w} = 0. \tag{2.97}$$

It is also convenient to assume that the cell is in an infinite bath, so that ionic currents do not change the external concentrations, and therefore the external concentrations are assumed to be fixed and known.

The differential equations (2.91), (2.92), (2.93), and (2.95) together with the electrostatic balance laws (2.97) describe the changes of cell volume and membrane potential as functions of time. Even though we formulated this model as a system of differential equations, we are interested, for the moment, only in their steady-state solution. Time-dependent currents and potentials become important in Chapter 4 for the discussion of excitability.

To understand these equations, we first introduce the nondimensional variables $v = \frac{FV}{RT}, P = \frac{pFq}{RTg_{\mathrm{Na}}}, \mu = \frac{w}{X}[\mathrm{Cl}^-]_e$ and set $y = e^{-v}$. Then the equation of intracellular electroneutrality becomes

$$\alpha y - \frac{1}{y} + \frac{z_x}{\mu} = 0, \tag{2.98}$$

and the equation of osmotic pressure balance becomes

$$\alpha y + \frac{1}{y} + \frac{1}{\mu} - 2 = 0, \tag{2.99}$$

where $\alpha = \frac{[\mathrm{Na}^+]_e e^{-3P} + [\mathrm{K}^+]_e e^{2P\gamma}}{[\mathrm{Na}^+]_e + [\mathrm{K}^+]_e}$ and $\gamma = g_{\mathrm{Na}}/g_{\mathrm{K}}$. In terms of these nondimensional variables, we find the ion concentrations to be

$$\frac{[\mathrm{Na}^+]_i}{[\mathrm{Na}^+]_e} = e^{-3P}y, \tag{2.100}$$

$$\frac{[\mathrm{K}^+]_i}{[\mathrm{K}^+]_e} = e^{2P\gamma}y, \tag{2.101}$$

$$\frac{[\text{Cl}^-]_i}{[\text{Cl}^-]_e} = \frac{1}{y}. \tag{2.102}$$

Solving (2.98) for its unique positive root, we obtain

$$y = \frac{-z_x + \sqrt{z_x^2 + 4\alpha\mu^2}}{2\alpha\mu}, \tag{2.103}$$

and when we substitute for $y$ back into (2.99), we find the quadratic equation for $\mu$:

$$4(1 - \alpha)\mu^2 - 4\mu + 1 - z_x^2 = 0. \tag{2.104}$$

For $z_x \leq -1$, this quadratic equation has one positive root if and only if $\alpha < 1$. Expressed in terms of concentrations, the condition $\alpha < 1$ is

$$\rho(P) = \frac{[\text{Na}^+]_e e^{-3P} + [\text{K}^+]_e e^{2P\gamma}}{[\text{Na}^+]_e + [\text{K}^+]_e} < 1. \tag{2.105}$$

One can easily see that $\rho(0) = 1$ and that for large $P$, $\rho(P)$ is exponentially large and positive. Thus, the only hope for $\rho(P)$ to be less than one is if $\rho'(0) < 0$. This occurs if and only if

$$\frac{3[\text{Na}^+]_e}{g_{\text{Na}}} > \frac{2[\text{K}^+]_e}{g_{\text{K}}}, \tag{2.106}$$

in which case there is a range of values of $P$ for which a finite, positive cell volume is possible and for which there is a corresponding nontrivial membrane potential.

To decide if this condition is ever satisfied we must determine "typical" values for $g_{\text{Na}}$ and $g_{\text{K}}$. This is difficult to do, because, as we will see, excitability of nerve tissue depends strongly on the fact that conductances are voltage dependent and can vary rapidly over a large range of values. However, at rest, in squid axon, reasonable values are $g_{\text{K}} = 0.367$ mS/cm$^2$ and $g_{\text{Na}} = 0.01$ mS/cm$^2$. For these values, and at the extracellular concentrations of 437 and 20 mM for sodium and potassium, respectively, the condition (2.106) is readily met.

One important property of the model is that the resting value of $V$ is equal to the Nernst potential of Cl$^-$, as can be seen from (2.93) or (2.102). Thus, the membrane potential is set by the activity of the Na$^+$–K$^+$ pump, and the intracellular Cl$^-$ concentration is set by the membrane potential.

In Figs. 2.13 and 2.14 the volume $\mu$ and the potential $V$ (assuming $RT/F = 25.8$ mV) are plotted as functions of the pump rate $P$. In addition, in Fig. 2.14 are shown the sodium and potassium equilibrium potentials. For these plots, $\gamma$ was chosen to be 0.11, and $z_x = -1$. Then, at $P = 1.6$, the sodium and potassium equilibrium potentials and the membrane potentials are close to their observed values for squid axon, of 56, $-77$ and $-68$ mV, respectively.

From these plots we can see the effect of changing pump rate on cell volume and membrane potential. At zero pump rate, the membrane potential is zero and the cell volume is infinite (dead cells swell). As the pump rate increases from zero, the cell volume and membrane potential rapidly decrease to their minimal values and then

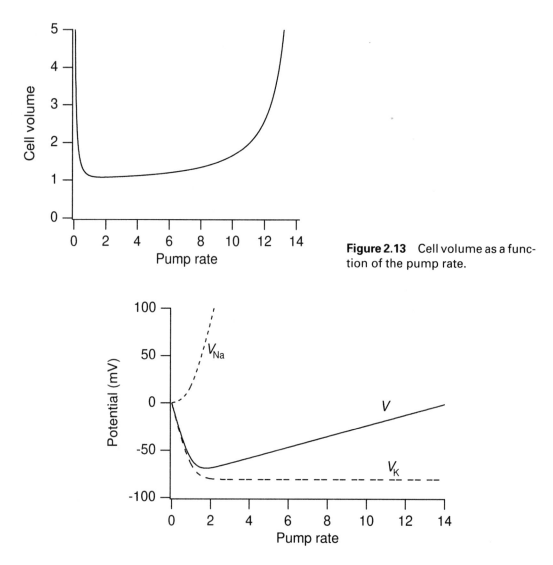

**Figure 2.13** Cell volume as a function of the pump rate.

**Figure 2.14** Membrane potential, sodium equilibrium potential, and potassium equilibrium potential as functions of the pump rate.

gradually increase until at some upper limit for pump rate, the volume and potential become infinite. The potassium equilibrium potential is seen to decrease rapidly as a function of pump rate until it reaches a plateau at a minimum value. The sodium equilibrium potential increases monotonically.

Physically realistic values of the membrane potential are achieved fairly close to the local minimum. Clearly, there is little advantage for a higher pump rate, and since the pump rate is proportional to energy expenditure, it would seem that the pump rate is chosen approximately to minimize cell volume, membrane potential, and en-

**Table 2.3** Resting potentials in some typical excitable cells.

| Cell Type | Resting Potential (mV) |
| --- | --- |
| Neuron | −70 |
| Skeletal muscle (mammalian) | −80 |
| Skeletal muscle (frog) | −90 |
| Cardiac muscle (atrial and ventricular) | −80 |
| Cardiac Purkinje fiber | −90 |
| Atrioventricular nodal cell | −65 |
| Sinoatrial nodal cell | −55 |
| Smooth muscle cell | −55 |

ergy expenditure. However, no mechanism for the regulation of energy expenditure is suggested.

### Generalizations

While the above model for control of volume and membrane potential is useful and gives some insight into the control mechanisms, as with most models there are important features that have been ignored but that might lead to substantially different behavior.

There are (at least) two significant simplifications in the model presented here. First, the conductances $g_{Na}$ and $g_K$ were treated as constants. In Chapter 4 we will see that the ability of cells to generate an electrical signal results from voltage and time dependence of the conductances. In fact, the discovery that ion channels have differing properties of voltage sensitivity was of fundamental importance to the understanding of neurons. The second simplification relates to the operation of the ion exchange pump. Figure 2.14 suggests that the minimal membrane potential is achieved at a particular pump rate and suggests the need for a tight control of pump rate that maintains the potential near this minimum. If indeed, such a tight control is required, it is natural to ask what that control mechanism might be.

A different model of the pump activity might be beneficial. Recall from (2.48) that with each "stroke" of the ion exchange pump, three intracellular sodium ions are exchanged for two extracellular potassium ions. The analysis of the $Na^+$–$K^+$ pump suggests that at low internal sodium concentrations, the pump rate can be represented in nondimensional variables as

$$P = \rho u^3, \tag{2.107}$$

where $u = [Na^+]_i/[Na^+]_e$. This representation is appropriate at high pump rates, where effects of saturation are of no concern. Notice that $P$ is proportional to the rate of ATP hydrolysis, and hence to energy consumption. Thus, as $u$ decreases, so also does the rate of energy consumption. With this change, the equation for the sodium concentration becomes

$$u \exp(3\rho u^3) = y, \tag{2.108}$$

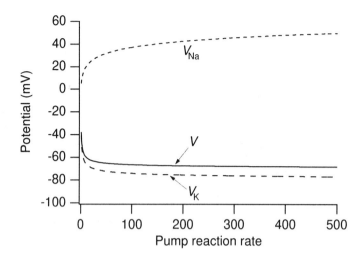

**Figure 2.15** Membrane potential, sodium equilibrium potential, and potassium equilibrium potential as functions of the pump rate.

and this must be solved together with the quadratic polynomials (2.98) and (2.99) (replacing (2.87) for $y$ and $\mu$).

In Fig. 2.15 are shown the membrane potential, and the sodium and potassium equilibrium potentials, plotted as functions of the nondimensional reaction rate $\rho$. Here we see something qualitatively different from what is depicted in Fig. 2.14. There the membrane potential had a noticeable local minimum and was sensitive to changes in pump rate. In this modified model, the membrane potential is insensitive to changes in the pump rate. The reason for this difference is clear. Since the effectiveness of the pump depends on the internal sodium concentration, increasing the speed of the pumping rate has little effect when the internal sodium is depleted, because of the diminished number of sodium ions available to be pumped.

While the pump rate is certainly ATP dependent, there are a number of drugs and hormones that affect the pump rate. Catecholamines rapidly increase the activity of the pump in skeletal muscle, thereby preserving proper $K^+$ during strenuous exercise. Within minutes, insulin stimulates pump activity in the liver, muscle, and fat tissues, whereas over a period of hours, aldosterone and corticosterones increase activity in the intestine.

On the other hand, digitalis (clinically known as digoxin) is known to suppress pump activity. Digitalis is an important drug used in the treatment of congestive heart failure and during the 1980s was the fourth most widely prescribed drug in the United States. At therapeutic concentrations, digitalis inhibits a moderate fraction (say, 30–40%) of the $Na^+$–$K^+$ ATPase, by binding with the sodium binding site on the extracellular side. This causes an increase in internal sodium, which has an inhibitory effect on the sodium–calcium antiport exchanger, slowing down the rate by which calcium

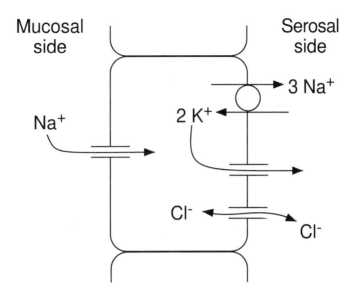

**Figure 2.16** Schematic diagram of the model of a $Na^+$-transporting epithelial cell, based on the model of Koefoed-Johnsen and Ussing (1958).

exits the cells. Increased levels of calcium result in increased myocardial contractility, a positive and useful effect. However, it is also clear that at higher levels, the effect of digitalis is toxic.

### 2.8.2  Volume Regulation and Ionic Transport

Many cells have a more difficult problem to solve, that of maintaining their cell volume in widely varying conditions, while transporting large quantities of ions through the cell. Here we present a simplified model for transport and volume regulation in a $Na^+$-transporting epithelial cell.

As are virtually all models of transporting epithelia, the model is based on that of Koefoed-Johnsen and Ussing (1958), the so-called KJU model. In the KJU model, an epithelial cell is modeled as a single cell layer separating a mucosal solution from the serosal solution (Fig. 2.16). (The mucosal side of an epithelial cell is that side on which mucus is secreted and from which various chemicals are withdrawn, for example, from the stomach. The serosal side is the side of the epithelial cell facing the interstitium, wherein lie capillaries, etc.) $Na^+$ transport is achieved by separating the $Na^+$ pumping machinery from the channels that allow $Na^+$ entry into the cell. Thus, the mucosal membrane contains $Na^+$ channels that allow $Na^+$ to diffuse down its concentration gradient into the cell, while the serosal membrane contains the $Na^+$-

K$^+$ ATPases, which remove Na$^+$ from the cell. The overall result is the transport of Na$^+$ from the mucosal side of the cell to the serosal side. The important question is whether the cell can maintain a steady volume under widely varying concentrations of Na$^+$ on the mucosal side.

We begin by letting $N$, $K$, and $C$ denote Na$^+$, K$^+$, and Cl$^-$ concentrations respectively, and letting subscripts $m$, $i$, and $s$ denote mucosal, intracellular and serosal concentrations. Thus, for example, $N_i$ is the intracellular Na$^+$ concentration, and $N_m$ is the mucosal Na$^+$ concentration. We now write down the conservation equations for Na$^+$, K$^+$, and Cl$^-$ at steady state. The conservation equations are the same as those of the pump–leak model with some minor exceptions. First, instead of the linear $I$–$V$ curve used in the pump–leak model, we use the GHK formulation to represent the ionic currents. This makes little qualitative change to the results and is more convenient because it simplifies the analysis that follows. Second, we assume that the rate of the Na$^+$–K$^+$ pump is proportional to the intracellular Na$^+$ concentration, $N_i$, rather than $N_i^3$, as was assumed in the generalized version of the pump–leak model. Thus,

$$P_{\text{Na}}v\frac{N_i - N_m e^{-v}}{1 - e^{-v}} + 3qpN_i = 0, \tag{2.109}$$

$$P_{\text{K}}v\frac{K_i - K_s e^{-v}}{1 - e^{-v}} - 2qpN_i = 0, \tag{2.110}$$

$$P_{\text{Cl}}v\frac{C_i - C_s e^{v}}{1 - e^{v}} = 0. \tag{2.111}$$

Note that the voltage, $v$, has been scaled by $F/(RT)$ and that the rate of the Na$^+$–K$^+$ pump is $pN_i$. Also note that the inward Na$^+$ current is assumed to enter from the mucosal side, and thus $N_m$ appears in the GHK current expression, but that no other ions enter from the mucosa. Here the membrane potential is assumed to be the same across the lumenal membrane and across the basal membrane. This is not quite correct, as the potential across the lumenal membrane is typically $-67$ mV while across the basal membrane it is about $-70$ mV.

There are two further equations to describe the electroneutrality of the intracellular space and the osmotic balance. In steady state, these are, respectively,

$$w(N_i + K_i - C_i) + z_x X = 0, \tag{2.112}$$

$$N_i + K_i + C_i + \frac{X}{w} = N_s + K_s + C_s, \tag{2.113}$$

where $X$ is the number of moles of protein, each with a charge of $z_x \leq -1$, that are trapped inside the cell, and $w$ is the cell volume. Finally, the serosal solution is assumed to be electrically neutral, and so in specifying $N_s, K_s$, and $C_s$ we must ensure that

$$N_s + K_s = C_s. \tag{2.114}$$

Since the mucosal and serosal concentrations are assumed to be known, we now have a system of 5 equations to solve for the 5 unknowns, $N_i, K_i, C_i, v$, and $\mu = w/X$. First, notice that (2.109), (2.110), and (2.111) can be solved for $N_i, K_i$, and $C_i$, respectively, to get

$$N_i(v) = \frac{vN_m e^{-v}}{v + 3\rho_n(1 - e^{-v})}, \tag{2.115}$$

$$K_i(v) = 2\rho_k N_i(v)\frac{1 - e^{-v}}{v} + K_s e^{-v}, \tag{2.116}$$

$$C_i(v) = C_s e^v, \tag{2.117}$$

where $\rho_n = pq/P_{\text{Na}}$ and $\rho_k = pq/P_{\text{K}}$.

Next, eliminating $N_i + K_i$ between (2.112) and (2.113), we find that

$$2\mu(C_i - C_s) = z_x - 1. \tag{2.118}$$

We now use (2.117) to find that

$$z_x - 1 = 2\mu C_s(e^v - 1), \tag{2.119}$$

and thus, using (2.119) to eliminate $\mu$ from (2.112), we get

$$N_i(v) + K_i(v) = \frac{C_s}{1 - z_x}[-2z_x + e^v(1 + z_x)] \equiv \phi(v). \tag{2.120}$$

Since $z_x - 1 < 0$, it must be (from (2.119)) that $v < 0$, and as $v \to 0$, the cell volume becomes infinite. Thus, we wish to find a negative solution of (2.120), with $N_i(v)$ and $K_i(v)$ specified by (2.115) and (2.116).

It is instructive to consider when solutions for $v$ (with $v < 0$) exist. First, notice that $\phi(0) = C_s$. Further, since $z_x \leq -1$, $\phi$ is a decreasing function of $v$, bounded above, with decreasing slope (i.e., concave down), as sketched in Fig. 2.17. Next, from (2.115) and (2.116) we determine that $N_i(v) + K_i(v)$ is a decreasing function of $v$ that approaches $\infty$ as $v \to -\infty$ and approaches zero as $v \to \infty$. It follows that a negative solution for $v$ exists if $N_i(0) + K_i(0) < C_s$, i.e., if

$$\frac{N_m}{1 + 3\rho_n} + \frac{2\rho_k N_m}{1 + 3\rho_n} + K_s < C_s. \tag{2.121}$$

Since $K_s + N_s = C_s$, this becomes

$$\frac{N_m}{N_s} < \frac{1 + 3\rho_n}{1 + 2\rho_k}. \tag{2.122}$$

This condition is sufficient for the existence of a solution, but not necessary. That is, if this condition is satisfied, we are assured that a solution exists, but if this condition fails to hold, it is not certain that a solution fails to exist. The problem, of course, is that negative solutions are not necessarily unique, nor is it guaranteed that increasing $N_m$ through $N_s\frac{1+3\rho_n}{1+2\rho_k}$ causes a negative solution to disappear. It is apparent from (2.115) and (2.116) that $N_i(v)$ and $K_i(v)$ are monotone increasing functions of the parameter $N_m$, so that no negative solutions exist for $N_m$ sufficiently large. However, for $N_m = N_s\frac{1+3\rho_n}{1+2\rho_k}$ to be the value at which the cell bursts by *increasing* $N_m$, it must also be true that

$$N_i'(0) + K_i'(0) < \phi'(0), \tag{2.123}$$

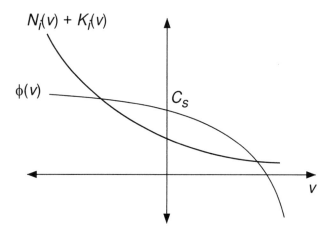

**Figure 2.17** Sketch (not to scale) of the function $\phi(v)$, defined as the right-hand side of (2.120), and of $N_i(v) + K_i(v)$, where $N_i$ and $K_i$ are defined by (2.115) and (2.116). $\phi(v)$ is sketched for $z_x < -1$.

or that

$$4(1 + 3\rho_n)C_s + N_s(1 - z_x)\frac{3\rho_n - 2\rho_k}{1 + 2\rho_k} > 0. \tag{2.124}$$

For the remainder of this discussion we assume that this condition holds, so that the failure of (2.122) also implies that the cell bursts.

According to (2.122), a transporting epithelial cell can maintain its cell volume, provided that the ratio of mucosal to serosal concentrations is not too large. When $N_m/N_s$ becomes too large, $\mu$ becomes unbounded, and the cell bursts. Typical solutions for the cell volume and membrane potential, as functions of the mucosal $Na^+$ concentration, are shown in Fig. 2.18.

Obviously, this state of affairs is unsatisfactory. In fact, some epithelial cells, such as those in the loop of Henle in the nephron (Chapter 20), must work in environments with extremely high mucosal sodium concentrations. To do so, these $Na^+$-transporting epithelial cells have mechanisms to allow operation over a much wider range of mucosal $Na^+$ concentrations than suggested by this simple model.

From (2.122) we can suggest some mechanisms by which a cell might avoid bursting at high mucosal concentrations. For example, the possibility of bursting is decreased if $\rho_n$ is increased or if $\rho_k$ is decreased. The reasons for this are apparent from (2.115) and (2.116), since $N_i(v) + K_i(v)$ is a decreasing function of $\rho_n$ and an increasing function of $\rho_k$. From a physical perspective, increasing $N_m$ causes an increase in $N_i$, which increases the osmotic pressure, inducing swelling. Decreasing the conductance of sodium ions from the mucosal side helps to control this swelling. Similarly, increasing the conductance of potassium ions allows more potassium ions to flow out of the cell, thereby decreasing the osmotic pressure from potassium ions and counteracting the tendency to swell.

It has been conjectured for some time that epithelial cells use both of these mechanisms to control their volume (Schultz, 1981; Dawson and Richards, 1990; Beck et al., 1994). There is evidence that as $N_i$ increases, epithelial cells decrease the $Na^+$ conduc-

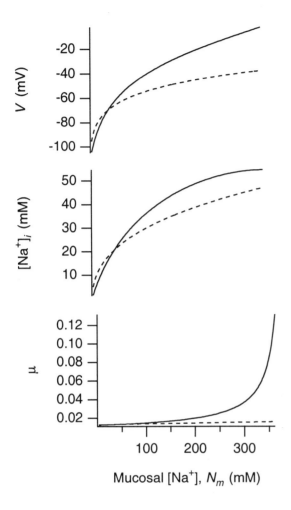

**Figure 2.18** Numerical solutions of the model for epithelial cell volume regulation and $Na^+$ transport. The membrane potential $V$, the scaled cell volume $\mu$, and the intracellular $Na^+$ concentration $[Na^+]_i$ are plotted as functions of the mucosal $Na^+$ concentration. The solid lines are the solutions of the simpler version of the model, where $P_{Na}$ and $P_K$ are assumed to be constant. The dashed lines are the solutions of the model when $P_{Na}$ is assumed to be a decreasing function of $N_i$, and $P_K$ is assumed to be an increasing function of $w$, as described in the text. Parameter values are $K_s = 2.5$, $N_s = 120$, $C_s = 122.5$, $P = 2$, $\gamma = 0.3$, $z_x = -2$. All concentrations are in mM.

tance on the mucosal side of the cell, thus restricting $Na^+$ entry. There is also evidence that as the cell swells, the $K^+$ conductance is increased, possibly by means of stretch-activated $K^+$ channels (Ussing, 1982. This assumption was used in the modeling work of Strieter et al., 1990).

To investigate the effects of these mechanisms in our simple model, we replace $P_{Na}$ by $P_{Na}20/N_i$ (20 is a scale factor, so that when $N_i = 20$ mM, $P_{Na}$ has the same value as in the original version of the model) and replace $P_K$ by $P_Kw/w_0$, where $w_0$ is the volume of the cell when $N_m = 100$ mM. As before, we can solve for $v$ and $\mu$ as functions of $N_m$, and the results are shown in Fig. 2.18. Clearly the incorporation of these mechanisms decreases the variation of cell volume and allows the cell to survive over a much wider range of mucosal $Na^+$ concentrations.

The model for control of ion conductance used here is extremely simplistic, as for example, there is no parametric control of sensitivity, and the model is heuristic, not mechanistic. More realistic and mechanistic models have been constructed and analyzed in detail (Lew et al., 1979; Civan and Bookman, 1982; Strieter et al., 1990; Weinstein, 1992, 1994, 1996; Tang and Stephenson, 1996).

## 2.9  EXERCISES

1.  Find the maximal enhancement for diffusion of carbon dioxide via binding with myoglobin using $D_s = 9 \times 10^{-4} \, cm^2/s$, $k_+ = 2 \times 10^8 \, cm^3/M \cdot s$, $k_- = 1.7 \times 10^{-2} \sigma$. Compare the amount of facilitation of carbon dioxide transport with that of oxygen at similar concentration levels. (Hint: For oxygen $\rho = 14$, whereas for carbon dioxide, $\rho = 2$.)

2.  Devise a model to determine the rate of production of product for a spherical enzyme capsule of radius $R_0$ in a bath of substrate at concentration $S_0$. Assume that the enzyme cannot diffuse within the capsule but that the substrate and product can freely diffuse into, within, and out of the capsule. Show that spheres of small radius have a larger rate of production than spheres of large radius.
    Hint: Reduce the problem to the nondimensional boundary value problem

    $$\frac{1}{y^2}(y^2\sigma')' - \alpha^2 \frac{\sigma}{\sigma + 1} = 0, \tag{2.125}$$

    $$\sigma'(0) = 0, \tag{2.126}$$

    $$\sigma(1) = \sigma_0, \tag{2.127}$$

    and solve numerically as a function of $\alpha$. How does the radius of the sphere enter the parameter $\alpha$?

3.  Suppose a membrane that contains water-filled pores separates two solutions.

    (a)  Suppose that the solution on either side of the membrane contains an impermeant solute. Show that the hydrostatic pressure of the water within the pores must be less than the hydrostatic pressure of the solutions on either side of the membrane.

    (b)  Show that if the solute can permeate the pore freely, no such drop in hydrostatic pressure exists.

    (c)  Show that if the solution on one side of the membrane contains both permeant and impermeant solutes, while the solution on the other side contains only the permeant solute, it is possible for water to flow against its chemical potential gradient (at least temporarily). This problem of wrong-way water flow has been observed experimentally, and it is discussed in detail by Dawson (1992).

4.  Red blood cells have a passive exchanger that exchanges a single $Cl^-$ ion for a bicarbonate ($HCO_3^-$) ion. Develop a model for this exchanger and find the flux.

5.  Modify (2.52) to take into account the fact that the concentration of ATP affects the rate of reaction (since pumping should stop if there is no ATP).

6.  Generalize (2.53) to account for the fact that with each turn of the sodium–potassium pump three sodium ions are exchanged for two potassium ions.

7.  Almost immediately upon entering a cell, glucose is phosphorylated in the first reaction step of glycolysis. How does this rapid and nearly unidirectional reaction affect the trans-membrane flux of glucose as represented by (2.41)? How is this reaction affected by the concentration of ATP?

8.  How does the concentration of ATP affect the rate of the sodium–potassium pump?

9.  The process by which calcium is taken up into the sarcoplasmic reticulum (SR) in muscle and cardiac cells is similar to the sodium–potassium ATPase, but simpler. Two intracellular calcium ions bind with a carrier protein with high affinity for calcium. ATP is dephosphorylated, with the phosphate bound to the carrier. There is a conformational change of the carrier protein that exposes the calcium to the interior of the SR and reduces the affinity of the binding sites, thereby releasing the two ions of calcium. The phosphate is released and the conformation changed so that the calcium binding sites are once again exposed to the intracellular space.
    Formalize this reaction and find the rate of calcium uptake by this pump.

10. A 1.5 oz bag of potato chips (a typical single serving) contains about 200 mg of sodium. When eaten and absorbed into the body, how many osmoles does this bag of potato chips represent?

11. Generalize formula (2.86) to take into account that the two fluids have different densities and to allow the columns to have different cross-sectional areas.

12. Two columns with cross-sectional area 1 cm$^2$ are initially filled to a height of one meter with water at $T = 300K$. Suppose 1 gm of sugar is dissolved in one of the two columns. How high will the sugary column be when equilibrium is reached? Hint: The weight of a sugar molecule is $3 \times 10^{-22}$ gm, and the force of gravity on 1cm$^3$ of water is 980 dynes.

13. Suppose an otherwise normal cell is placed in a bath of high extracellular potassium. What happens to the cell volume and resting potentials?

14. Based on what you know about glycolysis from Chapter 1, how would you expect anoxia (insufficient oxygen) to affect the volume of the cell? How might you incorporate this into a model of cell volume? Hint: Lactic acid does not diffuse out of a cell as does carbon dioxide.

15. Suppose 90% of the sodium in the bath of a squid axon is replaced by inert choline, preserving electroneutrality. What happens to the equilibrium potentials and membrane potentials?

16. Determine the effect of temperature (through the Nernst equation) on cell volume and membrane potential.

17. Write and analyze the balance equations for a cell in a finite bath. Hint: In a finite bath the total volume is conserved as are the total number of sodium, potassium, and chloride ions.

18. Simulate the time-dependent differential equations governing cell volume and ionic concentrations.

19. Many animal cells swell and burst when treated with the drug ouabain. Why? Hint: Ouabain competes with $K^+$ for external potassium binding sites of the $Na^+$–$K^+$ ATPase. How would you include this effect in a model of cell volume control?

# CHAPTER 3

# Membrane Ion Channels

Every cell membrane contains ion channels, macromolecular pores that allow specific ions to travel through the channels by a passive process, driven by their concentration gradient and the membrane potential. One of the most extensively studied problems in physiology is the regulation of such ionic currents. Indeed, in practically every chapter of this book we see examples of how the control of ionic current is vital for cellular function. Already we have seen how the cell membrane uses ion channels and pumps to maintain an intracellular environment that is different from the extracellular environment, and we have seen how such ionic separation results in a membrane potential. In subsequent chapters we will see that modulation of the membrane potential is one of the most important ways in which cells control their behavior or communicate with other cells. However, to understand the role played by ion channels in the control of membrane potential, it is first necessary to understand how membrane ionic currents depend on the voltage and ionic concentrations.

There is a vast literature, both theoretical and experimental, on the properties of ion channels. One of the best books on the subject is that of Hille (1992), to which the reader is referred for a more detailed presentation than that given here. The bibliography given there will also serve as a starting point for more detailed studies.

## 3.1 Current–Voltage Relations

Before we discuss specific models for ion channels, we emphasize an important fact that can be a source of confusion to the novice. Although the Nernst equation (2.54) for the equilibrium voltage generated by ionic separation can be derived from thermodynamic considerations and is thus universally applicable, there is no universal expression for the ionic current. An expression for, say, the $Na^+$ current cannot be derived from thermodynamic first principles and depends on the particular model used to describe membrane $Na^+$ channels. Already we have seen two different models for ionic currents. In the previous chapter we discussed two common models for $Na^+$ current as a function of the membrane potential and the internal and external $Na^+$ concentrations. In the simpler model, we assumed that the $Na^+$ current across the cell membrane was a linear function of the membrane potential, with a driving force given by the $Na^+$ Nernst potential. Thus,

$$I_{Na} = g_{Na}(V - V_{Na}), \tag{3.1}$$

where $V_{Na} = (RT/F)\ln([Na^+]_e/[Na^+]_i)$ is the Nernst potential of $Na^+$. (As usual, a subscript $e$ denotes the external concentration, while a subscript $i$ denotes the internal concentration.) Note that the $Na^+$ current is zero when $V$ is the Nernst potential, as must be the case. However, we also discussed an alternative model, where integration of the Nernst–Planck equation (2.59), assuming a constant electric field, gave the Goldman–Hodgkin–Katz (GHK), or constant-field, current equation:

$$I_{Na} = P_{Na}\frac{F^2}{RT}V\left[\frac{[Na^+]_i - [Na^+]_e \exp\left(\frac{-VF}{RT}\right)}{1 - \exp\left(\frac{-VF}{RT}\right)}\right]. \tag{3.2}$$

As before, the $Na^+$ current is zero when $V$ equals the Nernst potential, but here the current is a nonlinear function of the voltage. In Fig. 3.1A we compare the linear and GHK $I–V$ curves when there is only a single ion present.

There is no one "correct" expression for the $Na^+$ current, or any other ionic current for that matter. Different cells have different types of ion channels, each of which may have a current–voltage relation different from the rest. The challenge is to determine the current–voltage, or $I–V$, curve for a given ion channel and relate it to underlying biophysical mechanisms.

Our choice of these two models as examples was not coincidental, as they are the two most commonly used in theoretical models of cellular electrical activity. Not only are they relatively simple (at least compared to some of the other models we discuss later in this chapter), they also provide good quantitative descriptions of many ion channels. For example, the $I–V$ curves of open $Na^+$ and $K^+$ channels in the squid giant axon are approximately linear, and thus the linear model was used by Hodgkin and Huxley in their classic model of the squid giant axon (discussed in detail in Chapter 4). However, the $I–V$ curves of open $Na^+$ and $K^+$ channels in vertebrate axons are better described by the GHK equation, and so nonlinear $I–V$ curves are used for vertebrate models (Frankenhaeuser, 1960a,b, 1963; Campbell and Hille, 1976).

Because of the importance of these two models, we illustrate another way in which they differ. This also serves to illustrate the fact that although the Nernst potential is universal when there is only one ion present, the situation is more complicated when two or more species of ion can pass through the membrane. If both $Na^+$ and $K^+$ ions are present and both obey the GHK current equation, we showed in (2.72) that the reversal potential $V_r$ at which there is no net current flow is given by

$$V_r = \frac{RT}{F} \ln \left( \frac{P_{\text{Na}}[\text{Na}^+]_e + P_{\text{K}}[\text{K}^+]_e}{P_{\text{Na}}[\text{Na}^+]_i + P_{\text{K}}[\text{K}^+]_i} \right). \tag{3.3}$$

However, if we assume instead that the $I$–$V$ curves for $\text{Na}^+$ and $\text{K}^+$ are linear, then the reversal potential is given by

$$V_r = \frac{g_{\text{Na}} V_{\text{Na}} + g_{\text{K}} V_{\text{K}}}{g_{\text{Na}} + g_{\text{K}}}, \tag{3.4}$$

where $V_{\text{K}}$ is the Nernst potential of $\text{K}^+$. Clearly, the reversal potential is model-dependent. This is due to the fact that at the reversal potential the net current flow is zero, but the individual $\text{Na}^+$ and $\text{K}^+$ currents are not. Thus, the equilibrium arguments used to derive the Nernst equation do not apply, and a universal form for the reversal potential does not exist. As an illustration of this, in Fig. 3.1B we plot the reversal potentials $V_r$ from (3.3) and (3.4) as functions of $[\text{K}^+]_e$. Although the linear and GHK $I$–$V$ curves predict different reversal potentials, the overall qualitative behavior is similar, making it difficult to distinguish between a linear and a GHK $I$–$V$ curve on the basis of reversal potential measurements alone.

### 3.1.1  Steady-State and Instantaneous Current–Voltage Relations

Measurement of $I$–$V$ curves is complicated by the fact that ion channels can open or close in response to changes in the membrane potential. Suppose that in a population of ion channels, $I$ increases as $V$ increases. This increase could be the result of two different factors. One possibility is that more channels open as $V$ increases while the current through an individual channel remains unchanged. It is also possible that the same number of channels remain open but the current through each one increases. To understand how each channel operates, it is necessary to separate these two factors to determine the $I$–$V$ curve of a single open channel. This has motivated the definition of *steady-state* and *instantaneous* $I$–$V$ curves.

    If channels open or close in response to a change in voltage, but this response is slower than the change in current in an already open channel, it should be possible to measure the $I$–$V$ curve of a single open channel by changing the voltage quickly and measuring the channel current soon after the change. Presumably, if the measurement is performed fast enough, no channels in the population have time to open or close in response to the voltage change, and thus the observed current change reflects the current change through the open channels. Of course, this relies on the assumption that the current through each open channel changes instantaneously. The $I$–$V$ curve measured in this way (at least in principle) is called the instantaneous $I$–$V$ curve and reflects properties of the individual open channels. If the current measurement is performed after channels have had time to open or close, then the current change reflects the $I$–$V$ curve of a single channel as well as the proportion of open channels. In this way one obtains a steady-state $I$–$V$ curve.

**A**

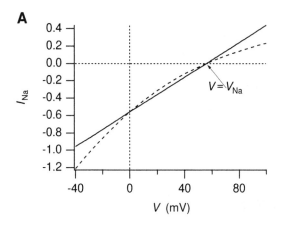

**B**

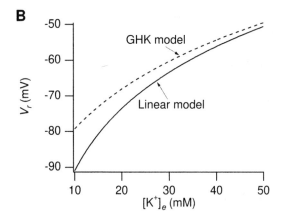

**Figure 3.1** A: $I$–$V$ curves of the linear and GHK models for $Na^+$ flux through a membrane. Both curves have the same reversal potential as expected, but the GHK model (dashed curve) gives a nonlinear $I$–$V$ curve. Typical concentrations and conductances of the squid axon were used: $[Na^+]_i = 50$ mM, $[Na^+]_e = 437$ mM, and $g_{Na} = 0.01$ mS/cm$^2$. $P_{Na}$ was chosen so that the GHK $I$–$V$ curve intersects the linear $I$–$V$ curve at $V = 0$. B: Reversal potentials of the linear and GHK models as functions of $[K^+]_e$. The membrane is permeable to both $Na^+$ and $K^+$. The same parameters as A, with $[K^+]_i = 397$ mM and $g_K = 0.367$ mS/cm$^2$. $P_K$ was chosen so that the GHK $I$–$V$ curve for $K^+$, with $[K^+]_e = 20$ mM, intersects the linear $I$–$V$ curve for $K^+$ at $V = 0$.

There are two basic types of model that are used to describe ion flow through open channels, and we discuss simple versions of each. In the first type of model, the channel is described as a continuous medium, and the ionic current is determined by the Nernst–Planck electrodiffusion equation, coupled to the electric field by means of the Poisson equation. In more complex models of this type, channel geometry and the effects of induced charge on the channel wall are incorporated. In the second type of model the channel is modeled as a sequence of binding sites, separated by barriers that impede the ion's progress: the passage of an ion through the channel is described as a process of "hopping" over barriers from one binding site to another. The height of each barrier is determined by the properties of the channel, as well as by the membrane potential. Thus, the rate at which an ion traverses the channel is a function both of the membrane potential and of the channel type. An excellent summary of the advantages and disadvantages of the two model types is given by Dani and Levitt (1990).

Finally, we discuss simple models for the kinetics of channel gating. These models will be of fundamental importance in Chapter 4, where we use an early model for the voltage-dependent gating of ion channels proposed by Hodgkin and Huxley as part of their model for the action potential in the squid giant axon. More detailed recent models for channel gating are not discussed at any length. The interested reader is referred to Hille (1992), Armstrong (1981), Armstrong and Bezanilla (1973, 1974, 1977), Aldrich et al. (1983), and Finkelstein and Peskin (1984) for a selection of models of how channels can open and close in response to changes in membrane potential. An important question that we do not consider here is how channels can discriminate between different ions. Detailed discussions of this and related issues are in Hille (1992) and the references therein.

## 3.2   Independence, Saturation, and the Ussing Flux Ratio

One of the most fundamental questions to be answered about an ion channel is whether the passage of an ion through the channel is independent of other ions. If so, the channel is said to obey the *independence principle*.

Suppose a membrane separates two solutions containing an ion species S with external concentration $c_e$ and internal concentration $c_i$. If the independence principle is satisfied, the flow of S is proportional to its local concentration, independent of the concentration on the opposite side of the membrane, and thus the flux from outside to inside, $J_{in}$, is

$$J_{in} = k_e c_e, \tag{3.5}$$

for some constant $k_e$. Similarly, the outward flux is given by

$$J_{out} = k_i c_i, \tag{3.6}$$

where in general, $k_e \neq k_i$. We let $V_S$ denote the Nernst potential of the ion S, and let $V$ denote the potential difference across the membrane. Now we introduce a hypothetical concentration $c_e^*$ defined as that external concentration necessary to maintain a Nernst potential $V$. Thus

$$\frac{c_e}{c_i} = \exp\left(\frac{V_S F}{RT}\right), \tag{3.7}$$

and

$$\frac{c_e^*}{c_i} = \exp\left(\frac{VF}{RT}\right). \tag{3.8}$$

When the external concentration is $c_e^*$ and the internal concentration is $c_i$, then the voltage is $V$, and there is no net flux across the membrane; i.e., the outward flux equals the inward flux, and so

$$k_e c_e^* = k_i c_i. \tag{3.9}$$

It follows that the flux ratio is given by

$$
\begin{aligned}
\frac{J_{\text{in}}}{J_{\text{out}}} &= \frac{k_e c_e}{k_i c_i} \\
&= \frac{k_e c_e}{k_e c_e^*} \\
&= \frac{c_e}{c_e^*} \\
&= \frac{\exp\left(\frac{V_S F}{RT}\right)}{\exp\left(\frac{VF}{RT}\right)} \\
&= \exp\left[\frac{(V_S - V)F}{RT}\right].
\end{aligned}
\tag{3.10}
$$

This expression for the ratio of the inward to the outward flux is usually called the *Ussing flux ratio*. It was first derived by Ussing (1949), although the derivation given here is due to Hodgkin and Huxley (1952a). Alternatively, the Ussing flux ratio can be written as

$$
\frac{J_{\text{in}}}{J_{\text{out}}} = \frac{c_e}{c_i} \exp\left(\frac{-VF}{RT}\right).
\tag{3.11}
$$

Note that when $V = 0$, the ratio of the fluxes is equal to the ratio of the concentrations, as might be expected intuitively.

As an illustration of the application of the Ussing flux ratio, suppose the $Na^+$ current is measured when the cell is immersed in a high $Na^+$ solution and then compared to the $Na^+$ current measured in a low $Na^+$ solution. The membrane potential and the internal $Na^+$ concentration are assumed to be the same in both cases. We let a prime denote quantities measured in the high $Na^+$ solution, and then

$$
\frac{I'_{Na}}{I_{Na}} = \frac{J'_{\text{out}} - J'_{\text{in}}}{J_{\text{out}} - J_{\text{in}}}.
\tag{3.12}
$$

Since the internal concentrations are the same, it follows from (3.6) that $J_{\text{out}} = J'_{\text{out}}$, and from (3.5) we find $J'_{\text{in}}/J_{\text{in}} = [Na^+]'_e/[Na^+]_e$. Substituting these into (3.12) and using the Ussing flux ratio, we find

$$
\frac{I'_{Na}}{I_{Na}} = \frac{([Na^+]'_e/[Na^+]_e)\exp[\frac{(V_{Na}-V)F}{RT}] - 1}{\exp[\frac{(V_{Na}-V)F}{RT}] - 1}.
\tag{3.13}
$$

Alternatively, this can be written as

$$
\frac{I'_{Na}}{I_{Na}} = \frac{[Na^+]_i - [Na^+]'_e \exp(\frac{-VF}{RT})}{[Na^+]_i - [Na^+]_e \exp(\frac{-VF}{RT})}.
\tag{3.14}
$$

By measuring the current ratio as a function of membrane potential, the $Na^+$ channel can thus conveniently be tested for independence.

Although many ion channels follow the independence principle approximately over a range of ionic concentrations, most show deviations from independence when the ionic concentrations are sufficiently large. This has motivated the development of models that show saturation at high ionic concentrations. For example, one could assume that ion flow through the channel can be described by a barrier-type model, in which the ion jumps from one binding site to another as it moves through the channel. If there are only a limited number of binding sites available for ion passage through the channel, and each binding site can bind only one ion, then as the ionic concentration increases there are fewer binding sites available, and so the flux is not proportional to the concentration. Equivalently, one could say that each channel has a single binding site for ion transfer, but there are only a limited number of channels. However, in many of these models the Ussing flux ratio is still obeyed, even though independence is not. Hence, although any ion channel obeying the independence principle must also satisfy the Ussing flux ratio, the converse is not true. We discuss saturating models later in this chapter.

Another way in which channels show deviations from independence is in flux-coupling. If ions can interact within a channel so that, for example, a group of ions must move through the channel together, then the Ussing flux ratio is not satisfied. The most common type of model used to describe such behavior is the so-called *multi-ion model*, in which it is assumed that there are a number of binding sites within a single channel and that the channel can bind multiple ions at the same time. The consequent interactions between the ions in the channel can result in deviations from the Ussing flux ratio. A more detailed consideration of multi-ion models is given later in this chapter. However, it is instructive to consider how the Ussing flux ratio is modified by a simple multi-ion channel mechanism in which the ions progress through the channel in single file (Hodgkin and Keynes, 1955).

Suppose a membrane separates two solutions, the external one (on the right) containing an ion S at concentration $c_e$, and the internal one (on the left) at concentration $c_i$. To keep track of where each S ion has come from, all the S ions on the left are labeled A, while those on the right are labeled B. Suppose also that the membrane contains $n$ binding sites and that S ions traverse the membrane by binding sequentially to the binding sites and moving across in single file. For simplicity we assume that there are no vacancies in the chain of binding sites. It follows that the possible configurations of the chain of binding sites are $[A_r, B_{n-r}]$, for $r = 0, \ldots, n$, where $[A_r, B_{n-r}]$ denotes the configuration such that the $r$ leftmost sites are occupied by A ions, while the rightmost $n - r$ sites are occupied by B ions. Notice that the only configuration that can result in the transfer of an A ion to the right-hand side is $[A_n B_0]$, i.e., if the chain of binding sites is completely filled with A ions.

Now we let $\alpha$ denote the total rate at which S ions are transferred from left to right. Since $\alpha$ denotes the total rate, irrespective of labeling, it does not take into account whether an A ion or a B ion is moved out of the channel from left to right. For this reason, $\alpha$ is not the same as the flux of labeled ions. Similarly, let $\beta$ denote the total flux of S ions, irrespective of labeling, from right to left. It follows that the rate at which $[A_r B_{n-r}]$ is converted to $[A_{r+1} B_{n-r-1}]$ is $\alpha[A_r B_{n-r}]$, and the rate of the reverse conversion is $\beta[A_{r+1} B_{n-r-1}]$. According to Hodgkin and Keynes, it is reasonable to assume that if there is a potential difference $V$ across the membrane, then the total flux ratio obeys the Ussing flux ratio,

$$\frac{\alpha}{\beta} = \frac{c_e}{c_i} \exp\left(\frac{-VF}{RT}\right).$$ (3.15)

This assumption is justified by the fact that a flux of one ion involves the movement of a single charge through the membrane (as in the independent case treated above) and thus should have the same voltage dependence. We emphasize that $\alpha/\beta$ is not the flux ratio of labeled ions, but the total flux ratio.

To obtain the flux ratio of labeled ions, notice that the rate at which A ions are transferred to the right-hand side is $\alpha[A_n B_0]$, and the rate at which B ions are transferred to the left hand side is $\beta[A_0 B_n]$. Thus, the flux ratio of labeled ions is

$$\frac{J_{\text{in}}}{J_{\text{out}}} = \frac{\alpha \, [A_n B_0]}{\beta \, [A_0 B_n]}.$$ (3.16)

At steady state there can be no net change in the distribution of configurations, so that

$$\frac{[A_{r+1} B_{n-r-1}]}{[A_r B_{n-r}]} = \frac{\alpha}{\beta}.$$ (3.17)

Thus,

$$\frac{J_{\text{in}}}{J_{\text{out}}} = \frac{\alpha \, [A_n B_0]}{\beta \, [A_0 B_n]} = \left(\frac{\alpha}{\beta}\right)^2 \frac{[A_{n-1} B_1]}{[A_0 B_n]} = \cdots = \left(\frac{\alpha}{\beta}\right)^{n+1},$$ (3.18)

so that

$$\frac{J_{\text{in}}}{J_{\text{out}}} = \left[\frac{c_e}{c_i} \exp\left(\frac{-VF}{RT}\right)\right]^{n+1}.$$ (3.19)

A similar argument, taking into account the fact that occasional vacancies in the chain arise when ions at the two ends dissociate and that these vacancies propagate through the chain, gives

$$\frac{J_{\text{in}}}{J_{\text{out}}} = \left[\frac{c_e}{c_i} \exp\left(\frac{-VF}{RT}\right)\right]^{n}.$$ (3.20)

Experimental data confirm this theoretical prediction (although historically, the theory was motivated by the experimental result, as is often the case). Hodgkin and Keynes (1955) showed that flux ratios in the $K^+$ channel of the *Sepia* giant axon could be described by the Ussing flux ratio raised to the power 2.5. Their result, as presented in modified form by Hille (1992), is shown in Fig. 3.2. Unidirectional $K^+$ fluxes were measured with radioactive $K^+$, and the ratio of the outward to the inward flux was plotted as a function of $V - V_K$. The best-fit line on a semilogarithmic plot has a slope of 2.5, which suggests that at least 2 $K^+$ ions traverse the $K^+$ channel simultaneously.

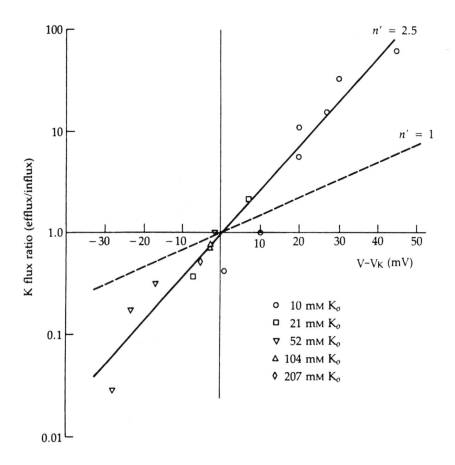

**Figure 3.2**   $K^+$ flux ratios as measured by Hodgkin and Keynes (1955), Fig. 7. Slightly modified into modern conventions by Hille (1992), page 375. $K_o$ is the external $K^+$ concentration, and $n'$ is the flux-ratio exponent, denoted by $n$ in (3.20). (Hille, 1992, Fig. 7, p. 375.)

## 3.3   Electrodiffusion Models

Most early work on ion channels was based on the theory of electrodiffusion. We saw in Chapter 2 that the movement of ions in response to a concentration gradient and an electric field is described by the Nernst–Planck equation,

$$J = -D\left(\frac{dc}{dx} + \frac{zF}{RT}c\frac{d\phi}{dx}\right),\tag{3.21}$$

where $J$ denotes the flux density, $c$ is the concentration of the ion under consideration, and $\phi$ is the electrical potential. If we make the simplifying assumption that the field $d\phi/dx$ is constant through the membrane, then (3.21) can be solved to give the Goldman–Hodgkin–Katz current and voltage equations (2.68) and (2.71). However, in general there is no reason to believe that the potential has a constant gradient in the membrane.

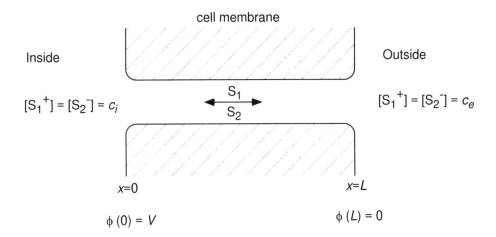

**Figure 3.3** Schematic diagram of the electrodiffusion model for current through an ionic channel. Each side of the channel is electrically neutral, and both ion types can diffuse through the channel.

Ions moving through the channel affect the local electric field, and this local field in turn affects ionic fluxes. Thus, to determine the electric field and consequent ionic fluxes, one must solve a coupled problem.

### 3.3.1 Multi-ion Flux: The Poisson–Nernst–Planck Equations

Suppose we have two types of ions, $S_1$ and $S_2$, with concentrations $c_1$ and $c_2$, passing through an ion channel, as shown schematically in Fig. 3.3. For convenience we assume that the valence of the first ion is 1 and that of the second is $-1$. Then the potential in the channel $\phi(x)$ must satisfy Poisson's equation,

$$\frac{d^2\phi}{dx^2} = -\frac{q}{\epsilon}(c_1 - c_2), \tag{3.22}$$

where $q$ is the unit electric charge and $\epsilon$ is the dielectric constant of the channel medium (usually assumed to be an aqueous solution). The flux densities $J_1$ and $J_2$ of $S_1$ and $S_2$ satisfy the Nernst–Planck equation, and at steady state $dJ_1/dx$ and $dJ_2/dx$ must both be zero to prevent any charge buildup within the channel. Hence, the steady-state flux through the channel is described by (3.22) coupled with

$$J_1 = -D_1\left(\frac{dc_1}{dx} + \frac{F}{RT}c_1\frac{d\phi}{dx}\right), \tag{3.23}$$

$$J_2 = -D_2\left(\frac{dc_2}{dx} - \frac{F}{RT}c_2\frac{d\phi}{dx}\right), \tag{3.24}$$

where $J_1$ and $J_2$ are constants. To complete the specification of the problem, it is necessary to specify boundary conditions for $c_1, c_2$, and $\phi$. We assume that the channel has length $L$, and that $x = 0$ denotes the left border, or inside, of the membrane. Then,

$$c_1(0) = c_i, \qquad c_1(L) = c_e,$$
$$c_2(0) = c_i, \qquad c_2(L) = c_e, \tag{3.25}$$
$$\phi(0) = V, \qquad \phi(L) = 0.$$

Note that we have specified that the solutions on both sides of the membrane are electrically neutral. $V$ is the potential difference across the membrane, defined, as usual, as the internal potential minus the external potential. While at first glance it might appear that there are too many boundary conditions for the differential equations, this is in fact not so, as the constants $J_1$ and $J_2$ are additional unknowns to be determined.

In general, it is not possible to obtain an exact solution to the Poisson–Nernst–Planck (PNP) equations (3.22)–(3.25). However, some simplified cases can be solved approximately. A great deal of work on the PNP equations has been done by Eisenberg and his colleagues (Chen, Barcilon, and Eisenberg, 1992; Barcilon, 1992; Barcilon, Chen, and Eisenberg, 1992; Chen and Eisenberg, 1993). Here we present simplified versions of their models, ignoring, for example, the charge induced on the channel wall by the presence of ions in the channel, and considering only the movement of two ion types, rather than three, through the channel. Similar models have also been discussed by Peskin (1991).

It is convenient first to nondimensionalize the PNP equations. We let $x^* = x/L$, $\phi^* = \phi F/RT$, $v = VF/RT$, $c_1^* = c_1/\tilde{c}$, and similarly for $c_2, c_i$, and $c_e$, where $\tilde{c} = c_e + c_i$. Substituting into (3.22)–(3.24) and dropping the stars, we find

$$-\bar{J}_1 = \frac{dc_1}{dx} + c_1 \frac{d\phi}{dx}, \tag{3.26}$$

$$-\bar{J}_2 = \frac{dc_2}{dx} - c_2 \frac{d\phi}{dx}, \tag{3.27}$$

$$\frac{d^2\phi}{dx^2} = -\lambda^2(c_1 - c_2), \tag{3.28}$$

where $\lambda^2 = L^2 q F \tilde{c}/(\epsilon RT)$, $\bar{J}_1 = J_1 L/(\tilde{c} D_1)$, and similarly for $\bar{J}_2$. The boundary conditions are

$$c_1(0) = c_i, \qquad c_1(1) = c_e,$$
$$c_2(0) = c_i, \qquad c_2(1) = c_e,$$
$$\phi(0) = v, \qquad \phi(1) = 0.$$

## The short-channel or low concentration limit

If the channel is short or the ionic concentrations on either side of the membrane are small, so that $\lambda \ll 1$, we can find an approximate solution to the PNP equations by setting $\lambda = 0$. This gives

$$\frac{d^2\phi}{dx^2} = 0, \tag{3.29}$$

and thus

$$\frac{d\phi}{dx} = -v. \tag{3.30}$$

Hence, $\lambda \approx 0$ implies that the electric potential has a constant gradient in the membrane, which is exactly the constant field assumption that was made in the derivation of the GHK equations (Chapter 2). The equation for $c_1$ is then

$$\frac{dc_1}{dx} - vc_1 = -\bar{J}_1, \tag{3.31}$$

and thus

$$c_1 = \frac{\bar{J}_1}{v} + K_1 e^{vx}. \tag{3.32}$$

From the boundary conditions $c_1(0) = c_i, c_1(1) = c_e$ it follows that

$$\bar{J}_1 = v \cdot \frac{c_i - c_e e^{-v}}{1 - e^{-v}}. \tag{3.33}$$

In dimensional form, this is

$$I_1 = FJ_1 = \frac{D_1}{L}\frac{F^2}{RT} \cdot V \cdot \left( \frac{c_i - c_e \exp(\frac{-VF}{RT})}{1 - \exp(\frac{-VF}{RT})} \right), \tag{3.34}$$

which is, as expected, the GHK current equation. Graphs of the concentration and voltage profiles through the membrane are shown in Fig. 3.4. It is reassuring that the widely used GHK equation for the ionic flux can be derived as a limiting case of a more general model.

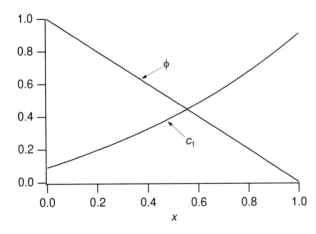

Figure 3.4 Graphs of the concentration and potential profiles for the short-channel limit of the Poisson–Nernst–Planck equations. Dimensionless parameters were set arbitrarily at $c_i = 50/550 = 0.091$, $c_e = 500/550 = 0.909$, $v = 1$. In this limit the electric field is constant through the channel (the potential has a constant slope), the concentration profile is nonlinear, and the GHK $I$–$V$ curve is obtained.

## The long-channel limit

Another interesting limit is obtained by letting the length of the channel go to infinity. If we let $\eta = 1/\lambda$ denote a small parameter, the model equations are

$$-\bar{J}_1 = \frac{dc_1}{dx} + c_1 \frac{d\phi}{dx}, \tag{3.35}$$

$$-\bar{J}_2 = \frac{dc_2}{dx} - c_2 \frac{d\phi}{dx}, \tag{3.36}$$

$$-\eta^2 \frac{d^2\phi}{dx^2} = (c_1 - c_2). \tag{3.37}$$

Since there is a small parameter multiplying the highest derivative, this is a singular perturbation problem. The solution obtained by setting $\eta = 0$ does not, in general, satisfy all the boundary conditions, as the degree of the differential equation has been reduced, resulting in an overdetermined system. In the present case, however, this reduction of order is not a problem.

Setting $\eta = 0$ in (3.37) gives $c_1 = c_2$, which happens to satisfy both the left and right boundary conditions. Thus, $c_1$ and $c_2$ are identically equal throughout the channel. From (3.35) and (3.36) it follows that

$$\frac{d}{dx}(c_1 + c_2) = -\bar{J}_1 - \bar{J}_2. \tag{3.38}$$

Since both $\bar{J}_1$ and $\bar{J}_2$ are constants, it follows that $dc_1/dx$ is a constant, and hence, from the boundary conditions,

$$c_1 = c_2 = c_i + (c_e - c_i)x. \tag{3.39}$$

We are now able to solve for $\phi$. Subtracting (3.37) from (3.36) gives

$$2c_1 \frac{d\phi}{dx} = 2\tilde{J}, \tag{3.40}$$

where $2\tilde{J} = \bar{J}_2 - \bar{J}_1$, and hence

$$\phi = \frac{\tilde{J}}{c_e - c_i} \ln[c_i + (c_e - c_i)x] + K, \tag{3.41}$$

for some other constant $K$. Applying the boundary conditions $\phi(0) = v, \phi(1) = 0$ we determine $\tilde{J}$ and $K$, with the result that

$$\phi = -\frac{v}{v_1} \ln\left[\frac{c_i}{c_e} + \left(1 - \frac{c_i}{c_e}\right)x\right], \tag{3.42}$$

where $v_1 = \ln(c_e/c_i)$ is the dimensionless Nernst potential of ion $S_1$. The flux density of one of the ions, say $S_1$, is obtained by substituting the expressions for $c_1$ and $\phi$ into (3.35) to get

$$\bar{J}_1 = \frac{c_e - c_i}{v_1}(v - v_1), \tag{3.43}$$

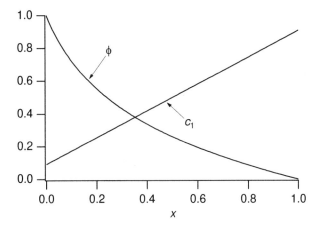

**Figure 3.5** Graphs of the concentration and potential profiles for the long-channel limit of the Poisson–Nernst–Planck equations. Dimensionless parameters were set arbitrarily at $c_i = 50/550 = 0.091$, $c_e = 500/550 = 0.909$, $v = 1$. In this limit the concentration profile has a constant slope, the potential profile is nonlinear, and the linear $I$–$V$ curve is obtained.

which is the linear $I$–$V$ curve that we met previously. Graphs of the corresponding concentration and voltage profiles through the channel are shown in Fig. 3.5.

In summary, by taking two different limits of the PNP equations we obtain either the GHK $I$–$V$ curve or a linear $I$–$V$ curve. In the short-channel limit, $\phi$ has a constant gradient through the membrane, but the concentration does not. In the long-channel limit the reverse is true, with a constant gradient for the concentration through the channel, but not for the potential. It is left as an exercise to prove that although the GHK equation obeys the independence principle and the Ussing flux ratio, the linear $I$–$V$ curve obeys neither. Given the above derivation of the linear curve, this is not surprising. A linear $I$–$V$ curve is obtained when either the channel is very long or the ionic concentrations on either side of the channel are very high. In either case, one does not expect the movement of each ion through the channel to be independent of other ions, and so one expects the independence principle to fail. Conversely, the GHK equation is obtained in the limit of low ionic concentrations or short channels, in which case the independent movement of ions is not unexpected.

## 3.4 Barrier Models

The second type of model that has been widely used to describe ion channels is based on the assumption that the movement of an ion through the channel can be modeled as the jumping of an ion over a discrete number of free-energy barriers (Eyring et al., 1949; Woodbury, 1971; Läuger, 1973). It is assumed that the potential energy of an ion passing through a channel is described by a potential energy profile of the general form shown in Fig. 3.6. The peaks of the potential energy profile correspond to barriers that impede the ion flow, while the local minima correspond to binding sites within the channel.

To traverse the channel the ion must hop from one binding site to another. According to the theory of chemical reaction rates, the rate at which an ion jumps from one

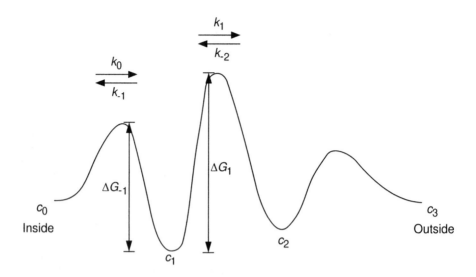

**Figure 3.6** General potential energy profile for barrier models. The local minima correspond to binding sites within the channel, and the local maxima are barriers that impede the ion flow. An ion progresses through the channel by hopping over the barriers from one binding site to another.

binding site to the next is an exponential function of the height of the potential energy barrier that it must cross. Thus, in the notation of the diagram,

$$k_j = \kappa \exp\left(\frac{-\Delta G_j}{RT}\right), \tag{3.44}$$

for some factor $\kappa$ with units of 1/time. One of the most difficult questions in the use of this expression is deciding on the precise form of the factor. According to Eyring rate theory (as used in this context by Hille (1992), for example), $\kappa = kT/h$, where $k$ is Boltzmann's constant, $T$ is the temperature, and $h$ is Planck's constant. The derivation of this expression for $\kappa$ relies on the quantization of the energy levels of the ion in some transition state as it binds to the channel binding sites. However, it is not clear whether at room temperature energy quantization has an important effect on ionic flows. Using methods from nonequilibrium statistical thermodynamics, an alternative form of the factor has been derived by Kramers (1940), and discussions of this, and other, alternatives may be found in McQuarrie (1967) and Laidler (1969). We do not enter this debate here, as it is unnecessary for our purposes. All we require is some factor, of plausible value, that can be used to fit the rate expressions to experimental data.

For simplicity, we assume that each local maximum occurs halfway between the local minima on each side. Barriers with this property are called *symmetrical*. An electric field in the channel also affects the rate constants. If the potential difference across the cell membrane is positive (so that the inside is more positive than the outside), it is easier for positive ions to cross the barriers in the outward direction but more difficult for positive ions to enter the cell. Thus, the heights of the barriers in the outward direction are reduced, while the heights in the inward direction are increased. If there is a potential difference of $\Delta V_j$ over the $j$th barrier, then

$$k_j = \kappa \exp\left[\frac{1}{RT}(-\Delta G_j + zF\Delta V_{j+1}/2)\right], \tag{3.45}$$

$$k_{-j} = \kappa \exp\left[\frac{1}{RT}(-\Delta G_{-j} - zF\Delta V_j/2)\right]. \tag{3.46}$$

The factor 2 appears because the barriers are assumed to be symmetrical, so that the maxima are lowered by $zF\Delta V_j/2$. A simple illustration of this is given in Fig. 3.7A and B and is discussed in detail in the next section.

In addition to symmetry, the barriers are assumed to have another important property, namely, that in the absence of an electric field the ends of the energy profile are at the same height, and thus

$$\sum_{j=0}^{n-1} \Delta G_j - \sum_{j=1}^{n} \Delta G_{-j} = 0. \tag{3.47}$$

If this were not so, then in the absence of an electric field and with equal concentrations on either side of the membrane, there would be a nonzero flux through the membrane, a situation that is clearly unphysiological.

A number of different models have been constructed along these general lines. First, we consider the simplest type of barrier model, in which the ionic concentration in the channel can become arbitrarily large, i.e., the channel does not saturate. This is similar to the continuous models discussed above and can be thought of as a discrete approximation to the constant field model. Because of this, nonsaturating models give the GHK $I$–$V$ curve in the limit of a homogeneous membrane. We then discuss saturating barrier models and multi-ion models. Before we do so, however, it is important to note that although barrier models can provide good quantitative descriptions of some experimental data, they are phenomenological. In other words, apart from the agreement between theory and experiment, there is often no reason to suppose that the potential energy barrier used to describe the channel corresponds in any way to physical properties of the channel. Thus, although their relative simplicity has led to their widespread use, mechanistic interpretations of the models should be made only with considerable caution. Of course, this does not imply that barrier models are inferior to continuous models such as the constant field model or the Poisson–Nernst–Planck equations, which suffer from their own disadvantages (Dani and Levitt, 1990).

## 3.4.1 Nonsaturating Barrier Models

In the simplest barrier model (Eyring et al., 1949; Woodbury, 1971), the potential energy barrier has the general form shown in Fig. 3.7A, and it is assumed that the movement of an ion S over a barrier is independent of the ionic concentrations at the neighboring

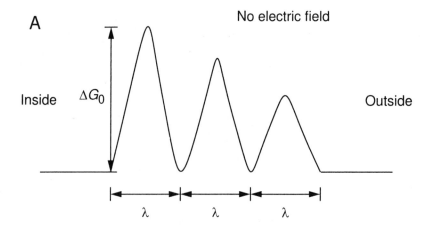

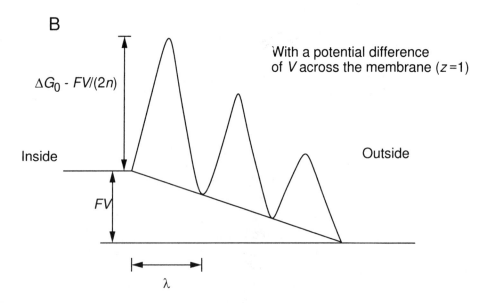

**Figure 3.7** The potential energy diagram used in the nonsaturating model of Woodbury (1971). There is an equal distance between the binding sites, and the barriers are symmetrical. A. In the absence of an electric field the barrier height decreases linearly through the membrane. B. The presence of a constant electric field skews the energy profile, bringing the outside end down relative to the inside. This increases the rate at which positive ions traverse the channel from inside to out and decreases their rate of entry.

barriers. This is equivalent to assuming that the concentration of S at any particular binding site can be arbitrarily large.

The internal concentration of S is denoted by $c_0$, while the external concentration is denoted by $c_n$. There are $n - 1$ binding sites (and thus $n$ barriers) in the membrane, and the concentration of S at the $j$th binding site is denoted by $c_j$. Note the slight change in notation from above. Instead of using $c_e$ and $c_i$ to denote the external and internal concentrations of S, we use $c_n$ and $c_0$. This allows the labeling of the concentrations on either side of the membrane to be consistent with the labeling of the concentrations at the binding sites. There is an equal voltage drop across each barrier, and thus the electrical distance between each binding site, denoted by $\lambda$, is the same. For convenience, we assume the stronger condition, that the physical distance between the binding sites is the same also, which is equivalent to assuming a constant electric field in the membrane. In the absence of an electric field, we assume that the heights of the energy barriers decrease linearly through the membrane, as in Fig. 3.7, with

$$\Delta G_j = \Delta G_0 - j\delta G, \tag{3.48}$$

for some constant increment $\delta G$. Finally, it is assumed that the flux from left to right, say, across the $j$th barrier, is proportional to $c_{j-1}$, and similarly for the flux in the opposite direction. Thus, the flux over the $j$th barrier, $J$, is given by

$$J = \lambda(k_{j-1}c_{j-1} - k_{-j}c_j). \tag{3.49}$$

Note that the units of $J$ are concentration $\times$ distance/time, or moles per unit area per time, so $J$ is a flux density. As usual, a flux from inside to outside (i.e., left to right) is defined as a positive flux.

At steady state the flux over each barrier must be the same, in which case we get a system of linear equations,

$$k_0c_0 - k_{-1}c_1 = k_1c_1 - k_{-2}c_2 = \cdots = k_{n-1}c_{n-1} - k_{-n}c_n = M, \tag{3.50}$$

where $M = J/\lambda$ is a constant. Hence

$$k_0c_0 = (k_1 + k_{-1})c_1 - k_{-2}c_2, \tag{3.51}$$

$$k_1c_1 = (k_2 + k_{-2})c_2 - k_{-3}c_3, \tag{3.52}$$

$$k_2c_2 = (k_3 + k_{-3})c_3 - k_{-4}c_4, \tag{3.53}$$

$$\vdots$$

Eventually, we need to determine $J$ in terms of the concentrations on either side of the membrane, $c_0$ and $c_n$. Solving (3.52) for $c_1$ and substituting into (3.51) gives

$$k_0c_0 = c_2k_2\left(1 + \frac{k_{-1}}{k_1} + \frac{k_{-1}k_{-2}}{k_1k_2}\right) - c_3k_{-3}\left(1 + \frac{k_{-1}}{k_1}\right), \tag{3.54}$$

and then solving (3.53) for $c_2$ and substituting into (3.54) gives

$$k_0c_0 = c_3k_3\left(1 + \frac{k_{-1}}{k_1} + \frac{k_{-1}k_{-2}}{k_1k_2} + \frac{k_{-1}k_{-2}k_{-3}}{k_1k_2k_3}\right) - c_4k_{-4}\left(1 + \frac{k_{-1}}{k_1} + \frac{k_{-1}k_{-2}}{k_1k_2}\right). \tag{3.55}$$

Repeating this process of sequential substitutions, and letting

$$\phi_n = 1 + \frac{k_{-1}}{k_1} + \frac{k_{-1}k_{-2}}{k_1k_2} + \cdots + \frac{k_{-1}\cdots k_{-n}}{k_1\cdots k_n}, \tag{3.56}$$

we find that

$$k_0 c_0 = k_{n-1} c_{n-1} \phi_{n-1} - c_n k_{-n} \phi_{n-2}. \tag{3.57}$$

Since

$$c_{n-1} = \frac{M + k_{-n} c_n}{k_{n-1}}, \tag{3.58}$$

it follows that

$$k_0 c_0 = \phi_{n-1}(M + k_{-n} c_n) - c_n k_{-n} \phi_{n-2}, \tag{3.59}$$

and hence

$$J = \lambda M = \frac{\lambda k_0 \left( c_0 - c_n \frac{k_{-1} \cdots k_{-n}}{k_0 \cdots k_{n-1}} \right)}{1 + \frac{k_{-1}}{k_1} + \frac{k_{-1} k_{-2}}{k_1 k_2} + \cdots + \frac{k_{-1} \cdots k_{-(n-1)}}{k_1 \cdots k_{n-1}}}. \tag{3.60}$$

It remains to express the rate constants in terms of the membrane potential. If there is a potential difference $V$ across the membrane (as shown in Fig. 3.7B), the constant electric field adds $FzV/(2n)$ to the barrier when moving from right to left, and $-FzV/(2n)$ when moving in the opposite direction. Hence

$$\Delta G_j = \Delta G_0 - j \delta G - \frac{FzV}{2n}, \tag{3.61}$$

$$\Delta G_{-j} = \Delta G_0 - (j - 1) \delta G + \frac{FzV}{2n}. \tag{3.62}$$

Now we use (3.44) to get

$$\frac{k_{-j}}{k_{j-1}} = \exp(-v/n), \qquad \frac{k_{-j}}{k_j} = \exp(-g - v/n), \tag{3.63}$$

where $g = \delta G/(RT)$ and $v = FzV/(RT)$. Hence

$$J = \frac{k_0 \lambda (c_0 - c_n e^{-v})}{1 + e^{-(g+v/n)} + e^{-2(g+v/n)} + \cdots + e^{-(n-1)(g+v/n)}},$$

$$= k_0 \lambda (c_0 - c_n e^{-v}) \frac{e^{-(g+v/n)} - 1}{e^{-n(g+v/n)} - 1}. \tag{3.64}$$

As expected, (3.64) satisfies both the independence principle and the Ussing flux ratio. Also, the flux is zero when $v$ is the Nernst potential of the ion.

### The homogeneous membrane simplification

One useful simplification of the nonsaturating barrier model is obtained if it is assumed that the membrane is homogeneous. We model a homogeneous membrane by setting $g = \delta G/(RT) = 0$ and letting $n \to \infty$. Thus, there is no increase in barrier height through the membrane, and the number of barriers approaches infinity. In this limit, keeping $n\lambda = L$ fixed,

$$J = \frac{k_0 0 \lambda^2}{L} \cdot v \cdot \frac{c_0 - c_n e^{-v}}{1 - e^{-v}}, \tag{3.65}$$

where $k_{00}$ is the value of $k_0$ at $V = 0$, $L$ is the width of the membrane, and $k_{00}\lambda^2$ is the diffusion coefficient of the ion over the first barrier in the absence of an electric field.

It follows that in the homogeneous membrane case,

$$J = \frac{D_S}{L} \cdot v \cdot \frac{c_0 - c_n e^{-v}}{1 - e^{-v}},$$

$$= P_S \cdot v \cdot \frac{c_0 - c_n e^{-v}}{1 - e^{-v}}, \tag{3.66}$$

which is exactly the GHK current equation (2.67) derived previously.

## 3.4.2 Saturating Barrier Models: One-Ion Pores

If an ion channel satisfies the independence principle, the flux of S is proportional to [S], even when [S] gets large. However, this is not usually found to be true experimentally. It is more common for the flux to saturate as [S] increases, reaching some maximum value as [S] gets large. This has motivated the development of models in which the flux is not proportional to [S] but is a nonlinear, saturating, function of [S]. As we will see, equations for such models are similar to those of enzyme kinetics.

The basic assumptions behind saturating barrier models are that to pass through the channel, ions must bind to binding sites in the channel, but that each binding site can hold only a single ion (Läuger, 1973; Hille, 1992). Hence, if all the binding sites are full, an increase in ionic concentration does not increase the ionic flux—the channel is saturated. Saturating barrier models can be further subdivided into one-ion pore models, in which each channel can bind only a single ion at any one time, and multi-ion pore models, in which each channel can bind multiple ions simultaneously. The theory of one-ion pores is considerably simpler than that of multi-ion pores, and so we discuss those models first.

### The simplest one-ion saturating model

We begin by considering the simplest one-ion pore model, with a single binding site. If we let $S_e$ denote the ion outside, $S_i$ the ion inside, and X the binding site, the passage of an ion through the channel can be described by the kinetic scheme

$$X + S_i \underset{k_{-1}}{\overset{k_0}{\rightleftarrows}} XS \underset{k_{-2}}{\overset{k_1}{\rightleftarrows}} X + S_e. \tag{3.67}$$

Essentially, the binding site acts like an enzyme that transfers the ion from one side of the membrane to the other, such as was encountered in Chapter 2 for the transport of glucose across a membrane. Following the notation of the previous section, we let $c_0$ denote [$S_i$] and $c_2$ denote [$S_e$]. However, instead of using $c_1$ to denote the concentration of $S$ at the binding site, it is more convenient to let $c_1$ denote the probability that the binding site is occupied. (In a population of channels, $c_1$ denotes the proportion of channels that have an occupied binding site.) Then, at steady state,

$$k_0 c_0 x - k_{-1} c_1 = k_1 c_1 - k_{-2} c_2 x, \tag{3.68}$$

where $x$ denotes the probability that the binding site is empty. Note that (3.68) is similar to the corresponding equation for the nonsaturating pore, (3.50), with the only difference that $x$ appears in the saturating model. In addition, we have a conservation equation for $x$,

$$x + c_1 = 1. \tag{3.69}$$

Solution of (3.68) and (3.69) gives the flux $J$ as

$$J = k_0 c_0 x - k_{-1} c_1 = \frac{k_0 k_1 c_0 - k_{-1} k_{-2} c_2}{k_0 c_0 + k_{-2} c_2 + k_{-1} + k_1}. \tag{3.70}$$

It is important to note that $J$, as defined by (3.70), does not have the same units (concentration $\times$ distance/time) as we used previously, but instead has units of number of ions crossing the membrane per unit time. The corresponding transmembrane current, $I$, is given by $I = zqJ$, where $q$ is the unit charge, and has the usual units of number of charges crossing the membrane per unit time. A plot of $J$ as a function of $c_0$ is shown in Fig. 3.8. When $c_0$ is small, $J$ is approximately a linear function of $c_0$, but as $c_0$ increases, $J$ saturates at the maximum value $k_1$.

We now use (3.44) to express the rate constants in terms of the membrane potential. As before, we assume that the local maxima of the energy profile occur midway between the local minima; i.e., we assume that the barriers are symmetrical. However, we no longer assume that the barriers are equally spaced through the channel. If the local minimum occurs at an electrical distance $\delta$ from the left-hand side, it follows that

$$k_0 = \kappa \exp\left[\frac{1}{RT}(-\Delta G_0 + \delta FV/2)\right], \tag{3.71}$$

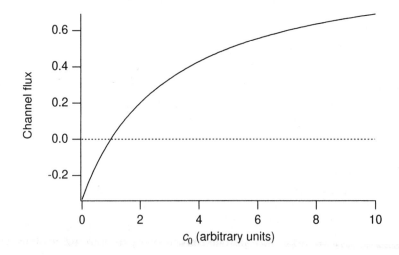

**Figure 3.8** Plot of $J$ against $c_0$ for the simplest saturating model with one binding site. When $c_0$ is small, the flux is approximately a linear function of $c_0$, but as $c_0$ increases, the flux saturates to a maximum value.

$$k_1 = \kappa \exp\left[\frac{1}{RT}(-\Delta G_1 + (1-\delta)FV/2)\right], \tag{3.72}$$

$$k_{-1} = \kappa \exp\left[\frac{1}{RT}(-\Delta G_{-1} - \delta FV/2)\right], \tag{3.73}$$

$$k_{-2} = \kappa \exp\left[\frac{1}{RT}(-\Delta G_{-2} - (1-\delta)FV/2)\right]. \tag{3.74}$$

Because $\delta$ denotes an electrical, not a physical, distance, it is not necessary to assume that the electric field in the membrane is constant, only that there is a drop of $\delta V$ over the first barrier and $(1-\delta)V$ over the second. In general, the energy profile of any particular channel is unknown. However, the number and positions of the binding sites and the values of the local maxima and minima can, in principle at least, be determined by fitting to experimental data. We consider an example of this procedure (for a slightly more complicated model) below.

## The Ussing flux ratio

Earlier in this chapter we stated that it is possible for a model to obey the Ussing flux ratio but not the independence principle. Single-ion saturating models provide a simple example of this. First, note that they cannot obey the independence principle, since the flux is not linearly proportional to the ionic concentration. This nonlinear saturation effect is illustrated in Fig. 3.8.

To see that the model obeys the Ussing flux ratio, it is necessary to set up the model in a slightly different form. Suppose we have two isotopes, S and $\bar{\text{S}}$, similar enough so that they have identical energy profiles in the channel. Then, we suppose that a channel has only S on the left-hand side and only $\bar{\text{S}}$ on the right. We let $c$ denote [S] and $\bar{c}$ denote [$\bar{\text{S}}$]. Since S and $\bar{\text{S}}$ have identical energy profiles in the channel, the rate constants for the passage of $\bar{\text{S}}$ through the channel are the same as those for S. From the kinetic schemes for S and $\bar{\text{S}}$ we get

$$k_0 c_0 x - k_{-1} c_1 = k_1 c_1 - k_{-2} c_2 x = J_S, \tag{3.75}$$
$$k_0 \bar{c}_0 x - k_{-1} \bar{c}_1 = k_1 \bar{c}_1 - k_{-2} \bar{c}_2 x = J_{\bar{S}}, \tag{3.76}$$

but here the conservation equation for $x$ is

$$x + \bar{c}_1 + c_1 = 1. \tag{3.77}$$

To calculate the individual fluxes of S and $\bar{\text{S}}$ it is necessary to eliminate $x$ from (3.75) and (3.76) using the conservation equation (3.77). However, to calculate the flux ratio this is not necessary. Solving (3.75) for $J_S$ in terms of $x$, $c_0$, and $c_2$, we find

$$J_S = x\left(\frac{k_0 c_0 - \dfrac{k_{-1}k_{-2}}{k_1}c_2}{1 + k_{-1}/k_1}\right), \tag{3.78}$$

and similarly,

$$J_{\bar{S}} = x \left( \frac{k_0 \bar{c}_0 - \dfrac{k_{-1} k_{-2}}{k_1} \bar{c}_2}{1 + k_{-1}/k_1} \right). \tag{3.79}$$

The variable $x$ cancels when we calculate the flux ratio, and so we do not need to use the conservation equation. If S is present only on the left-hand side and $\bar{S}$ only on the right, we then have $c_2 = 0$ and $\bar{c}_0 = 0$, in which case

$$\frac{J_S}{J_{\bar{S}}} = -\frac{k_0 k_1}{k_{-1} k_{-2}} \cdot \frac{c_0}{\bar{c}_2}. \tag{3.80}$$

The minus sign on the right-hand side appears because the fluxes are in different directions. Now we substitute for the rate constants, (3.71) to (3.74), and use the fact that the ends of the energy profile are at the same height (and thus $\Delta G_0 + \Delta G_1 - \Delta G_{-1} - \Delta G_{-2} = 0$) to find

$$\left| \frac{J_S}{J_{\bar{S}}} \right| = \exp\left( \frac{VF}{RT} \right) \cdot \frac{c_0}{\bar{c}_2}, \tag{3.81}$$

which is the Ussing flux ratio, as proposed.

### Multiple binding sites

When there are multiple binding sites within the channel, the analysis is essentially the same as the simpler case discussed above, but the details are more complicated. When there are $n$ barriers in the membrane (and thus $n - 1$ binding sites), the steady-state equations are

$$k_0 c_0 x - k_{-1} c_1 = k_1 c_1 - k_{-2} c_2 = \cdots = k_{n-1} c_{n-1} - k_{-n} c_n x = J, \tag{3.82}$$

where $x$ is the probability that all of the binding sites are empty and $c_j$ is the probability that the ion is bound to the $j$th binding site. Because the channel must be in either state $x$ or one of the states $c_1, \ldots, c_{n-1}$ (since there is only one ion in the channel at a time), it follows that

$$x = 1 - \sum_{i=1}^{n-1} c_i. \tag{3.83}$$

For convenience we define

$$\pi_j = \frac{k_{-1} \cdots k_{-j}}{k_1 \cdots k_j}, \qquad \pi_0 = 1, \tag{3.84}$$

$$\phi_j = \sum_{i=0}^{j} \pi_i. \tag{3.85}$$

Note that this definition for $\phi_j$ is the same as that given in (3.56). It is left as an exercise to show that

$$J = \frac{k_0 c_0 - k_{-n} c_n \pi_{n-1}}{\phi_{n-1} + \beta k_0 c_0 + k_{-n} c_n (\alpha \phi_{n-1} - \beta \phi_{n-2})}, \tag{3.86}$$

where

$$\alpha = \sum_{j=1}^{n-1} \frac{\phi_{n-2} - \phi_{j-1}}{k_j \pi_j}, \tag{3.87}$$

$$\beta = \sum_{j=1}^{n-1} \frac{\phi_{n-1} - \phi_{j-1}}{k_j \pi_j}. \tag{3.88}$$

Equation (3.86) does not satisfy the independence principle, but it does satisfy the Ussing flux ratio. However, the details are left as an exercise (Exercise 5).

### A model for sodium channels

Saturable one-ion barrier models have been used by a large number of authors to describe ion channels that show deviations from the independence principle (see, for example, Hille, 1975; Ciani and Ribalet, 1988; Robello et al., 1987). Here we focus our attention on the model of Hille (1975).

An example of Hille's experimental results is given in Fig. 3.9. A frog myelinated nerve was voltage clamped, held at $-80$ mV, and then depolarized in brief steps. The peak of the Na$^+$ current is plotted as a function of the voltage. If the Na$^+$ channels

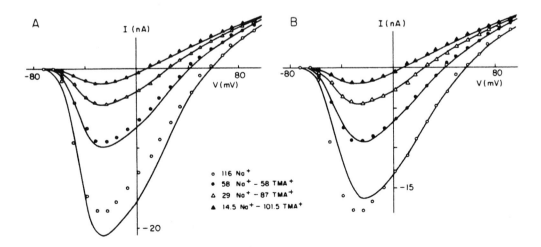

**Figure 3.9** Peak current plotted against voltage, for a range of external Na$^+$ concentrations (given in mM in the legend). For the model calculations, the internal Na$^+$ concentration was assumed to be 11.2 mM. A: The smooth curves are the predictions from the independence relation. B: the smooth curves are the predictions from the Hille four-barrier model, showing saturation of the flux at high Na$^+$ concentrations. (Hille, 1975, Fig. 5.)

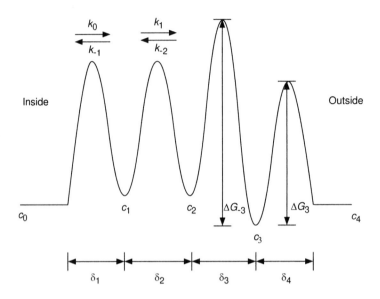

**Figure 3.10** Schematic diagram of the potential energy profile of the Hille four-barrier model of the Na$^+$ channel.

obeyed the independence principle, the resultant currents would then be predicted by (3.14). Thus, if the external Na$^+$ concentration is changed from $[Na^+]_e$ to $[Na^+]'_e$ while keeping the same internal $[Na^+]_i$, we expect

$$\frac{I'_{Na}}{I_{Na}} = \frac{[Na^+]_i - [Na^+]'_e \exp(\frac{-VF}{RT})}{[Na^+]_i - [Na^+]_e \exp(\frac{-VF}{RT})}. \tag{3.89}$$

The smooth curves in Fig. 3.9A are drawn from this relation, using the curve for $[Na^+]_e = 14.5$ mM as a reference curve. It is necessary to use one of the curves as a reference curve, because (3.89) determines only the ratio of the currents, not their absolute values. Clearly, the curves predicted from the independence principle agree with the data at low Na$^+$ concentrations, but do not agree with the data at high $[Na^+]$. In the latter case the observed Na$^+$ currents are smaller than predicted, suggesting that the Na$^+$ channel is saturated.

To explain the observed deviation from independence, Hille proposed a 4-barrier, 3 binding site model, sketched schematically in Fig. 3.10. As usual, each rate constant is described in terms of the free-energy profile and the voltage drop across the barrier, as in (3.71)–(3.74). For example,

$$k_0 = \kappa \exp\left[\frac{1}{RT}(-\Delta G_0 + \delta_1 FV/2)\right], \tag{3.90}$$

$$k_1 = \kappa \exp\left[\frac{1}{RT}(-\Delta G_1 + \delta_2 FV/2)\right], \tag{3.91}$$

**Table 3.1** Standard parameter set for the Hille model of the $Na^+$ channel (Hille, 1975). The $\Delta G_i$ and $\Delta G_{-i}$ are in units of $RT$. Note that $\sum_{i=1}^{4}(\Delta G_{i-1} - \Delta G_{-i}) = 0$, since the ends of the energy profile are at the same level, arbitrarily set to be $G = 0$. Also, $\sum_{i=1}^{4} \delta_i = 1$, since $\delta_i$ represents the fraction of the total voltage drop over the $i$th barrier. To determine these parameters, it was assumed that $\kappa = kT/h$, where $k$ is Boltzmann's constant, and $h$ is Planck's constant.

| | | |
|---|---|---|
| $\Delta G_0 = 7$ | $\Delta G_{-1} = 6.5$ | $\delta_1 = 0.25$ |
| $\Delta G_1 = 6.5$ | $\Delta G_{-2} = 6.5$ | $\delta_2 = 0.25$ |
| $\Delta G_2 = 8.5$ | $\Delta G_{-3} = 10$ | $\delta_3 = 0.23$ |
| $\Delta G_3 = 7$ | $\Delta G_{-4} = 6$ | $\delta_4 = 0.27$ |

$$k_{-1} = \kappa \exp\left[\frac{1}{RT}(-\Delta G_{-1} - \delta_1 FV/2)\right], \tag{3.92}$$

$$k_{-2} = \kappa \exp\left[\frac{1}{RT}(-\Delta G_{-2} - \delta_2 FV/2)\right], \tag{3.93}$$

and similarly for the other rate constants. The model is completely specified by the parameters $\Delta G_i$, $\Delta G_{-i}$, and $\delta_i$. All but two of these parameters were fixed at reasonable values, and the values of $\Delta G_3$ and $\Delta G_{-3}$ were varied to obtain agreement with experimental data. The standard parameter set used to plot the curves in Figs. 3.9B and 3.11 is given in Table 3.1.

With these parameters, the channel flux and current ratios are calculated using the methods of the previous two sections. Plotting the current ratio against the voltage gives the smooth curves shown in Fig. 3.9B. Clearly, the saturating model shows better agreement with the experimental data at high $[Na^+]$.

A typical $I$–$V$ curve of the model with the standard parameter set is shown in Fig. 3.11. A linear $I$–$V$ curve and the GHK current equation are also given for the sake of comparison. The $I$–$V$ curve of the Hille model has characteristics reminiscent of both the linear and the GHK curves. At large negative voltages the Hille model is similar to the GHK curve, but at large positive voltages it is more similar to the linear $I$–$V$ curve. All three curves have the same reversal potential, as indeed they must, since the Nernst potential is model-independent.

### 3.4.3 Saturating Barrier Models: Multi-Ion Pores

We showed above that single-ion models obey the Ussing flux ratio, even though they do not obey the independence principle. This means that to model the type of channel described in Fig. 3.2 it is necessary to use models that show flux coupling as predicted by Hodgkin and Keynes (1955). Such flux coupling arises in models in which more than one ion can be in the channel at any one time. Although the equations for such multi-ion models are essentially the same as the equations for the single-ion models

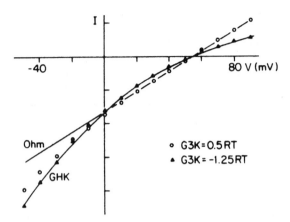

**Figure 3.11** The *I–V* curve of the Hille model (open circles), compared to a linear *I–V* curve (labeled Ohm), and the GHK *I–V* curve. The current scale marks correspond to 4–5 pA per channel. (Hille, 1975, Fig. 4.)

described in the previous section, the analysis is complicated considerably by the fact that there are many more possible channel states. Hence, numerical techniques are the most efficient for studying such models. A great deal has been written about multi-ion models (e.g., Hille and Schwartz, 1978; Begenisich and Cahalan, 1980; Schumaker and MacKinnon, 1990; Urban and Hladky, 1979; Kohler and Heckmann, 1979). We do not have space for a detailed discussion of the properties of these models, but present only a brief discussion of the simplest model. Hille and Schwarz (1978) and Hille (1992) give more detailed discussions.

Multi-ion models are based on assumptions similar to one-ion models. It is assumed that the passage of an ion through the channel can be described as the jumping of an ion over energy barriers, from one binding site to another. In one-ion models each binding site can either have an ion bound or not, and thus a channel with $n$ binding sites can be in one of $n$ independent states (i.e., the ion can be bound to any one of the binding sites). Hence, the steady-state ion distribution is found by solving a system of $n$ linear equations, treating the concentrations on either side of the membrane as known. If more than one ion can be present simultaneously in the channel, the situation is more complicated. Each binding site can be in one of two states: binding an ion or empty. Therefore, a channel with $n$ binding sites can be in any of $2^n$ states (at least; more states are possible if there is more than one ion type passing through the channel), and the steady-state probability distribution must be found by solving a large system of linear equations.

The simplest possible multi-ion model has three barriers and two binding sites, and so the channel can be in one of 4 possible states (Fig. 3.12). Arbitrary movements from one state to another are not possible. For example, the state OO (where both binding sites are empty) can change to OS or SO but cannot change to SS in a single step, as this would require two ions entering the channel simultaneously. We number the

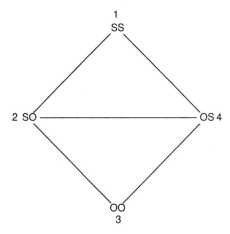

**Figure 3.12** State diagram for a multi-ion barrier model with two binding sites and a single ion.

states as in Fig. 3.12 and let $k_{ij}$ denote the rate of conversion of state $i$ to state $j$. Also, let $P_j$ denote the probability that the channel is in the $j$th state, and let $c_e$ and $c_i$ denote the external and internal ion concentrations, respectively. Then, the equations for the probabilities follow from the law of mass action; they are

$$\frac{dP_1}{dt} = -(k_{12} + k_{14})P_1 + k_{21}c_eP_2 + k_{41}c_iP_4, \tag{3.94}$$

$$\frac{dP_2}{dt} = -(k_{21}c_e + k_{23} + k_{24})P_2 + k_{12}P_1 + c_ik_{32}P_3 + k_{42}P_4, \tag{3.95}$$

$$\frac{dP_3}{dt} = -(c_ik_{32} + c_ek_{34})P_3 + k_{43}P_4 + k_{23}P_2, \tag{3.96}$$

$$\frac{dP_4}{dt} = -(k_{41}c_i + k_{42} + k_{43})P_4 + k_{14}P_1 + k_{24}P_2 + c_ek_{34}P_3. \tag{3.97}$$

The probabilities must also satisfy the conservation equation

$$\sum_{i=1}^{4} P_i = 1. \tag{3.98}$$

Using the conservation equation in place of the equation for $P_4$, the steady-state probability distribution is given by the linear system

$$\begin{pmatrix} -k_{12} - k_{14} & k_{21} & 0 & k_{41} \\ k_{12} & -k_{21} - k_{23} - k_{24} & c_ek_{32} & k_{42} \\ 0 & k_{23} & -c_ek_{32} - c_ik_{34} & k_{43} \\ 1 & 1 & 1 & 1 \end{pmatrix} \begin{pmatrix} P_1 \\ P_2 \\ P_3 \\ P_4 \end{pmatrix} = \begin{pmatrix} 0 \\ 0 \\ 0 \\ 1 \end{pmatrix}. \tag{3.99}$$

Since each rate constant is determined as a function of the voltage in the same way as one-ion models (as in, for example, (3.90)–(3.93)), solution of (3.98) gives each $P$ as a function of voltage and the ionic concentrations on each side of the membrane.

Finally, the membrane fluxes are calculated as the net rate of ions crossing any one barrier, and so, choosing the middle barrier arbitrarily, we have

$$J = P_2 k_{24} - P_4 k_{42}. \tag{3.100}$$

Although it is possible to solve such linear systems exactly (particularly with the help of symbolic manipulators such as Maple or Mathematica), it is often as useful to solve the equations numerically for a given energy profile. It is left as an exercise to show that the Ussing flux ratio is not obeyed by a multi-ion model with two binding sites and to compare the I–V curves of multi-ion and one-ion models.

### 3.4.4  Protein Ion Exchangers

We are now ready to examine the effect that the membrane potential has on ion exchangers. In Chapter 2 it was suggested that a protein ion exchanger could be modeled by the reaction mechanism

$$S_i + C_i \underset{k_-}{\overset{k_+}{\rightleftarrows}} P_i \underset{k^p_-}{\overset{k^p_+}{\rightleftarrows}} P_e \underset{k_+}{\overset{k_-}{\rightleftarrows}} S_e + C_e, \tag{3.101}$$

$$C_i \underset{k^c_-}{\overset{k^c_+}{\rightleftarrows}} C_e. \tag{3.102}$$

where C is the carrier protein, P is the complex of carrier protein and the ion to be transported. Subscripts $e$ and $i$ denote extracellular and intracellular concentrations, respectively.

A conformational change of the protein requires that a free-energy barrier be crossed. Thus,

$$k^\alpha_+ = \kappa \exp\left[\frac{1}{RT}(-\Delta G_+ + z^\alpha FV/2)\right], \tag{3.103}$$

$$k^\alpha_- = \kappa \exp\left[\frac{1}{RT}(-\Delta G_- - z^\alpha FV/2)\right], \tag{3.104}$$

where $V$ is the membrane potential, the superscript $\alpha$ can take on values $p$ or $c$, and $z^c$ and $z^p$ are the charges on the protein and the complexed protein, respectively. Since the transported quantity is an ion, $z^p = z^c + z$, where $z$ is the charge on the ion.

The analysis of this reaction is identical to that of a uniport given in Chapter 2, except that here we do not assume that the rates $k^p_\pm$ and $k^c_\pm$ are identical. The result of the calculation is that the flux of the ion through the port is given by

$$J = k_+ k_- C_0 \frac{k^p_- k^c_+ [S_e] - k^c_- k^p_+ [S_i]}{[S_i][S_e]K_{ei} + [S_e]K_e + [S_i]K_i + K}, \tag{3.105}$$

where $C_0$ is the total amount of the carrier protein, $K_{ei} = k^2_+(k^p_- + k^p_+)$, $K_e = k_+(k^p_-(k^c_+ + k_-) + k^c_+(k^p_+ + k_-))$, $K_i = k_+(k^c_-(k^p_- + k_-) + k^p_+(k^c_- + k_-))$, and $K = k_-(k^c_- + k^c_+)(k_- + k^p_+ + k^p_-)$. The important feature of this expression is that it is voltage dependent since the rates of conformational change are voltage-dependent. Specifically, the flux is inward ($J > 0$) whenever

$$\frac{[S_e]}{[S_i]} > \exp\left(\frac{zVF}{RT}\right). \qquad (3.106)$$

Notice that if $z > 0$ and if the membrane potential $V$ is negative, then it is possible to drive ions against their gradient. The term $\exp(\frac{zVF}{RT})$ is the *driving force* of the port.

To apply these ideas to a specific example, consider the sodium–calcium exchanger. This exchanger exchanges three ions of sodium for one of calcium, a net current with $z = 1$. We find the flux of the exchanger by replacing $S_e$ in (3.105) with $[Na^+]_e^3[Ca^{2+}]_i$ and $S_i$ with $[Na^+]_i^3[Ca^{2+}]_e$. Notice that there is a positive flux (sodium moving inward and calcium moving outward) whenever

$$\frac{[Ca^{2+}]_i}{[Ca^{2+}]_e} > \left(\frac{[Na^+]_i}{[Na^+]_e}\right)^3 \exp\left(\frac{VF}{RT}\right). \qquad (3.107)$$

For a cell with a resting potential of $-70$ mV, extracellular sodium of 440 mM, and intracellular sodium of 50 mM, this pump extracts calcium, provided that $\frac{[Ca^{2+}]_i}{[Ca^{2+}]_e} > 10^{-4}$. The advantage of having a net current with $z = 1$ is that the ability of the exchanger to extract calcium is improved by a factor of about 15 compared to an exchanger with no net current flow.

## 3.5 Channel Gating

So far in this chapter we have discussed how the current through a single open channel depends on the membrane potential and the ionic concentrations on either side of the membrane. However, it is probably of equally great interest to determine how ionic channels open and close in response to voltage. As we will see in Chapter 4, the opening and closing of ionic channels in response to changes in the membrane potential is the basis for electrical excitability and is thus of fundamental significance in neurophysiology.

Recall that there is an important difference between the instantaneous and steady-state *I–V* curves. In general, the current through a population of channels is the product of two terms,

$$I = g(V,t)\phi(V), \qquad (3.108)$$

where $\phi(V)$ is the *I–V* curve of a single open channel and $g(V,t)$ is the proportion of open channels in the population. In the previous sections we discussed electrodiffusion and barrier models for $\phi(V)$; in this section we discuss models for the dependence of $g$ on voltage and time.

Consider, for example, the curves in Fig. 3.13, which show typical responses of populations of $Na^+$ and $K^+$ channels. When the voltage is stepped from $-65$ mV to $-9$ mV, and held fixed at the new level, the $K^+$ conductance ($g_K$) slowly increases to a

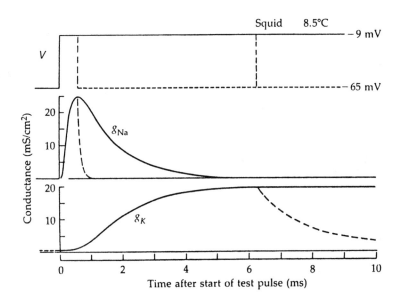

**Figure 3.13** Na$^+$ and K$^+$ conductances as a function of time after a step change in voltage from −65 mV to −9 mV. The dashed line shows how after repolarization $g_{Na}$ recovers quickly, and $g_K$ recovers more slowly. (Hille, 1992, Fig. 11, p. 40.)

new level, while the Na$^+$ conductance ($g_{Na}$) first increases and then decreases. From this data we can draw the following conclusions. First, as the voltage increases, the proportion of open K$^+$ channels increases. Second, although the proportion of open Na$^+$ channels initially increases, a second process is significant at longer times, as the Na$^+$ channel moves to an inactivated state. Thus, Na$^+$ channels first activate and then inactivate.

### 3.5.1 A Two-State K$^+$ Channel

The simplest model for the K$^+$ channel assumes that the channel can exist in either a closed state, C, or an open state, O, and that the rate of conversion from one state to another is dependent on the voltage. Thus,

$$C \underset{\beta(V)}{\overset{\alpha(V)}{\rightleftarrows}} O. \tag{3.109}$$

Letting $g$ denote the proportion of channels in the open state, we can write the differential equation for the rate of change of $g$ as

$$\frac{dg}{dt} = \alpha(V)(1-g) - \beta(V)g, \tag{3.110}$$

where we have used the fact that because channels are conserved, the proportion of closed channels is $1 - g$. Under voltage-clamp conditions (i.e., where the voltage is held fixed, as in Fig. 3.13), $\alpha$ and $\beta$ are constants, and thus we can readily solve for $g$ as a function of time. It is often convenient to write (3.110) as

$$\tau_g(V)\frac{dg}{dt} = g_\infty(V) - g, \tag{3.111}$$

where $g_\infty(V) = \alpha/(\alpha + \beta)$ is the steady-state value of $g$, and $\tau_g(V) = 1/(\alpha + \beta)$ is the time constant of approach to the steady state. From experimental data, such as that shown in Fig. 3.13, one can obtain values for $g_\infty$ and $\tau_g$, and thus $\alpha$ and $\beta$ can be unambiguously determined.

## 3.5.2 Multiple Subunits

An important generalization of the two-state model occurs when the channel is assumed to consist of multiple identical subunits, each of which can be in either the closed or open state. For example, suppose that the channel consists of two identical subunits, each of which can be closed or open. Then, the channel can take any of four possible states, $S_{00}, S_{10}, S_{01}$, or $S_{11}$, where the subscripts denote the different subunits, with 1 and 0 denoting open and closed subunits, respectively. A general model for this channel involves three differential equations (although there is a differential equation for each of the four variables, one equation is superfluous because of the conservation equation $S_{00} + S_{10} + S_{01} + S_{11} = 1$), but we can simplify the model by grouping the channel states with the same number of closed and open subunits. For example, because the subunits are identical, there should be no difference between $S_{10}$ and $S_{01}$, and thus they are amalgamated into a single variable.

So, we let $S_i$ denote the group of channels with exactly $i$ open subunits. Then, conversions between channel groups are governed by the reaction scheme

$$S_0 \underset{\beta}{\overset{2\alpha}{\rightleftarrows}} S_1 \underset{2\beta}{\overset{\alpha}{\rightleftarrows}} S_2. \tag{3.112}$$

The corresponding differential equations are

$$\frac{dx_0}{dt} = \beta x_1 - 2\alpha x_0, \tag{3.113}$$

$$\frac{dx_2}{dt} = \alpha x_1 - 2\beta x_2, \tag{3.114}$$

where $x_i$ denotes the proportion of channels in state $S_i$, and $x_0 + x_1 + x_2 = 1$. We make the change of variables $x_2 = n^2$, where $n$ satisfies the differential equation

$$\frac{dn}{dt} = \alpha(1 - n) - \beta n. \tag{3.115}$$

A simple substitution then shows that (3.113) and (3.114) are satisfied by $x_0 = (1 - n)^2$ and $x_1 = 2n(1 - n)$. Thus, (3.113) and (3.114) are equivalent to $x_0 = (1 - n)^2, x_1 = 2n(1 - n), x_2 = n^2$, where $n$ satisfies (3.115).

In fact, we can derive a stronger result. We let

$$x_0 = (1-n)^2 + y_0, \tag{3.116}$$

$$x_2 = n^2 + y_2, \tag{3.117}$$

so that of necessity, $x_1 = 2n(1-n) - y_0 - y_2$. It follows that

$$\frac{dy_0}{dt} = -2\alpha y_0 - \beta(y_0 + y_2), \tag{3.118}$$

$$\frac{dy_2}{dt} = -\alpha(y_0 + y_2) - 2\beta y_2. \tag{3.119}$$

This is a linear system of equations with eigenvalues $-(\alpha + \beta)$, $-2(\alpha + \beta)$, and so $y_0, y_2$ go exponentially to zero. This means that $x_0 = (1-n)^2, x_2 = n^2$ is an invariant stable manifold for the original system of equations; the solutions cannot leave this manifold, and with arbitrary initial data, the flow approaches this manifold exponentially. Notice that this is a stable invariant manifold even if $\alpha$ and $\beta$ are functions of time (so they can depend on voltage or other concentrations).

This argument generalizes to the case of $k$ identical independent binding sites where the invariant manifold for the flow is the binomial distribution with probability $n$ satisfying (3.115) (see Exercise 16). Thus, the channel conductance is proportional to $n^k$, where $n$ satisfies the simple equation (3.115). This multiple subunit model for channel gating provides the basis for the model of excitability that we examine in the next chapter.

### 3.5.3 The Sodium Channel

A more complex model is needed to explain the behavior of the $Na^+$ channel, which both activates and inactivates. The simplest approach is to extend the above analysis to the case of multiple subunits of two different types, $m$ and $h$, say, where each subunit can be either closed or open. To illustrate, we assume that the channel has one $h$ subunit and two $m$ subunits. The reaction diagram of such a channel is shown in Fig. 3.14. We let $S_{ij}$ denote the channel with $i$ open $m$ subunits and $j$ open $h$ subunits, and we let $x_{ij}$ denote the fraction of channels in state $S_{ij}$. As above, a simple substitution shows that the reaction scheme is equivalent to

$$x_{21} = m^2 h, \tag{3.120}$$

$$\frac{dm}{dt} = \alpha(1-m) - \beta m, \tag{3.121}$$

$$\frac{dh}{dt} = \gamma(1-h) - \delta h, \tag{3.122}$$

where the other variables are given by $x_{00} = (1-m)^2(1-h)$, $x_{10} = 2m(1-m)(1-h)$, $x_{20} = m^2(1-h)$, $x_{01} = (1-m)^2 h$, and $x_{11} = 2m(1-m)h$. Furthermore, the invariant manifold is again stable. A model of this type was used by Hodgkin and Huxley in their model of the nerve axon, which is discussed in detail in Chapter 4.

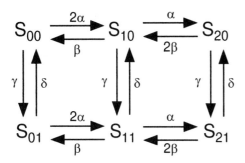

**Figure 3.14** Diagram of the possible states in a model of the Na$^+$ channel.

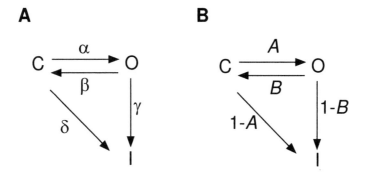

**Figure 3.15** A: Schematic diagram of the states of the Na$^+$ channel. C, O, and I denote the closed, open, and inactivated states, respectively. B: Time-independent transition probability diagram.

In an alternate model of the Na$^+$ channel (Aldrich et al., 1983; Peskin, 1991), it is assumed that the Na$^+$ channel can exist in three states, closed (C), open (O), or inactivated (I), and that once the channel is inactivated, it cannot return to either the closed or the open state (Fig. 3.15A). Transitions between states are described by

$$C \underset{\beta}{\overset{\alpha}{\rightleftharpoons}} O, \qquad O \overset{\gamma}{\longrightarrow} I, \qquad C \overset{\delta}{\longrightarrow} I. \tag{3.123}$$

Thus, the state I is absorbing. While this is clearly not true in general, it is a reasonable approximation at high depolarizations. As before, we let $g$ denote the proportion of open channels and let $c$ denote the proportion of closed channels. Then,

$$\frac{dc}{dt} = -(\alpha + \delta)c + \beta g, \tag{3.124}$$

$$\frac{dg}{dt} = \alpha c - (\beta + \gamma)g, \tag{3.125}$$

where as before, we have used the conservation of channels to eliminate the proportion of channels in the inactivated state. Initial conditions are $c(0) = 1, g(0) = 0$, i.e., all the channels are initially closed. By differentiating the equation for $s$ we eliminate $c$ to get

$$\frac{d^2g}{dt^2} + (\alpha + \beta + \gamma + \delta)\frac{dg}{dt} + [(\alpha + \delta)(\beta + \gamma) - \alpha\beta]g = 0, \tag{3.126}$$

which now has the initial conditions $g(0) = 0$, $g'(0) = \alpha$. This can be solved directly to give

$$g = a(e^{\lambda_1 t} - e^{\lambda_2 t}), \tag{3.127}$$

where $\lambda_2 < \lambda_1 < 0$ are the roots of

$$\lambda^2 + (\alpha + \beta + \gamma + \delta)\lambda + (\alpha + \delta)(\beta + \gamma) - \alpha\beta = 0, \tag{3.128}$$

and where

$$\alpha = a(\lambda_1 - \lambda_2) > 0. \tag{3.129}$$

As in the simple two-state model, $a$, $\lambda_1$, and $\lambda_2$ can be determined from experimental data. However, the rate constants cannot be determined uniquely. For since $\lambda_1$ and $\lambda_2$ are the roots of (3.128), it follows that

$$\alpha + \beta + \gamma + \delta = -\lambda_1 - \lambda_2, \tag{3.130}$$

$$(\alpha + \delta)(\beta + \gamma) - \alpha\beta = \lambda_1\lambda_2. \tag{3.131}$$

Here we have only three equations for the four unknowns, $\alpha, \beta, \gamma$, and $\delta$, so the system is underdetermined (see Exercise 18). This problem cannot be resolved using the macroscopic data that has been discussed so far, but requires data collected from a single channel, as described in the next section.

### Single-channel recordings

Since the late 1970s, the development of patch-clamp recording techniques has allowed the measurement of ionic current through a small piece of cell membrane, containing only a few, or even a single, ionic channel (Hamill et al., 1981; Sakmann and Neher, 1983; Neher and Sakmann received the 1991 Nobel Prize in physiology for their development of the patch-clamp technique). An example of an experimental record is given in Fig. 3.16. The current through an individual channel is stochastic (panel A) and cannot be predicted as a deterministic process. Nevertheless, the ensemble average over many experiments (panel B) is deterministic and reproduces the same properties that are seen in the macroscopic measurements of Fig. 3.13. However, the single-channel recordings contain more information than does the ensemble average.

To describe a channel with $n$ possible states (open, closed, inactivated, etc.), we introduce a stochastic variable $S(t) \in 1, 2, \ldots, n$ such that $S(t) = i$ if the channel is in state $i$ at time $t$. Further, if $k_{ij}$ (independent of time) is the rate constant for transitions from state $i$ to state $j$, then the probability that the channel changes from state $i$ to state $j$ in the time interval $(t, t + dt)$ is $k_{ij}dt$. In more condensed notation we write

$$P[S(t + dt) = j | S(t) = i] = k_{ij}dt. \tag{3.132}$$

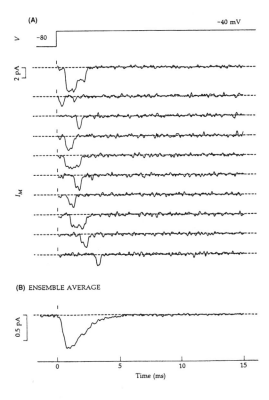

**Figure 3.16**   A: Na$^+$ currents from a single channel (or possibly two in the first trace) following a voltage step from $-80$ mV to $-40$ mV. B: Average open probability of the Na$^+$ channel, obtained by averaging over many traces of the type shown in A. (Hille, 1992, Fig. 6, p. 68.)

Note that this is valid only approximately and for small $dt$, since for large $dt$ and $k_{ij}$ fixed, this probability will exceed 1. Here, $P[x|y]$ is a conditional probability, meaning the probability of $x$ given $y$. Also, the probability that the channel does not change state in the time interval $(t, t + dt)$ is given by

$$P[S(t + dt) = i | S(t) = i] = 1 - K_i dt, \tag{3.133}$$

where $K_i = \sum_{j=1, j \neq i}^{n} k_{ij}$.

We now calculate the probability that the channel stays in state $i$ for time $t$. Let $M_i(t)$ be the logical random variable $\{S(\tau) = i, 0 < \tau < t\}$. If the interval $(0, t)$ is divided into $m$ subintervals, each of length $t/m$, then the channel stays in state $i$ only if it does not change state during any of the subintervals. Therefore,

$$P[M_i(t) | S(0) = i] = \left(1 - \frac{K_i t}{m}\right)^m. \tag{3.134}$$

Taking the limit $m \to \infty$, we find

$$P[M_i(t) | S(0) = i] = e^{-K_i t}. \tag{3.135}$$

To apply this theory to the determination of channel kinetics, we consider the probability diagram for the channel states (Fig. 3.15B). Here $A$ denotes the time-independent probability that a channel in the closed state moves to the open state (rather than to the inactivated state), and $B$ denotes the probability that a channel in the open state moves to the closed state. Comparing Fig. 3.15A with Fig. 3.15B, we see that

$$A = \frac{\alpha}{\alpha + \delta}, \qquad B = \frac{\beta}{\beta + \gamma}. \tag{3.136}$$

The probability $1 - A$ is easily determined experimentally, as it is the probability that a channel in the closed state inactivates without ever opening. Thus, $1 - A$ can be estimated by the proportion of experimental records in which no current is observed, even after the depolarizing stimulus was maintained for a long time.

Now let $T$ denote the time to first opening of the channel, often called the *latency* of the channel. $T$ is easily measured experimentally. Then,

$$P[T > t] = P[\text{First transition is to state I}]$$

$$+ P[\text{First transition is to state O and } T > t]$$

$$= 1 - A + P[\text{the transition is to state O}] \cdot P[\text{no transitions for time } t]$$

$$= 1 - A + A \cdot e^{-(\alpha + \delta)t}. \tag{3.137}$$

Thus, $P[T > t]$ is a decreasing exponential, and so $1 - A$ and $\alpha + \delta$ can be determined by fitting an exponential to the experimental measurements of $P[T > t]$. Hence, $\alpha$ and $\delta$ are unambiguously determined from the latency of the channel.

To determine the remaining two rate constants, let $N$ be the number of times the channel opens before it finally inactivates and determine the probability distribution for $N$. Clearly, $P[N = 0] = 1 - A$. Furthermore,

$$P[N = k] = P[N = k \text{ and channel enters I from O}]$$

$$+ P[N = k \text{ and channel enters I from C}]$$

$$= A^k B^{k-1}(1 - B) + A^k B^k(1 - A)$$

$$= (AB)^k \left( \frac{1 - AB}{B} \right). \tag{3.138}$$

Since $A$ can be determined from the latency, $B$ can be determined from an experimental plot of $P[N = k]$ vs. $k$. Finally, the distribution of open times is given by $\exp[-(\beta + \gamma)t]$, and so $\beta + \gamma$ can be determined from the open time distribution of the channel. This completes the characterization of the channel rate constants.

Since the work of Hodgkin and Huxley (described in Chapter 4), the traditional view of a $Na^+$ channel has been that it activates quickly and inactivates slowly. According to this view, the decreasing portion of the $g_{Na}$ curve in Fig. 3.13 is due entirely to inactivation of the channel. However, single-channel analysis has shown that this interpretation of macroscopic data is not always correct. It turns out that the rate of inactivation of some mammalian $Na^+$ channels is faster than the rate of activation. For example, Aldrich et al. (1983) found $\alpha = 1/ms$, $\beta = 0.4/ms$, $\gamma = 1.6/ms$ and $\delta = 1/ms$ at $V = 0$ for channels in a neuroblastoma cell line and a pituitary cell line. Although this reversal of activation and inactivation rates is not correct for all $Na^+$ channels in all species, the result does overturn some traditional ideas of how $Na^+$ channels work.

### 3.5.4  Drugs and Toxins

Many drugs act by blocking a specific ion channel. There are numerous specific channel blockers, such as sodium channel blockers, potassium channel blockers, calcium channel blockers, and so on. In fact, the discovery of site-specific and channel-specific blockers has been of tremendous benefit to the experimental study of ion channels. Examples of important channel blockers include verapamil (calcium-channel blocker), quinidine, sotolol, nicotine, DDT, various barbiturates (potassium-channel blockers), tetrodotoxin (TTX, the primary ingredient of puffer fish toxin), and scorpion toxins (sodium-channel blockers).

To include the effects of a drug or toxin like TTX in a model of a sodium channel is a relatively simple matter. We assume that a population $P$ of sodium channels is available for ionic conduction and that a population $B$ is blocked because they are bound by the toxin. Thus,

$$P + D \underset{k_-}{\overset{k_+}{\rightleftharpoons}} B, \tag{3.139}$$

where $D$ represents the concentration of the drug. Clearly, $P + B = P_0$, so that

$$\frac{dP}{dt} = k_-(P_0 - P) - k_+DP, \tag{3.140}$$

and the original channel conductance must be modified by multiplying by the percentage of unbound channels, $P/P_0$.

In steady state, we have

$$\frac{P}{P_0} = \frac{K_d}{K_d + D}. \tag{3.141}$$

The remarkable potency of TTX is reflected by its small equilibrium constant $K_d$, as $K_d \approx 1$–$5$ nM for sodium channels in nerve cells, and $K_d \approx 1$–$10$ $\mu$M for sodium channels in cardiac cells. By contrast, verapamil has $K_d \approx 140$–$940$ $\mu$M.

Other important drugs, such as lidocaine, flecainide, and encainide are so-called *use-dependent* sodium-channel blockers, in that they interfere with the sodium channel only when it is open. Thus, the more the channel is used, the more likely it will be blocked. Lidocaine is an important drug used in the treatment of cardiac arrhythmias. The folklore explanation of why it is useful is that because it is use-dependent, it helps prevent high-frequency firing of cardiac cells, which is commonly associated with cardiac arrhythmias. In fact, lidocaine, flecainide, and encainide are officially classified as antiarrhythmic drugs, even though it is now known that flecainide and encainide are proarrhythmic in certain postinfarction (after a heart attack) patients. A full explanation of this behavior is not known.

To keep track of the effect of a use-dependent drug on a two-state channel, we suppose that there are four classes of channels, those that are closed but unbound by the drug (C), those that are open and unbound by the drug (O), those that are closed and bound by the drug (CB), and those that are open and bound by the drug (OB) (but unable to pass a current). For this four-state model a reasonable reaction mechanism is

$$
C \underset{\beta}{\overset{\alpha}{\rightleftharpoons}} O, \qquad CB \underset{\beta}{\overset{\alpha}{\rightleftharpoons}} OB \tag{3.142}
$$

$$
CB \xrightarrow{k_+} C + D, \qquad O + D \underset{k_-}{\overset{k_+}{\rightleftharpoons}} OB. \tag{3.143}
$$

Notice that we have assumed that the drug does not interfere with the process of opening and closing, only with the actual flow of ionic current, and that the drug can bind the channel only when it is open. It is now a straightforward matter to find the differential equations governing these four states, and we leave this as an exercise.

This is not the only way that drugs might interfere with a channel. For example, for a channel with multiple subunits, the drug may bind only when certain of the subunits are in specific states. Indeed, the binding of drugs with channels can occur in many ways, and there are numerous unresolved questions concerning this complicated process.

## 3.6 EXERCISES

1. Derive the extended independence principle. Assume that there are more than one species of ion present, all with the same valence, and assume that the reversal potential is given by the GHK potential. Show that

$$
\frac{I'}{I} = \frac{\sum_j P_j[S_j]'_i - \sum_j P_j[S_j]'_e \exp(\frac{-VF}{RT})}{\sum_j P_j[S_j]_i - \sum_j P_j[S_j]_e \exp(\frac{-VF}{RT})}, \tag{3.144}
$$

where the sum over $j$ is over all the ionic species. Subscripts $i$ and $e$ denote internal and external concentrations, respectively.

2. Show that the GHK equation (3.2) satisfies both the independence principle and the Ussing flux ratio, but that the linear $I$–$V$ curve (3.1) satisfies neither.

3. In Section 3.3.1 we used the PNP equations to derive $I$–$V$ curves when two ions with opposite valence are allowed to move through a channel. Extend this analysis by assuming that two types of ions with positive valence and one type of ion with negative valence are allowed to move through the channel. Show that in the high concentration limit, although the negative ion still obeys a linear $I$–$V$ curve, the two positive ions do not. Details can be found in Chen, Barcilon, and Eisenberg (1992), equations (43)–(45).

4. (a) Show that (3.64) satisfies the independence principle and the Ussing flux ratio.

(b) Show that (3.64) can be made approximately linear by choosing $g$ such that

$$ng = \ln\left(\frac{c_n}{c_0}\right). \tag{3.145}$$

Although we know that a linear $I$–$V$ curve does not satisfy the independence principle, why does this result not contradict part (a)?

5. Show that (3.86) does not satisfy the independence principle, but does obey the Ussing flux ratio.

6. Derive (3.86) by solving the steady-state equations (3.82) and (3.83). First show that

$$J = x \cdot \frac{k_0 c_0 - k_{-n} c_n \pi_{n-1}}{\phi_{n-1}}. \tag{3.146}$$

Then show that

$$k_0 c_0 x = k_{n-1} c_{n-1} \phi_{n-1} - x k_{-n} c_n \phi_{n-2}, \tag{3.147}$$

$$k_j c_j = \frac{k_{n-1} c_{n-1}}{\pi_j} \cdot (\phi_{n-1} - \phi_{j-1}) - \frac{k_{-n} c_n x}{\pi_j} \cdot (\phi_{n-2} - \phi_{j-1}), \tag{3.148}$$

for $j = 1, \ldots, n-1$. Substitute these expressions into the conservation equation and solve for $x$.

7. Numerically plot some $I$–$V$ curves of the Hille $Na^+$ channel model for a selection of values for $[Na^+]_e$ and $[Na^+]_i$ and for a range of parameter values, not only those in Table 3.1. Compare to the linear and GHK $I$–$V$ curves.

8. By making a guess at the shape of the curve for $[Na^+] = 14.5$ mM in Fig. 3.9A, repeat the calculations to obtain the smooth curves in A and B of that figure. In other words, take an arbitrary curve of approximately the same shape as the $[Na^+] = 14.5$ mM curve and calculate the other smooth curves, first by using the independence principle, and second, by using the Hille model. (This is best done numerically.)

9. Write down state diagrams showing the channel states and the allowed transitions for a multi-ion model with two binding sites when the membrane is bathed with a solution containing:

(a) Only ion S on the left and only ion S' on the right.

(b) Ion S on both sides and ion S' only on the right.

(c) Ions S and S' on both the left and right.

In each case write down the corresponding system of linear equations that determine the steady-state ionic concentrations at the channel binding sites.

10. By using an arbitrary symmetric energy profile with two binding sites, show numerically that the Ussing flux ratio is not obeyed by a multi-ion model with two binding sites. (Note that since unidirectional fluxes must be calculated, it is necessary to treat the ions on each side of the membrane differently. Thus, an 8-state channel diagram must be used.) Hodgkin and Keynes predicted that the actual flux ratio is the Ussing ratio raised to the $(n + 1)$st power (cf. (3.19)). How does $n$ depend on the ionic concentrations on either side of the membrane, and on the energy profile?

11. Choose an arbitrary symmetric energy profile with two binding sites, and compare the $I$–$V$ curves of the one-ion and multi-ion models. Assume that the same ionic species is present on both sides of the membrane, so that only a 4-state multi-ion model is needed.

12. Suppose the sodium Nernst potential of a cell is 56 mV, its resting potential is −70 mV, and the extracellular calcium concentration is 1 mM. At what intracellular calcium concentration is the flux of a three-for-one sodium–calcium exchanger zero? (Use that $RT/F = 25.8$ mV at 27°C.)

13. Modify the pump–leak model of Chapter 2 to include a calcium current and the 3-for-1 sodium–calcium exchanger. What effect does this modification have on the relationship between pump rate and membrane potential?

14. Because there is a net current, the sodium–potassium pump current must be voltage dependent. Determine this dependence by including voltage dependence in the rates of conformational change in expression (2.53). How does voltage dependence affect the pump–leak model of Chapter 2?

15. Intestinal epithelial cells have a glucose–sodium symport that transports one sodium ion and one glucose molecule from the intestine into the cell. Model this transport process. Is the transport of glucose aided or hindered by the cell's negative membrane potential?

16. Suppose that a channel consists of $k$ identical, independent subunits, each of which can be open or closed, and that a current can pass through the channel only if all units are open.

    (a) Let $S_j$ denote the state in which $j$ subunits are open. Show that the conversions between states are governed by the reaction scheme

    $$S_0 \underset{\beta}{\overset{k\alpha}{\rightleftharpoons}} S_1, \ldots, S_{k-1} \underset{k\beta}{\overset{\alpha}{\rightleftharpoons}} S_k. \qquad (3.149)$$

    (b) Derive the differential equation for $x_j$, the proportion of channels in state $j$.

    (c) Show that $x_j = \binom{k}{j} n^j (1-n)^{k-j}$, where $\binom{k}{j} = \frac{k!}{j!(k-j)!}$ is the *binomial coefficient*, is a stable invariant manifold for the system of differential equations, provided that

    $$\frac{dn}{dt} = \alpha(1-n) - \beta n. \qquad (3.150)$$

17. Consider the model of the Na$^+$ channel shown in Fig. 3.14. Show that if $\alpha$ and $\beta$ are large compared to $\gamma$ and $\delta$, then $x_{21}$ is given (approximately) by

    $$x_{21} = \left(\frac{\alpha}{\alpha+\beta}\right)^2 h, \qquad (3.151)$$

    $$\frac{dh}{dt} = \gamma(1-h) - \delta h, \qquad (3.152)$$

    while conversely, if $\gamma$ and $\delta$ are large compared to $\alpha$ and $\beta$, then (approximately)

    $$x_{21} = m^2 \left(\frac{\gamma}{\gamma+\delta}\right), \qquad (3.153)$$

    $$\frac{dm}{dt} = \alpha(1-m) - \beta m. \qquad (3.154)$$

18. Show that (3.128) has two negative real roots. Show that when $\beta = 0$ and $a \le \frac{-\lambda_1}{\lambda_1 - \lambda_2}$, then (3.129)–(3.131) have two possible solutions, one with $\alpha + \delta = -\lambda_1$, $\gamma = -\lambda_2$, the other with $\alpha + \delta = -\lambda_2$, $\gamma = -\lambda_1$. In the first solution inactivation is faster than activation, while the reverse is true for the second solution.

19. Write a computer program to simulate the response of a stochastic three-state $Na^+$ channel (Fig. 3.15A) to a voltage step. Take the ensemble average of many runs to reproduce the macroscopic behavior of Fig. 3.13. Using the data from simulations, reconstruct the open-time distribution, the latency distribution, and the distribution of $N$, the number of times the channel opens. From these distributions calculate the rate constants of the simulation.

20. Find the differential equations describing the interaction of a two-state channel with a use-dependent blocker.

# CHAPTER 4

# Excitability

We have seen in previous chapters how the control of cell volume results in a potential difference across the cell membrane, and how this potential difference causes ionic currents to flow through channels in the cell membrane. Regulation of this membrane potential by control of the ionic channels is one of the most important cellular functions. Many cells, such as neurons and muscle cells, use the membrane potential as a signal, and thus the operation of the nervous system and muscle contraction (to name but two examples) are both dependent on the generation and propagation of electrical signals.

To understand electrical signaling in cells, it is helpful (and not too inaccurate) to divide all cell types into two groups: excitable cells and nonexcitable cells. Many cells maintain a stable equilibrium potential. For some, if currents are applied to the cell for a short period of time, the potential returns directly to its equilibrium value after the applied current is removed. Such cells are called nonexcitable, typical examples of which are the epithelial cells that line the walls of the gut. Photoreceptors are also nonexcitable, although in their case, membrane potential plays an extremely important signaling role nonetheless.

However, there are cells for which, if the applied current is sufficiently strong, the membrane potential goes through a large excursion, called an *action potential*, before eventually returning to rest. Such cells are called *excitable*. Excitable cells include cardiac cells, smooth and skeletal muscle cells, secretory cells, and most neurons. The most obvious advantage of excitability is that an excitable cell either responds in full to a stimulus or not at all, and thus a stimulus of sufficient amplitude may be reliably distinguished from background noise. In this way, noise is filtered out, and a signal is reliably transmitted.

There are many examples of excitability that occur in nature. A simple example of an excitable system is a household match. The chemical components of the match head are stable to small fluctuations in temperature, but a sufficiently large temperature fluctuation, caused, for example, by friction between the head and a rough surface, triggers the abrupt oxidation of these chemicals with a dramatic release of heat and light. The fuse of a stick of dynamite is a one-dimensional continuous version of an excitable medium, and a field of dry grass is its two-dimensional version. Both of these spatially extended systems admit the possibility of wave propagation. The field of grass has one additional feature that the match and dynamite fuse fail to have, and that is recovery. While it is not very rapid by physiological standards, given a few months of growth, a burned-over field of grass will regrow enough fuel so that another fire may spread across it.

Although the generation and propagation of signals have been extensively studied by physiologists for at least the past 100 years, the most important landmark in these studies is the work of Alan Hodgkin and Andrew Huxley, who developed the first quantitative model of the propagation of an electrical signal along a squid giant axon (deemed "giant" because of the size of the axon, *not* the size of the squid). Their model was originally used to explain the action potential in the long giant axon of a squid nerve cell, but the ideas have since been extended and applied to a wide variety of excitable cells. Hodgkin–Huxley theory is remarkable, not only for its influence on electrophysiology, but also for its influence, after some filtering, on applied mathematics. FitzHugh (in particular) showed how the essentials of the excitable process could be distilled into a simpler model upon which mathematical analysis could make some progress. Because this simplified model turned out to be of such great theoretical interest, it contributed enormously to the formation of a new field of applied mathematics, the study of excitable systems, a field that continues to stimulate a vast amount of research.

Because of the central importance of cellular electrical activity in physiology, because of the importance of the Hodgkin–Huxley model in the study of electrical activity, and because it forms the basis for the study of excitability, it is no exaggeration to say that the Hodgkin–Huxley model is the most important model in all of the physiological literature.

## 4.1 The Hodgkin–Huxley Model

In Chapter 2 we described how the cell membrane can be modeled as a capacitor in parallel with an ionic current, resulting in the equation

$$C_m \frac{dV}{dt} + I_{\text{ion}}(V,t) = 0, \tag{4.1}$$

where $V$, as usual, denotes the internal minus the external potential ($V = V_i - V_e$). In the squid giant axon, as in many neural cells, the principal ionic currents are the sodium current and the potassium current. Although there are other ionic currents, primarily

**Figure 4.1**  The infamous giant squid, having nothing to do with the work of Hodgkin and Huxley on squid giant axon. From *Dangerous Sea Creatures*, © 1976, 1977 Time-Life Films, Inc.

the chloride current, in the Hodgkin–Huxley theory they are small and lumped together into one current called the *leakage current*. Since the instantaneous $I$–$V$ curves of open $Na^+$ and $K^+$ channels in the squid giant axon are approximately linear, (4.1) becomes

$$C_m \frac{dV}{dt} = -g_{Na}(V - V_{Na}) - g_K(V - V_K) - g_L(V - V_L) + I_{app}, \qquad (4.2)$$

where $I_{app}$ is the applied current. During an action potential there is a measured influx of 3.7 pmoles/cm$^2$ of sodium and a subsequent efflux of 4.3 pmoles/cm$^2$ of potassium. These amounts are so small that it is realistic to assume that the ionic concentrations, and hence the equilibrium potentials, are constant and unaffected by an action potential. It is important to emphasize that our choice of linear $I$–$V$ curves for the three different channel types is dictated largely by experimental data. Axons in other species (such as vertebrates) have ionic channels that are better described by other $I$–$V$ curves, such as the GHK current equation (2.68). However, the qualitative nature of the results remains largely unaffected, and so the discussion in this chapter, which is mostly of a qualitative nature, remains correct for models that use more complex $I$–$V$ curves to describe the ionic currents.

Equation (4.2) is a first-order ordinary differential equation and can be rewritten in the form

$$C_m\frac{dV}{dt} = -g_{\text{eff}}(V - V_{\text{eq}}) + I_{\text{app}},\tag{4.3}$$

where $g_{\text{eff}} = g_{\text{Na}} + g_{\text{K}} + g_{\text{L}}$ and $V_{\text{eq}} = (g_{\text{Na}}V_{\text{Na}} + g_{\text{K}}V_{\text{K}} + g_{\text{L}}V_{\text{L}})/g_{\text{eff}}$. $V_{\text{eq}}$ is the membrane resting potential and is a balance between the reversal potentials for the three ionic currents. In fact, at rest, the sodium and leakage conductances are small compared to the potassium conductance, so that the resting potential is close to the potassium equilibrium potential.

The quantity $R_m = 1/g_{\text{eff}}$, the passive membrane resistance, is on the order of 1000 $\Omega\text{cm}^2$. The time constant for this equation is

$$\tau_m = C_m R_m,\tag{4.4}$$

on the order of 1 msec. It follows that with a steady applied current, the membrane potential should equilibrate quickly to

$$V = V_{\text{eq}} + R_m I_{\text{app}}.\tag{4.5}$$

For sufficiently small applied currents, this is indeed what happens. However, for larger applied currents, the response is quite different. Assuming that the model (4.2) is correct, the only possible explanation for these differences is that the conductances are not constant but depend in some way on the voltage. Historically, the key step to determining the conductances was being able to measure the individual ionic currents and from this to deduce the changes in conductances. This was brilliantly accomplished by Hodgkin and Huxley in 1952.

## 4.1.1 History of the Hodgkin–Huxley Equations

(This section is adapted from Rinzel, 1990.) In a series of five articles that appeared in the *Journal of Physiology* in 1952, Alan Lloyd Hodgkin and Andrew Fielding Huxley, along with Bernard Katz, who was a coauthor of the lead paper and a collaborator in several related studies, unraveled the dynamic ionic conductances that generate the nerve action potential (Hodgkin et al., 1952; Hodgkin and Huxley, 1952a,b,c,d). They were awarded the 1963 Nobel Prize in physiology and medicine (shared with John C. Eccles, for his work on potentials and conductances at motorneuron synapses).

Before about 1939, the membrane potential was believed to play an important role in the membrane's state, but there was no way to measure it. It was known that a cell's membrane separated different ionic concentrations inside and outside the cell.

Applying the Nernst equation, Bernstein (1902) was led to suggest that the resting membrane was semipermeable to potassium, implying that at rest, $V$ should be around $-70$ mV. He believed that during activity there was a breakdown in the membrane's resistance to all ionic fluxes, and potential differences would disappear, i.e., $V$ would approach zero.

In 1940, Cole and Curtis, using careful electrode placement coupled with biophysical and mathematical analysis, obtained the first convincing evidence for a substantial transient increase in membrane conductivity during passage of the action potential. While they estimated a large conductance increase, it was not infinite, so without a direct measurement of membrane potential it was not possible to confirm or nullify Bernstein's hypothesis. During a postdoctoral year in the U.S. in 1937–1938, Hodgkin established connections with Cole's group at Columbia and worked with them at Woods Hole in the summer. He and Curtis nearly succeeded in measuring $V$ directly by tunneling along the giant axon with a glass micropipette. When each succeeded later (separately, with other collaborators), they found, surprisingly, that $V$ rose transiently toward zero, but with a substantial overshoot. This finding brought into serious question the hypothesis of Bernstein and provided much food for thought during World War II, when Hodgkin, Huxley, and many other scientists were involved in the war effort.

By the time postwar experimental work was resuming in England, Cole and Marmont had developed the *space clamp technique*. This method allowed one to measure directly the total transmembrane current, uniform through a known area, rather than spatially nonuniform as generated by a capillary electrode. To achieve current control with space clamping, the axon was threaded with a metallic conductor (like a thin silver wire) to provide low axial resistance and thereby eliminate voltage gradients along the length of the axon. Under these conditions the membrane potential is no longer a function of distance along the axon, only of time. In addition, during the 1947 squid season, Cole and company made substantial progress toward controlling the membrane potential as well.

In 1948, Hodgkin went to visit Cole (then at Chicago) to learn directly of their methods. With some further developments of their own, Hodgkin, Huxley, and Katz applied the techniques with great success to record transient ionic fluxes over the physiological ranges of voltages. Working diligently, they collected most of the data for their papers in the summer of 1949. Next came the step of identifying the individual contributions of the different ion species. Explicit evidence that both sodium and potassium were important came from the work of Hodgkin and Katz (1949). This also explained the earlier puzzling observations that $V$ overshoots zero during an action potential, opposing the suggestion of Bernstein. Instead of supposing that there was a transient increase in permeability identical for all ions, Hodgkin and Katz realized that different changes in permeabilities for different ions could account for the $V$ time course, as $V$ would approach the Nernst potential for the ion to which the membrane was predominantly permeable, and this dominance could change with time. For example, at rest the membrane is most permeable to $K^+$, so that $V$ is close to $V_K$. However, if $g_K$ were to decrease and $g_{Na}$ were to increase, then $V$ would be pushed toward $V_{Na}$, which is positive, thus depolarizing the cell.

The question of how the changes in permeability were dynamically linked to $V$ was not completely stated until the papers of 1952. In fact, the substantial delay from data collection in 1949 until final publication in 1952 can be attributed to the considerable time devoted to data analysis, model formulation, and testing. Computer downtime was also a factor, as some of the solutions of the Hodgkin–Huxley equations were computed on a desktop, hand-cranked calculator. As Hodgkin notes, "The propagated action potential took about three weeks to complete and must have been an enormous labour for Andrew [Huxley]" (Hodgkin, 1976, p. 19).

The final paper of the 1952 series is a masterpiece of the scientific art. Therein they present their elegant experimental data, a comprehensive theoretical hypothesis, a fit of the model to the experimental data (obtained for fixed values of the membrane potential), and then, presto, a prediction (from their numerical computations) of the time course of the propagated action potential. In biology, where quantitatively predictive theories are rare, this work stands out as one of the most successful combinations of experiment and theory.

## 4.1.2 Voltage and Time Dependence of Conductances

The key step to sorting out the dynamics of the conductances came from the development of the *voltage clamp*. A voltage clamp fixes the membrane potential, usually by a rapid step from one voltage to another, and then measures the current that must be supplied in order to hold the voltage constant. Since the supplied current must equal the transmembrane current, the voltage clamp provides a way to measure the transient transmembrane current that results. The crucial point is that the voltage can be stepped from one constant level to another, and so the ionic currents can be measured at a constant, known, voltage. Thus, even when the conductances are functions of the voltage (as is actually the case), a voltage clamp eliminates any voltage changes and permits measurement of the conductances as functions of time only.

Hodgkin and Huxley found that when the voltage was stepped up and held fixed at a higher level, the total ionic current was initially inward, but at later times an outward current developed (Fig. 4.2). For a number of reasons, not discussed here, they argued that the initial inward current is carried almost entirely by $Na^+$ ions, while the outward current that develops later is carried largely by $K^+$ ions. With these assumptions, Hodgkin and Huxley were able to use a clever trick to separate the total ionic current into its constituent ionic parts. They replaced 90% of the extracellular sodium in the normal seawater bath with choline (a viscous liquid vitamin B complex found in many animal and vegetable tissues), which rendered the axon nonexcitable but changed the resting potential only slightly. Since it is assumed that immediately after the voltage has been stepped up, the ionic current is all carried by $Na^+$, it is possible to measure the initial $Na^+$ currents in response to a voltage step. Note that although the $Na^+$ currents can be measured directly immediately after the voltage step,

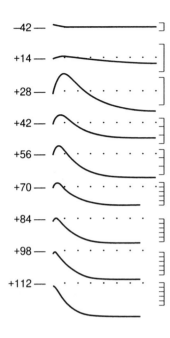

**Figure 4.2** Experimental results describing the total membrane current in response to a step depolarization. The numbers on the left give the final value of the membrane potential, in mV. The interval between dots on the horizontal scale is 1 ms, while one division on the vertical scale represents 0.5 mA/cm². (Hodgkin and Huxley, 1952a, Fig. 2a.)

they cannot be measured directly over a longer time period, as the total ionic current begins to include a contribution from the $K^+$ current. If we denote the $Na^+$ currents for the two cases of normal extracellular $Na^+$ and zero extracellular $Na^+$ by $I_{Na}^1$ and $I_{Na}^2$ respectively, then the ratio of the two currents,

$$I_{Na}^1 / I_{Na}^2 = K, \tag{4.6}$$

say, can be measured directly from the experimental data.

Next, Hodgkin and Huxley made two further assumptions. First, they assumed that the sodium current ratio $K$ is independent of time and is thus constant over the course of each voltage clamp experiment. In other words, the amplitude and direction of the $Na^+$ current may be affected by the low extracellular $Na^+$ solution, but its time course is not. Second, they assumed that the potassium channels are unaffected by the change in extracellular sodium concentration. There is considerable evidence that the sodium and potassium channels are independent. Tetrodotoxin (TTX) is known to block sodium currents while leaving the potassium currents almost unaffected, while tetraethylammonium (TEA) has the opposite effect of blocking the potassium current but not the sodium current. To complete the argument, since $I_{ion} = I_{Na} + I_K$, and $I_K^1 = I_K^2$, it follows that $I_{ion}^1 - I_{Na}^1 = I_{ion}^2 - I_{Na}^2$, and thus

$$I_{Na}^1 = \frac{K}{K-1}(I_{ion}^1 - I_{ion}^2), \tag{4.7}$$

$$I_K = \frac{I_{ion}^1 - K I_{ion}^2}{1 - K}. \tag{4.8}$$

Hence, given measurements of the total ionic currents in the two cases, and given the ratio $K$ of the $Na^+$ currents, it is possible to determine the complete time courses of both the $Na^+$ and $K^+$ currents.

Finally, from knowledge of the individual currents, one obtains the conductances as

$$g_{Na} = \frac{I_{Na}}{V - V_{Na}}, \qquad g_K = \frac{I_K}{V - V_K}. \tag{4.9}$$

Note that this result relies on the specific (linear) model used to describe the $I$–$V$ curve of the $Na^+$ and $K^+$ channels, but, as we discussed above, we assume throughout that the instantaneous $I$–$V$ curves of the $Na^+$ and $K^+$ channels are linear.

Samples of Hodgkin and Huxley's data are shown in Fig. 4.3. The plots show ionic conductances as functions of time following a step increase or decrease in the membrane potential. The important observation is that with voltages fixed, the conductances are time dependent. For example, when $V$ is stepped up and held fixed at a higher level, $g_K$ does not increase instantaneously, but instead increases over time to a final steady level. Both the time constant of the increase and the final value of $g_K$ are dependent on the value to which the voltage is stepped. Further, $g_K$ increases in a sigmoidal fashion, with a slope that first increases and then decreases (Fig. 4.3A and B). Following a step decrease in the voltage, $g_K$ falls in a simple exponential fashion (Fig. 4.3A). This particular feature of $g_K$ — a sigmoidal increase coupled with an exponential decrease — will be important when we model $g_K$. The behavior of $g_{Na}$ is more complex. Following a step increase in voltage, $g_{Na}$ first increases, but then decreases again, *all at the same fixed voltage* (Fig. 4.3C). Hence, the time dependence of $g_{Na}$ requires a more complex model than for that of $g_K$.

### The potassium conductance

From the experimental data shown in Fig. 4.3A and B, it is reasonable to expect that $g_K$ obeys some differential equation,

$$\frac{dg_K}{dt} = f(v, t), \tag{4.10}$$

say, where $v = V - V_{eq}$; i.e., $v$ is the difference between the membrane potential and the resting potential. (Of course, since $V_{eq}$ is a constant, $dv/dt = dV/dt$.) However, for $g_K$ to have the required sigmoidal increase and exponential decrease, Hodgkin and Huxley realized that it would be easier to write $g_K$ as some power of a different variable, $n$ say, where $n$ satisfies a first-order differential equation. Thus, they wrote

$$g_K = \bar{g}_K n^4, \tag{4.11}$$

for some constant $\bar{g}_K$. The fourth power was chosen not for physiological reasons, but because it was the smallest exponent that gave acceptable agreement with the experimental data. The secondary variable $n$ obeys the differential equation

$$\tau_n(v) \frac{dn}{dt} = n_\infty(v) - n, \tag{4.12}$$

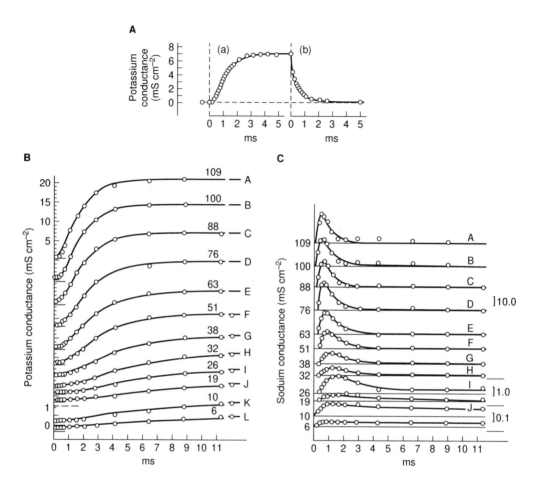

**Figure 4.3** Conductance changes as a function of time at different voltage clamps. A: The response of $g_K$ to a step increase in $V$ and then a step decrease. B: Responses of $g_K$ to step increases in $V$ of varying magnitudes. The number on each curve gives the depolarization in mV, and the smooth curves are calculated from solution of (4.11) and (4.12), with the initial condition $g_K(t=0) = 0.24$ mS/cm². The vertical scale is the same in curves A–J, but is increased by a factor of four in the lower two curves. For clarity, the baseline of each curve has been shifted up. C: Responses of $g_{Na}$ to step increases in $V$ of magnitudes given by the numbers on the left, in mV. The smooth curves are the model solutions. The vertical scales on the right are in units of mS/cm². (Hodgkin and Huxley, 1952d, Figs. 2, 3, and 6.)

for some functions $\tau_n(v)$ and $n_\infty(v)$ that must be determined from the experimental data in a manner that we describe soon. Equation (4.12) is often written in the form

$$\frac{dn}{dt} = \alpha_n(v)(1-n) - \beta_n(v)n, \tag{4.13}$$

where

$$n_\infty(v) = \frac{\alpha_n(v)}{\alpha_n(v) + \beta_n(v)}, \tag{4.14}$$

$$\tau_n(v) = \frac{1}{\alpha_n(v) + \beta_n(v)}. \tag{4.15}$$

At elevated potentials $n(t)$ increases monotonically and exponentially toward its resting value, thereby turning on, or *activating*, the potassium current. Since the Nernst potential is below the resting potential, the potassium current is an outward current at potentials greater than rest. The function $n(t)$ is called the *potassium activation*.

It is instructive to consider in detail how such a formulation for $g_K$ results in the required sigmoidal increase and exponential decrease. Suppose that at time $t = 0$, $v$ is increased from 0 to $v_0$ and then held constant, and suppose further that $n(0) = 0$. Solving (4.12) then gives

$$n(t) = n_\infty(v_0)\left[1 - \exp\left(\frac{-t}{\tau_n(v_0)}\right)\right], \tag{4.16}$$

which is an increasing curve (with monotonically decreasing slope) that approaches its maximum at $n_\infty(v_0)$. Raising $n$ to the fourth power gives a sigmoidally increasing curve as required. Higher powers of $n$ result in curves with a greater maximum slope at the point of inflection. In response to a step decrease in $v$, from $v_0$ to 0 say, the solution for $n$ is

$$n(t) = n_\infty(v_0)\exp\left(\frac{-t}{\tau_n(v_0)}\right), \tag{4.17}$$

in which case $n^4$ is exponentially decreasing, with no inflection point.

It remains to describe how the functions $n_\infty$ and $\tau_n$ are determined from the experimental data. For any given voltage step, the time constant $\tau_n$, and the final value of $n$, namely $n_\infty$, can be determined by fitting (4.16) to the experimental data. By this procedure one can determine $\tau_n$ and $n_\infty$ at a discrete set of values for $v$, i.e., those values used experimentally. Typical data points for $n_\infty$ are shown in Fig. 4.4 as symbols. To obtain a complete description of $g_K$, valid for all voltages and not only those used in the experiments, Hodgkin and Huxley fitted a smooth curve through the data points. The functional form of the smooth curve has no physiological significance, but is a convenient way of providing a continuous description of $n_\infty$. A similar procedure is followed for $\tau_n$. The continuous descriptions of $n_\infty$ and $\tau_n$ (expressed in terms of $\alpha_n$ and $\beta_n$) are given in (4.28) and (4.29) below.

### The sodium conductance

The time dependence for the sodium conductance is more difficult to unravel. From the experimental data it is suggested that there are two processes at work, one that turns on the sodium current and one that turns it off. Hodgkin and Huxley proposed that the sodium conductance is of the form

$$g_{Na}(v) = \bar{g}_{Na}m^3 h, \tag{4.18}$$

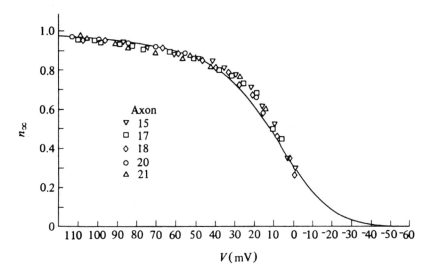

**Figure 4.4** Data points (symbols) of $n_\infty$, determined by fitting (4.16) to the experimental time courses. The smooth curve through the symbols provides a continuous description of $n_\infty$, and its functional form has no physiological significance. (Hodgkin and Huxley, 1952d, Fig. 5.)

and they fit the time-dependent behavior of $m$ and $h$ to exponentials with dynamics

$$\frac{dw}{dt} = \alpha_w(1-w) - \beta_w w, \tag{4.19}$$

where $w = m$ or $h$. Because $m$ is small at rest and first increases, it is called the *sodium activation*, and because $h$ shuts down, or inactivates, the sodium current, it is called the *sodium inactivation*. When $h = 0$, the sodium current is completely inactivated. The overall procedure is similar to that used in the specification of $g_K$. For any fixed voltage step, the unknown functions $\alpha_w$ and $\beta_w$ are determined by fitting to the experimental curves (Fig. 4.3C), and then smooth curves, with arbitrary functional forms, are fitted through the data points for $\alpha_w$ and $\beta_w$.

## Summary of the equations

In summary, the Hodgkin–Huxley equations for the space clamped axon are

$$C_m \frac{dv}{dt} = -\bar{g}_K n^4 (v - v_K) - \bar{g}_{Na} m^3 h (v - v_{Na}) - \bar{g}_L (v - v_L) + I_{app}, \tag{4.20}$$

$$\frac{dm}{dt} = \alpha_m(1-m) - \beta_m m, \tag{4.21}$$

$$\frac{dn}{dt} = \alpha_n(1-n) - \beta_n n, \tag{4.22}$$

$$\frac{dh}{dt} = \alpha_h(1-h) - \beta_h h. \tag{4.23}$$

The specific functions $\alpha$ and $\beta$ proposed by Hodgkin and Huxley were, in units of $(\text{ms})^{-1}$,

$$\alpha_m = 0.1 \frac{25 - v}{\exp\left(\frac{25-v}{10}\right) - 1}, \tag{4.24}$$

$$\beta_m = 4 \exp\left(\frac{-v}{18}\right), \tag{4.25}$$

$$\alpha_h = 0.07 \exp\left(\frac{-v}{20}\right), \tag{4.26}$$

$$\beta_h = \frac{1}{\exp\left(\frac{30-v}{10}\right) + 1}, \tag{4.27}$$

$$\alpha_n = 0.01 \frac{10 - v}{\exp\left(\frac{10-v}{10}\right) - 1}, \tag{4.28}$$

$$\beta_n = 0.125 \exp\left(\frac{-v}{80}\right). \tag{4.29}$$

For these expressions, the potential $v$ is the deviation from rest $(V = V_{\text{eq}} + v)$, measured in units of mV, current density is in units of $\mu A/cm^2$, conductances are in units of $mS/cm^2$, and capacitance is in units of $\mu F/cm^2$. The remaining constants are

$$\bar{g}_{Na} = 120, \qquad \bar{g}_K = 36, \qquad \bar{g}_L = 0.3, \tag{4.30}$$

with (adjusted) equilibrium potentials $v_{Na} = 115, v_K = -12$, and $v_L = 10.6$. In Fig. 4.5 are shown the steady-state functions, and the time constants are shown in Fig. 4.6.

In Chapter 3 we discussed simple models for the gating of $Na^+$ and $K^+$ channels and showed how the rate constants in simple kinetic schemes could be determined from whole-cell or single-channel data. We also showed how models of the form (4.20)–(4.23) can be derived by modeling the ionic channels as consisting of multiple subunits, each of which obeys a simple two-state model. In the Hodgkin–Huxley equations, it is assumed that the $Na^+$ channel consists of three "$m$" gates and one "$h$" gate, each of which can be either closed or open. If the gates operate independently, then the fraction

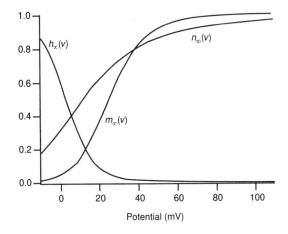

**Figure 4.5** Steady-state functions $m_\infty(v)$, $n_\infty(v)$ and $h_\infty(v)$.

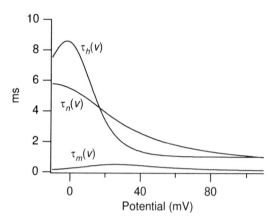

**Figure 4.6**   Time constants $\tau_m(v)$, $\tau_n(v)$, and $\tau_h(v)$.

of open $Na^+$ channels is $m^3h$, where $m$ and $h$ obey the equation of the two-state channel model. Similarly, if there are four "$n$" gates per potassium channel, all of which must be open for potassium to flow, then the fraction of open $K^+$ channels is $n^4$.

Now comes the most interesting challenge facing these equations. Having incorporated the measurements of conductance found from voltage-clamp experiments, one wonders whether these equations reproduce a realistic action potential, and if so, by what mechanism is the action potential produced? We can describe in qualitative terms how the Hodgkin–Huxley equations should work. If small currents are applied to a cell for a short period of time, the potential returns rapidly to its equilibrium $v = 0$ after the applied current is removed. The equilibrium potential is close to the potassium Nernst potential $v_K = -12$, because at rest, the sodium and leakage conductances are small. There is always competition among the three ionic currents to drive the potential to the corresponding resting potential. For example, if the potassium and leakage currents could be blocked or the sodium conductance dramatically increased, then the term $g_{Na}(V - V_{Na})$ should dominate (4.2), and as long as $v$ is below $v_{Na}$, an inward sodium current will drive the potential toward $v_{Na}$. Similarly, while $v$ is above $v_K$, the potassium current is outward in an attempt to drive $v$ toward $v_K$. Notice that since $v_K < v_L < v_{Na}$, $v$ is necessarily restricted to lie in the range $v_K < v < v_{Na}$.

If $g_{Na}$ and $g_K$ were constant, that would be the end of the story. The equilibrium at $v = 0$ would be a stable equilibrium, and following any stimulus, the potential would return exponentially to rest. But since $g_{Na}$ and $g_K$ can change, the different currents can exert their respective influences. The actual sequence of events is determined by the dynamics of $m, n$, and $h$. The most important observation for the moment is that $\tau_m(v)$ is much smaller than either $\tau_n(v)$ or $\tau_h(v)$, so that $m(t)$ responds much more quickly to changes in $v$ than either $n$ or $h$. We can now understand why the Hodgkin–Huxley system is an excitable system. As noted before, if the potential $v$ is raised slightly by a small stimulating current, the system returns to its stable equilibrium. However, during the period of time that the potential $v$ is elevated, the sodium activation $m$ is tracking $m_\infty(v)$. If the stimulating current is large enough to raise the potential and therefore

$m_\infty(v)$ to a high enough level (above its *threshold*), then before the system can return to rest, $m$ will increase sufficiently to change the sign of the net current, resulting in an autocatalytic inward sodium current. Now, as the potential rises, $m$ continues to rise, and the inward sodium current is increased, further adding to the rise of the potential.

If nothing further were to happen, the potential would be driven to a new equilibrium at $v_{Na}$. However, here is where the difference in time constants plays an important role. When the potential is at rest, the sodium inactivation $h$ is positive, about 0.6. As the potential increases, $h_\infty$ decreases toward zero, and as $h$ approaches zero, the sodium current is inactivated because $g_{Na}$ approaches zero. However, because the time constant $\tau_h(v)$ is much larger than $\tau_m(v)$, there is a considerable delay between turning on the sodium current when $m$ increases and turning off the sodium current when $h$ decreases. The net effect of the two different time scales on $m$ and $h$ is that the sodium current is at first turned on and later turned off, and this is seen as an initial increase of the potential, followed by a decrease toward rest.

At about the same time that the sodium current is inactivated, the outward potassium current is activated. This is because of the similarity of the time constants $\tau_n(v)$ and $\tau_h(v)$. Activation of the potassium current drives the potential below rest toward $v_K$. When $v$ is negative, $n$ declines, and the potential eventually returns to rest, and the whole process can start again. In Fig. 4.7 is shown a plot of the potential $v(t)$ during an action potential following a superthreshold stimulus. In Fig. 4.8, $m(t)$, $n(t)$, and $h(t)$ during the same action potential are shown.

There are four recognizable phases of an action potential: the *upstroke, excited, refractory*, and *recovery* phases. The refractory period is the period following the excited phase when additional stimuli evoke no substantial response, even though the potential is below or close to its resting value. There can be no response, since the sodium channels are still inactivated because $h$ is small. As $h$ gradually returns to its resting value, further responses once again become possible.

There are two ways that the Hodgkin–Huxley system can be made into an autonomous oscillator. The first is to inject a steady current of sufficient strength. Such

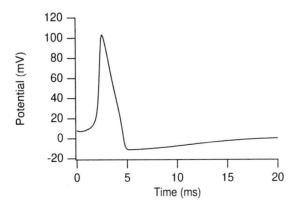

**Figure 4.7**  Action potential of the Hodgkin–Huxley equations.

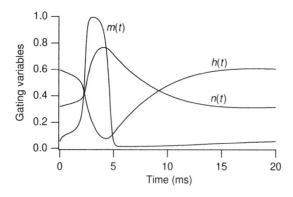

**Figure 4.8** The gating variables $m, n$, and $h$ during an action potential.

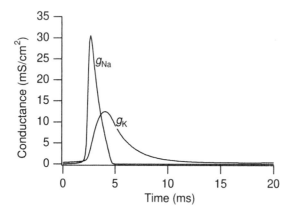

**Figure 4.9** Conductances $g_{Na}$ and $g_K$ during an action potential.

a current raises the resting potential above the threshold for an action potential, so that after the axon has recovered from an action potential, the potential rises to a superthreshold level at which another action potential is evoked.

Immersing the axon in a bath of high extracellular potassium has the same effect through a slightly different mechanism. An increase of extracellular potassium has the effect of increasing the potassium Nernst potential, effectively raising the rest potential (since the rest potential is close to the potassium Nernst potential). If this increase of the potassium Nernst potential is sufficiently large, the resting potential becomes superthreshold, and autonomous oscillations result. This mechanism of creating an autonomous oscillator out of normally excitable but nonoscillatory cells is important for certain cardiac arrhythmias.

### 4.1.3   Qualitative Analysis

FitzHugh (1960, 1961, 1969) has given a particularly elegant qualitative description of the Hodgkin–Huxley equations that allows a better understanding of the model's behavior. More detailed analyses have also been given by Rinzel (1978), Troy (1978), Cole et al. (1955), and Sabah and Spangler (1970). FitzHugh's approach is based on the fact that some of the model variables have fast kinetics, while others are much slower. In particular, $m$ and $v$ are fast variables (i.e., the Na$^+$ channel activates quickly, and the membrane potential changes quickly), while $n$ and $h$ are slow variables (i.e., Na$^+$ channels are inactivated slowly, and the K$^+$ channels are activated slowly). Thus, during the initial stages of the action potential, $n$ and $h$ remain essentially constant while $m$ and $v$ vary. This allows the full 4-dimensional phase space to be broken into smaller pieces by fixing the slow variables and considering the behavior of the model as a function only of the two fast variables. Although this description is accurate only for the initial stages of the action potential, it provides a useful way to study the process of excitation.

### The fast phase-plane

Thus motivated, we fix the slow variables $n$ and $h$ at their respective resting states, which we call $n_0$ and $h_0$, and consider how $m$ and $v$ behave in response to stimulation. The differential equations for the fast phase-plane are

$$C_m \frac{dv}{dt} = -\bar{g}_K n_0^4 (v - v_K) - \bar{g}_{Na} m^3 h_0 (v - v_{Na}) - \bar{g}_L (v - v_L), \tag{4.31}$$

$$\frac{dm}{dt} = \alpha_m (1 - m) - \beta_m m, \tag{4.32}$$

or, equivalently,

$$\tau_m \frac{dm}{dt} = m_\infty - m. \tag{4.33}$$

This is now a two-dimensional system and can be most easily studied in the $(m, v)$ phase-plane, a plot of which is given in Fig. 4.10. The curves defined by $dv/dt = 0$ and $dm/dt = 0$ are the $v$ and $m$ nullclines, respectively. The $m$ nullcline is the curve $m = m_\infty(v)$, which we have seen before (in Fig. 4.5), while the $v$ nullcline is defined by

$$v = \frac{\bar{g}_{Na} m^3 h_0 v_{Na} + \bar{g}_K n_0^4 v_K + \bar{g}_L v_L}{\bar{g}_{Na} m^3 h_0 + \bar{g}_K n_0^4 + \bar{g}_L}. \tag{4.34}$$

For the parameters of the Hodgkin–Huxley model, the $m$ and $v$ nullclines intersect in three places, corresponding to three steady states of the fast equations. Note that these three intersections are not steady states of the full model, only of the fast subsystem, and, to be precise, should be called pseudo-steady states. However, in the context of the fast phase-plane we continue to call them steady states. We label the three steady states $v_r$, $v_s$, and $v_e$ (for resting, saddle, and excited).

It is left as an exercise to show that $v_r$ and $v_e$ are stable steady states of the fast subsystem, while $v_s$ is a saddle point. Since $v_s$ is a saddle point, it has a one-dimensional stable manifold. This stable manifold divides the $(m, v)$ plane into two regions: any trajectory starting to the left of the stable manifold is prevented from reaching $v_e$ and must eventually return to the resting state, $v_r$. However, any trajectory starting to the right of the stable manifold is prevented from returning to the resting state and must eventually end up at the excited state, $v_e$. Hence, the stable manifold, in combination

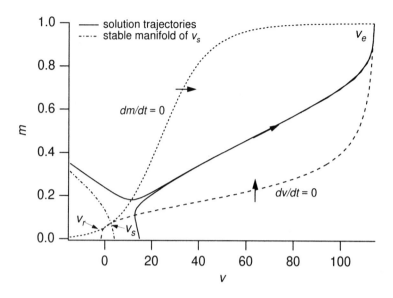

**Figure 4.10** The Hodgkin–Huxley fast phase-plane, showing the nullclines $dv/dt = 0$ and $dm/dt = 0$ (with $h_0 = 0.596$, $n_0 = 0.3176$), two sample trajectories and the stable manifold of the saddle point $v_s$.

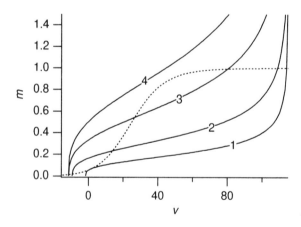

**Figure 4.11** The Hodgkin–Huxley fast phase-plane as a function of the slow variables, showing the $m$ nullcline (dashed) and the movement of the $v$ nullcline (solid) and the disappearance of the steady states. For these curves, parameter values are (1) $h_0 = 0.596$, $n_0 = 0.3176$; (2) $h_0 = 0.4$, $n_0 = 0.5$; (3) $h_0 = 0.2$, $n_0 = 0.7$; and (4) $h_0 = 0.1$, $n_0 = 0.8$.

with the two stable steady states, causes a threshold phenomenon. Any perturbation from the resting state that is not large enough to cross the stable manifold eventually dies away, but a perturbation that crosses the stable manifold results in a large excursion in the voltage, up to the excited state. Sample trajectories are sketched in Fig. 4.10.

If $m$ and $v$ were the only variables in the model, then $v$ would stay at $v_e$ indefinitely. However, as pointed out before, $v_e$ is not a steady state of the full model. Thus, to see what happens on a longer time scale, we must consider how slow variations in $n$ and

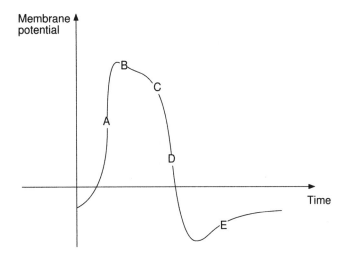

**Figure 4.12** Schematic diagram of a complete action potential. A: Superthreshold stimulus causes a fast increase of $v$ to the excited state. B: $v$ is sitting at the excited state, $v_e$, decreasing slowly as $n$ increases and $h$ decreases, i.e., as $v_e$ moves toward $v_s$. C: $v_e$ and $v_s$ disappear at a saddle-node bifurcation, and so, D: The solution must return to the resting state $v_r$. E: $n$ and $h$ slowly return to their resting states, and as they do so, $v_r$ slowly increases until the steady state of the full four-dimensional system is reached.

$h$ affect the qualitative nature of the fast phase-plane. First note that since $v_e > v_r$, it follows that $h_\infty(v_e) < h_\infty(v_r)$ and $n_\infty(v_e) > n_\infty(v_r)$. Hence, while $v$ is at the excited state, $h$ begins to decrease thus inactivating the Na$^+$ conductance, and $n$ starts to increase thus activating the K$^+$ conductance. Next note that although the $m$ nullcline in the fast phase-plane is independent of $n$ and $h$, the $v$ nullcline is not. In Fig. 4.10 the nullclines were drawn using the steady-state values for $n$ and $h$: different values of $n$ and $h$ change the shape of the $v$ nullcline. As $n$ increases and $h$ decreases, the $v$ nullcline moves to the left and up, as illustrated in Fig. 4.11. As the $v$ nullcline moves up and to the left, $v_e$ and $v_s$ move towards each other, while $v_r$ moves to the left. During this phase the voltage is at $v_e$ and thus decreases slowly. Eventually, $v_e$ and $v_s$ coalesce and disappear in a saddle-node bifurcation. When this happens $v_r$ is the only remaining steady state, and so the solution must return to the resting state. Note that since the $v$ nullcline has moved up and to the left, $v_r$ is not a steady state of the full system. However, when $v$ decreases to $v_r$, $n$ and $h$ both return to their steady states and as they do so, $v_r$ slowly increases until the steady state of the full system is reached and the action potential is complete. A schematic diagram of a complete action potential is shown in Fig. 4.12, and the important points are labeled for comparison with the phase-plane.

### The fast–slow phase-plane

In the above analysis, we simplified the four-dimensional phase space by taking a series of two-dimensional cross-sections, those with various fixed values of $n$ and $h$. However, by taking a different cross-section we can highlight other aspects of the action potential. In particular, by taking a cross-section involving one fast variable and one slow variable we obtain a description of the Hodgkin–Huxley model that has proven to be extraordinarily useful.

We extract a single fast variable by assuming that $m$ is an instantaneous function of $v$, and thus $m = m_\infty(v)$ at all times. This is equivalent to assuming that activation of the $Na^+$ conductance acts on a time scale even faster than that of the voltage. Next, FitzHugh noticed that during the course of an action potential, $h + n \approx 0.8$ (notice the approximate symmetry of $n(t)$ and $h(t)$ in Fig. 4.8), and thus $h$ can be eliminated by setting $h = 0.8 - n$. With these simplifications, the Hodgkin–Huxley model contains one fast variable $v$ and one slow variable $n$, and can be written as

$$-C_m \frac{dv}{dt} = \bar{g}_K n^4(v - v_K) + \bar{g}_{Na} m_\infty^3(v)(0.8 - n)(v - v_{Na}) + \bar{g}_L(v - v_L), \qquad (4.35)$$

$$\frac{dn}{dt} = \alpha_n(1 - n) - \beta_n n. \qquad (4.36)$$

For convenience we let $f(v,n)$ denote the right-hand side of (4.35), i.e.,

$$-f(v,n) = \bar{g}_K n^4(v - v_K) + \bar{g}_{Na} m_\infty^3(v)(0.8 - n)(v - v_{Na}) + \bar{g}_L(v - v_L). \qquad (4.37)$$

A plot of the nullclines of the fast–slow subsystem is given in Fig. 4.13. The $v$ nullcline is defined by $f(v,n) = 0$ and has a cubic shape, while the $n$ nullcline is $n_\infty(v)$ and is monotonically increasing. There is a single intersection (at least for the given parameter values) and thus a single steady state. Because $v$ is a fast variable and $n$ is a slow one, the solution trajectories are almost horizontal except where $f(v,n) \approx 0$. The curve $f(v,n) = 0$ is called the "slow manifold." Along the slow manifold the solution moves slowly in the direction determined by the sign of $dn/dt$, but away from the slow manifold the solution moves quickly in a horizontal direction. From the sign of $dv/dt$ it follows that the solution trajectories move away from the middle branch of

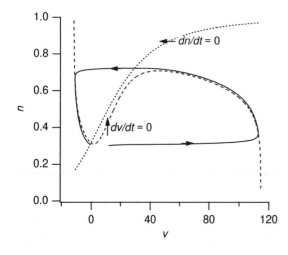

**Figure 4.13**  Fast–slow phase-plane of the Hodgkin–Huxley model.

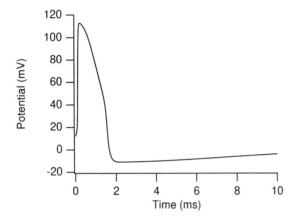

**Figure 4.14** Action potential for the reduced fast–slow Hodgkin–Huxley model.

the slow manifold and toward the left and right branches. Thus, the middle branch is termed the unstable branch of the slow manifold. This unstable branch acts as a threshold. Suppose a perturbation from the steady state is small enough so that $v$ does not cross the unstable manifold. Then, the trajectory moves horizontally towards the left and returns to the steady state. However, if the perturbation is large enough so that $v$ crosses the unstable manifold, then the trajectory moves to the right until it reaches the right branch of the slow manifold, which corresponds to the excited state. On this right branch $dn/dt > 0$, and so the solution moves slowly up the slow manifold until the turning point is reached. At the turning point, $n$ cannot increase any further, as the right branch of the slow manifold ceases to exist, and so the solution moves over to the left branch of the slow manifold. On this left branch $dn/dt < 0$, and so the solution moves down the left branch until the steady state is reached, completing the action potential (Fig. 4.13). A plot of the potential as a function of time is shown in Fig. 4.14.

The variables $v$ and $n$ are usually called the excitation and recovery variables, respectively: excitation because it governs the rise to the excited state, and recovery because it causes the return to the steady state. In the absence of $n$ the solution would stay at the excited state indefinitely.

There is a close relationship between the fast phase-plane and the fast–slow phase-plane. Recall that in the fast phase-plane, the $v$ and $m$ nullclines have three intersection points when $n = n_0$ and $h = h_0$. These three intersections correspond to the three branches of the curve $f(v, n_0) = 0$. In other words, when $n$ is fixed at $n_0$, the equation $f(v, n_0) = 0$ has three possible solutions, corresponding to $v_r$, $v_s$ and $v_e$ in the fast phase-plane. However, consideration of Fig. 4.13 shows that, as $n$ increases, the two rightmost branches of the slow manifold coalesce and disappear. This is analogous to the merging and disappearance of $v_e$ and $v_s$ seen in the fast phase-plane. The fast–slow phase-plane is a convenient way of summarizing how $v_r$, $v_s$, and $v_e$ depend on the slow variables.

This representation of the Hodgkin–Huxley model in terms of two variables, one fast and one slow, is the basis of the FitzHugh–Nagumo model for excitability, and we discuss models of this generic type in some detail throughout this book.

## 4.2 Two-Variable Models

There is considerable value in studying systems of equations that are simpler than the Hodgkin–Huxley equations but that retain many of their qualitative features. This is the motivation for the FitzHugh–Nagumo equations and their variants. Basically, the FitzHugh–Nagumo model extracts the essential behavior of the Hodgkin–Huxley fast–slow phase-plane and presents it in a simplified form. Thus, the FitzHugh–Nagumo model has two variables, one fast ($v$) and one slow ($w$). The fast variable has a cubic nullcline and is called the excitation variable, while the slow variable is called the recovery variable and has a nullcline that is monotonically increasing. The nullclines have a single intersection point, which, without loss of generality, may be assumed to be at the origin. A schematic diagram of the phase-plane is given in Fig. 4.15, where we introduce some of the notation used later in this section.

The FitzHugh–Nagumo model can be derived from a simplified model of the cell membrane (Fig. 4.16). Here the cell (or membrane patch) consists of three components, a capacitor representing the membrane capacitance, a nonlinear current–voltage device for the fast current, and a resistor, inductor, and battery in series for the recovery current. In the 1960s Nagumo, a Japanese electrical engineer, built this circuit using

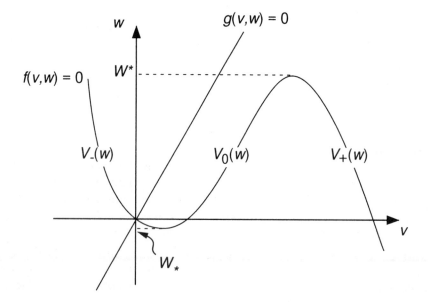

**Figure 4.15**   Schematic diagram of the generalized FitzHugh–Nagumo phase-plane.

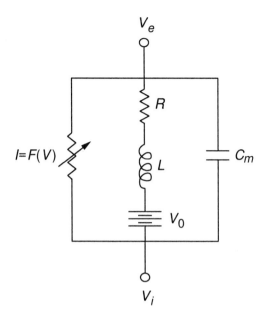

**Figure 4.16** Circuit diagram for the Fitz-Hugh–Nagumo equations.

a tunnel diode as the nonlinear element (Nagumo et al., 1964), thereby attaching his name to this system.

Using Kirchhoff's laws, we can write down equations for the behavior of this membrane circuit diagram. We find

$$C_m \frac{dV}{d\tau} + F(V) + i = -I_0, \tag{4.38}$$

$$L \frac{di}{d\tau} + Ri = V - V_0, \tag{4.39}$$

where $I_0$ is the applied external current, $i$ is the current through the resistor–inductor, $V = V_i - V_e$ is the membrane potential, and $V_0$ is the potential gain across the battery. Here $\tau$ is used to represent dimensional time because we will shortly introduce $t$ as a dimensionless time variable. The function $F(V)$ is assumed to be of "cubic" shape, having three zeros, of which the smallest $V = 0$ and largest $V = V_1$ are stable solutions of the differential equation $dV/d\tau = -F(V)$. We take $R_1$ to be the "passive" resistance of the nonlinear element, $R_1 = 1/F'(0)$. Now we introduce the dimensionless variables $v = V/V_1, w = R_1 i/V_1, f(v) = -R_1 F(V_1 v)/V_1$, and $t = L\tau/R_1$. Then (4.38) and (4.39) become

$$\epsilon \frac{dv}{dt} = f(v) - w - w_0, \tag{4.40}$$

$$\frac{dw}{dt} = v - \gamma w - v_0, \tag{4.41}$$

where $\epsilon = R_1^2 C_m/L, w_0 = R_1 I_0/V_1, v_0 = V_0/V_1$, and $\gamma = R/R_1$.

At this point we must specify $f(v)$. As we discussed above, the only requirement is that it have the general shape shown in Fig. 4.15. The classic choice is the cubic polynomial

$$f(v) = Av(v - \alpha)(1 - v) \qquad \text{with} \qquad 0 < \alpha < 1, \tag{4.42}$$

which gives the FitzHugh–Nagumo model. Other choices include the McKean model (McKean, 1970), for which

$$f(v) = H(v - \alpha) - v, \tag{4.43}$$

where $H$ is the Heaviside function. This choice recommends itself because then the model is piecewise linear, allowing explicit solutions of many interesting problems. Another piecewise linear model (also proposed by McKean, 1970) has

$$f(v) = \begin{cases} -v, & \text{for } v < \alpha/2, \\ v - \alpha, & \text{for } \dfrac{\alpha}{2} < v < \dfrac{1+\alpha}{2}, \\ 1 - v, & \text{for } v > \dfrac{1+\alpha}{2}. \end{cases} \tag{4.44}$$

A third piecewise linear model that has found widespread usage is the "Pushchino" model, so named because of its development in Pushchino (about 70 miles south of Moscow), by Krinksy, Panfilov, Pertsov, Zykov, and their coworkers. The details of the Pushchino model are described in Exercise 13.

An important variant of the FitzHugh–Nagumo equations is the *van der Pol oscillator*. An electrical engineer, van der Pol built the circuit using triodes because it exhibits stable oscillations. As there was little interest in oscillatory circuits at the time, he proposed his circuit as a model of an oscillatory cardiac pacemaker (van der Pol and van der Mark, 1928). Since then it has become a classic example of a system with limit cycle behavior and relaxation oscillations, included in almost every textbook on oscillations (see, for example, Stoker, 1950, or Minorsky, 1962).

If we eliminate the resistor $R$ from the circuit (Fig. 4.16), differentiate (4.38), and eliminate the current $i$, we get the second-order differential equation

$$C_m \frac{d^2 V}{d\tau^2} + F'(V)\frac{dV}{d\tau} + \frac{V}{L} = \frac{V_0}{L}. \tag{4.45}$$

Following rescaling, and setting $F(v) = A(v^3/3 - v)$, we arrive at the *van der Pol equation*

$$v'' + a(v^2 - 1)v' + v = 0. \tag{4.46}$$

From now on, by the *generalized FitzHugh–Nagumo equations* we mean the system of equations

$$\epsilon \frac{dv}{dt} = f(v, w) + I, \tag{4.47}$$

$$\frac{dw}{dt} = g(v, w), \tag{4.48}$$

where the nullcline $f(v, w) = 0$ is of "cubic" shape. By this we mean that for a finite range of values of $w$, there are three solutions $v = v(w)$ of the equation $f(v, w) = 0$. These we will denote by $v = V_-(w), v = V_0(w)$, and $v = V_+(w)$, and where comparison is possible (since these functions need not all exist for the same range of $w$),

$$V_-(w) \leq V_0(w) \leq V_+(w). \tag{4.49}$$

We denote the minimal value of $w$ for which $V_-(w)$ exists by $W_*$, and the maximal value of $w$ for which $V_+(w)$ exists by $W^*$. For values of $w$ above the nullcline $f(v, w) = 0, f(v, w) < 0$, and below the nullcline, $f(v, w) > 0$ (in other words, $f_w(v, w) < 0$).

The nullcline $g(v, w) = 0$ is assumed to have precisely one intersection with the curve $f(v, w) = 0$. Increasing $v$ beyond the curve $g(v, w) = 0$ makes $g(v, w)$ positive (i.e., $g_v(v, w) > 0$), and decreasing $w$ below the curve $g(v, w) = 0$ increases $g(v, w)$ (hence $g_w(v, w) < 0$). The nullclines $f$ and $g$ are illustrated in Fig. 4.15.

## 4.2.1  Phase-Plane Behavior

One attractive feature of the FitzHugh–Nagumo model is that because it is a two-variable system, it can be studied using phase-plane techniques. (For an example of a different approach, see Troy, 1976.) There are two characteristic phase portraits possible (shown in Figs. 4.17 and 4.19). By assumption, there is only one steady state, at $v = v^*, w = w^*$, with $f(v^*, w^*) = g(v^*, w^*) = 0$. Without loss of generality, we may assume that this steady state occurs at the origin, as this involves only a shift of the variables. Furthermore, it is typical that the parameter $\epsilon$ is a small number. For small $\epsilon$, if the steady state lies on either the left or right solution branch of $f(v, w) = 0$, i.e., the curves $v = V_\pm(w)$, it is linearly stable. Somewhere on the middle solution branch $v = V_0(w)$, near the extremal values of the curve $f(v, w) = 0$, there is a Hopf bifurcation point. That is, if parameters are varied so that the steady state passes through this point, a periodic orbit arises as a continuous solution branch and bifurcates into a stable limit cycle oscillation.

When the steady state is on the leftmost branch, but close to the minimum (Fig. 4.17), the system is excitable. This is because even though the steady state is linearly stable, a sufficiently large perturbation from the steady state sends the state variable on a trajectory that runs away from the steady state before eventually returning to rest. Such a trajectory goes rapidly to the rightmost branch, which it hugs as it gradually creeps upward, whence upon reaching the maximum, it goes rapidly to the leftmost branch and then gradually returns to rest, staying close to this branch as it does. Plots of the variables $v$ and $w$ are shown as functions of time in Fig. 4.18.

The mathematical description of these events follows from singular perturbation theory. With $\epsilon \ll 1$, the variable $v$ is a fast variable and the variable $w$ is a slow variable. This means that if possible, $v$ is adjusted rapidly to maintain a pseudo-equilibrium at $f(v, w) = 0$. In other words, if possible, $v$ clings to the stable branches of $f(v, w) = 0$, namely $v = V_\pm(w)$. Along these branches the dynamics of $w$ are governed by the reduced

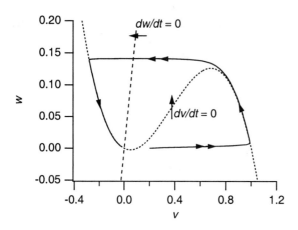

**Figure 4.17** Phase portrait for a Fitz-Hugh–Nagumo system with $f(v, w) = v(v - 0.1)(1 - v) - w$, $g(v, w) = v - 0.5w$, $\epsilon = 0.01$. For these parameter values the system has a unique globally stable rest point, but is excitable.

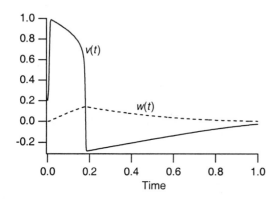

**Figure 4.18** Solutions of a FitzHugh–Nagumo system with $f(v, w) = v(v - 0.1)(1 - v) - w$, $g(v, w) = v - 0.5w$, $\epsilon = 0.01$.

dynamics

$$\frac{dw}{dt} = g(V_\pm(w), w) = G_\pm(w). \tag{4.50}$$

When it is not possible for $v$ to be in quasi-equilibrium, the motion is governed approximately by the differential equations,

$$\frac{dv}{d\tau} = f(v, w), \qquad \frac{dw}{d\tau} = 0, \tag{4.51}$$

found by making the change of variables to the fast time scale $t = \epsilon\tau$, and then setting $\epsilon = 0$. On this time scale, $w$ is constant, while $v$ equilibrates to a stable solution of $f(v, w) = 0$.

The evolution of $v$ and $w$ starting from specified initial conditions $v_0$ and $w_0$ can now be described. Suppose $v_0$ is greater than the rest value $v^*$. If $v_0 < V_0(w)$, then $v$ returns directly to the steady state. If $v_0 > V_0(w)$, then $v$ goes rapidly to the upper branch $V_+(w)$ with $w$ remaining nearly constant at $w_0$. The curve $v = V_0(w)$ is a *threshold curve*.

While $v$ remains on the upper branch, $w$ increases according to

$$\frac{dw}{dt} = G_+(w) \tag{4.52}$$

as long as possible. However, in the finite time

$$T_e = \int_{w_0}^{W^*} \frac{dw}{G_+(w)}, \tag{4.53}$$

$w$ reaches the "knee" of the nullcline $f(v, w) = 0$. This period of time constitutes the *excited phase* of the action potential.

When $w$ reaches $W^*$ it is no longer possible for $v$ to stay on the excited branch, so it must return to the lower branch $V_-(w)$. Once on this branch, $w$ decreases following the dynamics

$$\frac{dw}{dt} = G_-(w). \tag{4.54}$$

If the rest point lies on the lower branch, then $G_-(w^*) = 0$, and $w$ gradually returns to rest on the lower branch.

If the rest point lies on the middle branch $V_0(w)$ (depicted in Fig. 4.19), it is unstable. Instead of returning to rest after one excursion on the excited branch, the trajectory alternates periodically between the upper and lower branches, with $w$ varying between $W_*$ and $W^*$. This periodic limit cycle behavior (with the solution plotted as a function of time shown in Fig. 4.20) is called a *relaxation oscillation*. The period of the oscillation is approximately

$$T = \int_{W_*}^{W^*} \left( \frac{1}{G_+(w)} - \frac{1}{G_-(w)} \right) dw. \tag{4.55}$$

This number is finite because $G_+(w) > 0$, and $G_-(w) < 0$ for all appropriate $w$.

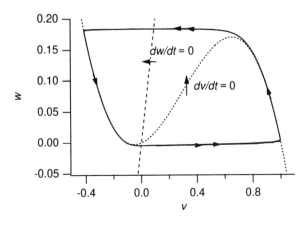

**Figure 4.19** Phase portrait for a Fitz-Hugh–Nagumo system with $f(v, w) = v(v+0.1)(1-v)-w$, $g(v, w) = v-0.5w$, $\epsilon = 0.01$. For these parameter values, the unique rest point is unstable and there is a globally stable periodic orbit.

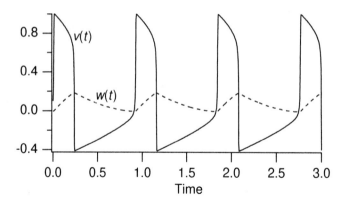

**Figure 4.20** Solutions of a FitzHugh–Nagumo system with $f(v, w) = v(v + 0.1)(1 - v) - w$, $g(v, w) = v - 0.5w$, $\epsilon = 0.01$.


# 4.3  Appendix: Cardiac Cells

Excitable cells come in many varieties. Some excitable cells that have been subjected to a substantial amount of study are the cardiac cells. The primary cell types of cardiac cells are nodal cells (the sinoatrial (SA) and atrioventricular (AV) nodes), Purkinje fiber cells, and myocardial cells, each with a slightly different function. For an introduction to the cardiac conduction system, see Chapter 11.

The primary function of SA nodal cells is to provide a pacemaker signal for the rest of the heart. AV nodal cells transmit the electrical signal from atria to ventricles with a delay. Purkinje fiber cells are primarily for fast conduction, to activate the myocardium, and myocardial cells, both atrial and ventricular, are muscle cells and so are contractile as well as excitable.

Because of these different functions, these cell types have slightly different action potential shapes. The action potential for SA nodal cells is the shortest, similar to the Hodgkin–Huxley action potential, while both Purkinje fiber cells and myocardial cells have substantially prolonged action potentials, facilitating muscular contraction. Typical action potentials for these are shown in Figs. 4.21, 4.22, and 4.24, and typical ionic concentrations are given in Table 4.1.

Subsequent to the work of Hodgkin and Huxley, there was substantial work done to apply their model to many different cell types, including cardiac cells. The quantitative models of cardiac cells are distinctively of Hodgkin–Huxley type, serving the purpose of reproducing the details of the action potential shape (which they do quite well) while attempting to give reasonable mechanistic explanations (via ionic channels and currents) of their behavior. The primary difficulty with cardiac cells is that there are many different cell types and many different types of ionic channels. Even within a single cell type, there is substantial variation. For example, in the ventricles, epicardial, midmyocardial, and endocardial cells have noticeable differences in action potential duration. AV nodal cells vary substantially, to the extent that they are sometimes classified into several different subtypes. In many ways, the squid giant axon worked on by Hodgkin and Huxley was an ideal first candidate because the number and types of channels are so few. One wonders whether they would have been as successful if they had begun instead with a cardiac cell.

Because of this increased complexity of cardiac cell structure, the models of their behavior, even though quantitative, are far from complete, and they retain a qualitative sense. That is, the details are not nearly as precise as one is led to believe. Indeed, these models reflect the artistry (or lack thereof) of modeling, in that some effects are retained, others are discarded, and still others are averaged together into conglomerates. There is continual competition between two forces, the one that would like all channels and all effects to be specified with great detail, and the other that recognizes that some details are inconsequential to the final behavior of the model.

The purpose of this appendix is to give a brief overview of the most important models of cardiac cell behavior and to give some (albeit limited) insight into the primary differences between cell types. This material is in an appendix because while the quantitative details of the models vary, there are no substantially new ideas used to develop the models.

## 4.3.1 Purkinje Fibers

### A phenomenological approach

The first model describing the action potential of a cardiac cell was proposed by Noble (1962) for Purkinje fiber cells. The primary purpose of the model was to show that the action potential of a Purkinje fiber cell, which is noticeably different from a squid axon action potential, could be captured by a model of Hodgkin–Huxley type. The Purkinje fiber cell is self-oscillatory, with a sharp upstroke that overshoots and returns to a prolonged plateau (300–400 ms compared to 3 ms for the squid axon) before recovering.

The Noble model is of Hodgkin–Huxley type, expressed in terms of ionic currents and conductances. In this model there are three currents, identified as an inward sodium current, an outward potassium current, and a chloride leak current, all of

**Table 4.1**  Ion concentrations in most cardiac cells.

| Ion | Extracellular (mM) | Intracellular (mM) | Nernst Potential (mV) |
|---|---|---|---|
| $Na^+$ | 145 | 15 | 60 |
| $Cl^-$ | 100 | 5 | −80 |
| $K^+$ | 4.5 | 160 | −95 |
| $Ca^{2+}$ | 1.8 | 0.0001 | 130 |
| $H^+$ | 0.0001 | 0.0002 | −18 |

which are assumed to satisfy a linear instantaneous $I$–$V$ relation,

$$I = g(V - V_{\text{eq}}). \tag{4.56}$$

For the Noble model, all changes in conductances that were measured in sodium-deficient solutions were assumed to be for currents carried by potassium ions and are therefore called potassium currents (thus, the chloride current is taken to be zero). There is no certainty that these are indeed carried by potassium, but the nomenclature is convenient.

Following the usual Hodgkin–Huxley formulation, the balance of transmembrane currents is expressed by the conservation law

$$C_m \frac{dV}{dt} + g_{\text{Na}}(V - V_{\text{Na}}) + (g_{K_1} + g_{K_2})(V - V_K) + g_{\text{an}}(V - V_{\text{an}}) = I_{\text{app}}, \tag{4.57}$$

where $g_{\text{an}} = 0$ and $V_{\text{an}} = -60$ are the conductance and equilibrium potential, respectively, for the anion current of the chloride ion (which therefore is not needed). In addition, $V_{\text{Na}} = 40$ and $V_K = -100$.

The Noble model assumes two different types of potassium channels, an instantaneous, voltage-dependent, channel, and a time-dependent channel. The time-dependent potassium channel has a similar form to the Hodgkin–Huxley potassium channel, except that it is about 100 times slower in its response, in order to prolong the action potential plateau. This current is sometimes called the *delayed rectifier current*, because it is delayed and because it is primarily an outward (rectified) current. The conductance for this channel, $g_{K_2}$, depends on a time-dependent potassium activation variable $n$ through

$$g_{K_2} = 1.2n^4. \tag{4.58}$$

The conductance for the instantaneous channel is described empirically by

$$g_{K_1} = 1.2 \exp\left(-\frac{V + 90}{50}\right) + 0.015 \exp\left(\frac{V + 90}{60}\right). \tag{4.59}$$

The sodium conductance for the Noble model is of a form similar to the Hodgkin–Huxley model, being

$$g_{\text{Na}} = 400m^3h + g_i, \tag{4.60}$$

where $g_i = 0.14$, with the fixed inward bias from $g_i$ enabling a prolonged action potential without necessitating major reworking of the dynamics of $h$ and $m$.

The time dependence of the variables $m, n,$ and $h$ is of the form

$$\frac{dw}{dt} = \alpha_w(1 - w) - \beta_w w, \tag{4.61}$$

with $w = m, n,$ or $h$, where $\alpha_w$ and $\beta_w$ are all of the form

$$\frac{C_1 \exp(\frac{V - V_0}{C_2}) + C_3(V - V_0)}{1 + C_4 \exp(\frac{V - V_0}{C_5})}. \tag{4.62}$$

**Table 4.2**  Defining values for rate constants $\alpha$ and $\beta$ for the Noble model.

|            | $C_1$ | $C_2$ | $C_3$  | $C_4$ | $C_5$ | $V_0$ |
|------------|-------|-------|--------|-------|-------|-------|
| $\alpha_m$ | 0     | —     | 0.1    | −1    | −15   | −48   |
| $\beta_m$  | 0     | —     | −0.12  | −1    | 5     | −8    |
| $\alpha_h$ | 0.17  | −20   | 0      | 0     | —     | −90   |
| $\beta_h$  | 1     | ∞     | 0      | 1     | −10   | −42   |
| $\alpha_n$ | 0     | —     | 0.0001 | −1    | −10   | −50   |
| $\beta_n$  | 0.002 | −80   | 0      | 0     | —     | −90   |

The constants $C_1, \ldots, C_5$ and $V_0$ are displayed in Table 4.2.

In the Noble model, $C_m = 12$, which is unrealistically large. This value was used because it gives a correct time scale for the length of the action potential. The choice was justified by arguing that the effective capacitance for a small bundle of cylindrical cells, for which the data were obtained, should be larger than for a single cylindrical cell, the surface area of which is only a small fraction of the total cell membrane area.

Numerical simulations show that the Noble model produces an action potential that has correct features, seen in Fig. 4.21. The sharp upstroke comes from a large, fast, inward sodium current, and the plateau is maintained by a continued inward sodium current (with conductance $g_i$), which nearly counterbalances the instantaneous outward potassium current. Gradually, the slow outward potassium current is activated, causing repolarization. A small inward sodium leak, called the *pacemaker current*, also allows the potential to creep upward, eventually initiating another action potential.

Because of the sharp spike at the beginning of the action potential, it is not possible to reproduce the Purkinje fiber action potential with a two-variable FitzHugh–Nagumo model. However, by setting $m = m_\infty(V)$, the Noble model can be reduced to a three-variable model that retains the primary qualitative features of the original model.

### A physiological approach

While the Noble model succeeds in reproducing the Purkinje fiber action potential with a model of Hodgkin–Huxley type, the underlying physiology is incorrect, primarily because the model was constructed before data on the ionic currents were available.

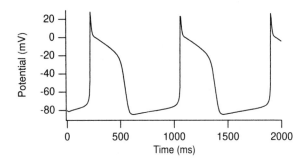

**Figure 4.21**  Action potential for the Noble model.

This lack of data was mostly because the voltage-clamp technique was not successfully applied to cardiac membrane until 1964.

The weakness of the physiology in the Noble model is exemplified by the fact that there is no current identified with calcium ions, and the inward sodium current was given the dual role of generating the upstroke and maintaining the plateau.

In 1975, McAllister, Noble, and Tsien (MNT) presented an improved model for the action potential of Purkinje fibers. This model is based on a "mosaic of experimental results," because unlike the Hodgkin–Huxley model, the required information was not obtained from a single experimental preparation. Furthermore, the model is known to have an inadequate description of the sodium current, so that the upstroke velocity is not accurate.

The MNT model is similar to the Noble model in that it is based on a description of transmembrane ionic currents. It is substantially more complicated than most other models, having 9 ionic currents and 9 gating variables. There are two inward currents, $I_{Na}$ and $I_{si}$ (called the "slow inward" current). The current $I_{Na}$ resembles the Hodgkin–Huxley sodium current and is represented as

$$I_{Na} = \bar{g}_{Na}m^3h(V - V_{Na}),  \tag{4.63}$$

where $m$ and $h$ are activation and inactivation gating variables, respectively, and $V_{Na} = 40$ mV. The inward current $I_{si}$ has slower kinetics than $I_{Na}$ and is carried, at least partly, by calcium ions. This current $I_{si}$ has two components and is given by

$$I_{si} = (0.8df + 0.04d')(V - V_{si}),  \tag{4.64}$$

where $V_{si} = 70$ mV. The variables $d$ and $f$ are time-dependent activation and inactivation variables, respectively, while $d'$ is only voltage-dependent, being

$$d' = \frac{1}{1 + \exp(-0.15(V + 40))}.  \tag{4.65}$$

In the MNT model, there are three time-dependent outward potassium currents, denoted by $I_{K_2}, I_{x_1}$, and $I_{x_2}$. None of these resemble the squid potassium current from a quantitative point of view, although all are described using an activation variable and no inactivation variable. The current $I_{K_2}$ is called the *pacemaker current* because it is responsible for periodically initiating an action potential, and it is given by

$$I_{K_2} = 2.8\bar{I}_{K_2}s,  \tag{4.66}$$

where

$$\bar{I}_{K_2} = \frac{\exp(0.04(V + 110)) - 1}{\exp(0.08(V + 60)) + \exp(0.04(V + 60))}.  \tag{4.67}$$

The currents $I_{x_1}$ and $I_{x_2}$ are called *plateau currents* and are governed by

$$I_{x_1} = 1.2x_1\frac{\exp(0.04(V + 95)) - 1}{\exp(0.04(V + 45))},  \tag{4.68}$$

$$I_{x_2} = x_2(25 + 0.385V).  \tag{4.69}$$

There is also a time-dependent outward current $I_{Cl}$ carried by chloride ions, which is described by

$$I_{Cl} = 2.5qr(V - V_{Cl}), \tag{4.70}$$

where $q$ and $r$ are activation and inactivation variables, and $V_{Cl} = -70$ mV.

Finally, there are several background (leak) currents that are time independent. There is an outward background current of potassium ions, described by

$$I_{K_1} = \bar{I}_{K_2} + 0.2\frac{V + 30}{1 - \exp(-0.04(V + 30))}, \tag{4.71}$$

where $\bar{I}_{K_2}$ is given by (4.67). There is an inward background sodium current described by

$$I_{Na,b} = 0.105(V - 40), \tag{4.72}$$

and, finally, a background chloride current, given by

$$I_{Cl,b} = 0.01(V + 70). \tag{4.73}$$

All of the conductances are specified in units of mS/cm$^2$, and voltage is in mV. The 9 gating variables $m, d, s, x_1, x_2, q, h, f$, and $r$ all satisfy first-order differential equations of the form (4.61), where $\alpha_w$ and $\beta_w$ are of the form (4.62). The constants $C_1, \ldots, C_5$ and $V_0$ are listed in Table 4.3.

The action potential for the MNT model is essentially the same as that for the Noble model, so the advantage of the MNT model is that it better isolates and depicts the

**Table 4.3**  Defining values for $\alpha$ and $\beta$ for the MNT model.

| | $C_1$ | $C_2$ | $C_3$ | $C_4$ | $C_5$ | $V_0$ |
|---|---|---|---|---|---|---|
| $\alpha_m$ | 0 | — | 1 | −1 | −10 | −47 |
| $\beta_m$ | 40 | −17.86 | 0 | 0 | — | −72 |
| $\alpha_h$ | 0.0085 | −5.43 | 0 | 0 | — | −71 |
| $\beta_h$ | 2.5 | $\infty$ | 0 | 1 | −12.2 | −10 |
| $\alpha_d$ | 0 | — | 0.002 | −1 | −10 | −40 |
| $\beta_d$ | 0.02 | −11.26 | 0 | 0 | — | −40 |
| $\alpha_f$ | 0.000987 | −25 | 0 | 0 | — | −60 |
| $\beta_f$ | 1 | $\infty$ | 0 | 1 | −11.49 | −26 |
| $\alpha_q$ | 0 | — | 0.008 | −1 | −10 | 0 |
| $\beta_q$ | 0.08 | −11.26 | 0 | 0 | — | 0 |
| $\alpha_r$ | 0.00018 | −25 | 0 | 0 | — | −80 |
| $\beta_r$ | 0.02 | $\infty$ | 0 | 1 | −11.49 | −26 |
| $\alpha_s$ | 0 | — | 0.001 | −1 | −5 | −52 |
| $\beta_s$ | $5.0 \times 10^{-5}$ | −14.93 | 0 | 0 | — | −52 |
| $\alpha_{x_1}$ | 0.0005 | 12.1 | 0 | 1 | 17.5 | −50 |
| $\beta_{x_1}$ | 0.0013 | −16.67 | 0 | 1 | −25 | −20 |
| $\alpha_{x_2}$ | $1.27 \times 10^{-4}$ | $\infty$ | 0 | 1 | −5 | −19 |
| $\beta_{x_2}$ | 0.0003 | −16.67 | 0 | 1 | −25 | −20 |

activity of different channels during an action potential. Of course, because it is much more complicated than the Noble model, it is also much harder to understand the model from a qualitative perspective. It therefore illustrates nicely the modeler's dilemma, the constant struggle to balance the demand for quantitative detail and qualitative understanding.

## 4.3.2  Sinoatrial Node

The primary function of SA nodal cells is to initiate a regular heartbeat. They have little contractile function and therefore little calcium. The most widely used model of action potential behavior for SA nodal cells is due to Yanagihara et al. (1980). As with all cardiac cell models, the YNI model is of Hodgkin–Huxley type. The YNI model includes four time-dependent currents. These are the sodium current $I_{Na}$, which is a fast inward current, and the potassium current $I_K$, both of which are similar to the Hodgkin–Huxley currents, as well as a slow inward current $I_s$, and a delayed inward current activated by hyperpolarization $I_h$. Finally, there is a time-independent leak current $I_l$.

The conservation of transmembrane current takes the form

$$C_m \frac{dV}{dt} + I_{Na} + I_K + I_l + I_s + I_h = I_{app}, \tag{4.74}$$

where

$$I_{Na} = 0.5 m^3 h (V - 30), \tag{4.75}$$

$$I_K = 0.7 p \frac{\exp(0.0277(V + 90)) - 1}{\exp(0.0277(V + 40))}, \tag{4.76}$$

$$I_l = 0.8 \left( 1 - \exp\left( -\frac{V + 60}{20} \right) \right), \tag{4.77}$$

$$I_s = 12.5(0.95d + 0.05)(0.95f + 0.05) \left( \exp\left( \frac{V - 10}{15} \right) - 1 \right), \tag{4.78}$$

$$I_h = 0.4 q (V + 45). \tag{4.79}$$

As usual, the 6 gating variables $m, h, p, d, f$, and $q$ satisfy first-order differential equations of the form (4.61). Some of the constants $\alpha_w$ and $\beta_w$ can be written in the form (4.62) with constant values as shown in Table 4.4. Those that do not fit this form are

$$\alpha_p = 9 \times 10^{-3} \frac{1}{1 + \exp(-\frac{V + 3.8}{9.71})} + 6 \times 10^{-4}, \tag{4.80}$$

$$\alpha_q = 3.4 \times 10^{-4} \frac{(V + 100)}{\exp(\frac{V + 100}{4.4}) - 1} + 4.95 \times 10^{-5}, \tag{4.81}$$

$$\beta_q = 5 \times 10^{-4} \frac{(V + 40)}{1 - \exp(-\frac{V + 40}{6})} + 8.45 \times 10^{-5}, \tag{4.82}$$

$$\alpha_d = 1.045 \times 10^{-2} \frac{(V + 35)}{1 - \exp(-\frac{V + 35}{2.5})} + 3.125 \times 10^{-2} \frac{V}{1 - \exp(-\frac{V}{4.8})}, \tag{4.83}$$

**Table 4.4** Defining values for $\alpha$ and $\beta$ for the YNI model.

| | $C_1$ | $C_2$ | $C_3$ | $C_4$ | $C_5$ | $V_0$ |
|---|---|---|---|---|---|---|
| $\alpha_m$ | 0 | — | 1 | −1 | −10 | −37 |
| $\beta_m$ | 40 | −17.8 | 0 | 0 | — | −62 |
| $\alpha_h$ | $1.209 \times 10^{-3}$ | −6.534 | 0 | 0 | — | −20 |
| $\beta_h$ | 1 | $\infty$ | 0 | 1 | −10 | −30 |
| $\beta_p$ | 0 | — | $-2.25 \times 10^{-4}$ | −1 | 13.3 | −40 |
| $\beta_d$ | 0 | — | $-4.21 \times 10^{-3}$ | −1 | 2.5 | 5 |
| $\alpha_f$ | 0 | — | $-3.55 \times 10^{-4}$ | −1 | 5.633 | −20 |

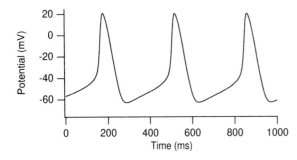

**Figure 4.22** Membrane potential for the YNI model of SA nodal behavior.

$$\beta_f = 9.44 \times 10^{-4} \frac{(V + 60)}{1 + \exp(-\frac{V+29.5}{4.16})}. \tag{4.84}$$

The behavior of the YNI equations is depicted in Fig. 4.22. The action potential is shaped similarly to the Hodgkin–Huxley action potential but is periodic in time and slower. The sodium current is a fast current, and there is little loss in accuracy in replacing $m(t)$ with $m_\infty(V)$. The most significant current in the YNI model is the slow inward current $I_s$. Not only does this current provide for most of the upstroke, it is also responsible for the oscillation, in that after repolarization by the potassium current, the slow inward current gradually depolarizes the node until threshold is reached and an action potential is initiated.

Because the action potential of the SA node has no initial spike, it is relatively easy to replicate it using a two-variable FitzHugh–Nagumo model. In Fig. 4.23 is shown the periodic activity of a cubic FitzHugh–Nagumo model with action potential spikes similar to those of the YNI model.

### 4.3.3 Ventricular Cells

The Beeler–Reuter equations (Beeler and Reuter, 1977) appeared shortly after the MNT equations and model the electrical behavior in ventricular myocardial cells. Like the models previously described, this model is based on data obtained from voltage-clamp experiments. The Beeler–Reuter equations are less complicated than the MNT equa-

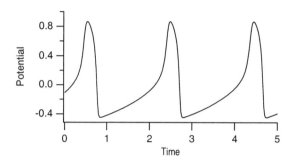

**Figure 4.23** The potential $v(t)$ for the FitzHugh–Nagumo model with $f(v, w) = v(1 - v)(v - \alpha) - w$, $g(v, w) = w - \gamma v$, with $\epsilon = 0.02$, $\alpha = -0.05$, $\gamma = -0.6$.

tions, as there are only 4 transmembrane currents that are described, two inward currents, one fast and one slow, and two outward currents, one time independent and one time dependent.

As usual, there is the inward sodium current

$$I_{Na} = (4m^3hj + 0.003)(V - 50), \tag{4.85}$$

which is gated by the variables $m, h$, and $j$. Here, Beeler and Reuter found it necessary to include the reactivation variable $j$, because the reactivation process is much slower than inactivation and cannot be accurately modeled with the single variable $h$. Thus, the sodium current is activated by $m$, inactivated by $h$, and reactivated by $j$, the slowest of the three variables. The functions $h_\infty$ and $j_\infty$ are identical; it is their time constants that differ. Notice also the inclusion of a sodium leak current; a similar sodium leak was included in the Noble and MNT models.

The potassium current has two components, a time-independent current

$$I_K = 1.4 \frac{\exp(0.04(V + 85)) - 1}{\exp(0.08(V + 53)) + \exp(0.04(V + 53))} + 0.07 \frac{V + 23}{1.0 - \exp(-0.04(V + 23))} \tag{4.86}$$

and a time-activated outward current

$$I_x = 0.8x \frac{\exp(0.04(V + 77)) - 1}{\exp(0.04(V + 35))}. \tag{4.87}$$

The pacemaker potassium current used in the MNT model is not active in myocardial tissue, which is not spontaneously oscillatory.

The primary difference between a ventricular cell and a Purkinje or SA nodal cell is the presence of calcium, which is needed to activate the contractile machinery. For the Beeler–Reuter equations, the calcium influx is modeled by the slow inward current

$$I_s = 0.09fd(V + 82.3 + 13.0287 \ln[Ca]_i), \tag{4.88}$$

activated by $d$ and inactivated by $f$. Since the reversal potential for $I_s$ is calcium dependent, the internal calcium concentration must be tracked, via

$$\frac{dc}{dt} = 0.07(1 - c) - I_s, \tag{4.89}$$

**Table 4.5** Defining values for $\alpha$ and $\beta$ for the Beeler–Reuter model.

|            | $C_1$  | $C_2$   | $C_3$ | $C_4$ | $C_5$   | $V_0$  |
|------------|--------|---------|-------|-------|---------|--------|
| $\alpha_m$ | 0      | —       | 1     | −1    | −10     | −47    |
| $\beta_m$  | 40     | −17.86  | 0     | 0     | —       | −72    |
| $\alpha_h$ | 0.126  | −4      | 0     | 0     | —       | −77    |
| $\beta_h$  | 1.7    | $\infty$| 0     | 1     | −12.2   | −22.5  |
| $\alpha_j$ | 0.055  | −4      | 0     | 1     | −5      | −78    |
| $\beta_j$  | 0.3    | $\infty$| 0     | 1     | −10     | −32    |
| $\alpha_d$ | 0.095  | −100    | 0     | 1     | −13.9   | 5      |
| $\beta_d$  | 0.07   | −58.5   | 0     | 1     | 20      | −44    |
| $\alpha_f$ | 0.012  | −125    | 0     | 1     | 6.67    | −28    |
| $\beta_f$  | 0.0065 | −50     | 0     | 1     | −5      | −30    |
| $\alpha_x$ | 0.0005 | 12      | 0     | 1     | 17.5    | −50    |
| $\beta_x$  | 0.0013 | −16.67  | 0     | 1     | −25     | −20    |

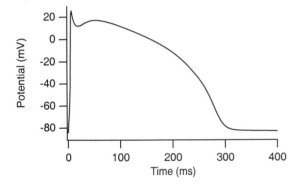

**Figure 4.24** Action potential for the Beeler–Reuter equations.

where $c = 10^7[\text{Ca}]_i$. Since currents are taken as positive outward, the intracellular source of calcium is $-I_s$.

The gating variables follow dynamics (4.61) where $\alpha_w$ and $\beta_w$ are of the form (4.62) with constants as displayed in Table 4.5. For these equations, units of $V$ are in mV, conductances are in units of mS/cm$^2$, and time is measured in milliseconds (ms). A plot of the Beeler–Reuter action potential is shown in Fig. 4.24. The long plateau of the Beeler–Reuter action potential is maintained by the slow inward (calcium) current, and the return to the resting potential is mediated by the slow outward potassium current $I_{x_1}$.

## 4.3.4 Summary

To the novice, these models may appear as a bunch of numbers and equations pulled magically out of a hat. To summarize and help give some perspective on the structure of these models, in Table 4.6 is presented an overview of the currents and gating variables used in each of the four models presented above.

**Table 4.6**  Summary of currents for different ionic models.

| Noble | Purkinje Fiber | | |
|---|---|---|---|
| | Current | Ion | Gating Variables |
| | $I_{Na}$ | $Na^+$ | $m, h$ |
| | $I_{K_1}$ | $K^+$ | |
| | $I_{K_2}$ | $K^+$ | $n$ |
| | $I_{An}$ | $Cl^-$ | |
| MNT | Purkinje fiber | | |
| | Current | Ion | Gating Variables |
| | $I_{Na}$ | $Na^+$ | $m, h$ |
| | $I_{si}$ | $Ca^{2+}$ | $d, f$ |
| | $I_{K_1}$ | $K^+$ | |
| | $I_{K_2}$ | $K^+$ | $s$ |
| | $I_{x_1}$ | $K^+$ | $x_1$ |
| | $I_{x_2}$ | $K^+$ | $x_2$ |
| | $I_{Cl}$ | $Cl^-$ | $q, r$ |
| | $I_{Na,b}$ | $Na^+$ | |
| | $I_{Cl,b}$ | $Cl^-$ | |
| YNI | SA Nodal cells | | |
| | Current | Ion | Gating Variables |
| | $I_{Na}$ | $Na^+$ | $m, h$ |
| | $I_s$ | | $d, f$ |
| | $I_K$ | $K^+$ | $p$ |
| | $I_l$ | leak | |
| | $I_h$ | | $q$ |
| BR | myocardial cells | | |
| | Current | Ion | Gating Variables |
| | $I_{Na}$ | $Na^+$ | $m, h, j$ |
| | $I_s$ | $Ca^{2+}$ | $d, f$ |
| | $I_K$ | $K^+$ | |
| | $I_x$ | $K^+$ | $x$ |

## 4.3.5  Further Developments

All of the foregoing models for cardiac cell electrical activity represent various approximations and simplifications. Consequently, there has been considerable activity to improve these models to better represent the correct physiology. Here we mention a few of these models.

### Pacemaker activity

Both SA node and Purkinje fiber cells exhibit autonomous oscillatory behavior. Examples of improved ionic models that reconstruct this behavior are given by DiFrancesco and Noble (1985) for Purkinje fiber cells and by Noble and Noble (1984) for the SA node.

## The fast sodium current

At the time the Beeler–Reuter equations were published, it was not possible to measure accurately the fast sodium inward current, because it activates so rapidly. As a result, all the early models (Noble, MNT, BR) used the Hodgkin–Huxley formulation of the sodium current. However, it is known that this does not give a sufficiently rapid upstroke for the action potential. This has little effect on the space-clamped action potential, but it has an important effect on the propagation speed for propagated action potentials.

Once appropriate data became available, it was possible to suggest an improved description of the sodium current. Thus, a modification of the sodium current was proposed by Ebihara and Johnson (1980) (EJ), which has since become the standard for most myocardial simulations. For the EJ model, the sodium current is given by

$$I_{Na} = 23m^3h(V - 29), \tag{4.90}$$

with

$$\alpha_m = 0.32\frac{V + 47.13}{1 - \exp(-(V + 47.13))}, \tag{4.91}$$

$$\beta_m = \exp\left(-\frac{V}{11}\right), \tag{4.92}$$

$$\alpha_h = \begin{cases} 0 & \text{if } V < -40, \\ 0.135\exp(-0.147(V + 80)) & \text{if } V > -40, \end{cases} \tag{4.93}$$

$$\beta_h = \begin{cases} \dfrac{7.69}{\exp(-0.09(V + 10.66)) + 1} & \text{if } V < -40, \\ 3.56\exp(0.079V) + 3.1 \times 10^5\exp(0.35V) & \text{if } V > -40. \end{cases} \tag{4.94}$$

## Calcium

The final form for a myocardial ionic model has not yet been determined, as there are continual suggestions for improvements and modifications. A major difficulty with the Beeler–Reuter model is with the calcium current and internal calcium concentration. This is not unexpected, as at the time the model was formulated, little was known about the mechanisms of calcium release and uptake.

While the inclusion of proper calcium kinetics into a myocardial ionic model is a topic of active research, one recent model deserves mention. This model is known as the Luo–Rudy (LR) model (Luo and Rudy, 1994a,b). An earlier model (Luo and Rudy, 1991) was a generalization of the Beeler–Reuter model. There is still significant debate about many of the details of the LR models, so it is unlikely that the LR models are the final word.

## 4.4 Exercises

1. Show that, if $k > 1$, then $(1 - e^{-x})^k$ has an inflection point, but $(e^{-x})^k$ does not.

2. Use the independence principle of Chapter 3 to derive an expression for $K$ in (4.6). Show that $K$ is independent of time, as assumed by Hodgkin and Huxley. How can this expression

be used to check the accuracy of the assumption that all the initial current in response to a voltage step is carried by $Na^+$ ions? Derive another expression for $K$ by assuming that the $Na^+$ channel has a linear $I$–$V$ curve (which, as we discussed in Chapter 3, does not obey the independence principle).

3. Explain why replacing the extracellular sodium with choline has little effect on the resting potential of an axon. Calculate the new resting potential with 90% of the extracellular sodium removed. Why is the same not true if potassium is replaced?

4. Plot the nullclines of the Hodgkin–Huxley fast subsystem. Show that $v_r$ and $v_e$ in the Hodgkin–Huxley fast subsystem are stable steady states, while $v_s$ is a saddle point. Compute the stable manifold of the saddle point and compute sample trajectories in the fast phase-plane, demonstrating the threshold effect.

5. Show how the Hodgkin–Huxley fast subsystem depends on the slow variables; i.e., show how the $v$ nullcline moves as $n$ and $h$ are changed, and demonstrate the saddle-node bifurcation in which $v_e$ and $v_s$ disappear.

6. Plot the nullclines of the fast–slow Hodgkin–Huxley phase-plane and compute a complete action potential. How does the fast–slow phase-plane behave in the presence of an applied current? How much applied current is needed to generate oscillations?

7. Suppose that in the Hodgkin–Huxley fast–slow phase-plane, $v$ is slowly decreased to $v^* < v_0$ (where $v_0$ is the steady state), held there for a considerable time, and then released. Describe what happens in qualitative terms, i.e., without actually computing the solution. This is called *anode break excitation* (Hodgkin and Huxley, 1952d. Also see Peskin, 1991). What happens if $v$ is instantaneously decreased to $v^*$ and then released immediately? Why do these two solutions differ?

8. Solve the full Hodgkin–Huxley system numerically with a variety of constant current inputs. For what range of inputs are there self-sustained oscillations? Why should one expect self-sustained oscillations for some current inputs?

9. The Hodgkin–Huxley equations are for the squid axon at $6.3°C$. Using that the absolute temperature enters the equations through the Nernst equation, determine how changes in temperature affect the behavior of the equations. In particular, simulate the equations at $0°C$ and $30°C$ to determine whether the equations become more or less excitable with an increase in temperature.

10. Show that a Hopf bifurcation occurs in the generalized FitzHugh–Nagumo model when $f_v(v^*, w^*) = -\epsilon g_w(v^*, w^*)$, assuming that
$$f_v(v^*, w^*)g_w(v^*, w^*) - g_v(v^*, w^*)f_w(v^*, w^*) > 0.$$

11. Morris and Lecar (1981) proposed the following two-variable model of membrane potential for a barnacle muscle fiber:
$$C_m \frac{dV}{dT} + I_{ion}(V, W) = I_{app}, \tag{4.95}$$

$$\frac{dW}{dT} = \phi \Lambda(V)[W_\infty(V) - W], \tag{4.96}$$

where $V$ = membrane potential, $W$ = fraction of open $K^+$ channels, $T$ = time, $C_m$ = membrane capacitance, $I_{app}$ = externally applied current, $\phi$ = maximum rate for closing

Table 4.7  Typical parameter values for the Morris–Lecar model.

| | | | |
|---|---|---|---|
| $C_m$ | $= 20\ \mu F/cm^2$ | $I_{app}$ | $= 0.06\ mamp/cm^2$ |
| $g_{Ca}$ | $= 4.4\ mS/cm^2$ | $g_K$ | $= 8\ mS/cm^2$ |
| $g_L$ | $= 2\ mS/cm^2$ | $\phi$ | $= 0.040\ (ms)^{-1}$ |
| $V_1$ | $= -1\ mV$ | $V_2$ | $= 15\ mV$ |
| $V_3$ | $= 0$ | $V_4$ | $= 30\ mV$ |
| $V_{Ca}^0$ | $= 100\ mV$ | $V_K^0$ | $= -70\ mV$ |
| $V_L$ | $= -50\ mV$ | | |

$K^+$ channels, and

$$I_{ion}(V, W) = g_{Ca}M_\infty(V)(V - V_{Ca}^0) + g_K W(V - V_K^0) + g_L(V - V_L^0), \tag{4.97}$$

$$M_\infty(V) = \frac{1}{2}\left(1 + \tanh\left(\frac{V - V_1}{V_2}\right)\right), \tag{4.98}$$

$$W_\infty(V) = \frac{1}{2}\left(1 + \tanh\left(\frac{V - V_3}{V_4}\right)\right), \tag{4.99}$$

$$\Lambda(V) = \cosh\left(\frac{V - V_3}{2V_4}\right). \tag{4.100}$$

Typical rate constants in these equations are shown in Table 4.7.

(a) Find a nondimensional representation of the Morris–Lecar equations in terms of the variables $v = \frac{V}{V_{Ca}^0}, t = \frac{g_K T}{2C_m}, w = W$.

(b) Sketch the phase portrait of the nondimensional Morris–Lecar equations. Show that there is a unique steady state at $v = -0.3173, w = 0.1076$ and determine its stability.

(c) Show that the Morris–Lecar equations can be reasonably well approximated by the cubic FitzHugh–Nagumo system

$$\frac{dv}{dt} = -k\left[(v - a)(v - b)(v - c) + \alpha(w - w_s)\right], \tag{4.101}$$

$$\frac{dw}{dt} = \phi(v - \beta w + \gamma), \tag{4.102}$$

where $a = -0.317, b = -0.18, c = 0.467, w_s = 0.1076, k = 10, \alpha = 0.2, \beta = 0.8333, \gamma = 0.47, \phi = 0.3$.

12. Does the Morris–Lecar model exhibit anode break excitation (see Exercise 7)? If not, why not?

13. The Pushchino model is a piecewise linear model of FitzHugh–Nagumo type proposed as a model for the ventricular action potential. The model has

$$f(v, w) = F(v) - w, \tag{4.103}$$

$$g(v, w) = \frac{1}{\tau(v)}(v - w), \tag{4.104}$$

where

$$F(v) = \begin{cases} -30v, & \text{for } v < v_1, \\ \gamma v - 0.12, & \text{for } v_1 < v < v_2, \\ -30(v-1), & \text{for } v > v_2, \end{cases} \qquad (4.105)$$

$$\tau(v) = \begin{cases} 2, & \text{for } v < v_1, \\ 16.6, & \text{for } v > v_1, \end{cases} \qquad (4.106)$$

with $v_1 = 0.12/(30 + \gamma)$ and $v_2 = 30.12/(30 + \gamma)$.
Simulate the action potential for this model. What is the effect on the action potential of changing $\tau(v)$?

14. Perhaps the most important example of a nonphysiological excitable system is the Belousov–Zhabotinsky reaction. This reaction denotes the oxidation of malonic acid by bromate in acidic solution in the presence of a transition metal ion catalyst. Kinetic equations describing this reaction are (Tyson and Fife, 1980)

$$\epsilon \frac{du}{dt} = -fv\frac{u-q}{u+q} + u - u^2, \qquad (4.107)$$

$$\frac{dv}{dt} = u - v, \qquad (4.108)$$

where $u$ denotes the concentration of bromous acid and $v$ denotes the concentration of the oxidized catalyst metal. Typical values for parameters are $\epsilon \approx 0.01, f = 1, q \approx 10^{-4}$. Describe the phase portrait for this system of equations.

15. It is not particularly difficult to build an electrical analogue of the FitzHugh–Nagumo equations with inexpensive and easily obtained electronic components. The parts list for one "cell" (shown in Fig. 4.27) includes two op-amps (operational amplifiers), two power supplies, a few resistors, and two capacitors, all readily available from any consumer electronics store (Keener, 1983).
The key component is an operational amplifier (Fig. 4.25). An op-amp is denoted in a circuit diagram by a triangle with two inputs on the left and a single output from the vertex on the right. Only three circuit connections are shown on a diagram, but two more are assumed, being necessary to connect with the power supply to operate the op-amp. Corresponding to the supply voltages $V_{s-}$ and $V_{s+}$, there are voltages $V_{r-}$ and $V_{r+}$, called the *rail voltages*, which determine the operational range for the output of an op-amp. The job of an op-amp is to compare the two input voltages $v_+$ and $v_-$, and if $v_+ > v_-$, to set (if possible) the output voltage $v_0$ to the high rail voltage $V_{r+}$, whereas if $v_+ < v_-$, then $v_0$ is set to $V_{r-}$. With reliable electronic components it is a good first approximation to assume that the input draws no current, while the output $v_0$ can supply whatever current is necessary to maintain the required voltage level.

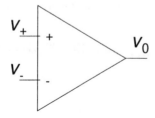

**Figure 4.25** Diagram for an operational amplifier (op-amp).

The response of an op-amp to changes in input is not instantaneous, but is described reasonably well by the differential equation

$$\epsilon_s \frac{dv_0}{dt} = g(v_+ - v_-) - v_0. \tag{4.109}$$

The function $g(v)$ is continuous, but quite close to the piecewise constant function

$$g(v) = V_{r+}H(v) + V_{r-}H(-v), \tag{4.110}$$

with $H(v)$ the Heaviside function. The number $\epsilon_s$ is small, and is the inverse of the *slew-rate*, which is typically on the order of $10^6$–$10^7$ V/sec. For all of the following circuit analysis, take $\epsilon_s \to 0$.

(a) Show that the simple circuit shown in Fig. 4.26 is a linear amplifier, with

$$v_0 = \frac{R_1 + R_2}{R_2} v_+, \tag{4.111}$$

provided that $v_0$ is within the range of the rail voltages.

(b) Show that if $R_1 = 0, R_2 = \infty$, then the device in Fig. 4.26 becomes a *voltage follower* with $v_0 = v_+$.

(c) Find the governing equations for the circuit in Fig. 4.27, assuming that the rail voltages for op-amp 2 are well within the range of the rail voltages for op-amp 1.
Show that

$$C_1 \frac{dv}{dt} + i_2 \left( 1 - \frac{R_4}{R_5} \right) + \frac{F(v)}{R_3} + \frac{v - v_g}{R_5} = 0, \tag{4.112}$$

$$C_2 R_4 R_5 \frac{di_2}{dt} + R_4 i_2 = v - v_g, \tag{4.113}$$

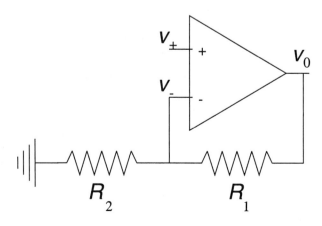

**Figure 4.26** Linear amplifier using an op-amp.

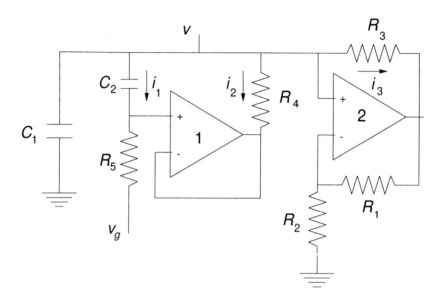

**Figure 4.27**   FitzHugh–Nagumo circuit using op-amps.

where $F(v)$ is the piecewise linear function

$$F(v) = \begin{cases} v - V_{r_+}, & \text{for } v > \alpha V_{r_+}, \\ -\dfrac{R_1}{R_2}v, & \text{for } \alpha V_{r_-} \leq v \leq \alpha V_{r_+}, \\ v - V_{r_-}, & \text{for } v < \alpha V_{r_-}, \end{cases} \tag{4.114}$$

and $\alpha = \frac{R_2}{R_1 + R_2}$.

(d)  Sketch the phase portrait for these circuit equations. Show that this is a piecewise linear FitzHugh–Nagumo system.

(e)  Use the singular perturbation approximation (4.55) to estimate the period of oscillation for the piecewise linear analog FitzHugh–Nagumo circuit in Fig. 4.27.

16.  Simulate the Noble equations with different values of $g_{an} = 0.0, 0.075, 0.18, 0.4$ mS/cm². Explain the results in qualitative terms.

---

**Table 4.8**   Parts list for the FitzHugh–Nagumo analog circuit.

| 2 LM 741 op-amps (National Semiconductor) | |
|---|---|
| $R_1 = R_2 = 100\text{k}\Omega$ | $R_3 = 2.4\Omega$ |
| $R_4 = 1\text{k}\Omega$ | $R_5 = 10\text{k}\Omega$ |
| $C_1 = 0.01\mu\text{F}$ | $C_2 = 0.5\mu\text{F}$ |
| Power supplies: | |
| $\pm15$V for op-amp #1 | $\pm12$V for op-amp #2 |

17. Simulate the MNT equations, and explain why the currents $I_{si}, I_{K_2}, I_{x_1}$, and $I_{x_2}$ are called the slow inward, pacemaker, and plateau currents, respectively.

18. Simulate the YNI model for the SA nodal action potential. Find parameter values for the FitzHugh–Nagumo cubic model that duplicate this behavior as best possible.

19. (a) Simulate the Beeler–Reuter equations and plot each of the currents and the calcium concentration. What terms are mostly responsible for the prolongation of the action potential?

    (b) Do the Beeler–Reuter equations exhibit anode break excitation?

# CHAPTER 5

# Calcium Dynamics

Calcium is critically important for a vast array of cellular functions, as can be seen by a quick look through any physiology book. For example, in this book we discuss the role that $Ca^{2+}$ plays in muscle mechanics, cardiac electrophysiology, bursting oscillations and secretion, hair cells, and adaptation in photoreceptors, among other things. Clearly, the mechanisms by which a cell controls its $Ca^{2+}$ concentration are of central interest in cell physiology.

There are a number of $Ca^{2+}$ control mechanisms operating on different levels, all designed to ensure that $Ca^{2+}$ is present in sufficient quantity to perform its necessary functions, but not in too great a quantity in the wrong places. Prolonged high concentrations of $Ca^{2+}$ are toxic. For example, since calcium causes contraction of muscle cells, failure to remove calcium can keep a muscle cell in a state of constant tension (as in rigor mortis).

In vertebrates, the majority of body $Ca^{2+}$ is stored in the bones, from where it can be released by hormonal stimulation to maintain an extracellular $Ca^{2+}$ concentration of around 1 mM, while intracellular $[Ca^{2+}]$ is kept at around 0.1 $\mu$M. Since the internal concentration is low, there is a steep concentration gradient from the outside of a cell to the inside. This disparity has the advantage that cells are able to raise their $[Ca^{2+}]$ quickly, by opening $Ca^{2+}$ channels and relying on passive flow down a steep concentration gradient, but it has the disadvantage that energy must be expended to keep the cytosolic $Ca^{2+}$ concentration low. Thus, cells have finely tuned mechanisms to control the influx and removal of cytosolic $Ca^{2+}$.

Calcium is removed from the cytoplasm in two principal ways: it is pumped out of a cell, and it is sequestered into internal membrane-bound compartments such as the mitochondria, the endoplasmic reticulum (ER) or sarcoplasmic reticulum (SR), and secretory granules. Since the $Ca^{2+}$ concentration in the cytoplasm is much lower than either the extracellular concentration or the concentration inside the internal compartments, both methods of $Ca^{2+}$ removal require expenditure of energy. Some of this is by a $Ca^{2+}$ ATPase, similar to the $Na^+$–$K^+$ ATPase discussed in Chapter 2, that uses energy stored in ATP to pump $Ca^{2+}$ out of the cell or into an internal compartment. There is also a $Na^+$–$Ca^{2+}$ exchanger in the cell membrane that uses the energy of the $Na^+$ electrochemical gradient to remove $Ca^{2+}$ from the cell at the expense of $Na^+$ entry (also discussed in Chapters 2 and 3).

Calcium influx also occurs via two principal pathways: inflow from the extracellular medium through $Ca^{2+}$ channels in the surface membrane and release from internal stores. The surface membrane $Ca^{2+}$ channels are of several different types: voltage-controlled channels that open in response to depolarization of the cell membrane, receptor-operated channels that open in response to the binding of an external ligand, second-messenger-operated channels that open in response to the binding of a cellular second messenger, and mechanically operated channels that open in response to mechanical stimulation. Voltage-controlled $Ca^{2+}$ channels are of great importance in other chapters of this book (in particular, when we consider models of bursting oscillations or cardiac cells), and we consider them in detail there. We also omit the consideration of the other surface membrane channels to concentrate on the properties of $Ca^{2+}$ release from internal stores.

Calcium release from internal stores such as the ER is the second major $Ca^{2+}$ influx pathway, and this is mediated principally by two types of $Ca^{2+}$ channels that are also receptors: the ryanodine receptor and the inositol (1,4,5)-trisphosphate ($IP_3$) receptor. The ryanodine receptor, so-called because of its sensitivity to the plant alkaloid ryanodine, plays an integral role in excitation–contraction coupling in skeletal and cardiac muscle cells, and is believed to underlie $Ca^{2+}$-induced $Ca^{2+}$ release, whereby a small amount of $Ca^{2+}$ entering the cardiac cell through voltage-gated $Ca^{2+}$ channels initiates an explosive release of $Ca^{2+}$ from the sarcoplasmic reticulum (Fig. 5.1, lower panel). Ryanodine receptors are also found in a variety of nonmuscle cells such as neurons, pituitary cells, and sea urchin eggs. The $IP_3$ receptor, although similar in structure to the ryanodine receptor, is found predominantly in nonmuscle cells, and is sensitive to the second messenger $IP_3$. The binding of an extracellular agonist such as a hormone or a neurotransmitter to a receptor in the surface membrane can cause, via a G-protein link to phospholipase C (PLC), the cleavage of phosphotidylinositol (4,5)-bisphosphate ($PIP_2$) into diacylglycerol (DAG) and $IP_3$ (Fig. 5.1, upper panel). The water-soluble $IP_3$ is free to diffuse through the cell cytoplasm and bind to $IP_3$ receptors situated on the ER membrane, leading to the opening of these receptors and subsequent release of $Ca^{2+}$ from the ER. Similarly to ryanodine receptors, $IP_3$ receptors are modulated by the cytosolic $Ca^{2+}$ concentration, with $Ca^{2+}$ both activating and inactivating $Ca^{2+}$ release, but at different rates.

As an additional control for the cytosolic $Ca^{2+}$ concentration, $Ca^{2+}$ is heavily buffered (i.e., bound) by large proteins, with estimates that approximately 99% of the total cytoplasmic $Ca^{2+}$ is bound to buffers. The $Ca^{2+}$ in the internal stores is also heavily buffered.

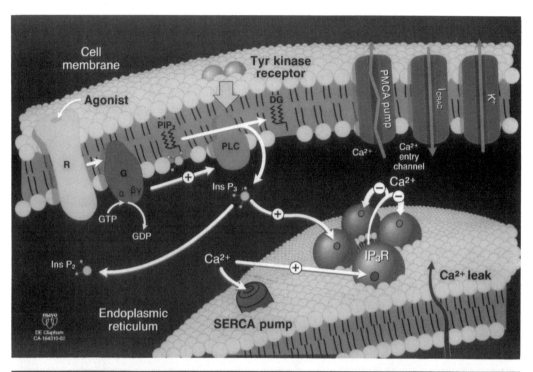

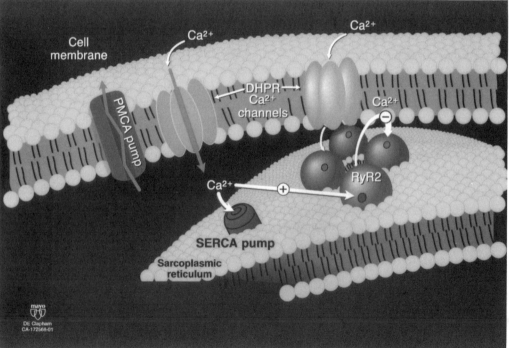

**Figure 5.1**  Diagram of the pathways involved in the control of cytoplasmic Ca²⁺ concentration. A: Via the production of IP₃. B: Via the ryanodine receptor. (Clapham, 1995, Fig. 1. We thank David Clapham for providing the original of this figure.)

# 5.1  Calcium Oscillations

In response to agonists such as hormones or neurotransmitters, many cell types exhibit oscillations in intracellular [$Ca^{2+}$]. These oscillations can be grouped into two major types: those that are dependent on periodic fluctuations of the cell membrane potential and the associated periodic entry of $Ca^{2+}$ through voltage-gated $Ca^{2+}$ channels, and those that occur in the presence of a voltage clamp. Our focus here is on the latter type, within which group further distinctions can be made by whether the oscillatory calcium flux is through ryanodine or $IP_3$ receptors. We consider models of both types here.

The period of $IP_3$-dependent oscillations ranges from a few seconds to a few minutes (Fig. 5.2). There is a great deal of evidence that in many cell types, these oscillations occur at constant [$IP_3$] and are therefore not driven by oscillations in [$IP_3$]. Although it is risky to generalize, some overall trends can be seen; as [$IP_3$] increases, the steady state [$Ca^{2+}$] also increases, the oscillation frequency increases, and the amplitude of the oscillations remains approximately constant. Calcium oscillations usually occur only when [$IP_3$] is greater than some critical value and disappear again when [$IP_3$] gets too large. Thus, there is an intermediate range of $IP_3$ concentrations that generate $Ca^{2+}$ oscillations.

Although it is known that $Ca^{2+}$ controls many cellular processes, the exact significance of $Ca^{2+}$ oscillations is not completely understood in most cell types. It is widely believed that the oscillations are a frequency-encoded signal that allows a cell to use $Ca^{2+}$ as a second messenger while avoiding the toxic effects of prolonged high [$Ca^{2+}$]. However, there are still relatively few examples where the signal carried by a $Ca^{2+}$ oscillation has been unambiguously decoded.

# 5.2  The Two-Pool Model

One of the earliest models for $IP_3$-dependent $Ca^{2+}$ release assumes the existence of two distinct internal $Ca^{2+}$ stores, one of which is sensitive to $IP_3$, the other of which is sensitive to $Ca^{2+}$ (Kuba and Takeshita, 1981; Goldbeter et al., 1990; Goldbeter, 1996). Agonist stimulation leads to the production of $IP_3$, which releases $Ca^{2+}$ from the $IP_3$-sensitive store through $IP_3$ receptors. The $Ca^{2+}$ that is thereby released stimulates the release of further $Ca^{2+}$ from the $Ca^{2+}$-sensitive store, possibly via ryanodine receptors. A crucial assumption of the model is that the concentration of $Ca^{2+}$ in the $IP_3$-sensitive store remains constant, as the store is quickly refilled from the extracellular medium. A schematic diagram of the model is given in Fig. 5.3.

Recent work by Dupont and Goldbeter (1993, 1994) has shown that the model does not depend on the existence of two separate pools of $Ca^{2+}$; the model equations can equally well be used to describe the release of $Ca^{2+}$ from a single pool, with the release modulated by both $IP_3$ and $Ca^{2+}$. Nevertheless, for convenience, we persist in calling this the two-pool model.

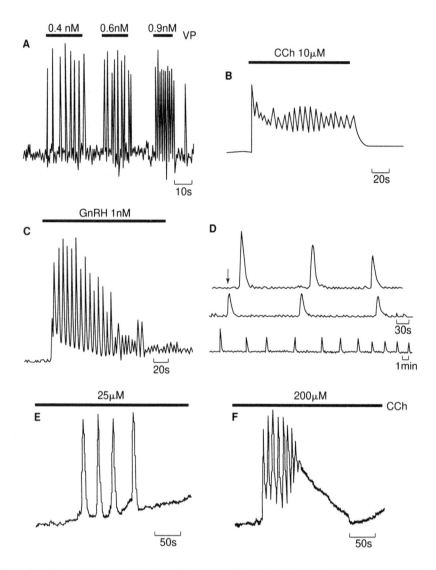

**Figure 5.2** Typical calcium oscillations from a variety of cell types. A: Hepatoctyes stimulated with vasopressin (VP). B: Rat parotid gland stimulated with carbachol (CCh). C: Gonadotropes stimulated with gonadotropin-releasing hormone (GnRH). D: Hamster eggs after fertilization. The time of fertilization is denoted by the arrow. E and F: Insulinoma cells stimulated with two different concentrations of carbachol. (Berridge and Galione, 1988, Fig. 2.)

Let $c$ denote the concentration of $Ca^{2+}$ in the cytoplasm, and $c_s$ the concentration of $Ca^{2+}$ in the $Ca^{2+}$-sensitive pool. We assume that $IP_3$ causes a steady flux $r$ of $Ca^{2+}$ into the cytosol, and that $Ca^{2+}$ is pumped out of the cytoplasm at the rate $-kc$. Then

$$\frac{dc}{d\tau} = r - kc - \tilde{f}(c, c_s), \tag{5.1}$$

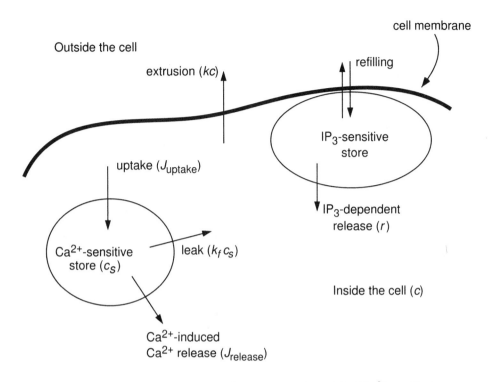

**Figure 5.3** Schematic diagram of the two-pool model of $Ca^{2+}$ oscillations.

$$\frac{dc_s}{d\tau} = \tilde{f}(c, c_s), \tag{5.2}$$

$$\tilde{f}(c, c_s) = J_{uptake} - J_{release} - k_f c_s, \tag{5.3}$$

where

$$J_{uptake} = \frac{V_1 c^n}{K_1^n + c^n}, \tag{5.4}$$

$$J_{release} = \left(\frac{V_2 c_s^m}{K_2^m + c_s^m}\right)\left(\frac{c^p}{K_3^p + c^p}\right), \tag{5.5}$$

and $\tau$ denotes time. The flux of $Ca^{2+}$ from the cytoplasm into the $Ca^{2+}$-sensitive pool is given by $\tilde{f}$; $J_{uptake}$ is the rate at which $Ca^{2+}$ is pumped from the cytosol into the $Ca^{2+}$-sensitive pool by an active process, and $J_{release}$ is the rate at which $Ca^{2+}$ is released from the $Ca^{2+}$-sensitive pool. Note that as $c$ increases, so does $J_{release}$. Thus, $Ca^{2+}$ stimulates its own release through positive feedback, usually called $Ca^{2+}$-induced $Ca^{2+}$ release, or CICR (Endo et al., 1970; Fabiato, 1983). It is this positive feedback that is central to the model's behavior. Finally, the rate at which $Ca^{2+}$ leaks from the $Ca^{2+}$-sensitive pool into the cytosol is $k_f c_s$. In the model, $r$ is constant for constant [IP$_3$] and is treated as a control parameter. Thus, the behavior of the model at different constant IP$_3$ concentrations can be studied by varying $r$.

**Table 5.1**  Typical parameter values for the two-pool model of $Ca^{2+}$ oscillations. (Goldbeter et al., 1990.)

| | | | |
|---|---|---|---|
| $k$ | $= 10\ s^{-1}$ | $K_1$ | $= 1\ \mu M$ |
| $K_2$ | $= 2\ \mu M$ | $K_3$ | $= 0.9\ \mu M$ |
| $V_1$ | $= 65\ \mu Ms^{-1}$ | $V_2$ | $= 500\ \mu Ms^{-1}$ |
| $k_f$ | $= 1\ s^{-1}$ | $m$ | $= 2$ |
| $n$ | $= 2$ | $p$ | $= 4$ |

For convenience we nondimensionalize the model equations. Let $u = c/K_1$, $t = \tau k$, $v = c_s/K_2$, $\alpha = K_3/K_1$, $\beta = V_1/V_2$, $\gamma = K_2/K_1$, $\delta = k_f K_2/V_2$, $\mu = r/(kK_1)$, and $\epsilon = kK_2/V_2$, to get

$$\frac{du}{dt} = \mu - u - \frac{\gamma}{\epsilon} f(u,v), \tag{5.6}$$

$$\frac{dv}{dt} = \frac{1}{\epsilon} f(u,v), \tag{5.7}$$

$$f(u,v) = \beta \left( \frac{u^n}{u^n + 1} \right) - \left( \frac{v^m}{v^m + 1} \right) \left( \frac{u^p}{\alpha^p + u^p} \right) - \delta v. \tag{5.8}$$

If the exchange of $Ca^{2+}$ between the cytosol and the $Ca^{2+}$-sensitive pool is fast (i.e., $V_1$ and $V_2$ are large), then $\epsilon$ is a small parameter. A table of typical parameter values in the model is given in Table 5.1. For these values, $\epsilon \approx 0.04$.

## 5.2.1  Excitability and Oscillations

The two-pool model can be put into the form of a generalized FitzHugh–Nagumo model (Chapter 4) by a simple change of variables. If we let $w = u + \gamma v$, then the two-pool model becomes

$$\frac{dw}{dt} = \mu - (w - \gamma v), \tag{5.9}$$

$$\frac{dv}{dt} = \frac{1}{\epsilon} f(w - \gamma v, v) = \frac{1}{\epsilon} F(w,v). \tag{5.10}$$

The nullclines of the transformed equations are shown in Fig. 5.4, where it can be seen that one nullcline is N-shaped, while the other is a straight line, as in the FitzHugh–Nagumo model. Thus, the analysis of the FitzHugh–Nagumo model can be applied, essentially without change, to the temporal behavior of the two-pool model. It is therefore not surprising that the two-pool model is excitable and exhibits oscillatory behavior. When $\mu$ is slightly below the lower Hopf bifurcation point (calculated below), a subthreshold addition of $Ca^{2+}$ gives a small response, while a superthreshold addition causes a large transient before the return to steady state.

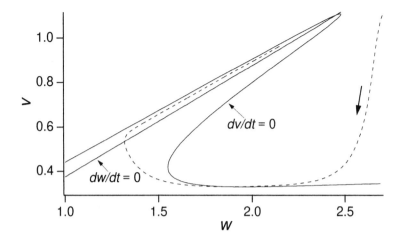

**Figure 5.4** Nullclines and sample trajectory of the two-pool model transformed to $w, v$ coordinates (Sneyd et al., 1993; Fig. 4.)

The steady state of the two-pool model, $(u_0, v_0)$, is given by

$$u_0 = \mu, \tag{5.11}$$

$$f(\mu, v_0) = 0, \tag{5.12}$$

and the stability of the steady state is determined by the roots of the characteristic equation

$$\lambda^2 + H\lambda - \frac{f_v}{\epsilon} = 0, \tag{5.13}$$

where

$$H = \frac{\gamma f_u(u_0, v_0)}{\epsilon} - \frac{f_v(u_0, v_0)}{\epsilon} + 1. \tag{5.14}$$

Since $f_v < 0$, the roots of (5.13) have negative real part (and thus the steady state is stable) if $H > 0$, and they have positive real part if $H < 0$. At $H = 0$ the steady state changes stability through a Hopf bifurcation, and at these points a branch of periodic orbits appears. The amplitude and period of these periodic orbits as a function of the bifurcation parameter $\mu$ can be tracked with the use of the software package AUTO (Doedel, 1986), and the results are shown in Fig. 5.5.

As $\mu$ is increased, oscillations appear at a Hopf bifurcation and disappear in the same manner. Both Hopf bifurcations are supercritical, and the two bifurcation points are connected by a branch of stable periodic orbits. Oscillations occur for a constant value of $\mu$, i.e., constant [IP$_3$]. Thus, the two-pool model shows that Ca$^{2+}$-induced Ca$^{2+}$ release is sufficient to produce oscillations in the absence of IP$_3$ oscillations. The function of IP$_3$ is to produce a steady influx of Ca$^{2+}$ into the cytosol from the IP$_3$-sensitive pool, and this steady influx drives the Ca$^{2+}$ oscillations. Many features of Ca$^{2+}$

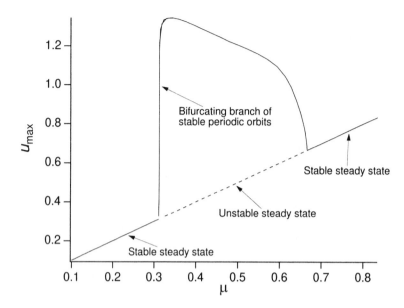

**Figure 5.5** Bifurcation diagram of the two-pool model, using the parameter values given in Table 5.1. Only the maximum of $u$ over the periodic orbit is shown here. (Sneyd et al., 1993, Fig. 2b.)

oscillations are reproduced well by the model. For example, the amplitude of the oscillations stays approximately constant (or increases slightly), and the period increases, as [IP$_3$] is decreased, as observed in many cell types. Furthermore, the oscillations show pronounced spike-like behavior, again in good agreement with many experiments. A detailed discussion of this model, and of many experimental results, can be found in Goldbeter (1995).

## 5.3 The Mechanisms of Calcium Release

### 5.3.1 IP$_3$ Receptors

Although the two-pool model reproduces experimental data extremely well, both qualitatively and quantitatively, recent experimental evidence indicates that the role of Ca$^{2+}$ is more complicated than was assumed in this model. In the two-pool model Ca$^{2+}$ stimulates its own release (thus the term $c^p/(K_3^p + c^p)$ in (5.5) ), while the flow of Ca$^{2+}$ from the internal store is terminated when the concentration of Ca$^{2+}$ in the internal store becomes too low (thus the term $c_s^m/(K_2^m + c_s^m)$). However, it now appears that not only does Ca$^{2+}$ stimulate its own release, it also inhibits it, but on a slower time scale (Parker and Ivorra, 1990; Finch et al., 1991; Bezprozvanny et al., 1991; Parys et al., 1992). It is hypothesized that this sequential activation and inactivation of the IP$_3$ receptor by Ca$^{2+}$ is the fundamental mechanism underlying IP$_3$-dependent Ca$^{2+}$ oscillations and waves, and a number of models incorporating this hypothesis have appeared (reviewed by Sneyd et al., 1995b and Tang et al., 1996).

### A detailed IP$_3$ receptor model

One approach to determining whether sequential activation and inactivation of the IP$_3$ receptor by Ca$^{2+}$ can produce Ca$^{2+}$ oscillations is to construct a detailed model of the IP$_3$ receptor, including all the possible receptor states and transitions between them (De Young and Keizer, 1992). To do so, we assume that the IP$_3$ receptor consists of three equivalent and independent subunits, all of which must be in a conducting state before the receptor allows Ca$^{2+}$ flux. Each subunit has an IP$_3$ binding site, an activating Ca$^{2+}$ binding site, and an inactivating Ca$^{2+}$ binding site, each of which can be either occupied or unoccupied, and thus each subunit can be in one of eight states. Each state of the subunit is labeled S$_{ijk}$, where $i, j$, and $k$ are equal to 0 or 1, with a 0 indicating that the binding site is unoccupied and a 1 indicating that it is occupied. The first index refers to the IP$_3$ binding site, the second to the Ca$^{2+}$ activation site, and the third to the Ca$^{2+}$ inactivation site. This is illustrated in Fig. 5.6. Although a fully general model would include 24 rate constants, we make two simplifying assumptions. First, the rate constants are assumed to be independent of whether activating Ca$^{2+}$ is bound or not. Second, the kinetics of Ca$^{2+}$ activation are assumed to be independent of IP$_3$ binding and Ca$^{2+}$ inactivation. This leaves only 10 rate constants, $k_1, \ldots, k_5$ and $k_{-1}, \ldots, k_{-5}$.

The fraction of subunits in the state S$_{ijk}$ is denoted by $x_{ijk}$. The differential equations for these are based on mass-action kinetics, and thus, for example,

$$\frac{dx_{000}}{dt} = -(V_1 + V_2 + V_3), \tag{5.15}$$

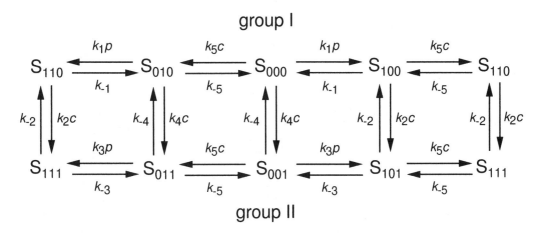

**Figure 5.6** The binding diagram for the IP$_3$ receptor model. Here, $c$ denotes [Ca$^{2+}$], and $p$ denotes [IP$_3$].

where

$$V_1 = k_1 p x_{000} - k_{-1} x_{100}, \tag{5.16}$$

$$V_2 = k_4 c x_{000} - k_{-4} x_{001}, \tag{5.17}$$

$$V_3 = k_5 c x_{000} - k_{-5} x_{010}, \tag{5.18}$$

where $p$ denotes [IP$_3$] and $c$ denotes [Ca$^{2+}$]. $V_1$ describes the rate at which IP$_3$ binds to and leaves the IP$_3$ binding site, $V_2$ describes the rate at which Ca$^{2+}$ binds to and leaves the inactivating site, and similarly for $V_3$. Since experimental data indicate that the receptor subunits act in a cooperative fashion, the model assumes that the IP$_3$ receptor passes Ca$^{2+}$ current only when three subunits are in the state $S_{110}$ (i.e., with one IP$_3$ and one activating Ca$^{2+}$ bound), and thus the open probability of the receptor is $x_{110}^3$.

The set of seven differential equations for the receptor states (there are eight receptor states, but only seven are independent, as the $x_{ijk}$s must sum to one) are combined with a differential equation for Ca$^{2+}$ transport to obtain the full model

$$\frac{dc}{dt} = \overbrace{(r_1 x_{110}^3 + r_2)(c_s - c)}^{\text{receptor flux}} - \overbrace{\frac{r_3 c^2}{c^2 + k_p^2}}^{\text{pumping}}, \tag{5.19}$$

where $c_s$ denotes the concentration in the ER, and $r_1$ and $r_2$ are constants. The first term is the Ca$^{2+}$ flux through the IP$_3$ receptor, and it is proportional to the concentration difference between the ER and the cytoplasm. It includes an IP$_3$-independent leak ($r_2$) from the ER into the cytoplasm. The second term, similar to the pump term in the two-pool model, describes the action of the Ca$^{2+}$ ATPases that pump Ca$^{2+}$ from the cytoplasm into the ER, and is based on experimental data that show that the Ca$^{2+}$ ATPase is cooperative, with a Hill coefficient of 2. For simplicity, we assume that the cell is closed, i.e., that there is no Ca$^{2+}$ exchange between the inside and outside of the cell. In this case, $c_s$ is determined by constraining the total amount of intracellular Ca$^{2+}$ to be constant, and thus

$$c_{\text{avg}} = v_c c_s + c, \tag{5.20}$$

where $v_c$ is the ratio of the ER volume to the cytoplasmic volume. However, this artificial constraint can be removed without affecting model behavior greatly. If Ca$^{2+}$ exchange between the inside and the outside of the cell is much slower than exchange between the cytoplasm and the ER, we can apply a quasi-steady-state hypothesis and assume that [Ca$^{2+}$] is constant on a fast time scale.

In Fig. 5.7 we show the open probability of the IP$_3$ receptor as a function of [Ca$^{2+}$], which is some of the experimental data upon which the model is based. Bezprozvanny et al. (1991) showed that this open probability is a bell-shaped function of [Ca$^{2+}$]. Thus, at low [Ca$^{2+}$], an increase in [Ca$^{2+}$] increases the open probability of the receptor, while at high [Ca$^{2+}$] an increase in [Ca$^{2+}$] decreases the open probability. Parameters in the model were chosen to obtain agreement with this steady-state data. The kinetic properties of the IP$_3$ receptor are equally important: the receptor is activated quickly by

**Table 5.2**   Parameters of the receptor model (De Young and Keizer, 1992) for $Ca^{2+}$ oscillations.

| | | | | | |
|---|---|---|---|---|---|
| $k_1$ | = | $400\ \mu M^{-1}s^{-1}$ | $k_{-1}$ | = | $52\ s^{-1}$ |
| $k_2$ | = | $0.2\ \mu M^{-1}s^{-1}$ | $k_{-2}$ | = | $0.21\ s^{-1}$ |
| $k_3$ | = | $400\ \mu M^{-1}s^{-1}$ | $k_{-3}$ | = | $377.2\ s^{-1}$ |
| $k_4$ | = | $0.2\ \mu M^{-1}s^{-1}$ | $k_{-4}$ | = | $0.029\ s^{-1}$ |
| $k_5$ | = | $20\ \mu M^{-1}s^{-1}$ | $k_{-5}$ | = | $1.64\ s^{-1}$ |
| $c_{avg}$ | = | $2\ \mu M$ | $r_1$ | = | $1.11\ s^{-1}$ |
| $v_c$ | = | $0.185$ | $r_2$ | = | $0.02\ s^{-1}$ |
| $k_p$ | = | $0.1\ \mu M$ | $r_3$ | = | $0.9\ \mu M^{-1}s^{-1}$ |

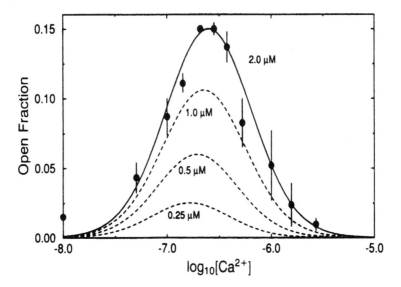

**Figure 5.7**   The steady-state open probability of the $IP_3$ receptor, as a function of $[Ca^{2+}]$. The symbols are the experimental data of Bezprozvanny et al. (1991), and the smooth curves are from the receptor model (calculated at four different $IP_3$ concentrations). (De Young and Keizer, 1992, Fig. 2A.)

$Ca^{2+}$, but inactivated by $Ca^{2+}$ on a slower time scale. In the model, this is incorporated in the magnitude of the rate constants and is the basis of a simplification of the model that we discuss below.

As $[IP_3]$ is increased, periodic orbits appear via a supercritical Hopf bifurcation and disappear in the same manner. For many parameters in the physiological range, the two Hopf bifurcations are connected by a branch of periodic orbits (Fig. 5.8), and

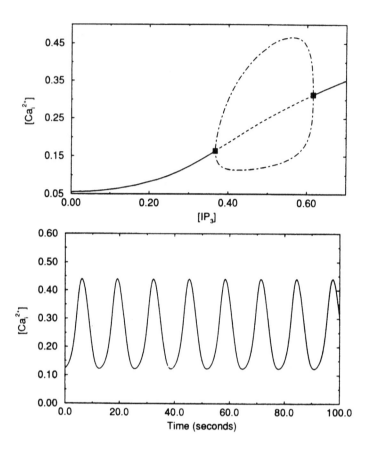

**Figure 5.8** Oscillations in the receptor model as a function of the $IP_3$ concentration. (De Young and Keizer, 1992, Figs. 3A and 4A.) A: Bifurcation diagram, showing the stable steady state (solid curve), the unstable steady state (dashed curve), and the maximum and minimum of the oscillations (dot-dashed curve). B: A typical periodic orbit in the receptor model, calculated for $[IP_3] = 0.5 \ \mu M$.

the period of the orbits is a decreasing function of $[IP_3]$, as observed experimentally. This behavior, similar to that seen in the two-pool model, seems to be typical of many models of $Ca^{2+}$ oscillations. However, in many respects the behavior of the receptor model does not agree as well with experimental data as does that of the two-pool model, even though it is based on more realistic assumptions. For example, the amplitude of the oscillations is not constant as $[IP_3]$ varies, and neither are the oscillations very spike-like. The receptor model is sufficient to explain some data, but certainly not all.

### Reduction of the detailed $IP_3$ receptor model

The complexity of the receptor model (eight differential equations and numerous parameters) provides ample motivation to seek a simpler model that retains its essential properties. Since $IP_3$ binds quickly to its binding site and $Ca^{2+}$ binds quickly to the activating site, we can dispense with the transient details of these binding processes and assume instead that the receptor is in quasi-steady state with respect to IP$_3$ binding and Ca$^{2+}$ activation (De Young and Keizer, 1992; Keizer and De Young, 1994; Li and Rinzel, 1994; Tang et al., 1996). Notice that this is implied by the parameter values for the detailed receptor model shown in Table 5.2, where $k_1, k_3$, and $k_5$ are substantially larger than $k_2$ and $k_4$, and $k_{-1}, k_{-3}$, and $k_{-5}$ are also larger than $k_{-2}$ and $k_{-4}$. The process by which fast binding can be used to simplify a complicated model is identical in spirit to the reductions for enzyme kinetics used in Chapter 1, and so most of the details are left as an exercise (Exercise 5).

As shown in Fig. 5.6, the receptor states are arranged into two groups: those without Ca$^{2+}$ bound to the inactivating site ($S_{000}, S_{010}, S_{100}$, and $S_{110}$, shown in the upper line of Fig. 5.6; called group I states), and those with Ca$^{2+}$ bound to the inactivating site ($S_{001}, S_{011}, S_{101}$, and $S_{111}$, shown in the lower line of Fig. 5.6; called group II states). Because the binding of IP$_3$ and the binding of Ca$^{2+}$ to the activating site are assumed to be fast processes, it follows that within each group the binding states are at quasi-steady state with respect to transitions within the group. However, the transitions between group I and group II (between top and bottom in Fig. 5.6), due to binding or unbinding of the inactivating site, are slow, and so the group I states are not in equilibrium with the group II states.

To carry out this calculation, we write the differential equations governing the states in group I, say, to get

$$\frac{dx_{000}}{dt} = -x_{000}(k_5 c + k_1 p + k_4 c) + k_{-5} x_{010} + k_{-1} x_{100} + k_{-4} x_{001}, \tag{5.21}$$

$$\frac{dx_{100}}{dt} = -x_{100}(k_5 c + k_{-1} + k_2 c) + k_{-5} x_{110} + k_1 p x_{000} + k_{-2} x_{101}, \tag{5.22}$$

$$\frac{dx_{010}}{dt} = -x_{010}(k_{-5} + k_1 p + k_4 c) + k_5 c x_{000} + k_{-1} x_{110} + k_{-4} x_{011}. \tag{5.23}$$

The differential equation for the fourth receptor state, $x_{110}$, is superfluous, as we have the constraint

$$x_{000} + x_{010} + x_{100} + x_{110} = 1 - y, \tag{5.24}$$

where

$$y = x_{001} + x_{011} + x_{101} + x_{111}. \tag{5.25}$$

The mathematically "proper" way to reduce these equations is to introduce appropriate nondimensional variables, determine which parameters are small and then find approximate equations by setting the small parameters to zero. However, it is not necessary to go through this formal procedure. Since we are assuming that the group I binding sites are all in quasi-steady state, the quasi-steady-state equations may be obtained by setting the fast terms on the right-hand sides of (5.21)–(5.23) to zero. Thus,

$$x_{000}(k_5 c + k_1 p) = k_{-5} x_{010} + k_{-1} x_{100}, \tag{5.26}$$

$$x_{100}(k_5 c + k_{-1}) = k_{-5} x_{110} + k_1 p x_{000}, \tag{5.27}$$

$$x_{010}(k_{-5} + k_1 p) = k_5 c x_{000} + k_{-1} x_{110}. \tag{5.28}$$

We solve these together with the constraint (5.24) to find the group I state probabilities. This gives, for example,

$$x_{000} = \frac{K_1 K_5 (1 - y)}{(p + K_1)(c + K_5)}, \tag{5.29}$$

where $K_i = k_{-i}/k_i$. An identical procedure applied to the group II receptor states gives the quasi-steady-state equations for the group II states.

It now remains to derive a differential equation for $y$. Notice that $y$ changes only on a slow time scale, since any changes in $y$ involve $Ca^{2+}$ leaving or binding to the inactivating site, a process that is assumed to be slow. Thus we write the differential equations for the group II sites, taking care to include the transitions between the group I and group II sites, add the four equations, and substitute all the quasi-steady-state expressions to get, finally,

$$\frac{dy}{dt} = \left[ \frac{(k_{-4} K_2 K_1 + k_{-2} p K_4) c}{K_4 K_2 (p + K_1)} \right] (1 - y) - \left( \frac{k_{-4} p + k_{-2} K_3}{p + K_3} \right) y. \tag{5.30}$$

This can be written in the form

$$\tau_y(c, p) \frac{dy}{dt} = y_\infty(c, p) - y, \tag{5.31}$$

which is useful for comparison with other models.

It is now a relatively simple matter to show how the reduced receptor model can be used to construct a simpler model of $Ca^{2+}$ oscillations. First, recall that the equation governing the $Ca^{2+}$ dynamics is (5.19)

$$\frac{dc}{dt} = (r_1 x_{110}^3 + r_2)(c_s - c) - \frac{r_3 c^2}{c^2 + k_p^2}. \tag{5.32}$$

Into this equation we substitute the expression for $x_{110}$,

$$x_{110} = \frac{pc(1 - y)}{(p + K_1)(c + K_5)}, \tag{5.33}$$

and the differential equation for $y$, to get a model of $Ca^{2+}$ oscillations consisting of two differential equations rather than the original eight.

Note that $1 - y$, which is the proportion of receptors that are not inactivated by $Ca^{2+}$, plays the role of an inactivation variable, similar in spirit to the variable $h$ in the Hodgkin–Huxley equations (Chapter 4). To emphasize this similarity, the reduced model can be written in the form

$$x_{110} = \frac{pc}{(p + K_1)(c + K_5)} h, \tag{5.34}$$

$$\tau_h(c, p) \frac{dh}{dt} = h_\infty(c, p) - h, \tag{5.35}$$

where $h = 1 - y$, and $\tau_h$ and $h_\infty$ are readily calculated from the corresponding differential equation for $y$.

### A heuristic model of the $IP_3$ receptor

An alternative approach to modeling $Ca^{2+}$ release assumes that $Ca^{2+}$ inactivates the $IP_3$ receptor in a cooperative manner (Atri et al., 1993). We assume first that the $IP_3$ receptor consists of three binding domains, the first of which binds $IP_3$, the other two binding $Ca^{2+}$, and second, that the receptor passes $Ca^{2+}$ current only when $IP_3$ is bound to domain 1, $Ca^{2+}$ is bound to domain 2, but $Ca^{2+}$ is *not* bound to domain 3. Thus, $Ca^{2+}$ activates the receptor by binding to domain 2 and inactivates the receptor by binding to domain 3. Each binding domain consists of a number of binding sites, grouped on the basis of functionality. If $p_1$ is the probability that $IP_3$ is bound to domain 1, $p_2$ is the probability that $Ca^{2+}$ is bound to domain 2, and $1 - p_3$ is the probability that $Ca^{2+}$ is bound to domain 3, then, assuming independence of the domains, it follows that the steady-state $Ca^{2+}$ flux through the $IP_3$ receptor, $J_{channel}$, is given by

$$J_{channel} = k_f p_1 p_2 p_3, \tag{5.36}$$

for some constant $k_f$. The probabilities $p_i, i = 1\text{–}3$ are chosen such that $J_{channel}$ agrees with the steady-state experimental data of Parys et al. (1992) from *Xenopus* oocytes. Good agreement with data is obtained by choosing

$$p_1 = \mu_0 + \frac{\mu_1 p}{p + k_\mu}, \tag{5.37}$$

$$p_2 = b + \frac{(1 - b)c}{k_1 + c}, \tag{5.38}$$

$$p_3 = \frac{k_2^2}{k_2^2 + c^2}, \tag{5.39}$$

where $p$ denotes $[IP_3]$, $c$ denotes $[Ca^{2+}]$, and where $\mu_0, \mu_1, b, k_1$, and $k_2$ are constants. The steady-state open probabilities of the $IP_3$ receptor in the model and experiment are shown in Fig. 5.9. Note that the expression for $p_3$ assumes that $Ca^{2+}$ binds to the inactivating domain in a cooperative manner, with Hill coefficient of 2 (cf. Chapter 1 and Exercise 8).

To complete the model, it is assumed that $p_1$ and $p_2$ are instantaneous functions of $[Ca^{2+}]$ and $[IP_3]$, but that $p_3$ acts on a slower time scale, so that

$$J_{channel} = k_f p_1 p_2 h, \tag{5.40}$$

where $h$ is a time-dependent inactivation variable satisfying the differential equation

$$\tau_h \frac{dh}{dt} = \frac{k_2^2}{k_2^2 + c^2} - h. \tag{5.41}$$

Thus,

$$\frac{dc}{dt} = \overbrace{k_f \left( \mu_0 + \frac{\mu_1 p}{p + k_\mu} \right) \left( b + \frac{(1 - b)c}{k_1 + c} \right) h}^{\text{receptor flux}} - \overbrace{\frac{\gamma c}{k_\gamma + c}}^{\text{pumping}} + \beta. \tag{5.42}$$

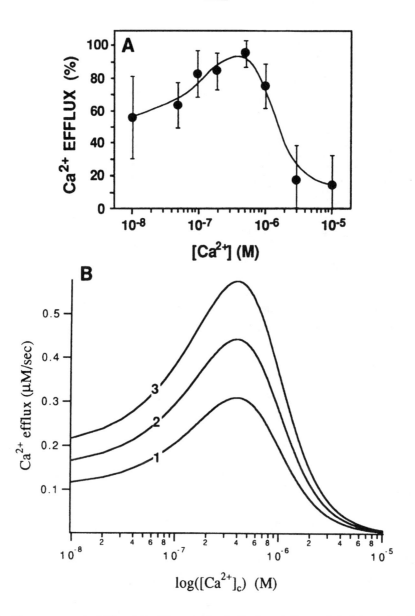

**Figure 5.9**  The open probability of the $IP_3$ receptor in the heuristic model, determined by fitting to the experimental data of Parys et al. (1992). A: The experimental data of Parys et al. (1992) from *Xenopus* oocytes. B: The model results, for three different $IP_3$ concentrations. Curves 1, 2, and 3 are in order of increasing $IP_3$ concentration. (Atri et al., 1993, Fig. 2.)

In fashion similar to the De Young–Keizer model, the term $\gamma c/(k_\gamma + c)$ represents pumping of $Ca^{2+}$ out of the cytoplasm, and $\beta$ represents a constant leak into the cytoplasm. The parameters of the model are given in Chapter 12, Table 12.1, when waves in the model are discussed.

It is important to note some features of the heuristic model. First, the model equations are of the same form as the reduced receptor model discussed above, and of the same form as the Hodgkin–Huxley and FitzHugh–Nagumo models. Obviously, this combination of fast variables and slow variables, with inactivation acting on a slow time scale, is a feature common to many physiological systems. Second, the functions $\tau_h$ and $h_\infty$ have a simpler form in the heuristic model. However, in attaining greater simplicity, some accuracy has been lost. For example, in the heuristic model $\tau_h$ is a constant, independent of $c$ and $p$; the more detailed receptor model, and its reduced version, find $\tau_h$ to be a function of $c$ and $p$; i.e., the kinetics of receptor inactivation are affected by the concentrations of $Ca^{2+}$ and $IP_3$. In a more realistic model of cooperativity at the inactivating site, $\tau_h$ would be a nonconstant function of $Ca^{2+}$ (Exercise 8), but such complications are ignored in this model for the sake of simplicity.

The heuristic model exhibits oscillations in a manner similar to the models discussed above. For a wide range of parameter values, two Hopf bifurcation points exist and are connected by a branch of stable periodic orbits. This is behavior typical of models of $Ca^{2+}$ oscillations, and we have seen similar bifurcation diagrams previously (cf. Figs. 5.5 and 5.8).

A number of other points about the heuristic model are worth noting. First, it does not include the factor $c_s - c$ in the term describing the $IP_3$-sensitive $Ca^{2+}$ current. Thus, it assumes that the concentration of $Ca^{2+}$ in the ER is so high, and so well buffered, that depletion of the ER has only a negligible effect on intracellular $Ca^{2+}$ dynamics for most of the physiological regime. Because of this omission, the structure of the heuristic model is different from the usual FitzHugh–Nagumo system, and the model becomes unphysiological when $\tau_h$ is too large. Nevertheless, in the physiological regime the heuristic model agrees well with experimental data. Also, the form of the pumping term is different from that in the detailed receptor model, which uses a Hill equation with coefficient 2. There is experimental evidence that the form used in the receptor model is a more accurate description of the $Ca^{2+}$ ATPase found in a variety of cell types (Lytton et al., 1992), but it is not clear how this change in the pumping term affects model behavior. The differences and similarities between the detailed receptor model and the heuristic model underline the fact that there are many choices to make in the construction of even the simplest model. The obvious question to ask is to what extent the choice of assumptions affects the final results. In other words, how sensitive are the model predictions to the underlying assumptions? How much complication is vital, and how much is a waste of time and money? In general these are difficult questions to answer, and they can be answered completely only after detailed (and time-consuming) comparisons between the models have been made. For this reason such comparisons are rarely performed, or at least not before experimental data indicates that the comparison will lead to useful distinctions. Despite these difficulties however, the similarities between these two models suggest strongly that fast activation and slow inactivation of the $IP_3$ receptor by $Ca^{2+}$ are some of the most significant mechanisms underlying $Ca^{2+}$ oscillations.

Further complications have been introduced recently by the experimental measurements of Finch et al. (1991), Parker et al. (1996), and Dufour et al. (1997), who have shown that the value used for $\tau_h$ is an order of magnitude too large, at least for *Xenopus* oocytes and rat brain synaptosomes. All of the models discussed above use a value of around 2 seconds (or greater) for $\tau_h$, but a more realistic value is apparently around 0.2 seconds or less. We conclude that although there is strong evidence that modulation of the $IP_3$ receptor by both $IP_3$ and $Ca^{2+}$ can explain many features of $Ca^{2+}$ oscillations and waves, at least qualitatively, the properties of the $IP_3$ receptor are not sufficiently well understood to give a fully quantitative theory.

## 5.3.2  Ryanodine Receptors

The second principal way in which $Ca^{2+}$ can be released from intracellular stores is through ryanodine receptors, which are found in a variety of cells, including cardiac cells, smooth muscle, skeletal muscle, chromaffin cells, pituitary cells, neurons, and sea urchin eggs. Ryanodine receptors share many structural and functional similarities with $IP_3$ receptors, particularly in their sensitivity to $Ca^{2+}$. Just as $Ca^{2+}$ can activate $IP_3$ receptors and increase the $Ca^{2+}$ flux, so too can $Ca^{2+}$ trigger $Ca^{2+}$-induced $Ca^{2+}$ release (CICR) from the sarcoplasmic or endoplasmic reticulum through ryanodine receptors (Endo et al., 1970; Fabiato, 1983). Calcium can also inactivate ryanodine receptors, although the physiological significance of such inactivation is unclear. Ryanodine receptors are so named because of their sensitivity to ryanodine, which decreases the open probability of the channel. On the other hand, caffeine increases the open probability of ryanodine receptors.

### Calcium oscillations in bullfrog sympathetic neurons

Sympathetic neurons respond to caffeine, or mild depolarization, with robust and reproducible $Ca^{2+}$ oscillations. Although these oscillations are dependent on external $Ca^{2+}$, they occur at a fixed membrane potential and involve the release of $Ca^{2+}$ from the ER via ryanodine receptors, as is indicated by the fact that they are abolished by ryanodine. Typical oscillations are shown in Fig. 5.10.

A particularly simple model of CICR (Friel, 1995) provides an excellent quantitative description of the behavior of these oscillations in the bullfrog sympathetic neuron. Despite the model's simplicity (or perhaps because of it), it is a superb example of how theory can supplement experiment, providing an interpretation of experimental results as well as quantitative predictions that can subsequently be tested.

Initially, we construct a linear model and determine the kinetic parameters by fitting the model to the responses following small perturbations. A schematic diagram of the model is given in Fig. 5.11. A single intracellular $Ca^{2+}$ store exchanges $Ca^{2+}$ with the cytoplasm (with fluxes $J_{L2}$ and $J_{P2}$), which in turn exchanges $Ca^{2+}$ with the external medium ($J_{L1}$ and $J_{P1}$). Thus,

$$\frac{dc}{dt} = J_{L1} - J_{P1} + J_{L2} - J_{P2}, \tag{5.43}$$

$$\frac{dc_s}{dt} = -J_{L2} + J_{P2}, \tag{5.44}$$

where $c$ denotes $[Ca^{2+}]$ in the cytoplasm and $c_s$ denotes $[Ca^{2+}]$ in the intracellular store. The fluxes are chosen in a simple way, as linear functions of the concentrations:

$$J_{L1} = k_1(c_e - c), \quad Ca^{2+} \text{entry}, \tag{5.45}$$

$$J_{P1} = k_2 c, \quad Ca^{2+} \text{extrusion}, \tag{5.46}$$

$$J_{L2} = k_3(c_s - c), \quad Ca^{2+} \text{release}, \tag{5.47}$$

$$J_{P2} = k_4 c, \quad Ca^{2+} \text{uptake}, \tag{5.48}$$

where $c_e$ denotes the external $[Ca^{2+}]$, which is assumed to be fixed. (For instance, in the simulations of Fig. 5.10 $c_e$ was fixed at 1, 0.5 and 0.7 mM). Depolarization induced by the application of high external $K^+$ can be modeled as an increase in $k_1$, the rate of $Ca^{2+}$ entry from the outside, while the application of caffeine (which increases the rate of $Ca^{2+}$ release from the internal store) can be modeled by an increase in $k_3$. If these changes are small enough, the cell responds in a linear fashion, with responses described by the exponential solutions of (5.43) and (5.44). By fitting these exponential solutions to the data, the kinetic constants $k_1, \ldots, k_4$ can be determined.

We now extend the linear model to account for the observed $Ca^{2+}$ oscillations. We model CICR in a simple way by making $k_3$ an increasing function of $c$, i.e.,

$$k_3 = \kappa_1 + \frac{\kappa_2 c^n}{K_d^n + c^n}, \tag{5.49}$$

and then, using the parameters determined from the linear fit as a starting point, determine the parameters of the nonlinear model by fitting to the time course of an oscillation (Table 5.3). A typical result is shown in Fig. 5.12.

Not only does this model provide an excellent quantitative description of the $Ca^{2+}$ oscillation, it also predicts the fluxes that should be observed over the oscillatory cycle. Subsequent measurement of these fluxes confirmed the model predictions, as seen in the lower panel of Fig. 5.12. It thus appears that CICR (at least in bullfrog sympathetic neurons) can be well described by a relatively simple model. It is necessary only for the ryanodine receptors to be activated by $Ca^{2+}$ to generate physiological oscillations — inactivation by $Ca^{2+}$ is not necessary.

---

**Table 5.3** Parameters of the model of $Ca^{2+}$ oscillations in sympathetic neurons, determined by fitting to the time course of an oscillation.

| | | | | | |
|---|---|---|---|---|---|
| $k_1$ | = | $5 \times 10^{-6}$ s$^{-1}$ | $\kappa_2$ | = | 2.4 s$^{-1}$ |
| $k_2$ | = | 0.132 s$^{-1}$ | $K_d$ | = | 1 $\mu$M |
| $k_4$ | = | 3.78 s$^{-1}$ | $n$ | = | 3 |
| $\kappa_1$ | = | 0.054 s$^{-1}$ | | | |

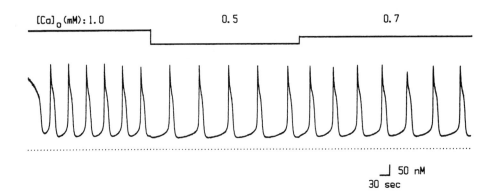

**Figure 5.10** Caffeine-induced Ca$^{2+}$ oscillations in sympathetic neurons, and their dependence on the extracellular Ca$^{2+}$ concentration. [Ca]$_0$ stands for $c_e$. (Friel, 1995, Fig. 5a.)

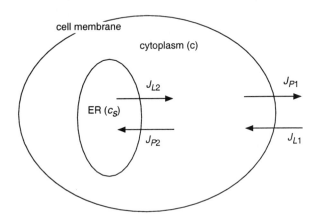

**Figure 5.11** Schematic diagram of the CICR model of Ca$^{2+}$ oscillations in bullfrog sympathetic neurons.

## Excitation–contraction coupling in cardiac cells

Although this simple CICR model does a good job of describing the behavior of one cell type, the similarities between IP$_3$ and ryanodine receptors suggest that it should be possible to construct more detailed models of the ryanodine receptor along the lines of the detailed model of the IP$_3$ receptor. The goal is to explain a wider range of behaviors than can be explained by the simpler model of CICR.

CICR is of particular importance in cardiac cells. In these cells, membrane depolarization causes a small influx of Ca$^{2+}$ through a voltage-sensitive membrane channel, which, in turn, initiates the release of Ca$^{2+}$ from the sarcoplasmic reticulum (SR) through the ryanodine receptor, leading to muscle contraction. Although there is general agreement on the outline of the process, there is little agreement on the details. For example, it is not yet clear whether inactivation of ryanodine receptors by Ca$^{2+}$ plays any role in intact cardiac cells, or how many functionally distinct intracellular

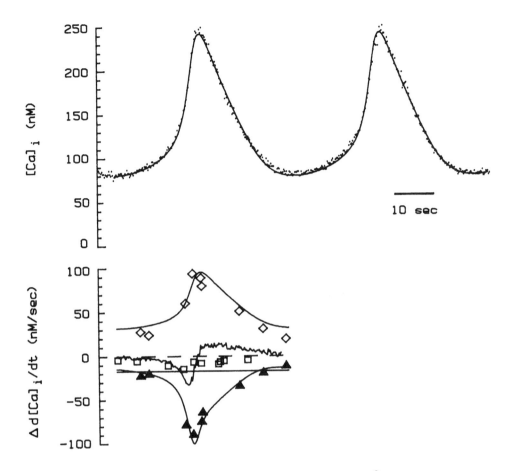

**Figure 5.12** The top panel shows an experimentally measured $Ca^{2+}$ oscillation (dots) and a model oscillation (smooth curve). The model parameters were determined by fitting the model to the oscillation time course, which explains the excellent agreement between model and data. The lower panel shows the predicted and measured $Ca^{2+}$ fluxes. Open squares are $J_{L1}$, open diamonds are $J_{P1} + J_{P2}$, solid triangles are $J_{L2}$. (Adapted from Friel, 1995, Fig. 9.)

$Ca^{2+}$ pools contribute to CICR. The question is complicated by the spatial positioning of the ryanodine receptors; the release of $Ca^{2+}$ into a confined space between the cell membrane and the SR can result in a much higher local $[Ca^{2+}]$ than is predicted by a spatially homogeneous model.

One model of CICR in cardiac cells (Fabiato, 1992; Tang and Othmer, 1994) is based on the assumption (similar to that made for $IP_3$ receptors) that $Ca^{2+}$ can both activate and inactivate the ryanodine receptor. It is assumed the ryanodine receptor can exist in four different states (Fig. 5.13): $S_{00}$, the bare ryanodine receptor; $S_{01}$, the receptor with one $Ca^{2+}$ bound to an inactivating site; $S_{10}$, the receptor with one $Ca^{2+}$ bound to an activating site; and $S_{11}$, with two $Ca^{2+}$ bound. $S_{10}$ is the open state of the receptor.

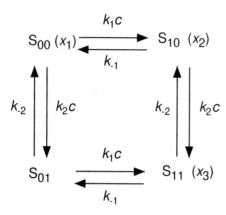

**Figure 5.13**  State diagram of the model of CICR in cardiac cells.

For simplicity, we use a slightly different notation from that used for the model of the $IP_3$ receptor. We let $x_1$, $x_2$, and $x_3$ denote the fraction of receptors in the states $S_{00}$, $S_{10}$, and $S_{11}$, respectively. With $c$ denoting $[Ca^{2+}]$, the equations for the receptor states are

$$\frac{dx_1}{dt} = k_{-1}x_2 + k_{-2}\left(1 - \sum_{i=1}^{3} x_i\right) - (k_1 + k_2)x_1 c, \qquad (5.50)$$

$$\frac{dx_2}{dt} = -k_{-1}x_2 + k_{-2}x_3 + (k_1 x_1 - k_2 x_2)c, \qquad (5.51)$$

$$\frac{dx_3}{dt} = \left[k_2 x_2 + k_1\left(1 - \sum_{i=1}^{3} x_i\right)\right]c - (k_{-2} + k_{-1})x_3, \qquad (5.52)$$

where we have used the fact that the fraction of receptors in state $S_{01}$ is $1 - \sum_{i=1}^{3} x_i$. It is also assumed that the rate of $Ca^{2+}$ binding to the activating site is independent of whether or not $Ca^{2+}$ is already bound to the inactivating site (i.e., the rate at which $S_{00} \to S_{10}$ is the same as the rate at which $S_{01} \to S_{11}$), and vice versa.

To these equations we add two more: one for $c$, and one for $c_s$, the concentration of $Ca^{2+}$ in the sarcoplasmic reticulum. To get these equations, we assume

1. that $Ca^{2+}$ leaks into the cell from the outside at the rate $g_2(c_e - c)$, where $c_e$ is the external $Ca^{2+}$ concentration, $c$ is the cytoplasmic $Ca^{2+}$ concentration, and $g_2$ is a constant,
2. that $Ca^{2+}$ leaks into the cell from the SR at the rate $g_1(c_s - c)$, where $c_s$ is the concentration of $Ca^{2+}$ in the SR, and $g_1$ is a constant,
3. that $Ca^{2+}$ is pumped out of the cell at the rate $q_1 c^2/(c^2 + q_2^2)$,
4. that $Ca^{2+}$ is pumped from the cytoplasm into the SR at a rate $p_1 c^2/(c^2 + p_2^2)$,
5. and finally, that the rate of $Ca^{2+}$ release from the SR through ryanodine receptors is $k_f x_2(c_s - c)$ for some constant $k_f$.

**Table 5.4**  Parameters of the model of CICR in cardiac cells.

| | | | | |
|---|---|---|---|---|
| $k_1$ | $=$ | $15\ \mu M^{-1}s^{-1}$ | $p_1$ | $=$ | $1038\ \mu M^{-1}s^{-1}$ |
| $k_{-1}$ | $=$ | $7.6\ s^{-1}$ | $p_2$ | $=$ | $0.12\ \mu M$ |
| $k_2$ | $=$ | $0.8\ \mu M^{-1}s^{-1}$ | $q_1$ | $=$ | $19\ \mu M^{-1}s^{-1}$ |
| $k_{-2}$ | $=$ | $0.84\ s^{-1}$ | $q_2$ | $=$ | $0.06\ \mu M$ |
| $k_f$ | $=$ | $80\ s^{-1}$ | $g_1$ | $=$ | $0.4\ s^{-1}$ |
| $c_e$ | $=$ | $1.5\ mM$ | $g_2$ | $=$ | $0.01\ s^{-1}$ |
| $v_c$ | $=$ | $0.185$ | | | |

Combining these assumptions and letting $v_c$ denote the ratio of SR volume to cytoplasm volume, we get

$$\frac{dc}{dt} = v_c \left[ (k_f x_2 + g_1)(c_s - c) - \frac{p_1 c^2}{p_2^2 + c^2} \right] + g_2(c_e - c) - \frac{q_1 c^2}{q_2^2 + c^2} + J(t), \qquad (5.53)$$

$$\frac{dc_s}{dt} = -(k_f x_2 + g_1)(c_s - c) + \frac{p_1 c^2}{p_2^2 + c^2}. \qquad (5.54)$$

$J(t)$ is a specified flux that models the $Ca^{2+}$ influx resulting from the opening of voltage-gated $Ca^{2+}$ channels in the sarcolemma. $J$ is modeled as a square pulse lasting for 240 ms and with height $A_0$. $A_0$ is a variable parameter that determines the size of the initial $Ca^{2+}$ stimulus. The parameters of the model are given in Table 5.4.

It is worth noting some of the similarities and differences between this model and some of the models discussed previously. First, the ryanodine receptor model assumes that the receptor flux is proportional to $x_2$, not $x_2^3$ as was assumed for the $IP_3$ receptor. In this sense, the ryanodine receptor model is closer to the heuristic model of the $IP_3$ receptor. However, the various leaks and pumps are modeled in a more detailed fashion, with two $Ca^{2+}$ ATPases included explicitly, one pumping $Ca^{2+}$ into the SR, the other pumping $Ca^{2+}$ out of the cell.

As is seen from Table 5.4, the binding of calcium to the activating site is a much faster process than binding to the inactivating site, as $k_1$ is much larger than $k_2$, and $k_{-1}$ is much larger than $k_{-2}$. This suggests that (5.50), (5.51), and (5.52) can be reduced by quasi-steady-state analysis to a simpler equation for the inactivation variable $y$. Indeed, following standard arguments, we find that

$$x_2 = \frac{k_1 c}{k_{-1} + k_1 c}(1 - y), \qquad (5.55)$$

where

$$\frac{dy}{dt} = k_2 c \left( \frac{k_1 c}{k_{-1} + k_1 c} \right)(1 - y) - k_{-2} y. \qquad (5.56)$$

The ryanodine receptor model agrees well with experimental data from cardiac cells. The steady-state fraction of open channels as a function of $c$ is a bell-shaped curve, with a maximum at about 1 $\mu$M. This is in good qualitative agreement with the results of Fabiato (1985), who showed that the amplitude of the tension transient in skinned cardiac cells was a bell-shaped function of the triggering $Ca^{2+}$ concentration. Further, the model incorporates the results of Györke and Fill (1993), who showed that ryanodine receptors in lipid bilayers adapted to a maintained $Ca^{2+}$ stimulus. In response to a step increase in $[Ca^{2+}]$, the open probability of ryanodine receptors in bilayers peaks rapidly, subsequently declining to a lower plateau. These results underline the similarities between ryanodine receptors and IP$_3$ receptors. Both have bell-shaped open probabilities as a function of $[Ca^{2+}]$, and both adapt to a maintained $Ca^{2+}$ stimulus. A number of other experimental results are also reproduced by the model. First, it shows a scaled response to a graded series of $Ca^{2+}$ stimuli; as is observed experimentally, the peak of the response increases as the stimulus increases. Second, experiments show that under conditions where $Ca^{2+}$ leakage from the outside into the cell is increased (by damage to the sarcolemma, for example), cardiac cells can exhibit spontaneous oscillations, with frequency dependent on the magnitude of the leak, a feature that is also reproduced by the model. Finally, the model exhibits traveling waves that annihilate upon intersection and that travel at approximately 80 $\mu$ms$^{-1}$, in the physiological range.

Far more complex models of CICR have been constructed, incorporating multiple compartments and detailed descriptions of $Ca^{2+}$ fluxes (for example, Wong et al., 1992). Although they can do an excellent job of reproducing experimental data, they are too complicated to be discussed here, and so we content ourselves with the simpler model discussed above. A model similar to the one above has been constructed by Keizer and Levine (1996), who study adaptation of the ryanodine receptor in detail.

## Spatial effects

The above model for CICR in cardiac cells shows that many features of the ryanodine receptor can be explained by a model that is similar to models of the IP$_3$ receptor. Although this emphasizes the behavioral similarities between the receptors, there are also major differences. For example, inactivation of ryanodine receptors by $Ca^{2+}$ has not yet been observed in intact cells (as opposed to lipid bilayers), and there is not yet general agreement that such inactivation is physiologically important. Another way in which ryanodine receptors may differ from IP$_3$ receptors is their position in the cell. There is a close relationship between the voltage-sensitive $Ca^{2+}$ channel in the sarcolemma that lets in the initial $Ca^{2+}$ influx and the ryanodine receptor in the sarcoplasmic reticulum that initiates CICR (Stern, 1992). Thus, $Ca^{2+}$ entering through the voltage-sensitive channel may have a much greater effect on the ryanodine receptor than $Ca^{2+}$ in the cytoplasm. Further, if $Ca^{2+}$ is released from the SR into a confined space between the SR and the sarcolemma, the ryanodine receptor could have a microenvironment radically different from that experienced by the rest of the cell (Stern, 1992; Kargacin, 1994; Stern et al., 1997).

Stern (1992) has constructed a series of *local control*, or *calcium synapse*, models that incorporate the effects of the close spatial relationship between the ryanodine receptors and the voltage-sensitive $Ca^{2+}$ channels. However, since these models are complex, we do not discuss them in detail here. Spatial issues have also been discussed by Peskoff et al. (1992), Langer and Peskoff (1996), Wang et al. (1996), and Peskoff and Langer (1998).

## 5.4  EXERCISES

1. Murray (1989) discusses a simple model of CICR that has been used by a number of modelers (Cheer et al., 1987; Lane et al., 1987). In the model, $Ca^{2+}$ release from the ER is an increasing sigmoidal function of $Ca^{2+}$, and $Ca^{2+}$ is removed from the cytoplasm with linear kinetics. Thus,

$$\frac{dc}{dt} = L + \frac{k_1 c^2}{1 + c^2} - k_2 c,$$

   where $L$ is a constant leak of $Ca^{2+}$ from the ER into the cytoplasm.

   (a) Show that when $L = 0$ and $k_1 > 2k_2$, there are two positive steady states and determine their stability. For the rest of this problem assume that $k_1 > 2k_2$.

   (b) How does the nullcline $dc/dt = 0$ vary as the leak from the internal store increases? Show that there is a critical value of $L$, $L_c$ say, such that when $L > L_c$, only one positive solution exists.

   (c) Fix $L < L_c$ and suppose the solution is initially at the lowest steady state. How does $c$ behave when small perturbations are applied to $c$? How does $c$ behave when large perturbations are applied? How does $c$ behave when $L$ is raised above $L_c$ and then decreased back to zero? Plot the bifurcation diagram in the $L, c$ phase-plane, indicating the stability (or otherwise) of the branches. Why is this behavior called a biological switch? Is there hysteresis in this model?

2. Show that in a closed cell (i.e., one without any interaction with the extracellular environment) the two-pool model cannot exhibit $Ca^{2+}$ oscillations.

3. Test various simplifications of the two-pool model to determine whether or not oscillations occur. Do this by looking for Hopf bifurcations analytically. For example, consider the cases

   (a) $n = 1, m = 1, p = 0$.

   (b) $n = p = m = 1$, with $c_1 \ll K_1, c_1 \ll K_3, c_2 \ll K_2$.

   What is the simplest version that supports oscillations?

4. Show that in a general model of intracellular $Ca^{2+}$ dynamics, the resting level of intracellular $Ca^{2+}$ is independent of $Ca^{2+}$ exchange with the internal pools.

5. Complete the details of the reduction of the receptor model (Section 5.3.1). First, derive the quasi-steady-state equations,

$$\frac{x_{000}}{K_1 K_5} = \frac{x_{010}}{K_1 c} = \frac{x_{100}}{K_5 p} = \frac{x_{110}}{pc} = \frac{(1 - y)}{(p + K_1)(c + K_5)}, \tag{5.57}$$

$$\frac{x_{001}}{K_3 K_5} = \frac{x_{011}}{K_3 c} = \frac{x_{101}}{K_5 p} = \frac{x_{111}}{pc} = \frac{y}{(p + K_3)(c + K_5)}, \tag{5.58}$$

where $K_i = k_{-i}/k_i$. Next, write the differential equations for the group II sites, and add the four equations, to get

$$\frac{dy}{dt} = k_4 c(x_{000} + x_{010}) + k_2 c(x_{100} + x_{110}) - k_{-4}(x_{001} + x_{011}) - k_{-2}(x_{101} + x_{111}). \tag{5.59}$$

Finally, substitute in the quasi-steady-state solutions to get (5.30).

6.  Show that (5.30) can be written in the form (5.31), and interpret the qualitative behavior of the functions $\tau_y$ and $y_\infty$ in terms of the underlying physiology.

7.  Write down the equations for the reduced receptor model (Section 5.3.1) when $k_4 = k_2$ and $k_{-4} = k_{-2}$. Let $h = 1 - y$. What is the differential equation for $h$? Write it in the form

$$\tau_h \frac{dh}{dt} = h_\infty - h. \tag{5.60}$$

Derive this simplified model directly from the state diagram in Fig. 5.6.

8.  Write down a reaction scheme like that of Fig. 5.6, but assuming that 2 $Ca^{2+}$ ions inactivate the receptor in a cooperative fashion. Assume a simple model of cooperativity,

$$S_{ij0} \underset{k_{-2}}{\overset{c^2 k_2}{\rightleftarrows}} S_{ij1}, \tag{5.61}$$

for $i, j = 0$ or 1, and assume that the group I and group II states are each in quasi-steady state (as was done to obtain the reduced receptor model). Derive a version of the heuristic model (Section 5.3.1) in which $\tau_h$ is a function of $Ca^{2+}$.

9.  Without assuming that two $Ca^{2+}$ ions bind simultaneously, repeat the previous derivation. Assume that the first $Ca^{2+}$ ion binds slowly, while the second binds quickly.

10. Check that the binding diagram in Fig. 5.6 satisfies the principle of detailed balance (Chapter 1, Exercise 3). (In fact, the model parameters were chosen to ensure this.)

11. Plot the nullclines of the heuristic model of the $IP_3$ receptor (Section 5.3.1). Show that it has a form different from that of the FitzHugh–Nagumo model of oscillations (Chapter 4). Modify the model by assuming that $Ca^{2+}$ release through the $IP_3$ receptor is proportional to $c_s - c$, where $c_s$ is the concentration of $Ca^{2+}$ in the ER. How does this change the nullclines and the basic model structure? How do these two versions of the model behave as $\tau_h \to 0$?

12. Using AUTO, show that for the parameters given in Table 12.1, the bifurcation diagram of the heuristic model of the $IP_3$ receptor (Section 5.3.1) becomes considerably more complicated as the steady-state curve folds up and breaks the branch of periodic orbits into two separate branches (Fig. 5.14). Describe and sketch the series of bifurcations that occurs as $\mu$ is decreased.

13. Assuming $k_3$ to be a constant, calculate the solution to (5.43)–(5.48). Stimulation by high external $K^+$ can be modeled as a step increase in $k_1$ (as the cell depolarizes), and stimulation by external caffeine can be modeled as a step increase in $k_3$. What are the solutions for these two cases?

14. Plot the bifurcation diagram of the model of CICR in bullfrog sympathetic neurons (Section 5.3.2) using the external $Ca^{2+}$ concentration $c_e$ as the main bifurcation parameter. Verify the behavior seen in Fig. 5.10, that the period but not the amplitude of the oscillations is sensitive to $c_e$.

15. Use quasi-steady-state analysis to reduce the detailed ryanodine receptor model equations (5.50), (5.51), and (5.52) to (5.55) and (5.56).

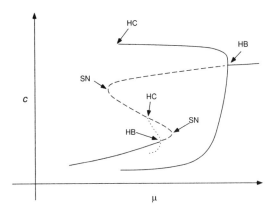

**Figure 5.14** Typical bifurcation diagram of the heuristic model of Ca$^{2+}$ oscillations, not drawn to scale (see Exercise 11). HB denotes a Hopf bifurcation, SN denotes a saddle-node bifurcation, HC denotes a homoclinic bifurcation.

16. Calculate the steady-state fraction of open channels in the model of CICR in cardiac cells (Section 5.3.2) and show that it is a bell-shaped function of $c$. Plot the bifurcation diagram of the model. Find the Hopf bifurcations and plot some representative oscillations.

# CHAPTER 6

# Bursting Electrical Activity

Neurons communicate by firing and transmitting action potentials. Commonly, action potentials occur in periodic fashion, as in response to a constant applied current of sufficient magnitude. For example, in both the Hodgkin–Huxley and FitzHugh–Nagumo models, a constant applied current can cause the repetitive firing of action potentials. Many cell types exhibit more complex behavior, characterized by brief bursts of oscillatory activity interspersed with quiescent periods during which the membrane potential changes only slowly. This behavior is called *bursting*, and typical experimental results from a number of different cell types are shown in Fig. 6.1.

Although bursting has been studied extensively for many years, most mathematical studies are based on the pioneering work of Rinzel (1985, 1987), which was in turn based on one of the first biophysical models of a pancreatic $\beta$-cell (Chay and Keizer, 1983). Rinzel's interpretation of bursting in terms of nonlinear dynamics is one of the recent success stories of mathematical physiology and provides an excellent example of how mathematics can be used to understand complex biological dynamical systems. However, despite the extensive studies, there is yet no consensus on the underlying mechanisms that cause bursting. In fact, in many cell types it is not even clear whether bursting is caused by cellular-level processes or is generated by the coupling of many cells into an electrical network. For example, bursting in pancreatic $\beta$-cells, one of the classic examples of bursting, does not usually occur in isolated cells, but only in intact islets or in groups of $\beta$-cells coupled by gap junctions. Thus, although there is an elegant interpretation of many experimental results in terms of nonlinear dynamical systems, the phenomenon of bursting is not completely understood.

Models for electrical bursting can be divided into two major groups (well summarized by De Vries, 1995). Earlier models were generally based on the assumption that bursting was caused by an underlying slow oscillation in the intracellular $Ca^{2+}$ concen-

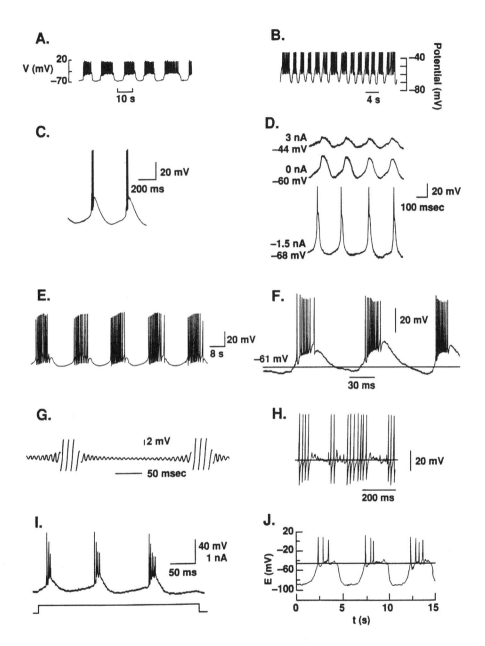

**Figure 6.1** Electrical bursting in a range of different cell types. A: Pancreatic $\beta$-cell. B: Dopamine-containing neurons in the rat midbrain. C: Cat thalamocortical relay neuron. D: Guinea pig inferior olivary neuron. E: *Aplysia* R15 neuron. F: Cat thalamic reticular neuron. G: *Sepia* giant axon. H: Rat thalamic reticular neuron. I: Mouse neocortical pyramidal neuron. J: Rat pituitarygonadotropin-releasing cell. (Wang and Rinzel, 1995, Fig. 2.)

tration (Chay, 1986, 1987; Chay and Cook, 1988; Chay and Kang, 1987; Himmel and Chay, 1987; Keizer and Magnus, 1989). In light of more recent experimental evidence showing that $Ca^{2+}$ is probably not the slow variable underlying bursting, more recent models have modified this assumption, relying on alternative mechanisms to produce the underlying slow oscillation (Keizer and Smolen, 1991; Smolen and Keizer, 1992). In this chapter we focus our attention on two early models, discussing how they fit into the general classification scheme proposed by Rinzel (1987). Later models, being similar in mathematical structure to the early models, are not discussed in any detail. We then discuss some mathematical properties of bursting models and finally show how the important properties can be incorporated into simpler polynomial models, in much the same way that the FitzHugh–Nagumo model provides a simplification of the Hodgkin–Huxley model.

## 6.1   Bursting in the Pancreatic $\beta$-Cell

In response to glucose, $\beta$-cells of the pancreatic islet secrete insulin, which causes the increased use or uptake of glucose in target tissues such as muscle, liver, and adipose tissue. When blood levels of glucose decline, insulin secretion stops, and the tissues begin to use their energy stores instead. Interruption of this control system results in diabetes, a disease that if left uncontrolled can result in kidney failure, heart disease, and death. It is believed that electrical bursting, a typical example of which is shown in Fig. 6.1A, plays an important (but not exclusive) role in the release of insulin from the cell. Other aspects of insulin secretion and the control of blood glucose are discussed in Chapter 19. In this chapter we focus on models for the bursting electrical activity observed in single cells and cell clusters.

One of the first models for bursting was proposed by Atwater et al. (1980). It was based on extensive experimental data, incorporating the important cellular mechanisms that were thought to underlie bursting, and was later developed into a mathematical model by Chay and Keizer (1983). Although the mathematical model includes only those processes believed to be essential to the bursting process and thus omits many features of the cell, it is able to reproduce many of the basic properties of bursting. The ionic currents in the model are:

1. A $Ca^{2+}$-activated $K^+$ channel with conductance an increasing function of $c = [Ca^{2+}]$ of the form

$$g_{K,Ca} = \bar{g}_{K,Ca} \frac{c}{K_d + c},  \tag{6.1}$$

   for some constant $\bar{g}_{K,Ca}$.

2. A voltage-gated $K^+$ channel modeled in the same way as in the Hodgkin–Huxley model, with

$$g_K = \bar{g}_K n^4,  \tag{6.2}$$

where $n$ obeys the same differential equation as in the Hodgkin–Huxley model (Chapter 4), except that the voltage is shifted by $V^*$, so that $V$ in (4.28) and (4.29) is replaced by $V + V^*$. For example, $\beta_n(V) = 0.125 \exp[(-V - V^*)/80]$.

3. A voltage-gated $Ca^{2+}$ channel, with conductance

$$g_{Ca} = \bar{g}_{Ca} m^3 h, \tag{6.3}$$

where again $m$ and $h$ satisfy Hodgkin–Huxley-type differential equations, shifted along the voltage axis by an amount $V'$. In effect, the inward $Na^+$ current of the Hodgkin–Huxley model is replaced by an identical inward $Ca^{2+}$ current.

Combining these ionic currents and adding the usual leak current gives

$$C_m \frac{dV}{dt} = -(g_{K,Ca} + g_K)(V - V_K) - 2g_{Ca}(V - V_{Ca}) - g_L(V - V_L), \tag{6.4}$$

where $C_m$ is the membrane capacitance.

To complete the model, there is an equation for the regulation of intracellular $Ca^{2+}$, where it is assumed that glucose can regulate the removal of $Ca^{2+}$ from the cytoplasm; i.e., glucose acts by lowering $[Ca^{2+}]$, leading to bursting oscillations and subsequent insulin release. Hence,

$$\frac{dc}{dt} = f(-k_1 I_{Ca} - k_c c), \tag{6.5}$$

where the $Ca^{2+}$ current is $I_{Ca} = \bar{g}_{Ca} m^3 h(V - V_{Ca})$ and where $k_1$ and $k_c$ are constants. The constant $f$ is a scale factor relating total changes in $[Ca^{2+}]$ to the changes in free $[Ca^{2+}]$ (as discussed in the section on calcium buffering in Chapter 12) and is usually a small number. Although $k_c$ is an increasing function of glucose concentration, the concentration of glucose is not a dynamic variable in the model. Thus, $k_c$ can be regarded as fixed, and the behavior of the model can be studied for a range of values of $k_c$.

As shown in Fig. 6.2, the model exhibits bursts that bear a qualitative resemblance to those seen experimentally. Further, as glucose is increased (i.e., as $k_c$ is increased), the length of the bursts increases until at the level $k_c = 0.06$, bursting is continuous.

Consideration of the $Ca^{2+}$ concentration as a function of time (Fig. 6.2) shows that there is a slow oscillation in $c$ underlying the bursts, with bursting occurring during the peak of the $Ca^{2+}$ oscillation. The fact that $Ca^{2+}$ oscillations occur on a slower time scale is built into the $Ca^{2+}$ equation explicitly by means of the parameter $f$. As $f$ becomes smaller, the $Ca^{2+}$ equation evolves more slowly, and thus the relative speeds of the voltage and $Ca^{2+}$ equations can be directly controlled. It therefore appears that there are two oscillatory processes interacting to give bursting, with a fast oscillation in $V$ superimposed on a slower oscillation in $c$. This fact is the basis of the phase-plane analysis that we consider next.

## 6.1.1  Phase-Plane Analysis

The $\beta$-cell model can be simplified by ignoring the dynamics of $m$ and $h$, thus removing the time dependence of the $Ca^{2+}$ current (Rinzel and Lee, 1986). The simplified model

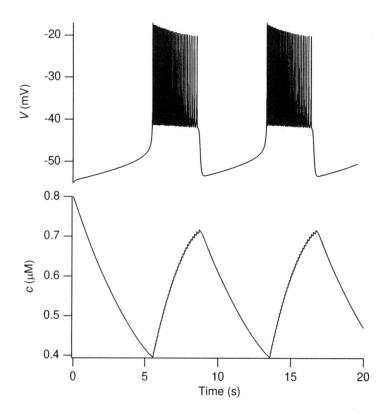

**Figure 6.2** Bursting oscillations in the $\beta$-cell model, calculated using the parameter values in Table 6.1.

equations are

$$C_m \frac{dV}{dt} = -I_{Ca}(V) - \left( \bar{g}_K n^4 + \frac{\bar{g}_{K,Ca} c}{K_d + c} \right) (V - V_K) - \bar{g}_L (V - V_L), \qquad (6.6)$$

$$\tau_n(V) \frac{dn}{dt} = n_\infty(V) - n, \qquad (6.7)$$

$$\frac{dc}{dt} = f(-k_1 I_{Ca}(V) - k_c c), \qquad (6.8)$$

where $I_{Ca} = \bar{g}_{Ca} m_\infty^3(V) h_\infty(V)(V - V_{Ca})$.

This separates the $\beta$-cell model into a fast subsystem (the $V$ and $n$ equations) and a slow equation for $c$. The advantage of this simplification is that the fast subsystem can be studied using phase-plane methods. Treating $c$ as a constant parameter, we first consider the bifurcation structure of the fast subsystem as a function of $c$. It is then easier to see how the qualitative behavior of the fast phase-plane changes as $c$ is varied slowly.

**Table 6.1**   Parameters of the model for electrical bursting in pancreatic $\beta$-cells.

| | | | | | |
|---|---|---|---|---|---|
| $C_m$ | $=$ | $1\ \mu F/cm^2$ | $\bar{g}_{K,Ca}$ | $=$ | $0.02\ mS/cm^2$ |
| $\bar{g}_K$ | $=$ | $3\ mS/cm^2$ | $\bar{g}_{Ca}$ | $=$ | $3.2\ mS/cm^2$ |
| $\bar{g}_L$ | $=$ | $0.012\ mS/cm^2$ | $V_K$ | $=$ | $-75\ mV$ |
| $V_{Ca}$ | $=$ | $100\ mV$ | $V_L$ | $=$ | $-40\ mV$ |
| $V^*$ | $=$ | $30\ mV$ | $V'$ | $=$ | $50\ mV$ |
| $K_d$ | $=$ | $1\ \mu M$ | $f$ | $=$ | $0.007$ |
| $k_1$ | $=$ | $0.0275\ \mu M\ cm^2/nC$ | $k_c$ | $=$ | $0.02\ ms^{-1}$ |

A

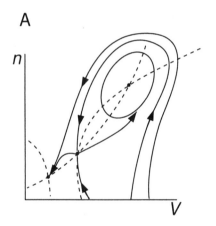

B

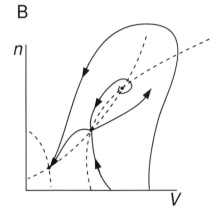

**Figure 6.3**   Phase-planes of the fast subsystem of the $\beta$-cell model, for two different values of $c$, both in the intermediate range. The phase-planes are sketched, not drawn to scale. Nullclines are denoted by dashed lines, and the intersections of the nullclines show the positions of the fixed points. For both values of $c$ there are three fixed points, of which the middle one is a saddle point. However, in A (with $c_{hb} < c < c_{hc}$; see Fig. 6.4) the unstable node is surrounded by a stable limit cycle, while in B (corresponding to $c > c_{hc}$) the limit cycle has disappeared via a homoclinic bifurcation.

When $c$ is low, the $Ca^{2+}$-activated $K^+$ channel is not activated, and the fast subsystem has a unique fixed point where $V$ is high. Conversely, when $c$ is high, the $Ca^{2+}$-activated $K^+$ channel is fully activated, and the fast subsystem has a unique fixed point where $V$ is low, as the high conductance of the $Ca^{2+}$-activated $K^+$ channels pulls the membrane potential closer to the Nernst potential of $K^+$, which is about $-75$ mV. However, for intermediate values of $c$ there are three fixed points, and the phase-plane is much more interesting and intricate. Phase-planes of the $V, n$ subsystem for two different intermediate values of $c$ are shown in Fig. 6.3.

**A**

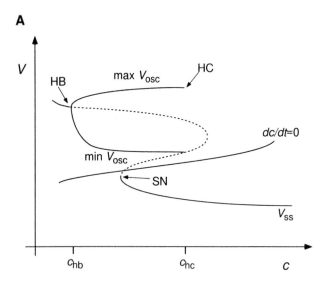

**B**

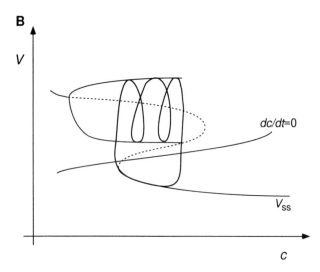

**Figure 6.4** A: Sketch of the bifurcation diagram of the simplified $\beta$-cell model, with $c$ as the bifurcation parameter. $V_{ss}$ denotes the curve of steady states of $V$ as a function of $c$. A solid line indicates a stable steady state; a dashed line indicates an unstable steady state. The two branches of $V_{osc}$ denote the maximum and minimum of $V$ over one oscillatory cycle. HB denotes a Hopf bifurcation, HC denotes a homoclinic bifurcation, and SN denotes a saddle-node bifurcation. B: A burst cycle projected onto the $(V, c)$ plane. (Adapted from Rinzel and Lee, 1986, Fig. 3.)

In both cases, the lower fixed point is stable, the middle fixed point is a saddle point, and the upper fixed point is unstable. For some values of $c$ the upper fixed point is surrounded by a stable limit cycle, which in turn is surrounded by the stable manifold of the saddle point (Fig 6.3A). However, as $c$ increases (still in the intermediate range), the limit cycle "hits" the saddle point and forms a homoclinic connection (a homoclinic bifurcation). Increasing $c$ further breaks the homoclinic connection, and the stable manifold of the saddle point forms a heteroclinic connection with the upper, unstable, critical point (Fig. 6.3B). There is now no limit cycle.

This sequence of bifurcations is easiest to understand in a bifurcation diagram, with $V$ plotted against the control parameter $c$ (Fig. 6.4A). The Z-shaped curve is the curve of fixed points, and as usual, the stable oscillation around the upper steady state is denoted by the maximum and minimum of $V$ over one cycle. As $c$ increases, oscillations appear via a Hopf bifurcation ($c_{hb}$) and disappear again via a homoclinic bifurcation ($c_{hc}$). For a range of values of $c$ the fast subsystem is bistable, with a lower stable fixed point and an upper stable periodic orbit. This bistability is crucial to the appearance of bursting.

We now couple the dynamics of the fast subsystem to the slower dynamics of $c$. Included in Fig. 6.4A is the curve defined by $dc/dt = 0$, i.e., the $c$ nullcline. When $V$ is above the $c$ nullcline, $dc/dt > 0$, and so $c$ increases, but when $V$ is below the $c$ nullcline, $c$ decreases. Now suppose $V$ starts on the lower fixed point for a value of $c$ that is greater than $c_{hc}$. Since $V$ is below the $c$ nullcline, $c$ starts to decrease, and $V$ follows the lower branch of fixed points. However, when $c$ becomes too small, this lower branch of fixed points disappears in a saddle-node bifurcation (SN), and so $V$ must switch to the upper branch of the Z-shaped curve. Since this upper branch is unstable and surrounded by a stable limit cycle, $V$ begins to oscillate. However, since $V$ now lies entirely above the $c$ nullcline, $c$ begins to increase. Eventually, $c$ increases enough to cross the homoclinic bifurcation at $c_{hc}$, the stable limit cycles disappear, and $V$ switches back to the lower branch, completing the cycle. Repetition of this process causes bursting. The quiescent phase of the bursting cycle is when $V$ is on the lower branch of the Z-shaped curve, and during this phase $V$ increases slowly. A burst of oscillations occurs when $V$ switches to the upper branch, and disappears again after passage through the homoclinic bifurcation. Clearly, in this scenario, bursting relies on the coexistence of both a stable fixed point and a stable limit cycle, and the bursting cycle is a hysteresis loop that switches between branches of the Z-shaped curve. Bursting also relies on the $c$ nullcline intersecting the Z-shaped curve in the right location. For example, if the $c$ nullcline intersects the Z-shaped curve on its lower branch, there is a unique stable fixed point for the whole system, and bursting does not occur. A projection of the bursting cycle on the $(V, c)$ phase-plane is shown in Fig. 6.4B. The periods of the oscillations in the burst increase through the burst, as the limit cycles get closer to the homoclinic trajectory, which has infinite period.

The relationship between bursting patterns and glucose concentration can also be deduced from Fig. 6.4. Notice that the $\frac{dc}{dt} = 0$ nullcline, given by $c = -\frac{k_1}{k_c} I_{Ca}(V)$, is inversely scaled by $k_c$. Thus, when $k_c$ is small, the nullcline intersects the lower branch of the $V$ nullcline. On the other hand, if $k_c$ is extremely large, the $c$ nullcline intersects the upper branch of the $V$ nullcline, possibly to the left of $c_{hb}$. At intermediate values of $c$, the $c$ nullclines intersects the middle branch of the $V$ nullcline.

Under the assumption that $k_c$ is related to the glucose concentration, we see that when the glucose concentration is low, the system is at a stable rest point on the lower $V$ nullcline; there is no bursting. If glucose is increased so that the $c$ nullcline intersects the middle $V$ nullcline with $c < c_{hc}$, there is bursting. However, the length of the bursting phase increases and the length of the resting phase decreases with increasing glucose, simply because calcium increases at a slower rate and decreases at a faster rate when $k_c$ is increased. For large enough $k_c$ the bursting is sustained with no rest phase, as $c$

becomes stalled below $c_{hc}$. Finally, at extremely high $k_c$ values, bursting is replaced by a permanent high membrane potential, with $c < c_{hb}$. This scenario of the dependence of the bursting phase on glucose is confirmed by experiments.

## 6.2   Parabolic Bursting

Another well-studied example of bursting is found in the Aplysia R-15 neuron (Fig. 6.1E). Analysis of a detailed model by Plant (1981) shows that the mathematical structure of this bursting oscillator is different from that in the $\beta$-cell model (Rinzel and Lee, 1987). The $\beta$-cell model has two fast variables, one slow variable, bistability, and a hysteresis loop. At the end of a burst, a homoclinic bifurcation is crossed, leading to an increasing period through the burst. Plant's model, on the other hand, has no bistability, with bursting arising from the presence of two slow variables with their own oscillation. A homoclinic bifurcation is crossed at the beginning and the end of the burst, and so the instantaneous period of the burst oscillations starts high, decreases, and then increases again. The fact that the period is roughly a parabolic function of time has led to the name *parabolic* bursting.

Plant's parabolic bursting model is similar in some respects to the $\beta$-cell model, incorporating $Ca^{2+}$-activated $K^+$ channels and voltage-dependent $K^+$ channels. However, it also includes a voltage-dependent $Na^+$ channel that activates and inactivates in typical Hodgkin–Huxley fashion and a slowly activating $Ca^{2+}$ current. The $Na^+$, $K^+$, and leak currents form the fast subsystem

$$C_m \frac{dV}{dt} = -\bar{g}_{Na} m_\infty^3(V) h (V - V_{Na}) - \bar{g}_{Ca} x (V - V_{Ca})$$

$$- \left( \bar{g}_K n^4 + \frac{\bar{g}_{K,Ca} c}{0.5 + c} \right) (V - V_K) - \bar{g}_L (V - V_L), \tag{6.9}$$

$$\tau_h(V) \frac{dh}{dt} = h_\infty(V) - h, \tag{6.10}$$

$$\tau_n(V) \frac{dn}{dt} = n_\infty(V) - n, \tag{6.11}$$

while the $Ca^{2+}$ current and its activation $x$ form the slow subsystem

$$\tau_x \frac{dx}{dt} = x_\infty(V) - x, \tag{6.12}$$

$$\frac{dc}{dt} = f(k_1 x (V_{Ca} - V) - c). \tag{6.13}$$

Full details of the model are specified in Exercise 3, as described in the appendix of Rinzel and Lee (1987).

For a fixed $x$, the bifurcation diagram of the fast subsystem as $c$ varies is shown in Fig. 6.5. Note that in general, $V_{ss}$ is a function of both $c$ and $x$, and therefore the fast subsystem is properly described by a bifurcation surface. However, since surfaces are more difficult to draw and understand, we first examine a cross-section of the bifurcation surface for fixed $x$ and then discuss how the important points behave as $x$ varies. As before, $c_{hb}$ denotes the value of $c$ at which a Hopf bifurcation to periodic orbits occurs, and $c_{hc}$ denotes the value of $c$ where the periodic solutions disappear in a homoclinic bifurcation. The bifurcation diagram is similar to that of the fast subsystem of the $\beta$-cell model (Fig. 6.4), except that the branch of periodic solutions around the upper branch of the Z-shaped curve does not extend past the lower "knee" of the Z-shaped curve. In fact, $c_{hc}$ appears to coincide with the saddle-node bifurcation at the knee of the Z-shaped curve. Hence, there is no bistability in the model, and a simple one-variable slow subsystem is insufficient to give bursting, because it is unable to move the fast subsystem in and out of the region where the oscillations occur, as in the $\beta$-cell model. However, because the parabolic bursting model has two slow variables, $x$ and $c$, a slow oscillation in these variables moves $c$ backward and forward across the homoclinic bifurcation, leading to bursts of fast oscillations during one portion of the slow oscillations.

In Fig. 6.6A we plot $c_{hc}$ and $c_{hb}$ as $x$ varies. Note that the curves in Fig. 6.5 correspond to taking a cross-section of Fig. 6.6 for a fixed value of $x$ (the dashed line). In region $J_{ss}$ there is a single stable fixed point of the fast subsystem, while in region $J_{osc}$ the fast subsystem has stable oscillations. Now suppose that the slow subsystem has a periodic solution. In the parabolic bursting model, slow oscillations do not occur independently of the fast subsystem but rely on the interaction between the fast and slow variables,

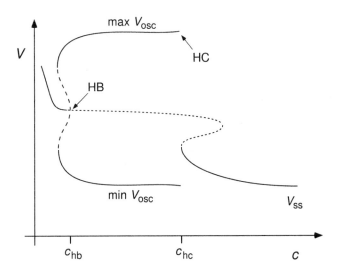

**Figure 6.5**  Sketch (not to scale) of the bifurcation diagram of the fast subsystem of the parabolic bursting model, with $x = 0.7$. The notation is similar to that used in Fig. 6.4. Note that the homoclinic bifurcation coincides with the lower knee of the Z-shaped curve of steady states and that the oscillations arise via a subcritical Hopf bifurcation instead of a supercritical one as was seen in the $\beta$-cell model.

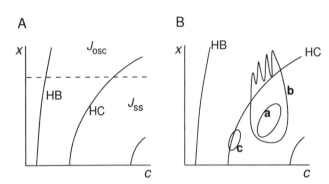

**Figure 6.6** A: The Hopf and homoclinic bifurcations in the $(x, c)$ plane, i.e., a projection of the bifurcation surface onto the $(x, c)$ plane. The bifurcation diagram in the previous figure corresponds to a cross-section of this figure for a fixed value of $x$, as denoted by the dashed line. In the region (labeled $J_{osc}$) between the two bifurcations, a stable limit cycle exists, while to the right of the homoclinic bifurcation ($J_{ss}$) there is a unique stable steady state. The rightmost (unlabeled) curve corresponds to where the middle branch of the Z-shaped curve meets the upper branch. B: Three typical slow oscillations in $x$ and $c$.

the details of which do not concern us here. Similar results are obtained when the slow variables oscillate independently of the fast variables, acting as a periodic driver of the fast subsystem (Kopell and Ermentrout, 1986). In any case, these oscillations correspond to closed curves in the $(x, c)$ phase-plane, three possible examples of which are shown in Fig. 6.6B. In case **a**, the dynamics of $x$ and $c$ are such that the slow periodic solution lies entirely within the region $J_{ss}$; i.e., the fast subsystem "lives" entirely on the lower branch of the Z-shaped curve and does not oscillate. In case **b**, the slow oscillation crosses the line of homoclinic bifurcations into the region $J_{osc}$, in which region the fast subsystem oscillates rapidly. In case **c**, the slow oscillations are so arranged that the system spends only a small amount of time in region $J_{osc}$, but not long enough to generate a burst of oscillations. Thus, by tuning the parameters of the underlying slow oscillation, different patterns of bursts can be obtained from one model. We have already mentioned the parabolic nature of the period of the fast oscillations. This is easily understood in terms of the foregoing analysis. For a burst to occur, the slow oscillation must cross the line of homoclinic bifurcations both at the beginning and at the end of the burst. Since the period of the limit cycle tends to infinity at the homoclinic bifurcation, the interspike interval is large at both the beginning and end of the burst. However, as the slow periodic orbit penetrates further into region $J_{osc}$, away from the homoclinic bifurcation, the interspike interval decreases.

## 6.3  A Classification Scheme for Bursting Oscillations

The above examples are only two of a range of different mechanisms that give rise to bursting. A classification scheme for the different mechanisms was proposed by Rinzel

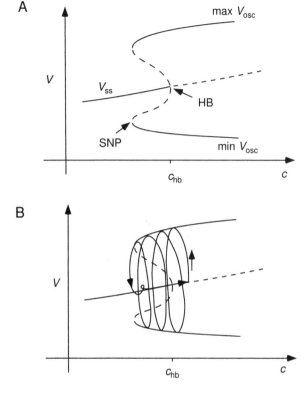

**Figure 6.7** A: Sketch of the bifurcation diagram for the fast subsystem of a type III burster. $V_{ss}$ denotes the curves of steady states. The steady state loses stability at $c_{hb}$ in a Hopf bifurcation (HB), and a branch of unstable periodic orbits is formed. This merges with a branch of stable periodic orbits in a saddle node of periodics bifurcation (SNP). B: Sketch of a typical bursting trajectory.

(1987) and extended by Bertram et al. (1995). Originally, bursting oscillations were grouped into three classes: type I, with bursts arising from hysteresis and bistability as in the $\beta$-cell model; type II, with bursts arising from an underlying slow oscillation, as in the parabolic bursting model; and type III, which arises from a subcritical Hopf bifurcation.

## 6.3.1 Type III Bursting

Suppose that the fast subsystem exhibits a subcritical Hopf bifurcation at some value of the slow variable (there need be only one slow variable for type III bursting). For convenience, we continue to label the slow and fast variables $c$ and $V$ respectively, and we denote the Hopf bifurcation point by $c_{hb}$ as shown in Fig. 6.7A.

Immediately below $c_{hb}$ the fast subsystem is bistable, with a stable fixed point and a stable periodic orbit. Suppose further that on the branch of fixed points to the left of $c_{hb}$, $c$ is slowly increasing, while on the branch of stable periodic orbits the dynamics are such that $c$ is slowly decreasing. The burst cycle, illustrated in Fig. 6.7B, is as follows: starting with $c < c_{hb}$, $c$ increases past the Hopf bifurcation, the fixed point loses stability, and $V$ switches to the branch of periodic orbits (the only stable solutions for $c > c_{hb}$) and begins to oscillate. On this upper branch $c$ decreases. Once $c$

has decreased far enough, $V$ "falls off" the branch of periodic orbits, and returns to the branch of fixed points, completing the cycle. The active period (i.e., the period where the fast oscillations occur) ends where the branch of stable periodic solutions meets the branch of unstable periodic solutions. Such a bifurcation point, where a stable and an unstable periodic orbit appear, is called a saddle node of periodics, or SNP. Although type III bursting relies on bistability and hysteresis as does type I bursting, it does not involve passage through a homoclinic bifurcation, or a Z-shaped curve of fixed points in the fast subsystem, leading to the different classification. One important qualitative difference between type I and type III bursting arises from the different bifurcations that end the active phase. In type I the active phase ends at a homoclinic bifurcation, leading to an increase in the period of the oscillations toward the end of a burst. In type III no such homoclinic bifurcation appears, and the active phase ends at an SNP bifurcation. Thus, in general, the periods of the oscillations in the active phase follow no particular pattern; the spike frequency is indeterminate.

### 6.3.2   Type Ib Bursting

The final type of bursting that we discuss (although there are more) is a subclass of type I. It is important because although the models for this subclass of bursting have a similar underlying bifurcation structure to those for type I, the bursts nevertheless can behave quite differently.

In type Ib bursting (Fig. 6.8), the stable limit cycle surrounds all three fixed points, and the burst cycle is similar to that of type I (which we call type Ia from now on). When $V$ is on the lower branch of the Z-shaped curve, $c$ decreases, as $V$ lies below the $c$ nullcline. As $c$ decreases, the solution crosses the saddle-node bifurcation at the lower knee of the Z-shaped curve and jumps to the branch of stable periodic orbits. Although $c$ does not increase monotonically during each oscillation, the average value of $V$ is

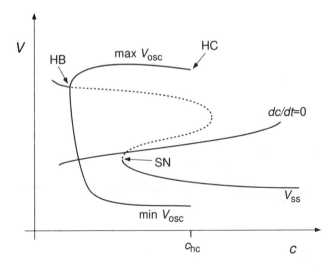

**Figure 6.8** Bifurcation diagram of the fast subsystem for a type Ib burster. HB denotes a Hopf bifurcation, HC denotes a homoclinic bifurcation, and SN denotes a saddle-node bifurcation. The position of the homoclinic bifurcation is denoted by $c_{hc}$. Type Ib is similar to type Ia, with the major difference that the stable periodic orbit surrounds all three steady states.

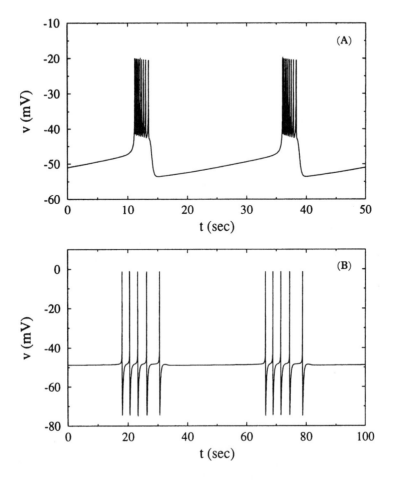

**Figure 6.9** Comparison of type Ia (A) and type Ib (B) bursting in the Chay–Cook model. Although this model is not discussed in detail in the text, these numerical solutions of Bertram et al. (1995) provide an excellent comparison of the two bursting types. (Bertram et al., 1995, Fig. 3.)

high enough to cause a net increase in $c$ over each cycle. Thus, the solution moves to the right until the branch of periodic orbits disappears at a homoclinic bifurcation, at which time the solution reverts to the lower branch of the Z-shaped curve, completing the burst cycle. In Fig. 6.9 we compare type Ia and Ib bursting patterns. The numerical simulations are from a model not discussed here (Chay and Cook, 1988), but the figure serves as an excellent comparison of the bursting types. In type Ia, the burst pattern is superimposed on a high-voltage baseline, forming a square-wave pattern as seen in pancreatic $\beta$-cells. In type Ib, the minimum of the fast oscillation lies below the lower branch of the Z-shaped curve, and thus the minimum of the fast oscillation lies below the quiescent phase.

### 6.3.3　Summary of Types I, II, and III

For convenience, we summarize the properties of the three types of bursting.

**Type I.** The active phase begins at a saddle-node bifurcation, i.e., at the knee of the Z-shaped curve, and ends at a homoclinic bifurcation. The fast subsystem is bistable, and only one slow variable is needed. The spike period tends to increase monotonically through the active phase. In types Ia and Ib, the minimum of the burst lies, respectively, above and below the quiescent phase.

**Type II.** The active phase begins and ends at a homoclinic bifurcation. In the parabolic bursting model, this homoclinic bifurcation occurs at a saddle-node bifurcation. The fast subsystem is monostable, and two slow variables are necessary for bursting. The spike period is parabolic.

**Type III.** The active phase begins at a Hopf bifurcation and ends at a saddle node of periodics. The fast subsystem is bistable, and only one slow variable is needed. The spike frequency is indeterminate.

These three types are illustrated in Fig. 6.10.

## 6.4　Bursting in Clusters

In the discussion so far, we have ignored the inconvenient fact that isolated pancreatic $\beta$-cells usually do not burst in a regular fashion. It is not until several thousand cells are grouped into an islet and electrically coupled by gap junctions that regular spiking is seen in any cell. An isolated $\beta$-cell behaves in a much more irregular fashion, with no discernible pattern of bursting. Indeed, blockage of gap junctions in an islet greatly reduces insulin secretion, and thus intercellular mechanisms that control bursting are of great physiological importance. Figure 6.11 shows how the behavior of an individual cell changes as a function of the number of other cells to which it is coupled.

### 6.4.1　Channel-Sharing

In 1983, Atwater et al. proposed a qualitative mechanism to account for the difference between the single-cell behavior and the behavior of the cell in a cluster. They proposed that an individual cell contains a small number of calcium-sensitive $K^+$ (K–Ca) channels each of which has a high conductance. At resting $V$ and $[Ca^{2+}]$, a K–Ca channel is open only infrequently, but the opening of a single channel passes enough current to cause a significant perturbation of the cell membrane potential. Thus, stochastic channel opening and closing causes the observed random fluctuations in $V$. However, when the cells are electrically coupled in a cluster, each K–Ca channel has a much smaller effect on the potential of each individual cell, as the channel current is spread over the network of cells. Each cell is integrating the effects of a large number of K–Ca channels, each of which has only a small influence. The tighter the electrical coupling between

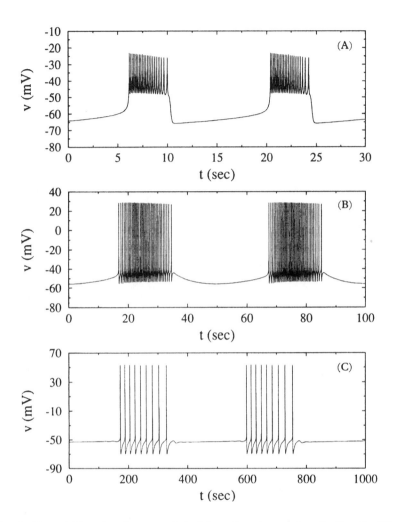

**Figure 6.10**    A: Type I bursting in a model of bursting in pancreatic $\beta$-cells (Sherman and Rinzel, 1992). B: Type II bursting in the parabolic bursting model (Rinzel and Lee, 1987). C: Type III bursting in a model of bursting in cardiac ganglion cells of the lobster (Av-Ron et al., 1993). This model is not discussed in the text, but is used here for purposes of comparison. (Bertram et al., 1995, Fig. 1.)

cells, the better each cell is able to integrate the effects of all the K–Ca channels in the cluster, and the more regular and deterministic is the overall behavior.

We can use this qualitative explanation as the basis for a quantitative model (Sherman et al., 1988; Chay and Kang, 1988). Initially, we assume infinitely tight coupling of the cells in the cluster, calling the cluster a "supercell," and show how the bursting becomes more regular as the size of the cluster increases.

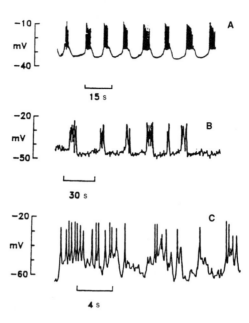

**Figure 6.11** Bursting in clusters of $\beta$-cells, compared to the electrical activity of an isolated $\beta$-cell (Sherman et al., 1988, Fig. 1). A: Recording from a $\beta$-cell in an intact cluster (Atwater and Rinzel, 1986), B: Recording from a cluster of $\beta$-cells, with a radius of 70 $\mu$m (Rorsman and Trube, 1986, Fig. 1a). C: Recording from an isolated $\beta$-cell (Rorsman and Trube, 1986, Fig. 2c).

The equations of the supercell model are similar to those of the $\beta$-cell model. Recall that in the $\beta$-cell model the conductance of the K–Ca channel, $g_{K,Ca}$, is given by

$$g_{K,Ca} = \bar{g}_{K,Ca} \frac{c}{K_d + c}, \tag{6.14}$$

where $c$, as usual, denotes [Ca$^{2+}$]. This can be derived from a simple channel model in which the channel has one closed state and one open state, switching from closed to open upon the binding of a single Ca$^{2+}$ ion. Thus

$$C + Ca^{2+} \underset{k_-}{\overset{k_+}{\rightleftarrows}} O, \tag{6.15}$$

where C is a closed channel and O is an open one. If the rate constants $k_+$ and $k_-$ are both large compared to the other kinetic parameters of the model, then

$$[O] = \frac{k_+[Ca^{2+}]}{k_-}[C] \tag{6.16}$$

$$= \frac{[Ca^{2+}]}{K_d}(1 - [O]), \tag{6.17}$$

where $K_d = k_-/k_+$. Hence $[O] = [Ca^{2+}]/(K_d + [Ca^{2+}])$, from which (6.14) follows.

In the supercell model, (6.15) is interpreted as a Markov rather than a deterministic process, with $k_+$ denoting the probability per unit time that the closed channel switches to the open state, and similarly for $k_-$. Thus, the mean open and closed times are, respectively, $1/k_-$ and $1/k_+$. If we let $\langle N_o \rangle$ and $\langle N_c \rangle$ denote the mean number of open channels or closed channels respectively, then at equilibrium we have

$$\frac{\langle N_o \rangle}{\langle N_c \rangle} = \frac{k_+}{k_-}. \tag{6.18}$$

To incorporate the $Ca^{2+}$ dependence of the channel we make

$$\frac{k_+}{k_-} = \frac{[Ca^{2+}]}{K_d}, \tag{6.19}$$

which gives the steady-state mean proportion of open channels as

$$\langle p \rangle = \frac{\langle N_o \rangle}{\langle N_o \rangle + \langle N_c \rangle} = \frac{[Ca^{2+}]}{K_d + [Ca^{2+}]} = \frac{c}{K_d + c}. \tag{6.20}$$

The associated stochastic model is similar to (6.6)–(6.8), with the major difference being that the $Ca^{2+}$-sensitive $K^+$ current is governed by the above stochastic process. In other words, $p$ is a random variable denoting the proportion of open channels in a single cell, and is calculated (as a function of time) by numerical simulation of the Markov process described by (6.15). Thus,

$$C_m \frac{dV}{dt} = -I_{Ca} - I_K - \bar{g}_{K,Ca} p (V - V_K), \tag{6.21}$$

$$\tau_n(V) \frac{dn}{dt} = \lambda(n_\infty(V) - n), \tag{6.22}$$

$$\frac{dc}{dt} = f(-\alpha I_{Ca} - k_c c), \tag{6.23}$$

where $I_{Ca} = \bar{g}_{Ca} m_\infty(V) h(V)(V - V_{Ca})$ and $I_K = \bar{g}_K n (V - V_K)$. The functions appearing in the supercell model are

$$m_\infty(V) = \frac{1}{1 + \exp\left[\frac{4-V}{14}\right]}, \tag{6.24}$$

$$n_\infty(V) = \frac{1}{1 + \exp\left[\frac{-15-V}{5.6}\right]}, \tag{6.25}$$

$$\tau_n(V) = \frac{\bar{\tau}_n}{\exp\left[\frac{V+75}{65}\right] + \exp\left[\frac{-V-75}{20}\right]}, \tag{6.26}$$

$$h(V) = \frac{1}{1 + \exp\left[\frac{V+10}{10}\right]}. \tag{6.27}$$

The other parameters of the model are summarized in Table 6.2. Note that although the form of the model is similar to that of the $\beta$-cell model, the details have been changed to agree with more recent experimental data. In particular, $m_\infty$ appears only to the first power instead of the third, while $n$ also appears to the first power. Further, the $I$–$V$

**Table 6.2** Parameters of the supercell model for electrical bursting in clusters of pancreatic $\beta$-cells.

| | | | | |
|---|---|---|---|---|
| $C_m$ | = | 5.3 pF | $\bar{g}_{K,Ca}$ | = | 30 nS |
| $\bar{g}_K$ | = | 2.5 nS | $\bar{g}_{Ca}$ | = | 1.4 nS |
| $V_{Ca}$ | = | 110 mV | $V_K$ | = | −75 mV |
| $k_c$ | = | 0.03 ms$^{-1}$ | $f$ | = | 0.001 |
| $\alpha$ | = | 4.5 $\mu$M/C | $K_d$ | = | 100 $\mu$M |
| $\bar{\tau}_n$ | = | 60 ms | | | |

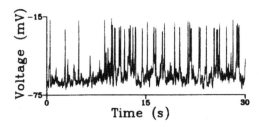

**Figure 6.12** Behavior of an isolated cell in the supercell stochastic model. Because of the stochastic nature of each high-conductance K–Ca channel, and because there are few channels in a single cell, no organized bursting appears. (Sherman et al., 1988, Fig. 6).

curve of the open $Ca^{2+}$ channel is assumed to be of the form $h(V)(V - V_{Ca})$, and $h(V)$ is chosen to fit the experimentally observed $I$–$V$ curve. This has an effect similar to that of the function $h_\infty(V)$ that was used in the $\beta$-cell model.

One of the most important features of the stochastic model is that the conductance of a single K–Ca channel is an order of magnitude greater than the conductances of the other two channels (Table 6.2). However, each cell contains only a small number of K–Ca channels. Thus, the opening of a single K–Ca channel in an isolated cell has a disproportionately large effect on the membrane potential of the cell; the stochastic nature of each K–Ca channel then causes seemingly random fluctuations in membrane potential (Fig. 6.12).

However, when identical cells are coupled by gap junctions with zero resistance, different behavior emerges. First, since the gap junctions are assumed to have zero resistance, and thus the entire group of cells has the same membrane potential, it is not necessary to treat each cell explicitly. Second, the membrane capacitance and the ionic currents depend on the surface area of the cluster and are therefore proportional to the number of cells in the cluster, $N_{cell}$. Finally, the total number of K–Ca channels is proportional to the total number of cells in the cluster, but the effect of each individual channel on the membrane potential of the cluster is proportional to $1/N_{cell}$. It follows that the cluster of cells behaves in the same way as a very large single cell with many K–Ca channels, each with a smaller conductance. We expect the cluster of cells to behave in the same manner as a deterministic single-cell model in the limit as $N_{cell} \longrightarrow \infty$.

To see that this is what happens, we let $\hat{g}$ denote the conductance of a single K–Ca channel and let $N_o^i$ denote the number of open K–Ca channels in the $i$th cell. Then

$$N_{cell}C_m\frac{dV}{dt} = -N_{cell}(I_{Ca} + I_K) - \hat{g}\sum_{i=1}^{N_{cell}}N_o^i(V - V_K),\tag{6.28}$$

and so

$$C_m\frac{dV}{dt} = -(I_{Ca} + I_K) - \hat{g}\bar{N}\frac{1}{N_{cell}\bar{N}}\sum_{i=1}^{N_{cell}}N_o^i(V - V_K)\tag{6.29}$$

$$= -(I_{Ca} + I_K) - \bar{g}_{K,Ca}p(V - V_K),\tag{6.30}$$

where $\bar{N}$ is the number of K–Ca channels per cell, or the channel density, and where, as before, $\bar{g}_{K,Ca} = \hat{g}\bar{N}$ is the total K–Ca conductance per cell. Note that in the supercell model $p = \frac{1}{N_{cell}\bar{N}}\sum_{i=1}^{N_{cell}}N_o^i$ must be interpreted as the fraction of open channels in the cluster, rather than the fraction of open channels in a single cell. The mean of $p$ is the same in both these cases, but as $N_{cell}$ increases, the standard deviation of $p$ decreases, leading to increasingly regular behavior. As before, $p$ must be obtained by direct simulation of the Markov process. Simulations for different numbers of cells are shown in Fig. 6.13. Clearly, as the size of the cluster increases, bursting becomes more regular.

One obvious simplification in the supercell model is the assumption that the gap junctions have zero resistance and thus that every cell in the cluster has the same membrane potential. We can relax this assumption by modeling the cluster as individual cells coupled by gap junctions with finite conductance (Sherman and Rinzel, 1991). An individual cell, cell $i$ say, satisfies a voltage equation of the form

$$C_m\frac{dV_i}{dt} = -I_{Ca}(V_i) - I_K(V_i, n_i) - \bar{g}_{K,Ca}p_i(V_i - V_K) - g_c\sum_j d_{ij}(V_i - V_j),\tag{6.31}$$

where $g_c$ is the gap junction conductance and where $d_{ij}$ are coupling coefficients, with value one if cells $i$ and $j$ are coupled, and zero otherwise. As $g_c \rightarrow \infty$, the sum $\sum_j d_{ij}(V_i - V_j)$ must approach zero for every cell in the cluster. If all the cells are connected by some connecting path (so that there are no isolated cells or subclusters), then every cell must have the same voltage (see Exercise 6). Thus, in the limit of infinite conductance,

$$V_i \longrightarrow \bar{V} = \frac{1}{N_{cell}}\sum_{j=1}^{N_{cell}}V_j.\tag{6.32}$$

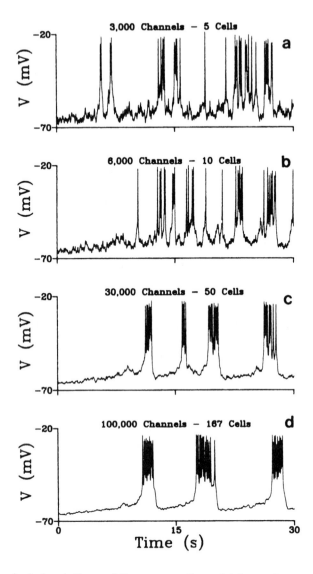

**Figure 6.13** Numerical simulations of the supercell model for a cluster of cells ranging in size from 5 to 167 cells. As the size of the cluster increases, more organized bursting appears. (Sherman et al., 1988, Fig. 8.)

For large but finite coupling, $V_i = \bar{V} + O(\frac{1}{g_c})$. If we now sum (6.31) over all the cells in the cluster and divide by $N_{\text{cell}}$, we find

$$C_m \frac{d\bar{V}}{dt} = -I_{\text{Ca}}(\bar{V}) - I_{\text{K}}(\bar{V}, n) - \bar{g}_{\text{K,Ca}} \frac{1}{N_{\text{cell}}} \sum_{j=1}^{N_{\text{cell}}} p_j(\bar{V} - V_{\text{K}}) + O\left(\frac{1}{N_{\text{cell}}}\right), \quad (6.33)$$

$$= -I_{\text{Ca}}(\bar{V}) - I_{\text{K}}(\bar{V}, n) - \bar{g}_{\text{K,Ca}}\bar{p}(\bar{V} - V_{\text{K}}) + O\left(\frac{1}{N_{\text{cell}}}\right), \quad (6.34)$$

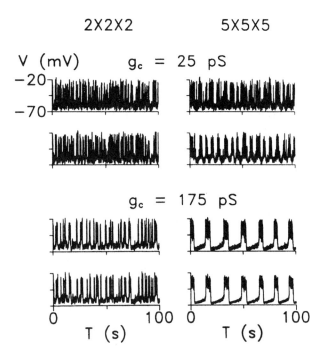

**Figure 6.14** Numerical simulations of the multicell model, in which the cells in the cluster are coupled by gap junctions with finite conductance. Results are shown for two cells (upper and lower traces in each pair) from two different cluster sizes ($2 \times 2 \times 2$ cells and $5 \times 5 \times 5$ cells) and two different junctional conductances. (Sherman and Rinzel, 1991, Fig. 3.)

where $\bar{p} = \frac{\sum_i p_i}{N_{cell}} = \frac{\sum_i p_i \bar{N}}{N_{cell} \bar{N}}$ is the proportion of open K–Ca channels in the cluster. Hence, the model with finite gap-junctional conductance (the *multicell* model) turns into the supercell model as the gap-junctional conductance and the number of cells in the cluster approaches infinity.

As expected, synchronized bursting appears as the number of cells in the cluster increases and the coupling strength increases. Both strong enough coupling and a large enough cluster size are required to achieve regular bursting. This is illustrated in Fig. 6.14, where we show numerical simulations for two different cluster sizes and two different coupling strengths. However, what is not expected (and is therefore particularly interesting) is that there is a coupling strength at which the length of the burst period is maximized. The reasons for this have been analyzed in depth in a simpler system consisting of two coupled cells (Sherman, 1994).

## 6.5 Qualitative Bursting Models

To a large extent, all of the models discussed above are based on biophysical mechanisms, with the parameters and functions in the models derived from experimental data. An alternative approach is to construct a polynomial model that retains the important qualitative features but is simpler to analyze and understand (Hindmarsh and Rose, 1984; Pernarowski, 1994). Such a model has the same relationship to the above models as the FitzHugh–Nagumo model does to the Hodgkin–Huxley model: it is phenomenological in nature and based on the fact that the underlying behavior of the biophysical models can be distilled into a simpler model containing only polynomials.

## 6.5.1 A Polynomial Model

We begin by presenting a modified version of the FitzHugh–Nagumo model that has the nice property of generating oscillations with a long interspike interval (Hindmarsh and Rose, 1982, 1984). We let $v$ denote the excitatory variable and $w$ the recovery variable (as in Chapter 4). Then, the model equations are

$$\frac{dv}{dt} = \alpha(\beta w - f(v) + I), \tag{6.35}$$

$$\frac{dw}{dt} = \gamma(g(v) - \delta w), \tag{6.36}$$

where $I$ is the applied current and $\alpha, \beta, \gamma,$ and $\delta$ are constants. As in the FitzHugh–Nagumo model $f(v)$ is cubic, but unlike the FitzHugh–Nagumo model $g(v)$ is not a linear function. In fact, as can be seen from Fig. 6.15, the $w$ nullcline curves around to lie close to the $v$ nullcline to the left of the oscillatory critical point. (Of course, this occurs only for a range of values of the applied current $I$ for which oscillations occur.) As a result, between the peaks of the oscillation, the limit cycle trajectory lies close to both nullclines, and thus both derivatives $\dot{v}$ and $\dot{w}$ are small over that portion of the cycle. It follows that the intervals between the spikes are large.

With only a slight change, this modified FitzHugh–Nagumo model can be used as a model for bursting. Following the discussions in this chapter, it should come as no surprise to learn that bursting can arise in this model when bistability is introduced. This can be done by deforming the $\dot{w} = 0$ nullcline so that it intersects the $\dot{v} = 0$ nullcline in three places rather than only one. Since the nullclines lie close to one

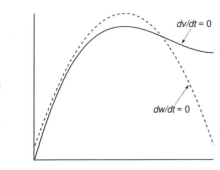

**Figure 6.15** Sketch of typical nullclines in the modified FitzHugh–Nagumo model.

another in the original model, only a slight deformation is required to create two new critical points. With this change, the new model equations are

$$\frac{dv}{dt} = \alpha(\beta w - f(v) + I), \tag{6.37}$$

$$\frac{dw}{dt} = \gamma(g(v) + h(v) - \delta w), \tag{6.38}$$

where $h(v)$ is chosen such that the nullclines now intersect in three places. For convenience, we scale and nondimensionalize the model by introducing the variables $T = \gamma\delta t$, $x = v$, and $y = \alpha\beta w/(\gamma\delta)$, in which variables the model becomes

$$\frac{dx}{dT} = y - \tilde{f}(x), \tag{6.39}$$

$$\frac{dy}{dT} = \tilde{g}(x) - y, \tag{6.40}$$

where $\tilde{f}(x) = \alpha f(x)/(\gamma\delta)$ and $\tilde{g}(x) = \alpha\beta[g(x)+h(x)]/(\gamma\delta^2)$. Note that the form of the model is the same as that of the FitzHugh–Nagumo model, although the functions appearing in the model are different. With appropriate choices for $\tilde{f}$ and $\tilde{g}$ the model exhibits bistability; we use the particular functions

$$\frac{dx}{dT} = y - x^3 + 3x^2 + I, \tag{6.41}$$

$$\frac{dy}{dT} = 1 - 5x^2 - y, \tag{6.42}$$

the phase-plane of which is shown in Fig. 6.16 for $I = 0$.

There are three critical points: a stable node to the left at $x = -\frac{1}{2}(1 + \sqrt{5})$, (the resting state), a saddle point in the middle at $x = -1$, and an unstable node to the right at $x = \frac{1}{2}(-1 + \sqrt{5})$, which is surrounded by a stable limit cycle. As in the models of bursting discussed above, the stable manifold of the saddle point acts as a threshold; if the perturbation from the resting state is large enough that the stable manifold is crossed, the trajectory approaches the stable limit cycle. Smaller perturbations die

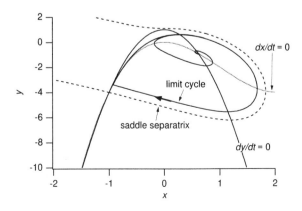

**Figure 6.16** Phase-plane of the polynomial bursting model.

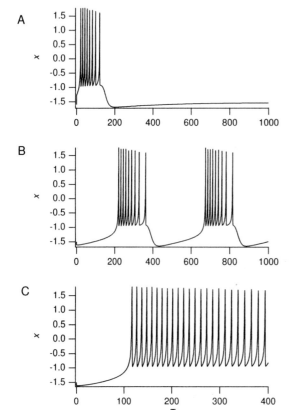

**Figure 6.17** Bursting in the polynomial model, calculated numerically from (6.43)–(6.45) for three values of the applied current. A: $I = 0.4$; B: $I = 2$; C: $I = 4$.

away to the resting state. This is called *triggered firing*. This bistable phase-plane is essentially the same as the phase-plane shown in Fig. 6.3A, and (6.41)–(6.42) are a simple realization of the qualitative theory developed by Rinzel and others.

To generate bursting in addition to bistability, it is also necessary to have a slow variable so that the voltage can be moved in and out of the bistable regime. This is accomplished in this model by introducing a third variable that modulates the applied current $I$ on a slower time scale. Thus,

$$\frac{dx}{dT} = y - x^3 + 3x^2 + I - z, \tag{6.43}$$

$$\frac{dy}{dT} = 1 - 5x^2 - y, \tag{6.44}$$

$$\frac{dz}{dT} = r[s(x - x_1) - z], \tag{6.45}$$

where $x_1 = -\frac{1}{2}(1 + \sqrt{5})$ is the $x$-coordinate of the resting state in the two-variable model (6.41)–(6.42), and where $I$, as before, is the applied current. When $r = 0.001$ and $s = 4$,

(6.43)–(6.45) exhibit type I bursting (Fig. 6.17), arising via the same mechanism as in the Rinzel–Lee simplification of the Chay–Keizer model.

## 6.6  EXERCISES

1. (a) Numerically simulate the system of differential equations

$$\frac{dv}{dt} = f(v) - w - gs(v - v_\theta),\tag{6.46}$$

$$5\frac{dw}{dt} = w_\infty(v) - w,\tag{6.47}$$

$$\frac{ds}{dt} = f_s(s) + \alpha_s(x - 0.3),\tag{6.48}$$

$$\frac{dx}{dt} = \beta_x\left((1 - x)H(v) - x\right),\tag{6.49}$$

where $f(v) = 1.35v(1 - \frac{1}{3}v^2), f_s(s) = -0.2(s - 0.05)(s - 0.135)(s - 0.21), w_\infty(v) = \tanh(5v)$, and $H(v) = \frac{3}{2}(1 + \tanh(5x - 2.5))$, and $v_\theta = -2, \alpha_s = 0.002, \beta_x = 0.00025, g = 0.73$.

   (b) Give a fast–slow analysis of this burster. Hint: The equations for $v, w$ comprise the fast subsystem, while those for $s, x$ comprise the slow subsystem.

   (c) Describe the bursting mechanism in this model. For what kind of burster might this be a reasonable model?

2. Compute some numerical solutions of the Rinzel–Lee simplification of the Chay–Keizer $\beta$-cell model. How does the value of $k_c$ affect the burst length? Can you find parameter values such that the model behaves like a type Ib burster?

3. Simulate the Plant model for parabolic bursting (the differential equations (6.9)–(6.13)) using the parameter values displayed in Table 6.3. The voltage dependence of the variables $\alpha_w$ and $\beta_w$ with $w = m, n,$ or $h$ is of the form

$$\frac{C_1\exp(\frac{V - V_0}{C_2}) + C_3(V - V_0)}{1 + C_4\exp(\frac{V - V_0}{C_5})},\tag{6.50}$$

in units of $ms^{-1}$, and the asymptotic values $w_\infty(V)$ and time constants $\tau_w(V)$ are of the form

$$w_\infty(V) = \frac{\alpha_w(\tilde{V})}{\alpha_w(\tilde{V}) + \beta_w(\tilde{V})}\tag{6.51}$$

$$\tau_w(V) = \frac{1}{\lambda\left(\alpha_w(\tilde{V}) + \beta_w(\tilde{V})\right)}\tag{6.52}$$

for $w = m, n,$ or $h$ (although $\tau_m(V)$ is not used), with $\tilde{V} = c_1 V + c_2, c_1 = 127/105, c_2 = 8265/105$. The constants $C_1, \ldots, C_5$ and $V_0$ are displayed in Table 6.4. Finally,

$$x_\infty(V) = \frac{1}{\exp\{-0.15(V + 50)\} + 1},\tag{6.53}$$

and $\tau_x = 235$ ms.

4. Determine the value of $I$ in the Hindmarsh–Rose fast subsystem (6.41), (6.42) for which the trajectory from the unstable node is also the saddle point separatrix.

5. (a) An equation of the form $\frac{d^2x}{dT^2} + F(x)\frac{dx}{dT} + x = 0$ is called an equation in the Liénard form (Minorsky, 1962; Stoker, 1955) and was important in the classical development of the

**Table 6.3**   Parameters of the Plant model for parabolic bursting.

| | | | | | |
|---|---|---|---|---|---|
| $C_m$ | = | $1\ \mu\text{F/cm}^2$ | $\bar{g}_{K,Ca}$ | = | $0.03\ \text{mS/cm}^2$ |
| $\bar{g}_{Ca}$ | = | $0.004\ \text{mS/cm}^2$ | $V_{Ca}$ | = | $140\ \text{mV}$ |
| $\bar{g}_{Na}$ | = | $4.0\ \text{mS/cm}^2$ | $V_{Na}$ | = | $30\ \text{mV}$ |
| $\bar{g}_K$ | = | $0.3\ \text{mS/cm}^2$ | $V_K$ | = | $-75\ \text{mV}$ |
| $\bar{g}_L$ | = | $0.003\ \text{mS/cm}^2$ | $V_L$ | = | $-40\ \text{mV}$ |
| $f$ | = | $0.0003\ \text{ms}^{-1}$ | $k_1$ | = | $0.0085\ \text{mV}^{-1}$ |
| $\lambda$ | = | $1/12.5$ | | | |

**Table 6.4**   Defining values for rate constants $\alpha$ and $\beta$ for the Plant parabolic bursting model.

| | $C_1$ | $C_2$ | $C_3$ | $C_4$ | $C_5$ | $V_0$ |
|---|---|---|---|---|---|---|
| $\alpha_m$ | 0 | — | 0.1 | −1 | −10 | 50 |
| $\beta_m$ | 4 | −18 | 0 | 0 | — | 25 |
| $\alpha_h$ | 0.07 | −20 | 0 | 0 | — | 25 |
| $\beta_h$ | 1 | 10 | 0 | 1 | 10 | 55 |
| $\alpha_n$ | 0 | — | 0.01 | −1 | −10 | 55 |
| $\beta_n$ | 0.125 | −80 | 0 | 0 | — | 45 |

theory of nonlinear oscillators. Show that the polynomial bursting model (6.43)–(6.45) can be written in the *generalized* Liénard form

$$\frac{d^2x}{dT^2} + F(x)\frac{dx}{dT} + G(x,z) = -\epsilon H(x,z), \tag{6.54}$$

$$\frac{dz}{dT} = \epsilon H(x,z), \tag{6.55}$$

where $\epsilon$ is a small parameter, and where $F$, $G$, and $H$ are the polynomial functions

$$F(u) = a[(u-\hat{u})^2 - \eta^2], \tag{6.56}$$
$$G(u,c) = c + u^3 - 3(u+1), \tag{6.57}$$
$$H(u,c) = \beta(u-\bar{u}) - c. \tag{6.58}$$

(b)   Construct the fast subsystem bifurcation diagram for this polynomial model for the following three different cases with $a = 0.25$, $\beta = 4$, $\hat{u} = -0.954$, and $\epsilon = 0.0025$. You may want to use a bifurcation tracking program such as AUTO (Doedel, 1986).

  i.   $\eta = 0.7$, $\hat{u} = 1.6$. Show that in this case the model exhibits square-wave bursting, of type Ia.

  ii.   $\eta = 0.7$, $\hat{u} = 2.1$. Show that the model exhibits tapered bursting, resulting from passage through a supercritical Hopf bifurcation.

  iii.   $\eta = 1.2$, $\hat{u} = 1.0$. Show that the model exhibits type Ib bursting.

Solve the model equations numerically to confirm the predictions from the bifurcation diagrams. A detailed analysis of this model has been performed by Pernarowski (1994) and de Vries (1995), with a perturbation analysis given by (Pernarowski et al., 1991, 1992).

6.  Prove that if $D$ is an irreducible matrix with nonnegative entries $d_{ij}$, then the only nontrivial solution of the system of equations $\sum_{j=1}^{n} d_{ij}(V_i - V_j) = 0, i = 1, \ldots, N$, is the constant vector. Remark: An irreducible nonnegative matrix is one for which some power of the matrix has no zero elements. To understand irreducibility, think of the elements of the matrix $D$ as providing connections between nodes of a graph, and $d_{ij}$ is positive if there is a path from node $i$ to node $j$, but zero if not. Such a matrix is irreducible if between any two points there is a connecting path, with possibly multiple intermediate points. The smallest power of the matrix that is strictly positive is the smallest number of connections that is certain to connect any node with any other node.

    Hint: Represent the system of equations as $Av = 0$ and show that some shift $A + \lambda I$ of the matrix $A$ is irreducible and has only nonnegative entries. Invoke the Perron–Frobenius theorem (a nonnegative irreducible matrix has a unique, positive, simple eigenvector with eigenvalue larger in magnitude than all other eigenvalues) to show that the null space of $A$ is one-dimensional, spanned by the constant vector.

# CHAPTER 7

# Intercellular Communication

For multicellular organisms to form and operate, cellular behavior must be vastly more complex than what is seen on the single-cell level. Cells must not only regulate their own growth and behavior, they must also communicate and interact with their neighbors to ensure the correct behavior of the entire organism. Intercellular communication occurs in a variety of ways, ranging from hormonal communication on the level of the entire body to localized interactions between individual cells. Our discussion in this chapter is limited to cellular communication processes that occur between cells or over a region of a small number of cells. Other forms of communication and control, such as hormone feedback systems, will be studied in other chapters.

There are two primary ways that cells communicate with neighbors. Many cells (muscle and cardiac cells for example) are connected to their immediate neighbors by gap junctions in the cell membrane that form a relatively nonselective, low-resistance, pore through which electrical current or chemical species can flow. Hence, a gap junction is also called an *electrical synapse*. The second means of communication is through a *chemical synapse*, in which the message is mediated by the release of a chemical from one cell and detected by receptors on its neighbor. Electrically active cells such as neurons typically communicate via chemical synapses, which are thus a crucial feature of the nervous system. Because chemical synapses form the basis for neuronal communication they have been studied in considerably more detail than electrical synapses.

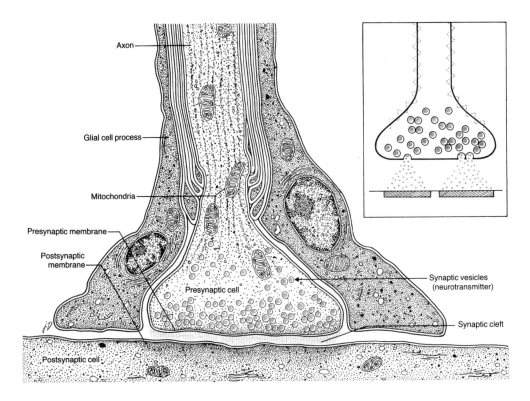

**Figure 7.1** Schematic diagram of a chemical synapse. (Davis et al., 1985, Fig. 8-11, p. 135.)

# 7.1 Chemical Synapses

At a chemical synapse (Fig. 7.1) the nerve axon and the postsynaptic cell are in close apposition, being separated by the *synaptic cleft*, which is about 500 angstroms wide. When an action potential reaches the nerve terminal, it opens voltage-gated $Ca^{2+}$ channels, leading to an influx of $Ca^{2+}$ into the nerve terminal. Increased $[Ca^{2+}]$ causes the release of a chemical neurotransmitter, which diffuses across the synaptic cleft, binds to receptors on the postsynaptic cell, and initiates changes in its membrane potential. The neurotransmitter is then removed from the synaptic cleft by diffusion and hydrolysis.

There are over 40 different types of synaptic transmitters, with differing effects on the postsynaptic membrane. For example, acetylcholine (ACh) binds to ACh receptors, which in skeletal muscle act as cation channels. Thus, when they open, the flow of the cation causes a change in the membrane potential, either depolarizing or hyperpolarizing the membrane. If the channel is a sodium channel, then the flow is inward and depolarizing, whereas if the channel is a potassium channel, then the flow is outward and hyperpolarizing. Other receptors, such as those for gamma-aminobutyric acid (GABA), open anion channels (mainly chloride), thus hyperpolarizing the post-synaptic membrane, rendering it less excitable. Synapses are classified as excitatory or inhibitory according to whether they depolarize or hyperpolarize the postsynaptic membrane. Other important neurotransmitters include epinephrine (adrenaline), norepinephrine, dopamine, glycine, glutamate, and serotonin. Of these, dopamine, glycine, and GABA are usually inhibitory, while glutamate and ACh are usually, but not always, excitatory.

The loss of specific neurotransmitter function corresponds to certain diseases. For example, *Huntington's disease*, a hereditary disease characterized by flicking movements at individual joints progressing to severe distortional movements of the entire body, is associated with the loss of certain GABA-secreting neurons in the brain. The resulting loss of inhibition is believed to allow spontaneous outbursts of neural activity leading to distorted movements.

Similarly, Parkinson's disease results from widespread destruction of dopamine-secreting neurons in the basal ganglia. The disease is associated with rigidity of much of the musculature of the body, involuntary tremor of involved areas, and a serious difficulty in initiating movement. Although the causes of these abnormal motor effects are uncertain, the loss of dopamine inhibition could lead to overexcitation of many muscles, hence rigidity, or to lack of inhibitory control of feedback circuits with high feedback gain, leading to oscillations, i.e., muscular tremor.

## 7.1.1 Quantal Nature of Synaptic Transmission

Chemical synapses are typically small and inaccessible, crowded together in very large numbers in the brain. However, neurons also make synapses with skeletal muscle cells, and these are usually much easier to isolate and study. For this reason, a great deal of the early experimental and theoretical work on synaptic transmission was performed on the neuromuscular junction, where the axon of a motorneuron forms a chemical synapse with a skeletal muscle fiber. The response of the muscle cell to a neuronal stimulus is called an *end-plate potential*, or epp.

In 1952 Fatt and Katz discovered that when the concentration of $Ca^{2+}$ in the synaptic cleft was very low, an action potential stimulated only a small end-plate potential (Fig. 7.2). Further, these miniature end-plate potentials appeared to consist of multiples of an underlying "minimum" epp of the same amplitude as an epp arising spontaneously, i.e., due to random activity other than an action potential. Their findings suggested that an epp is made up of a large number of identical building blocks each of which is of small amplitude.

It is now known that quantal synaptic transmission results from the packaging of ACh into discrete vesicles. Each nerve terminal contains a large number of synaptic vesicles that contain ACh. Upon stimulation, these vesicles fuse with the cell membrane, releasing their contents into the synaptic cleft. Even in the absence of stimulation, background random activity can cause vesicles to fuse with the cell membrane and release their contents. The epp seen in spontaneous activity results from the release of

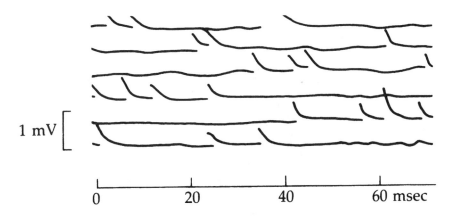

1 mV

| | | | |
|---|---|---|---|
| 0 | 20 | 40 | 60 msec |

**Figure 7.2** Miniature end-plate potentials (epps) in the frog neuromuscular junction. Each epp has an amplitude of around 1 mV and, as described in the text, results from the independent release of a single quantum of ACh. (Kuffler et al., 1984, p. 251, reproducing figure of Fatt and Katz, 1952.)

the contents of a single vesicle, while the miniature epps result from the fusion of a small integer number of vesicles, and thus appear in multiples of the spontaneous epp.

Based on their observations in frog muscle, del Castillo and Katz (1954) proposed a probabilistic model of ACh release. Their model was later applied to mammalian neuromuscular junctions by Boyd and Martin (1956). The model is based on the assumption that the synaptic terminal of the neuron consists of a large number, say $n$, of releasing units, each of which releases a fixed amount of ACh with probability $p$. If each releasing site operates independently, then the number of quanta of ACh that is released by an action potential is binomially distributed. The probability that $k$ releasing sites "fire" (i.e., release a quantum of ACh) is the probability that $k$ sites fire and the remaining sites do not, and so is given by $p^k(1-p)^{n-k}$. Since $k$ sites can be chosen from $n$ total sites in $n!/[(k!(n-k)!)]$ ways, it follows that

$$\text{Probability } k \text{ sites fire} = P(k) = \frac{n!}{k!(n-k)!}p^k(1-p)^{n-k}. \tag{7.1}$$

Under normal conditions, $p$ is large (and furthermore, the assumption of independent release sites is probably inaccurate). However, under conditions of low external $Ca^{2+}$ and high $Mg^{2+}$, $p$ is small. This is because $Ca^{2+}$ entry into the synapse is required for the release of a quantum of ACh. If only a small amount of $Ca^{2+}$ is able to enter (because the external $Ca^{2+}$ concentration is low), the probability of transmitter release is small. If $n$ is correspondingly large, while $np = m$ remains fixed, the binomial distribution can be approximated by a Poisson distribution. That is (Exercise 1a),

$$\lim_{n\to\infty} P(k) = \lim_{n\to\infty} \left[ \frac{n!}{k!(n-k)!} \left(\frac{m}{n}\right)^k \left(1-\frac{m}{n}\right)^{n-k} \right]$$

$$= \frac{m^k}{k!} \lim_{n\to\infty} \left[ \frac{n!}{n^k(n-k)!} \left(1 - \frac{m}{n}\right)^{n-k} \right]$$

$$= \frac{m^k}{k!} \lim_{n\to\infty} \left(1 - \frac{m}{n}\right)^n$$

$$= \frac{e^{-m}m^k}{k!}, \tag{7.2}$$

which is the Poisson distribution with mean, or expected value, $m$.

There are two ways to estimate $m$. First, notice that $P(0) = e^{-m}$, so that

$$e^{-m} = \frac{\text{number of action potentials with no epps}}{\text{total number of action potentials}}. \tag{7.3}$$

Second, according to the assumptions of the model, a spontaneous epp results from the release of a single quantum of ACh, and a miniature epp is a linear sum of spontaneous epps. Thus, $m$ can be calculated by dividing the mean amplitude of a miniature epp response by the mean amplitude of a spontaneous epp, giving the mean number of quanta in a miniature epp, which should be $m$. As del Castillo and Katz showed, these two estimates of $m$ agree very well, which confirms the model hypotheses.

The spontaneous epps are not of constant amplitude, because the amounts of ACh released from each vesicle are not identical. In the inset to Fig. 7.4 is shown the amplitude distribution of spontaneous epps. To a good approximation, the amplitudes of single-unit release, denoted by $A_1(x)$, are normally distributed (i.e., a Gaussian distribution), with mean $\mu$ and variance $\sigma^2$. From this it is possible to calculate the amplitude distribution of the miniature epps, as follows. We know that if $k$ vesicles are released, the amplitude distribution, denoted by $A_k(x)$, will be normally distributed with mean $k\mu$ and variance $k\sigma^2$, being the sum of $k$ independent, normal distributions each of mean $\mu$ and variance $\sigma^2$ (Fig. 7.3). Summing the distributions for $k = 1, 2, 3, \ldots$, and noting that the probability of $A_k(x)$ is $P(k)$, gives the amplitude distribution

$$A(x) = \sum_{k=1}^{\infty} P(k)A_k(x) \tag{7.4}$$

$$= \frac{1}{\sqrt{2\pi\sigma^2}} \sum_{k=1}^{\infty} \frac{e^{-m}m^k}{k!\sqrt{k}} \exp\left[\frac{-(x-k\mu)^2}{2k\sigma^2}\right], \tag{7.5}$$

which is graphed in Fig. 7.4. There are clear peaks corresponding to 1, 2, or 3 released quanta, but these peaks are smeared out and flattened by the normal distribution of amplitudes. There is excellent agreement between the theoretical prediction and the experimental observations, lending further support to the quantal model of synaptic transmission (del Castillo and Katz, 1954; Boyd and Martin, 1956).

## 7.1.2 Presynaptic Voltage-Gated Calcium Channels

Chemical synaptic transmission begins when an action potential reaches the nerve terminal and opens voltage-gated $Ca^{2+}$ channels, leading to an influx of $Ca^{2+}$ and con-

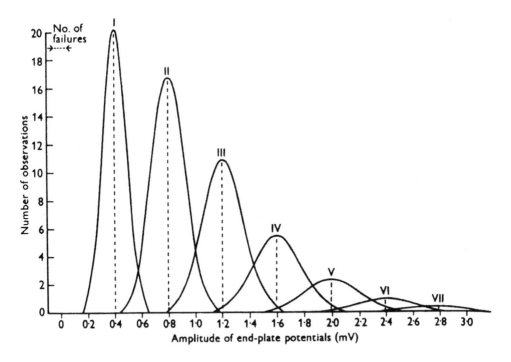

**Figure 7.3** Theoretical distributions for epps consisting of integer multiples of the spontaneous epp, which is the basic building block, or quantum, of the epp. Summation of these curves for all integral numbers of quanta gives the theoretical prediction for the overall amplitude distribution, (7.5), which is plotted in the next figure. (Boyd and Martin, 1956, Fig. 9.)

sequent neurotransmitter release. Based on voltage clamp data from the squid giant synapse, Llinás et al. (1976) constructed a model of the $Ca^{2+}$ current and its relation to synaptic transmission.

When the presynaptic voltage is stepped up and clamped at a constant level, the presynaptic $Ca^{2+}$ current $I_{Ca}$ increases in a sigmoidal fashion (Fig. 7.5). To model this data, we assume that the voltage-gated $Ca^{2+}$ channel consists of $n$ identical subunits, each of which can be in one of two states, S and O. Only when all subunits are in the state O does the channel admit $Ca^{2+}$ current. Hence

$$S \overset{k_1}{\underset{k_2}{\rightleftarrows}} O, \tag{7.6}$$

and the number of open channels is proportional to $o^n$ where $o$ is the number of open channels. To incorporate the voltage dependence of the channel, the opening and closing rate constants $k_1$ and $k_2$ are assumed to be functions of voltage of the form

$$k_1 = k_1^0 \exp\left(\frac{qz_1V}{kT}\right), \qquad k_2 = k_2^0 \exp\left(\frac{qz_2V}{kT}\right), \tag{7.7}$$

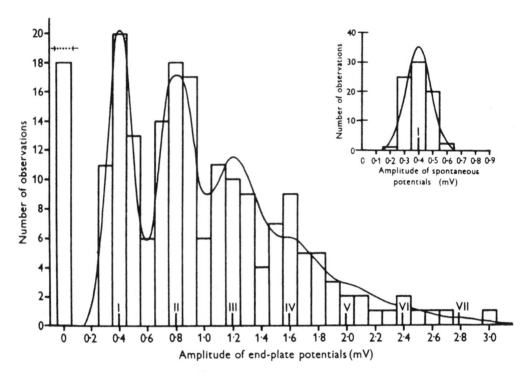

**Figure 7.4**   Amplitude distribution of miniature epps. The histogram gives the frequency of the miniature epp as a function of its amplitude, as measured experimentally. The smooth curve is a fit of the theoretical prediction (7.5). The inset shows the amplitude distribution of spontaneous epps: the smooth curve is a fit of the normal distribution to the data. (Boyd and Martin, 1956, Fig. 8.)

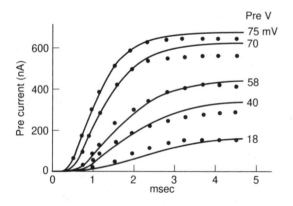

**Figure 7.5**   Presynaptic calcium currents in response to presynaptic voltage steps in the squid stellate ganglion. The continuous lines are the fits from the model described in the text. (Llinás et al., 1976, Fig. 1E.)

where $k$ is Boltzmann's constant, $T$ is the absolute temperature, $V$ is the membrane potential, $q$ is the positive elementary electric charge, $z_1$ and $z_2$ are the number of charges that move across the width of the membrane as S → O and O → S respectively, and $k_1^0$ and $k_2^0$ are constants. This is the same type of expression as that for the rate constant seen in Chapter 3 (for example, (3.45) and (3.46)). In Chapter 3, $z$ referred to the number of charges on each ion crossing the membrane by passing through a channel. In this model, $z_1$ and $z_2$ denote charges that cross the membrane as a result of a change in the conformation of the channel as it opens or closes. In either case, the result of $z$ charges crossing the membrane is the same, and we have a simple and plausible way to incorporate voltage dependence into the rate constants.

The unknown constants $k_1^0, k_2^0, z_1, z_2$, and $n$ are determined by fitting to the voltage clamp data shown in Fig. 7.5. From (7.6) it follows that, with $s$ denoting the number of shut channels,

$$\frac{do}{dt} = k_1 s - k_2 o = k_1(s_0 - o) - k_2 o, \tag{7.8}$$

where $s_0$ is the total number of subunits, assumed to be conserved. We assume the membrane potential jumps instantaneously from 0 to $V$ at time $t = 0$ and that $o(0) = 0$. Then we can solve the differential equation (7.8) to find that

$$o(t) = s_0 \frac{k_1}{k_1 + k_2}(1 - \exp[-(k_1 + k_2)t]). \tag{7.9}$$

Now we assume that the single-channel current for an open calcium channel, $j$, is given by the Goldman–Hodgkin–Katz current equation (Chapters 2 and 3). Hence, if $c_i$ and $c_e$ denote the internal and external $Ca^{2+}$ concentrations respectively, then

$$j = P_{Ca} \cdot \frac{2V}{kT} \cdot \frac{c_i - c_e \exp(\frac{-2qV}{kT})}{1 - \exp(\frac{-2qV}{kT})}, \tag{7.10}$$

where $P_{Ca}$ is the permeability of the $Ca^{2+}$ channel. Finally, $I_{Ca}$ is the product of the number of open channels with the single-channel current, and so

$$I_{Ca} = j\frac{s_0}{n}\left(\frac{o}{s_0}\right)^n, \tag{7.11}$$

since $\frac{s_0}{n}$ is the total number of channels and $(\frac{o}{s_0})^n$ is the percentage of open channels. By fitting curves of this form to the data shown in Fig. 7.5, Llinás et al. determined that the best-fit values for the unknowns are $n = 5$, $k_1^0 = 2$ ms$^{-1}$, $k_2^0 = 1$ ms$^{-1}$, $z_1 = 1$, and $z_2 = 0$. Other fixed parameters are $c_i = 0.1$ $\mu$M and $c_e = 40$ mM. These values were used to calculate the smooth curves in Fig. 7.5. (The internal $Ca^{2+}$ concentration is much smaller than the external concentration, so that the exact number used makes no essential difference to the result.) Hence, the best-fit parameters imply that the $Ca^{2+}$ channel consists of 5 independent subunits, that the conversion of O to S is independent of voltage ($z_2 = 0$), but that the conversion of S to O involves the movement of a single charge across the membrane ($z_1 = 1$) and is thus dependent on the membrane potential. Because the conversion of a closed channel to an open channel involves the movement of a charge across the cell membrane, there must be a current associated with channel opening, i.e., a gating current. This is generally the case when the rate constants for conformational changes of a channel protein are voltage dependent, and these gating currents have been measured experimentally. We do not discuss gating currents any further; the interested reader is referred to Hille (1992) and Armstrong and Bezanilla (1973, 1974, 1977).

## Synaptic suppression

At steady state, the percentage of open channels is

$$\left(\frac{o(t=\infty)}{s_0}\right)^5 = \left(\frac{k_1}{k_1+k_2}\right)^5, \tag{7.12}$$

which is an increasing function of $V$. However, the single-channel current (7.10) is a decreasing function of $V$. Thus, the steady state $I_{Ca}$, being a product of these two functions, is a bell-shaped function of $V$, as illustrated in Fig. 7.6. There are two time scales in the model; the single channel current depends instantaneously on the voltage, while the number of open channels is controlled by the voltage on a slower time scale. When the voltage is stepped up, the single-channel current decreases instantaneously. However, since there are so few channels open, the instantaneous decrease in the single-channel current has little effect on the total current. On a longer time scale, the channels gradually open in response to the increase in voltage, and this results in the slow monotonic responses to a positive step seen in Fig. 7.5. Of course, if the single-channel current has been reduced to zero, no increase in the current is seen as the channels begin to open.

In response to a step *decrease* in voltage, the single-channel $Ca^{2+}$ current increases instantaneously, but in contrast to the previous case where there were few channels open before the stimulus, there are now many open channels. Hence, the instantaneous increase in the single-channel current results in a large and fast increase in the total current. Over a longer time scale, the decrease in the voltage then leads to a slow decrease in the number of open channels, and thus a slow decrease in the total current.

These responses are illustrated in Fig. 7.7. When a small positive step is turned on and then off, the calcium current $I_{Ca}$ responds with a monotonic increase followed by a monotonic decrease (curve a). When the step is increased to 70 mV, the increase is still monotonic, but the decrease is preceded by a small "bump" as the current initially increases slightly (curve b). For a large step of 150 mV, the initial response is suppressed completely as the single-channel current is essentially zero, but when this suppression is released, a large voltage response is seen, which finally decreases to the resting state (curve c). This phenomenon is called *synaptic suppression*, and the theoretical results agree well with experimental observations.

## Response to an action potential

So far we have analytically calculated the response of the model to a voltage step. This was possible because under voltage clamp conditions, the voltage is piecewise constant. However, a more realistic stimulus would be a time-varying voltage corresponding to an action potential at the nerve terminal. It is easiest to find the solution

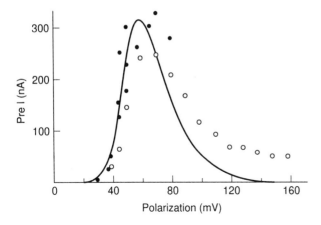

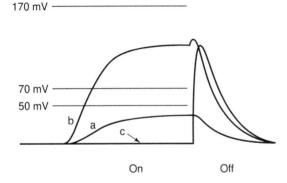

**Figure 7.6** Steady-state $I_{Ca}$ as a function of $V$. Symbols are experimental data. The smooth curve is a model simulation. (Llinás et al., 1976, Fig. 2B.)

**Figure 7.7** Numerical solutions of the Llinás model, demonstrating synaptic suppression of the calcium current. (Llinás et al., 1976, Fig. 2D.)

of (7.8) numerically, and this is shown in Fig. 7.8. Given an input $V(t)$ that looks like an action potential, the number of open channels (curve a) and the presynaptic $Ca^{2+}$ current (curve b) can be calculated. Figure 7.8 also includes theoretical predictions of the postsynaptic current (curve c) and the postsynaptic membrane potential (curve d). The postsynaptic current is obtained by assuming that it has the same form as the presynaptic current, delayed by 200 ms and amplified appropriately, assumptions that are justified by experimental evidence not discussed here. The postsynaptic membrane potential is obtained by using the postsynaptic current as an input into a model electrical circuit that we describe later.

Although the Llinás model provides a detailed description of the initial stages of synaptic transmission and the voltage-gated $Ca^{2+}$ channels, its picture of the postsynaptic response is oversimplified. There are a number of steps between the presynaptic $Ca^{2+}$ current and the postsynaptic current that in this model are assumed to be linearly related. Thus, a decrease in the postsynaptic current is the direct result of a decrease in the presynaptic $Ca^{2+}$ current, leading to a decrease in the concentration of neurotransmitter in the synaptic cleft. However, more detailed models of neurotransmitter kinetics show that at least at the neuromuscular junction, this is not an accurate description.

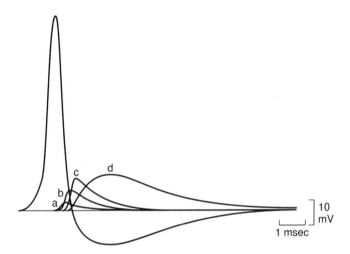

**Figure 7.8** Theoretical responses to an action potential. Using the experimentally measured action potential (the leftmost curve in the figure) as input, the model can be used to predict the time courses of (a) the proportion of open channels, (b) the $Ca^{2+}$ current, (c) the postsynaptic current, and (d) the postsynaptic potential. Details of how curve d is calculated are given in the text. Curve c is obtained by assuming that it has the same form as curve b, delayed by 200 ms and amplified. (Llinás et al., 1976, Fig. 2C.)

### 7.1.3 Calcium Diffusion, Binding, and Facilitation

One of the fundamental assumptions of the above model (and of the others in this chapter) is that neurotransmitter release is caused by the entry of $Ca^{2+}$, through voltage-sensitive channels, into the presynaptic neuron. However, although there is much evidence in favor of this hypothesis, there is also evidence that cannot be easily reconciled with this model. This has led to considerable controversy; some favor the *calcium hypothesis*, in which transmitter release is the direct result of the influx of $Ca^{2+}$ (Fogelson and Zucker, 1985; Zucker and Fogelson, 1986; Zucker and Landò, 1986; Yamada and Zucker, 1992), while others favor the *calcium–voltage hypothesis*, in which transmitter release can be triggered directly by the presynaptic membrane potential, with $Ca^{2+}$ playing a regulatory role (Parnas and Segel, 1980; Parnas et al., 1989; Aharon et al., 1994; and many other references). The major difference between the two hypotheses is in the role played by the voltage; the first group assumes that the only role of the voltage is to cause calcium influx, while the second group believes that voltage also has a direct effect on neurotransmitter release.

The calcium/voltage controversy is particularly interesting because of the role mathematical models have played. In 1985, Fogelson and Zucker proposed a model in which the diffusion of $Ca^{2+}$ from an array of single channels was used to explain the duration of transmitter release and the decay of facilitation. This model was later used as the basis for a large number of other modeling studies, some showing that $Ca^{2+}$ diffusion could not by itself explain all the experimental data, others showing how various refinements of the basic model could result in better agreement with experiment. Experimental and theoretical groups alike used the model as a basis for discussion, and thus, irrespective of the final verdict concerning its accuracy, the Fogelson–Zucker model is an excellent example of the value and use of modeling. However, here we do not repeat the many technical arguments both for and against the model; interested readers should consult the original literature referenced above.

Related to this controversy is the phenomenon of *facilitation*, which occurs when the amount of neurotransmitter release caused by an action potential is increased by an earlier action potential, provided that the time interval between the action potentials is not too great. One of the earliest hypotheses (Katz and Miledi, 1968) was that facilitation is caused by the buildup of free calcium at the transmitter release site, the so-called *residual free calcium* hypothesis. We consider here a more recent model that does not rely on diffusion arguments, but rather on the binding of $Ca^{2+}$ to transmitter release sites (Bertram et al., 1996). This model is based on recent experimental results showing that the minimum latency between $Ca^{2+}$ influx and the onset of transmitter release can be as short as 200 $\mu$s. Since the $Ca^{2+}$ binding site must thus be very close to the $Ca^{2+}$ channel, it is suggested that transmitter release is the result of $Ca^{2+}$ entering through a single channel, the so-called $Ca^{2+}$-domain hypothesis. If the $Ca^{2+}$ channels are far enough apart, or if only few open during each action potential, the $Ca^{2+}$ domains of individual channels are independent. Thus, each transmitter release site is assumed to be located very close to a single $Ca^{2+}$ channel. Our principal goal here is to provide a plausible explanation for the intriguing experimental result that facilitation increases in a step-like fashion as a function of the frequency of the conditioning action potential train.

We assume that $Ca^{2+}$ entering through the $Ca^{2+}$ channel is immediately available to bind to the transmitter release site, which itself consists of four independent gates, denoted by $S_1$ through $S_4$. Gate $S_j$ can be either closed (with probability $C_j$) or open (with probability $O_j$), and thus

$$Ca^{2+} + C_j \underset{k_{-j}}{\overset{k_{+j}}{\rightleftharpoons}} O_j. \tag{7.13}$$

Hence,

$$\frac{dO_j}{dt} = k_{+j}c - \frac{O_j}{\tau_j(c)}, \tag{7.14}$$

where $\tau_j(c) = 1/(k_{+j}c + k_{-j})$, and $c$ is the $Ca^{2+}$ concentration. Finally, the probability $R$ that the release site is activated is

$$R = O_1 O_2 O_3 O_4. \tag{7.15}$$

The rate constants were chosen to give good agreement with experimental data and are shown in Table 7.1. Note that the rates of closure of $S_3$ and $S_4$ are much greater than for $S_1$ and $S_2$, and thus $Ca^{2+}$ remains bound to $S_1$ and $S_2$ for a relatively long time, providing the possibility of facilitation.

---

**Table 7.1**  Parameter values for the binding model of synaptic facilitation (Bertram et al., 1996).

---

| | | | | | |
|---|---|---|---|---|---|
| $k_{+1}$ | $=$ | $3.75 \times 10^{-3}$ ms$^{-1}\mu$M$^{-1}$ | $k_{-1}$ | $=$ | $4 \times 10^{-4}$ ms$^{-1}$ |
| $k_{+2}$ | $=$ | $2.5 \times 10^{-3}$ ms$^{-1}\mu$M$^{-1}$ | $k_{-2}$ | $=$ | $1 \times 10^{-3}$ ms$^{-1}$ |
| $k_{+3}$ | $=$ | $5 \times 10^{-4}$ ms$^{-1}\mu$M$^{-1}$ | $k_{-3}$ | $=$ | $0.1$ ms$^{-1}$ |
| $k_{+4}$ | $=$ | $7.5 \times 10^{-3}$ ms$^{-1}\mu$M$^{-1}$ | $k_{-4}$ | $=$ | $10$ ms$^{-1}$ |

To demonstrate how facilitation works in this model, we suppose that a train of square pulses of Ca$^{2+}$ (each of width $t_p$ and amplitude $c_p$) arrives at the synapse. We want to calculate the level of activation at the end of each pulse and show that this is an increasing function of time. The reason for this increase is clear from the governing differential equation, (7.14). If a population of gates is initially closed, then a calcium pulse will begin to open them, but when calcium is absent, the gates will close. If the interval between pulses is sufficiently short and the decay time constant sufficiently large, then when the next pulse arrives, some gates are already open, so the new pulse achieves a larger percentage of open channels than the first, and so on.

To quantify this observation we define $t_n$ to be the time at the end of the $n$th pulse,

$$t_n = t_p + (n-1)T, \tag{7.16}$$

where $T = t_p + t_I$ is the period and $t_I$ is the interpulse interval. For any gate (temporarily omitting the subscript $j$) with $O(0) = 0$, the open probability at the end of the first pulse is

$$O(t_1) = O_\infty(1 - e^{-t_p/\tau_p}), \tag{7.17}$$

where $O_\infty = k_+ c_p \tau_p$ is the steady-state probability corresponding to a steady concentration of Ca$^{2+}$, $c_p$, and $\tau_p = \tau(c_p) = 1/(k_+ c_p + k_-)$.

Suppose that $O(t_{n-1})$ is the open probability at the end of the $(n-1)$st calcium pulse. During the interpulse period, $O$ decays with rate constant $\tau(0)$. Thus, at the start of the $n$th pulse,

$$O(t_{n-1} + t_I) = O(t_{n-1})e^{-t_I/\tau(0)}, \tag{7.18}$$

and so at the end of the $n$th pulse,

$$O(t_n) = O(t_{n-1})e^{-t_I/\tau(0)}e^{-t_p/\tau_p} + O_\infty(1 - e^{-t_p/\tau_p}) \tag{7.19}$$

$$= \alpha O(t_{n-1}) + O(t_1), \tag{7.20}$$

where $\alpha = \exp(-(t_I/\tau(0) + t_p/\tau_p)) = \exp(-k_-(T + t_p\frac{c_p}{K}))$ and $K = k_-/k_+$. This is a geometric series for $O(t_n)$, so that

$$\frac{O(t_n)}{O(t_1)} = \frac{1 - \alpha^n}{1 - \alpha}. \tag{7.21}$$

Notice that as the interpulse interval gets large ($t_I \to \infty$), $\alpha \to 0$, so that $O(t_n)$ is independent of $n$. On the other hand, $\alpha$ increases if the calcium pulses are shortened ($t_p$ is decreased).

Now we define facilitation as the ratio

$$F_n = \frac{R(t_n)}{R(t_1)} \tag{7.22}$$

and find that

$$F_n = \left(\frac{1 - \alpha_1^n}{1 - \alpha_1}\right)\left(\frac{1 - \alpha_2^n}{1 - \alpha_2}\right)\left(\frac{1 - \alpha_3^n}{1 - \alpha_3}\right)\left(\frac{1 - \alpha_4^n}{1 - \alpha_4}\right), \tag{7.23}$$

where $\alpha_j$ is the $\alpha$ corresponding to gate $j$. For the numbers shown in Table 7.1, $\alpha_4$ is nearly zero in the physiologically relevant range of frequencies, so it can be safely ignored. A plot of $F_n$ against the pulse train frequency shows a step-like function, as is observed experimentally. In Fig. 7.9 is shown the maximal facilitation,

$$F_{max} = \lim_{n \to \infty} F_n \tag{7.24}$$

$$= \left(\frac{1}{1 - \alpha_1}\right)\left(\frac{1}{1 - \alpha_2}\right)\left(\frac{1}{1 - \alpha_3}\right), \tag{7.25}$$

which also has a step-like appearance.

## 7.1.4  Neurotransmitter Kinetics

When the end-plate voltage is clamped and the nerve stimulated (so that the end-plate receives a stimulus of ACh, of undetermined form), the end-plate current first rises to a peak and then decays exponentially, with a decay time constant that is an exponential function of the voltage. Magleby and Stevens (1972) constructed a detailed

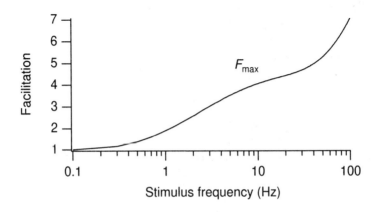

**Figure 7.9**  Facilitation as a function of stimulus frequency in the binding model for synaptic facilitation, calculated using $t_p c_p = 200 \ \mu M \, ms$. Here $F_{max} = \lim_{n \to \infty} F_n$ is the maximal facilitation produced by a pulse train.

model of end-plate currents in the frog neuromuscular junction that gives a mechanistic explanation of this observation and shows how a simple model of the receptor kinetics can quantitatively reproduce the observed end-plate currents.

First, Magleby and Stevens showed that the instantaneous end-plate current–voltage relationship is linear, and thus, for a fixed voltage, the end-plate current is proportional to the end-plate conductance. Because of this, it is sufficient to study the end-plate conductance rather than the end-plate current. Since the end-plate conductance is a function of the concentration of ACh, we restrict our attention to the kinetics of ACh in the synaptic cleft.

We assume that ACh reacts with its receptor, R, in enzymatic fashion,

$$\text{ACh} + \text{R} \underset{k_2}{\overset{k_1}{\rightleftharpoons}} \text{ACh} \cdot \text{R} \underset{\alpha}{\overset{\beta}{\rightleftharpoons}} \text{ACh} \cdot \text{R}^*, \tag{7.26}$$

and that the ACh-receptor complex passes current only when it is in the open state $\text{ACh} \cdot \text{R}^*$. We let $c = [\text{ACh}]$, $y = [\text{ACh} \cdot \text{R}]$, and $x = [\text{ACh} \cdot \text{R}^*]$, and then it follows from the law of mass action that

$$\frac{dx}{dt} = -\alpha x + \beta y, \tag{7.27}$$

$$\frac{dy}{dt} = \alpha x + k_1 c(N - x - y) - (\beta + k_2)y, \tag{7.28}$$

$$\frac{dc}{dt} = f(t) - k_e c - k_1 c(N - x - y) + k_2 y, \tag{7.29}$$

where $N$ is the total concentration of ACh receptor, which is assumed to be conserved, and ACh decays by a simple first-order process at rate $k_e$. The postsynaptic conductance is assumed to be proportional to $x$, and the rate of formation of ACh is some given function $f(t)$. One option for $f(t)$ would be to use the output of the single-domain/bound-calcium model described in the previous section (Exercise 9).

The model equations, as given, are too complicated to solve analytically, and so we proceed by making some simplifying assumptions. First, we assume that the kinetics of ACh binding to its receptor are much faster than the other reactions in the scheme, so that $y$ is in instantaneous equilibrium with $c$. To formalize this assumption, we introduce dimensionless variables $X = x/N, Y = y/N, C = k_1 c/k_2$, and $\tau = \alpha t$, in terms of which (7.28) becomes

$$\epsilon \frac{dY}{d\tau} = \epsilon X + C(1 - X - Y) - \left(\epsilon \frac{\beta}{\alpha} + 1\right) Y, \tag{7.30}$$

where $\epsilon = \alpha/k_2 \ll 1$. Upon setting $\epsilon$ to zero, we find the quasi-steady approximation

$$Y = \frac{C(1 - X)}{1 + C}, \tag{7.31}$$

or in dimensioned variables,

$$y = \frac{c(N - x)}{K + c}, \tag{7.32}$$

where $K = k_2/k_1$. Now we can eliminate $y$ from (7.27) to obtain

$$\frac{dx}{dt} = -\alpha x + \beta \frac{c(N - x)}{K + c}.$$  (7.33)

Next we observe that

$$\frac{dx}{dt} + \frac{dy}{dt} + \frac{dc}{dt} = f(t) - k_e c.$$  (7.34)

In dimensionless variables this becomes

$$\frac{N}{K}\left(\frac{dX}{d\tau} + \frac{dY}{d\tau}\right) + \frac{dC}{d\tau} = F(\tau) - K_e C,$$  (7.35)

where $F(\tau) = \frac{f(t)}{\alpha K}$ and $K_e = k_e/\alpha$ are assumed to be of order one. If we suppose that $N \ll K$, then setting $N/K$ to zero in (7.35), we find (in dimensioned variables)

$$\frac{dc}{dt} = f(t) - k_e c.$$  (7.36)

One further simplification is possible if we assume that $\beta \ll \alpha$. Notice from (7.33) that

$$\frac{dx}{dt} < -\alpha x + \beta(N - x),$$  (7.37)

so that

$$x(t) \le x(0)e^{-(\alpha+\beta)t} + \frac{\beta N}{\alpha + \beta}(1 - e^{-(\alpha+\beta)t}).$$  (7.38)

Once this process has been running for some time, so that effects of initial data can be ignored, $x$ is of order $\frac{\beta N}{\alpha+\beta}$. If $\beta \ll \alpha$, then $x \ll N$, and (7.33) simplifies to

$$\frac{dx}{dt} = -\alpha x + \beta \frac{cN}{K + c}.$$  (7.39)

For any given input $f(t)$, (7.36) can be solved for $c(t)$, and then (7.39) can be solved to give $x(t)$, the postsynaptic conductance.

As mentioned above, the decay of the postsynaptic current has a time constant that depends on the voltage. This could happen in two principal ways. First, if the conformational changes of the receptor were much faster than the decay of ACh in the synaptic cleft, $x$ would be in quasi-equilibrium, and we would have

$$x = \frac{\beta c(t)N}{\alpha[K + c(t)]}.$$  (7.40)

Thus, if $c$ is small, $x$ would be approximately proportional to $c$. In this case an exponential decrease of $c$ caused by the decay term $-k_e c$ would cause an exponential decrease in the postsynaptic conductance. An alternative possibility is that ACh degrades quickly in the synaptic cleft, so that $c$ quickly approaches zero, but that the decay of the endplate current is due to conformational changes of the ACh receptor. According to this hypothesis, the release of ACh into the cleft causes an increase in $x$, which then decays according to

$$\frac{dx}{dt} = -\alpha x \tag{7.41}$$

(since $c$ is nearly zero). In this case, the exponential decrease of end-plate current would be governed by the term $-\alpha x$.

Magleby and Stevens argued that the latter hypothesis is preferable. Assuming therefore that the rate-limiting step in the decay of the end-plate current is the decay of $x$, $\alpha$ can be estimated directly from experimental measurements of end-plate current decay to be

$$\alpha(V) = Be^{AV}, \tag{7.42}$$

where $A = 0.008$ mV$^{-1}$ and $B = 1.43$ ms$^{-1}$.

To calculate the complete time course of the end-plate current from (7.39), it remains to determine $c(t)$. In general this is not known, as it is not possible to measure synaptic cleft concentrations of ACh accurately.

A method to determine $c(t)$ from the experimental data was proposed by Magleby and Stevens. First, suppose that $\beta$ is also a function of $V$, as is expected, since $\alpha$ is a function of $V$. Then (7.39) can be written as

$$\frac{dx}{dt} = -\alpha(V)x + \beta(V)W(t), \tag{7.43}$$

where $W(t) = Nc(t)/[K + c(t)]$. Since for any fixed voltage the time course of $x$ can be measured experimentally (recall that the experiments were done under voltage clamp conditions), it follows that $\beta(V)W(t)$ can be determined from

$$\beta(V)W(t) = \frac{dx}{dt} + \alpha(V)x. \tag{7.44}$$

Although this requires numerical differentiation (which is a notoriously unstable procedure), the experimental records are smooth enough to permit a reasonable determination of $W$ from the time course of $x$. Since $W$ does not depend on $V$, it can be determined (up to an arbitrary scale factor) from a time course of $x$ obtained at any fixed voltage. Further, if the model is valid, then we expect the same result no matter what voltage is used to obtain $W$. A typical result for $W$, shown in Fig. 7.10, rises and falls in a way reminiscent of the responses calculated from the Llinás model described above (Fig. 7.8).

The final unknown is $\beta(V)$, the scale factor in the determination of $W$. Relative values of $\beta$ can be obtained by comparing time courses taken at different voltages. If $\beta(V_1)W(t)$ and $\beta(V_2)W(t)$ are time courses obtained from (7.44) at two different voltages, the ratio $\beta(V_1)/\beta(V_2)$ is obtained from the ratio of the time courses. However, because of experimental variability or invalid model assumptions, this ratio may not be constant as a function of time, in which case the ratio cannot be determined unambiguously. Magleby and Stevens used the ratio of the maximum amplitudes of the time courses, in which case $\beta(V_1)/\beta(V_2)$ can be obtained uniquely. They determined that $\beta$, like $\alpha$, is an exponential function of $V$,

$$\beta(V) = be^{aV}, \tag{7.45}$$

where $a = 0.00315$ mV$^{-1}$ and $b$ is an arbitrary scaling factor.

Equation (7.43) can now be solved numerically to determine the time course of the end-plate current for various voltages. Typical results are shown in Fig. 7.11. Although the model construction guarantees the correct peak response (because that is how $\beta$ was determined) and also guarantees the correct time course at one particular voltage (because that is how $W$ was determined), the model responses agree well with the experimental records over all times and voltages. This confirms the underlying assumption that $W$ is independent of voltage.

Although the approach of Magleby and Stevens of determining $W(t)$ directly from the data leads to excellent agreement with the experimental data, it suffers from the disadvantage that no mechanistic rationale is given for the function $W$. It would be preferable to have a derivation of the behavior of $W$ from fundamental assumptions about the kinetics of ACh release and degradation in the synaptic cleft, but such is not presently available.

## 7.1.5   The Postsynaptic Membrane Potential

Acetylcholine acts by opening ionic channels in the postsynaptic membrane that are permeable to Na$^+$ and K$^+$ ions. A schematic diagram of the electrical circuit model of the postsynaptic membrane is given in Fig. 7.12. This model is based on the usual assumptions (see, for example, Chapter 2) that the membrane channels can be modeled as ohmic resistors and that the membrane acts like a capacitor, with capacitance $C_m$.

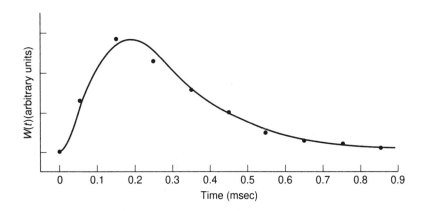

**Figure 7.10**   $W(t)$ calculated from the time course of $x$ using (7.44). (Magleby and Stevens, 1972, Fig. 4.)

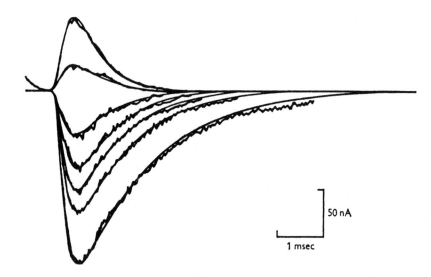

**Figure 7.11**  End-plate currents from the Magleby–Stevens model. Equation (7.43) was solved numerically, using as input the function plotted in Fig. 7.10. The functions $\alpha(V)$ and $\beta(V)$ are given in the text. The corresponding values for $V$ are, from top to bottom, 32, 20, −30, −56, −82, −106 and −161 mV. The wavy lines are the experimental data; the smooth curves are the model fit. (Magleby and Stevens, 1972, Fig. 6.)

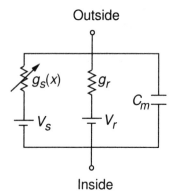

**Figure 7.12**  Electrical circuit model of the postsynaptic membrane.

The ACh-sensitive channels have a reversal potential $V_s$ of about −15 mV and a conductance that depends on the concentration of ACh. The effects of all the other ionic channels in the membrane are summarized by a resting conductance, $g_r$ and a resting potential $V_r$ of about −70 mV. In the usual way, the equation for the membrane potential $V$ is

$$C_m\frac{dV}{dt} + g_r(V - V_r) + g_s(V - V_s) = 0. \tag{7.46}$$

In general, $g_s$ is a function of the number of ACh receptors with ACh bound, i.e., in the notation of the previous section, $g_s = g_s(x)$. Since $x$ is a function of time, $g_s$ is also a function of time. Hence,

$$C_m \frac{dV}{dt} + [g_r + g_s(t)]V = g_r V_r + g_s(t)V_s. \tag{7.47}$$

This equation can be reduced to quadratures by using an integrating factor, and so in principle, the response of the postsynaptic membrane can be calculated from knowledge of the time course of $x$.

A simple solution is obtained when $g_s$ is taken to be proportional to $x$ and the input $f(t)$ to the Magleby and Stevens model is assumed to be $\gamma\delta(t)$ for some small $\gamma$. In this case,

$$g_s(t) = x(t) = \frac{\gamma\beta N}{K(\alpha - k_e)}\left(e^{-k_e t} - e^{-\alpha t}\right), \tag{7.48}$$

as derived in Exercise 5. Using this expression for $g_s$, we look for a solution for (7.47) that is close to $V_r$, i.e., $V = V_r + \gamma V_1$. This gives the linear differential equation for $V_1$,

$$C_m \frac{dV_1}{dt} + g_r V_1 + g_s(t)V_1 = \frac{g_s(t)}{\gamma}(V_s - V_r). \tag{7.49}$$

Solution of this linear equation is left as an exercise (Exercise 8).

## 7.1.6 Drugs and Toxins

The foregoing models are sufficient to piece together a crude model of synaptic transmission. However, many features were ignored, and there are many situations that can change the behavior of this system. Primary among these are drugs and toxins that affect specific events in the neurotransmission process. For example, the influx of calcium is reduced by divalent metal ions, such as $Pb^{++}$, $Cd^{++}$, $Hg^{++}$, and $Co^{++}$. By reducing the influx of calcium, these cations depress or abolish the action-potential-evoked transmitter release. Certain toxins, including tetanus and clostridial botulinus, are potent inhibitors of transmitter exocytosis, an action that is essentially irreversible. Botulinus neurotoxin is selective for cholinergic synapses and is one of the most potent neuroparalytic agents known. Tetanus toxin is taken up by spinal motor nerve terminals and transported retrogradely to the spinal cord, where it blocks release of glycine at inhibitory synapses. Spread of the toxin throughout the brain and spinal cord can lead to severe convulsions and death. The venom from black widow spider contains a toxin ($\alpha$-latrotoxin) that causes massive transmitter exocytosis and depletion of synaptic vesicles from presynaptic nerve terminals.

Agents that compete with the transmitter for receptor binding sites, thereby preventing receptor activation, are called receptor antagonists. An example of an antagonist of the ACh receptors of the skeletal neuromuscular junction is curare. By inhibiting ACh binding at receptor sites, curare causes progressive decrease in amplitude and shortening of epps. In severe curare poisoning, transmission is blocked.

Selective antagonists exist for most transmitter receptors. For example, bicuculline is an antagonist of GABA receptors, and is a well-known convulsant.

Agents that mimic the action of natural transmitters are known as receptor agonists. A well-known agonist of ACh receptors in neuromuscular junction is nicotine. Nicotine binds to the ACh receptor and activates it in the same manner as ACh. However, nicotine causes persistent receptor activation because it is not degraded, as is ACh, by ACh-esterase. On the other hand, diisopropylphosphofluoridate (commonly known as nerve gas) is an example of an anticholinesterase, because it inhibits the activity of ACh-esterase, so that ACh persists in the synaptic cleft. Similarly, one effect of cocaine is to prolong the activity of dopamine, by blocking the uptake of dopamine from the synaptic cleft.

Other agents interfere with receptor-gated permeabilities by interfering with the channel itself. Thus, picrotoxin, which blocks GABA-activated $Cl^-$ channels, and strychnine, which blocks glycine-activated $Cl^-$ channels, are potent blockers of inhibitory synapses and known convulsants.

## 7.2   Gap Junctions

Gap junctions are small nonselective channels (with diameters of about 1.2 nm) that form direct intercellular connections through which ions or other small molecules can flow. They are formed by the joining of two *connexons*, hexagonal arrays of *connexin* protein molecules (Fig. 7.13). Despite being called electrical synapses, in this chapter we concentrate on models in which membrane potential plays no role, focusing instead on the interaction between intracellular diffusion and intercellular permeability. Electrical aspects of gap junctions are important for the function of cardiac cells and are discussed in that context in Chapter 11. An example of how gap junctions are used for intercellular signaling via second messengers is detailed in Chapter 12, where we discuss a model of intercellular calcium wave propagation.

### 7.2.1   Effective Diffusion Coefficients

We first consider a one-dimensional situation where a species $u$ diffuses along a line of cells which are connected by gap junctions. Because of their relatively high resistance to flow (compared to cytoplasm), the gap junctions decrease the rate at which $u$ diffuses along the line. Since this is a one-dimensional problem, we assume that each intercellular membrane acts like a single resistive pore with a given permeability $F$. The effect of gap-junction distribution within the intercellular membrane is discussed later in this chapter.

We assume that Fick's law holds and thus the flux, $J$, of $u$ is proportional to the gradient of $u$; i.e.,

$$J = -D\frac{\partial u}{\partial x},$$

(7.50)

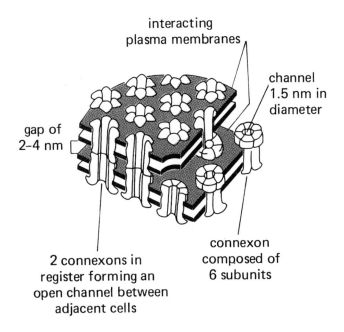

interacting
plasma membranes

channel
1.5 nm in
diameter

gap of
2–4 nm

connexon
composed of
6 subunits

2 connexons in
register forming an
open channel between
adjacent cells

**Figure 7.13** Diagram of a region of membrane containing gap junctions, based on electron microscope images and X-ray diffraction data. Each connexon is composed of six gap-junction proteins, called connexins, arranged hexagonally. Connexons in apposed membranes meet in the intercellular space to form the gap junction. (Alberts et al., 1994, Fig. 19-15.)

where $D$ is the diffusion coefficient for the intracellular space. From the conservation of $u$ it follows that in the interior of each cell,

$$\frac{\partial u}{\partial t} = D\frac{\partial^2 u}{\partial x^2}. \tag{7.51}$$

However, $u$ need not be continuous across the intercellular boundary. In fact, if there is a cell boundary at $x = x_b$, the flux through the boundary is assumed to be proportional to the concentration difference across the boundary. Then, conservation of $u$ across the boundary implies that

$$-D\frac{\partial u(x_b^-, t)}{\partial x} = -D\frac{\partial u(x_b^+, t)}{\partial x} = F[u(x_b^-, t) - u(x_b^+, t)], \tag{7.52}$$

for some constant $F$, called the *permeability coefficient*, with units of distance/time. The + and − superscripts indicate that the function values are calculated as limits from the right and left, respectively.

When the cells through which $u$ diffuses are short compared to the total distance that $u$ moves, the movement of $u$ can be described by an effective diffusion coefficient. The effective diffusion coefficient is defined and is measurable experimentally by assuming that the analogue of Ohm's law holds. Thus, in a preparation of $N$ cells, each of length $L$, with $u = U_0$ at $x = 0$ and $u = U_1$ at $x = NL$, the effective diffusion coefficient

$D_e$ is defined by

$$J = \frac{D_e}{NL}(U_0 - U_1), \tag{7.53}$$

where $J$ is the steady-state flux of $u$.

To calculate $D_e$, we look for a function $u(x)$ that satisfies $u_{xx} = 0$ when $x \neq (j + \frac{1}{2})L$ and satisfies (7.52) at $x = (j + \frac{1}{2})L, j = 0, \ldots, N-1$. Further, we require that $u(0) = U_0$, and $u(NL) = U_1$. Note that we are assuming that the cell boundaries occur at $L/2, 3L/2, \ldots$, and thus the boundary conditions at $x = 0$ and $x = NL$ occur halfway through a cell.

A typical solution $u$ that satisfies these conditions must be linear within each cell, and piecewise continuous with jumps at the cell boundaries. Suppose that the slope of $u$ within each cell is $-\lambda$, and that the jump in $u$ between cells is $u(x_b^+) - u(x_b^-) = -\Delta$. Then, since there are $N-1$ whole cells, two half cells (at the boundaries), and $N$ interior cell boundaries, we have

$$(N - 1)\lambda L + 2\lambda \left(\frac{L}{2}\right) + N\Delta = U_0 - U_1. \tag{7.54}$$

Furthermore, it follows from (7.52) that

$$D\lambda = F\Delta. \tag{7.55}$$

We find from (7.53) and (7.54) that

$$D\lambda = -D\frac{\partial u}{\partial x} = J = \frac{D_e}{NL}(U_0 - U_1) \tag{7.56}$$

$$= \frac{D_e}{L}(L\lambda + \Delta) \tag{7.57}$$

$$= \lambda D_e \left(1 + \frac{D}{FL}\right), \tag{7.58}$$

from which it follows that

$$\frac{1}{D_e} = \frac{1}{D} + \frac{1}{LF}. \tag{7.59}$$

## 7.2.2 Homogenization

The above calculation of the effective diffusion coefficient can be formalized by the process of *homogenization*. Homogenization is an important technique that we will see again in the context of cardiac propagation in Chapter 11. The point of homogenization is that there are two spatial scales, a microscopic and a macroscopic scale, and we are interested in knowing the behavior of the solution on the macroscopic scale while accounting for influences from the microscopic scale, without calculating the full details of the solution on the microscopic scale. An introduction to homogenization theory in the context of asymptotic methods can be found in Holmes (1995).

Here we illustrate the technique on the one-dimensional diffusion equation with gap junctions, but it can readily be extended to higher dimensions (see Exercise 12) and to nonlinear equations (Chapter 11). The basic assumption is that Fick's law holds, but that the resistance, $R$, is a rapidly varying function, so that the flux is

$$J = -\frac{1}{R(\frac{x}{\epsilon})}\frac{\partial u}{\partial x}. \tag{7.60}$$

The dimensionless parameter $\epsilon$ is small, indicating that the variations of $R$ are rapid compared to other spatial scales of the problem. The resistance $R$ is taken to be a periodic function of period one, which is of order one in the interior of cells and large at the gap junctions. It is important to note that since $R$ is of order one, the diffusion coefficient is also of order one, but periodic in $x$ with period $\epsilon$, which is small. Thus, cells are assumed to be short compared to the diffusion length scale.

As a practice problem, we solve the heat equation

$$\frac{\partial u}{\partial t} = \frac{\partial}{\partial x}\left(\frac{1}{R(\frac{x}{\epsilon})}\right)\frac{\partial u}{\partial x}. \tag{7.61}$$

To take into account that there are two space scales, we explicitly introduce two independent spatial variables, $z = x$ and $\sigma = \frac{x}{\epsilon}$, and then partial differentiation with respect to $x$ becomes

$$\frac{\partial}{\partial x} = \frac{\partial}{\partial z} + \frac{1}{\epsilon}\frac{\partial}{\partial \sigma}. \tag{7.62}$$

In terms of these new variables, the heat equation (7.61) becomes

$$\frac{\partial u}{\partial t} = \left(\frac{\partial}{\partial z} + \frac{1}{\epsilon}\frac{\partial}{\partial \sigma}\right)\left(\frac{1}{R(\sigma)}\left(\frac{\partial u}{\partial z} + \frac{1}{\epsilon}\frac{\partial u}{\partial \sigma}\right)\right). \tag{7.63}$$

To account for the fact that we expect the solution to exhibit behavior on the two space scales, we write the solution $u$ as

$$u(x,t) = u_0(z,t) + \epsilon u_1(\sigma,z,t) + \epsilon^2 u_2(\sigma,z,t) + O(\epsilon^3). \tag{7.64}$$

Because we want $u_0(z,t)$ to be the average, slowly varying solution, we require $u_1$ to be periodic and have zero mean in $\sigma$. We expand the governing equation in powers of $\epsilon$ and collect like powers into a hierarchy of equations, to be solved sequentially. Thus,

$$\frac{\partial}{\partial \sigma}\left(\frac{1}{R(\sigma)}\left(\frac{\partial u_0}{\partial z} + \frac{\partial u_1}{\partial \sigma}\right)\right) = 0, \tag{7.65}$$

and

$$\frac{\partial}{\partial \sigma}\left(\frac{1}{R(\sigma)}\frac{\partial u_2}{\partial \sigma}\right) = \frac{\partial u_0}{\partial t} - \frac{\partial}{\partial \sigma}\frac{1}{R(\sigma)}\frac{\partial u_1}{\partial z} - \frac{\partial}{\partial z}\frac{1}{R(\sigma)}\frac{\partial u_1}{\partial \sigma} - \frac{1}{R(\sigma)}\frac{\partial^2 u_0}{\partial z^2}. \tag{7.66}$$

From (7.65) it follows that

$$\frac{\partial u_1}{\partial \sigma} + \frac{\partial u_0}{\partial z} = R(\sigma)\phi(z), \tag{7.67}$$

for some function $\phi(z)$. Recalling that $u_1$ is periodic in $\sigma$, with period 1 and mean 0, we integrate (7.67) with respect to $\sigma$ from 0 to 1 to get

$$\frac{\partial u_0}{\partial z} = \phi(z) \int_0^1 R(\sigma)d\sigma = \bar{R}\phi, \tag{7.68}$$

where

$$\bar{R} = \int_0^1 R(s)\,ds \tag{7.69}$$

is the average resistance. It follows that

$$R\phi = \frac{R}{\bar{R}}\frac{\partial u_0}{\partial z}, \tag{7.70}$$

so that eliminating $R\phi$ from (7.67),

$$\frac{\partial u_1}{\partial \sigma} = \left(\frac{R}{\bar{R}} - 1\right)\frac{\partial u_0}{\partial z}. \tag{7.71}$$

Integrating (7.71) with respect to $\sigma$ then gives

$$u_1 = w(\sigma)\frac{\partial u_0}{\partial z}, \tag{7.72}$$

where $w(\sigma)$ is periodic with zero mean and satisfies the differential equation

$$\frac{dw}{d\sigma} = \frac{R}{\bar{R}} - 1. \tag{7.73}$$

The function $w(\sigma)$ determines the small-scale structure of the solution.

The next step is to observe that (7.66) can have a periodic solution for $u_2$ only if the right-hand side of this equation has zero average in $\sigma$. This implies that

$$\frac{\partial u_0}{\partial t} = \frac{\partial}{\partial z}\int_0^1 \frac{1}{R(\sigma)}\left(\frac{\partial u_1}{\partial \sigma} + \frac{\partial u_0}{\partial z}\right)d\sigma. \tag{7.74}$$

We use (7.72) to eliminate $u_1$ from (7.74) and find that

$$\frac{\partial u_0}{\partial t} = D_e \frac{\partial^2 u_0}{\partial z^2}, \tag{7.75}$$

where the effective diffusion coefficient is

$$D_e = \frac{1}{\bar{R}}. \tag{7.76}$$

Notice that now we are able to solve the diffusion equation (7.75) for $u_0$ on the macroscopic scale, yet we determine the solution on the microscopic scale through

$$u(x,t) = u_0 + \epsilon w\left(\frac{x}{\epsilon}\right)\frac{\partial u_0}{\partial x} + O(\epsilon^2). \tag{7.77}$$

To apply this technique to the specific problem of gap junctions, we take $R(x) = r_c + r_g \sum_k \delta(x - kL)$ to reflect the periodic occurrence of gap junctions with resistance $r_g$ evenly spaced at the ends of cells of length $L$. Notice that $D = 1/r_c$ is the diffusion coefficient for the intracellular space, while $F = 1/r_g$ is the intercellular permeability. It follows easily that

$$\bar{R} = r_c + \frac{r_g}{L}, \tag{7.78}$$

which is the same as (7.59).

### 7.2.3 Measurement of Permeabilities

Although an effective diffusion coefficient is useful when the species of interest diffuses through a large number of cells, in some experimental situations one is interested in how a dye molecule (or a second messenger such as $IP_3$) diffuses through a relatively small number of cells. In this case the effective diffusion coefficient approximation cannot always be used, and it is necessary to solve the equations with internal boundary conditions (Brink and Ramanan, 1985; Ramanan and Brink, 1990). By calculating exact solutions to the linear diffusion equation with internal boundary conditions (using transform methods, for example) and fitting them to experimental measurements on the movement of fluorescent probes, it is possible to obtain estimates of the intracellular diffusion coefficient as well as the permeability of the intercellular membrane.

The analytic solutions of Brink and Ramanan are useful only when the underlying equations are linear. In many cases, however, the species of interest are also reacting in a nonlinear way. This results in a system of nonlinear diffusion equations coupled by jump conditions at the gap junctions, a system that can only be solved numerically. Two groups have used numerical methods to study problems of this kind. Christ et al. (1994) studied the problem of diffusion through gap junctions, assuming that the diffusing species $u$ decreases the permeability of the gap junction in a nonlinear fashion. A similar model was used by Sneyd et al. (1995a) to study the spread of a calcium wave through a layer of cells coupled by gap junctions, and we discuss this model in Chapter 12.

### 7.2.4 The Role of Gap-Junction Distribution

It is not widely appreciated that the intercellular permeability is strongly influenced by the distribution of gap junctions in the intercellular membrane, although it is a common observation in introductory biology texts that there is a similar relationship between the distribution of stomata on leaves and the rate of evaporation of water through the leaf surface. Individual gap junctions are usually found in aggregates forming larger junctional plaques, as individual gap-junction particles are not easily distinguished from other nonjunctional particles. However, numerical simulations show that the permeability of the intercellular membrane decreases as the gap junction particles aggregate in larger groupings. This raises the intriguing possibility that intercellular permeability may be lowest when the gap-junctional plaques are easiest to see. This in turn provides a possible explanation for the fact that it has been difficult to establish a direct link between the number of recognizable gap junctions and the intercellular permeability.

Chen and Meng (1995) constructed a cubic lattice model of a two-cell system with a common border. A number of gap-junction particles, with varying degrees of aggregation, were placed on the border lattice points. Marker particles were placed in one of the cubes and followed a random walk over the lattice points of the cube. When they encountered a gap-junction lattice point on the boundary, there was an assigned probability that the marker particle would move across to the other cell. By measuring the time required for a certain percentage of marker particles to cross from one cell to the other, Chen and Meng obtained a quantitative estimate of the efficiency of intercellular transport as a function of gap-junction aggregation. Their results are summarized in Fig. 7.14. When the gap junctions are clumped together in a single junctional plaque, 10,000 time steps were required for the transfer of about 10% of the marker particles. However, when the gap-junction particles were randomly scattered, only 1,000 time steps were required for the same transfer. The magnitude of this discrepancy emphasizes the fact that gap junction distribution can have a huge effect on the rate of intercellular transport.

To get an analytical understanding of how the distribution of gap junctions affects the diffusion coefficient, we solve a model problem, similar to the one-dimensional problem solved in Section 7.2.1. We consider cells to be two-dimensional rectangles, with a portion of their ends open for diffusive transport (the gap junctions) and the remainder closed (Fig. 7.15A). The dashed lines in this figure are lines of symmetry across which there is no flux in a steady-state problem, so we can reduce the cell configuration to that shown in Fig. 7.15B.

To study diffusion in the $x$-coordinate direction, we assume that the vertical walls are separated by length $L$ and have regularly spaced openings of width $2\delta$ with centers separated by length $2l$. The fraction of the vertical separator that is open between cells is $\Delta = \delta/l$. To study how the distribution of gap junctions affects the diffusion coefficient, we hold $\Delta$ fixed while varying $l$. When $l$ is small, the gap junctions are small and uniformly distributed, while when $l$ is large, the gap junctions are clumped together into larger aggregates; in either case the same fraction ($\Delta$) of the intercellular membrane is occupied by gap junctions.

Suppose that there are a large number of cells (say $N$) each of length $L$ connected end to end. We impose a fixed concentration gradient across the array and use the definition (7.53) to define the effective diffusion coefficient for this array.

To find the flux, we solve Laplace's equation subject to no-flux boundary conditions on the horizontal lines $y = 0$ and $y = l$ and on the vertical lines $\delta < y < l, x = pL, p = 0, \ldots, N$. We further divide this region into two subregions, one for $y \geq \delta$ and one for $y \leq \delta$.

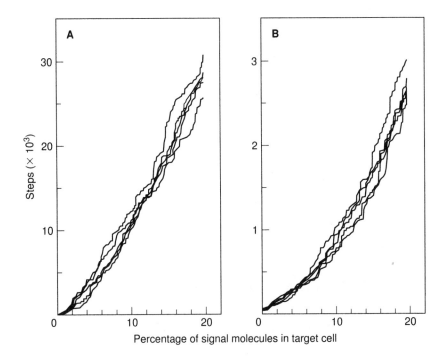

**Figure 7.14** Simulation results of the cubic-lattice gap-junction model on a $50 \times 50 \times 50$ lattice with 1000 signal molecules in the source cell at time 0. In A, 100 gap-junction particles are arranged in a compact junctional plaque, while in B they are scattered randomly on the intercellular interface. The random scattering of gap-junction particles results in a greatly increased intercellular transfer rate (note the different scales for the two panels). (Chen and Meng, 1995, Fig. 1.)

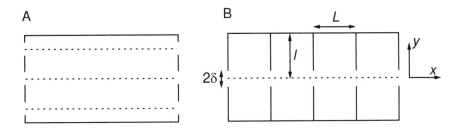

**Figure 7.15** A: Sketch of a single rectangular cell with gap-junctional openings in the end faces. B: Sketch of cell array, reduced by symmetry to a single "half-channel."

Consider first the solution on the upper region. The solution for a single cell $0 \leq x \leq L$ can be found by separation of variables to be

$$u(x,y) = \sum_{n=0}^{\infty} a_n \cos\left(\frac{n\pi x}{L}\right) \cosh\left(\frac{n\pi(y-l)}{L}\right). \tag{7.79}$$

This solution satisfies no-flux boundary conditions at $y = l$ and at $x = 0, L$. Notice also that this solution is periodic, so it is a contender for the solution for any cell.

Now recall that in the one-dimensional case, the solution is piecewise linear, with jumps at the cell boundaries, and that the slope of the solution within each cell is the same (Section 7.2.1). This suggests that the derivative of the solution in the two-dimensional case should be the same in each cell, or equivalently, the solution in each cell should be the same up to an additive constant. Thus,

$$u(x,y) = \sum_{n=1}^{\infty} A_n \frac{\cosh(n\pi(y-l)/L)}{\cosh(n\pi(\delta-l)/L)} \cos\left(\frac{n\pi}{L}(x - pL)\right) + \alpha_p \tag{7.80}$$

for $pL < x < (p+1)L$, $\delta < y < l$, and $p = 0, \ldots, N - 1$. We have scaled the unknown constants by $\cosh(n\pi(\delta - l)/L)$ for convenience.

On the lower region, a similar process gives

$$u(x,t) = (U_1 - U_0)\frac{x}{NL} + U_0 + \sum_{n=1}^{\infty} \frac{\cosh(2n\pi y/L)}{\cosh(2n\pi\delta/L)}\left(C_n \sin\frac{2n\pi x}{L}\right) \tag{7.81}$$

for $0 < x < NL, 0 < y < \delta$. Notice that this solution satisfies a no-flux boundary condition at $y = 0$ and has the correct overall concentration gradient.

Now, to make these into a smooth solution of Laplace's equation we require that $u(x,y)$ and $u_y(x,y)$ be continuous at $y = \delta$. This gives two conditions,

$$\sum_{n=1}^{\infty} A_n \cos\left(\frac{n\pi}{L}(x - pL)\right) = \sum_{n=1}^{\infty}\left(C_n \sin\frac{2n\pi x}{L}\right) + (U_1 - U_0)\frac{x}{nL} + U_0 - \alpha_p \tag{7.82}$$

and

$$\sum_{n=1}^{\infty} nA_n \tanh\left(\frac{n\pi}{L}(\delta - l)\right)\cos\left(\frac{n\pi}{L}(x - pL)\right) = \sum_{n=1}^{\infty} 2n \tanh\frac{2n\pi\delta}{L}C_n \sin\frac{2n\pi x}{L} \tag{7.83}$$

on the interval $pL < x < (p+1)L$.

We now determine $\alpha_p$ by averaging (7.82) over cell $p$. Integrating (7.82) from $x = (p-1)L$ to $x = pL$ gives

$$\alpha_p = \frac{1}{L}\int_{(p-1)L}^{pL} (U_1 - U_0)\frac{x}{NL}\,dx + U_0, \tag{7.84}$$

since all the trigonometric terms integrate to zero. Hence,

$$\alpha_p = U_0 - (U_0 - U_1)\frac{p - 1/2}{N}. \tag{7.85}$$

Finally, for convenience, we choose $U_0 = N/2$ and $U_1 = -N/2$, which gives $\alpha_p = p + (1 + N)/2$. Since this is a linear problem, the values chosen for $U_0$ and $U_1$ have no effect on the effective diffusion coefficient.

To obtain equations for the coefficients, we project each of these onto $\cos\frac{k\pi x}{L}$ by multiplying by $\cos\frac{k\pi x}{L}$ and integrating from 0 to $L$. We find that

$$A_k \frac{L}{2} = F_k + \sum_{n=1}^{\infty} C_n I_{2n,k} \tag{7.86}$$

and

$$k A_k \tanh\left(\frac{k\pi}{L}(\delta - l)\right)\frac{L}{2} = \sum_{n=1}^{\infty} 2n \tanh\frac{2n\pi\delta}{L}\left(C_n I_{2n,k}\right), \tag{7.87}$$

where

$$F_k = \int_0^L \left(\frac{x}{L} - \frac{1}{2}\right)\cos\frac{k\pi x}{L}dx = \frac{L}{n^2\pi^2}((-1)^k - 1), \tag{7.88}$$

$$I_{n,k} = \int_0^L \sin\frac{n\pi x}{L}\cos\frac{k\pi x}{L}dx = \frac{Ln}{\pi}\left(\frac{1 - (-1)^{n+k}}{n^2 - k^2}\right). \tag{7.89}$$

There is an immediate simplification. Notice that $I_{2n,k} = 0$ and $F_k = 0$ when $k$ is even. Thus, $A_k = 0$ for all even $k$. Now we eliminate the coefficients $A_k$ from (7.86) and (7.87) to obtain

$$\sum_{n=1}^{\infty} C_n \left(\frac{2n}{k}\frac{\tanh\frac{2n\pi l}{L}\Delta}{\tanh\frac{k\pi l}{L}(1 - \Delta)} + 1\right)\frac{n}{4n^2 - k^2} = \frac{1}{2\pi k^2} \tag{7.90}$$

for all odd $k$, with $\Delta = \delta/l$.

In these terms, the average flux is

$$J = \frac{D}{l}\int_0^\delta \left.\frac{\partial u}{\partial x}\right|_{x=0} dy = D\left(\frac{\Delta}{L} + \frac{1}{l}\sum_{n=1}^{\infty} C_n \tanh\frac{2n\pi l\Delta}{L}\right). \tag{7.91}$$

It follows that the effective diffusion coefficient is

$$D_e = D\left(\frac{L}{l}\sum_{n=1}^{\infty} C_n \tanh\frac{2n\pi l\Delta}{L} + \Delta\right). \tag{7.92}$$

Since $k$ can take on any odd positive integer value, (7.90) is an infinite set of equations for the coefficients $C_n$. Since the solution of the differential equation converges, we can truncate this system of equations and solve the resulting finite linear system numerically. Typical results are shown in Fig. 7.16A, where the ratio $D_e/D$ is shown plotted as a function of $l/L$ for different values of fixed $\Delta = \delta/l$, and in Fig. 7.16B, where $D_e/D$ is shown plotted as a function of $\Delta$ for fixed $l/L$.

There are a number of important observations that can be made. First, notice that in the limit $\Delta \to 1$, or $l/L \to \infty$, $\frac{L}{l}C_n \tanh\frac{2n\pi l\Delta}{L} \to 0$. Thus,

$$\lim_{\Delta \to 1} D_e = D, \tag{7.93}$$

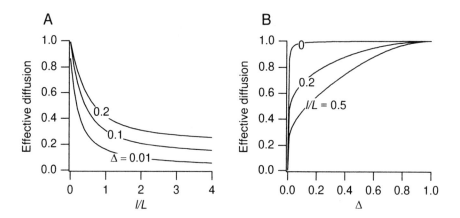

**Figure 7.16**  A: Effective diffusion ratio $D_e/D$ as a function of the distribution ratio $l/L$ for fixed gap-junction fraction $\Delta$. B: Effective diffusion ratio $D_e/D$ as a function of the gap-junction fraction $\Delta$ for fixed distribution ratio $l/L$.

and

$$\lim_{\frac{l}{L} \to \infty} D_e = D\Delta. \tag{7.94}$$

Finally, from the numerical solution, it can be seen that $D_e$ is a decreasing function of $l/L$. Thus, clumping of gap junctions lowers the effective diffusion coefficient compared with spreading them out uniformly.

From Fig. 7.16B we see that when gap junctions are small but uniformly spread, there is little decrease in the effective diffusion coefficient, unless $\Delta$ is quite small. Thus, for example, with $\Delta = 0.01$ (so that gap junctions comprise 1% of the membrane surface area) and $l = 0$, the effective diffusion coefficient is about 86% of the cytoplasmic diffusion. On the other hand, with only one large gap junction with $\Delta = 0.01$ in the end membrane of a square cell ($l = 0.5$), the effective diffusion coefficient is reduced to about 27% of the original.

It is interesting to relate these results to the one-dimensional solution. This two-dimensional problem becomes effectively one-dimensional in the limit $\frac{l}{L} \to 0$, with a piecewise linear profile in the interior of the cells and small boundary or corner layers at the end faces. In this limit, the effective diffusion coefficient satisfies

$$\frac{D}{D_e} = 1 + \mu \frac{1 - \Delta}{\Delta}, \tag{7.95}$$

with $\mu = 0.0016$. This formula was found by plotting the curve $\Delta(\frac{D}{D_e} - 1)$ against $\Delta$, which, remarkably, is numerically indistinguishable from the straight line $\mu(1 - \Delta)$. Comparing this with the one-dimensional result, we find that the end-face permeability can be related to the fraction of gap junctions $\Delta$ through

$$F = \frac{D}{L} \frac{\Delta}{\mu(1 - \Delta)}. \tag{7.96}$$

## 7.3  EXERCISES

1. (a) Verify the last step of (7.2). Hint: Use Stirling's formula $n! \approx n^{n+1}e^{-n}\sqrt{\frac{2\pi}{n}}$, and show that $\lim_{n \to \infty} \ln[(1 - \frac{x}{n})^n] = -x$.

   (b) Verify that the Poisson distribution $P(k) = \frac{e^{-m}m^k}{k!}$ has mean $m$, by verifying that $\langle k \rangle = \sum_{k=0}^{\infty} kP(k) = m$.

   (c) Verify that the sum of $k$ identical Gaussian distributions with mean $\mu$ and variance $\sigma^2$ is a Gaussian distribution with mean $k\mu$ and variance $k\sigma^2$.

2. This question is based on the model presented by Peskin (1991). Motivated by the smallness of $c_i$ with respect to $c_e$, simplify the Llinás model by setting $c_i \approx 0$. How much difference does this make to the $Ca^{2+}$ currents plotted in Fig. 7.5? Calculate the steady-state $Ca^{2+}$ current as a function of $V$ and show that it is bell-shaped. Solve for a general step in voltage from $V_1$ to $V_2$ and demonstrate synaptic suppression.

3. Calculate the analytic solution to (7.8) when $V$ is a given function of $t$.

4. Construct a simple function $F(t)$ with the same qualitative shape as the function $W(t)$ used in the Magleby and Stevens model, and calculate the analytic solution to (7.43) for that $F$. Compare to the numerical solutions shown in Fig. 7.11.

5. In the Magleby and Stevens model, a simple choice for the release function $f(t)$ results in end-plate conductances with considerable qualitative similarity with those in Fig. 7.11. Suppose there is a sudden release of ACh into the synaptic cleft at time $t = 0$. We take $f(t) = \gamma\delta(t)$, where $\delta$ is the Dirac delta function. Show that the resulting differential equation is

$$\frac{dc}{dt} = -k_e c, \qquad c(0) = \gamma, \tag{7.97}$$

   for $t \geq 0$. Solve for $c$, and, assuming that $\gamma$ is small, substitute this expression for $c$ into the differential equation for $x$, (7.39). Look for a solution for $x$ of order $\gamma$, and show that

$$x(t) = \frac{\gamma\beta N}{K(\alpha - k_e)}\left(e^{-k_e t} - e^{-\alpha t}\right), \tag{7.98}$$

   which is always positive. Sketch the solution.

6. Peskin (1991) presented a more complex version of the Magleby and Stevens model. His model is based on the reaction scheme

$$\longrightarrow ACh, \qquad \text{rate } r_T \text{ per unit volume}, \tag{7.99}$$

$$ACh + R \underset{k_2}{\overset{k_1}{\rightleftarrows}} ACh \cdot R \underset{\alpha}{\overset{\beta}{\rightleftarrows}} ACh \cdot R^*, \tag{7.100}$$

$$ACh + E \underset{k_4}{\overset{k_3}{\rightleftarrows}} ACh \cdot E \overset{\gamma}{\rightarrow} E, \tag{7.101}$$

   where E is some enzyme that degrades ACh in the synaptic cleft. (The assumption of enzymatic degradation of ACh is one of the ways in which the Peskin model differs from the Magleby and Stevens model. The other difference is that the Peskin model does not assume that the amount of ACh bound to its receptor is negligible.) Write down the equations for the 6 dependent variables. Use conservation laws to eliminate two of the equations.

Then assume that the reactions involving ACh with R, and ACh with E (with reaction rates $k_i, i = 1, 4$), are fast to obtain expressions for [R] and [E] in terms of the other variables. Substitute these expressions into the differential equations for [ACh · R*] and [ACh] + [R] − [E] to end up with two differential equations in [ACh · R*] and [ACh]. Solve these equations when the stimulus is a small sudden release of ACh (i.e., assume that $r_T = \epsilon\delta(t)$ and look for solutions of $O(\epsilon)$), and show that the solution has the same form as (7.98) but that the exponential coefficients are given by the roots of a quadratic polynomial. What is the rate of ACh degradation?

7. Solve the above exercise (and obtain the same solution!) by nondimensionalizing, finding a small parameter, and then solving in terms of an asymptotic expansion. Hint: One method is to nondimensionalize time by $\gamma$ and let $\gamma/k_2 = \epsilon$ be the small parameter. To lowest order in $\epsilon$ one gets only three equations for four unknowns, and so to solve the lowest-order problem completely it is necessary to go to the higher-order terms. The differential equations obtained at higher order must then be added in the appropriate manner (as in the previous question) so that unwanted terms cancel.

8. Solve (7.49) and plot the solution (Peskin, 1991). What is the solution as $\gamma \to 0$? What is the slope of the solution at $t = 0$? Compare to the curve d in Fig. 7.8.

9. By linking the output of the Llinás model to the input of the single-domain/bound-calcium model, and then linking this to the input of the Magleby and Stevens model (the rate of production of ACh) construct a unified model for the synaptic cleft that connects the presynaptic action potential to the postsynaptic voltage via the concentration of ACh in the synaptic cleft. Solve the model numerically and compare to the simpler model presented briefly in Fig. 7.8.

10. Incorporate the effects of nicotine into a model of ACh activation of receptors.

11. Calculate the effective diffusion coefficient $D_e$ for a periodic medium with periodic diffusion coefficient $D(x)$.
    Answer:
    $$\frac{1}{D_e} = \frac{1}{P}\int_0^P \frac{1}{D(x)}dx, \tag{7.102}$$
    where the period of $D(x)$ is $P$.

12. Use homogenization arguments to solve the Poisson equation
    $$\nabla \cdot D\left(\frac{x}{\epsilon}\right)\nabla u = f\left(x, \frac{x}{\epsilon}\right), \tag{7.103}$$
    assuming that $D$ is periodic with a basic spatial subunit $\Omega$ of total volume $V$. Show that $u = U(x) + \epsilon W(\frac{x}{\epsilon}) \cdot \nabla U(x) + O(\epsilon^2)$, where $U(x)$ satisfies the averaged Poisson equation
    $$\nabla \cdot D_e \nabla U = \frac{1}{V}\int_\Omega f(x, \sigma)d\sigma \tag{7.104}$$
    and where the effective diffusion coefficient is
    $$D_e = \frac{1}{V}\int_\Omega D(\sigma)(\nabla_\sigma W + I)d\sigma. \tag{7.105}$$
    Show that the partial differential equation governing $W(\sigma)$ is
    $$\nabla_\sigma \cdot D(\sigma)(\nabla_\sigma W + I) = 0, \tag{7.106}$$
    with $\sigma = x/\epsilon$.

# CHAPTER 8

# Passive Electrical Flow in Neurons

Neurons are among the most important and interesting cells in the body. They are the fundamental building blocks of the central nervous system and hence responsible for motor control, cognition, perception, and memory, among other things. Although our understanding of how networks of neurons interact to form an intelligent system is extremely limited, one prerequisite for an understanding of the nervous system is an understanding of how individual nerve cells behave.

There is a great deal of experimental data indicating that parts of neurons conduct electricity in a passive manner. Thus, there has been developed an extensive body of theory describing the flow of electricity in neurons using the theory of electrical flow in cables. A cable is any structure that provides a one-dimensional pathway for communication via an electrical signal. Neurons are among the most abundant cells with a cable structure, although skeletal muscle and cardiac cells also have cable-like features.

A typical neuron consists of three principal parts: the *dendrites*; the cell body, or *soma*; and the *axon*. The structure of some typical neurons is shown in Fig. 8.1. Dendrites are the input stage of a neuron and receive synaptic input from other neurons. The soma contains the necessary cellular machinery such as a nucleus and mitochondria, and the axon is the output stage. At the end of the axon (which may also be branched, as are the dendrites) are synapses, which are cellular junctions specialized for the transmission of an electrical signal (Chapter 7). Thus, a single neuron may receive input along its dendrites from a large number of other neurons, which is called *convergence*, and may similarly transmit a signal along its axon to many other neurons, called *divergence*.

The behaviors of the dendrites, axon, and synapse are all quite different. The spread of electrical current in a dendritic network is (mostly) a passive process that can be well described by the diffusion of electricity along a leaky cable. The axon, on the other

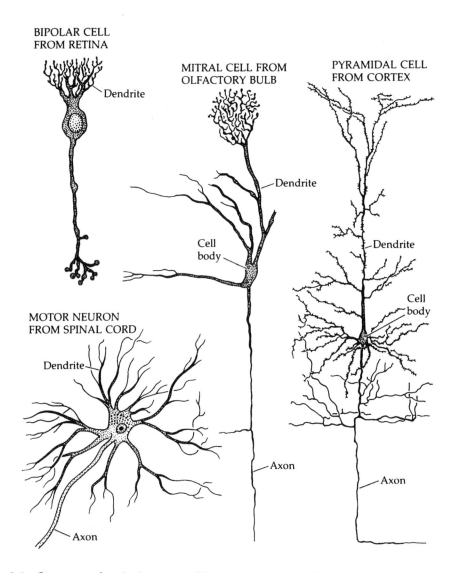

BIPOLAR CELL
FROM RETINA

Dendrite

MITRAL CELL FROM
OLFACTORY BULB

PYRAMIDAL CELL
FROM CORTEX

Dendrite

Cell
body

Dendrite

Cell
body

MOTOR NEURON
FROM SPINAL CORD

Dendrite

Axon

Axon

Axon

**Figure 8.1**  Structure of typical neurons. The motor neuron is from a mammalian spinal cord and was drawn by Dieters in 1869. The other cells were drawn by Ramón y Cajal. The pyramidal cell is from mouse cortex, and the mitral cell from the olfactory bulb of a cat. (Kuffler et al., 1984, Fig. 1, p. 10.)

hand, has an excitable membrane of the type described in Chapter 4, and thus can propagate an electrical signal actively. At the synapse (Chapter 7), the membrane is specialized for the release or reception of chemical neurotransmitters. In this chapter we discuss how to model the behavior of a cable, and then focus on the passive spread of current in a dendritic network; in the following chapter we show how an excitable membrane can actively propagate an electrical impulse, or action potential.

## 8.1 The Cable Equation

One of the first things to realize from the pictures in Fig. 8.1 is that it is unlikely that the membrane potential is the same at each point. In some cases spatial uniformity can be achieved experimentally (for instance, by threading a silver wire along the axon, as did Hodgkin and Huxley), but *in vivo*, the intricate branched structure of the neuron can create spatial gradients in the membrane potential. Although this seems clear to us now, it was not until the pioneering work of Wilfrid Rall in the 1950s and 1960s that the importance of spatial effects gained widespread acceptance.

To understand something of how spatial distribution affects the behavior of a cable, we derive the *cable equation*. The theory of the flow of electricity in a leaky cable dates back to the work of Lord Kelvin in 1855, who derived the equations to study the transatlantic telegraph cable then under construction. However, the application of the cable equation to neuronal behavior is mainly due to Hodgkin and Rushton (1946), and then a series of classic papers by Rall (1957, 1959, 1960, 1969; an excellent summary of much of Rall's work on electrical flow in neurons is given in Segev et al., 1995.)

We view the cell as a long cylindrical piece of membrane surrounding an interior of cytoplasm (called a cable). We suppose that everywhere along its length, the potential depends only on the length variable and not on radial or angular variables, so that the cable can be viewed as one-dimensional. This assumption is called the *core conductor assumption* (Rall, 1977). We now divide the cable into a number of short pieces of isopotential membrane each of length $dx$. In any cable section, all currents must balance, and there are only two types of current, namely, transmembrane cur-

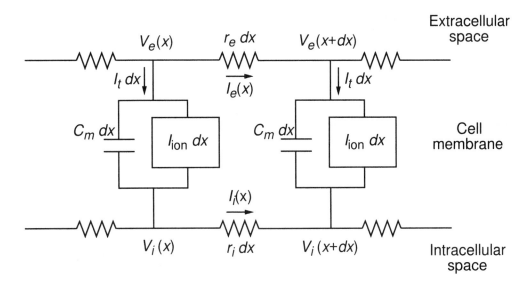

**Figure 8.2** Schematic diagram of a discretized cable, with isopotential circuit elements of length $dx$.

rent and axial current (Fig. 8.2). The axial current has intracellular and extracellular components, both of which we assume to be ohmic, i.e., linear functions of the voltage. Hence,

$$V_i(x + dx) - V_i(x) = -I_i(x)r_i dx, \tag{8.1}$$

$$V_e(x + dx) - V_e(x) = -I_e(x)r_e dx, \tag{8.2}$$

where $I_i$ and $I_e$ are the intracellular and extracellular axial currents respectively. The minus sign on the right-hand side appears because of the convention that positive current is a flow of positive charges from left to right (i.e., in the direction of increasing $x$). If $V_i(x + dx) > V_i(x)$, then positive charges flow in the direction of decreasing $x$, giving a negative current. In the limit $dx \to 0$,

$$I_i = -\frac{1}{r_i}\frac{\partial V_i}{\partial x}, \tag{8.3}$$

$$I_e = -\frac{1}{r_e}\frac{\partial V_e}{\partial x}. \tag{8.4}$$

The numbers $r_i$ and $r_e$ are the resistances per unit length of the intracellular and extracellular media, respectively. In general,

$$r_i = \frac{R_c}{A_i}, \tag{8.5}$$

where $R_c$ is the *cytoplasmic resistivity*, measured in units of Ohms-length, and $A_i$ is the cross-sectional area of the cylindrical cable. A similar expression holds for the extracellular space, so if the cable is in a bath with large (effectively infinite) cross-sectional area, the extracellular resistance $r_e$ is nearly zero.

Next, from Kirchhoff's laws, any change in extracellular or intracellular axial current must be due to a transmembrane current, and thus

$$I_i(x) - I_i(x + dx) = I_t dx = I_e(x + dx) - I_e(x), \tag{8.6}$$

where $I_t$ is the total transmembrane current (positive outward) per unit length of membrane. In the limit as $dx \to 0$, this becomes

$$I_t = -\frac{\partial I_i}{\partial x} = \frac{\partial I_e}{\partial x}. \tag{8.7}$$

In a cable with no additional current sources, the total axial current is $I_T = I_i + I_e$, so using that $V = V_i - V_e$, we find

$$- I_T = \frac{r_i + r_e}{r_i r_e}\frac{\partial V_i}{\partial x} - \frac{1}{r_e}\frac{\partial V}{\partial x}, \tag{8.8}$$

from which it follows that

$$\frac{1}{r_i}\frac{\partial V_i}{\partial x} = \frac{1}{r_i + r_e}\frac{\partial V}{\partial x} - \frac{r_e}{r_i + r_e}I_T. \tag{8.9}$$

On substituting (8.9) into (8.7), we obtain

$$I_t = \frac{\partial}{\partial x} \left( \frac{1}{r_i + r_e} \frac{\partial V}{\partial x} \right), \tag{8.10}$$

where we have used (8.3) and the fact that $I_T$ is constant. Finally, recall that the transmembrane current $I_t$ is a sum of the capacitive and ionic currents, and thus

$$I_t = p \left( C_m \frac{\partial V}{\partial t} + I_{\text{ion}} \right) = \frac{\partial}{\partial x} \left( \frac{1}{r_i + r_e} \frac{\partial V}{\partial x} \right), \tag{8.11}$$

where $p$ is the perimeter of the axon. Equation (8.11) is usually referred to as the cable equation. Note that $C_m$ has units of capacitance per unit area of membrane, and $I_{\text{ion}}$ has units of current per unit area of membrane. If a current $I_{\text{applied}}$, with units of current per unit area, is applied across the membrane (as before, taken positive in the outward direction), then the cable equation becomes

$$I_t = p \left( C_m \frac{\partial V}{\partial t} + I_{\text{ion}} + I_{\text{applied}} \right) = \frac{\partial}{\partial x} \left( \frac{1}{r_i + r_e} \frac{\partial V}{\partial x} \right). \tag{8.12}$$

It is useful to nondimensionalize the cable equation. To do so we define the *membrane resistivity* $R_m$ as the resistance of a unit square area of membrane, having units of $\Omega\,\text{cm}^2$. For any fixed $V_0$, $R_m$ is determined by measuring the change in membrane current when $V$ is perturbed slightly from $V_0$. In mathematical terms,

$$\frac{1}{R_m} = \frac{dI_{\text{ion}}}{dV} \bigg|_{V=V_0}. \tag{8.13}$$

Although the value of $R_m$ depends on the chosen value of $V_0$, it is typical to take $V_0$ to be the resting membrane potential to define $R_m$. Note that if the membrane is an ohmic resistor, then $I_{\text{ion}} = V/R_m$, in which case $R_m$ is independent of the value $V_0$.

Assuming that $r_i$ and $r_e$ are constant, the cable equation (8.11) can now be written in the form

$$\tau_m \frac{\partial V}{\partial t} + R_m I_{\text{ion}} = \lambda_m^2 \frac{\partial^2 V}{\partial x^2}, \tag{8.14}$$

where

$$\lambda_m = \sqrt{\frac{R_m}{p(r_i + r_e)}} \tag{8.15}$$

has units of distance and is called the cable *space constant*, and where

$$\tau_m = R_m C_m \tag{8.16}$$

has units of time and is called the membrane *time constant*. If we ignore the extracellular resistance, then

$$\lambda_m = \sqrt{\frac{R_m d}{4R_c}}, \tag{8.17}$$

**Table 8.1**   Typical parameter values for a variety of excitable cells.

| parameter | $d$ | $R_c$ | $R_m$ | $C_m$ | $\tau_m$ | $\lambda_m$ |
|---|---|---|---|---|---|---|
| units | $10^{-4}$ cm | $\Omega$ cm | $10^3\,\Omega$ cm$^2$ | $\mu$F/cm$^2$ | ms | cm |
| squid giant axon | 500 | 30 | 1 | 1 | 1 | 0.65 |
| lobster giant axon | 75 | 60 | 2 | 1 | 2 | 0.25 |
| crab giant axon | 30 | 90 | 7 | 1 | 7 | 0.24 |
| earthworm giant axon | 105 | 200 | 12 | 0.3 | 3.6 | 0.4 |
| marine worm giant axon | 560 | 57 | 1.2 | 0.75 | 0.9 | 0.54 |
| mammalian cardiac cell | 20 | 150 | 7 | 1.2 | 8.4 | 0.15 |
| barnacle muscle fiber | 400 | 30 | 0.23 | 20 | 4.6 | 0.28 |

where $d$ is the diameter of the axon (assuming circular cross-section). Finally, we rescale the ionic current by defining $I_{\text{ion}} = -f(V,t)/R_m$ for some $f$, which, in general, is a function of both voltage and time and has units of voltage, and we nondimensionalize space and time by defining new variables $X = x/\lambda_m$ and $T = t/\tau_m$. In the new variables the cable equation is

$$\frac{\partial V}{\partial T} = \frac{\partial^2 V}{\partial X^2} + f(V,T). \tag{8.18}$$

Although we write $f$ as a function of voltage and time, in many of the simpler versions of the cable equation, $f$ is a function of $V$ only (for example, (8.19) below). Typical parameter values for a variety of cells are shown in Table 8.1.

## 8.2   Dendritic Conduction

To complete the description of a spatially distributed cable, we must specify how the ionic current depends on voltage and time. In the squid giant axon, $f(V,t)$ is a function of $m, n, h$, and $V$ as described in Chapter 4. This choice for $f$ allows waves that propagate along the axon at constant speed and with a fixed profile, as we will see in the next chapter. They require the input of energy from the axon, which must expend energy to maintain the necessary ionic concentrations, and thus they are often called *active waves*.

Any electrical activity for which the approximation $f = -V$ is valid (i.e., if the membrane is an Ohmic resistor) is said to be *passive* activity. There are some cables, primarily in neuronal dendritic networks, for which this is a good approximation in the range of normal activity. For other cells, activity is passive only if the membrane potential is sufficiently small. For simplicity in a passive cable, we shift $V$ so that the resting potential is at $V = 0$. Thus,

$$\frac{\partial V}{\partial T} = \frac{\partial^2 V}{\partial X^2} - V, \tag{8.19}$$

which is called the *linear cable equation*. In the linear cable equation, current flows along the cable in a passive manner, leaking to the outside at a linear rate.

There is a vast literature on the application of the linear cable equation to dendritic networks. In particular, the books by Jack et al. (1975) and Tuckwell (1988) are largely devoted to this problem, and provide detailed discussions of the theory. Koch and Segev (1989) also provides an excellent introduction. In this chapter we provide only a brief introduction to this topic.

## 8.2.1 Boundary Conditions

To determine the behavior of a single dendrite, we must first specify initial and boundary conditions. Usually, it is assumed that at time $T = 0$, the dendritic cable is in its resting state, $V = 0$, and so

$$V(X, 0) = 0. \tag{8.20}$$

Boundary conditions can be specified in a number of ways. Suppose that $X = X_b$ is a boundary point.

1. Voltage-clamp boundary conditions: If the voltage is fixed (i.e., clamped) at $X = X_b$, then the boundary condition is of Dirichlet type,

$$V(X_b, T) = V_b, \tag{8.21}$$

where $V_b$ is the specified voltage level.

2. Short circuit: If the ends of the cable are short-circuited, so that the extracellular and intracellular potentials are the same at $X = X_b$, then

$$V(X_b, T) = 0. \tag{8.22}$$

This is a special case of the voltage clamp condition in which $V_b = 0$.

3. Current injection: Suppose a current $I(T)$ is injected at one end of the cable. Since

$$I_i = -\frac{1}{r_i}\frac{\partial V_i}{\partial x} = -\frac{1}{r_i \lambda_m}\frac{\partial V_i}{\partial X}, \tag{8.23}$$

the boundary condition (if we ignore extracellular resistance, so that the extracellular potential is uniform) is

$$\frac{\partial V(X_b, T)}{\partial X} = -r_i \lambda_m I(T). \tag{8.24}$$

If $X_b$ is at the left end, this corresponds to an inward current, while if it is on the right end, this is an outward current.

4. Sealed ends: If the end at $X = X_b$ is sealed to ensure that there is no current across the endpoint, then the boundary condition is the homogeneous Neumann condition,

$$\frac{\partial V(X_b, T)}{\partial X} = 0, \tag{8.25}$$

a special case of an injected current for which $I(T) = 0$.

## 8.2.2  Input Resistance

One of the most important simple solutions of the cable equation corresponds to the situation in which a steady current is injected at one end of a semi-infinite cable. This is a common experimental protocol (although never with a truly semi-infinite cable) that can be used to determine the cable parameters $R_m$ and $R_c$. Suppose the cable extends from $X = 0$ to $X = \infty$ and that a steady current $I_0$ is injected at $X = 0$. Then, the boundary condition at $X = 0$ is

$$\frac{dV(0)}{dX} = -r_i \lambda_m I_0. \tag{8.26}$$

Setting $\partial V/\partial T = 0$ and solving (8.19) subject to the boundary condition (8.19) gives

$$V(X) = \lambda_m r_i I_0 e^{-X} = V(0)e^{-X} = V(0)e^{-x/\lambda_m}. \tag{8.27}$$

Clearly, by measuring the rate at which the voltage decays along the cable, $\lambda_m$ can be determined from experimental data. The *input resistance* $R_{in}$ of the cable is defined to be the ratio $V(0)/I_0 = \lambda_m r_i$. Recall that when the extracellular resistance is ignored,

$$\lambda_m = \sqrt{\frac{R_m d}{4 R_c}}. \tag{8.28}$$

Combining this with (8.5) gives

$$R_{in} = \lambda_m r_i = \sqrt{\frac{4 R_m R_c}{\pi^2}} \frac{1}{d^{\frac{3}{2}}}. \tag{8.29}$$

Hence, the input resistance of the cable varies with the $-3/2$ power of the cable diameter, a fact that is of importance for the behavior of the cable equation in a branching structure. Since both the input resistance and the space constant of the cable can be measured experimentally, $R_m$ and $R_c$ can be calculated from experimental data.

Some solutions to the cable equation for various types of cable and boundary conditions are discussed in the exercises. Tuckwell (1988) gives the most detailed discussion of the various types of solutions and how they are obtained.

## 8.2.3  Branching Structures

The property of neurons that is most obvious from Fig. 8.1 is that they are extensively branched. While the procedure to find solutions on a branched cable network

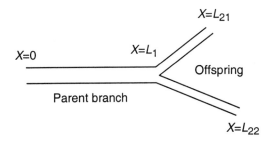

**Figure 8.3** Diagram of the simplest possible branched cable.

is straightforward in concept, it can be quite messy in application. Thus, in what follows, we emphasize the procedure for obtaining the solution on branching structures, without calculating specific formulae.

### The steady-state solution

It is useful first to consider the simplest branched cable, depicted in Fig. 8.3. The cable has a single branch point, or *node*, at $X = L_1$, and the two offspring branches extend to $L_{21}$ and $L_{22}$, respectively. For convenience we express all lengths in nondimensional form, with the reminder that nondimensional length does not correspond to physical length, as the distance variable $x$ along each branch of the cable is scaled by the length constant $\lambda_m$ appropriate for that branch, and each branch may have a different length constant.

We construct the solution in three parts: $V_1$ on cylinder 1, and $V_{21}$ and $V_{22}$ on the two offspring cylinders. At steady state each $V$ satisfies the differential equation $V'' = V$, and so we can immediately write the general solution as

$$V_1 = A_1 e^{-X} + B_1 e^X, \tag{8.30}$$

$$V_{21} = A_{21} e^{-X} + B_{21} e^X, \tag{8.31}$$

$$V_{22} = A_{22} e^{-X} + B_{22} e^X, \tag{8.32}$$

where the $A$s and $B$s are unknown constants. To determine the 6 unknown constants, we need 6 constraints, which come from the boundary and nodal conditions. For boundary conditions, we assume that a current $I_0$ is injected at $X = 0$ and that the terminal ends (at $X = L_{21}$ and $X = L_{22}$) are held fixed at $V = 0$. Thus,

$$\frac{dV_1(0)}{dX} = -r_i \lambda_m I_0, \tag{8.33}$$

$$V_{21}(L_{21}) = V_{22}(L_{22}) = 0. \tag{8.34}$$

The remaining three constraints come from conditions at the node. We require that $V$ be a continuous function and that current be conserved at the node. It follows that

$$V_1(L_1) = V_{21}(L_1) = V_{22}(L_1), \tag{8.35}$$

and

$$d_1^{3/2} \sqrt{\frac{\pi^2}{4R_m R_c}} \frac{dV_1(L_1)}{dX} = d_{21}^{3/2} \sqrt{\frac{\pi^2}{4R_m R_c}} \frac{dV_{21}(L_1)}{dX} + d_{22}^{3/2} \sqrt{\frac{\pi^2}{4R_m R_c}} \frac{dV_{22}(L_1)}{dX}. \tag{8.36}$$

If we make the natural assumption that each branch of the cable has the same physical properties (and thus have the same $R_m$ and $R_c$), although possibly differing in diameter, the final condition for conservation of current at the node becomes

$$d_1^{3/2} \frac{dV_1(L_1)}{dX} = d_{21}^{3/2} \frac{dV_{21}(L_1)}{dX} + d_{22}^{3/2} \frac{dV_{22}(L_1)}{dX}. \tag{8.37}$$

We thus have six linear equations for the six unknown constants; explicit solution of this linear system is left for Exercise 6.

## More general branching structures

For this method to work for more general branching networks, there must be enough constraints to solve for the unknown constants. The following argument shows that this is the case. First, we know that each branch of the tree contributes two unknown constants, and thus, if there are $N$ nodes, there are $1 + 2N$ individual cables with a total of $2 + 4N$ unknown constants. Each node contributes three constraints, and there are $2 + N$ terminal ends (including that at $X = 0$), each of which contributes one constraint, thus giving a grand total of $2 + 4N$ constraints. Thus, the resulting linear system is well-posed. Of course, a unique solution is guaranteed only if this system is invertible, which is not known a priori.

## Equivalent cylinders

One of the most important results in the theory of dendritic trees is due to Rall (1959), who showed that under certain conditions, the equations for passive electrical flow over a branching structure reduce to a single equation for electrical flow in a single cylinder, the so-called *equivalent cylinder*.

To see this reduction in a simple setting, consider again the branching structure of Fig. 8.3. To reduce this to an equivalent cylinder we need some additional assumptions. We assume, first, that the two offspring branches have the same dimensionless lengths, $L_{21} = L_{22}$, and that their terminals have the same boundary conditions. Since $V_{21}$ and $V_{22}$ obey the same differential equation on the same domain, obey the same boundary conditions at the terminals, and are equal at the node, it follows that they must be equal. That is,

$$\frac{dV_{21}(L_1)}{dX} = \frac{dV_{22}(L_1)}{dX}. \tag{8.38}$$

Substituting (8.38) into (8.37) we then get

$$d_1^{3/2} \frac{dV_1(L_1)}{dX} = (d_{21}^{3/2} + d_{22}^{3/2}) \frac{dV_{21}(L_1)}{dX}. \tag{8.39}$$

Finally (and this is the crucial assumption), if we assume that

$$d_{21}^{3/2} + d_{22}^{3/2} = d_1^{3/2}, \tag{8.40}$$

then

$$\frac{dV_1(L_1)}{dX} = \frac{dV_{21}(L_1)}{dX}. \tag{8.41}$$

Thus $V_1$ and $V_{21}$ have the same value and derivative at $L_1$ and obey the same differential equation. It follows that the composite function

$$V = \begin{cases} V_1(X), & 0 \leq X \leq L_1, \\ V_{21}(X), & L_1 \leq X \leq L_{21}, \end{cases} \tag{8.42}$$

is continuous with a continuous derivative on $0 < X < L_{21}$ and obeys the cable equation on that same interval. Thus, the simple branching structure is equivalent to a cable of length $L_{21}$ and diameter $d_1$.

More generally, if we have a branching structure that satisfies the following conditions:

1. $R_m$ and $R_c$ are the same for each branch of the cable;
2. At every node the cable diameters satisfy an equation analogous to (8.40). That is, if $d_0$ is the diameter of the parent branch, and $d_1, d_2, \ldots$ are the diameters of the offspring, then

$$d_0^{3/2} = d_1^{3/2} + d_2^{3/2} + \cdots; \tag{8.43}$$

3. The boundary conditions at the terminal ends are all the same;
4. Each terminal is the same dimensionless distance $L$ from the origin of the tree (at $X = 0$);

then the entire tree is equivalent to a cylinder of length $L$ and diameter $d_1$, where $d_1$ is the diameter of the cable at $X = 0$.

Using an inductive argument, it is not difficult to show that this is so (although a rigorous proof is complicated by the notation). Working from the terminal ends, one can condense the outermost branches into equivalent cylinders, then work progressively inwards, condensing the equivalent cylinders into other equivalent cylinders, and so on, until only a single cylinder remains. It is left as an exercise (Exercise 7) to show that during this process the requirements for condensing branches into an equivalent cylinder are never violated.

## 8.3 The Rall Model of a Neuron

When studying a model of a neuron, the item of greatest interest is often the voltage at the cell body, or soma. This is primarily because the voltage at the cell body can be measured experimentally with greater ease than can the voltage in the dendritic

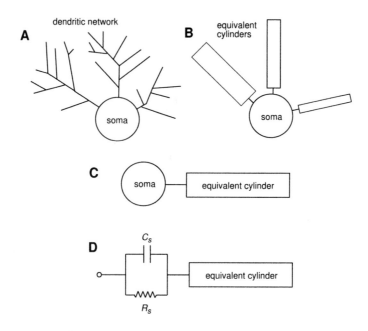

**Figure 8.4** Schematic diagram of the Rall lumped-soma model of the neuron. First, it is assumed that the dendritic network pictured in A is equivalent to the equivalent cylinders shown in B, and that these cylinders are themselves equivalent to a single cylinder as in C. The soma is assumed to be isopotential and to behave like a resistance and capacitance in parallel, as in D.

network, and further, it is the voltage at the soma that determines whether or not the neuron fires an action potential. Therefore, it is important to determine the solution of the cable equation on a dendritic network when one end of the network is connected to a soma. The most common approach to incorporating a soma into the model is due to Rall (1960), and is called the *Rall lumped-soma model.*

The three basic assumptions of the Rall model are, first, that the soma is isopotential (i.e., that the soma membrane potential is the same at all points), second, that the soma acts like a resistance ($R_s$) and a capacitance ($C_s$) in parallel, and, third, that the dendritic network can be collapsed into a single equivalent cylinder. This is illustrated in Fig. 8.4.

The potential $V$ satisfies the cable equation on the equivalent cylinder. The boundary condition must account for current flow within the soma and into the cable. Thus, if $I_0$ denotes an applied current at $X = 0$, then the boundary condition is

$$I_0 = -\frac{1}{r_i}\frac{\partial V(0,t)}{\partial x} + C_s\frac{\partial V(0,t)}{\partial t} + \frac{V(0,t)}{R_s}, \tag{8.44}$$

so that

$$R_s I_0 = -\gamma\frac{\partial V(0,T)}{\partial X} + \sigma\frac{\partial V(0,T)}{\partial T} + V(0,T), \tag{8.45}$$

where $\sigma = C_s R_s / \tau_m = \tau_s / \tau_m$ and $\gamma = R_s / (r_i \lambda_m)$. For convenience we assume that the time constant of the soma is the same as the membrane time constant, so that $\sigma = 1$.

## 8.3.1 A Semi-Infinite Neuron with a Soma

We first calculate the steady response of a semi-infinite neuron to a current $I_0$ injected at $X = 0$, as in Section 8.2.2. As before, we set the time derivative to zero to get

$$\frac{d^2 V}{dX^2} = V, \tag{8.46}$$

$$V(0) - \gamma \frac{dV(0)}{dX} = R_s I_0, \tag{8.47}$$

which can easily be solved to give

$$V(X) = \frac{R_s}{r_i \lambda_m + R_s} r_i \lambda_m I_0 e^{-X}. \tag{8.48}$$

This solution is nearly the same as the steady response of the equivalent cylinder without a soma to an injected current, except that $V$ is decreased by the constant factor $R_s / (R_s + r_i \lambda_m) < 1$. As $R_s \to \infty$, in which limit the soma carries no current, the solution to the lumped-soma model approaches the solution to the simple cable.

The input resistance $R_{\text{in}}$ of the lumped-soma model is

$$R_{\text{in}} = \frac{V(0)}{I_0} = \frac{r_i \lambda_m R_s}{r_i \lambda_m + R_s}, \tag{8.49}$$

and thus

$$\frac{1}{R_{\text{in}}} = \frac{1}{r_i \lambda_m} + \frac{1}{R_s}. \tag{8.50}$$

Since $r_i \lambda_m$ is the input resistance of the cylinder, the input conductance of the lumped-soma model is the sum of the input conductance of the soma and the input conductance of the cylinder. This is as expected, since the equivalent cylinder and the soma are in parallel.

## 8.3.2 A Finite Neuron and Soma

We now calculate the time-dependent response of a finite cable and lumped soma to a delta function current input at the soma, as this is readily observed experimentally.

We assume that the equivalent cylinder has finite length $L$. Then the potential satisfies

$$\frac{\partial V}{\partial T} = \frac{\partial^2 V}{\partial X^2} - V, \qquad 0 < X < L, T > 0, \tag{8.51}$$

$$V(X, 0) = 0, \tag{8.52}$$

with boundary conditions

$$\frac{\partial V(L,T)}{\partial X} = 0, \qquad (8.53)$$

$$\frac{\partial V(0,T)}{\partial T} + V(0,T) - \gamma\frac{\partial V(0,T)}{\partial X} = R_s\delta(T). \qquad (8.54)$$

Note that the boundary condition (8.54) is equivalent to

$$\frac{\partial V(0,T)}{\partial T} + V(0,T) - \gamma\frac{\partial V(0,T)}{\partial X} = 0, \qquad T > 0, \qquad (8.55)$$

together with the initial condition

$$V(0,0) = R_s. \qquad (8.56)$$

No single method such as Fourier series or Laplace transforms suffices to solve these equations. However, by combining aspects of both methods, the exact solution can be obtained (Tuckwell, 1988). We begin by finding a generalized Fourier series expansion of the solution. Using separation of variables, we find solutions of the form

$$V(X,T) = \phi(X)e^{-\mu^2 T}, \qquad (8.57)$$

where $\phi$ satisfies the differential equation

$$\phi'' - (1 - \mu^2)\phi = 0, \qquad (8.58)$$

with boundary conditions (when $T > 0$)

$$\phi'(L) = 0, \qquad (8.59)$$

$$\phi'(0) = \phi(0)\frac{1 - \mu^2}{\gamma}. \qquad (8.60)$$

Setting $\lambda^2 = \mu^2 - 1$, we solve for $\phi$ to get

$$\phi = A\cos(\lambda X) + B\sin(\lambda X), \qquad (8.61)$$

for some constants $A$ and $B$, and then apply the boundary conditions, finding

$$B = \frac{-\lambda A}{\gamma} \qquad (8.62)$$

and

$$\tan(\lambda L) = -\frac{\lambda}{\gamma}. \qquad (8.63)$$

The roots of (8.63) determine the eigenvalues. Although the eigenvalues cannot be found analytically, they can be determined numerically. A graph of (8.63), showing the location of the eigenvalues, is given in Fig. 8.5. There is an infinite number of discrete eigenvalues, and we label them $\lambda_n$, with $\lambda_0 = 0$. Expanding the solution in terms of the

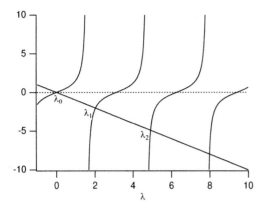

**Figure 8.5** The eigenvalues of (8.51)–(8.54) are determined by the intersections of the curves $\tan(\lambda L)$ and $-\lambda/\gamma$. In this figure, $L = \gamma = 1$.

eigenfunctions, we write

$$V(X,T) = \sum_{n=0}^{\infty} A_n \left( \cos(\lambda_n X) - \frac{\lambda_n}{\gamma} \sin(\lambda_n X) \right) e^{-(1+\lambda_n^2)T}. \tag{8.64}$$

Note that if $\lambda_n$ is an eigenvalue, so also is $-\lambda_n$, but the eigenfunction $\phi_n(X) = \cos(\lambda_n X) - \frac{\lambda_n}{\gamma} \sin(\lambda_n X)$ is an even function of $\lambda_n$, so it suffices to include only positive eigenvalues in the expansion.

We now strike a problem. The usual procedure is to expand the initial condition in terms of the eigenfunctions and thereby determine the coefficients $A_n$. However, in this case the eigenfunctions are not mutually orthogonal, and this procedure does not work easily. One way around this is to construct a nonorthogonal expansion of the initial condition, an approach used by Durand (1984). Here we use a different approach (Bluman and Tuckwell, 1987; Tuckwell, 1988) in which we calculate the Laplace transform of the solution, and then, by matching the two forms of the solution, obtain expressions for the unknown coefficients.

Taking the Laplace transform of (8.51)–(8.54) gives

$$\bar{V}'' = (s+1)\bar{V}, \tag{8.65}$$

$$\bar{V}'(L,s) = 0, \tag{8.66}$$

$$(s+1)\bar{V}(0,s) - \gamma\bar{V}'(0,s) = R_s, \tag{8.67}$$

where $\bar{V}(X,s)$ denotes the Laplace transform of $V$.

It is left as an exercise (Exercise 10) to show that the solution for $\bar{V}$ is

$$\bar{V}(X,s) = \frac{R_s \cosh[\sqrt{s+1}(X-L)]}{\sqrt{s+1}\{\sqrt{s+1}\cosh(\sqrt{s+1}L) + \gamma\sinh(\sqrt{s+1}L)\}}. \tag{8.68}$$

It now follows that

$$\bar{V}(0,s) = \frac{R_s}{s+1+\sqrt{s+1}\gamma\tanh(L\sqrt{s+1})}. \tag{8.69}$$

However, taking the Laplace transform of (8.64) we get

$$\bar{V}(0,s) = \sum_{n=0}^{\infty} \frac{A_n}{1+s+\lambda_n^2}. \tag{8.70}$$

Equating (8.69) and (8.70) gives

$$\frac{R_s}{s+1+\sqrt{s+1}\gamma \tanh(L\sqrt{s+1})} = \sum_{n=0}^{\infty} \frac{A_n}{1+s+\lambda_n^2}. \tag{8.71}$$

We can calculate the coefficients $A_n$ using contour integration. That is, if $C_n$ denotes a small circle in the complex $s$ plane, centered at $s = -1 - \lambda_n^2$, then integrating (8.70) around $C_n$ in the counterclockwise direction and using the residue theorem gives

$$A_n = \frac{1}{2\pi i} \int_{C_n} \bar{V}(0,s)ds. \tag{8.72}$$

Now we define

$$f(s) = s+1+\sqrt{s+1}\gamma \tanh(L\sqrt{s+1}) \tag{8.73}$$

and notice that $f(-1-\lambda_n^2) = 0$. This easily follows from the identity $\tanh(i\lambda_n) = i \tan(\lambda_n)$ and (8.63). Thus, by the residue theorem,

$$A_n = \frac{1}{2\pi i} \int_{C_n} \frac{R_s}{f(s)}ds \tag{8.74}$$

$$= \frac{R_s}{f'(-1-\lambda_n^2)}, \tag{8.75}$$

so that

$$\frac{1}{A_n} = \frac{d}{ds}\left(\frac{1}{\bar{V}(0,s)}\right)\Bigg|_{s=-1-\lambda_n^2}. \tag{8.76}$$

These coefficients are easily evaluated numerically.

### 8.3.3   Other Compartmental Models

The methods presented above give some idea of the difficulty of calculating analytical solutions to the cable equation on branching structures, with or without a soma termination. Since modern experimental techniques can determine the detailed structure of a neuron (for instance, by staining with horseradish peroxidase), it is clear that more experimental information can be obtained than can be incorporated into an analytical model (as is often the case). Thus, one common approach is to construct a large computational model of a neuron and then determine the solution by a numerical method.

In a numerical approach, a neuron is divided into a large number of small pieces, or compartments, each of which is assumed to be isopotential. Within each compartment the properties of the neuronal membrane are specified, and thus some compartments may have excitable kinetics, while others are purely passive. The compartments are then connected by an axial resistance, resulting in a large system of coupled ordinary differential equations, with the voltage specified at discrete places along the neuron.

Compartmental models, numerical methods for their solution, and software packages used for these kinds of models are discussed in detail in Koch and Segev (1989), to which the interested reader is referred.

## 8.4 Appendix: Transform Methods

To follow all of the calculations and complete all the exercises in this chapter, you will need to know about Fourier and Laplace transforms, generalized functions and the delta function, Green's functions, as well as some aspects of complex variable theory, including contour integration and the residue theorem. If you have made it this far into this book, then you are probably familiar with these classic techniques. However, should you need a reference for these techniques, there are many books with the generic title "Advanced Engineering Mathematics," from which to choose (see, for example, Kreyszig (1994), O'Neill (1983), or Kaplan (1981)). At an intermediate level one might consider Strang (1986) or Boyce and DiPrima (1997). Keener (1988) provides a more advanced coverage of this material.

## 8.5 Exercises

1. Calculate the input resistance of a cable with a sealed end at $X = L$. Determine how the length of the cable, and the boundary condition at $X = L$, affects the input resistance, and compare to the result for a semi-infinite cable.

2. (a) Find the fundamental solution $K$ of the linear cable equation satisfying

$$-\frac{d^2K}{dX^2} + K = \delta(X - \xi), \quad -\infty < X < \infty, \tag{8.77}$$

where $\delta(X - \xi)$ denotes an *inward* flow of positive current at the point $X = \xi$.
Answer:

$$K(X, \xi) = \frac{1}{2}e^{-|X-\xi|}. \tag{8.78}$$

(b) Use the fundamental solution to construct a solution of the cable equation with inhomogeneous current input

$$-\frac{d^2V}{dX^2} + V = I(X). \tag{8.79}$$

Answer:

$$V(X) = \int_{-\infty}^{\infty} I(\xi)K(X, \xi) \, d\xi = \frac{1}{2}\int_{-\infty}^{\infty} I(\xi)e^{-|X-\xi|} \, d\xi. \tag{8.80}$$

3. Solve

$$-\frac{d^2G(X)}{dx^2} + G(X) = \delta(X - \xi), \quad 0 < X, \xi < L, \tag{8.81}$$

subject to (i) sealed end, and (ii) short circuit, boundary conditions. The function $G$ is called the time-independent Green's function, and is similar to the fundamental solution. The only difference between the two is that the Green's function satisfies the given boundary conditions, while the fundamental solution does not.

4. (a) Use Laplace transforms to find the solution of the semi-infinite (time-dependent) cable equation with clamped voltage

$$V(X, 0) = 0 \tag{8.82}$$

and current input

$$\frac{\partial V(0, T)}{\partial X} = -r_i \lambda_m I_0 H(T), \tag{8.83}$$

where $H$ is the Heaviside function.
Answer:

$$V(X, T) = \frac{r_i \lambda_m I_0}{2} \left\{ e^{-X} \operatorname{erfc}\left( \frac{X}{2\sqrt{T}} - \sqrt{T} \right) - e^X \operatorname{erfc}\left( \frac{X}{2\sqrt{T}} + \sqrt{T} \right) \right\}, \tag{8.84}$$

where erfc is the complementary error function,

$$\operatorname{erfc}(x) = \frac{2}{\sqrt{\pi}} \int_x^\infty e^{-y^2} \, dy. \tag{8.85}$$

Hint: Use the identity

$$\frac{2}{s\sqrt{s+1}} = \frac{1}{s+1-\sqrt{s+1}} - \frac{1}{s+1+\sqrt{s+1}}, \tag{8.86}$$

and then use

$$\mathcal{L}^{-1}\left\{ \frac{e^{-a\sqrt{s}}}{s + b\sqrt{s}} \right\} = e^{b^2 T + ab} \operatorname{erfc}\left( \frac{a}{2\sqrt{T}} + b\sqrt{T} \right), \tag{8.87}$$

where $\mathcal{L}^{-1}$ denotes the inverse Laplace transform.

(b) Show that

$$V(X, T) \to r_i \lambda_m I_0 e^{-X} \tag{8.88}$$

as $T \to \infty$.

5. Calculate the time-dependent Green's function for a finite cylinder of length $L$ for (i) sealed end, and (ii) short circuit, boundary conditions. These may be calculated in two different ways, either using Fourier series or by constructing sums of fundamental solutions.

6. By solving for the unknown constants, calculate the solution of the cable equation on the simple branching structure of Fig. 8.3. Show explicitly that this solution is the same as the equivalent cylinder solution, as long as the necessary conditions are satisfied.

7. Show that if the conditions in Section 8.2.3 are satisfied, a branching structure can be condensed into a single equivalent cylinder.

8. Show that

$$\frac{\partial V}{\partial T} = \frac{\partial^2 V}{\partial X^2} - V + \delta(X)\delta(T), \qquad T \geq 0, \tag{8.89}$$

with $V(X, T) = 0$ for $T < 0$, is equivalent to

$$\frac{\partial V}{\partial T} = \frac{\partial^2 V}{\partial X^2} - V, \qquad T > 0, \tag{8.90}$$

with initial condition

$$V(X, 0) = \delta(X). \tag{8.91}$$

9.  Show that as $n \to \infty$, the eigenvalues $\lambda_n$ of (8.63) are approximately $(2n - 1)\pi/(2L)$.

10. Solve (8.65)–(8.67) for $\bar{V}(X, s)$.

11. Using the method of Section 8.3.2, find the Green's function for the finite cylinder and lumped soma; i.e., solve

$$\frac{\partial V}{\partial T} = \frac{\partial^2 V}{\partial X^2} - V + \delta(X - \xi)\delta(T), \tag{8.92}$$

$$V(X, 0) = 0, \tag{8.93}$$

with boundary conditions

$$\frac{\partial V(L, T)}{\partial X} = 0, \tag{8.94}$$

$$\frac{\partial V(0, T)}{\partial T} + V(0, T) - \gamma \frac{\partial V(0, T)}{\partial X} = 0. \tag{8.95}$$

Show that as $\xi \to 0$ the solution approaches that found in Section 8.3.2, scaled by the factor $\gamma/R_s$.

# CHAPTER 9

# Nonlinear Wave Propagation

The problem of current flow in the axon of a nerve is much more complicated than that of flow in dendritic networks. We saw in Chapter 4 how the voltage dependence of the ionic currents can lead to excitability and action potentials. In this chapter we show that when an excitable membrane is incorporated into a *nonlinear* cable equation, it can give rise to traveling waves of electrical excitation.

Indeed, this property of spatially distributed Hodgkin–Huxley theory is one of the reasons that the Hodgkin–Huxley model is so important. In addition to producing a realistic description of a space clamped action potential, Hodgkin and Huxley showed that this action potential should propagate along an axon with a fixed speed, which could be calculated. Their model spawned an entire cottage industry of nonlinear wave propagation in excitable media.

## 9.1  Brief Overview of Wave Propagation

There is a vast literature on wave propagation in biological systems. In addition to the books by Murray (1989), Britton (1986), and Grindrod (1991), there are numerous articles in journals and books, many of which we cite in this chapter. To avoid confusion, we emphasize at the outset that when we use the term *traveling wave*, we mean a solution that travels at constant velocity with fixed shape. On an infinite domain (a fictional object, of course), a traveling wave would travel at a constant velocity indefinitely.

There are many different kinds of waves in biological systems. There are traveling waves in excitable systems, waves that arise from the underlying excitability of the cell. An excitable wave acts as a model for, among other things, the propagation of an action potential along the axon of a nerve or the propagation of a grass fire on a

**A**                       **B**

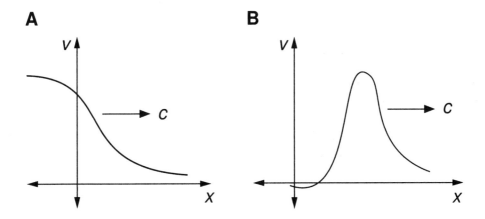

**Figure 9.1** Schematic diagram of A: Traveling front, B: Traveling pulse.

prairie. However, if the underlying kinetics are oscillatory but not excitable, and a large number of individual oscillatory units are coupled by diffusion, the resulting behavior is oscillatory waves and periodic wave trains. In this chapter we focus our attention on waves in excitable media, and delay consideration of the theory of coupled oscillators until later (Chapters 14 and 21).

It is also helpful to make a distinction between the two most important types of traveling waves in excitable systems. First, there is the wave that looks like a moving plateau. If we use $v$ to denote the wave variable, then in front of the wave, $v$ is steady at some low value, and behind the wave, $v$ is steady at a higher value (Fig. 9.1A). We call such waves *traveling fronts*. The second type of wave begins and ends at the same value of $v$ (Fig. 9.1B) and resembles a moving bump. We call this type of wave a *traveling pulse*.

These two wave types can be interpreted in the terminology of the Hodgkin–Huxley fast–slow phase-plane discussed in Chapter 4. A traveling front depends on the excitation variable $v$. We saw that when the recovery variable is fixed at the steady state, the fast–slow phase-plane has two stable steady states, $v_r$ and $v_e$ (i.e., it is bistable). Under appropriate conditions there exists a traveling front with $v = v_r$ in front of the wave and $v = v_e$ behind the wave. Thus, the traveling front acts like a zipper, changing the domain from the resting to the excited state. However, if the recovery variable $n$ is allowed to vary, the solution is eventually forced back to the resting state and the traveling front becomes a traveling pulse. The primary difference between the traveling front and the traveling pulse is that in the former case there is no recovery (or recovery is static), while in the latter case there is.

One of the simplest models for biological wave propagation is Fisher's equation. Although this equation is used extensively in population biology and ecology, it is much less relevant for the study of physiological waves, and so we do not discuss it here (see Exercise 13 and Fife, 1979).

The next level of complexity is the *bistable equation*. The bistable equation is so named because it has two stable rest points, and it is related to the FitzHugh–Nagumo model without recovery. For the bistable equation, one expects to find traveling fronts but not usually traveling pulses. Inclusion of the recovery variable leads to a more complex model, the spatially distributed FitzHugh–Nagumo model, for which one expects to find traveling pulses (among other types of waves). Wave propagation in the FitzHugh–Nagumo model is still not completely understood, especially in higher-dimensional domains. At the highest level of complexity are the spatially distributed models of Hodgkin–Huxley type, systems of equations that are resistant to analytical approaches.

## 9.2  Traveling Fronts

### 9.2.1  The Bistable Equation

The bistable equation is a special version of the cable equation (8.18), namely

$$\frac{\partial V}{\partial t} = \frac{\partial^2 V}{\partial x^2} + f(V), \tag{9.1}$$

where $f(V)$ has three zeros at $0, \alpha,$ and $1$, where $0 < \alpha < 1$. The values $V = 0$ and $V = 1$ are stable steady solutions of the ordinary differential equation $dV/dt = f(V)$. Notice that the variable $V$ may need to be rescaled so that $0$ and $1$ are zeros of $f(V)$. In the standard nondimensional form, $f'(0) = -1$. (Recall from (8.13) that the passive cable resistance was defined so that the ionic current has slope 1 at rest.) However, this restriction is often ignored.

An example of such a function can be found in the Hodgkin–Huxley fast–slow phase-plane. When the recovery variable $n$ is held fixed at its steady state, the Hodgkin–Huxley fast–slow model is bistable. Two other examples of functions that are often used in this context are the cubic polynomial

$$f(V) = aV(V - 1)(\alpha - V), \qquad 0 < \alpha < 1, \tag{9.2}$$

and the piecewise linear function

$$f(V) = -V + H(V - \alpha), \qquad 0 < \alpha < 1. \tag{9.3}$$

where $H(V)$ is the Heaviside function (Mckean, 1970). This piecewise linear function is not continuous, nor does it have three zeros, yet it is useful in the study of traveling wave solutions of the bistable equation because it is an analytically tractable model that retains many important qualitative features.

By a traveling wave solution, we mean a translation-invariant solution of (9.1) that provides a transition between two stable rest states (zeros of the nonlinear function $f(V)$) and travels with constant speed. In particular, if the traveling wave has the form of a traveling front, it provides a transition between two different zeros of $f$. That is, we seek a solution of (9.1) of the form

$$V(x,t) = U(x + ct) = U(\xi) \tag{9.4}$$

for some (yet to be determined) value of $c$. The new variable $\xi$, called the traveling wave variable, has the property that fixed values move in space–time with fixed speed $c$. When written as a function of $\xi$, the wave appears stationary. By substituting (9.4) into (9.1) it can be seen that any traveling wave solution must satisfy

$$U_{\xi\xi} - cU_\xi + f(U) = 0, \tag{9.5}$$

and this, being an ordinary differential equation, should be easier to analyze than the original partial differential equation. For $U(\xi)$ to provide a transition between rest points, it must be that $f(U(\xi)) \to 0$ as $\xi \to \pm\infty$.

To study (9.5) it is convenient to write it as two first order equations,

$$U_\xi = W, \tag{9.6}$$
$$W_\xi = cW - f(U). \tag{9.7}$$

To find traveling front solutions for the bistable equation, we look for a solution of (9.6) and (9.7) that connects the rest points $(U, W) = (0, 0)$ and $(U, W) = (1, 0)$ in the $(U, W)$ phase-plane. Such a trajectory, connecting two different steady states, is called a heteroclinic trajectory, and in this case is parametrized by $\xi$; the trajectory approaches $(0, 0)$ as $\xi \to -\infty$ and approaches $(1, 0)$ as $\xi \to +\infty$. The steady states at $U = 0$ and $U = 1$ are both saddle points, while for the steady state $U = \alpha$, the real part of both eigenvalues have the same sign, negative if $c$ is positive and positive if $c$ is negative, so that this is a node or a spiral point. Since the points at $U = 0$ and $U = 1$ are saddle points, our goal is to determine whether the parameter $c$ can be chosen such that the trajectory that leaves $U = 0$ at $\xi = -\infty$ can be made to connect with the saddle point $U = 1$ at $\xi = +\infty$. This mathematical procedure is called *shooting*, and some sample trajectories are shown in Fig. 9.2.

First of all, we can determine the sign of $c$. If a monotone increasing $(U_\xi > 0)$ connecting trajectory exists, we can multiply (9.5) by $U_\xi$ and integrate from $\xi = -\infty$ to $\xi = \infty$ with the result that

$$c \int_{-\infty}^{\infty} W^2 d\xi = \int_0^1 f(u) du. \tag{9.8}$$

In other words, if a traveling wave solution exists, then the sign of $c$ is the same as the sign of the area under the curve $f(u)$ between $u = 0$ and $u = 1$. If this area is positive, then the traveling solutions move the state variable $U$ from $U = 0$ to $U = 1$, and the state at $U = 1$ is said to be *dominant*. In both of the special cases (9.2) and (9.3), the state $U = 1$ is dominant if $\alpha < 1/2$.

Suppose $\int_0^1 f(u) du > 0$. We will try different values of $c$ to see what happens to the unstable trajectory that leaves the saddle point $U = 0, U_\xi = 0$. With $c = 0$, we can find an explicit expression for this trajectory by multiplying (9.5) by $U_\xi$ and integrating to get

$$\frac{W^2}{2} + \int_0^U f(u)du = 0. \tag{9.9}$$

If this trajectory were to reach $U = 1$ for some value of $W$, we would have

$$\frac{W^2}{2} + \int_0^1 f(u)du = 0, \tag{9.10}$$

in which case $\int_0^1 f(u)du < 0$. Since this contradicts our original assumption, we conclude that $U$ cannot reach 1. Neither can this trajectory stay in the first quadrant, as $W > 0$ implies that $U$ is always increasing there. Thus, this trajectory must intersect the $W = 0$ axis at some value of $U < 1$ (Fig. 9.2). It cannot be the connecting trajectory.

Next, suppose $c$ is large. In the $(U, W)$ phase-plane, the slope of the unstable trajectory leaving the rest point at $U = 0$ is the positive root of $\lambda^2 - c\lambda + f'(0) = 0$, which is always larger than $c$ (Exercise 1). Let $K$ be the smallest positive number for which $f(u)/u \leq K$ for all $u$ on the interval $0 < u \leq 1$ (Exercise: How do we know $K$ exists?), and let $\sigma$ be any fixed positive number. On the line $W = \sigma U$ the slope of trajectories satisfies

$$\frac{dW}{dU} = c - \frac{f(U)}{W} = c - \frac{f(U)}{\sigma U} \geq c - \frac{K}{\sigma}. \tag{9.11}$$

By picking $c$ large enough, we are assured that $c - K/\sigma > \sigma$, so that once trajectories are above the line $W = \sigma U$, they stay above it. We know that for large enough $c$, the trajectory leaving the saddle point $U = 0$ starts out above this curve. Thus, this trajectory always stays above the line $W = \sigma U$, and therefore passes above the rest point at $(U, W) = (1, 0)$.

Now we have two trajectories, one with $c = 0$, which misses the rest point at $U = 1$ by crossing the $W = 0$ axis at some point $U < 1$, and one with $c$ large, which misses this rest point by staying above it at $U = 1$. Since trajectories depend continuously on the

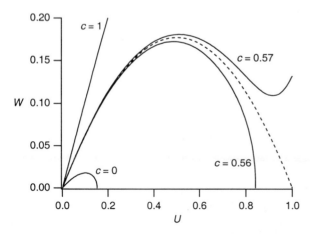

**Figure 9.2** Trajectories in the $(U, W)$ phase-plane leaving the rest point $U = 0$, $W = 0$ for the equation $U_{\xi\xi} - cU_\xi + U(U - 0.1)(1 - U) = 0$, with $c = 0.0$, 0.56, 0.57, and 1.0. Dashed curve shows the connecting heteroclinic trajectory.

parameters of the problem, there is a continuous family of trajectories depending on the parameter $c$ between these two special trajectories, and therefore there is at least one trajectory that hits the point $U = 1, W = 0$ exactly.

The value of $c$ for which this heteroclinic connection occurs is unique. To verify this statement, notice from (9.11) that the slope $dW/dU$ of trajectories in the $(U, W)$ plane is a monotone increasing function of the parameter $c$. Suppose at some value of $c = c_0$ there is known to be a connecting trajectory. For any value of $c$ that is larger than $c_0$, the trajectory leaving the saddle point at $U = 0$ must lie above the connecting curve for $c_0$. For the same reason, with $c > c_0$, the trajectory approaching the saddle point at $U = 1$ as $\xi \to \infty$ must lie below the connecting curve with $c = c_0$. A single curve cannot simultaneously lie above and below another curve, so there cannot be a connecting trajectory for $c > c_0$. By a similar argument, there cannot be a connecting trajectory for a smaller value of $c$, so the value $c_0$, and hence the connecting trajectory, is unique.

For most functions $f(V)$, it is necessary to calculate the speed of propagation of the traveling front solution numerically. However, in the two special cases (9.2) and (9.3) the speed of propagation can be calculated explicitly. In the piecewise linear case (9.3) one calculates directly that

$$c = \frac{1 - 2\alpha}{\sqrt{\alpha - \alpha^2}} \tag{9.12}$$

(see Exercise 4).

Suppose $f(u)$ is the cubic polynomial

$$f(u) = -A^2(u - u_0)(u - u_1)(u - u_2), \tag{9.13}$$

where the zeros of the cubic are ordered $u_0 < u_1 < u_2$. We want to find a heteroclinic connection between the smallest zero $u_0$, and the largest zero $u_2$, so we guess that

$$W = -B(U - u_0)(U - u_2). \tag{9.14}$$

We substitute this guess into the governing equation (9.5), and find that we must have

$$B^2(2U - u_0 - u_2) - cB - A^2(U - u_1) = 0. \tag{9.15}$$

This is a linear function of $U$ that can be made identically zero only if we choose $B = A/\sqrt{2}$ and

$$c = \frac{A}{\sqrt{2}}(u_2 - 2u_1 + u_0). \tag{9.16}$$

It follows from (9.14) that

$$U(\xi) = \frac{u_0 + u_2}{2} + \frac{u_2 - u_0}{2} \tanh\left(\frac{A}{\sqrt{2}} \frac{u_2 - u_0}{2}\xi\right), \tag{9.17}$$

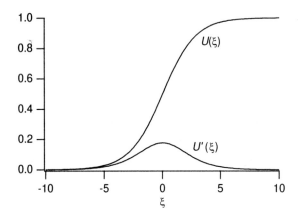

**Figure 9.3**  Profile of traveling wave solution of the bistable equation with $f(V) = V(1 - V)(0.1 - V)$.

which is independent of $u_1$. In the case that $u_0 = 0, u_1 = \alpha$, and $u_2 = 1$, the speed reduces to

$$c = \frac{A}{\sqrt{2}}(1 - 2\alpha), \tag{9.18}$$

showing that the speed is a decreasing function of $\alpha$ and the direction of propagation changes at $\alpha = 1/2$. The profile of the traveling wave in this case is

$$U(\xi) = \frac{1}{2}\left[1 + \tanh\left(\frac{A}{2\sqrt{2}}\xi\right)\right]. \tag{9.19}$$

A plot of this traveling wave profile is shown in Fig. 9.3.

Once the solution of the nondimensional cable equation (9.1) is known, it is a simple matter to express the solution in terms of physical parameters as

$$V(x,t) = U\left(\frac{x}{\lambda_m} + c\frac{t}{\tau_m}\right), \tag{9.20}$$

where $\lambda_m$ and $\tau_m$ are, respectively, the space and time constants of the cable, as described in Chapter 8. The speed of the traveling wave is

$$s = \frac{c\lambda_m}{\tau_m} = \frac{c}{2C_m}\sqrt{\frac{d}{R_m R_c}}, \tag{9.21}$$

which shows how the wave speed depends on capacitance, membrane resistance, cytoplasmic resistance, and axonal diameter. The dependence of the speed on ionic channel conductances is contained (but hidden) in $c$. According to empirical measurements, a good estimate of the speed for an axon is given by

$$s = \sqrt{\frac{d}{10^{-6}\text{m}}} \text{ m/sec.} \tag{9.22}$$

Using $d = 500\mu\text{m}$ for squid axon, this estimate gives $s = 22.4$ mm/ms, which compares favorably to the measured value of $s = 21.2$ mm/ms.

**Table 9.1** Sodium channel densities in selected excitable tissues.

| Tissue | Channel density (channels/$\mu$m$^2$) |
|---|---|
| Mammalian | |
|     Vagus nerve (nonmyelinated) | 110 |
|     Node of Ranvier | 2100 |
|     Skeletal muscle | 205–560 |
| Other animals | |
|     Squid giant axon | 166–533 |
|     Frog sartorius muscle | 280 |
|     Electric eel electroplax | 550 |
|     Garfish olfactory nerve | 35 |
|     Lobster walking leg nerve | 90 |

Scaling arguments can also be used to find the dependence of speed on certain other parameters. Suppose, for example, that a drug is applied to the membrane that blocks a percentage of all ion channels, irrespective of type. If $\rho$ is the fraction of remaining operational channels, then the speed of propagation is reduced by the factor $\sqrt{\rho}$. This follows directly by noting that the bistable equation with reduced ion channels

$$V'' - sV' + \rho f(V) = 0 \tag{9.23}$$

can be related to the original bistable equation (9.5) by taking $V(\xi) = U(\sqrt{\rho}\xi), s = c\sqrt{\rho}$.

## Thresholds and stability

There are many other features of the bistable equation, the details of which are beyond the scope of this book. Perhaps the most important of these features is that solutions of the bistable equation satisfy a comparison property: any two solutions of the bistable equation, say $u_1(x,t)$ and $u_2(x,t)$, that are ordered with $u_1(x,t_0) \leq u_2(x,t_0)$ at some time $t = t_0$ remain ordered for all subsequent times, i.e., $u_1(x,t) \leq u_2(x,t)$ for $t \geq t_0$.

With comparison arguments it is possible to prove a number of additional facts (Aronson and Weinberger, 1975). For example, the bistable equation exhibits threshold phenomena. Specifically, if initial data are sufficiently small, then the solution of the bistable equation approaches zero in the limit $t \to \infty$. However, there are initial functions with compact support lying between 0 and 1 for which the solution approaches 1 in the limit $t \to \infty$. Because of the comparison theorem any larger initial function also initiates a solution that approaches 1 in the limit $t \to \infty$. Such initial data are said to be *superthreshold*.

Furthermore, the traveling wave solution of the bistable equation is stable in a very strong way (Fife, 1979; Fife and McLeod, 1977), as follows. Starting from any initial data that lie between 0 and $\alpha$ in the limit $x \to -\infty$ and between $\alpha$ and 1 in the limit $x \to \infty$, the solution becomes arbitrarily close to some phase shift of the traveling wave solution for sufficiently large time.

## 9.3  Myelination

Most nerve fibers are coated with a lipid material called *myelin* with periodic gaps of exposure called *nodes of Ranvier*. The myelin sheath consists of a single cell, called a *Schwann cell*, which is wrapped many times (roughly 100 times) around the axonal membrane. This wrapping of the axon increases the effective membrane resistance by a factor of about 100 and decreases the membrane capacitance by a factor of about 100. Indeed, rough data are that $R_m$ is $10^3 \, \Omega \, \text{cm}^2$ for cell membrane and $10^5 \, \Omega \, \text{cm}^2$ for myelin sheath, and that $C_m$ is $10^{-6} \, \mu\text{F/cm}^2$ for cell membrane and $10^{-8} \, \mu\text{F/cm}^2$ for a myelinated fiber. The length of myelin sheath is typically 1 to 2 mm (close to 100 $d$, where $d$ is the fiber diameter), and the width of the gap is about 1 $\mu$m.

Propagation along myelinated fiber is faster than along nonmyelinated fiber. This is presumably caused by the fact that there is little transmembrane ionic current and little capacitive current in the myelinated section, allowing the axon to act as a simple resistor. An action potential does not propagate along the myelinated fiber but rather jumps from node-to-node. This node-to-node propagation is said to be *saltatory* (from the Latin word *saltare*, to leap or dance).

The pathophysiological condition of nerve cells in which damage of the myelin sheath impairs nerve impulse transmission in the central nervous system is called *mul-*

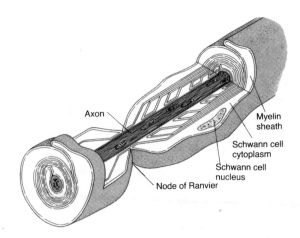

**Figure 9.4**  Schematic diagram of the myelin sheath. (Guyton and Hall, 1996, Fig. 5-16, p. 69.)

*tiple sclerosis* (MS). MS is a disease that usually affects young adults between the ages of 18 and 40, occurring slightly more often in females than males. MS attacks the white matter of the brain and spinal cord, causing demyelination of nerve fibers at various locations throughout the central nervous system, although the underlying nerve axons and cell bodies are not usually damaged. The loss of myelin slows or stops the transmission of action potentials, with the resultant symptoms of muscle fatigue and weakness or extreme "heaviness."

To model the electrical activity in a myelinated fiber we assume that the capacitive and transmembrane ionic currents are negligible, so that along the myelin sheath the axial currents

$$I_e = -\frac{1}{r_e}\frac{\partial V_e}{\partial x}, \qquad I_i = -\frac{1}{r_i}\frac{\partial V_i}{\partial x} \tag{9.24}$$

are constant, where we are using the same notation as in Chapter 8. We also assume that $V$ does not vary within each node of Ranvier (i.e., that the nodes are isopotential), and that the voltage at the $n$th node is given by $V_n$. Then the axial currents between node $n$ and node $n+1$ are

$$I_e = -\frac{1}{Lr_e}(V_{e,n+1} - V_{e,n}), \qquad I_i = -\frac{1}{Lr_i}(V_{i,n+1} - V_{i,n}), \tag{9.25}$$

where $L$ is the length of the myelin sheath between nodes. The total transmembrane current at a node is given by

$$\mu p \left( C_m \frac{\partial V_n}{\partial t} + I_{\text{ion}} \right) = I_{i,n} - I_{i,n+1}$$

$$= \frac{1}{L(r_i + r_e)}(V_{n+1} - 2V_n + V_{n-1}), \tag{9.26}$$

where $\mu$ is the length of the node.

We can introduce dimensionless time $\tau = \frac{t}{C_m R_m} = t/\tau_m$ (but not dimensionless space), to rewrite (9.26) as

$$\frac{dV_n}{d\tau} = f(V_n) + D(V_{n+1} - 2V_n + V_{n-1}), \tag{9.27}$$

where $D = \frac{R_m}{\mu L p (r_i + r_e)}$ is the coupling coefficient. We call this equation the *discrete* cable equation.

## 9.3.1 The Discrete Bistable Equation

The discrete bistable equation is the system of equations (9.27) where $f(V)$ has typical bistable form, as, for example, (9.2) or (9.3). The study of the discrete bistable equation is substantially more difficult than that of the continuous version (9.1). While the discrete bistable equation looks like a finite difference approximation of the continuous bistable equation, solutions of the two have significantly different behavior.

It is a highly nontrivial matter to prove that traveling wave solutions of the discrete system exist (Zinner, 1992). However, a traveling wave solution, if it exists, satisfies the special relationship $V_{n+1}(\tau) = V_n(\tau - \tau_d)$. In other words, the $(n+1)$st node experiences exactly the same time course as the $n$th node, but with time delay $\tau_d$. Furthermore, if $V_n(\tau) = V(\tau)$, it follows from (9.27) that $V(\tau)$ must satisfy the delay differential equation

$$\frac{dV}{d\tau} = D(V(\tau + \tau_d) - 2V(\tau) + V(\tau - \tau_d)) + f(V(\tau)). \tag{9.28}$$

If the function $V(\tau)$ is sufficiently smooth and if $\tau_d$ is sufficiently small, then we can approximate $V(\tau + \tau_d)$ with its Taylor series $V(\tau + \tau_d) = \sum_{n=0} \frac{1}{n!} V^{(n)}(\tau)\tau_d^n$, so that (9.28) is approximated by the differential equation

$$D\left(\tau_d^2 V_{\tau\tau} + \frac{\tau_d^4}{12} V_{\tau\tau\tau\tau}\right) - V_\tau + f(V) = 0, \tag{9.29}$$

ignoring terms of order $\tau_d^6$ and higher.

Now we suppose that $\tau_d$ is small. The leading-order equation is

$$D\tau_d^2 V_{\tau\tau} - V_\tau + f(V) = 0, \tag{9.30}$$

which has solution $V_0(\tau) = U(c\tau)$, provided that $D\tau_d^2 = 1/c^2$, where $U$ is the traveling front solution of the bistable equation (9.5) and $c$ is the dimensionless wave speed for the continuous equation. The wave speed $s$ is the internodal distance $L + \mu$ divided by the time delay $\tau_m \tau_d$, so that

$$s = \frac{L + \mu}{\tau_m \tau_d} = (L + \mu)c\frac{\sqrt{D}}{\tau_m}. \tag{9.31}$$

For myelinated nerve fiber we know that $D = \frac{R_m}{\mu L p(r_i + r_e)}$. If we ignore extracellular resistance, we find a leading order approximation for the velocity of

$$s = \frac{L + \mu}{\sqrt{\mu L}} \frac{c}{2C_m} \sqrt{\frac{d}{R_m R_c}}, \tag{9.32}$$

giving a change in velocity compared to nonmyelinated fiber by the factor $\frac{L+\mu}{\sqrt{\mu L}}$. If we estimate $L = 100\,d$ and take $\mu = 1\ \mu$m, this increase in velocity is by a factor of $10\sqrt{\frac{d}{10^{-6}\text{m}}}$, which is quite substantial. Empirically it is known that the improvement of velocity for myelinated fiber compared to nonmyelinated fiber is by a factor of about $6\sqrt{\frac{d}{10^{-6}\text{m}}}$.

## Higher-order approximation

We can find a higher-order approximation to the speed of propagation by using a standard regular perturbation argument. We set $\epsilon = 1/D$ and seek a solution of (9.29) of the form

$$V(\tau) = V_0(\tau) + \epsilon V_1(\tau) + \cdots, \tag{9.33}$$

$$\tau_d^2 = \frac{\epsilon}{c^2} + \epsilon^2 \tau_1 + \cdots. \tag{9.34}$$

We expand (9.29) into its powers of $\epsilon$ and set the coefficients of $\epsilon$ to zero. The first equation we obtain from this procedure is (9.30), and the second equation is

$$L[V_1] = \frac{1}{c^2}V_1'' - V_1' + f'(V_0)V_1 = -\frac{V_0''''}{12c^4} - \tau_1 V_0''. \tag{9.35}$$

Note that here we are using $L[\cdot]$ to denote a linear differential operator. The goal is to find solutions of (9.35) that are square integrable on the infinite domain, so that the solution is "close" to $V_0$. The linear operator $L[\cdot]$ is not an invertible operator in this space by virtue of the fact that $L[V_0'(\tau)] = 0$. (This follows by differentiating (9.30) once with respect to $\tau$.) Thus, it follows from the Fredholm alternative theorem (Keener, 1988) that a solution of (9.35) exists if and only if the right-hand side of the equation is orthogonal to the null space of the adjoint operator $L^*$. Here the adjoint differential operator is

$$L^*[V] = \frac{1}{c^2}V'' + V' + f'(V_0)V, \tag{9.36}$$

and the one element of the null space (a solution of $L^*[V] = 0$) is

$$V^*(\tau) = \exp\left(-c^2\tau\right)V_0'(\tau). \tag{9.37}$$

This leads to the solvability condition

$$\tau_1 \int_{-\infty}^{\infty} \exp\left(-c^2\tau\right)V_0'(\tau)V_0''(\tau)d\tau = -\frac{1}{12c^4}\int_{-\infty}^{\infty}\exp\left(-c^2\tau\right)V_0'(\tau)V_0''''(\tau)d\tau. \tag{9.38}$$

As a result, $\tau_1$ can be calculated (either analytically or numerically) by evaluating two integrals, and the speed of propagation is determined as

$$s = (L + \mu)\frac{c}{\tau_m}\sqrt{D}\left(1 - \frac{\tau_1 c^2}{2D} + O\left(\left(\frac{c^2}{D}\right)^2\right)\right). \tag{9.39}$$

This exercise is interesting from the point of view of numerical analysis, as it shows the effect of numerical discretization on the speed of propagation. This method can be applied to other numerical schemes for an equation with traveling wave solutions (Exercise 17).

## Propagation failure

The most significant difference between the discrete and continuous equations is that the discrete system has a coupling threshold for propagation, while the continuous model allows for propagation at all coupling strengths. It is readily seen from (9.21) that for the continuous cable equation, continuous changes in the physical parameters lead to continuous changes in the speed of propagation, and the speed cannot be driven to zero unless the diameter is zero or the resistances or capacitance are infinite. Such is not the case for the discrete system, and propagation may fail if the coupling coefficient is too small. This is easy to understand when we realize that if the coupling strength is very weak, so that the effective internodal resistance is large, the current flow from an excited node to an unexcited node may be so small that the threshold of the unexcited node is not exceeded, and propagation cannot continue.

We seek standing (time-independent, i.e., $dV_n/d\tau = 0$) solutions of the discrete equation (9.27). The motivation for this comes from the maximum principle and comparison arguments. One can show that if two sets of initial data for the discrete bistable equation are initially ordered, the corresponding solutions remain ordered for all time. It follows that if the discrete bistable equation has a monotone increasing stationary front solution, then there cannot be a traveling wave front solution.

A standing front solution of the discrete bistable equation is a sequence $\{V_n\}$ satisfying the finite difference equation

$$0 = D(V_{n+1} - 2V_n + V_{n-1}) + f(V_n) \tag{9.40}$$

for all integers $n$, for which $V_n \to 1$ as $n \to \infty$ and $V_n \to 0$ as $n \to -\infty$.

One can show (Keener, 1987) that for any bistable function $f$, there is a number $D^* > 0$ such that for $D \leq D^*$, the discrete bistable equation has a standing solution, that is, propagation fails. To get a simple understanding of the behavior of this coupling threshold, we solve (9.40) in the special case of piecewise linear dynamics (9.3). Since the discrete equation with dynamics (9.3) is linear, the homogeneous solution can be expressed as a linear combination of powers of some number $\lambda$ as

$$V_n = A\lambda^n + B\lambda^{-n}, \tag{9.41}$$

where $\lambda$ is a solution of the characteristic polynomial equation

$$\lambda^2 - \left(2 + \frac{1}{D}\right)\lambda + 1 = 0. \tag{9.42}$$

Note that this implies that

$$D = \frac{\lambda}{(\lambda - 1)^2}. \tag{9.43}$$

The characteristic equation has two positive roots, one larger and one smaller than 1. Let $\lambda$ be the root that is smaller than one. Then, taking the conditions at $\pm\infty$ into account, we write the solution as

$$V_n = \begin{cases} 1 + A\lambda^n, & \text{for } n \geq 0, \\ B\lambda^{-n}, & \text{for } n < 0. \end{cases} \tag{9.44}$$

This expression for $V_n$ must also satisfy the piecewise linear discrete bistable equation for $n = -1, 0$. Thus,

$$D(V_1 - 2V_0 + V_{-1}) = V_0 - 1, \tag{9.45}$$

$$D(V_0 - 2V_{-1} + V_{-2}) = V_{-1}, \tag{9.46}$$

where we have used the requirement that $V_n \geq \alpha$ for all $n \geq 0$, and $V_n \leq \alpha$ for all $n \leq 0$. Substituting in (9.43) for $D$, and solving for $A$ and $B$, then gives $B = A + 1 = \frac{1}{1+\lambda}$.

Finally, this is a solution for all $n$, provided that $V_0 \geq \alpha$. Since $V_0 = B = \frac{1}{1+\lambda}$, we need $\frac{1}{1+\lambda} \geq \alpha$, or $\lambda \leq \frac{1-\alpha}{\alpha}$. However, when $\lambda < 1$, $D$ is an increasing function of $\lambda$, and thus $\lambda \leq \frac{1-\alpha}{\alpha}$ whenever

$$D \leq D\left(\frac{1-\alpha}{\alpha}\right) = \frac{\alpha(1-\alpha)}{(2\alpha-1)^2} = D^*. \tag{9.47}$$

In other words, there is a standing wave, precluding propagation, whenever the coupling is small, with $D \leq D^*$. Since $\alpha$ is a measure of the excitability of this medium, we see that when the medium is weakly excitable ($\alpha$ is near 1/2), then $D^*$ is large and very little resistance is needed to halt propagation. On the other hand, when $\alpha$ is small, so that the medium is highly excitable, the resistance threshold is quite large, and propagation is relatively difficult to stop.

## 9.4 Traveling Pulses

A traveling pulse (often called a *solitary pulse*) is a traveling wave solution that starts and ends at the same steady state of the governing equations. Recall that a traveling front solution corresponds to a heteroclinic trajectory in the $(U, W)$ phase-plane, i.e., a trajectory, parametrized by $\xi$, that connects two different steady states of the system. A traveling pulse solution is similar, corresponding to a trajectory that begins and ends at the *same* steady state in the traveling wave coordinate system. Such trajectories are called *homoclinic orbits*.

There are three main approaches to finding traveling pulses for excitable systems. First, one can approximate the nonlinear functions with piecewise linear functions, and then find traveling pulse solutions as exact solutions of transcendental equations. Second, one can use perturbation methods exploiting the different time scales to find approximate analytical expressions. Finally, one can use numerical simulations to solve the governing differential equations. We illustrate each of these techniques in turn.

### 9.4.1 The FitzHugh–Nagumo Equations

To understand the structure of a traveling pulse it is helpful first to study traveling pulse solutions in the FitzHugh–Nagumo equations

$$\epsilon \frac{\partial v}{\partial t} = \epsilon^2 \frac{\partial^2 v}{\partial x^2} + f(v, w), \tag{9.48}$$

$$\frac{\partial w}{\partial t} = g(v, w), \tag{9.49}$$

where $\epsilon$ is assumed to be a small positive number. Without any loss of generality, space has been scaled so that the diffusion coefficient of $v$ is $\epsilon^2$. It is important to realize that this does not imply anything about the magnitude of the physical diffusion coefficient. We are simply scaling the space variable so that in the new coordinate system, the wave front appears steep, a procedure that facilitates the study of the wave as a whole. The variable $v$ is spatially coupled with diffusion, but the variable $w$ is not, owing to the fact that $v$ represents the membrane potential, while $w$ represents a slow ionic current or gating variable.

To study traveling waves, we first place the system of equations (9.48)–(9.49) in a traveling coordinate frame of reference. We define the traveling wave coordinate $\xi = x - ct$, where $c > 0$ is the wave speed, yet to be determined. Then the partial differential equations (9.48)–(9.49) become the ordinary differential equations

$$\epsilon^2 v_{\xi\xi} + c\epsilon v_\xi + f(v, w) = 0, \tag{9.50}$$

$$cw_\xi + g(v, w) = 0. \tag{9.51}$$

## A piecewise linear model

We begin by examining the simplest case, the piecewise linear dynamics (Rinzel and Keller, 1973)

$$f(v, w) = H(v - \alpha) - v - w, \tag{9.52}$$

$$g(v, w) = v. \tag{9.53}$$

Because the dynamics are piecewise linear, the exact solution can be constructed in the following manner. We look for solutions of the form sketched in Fig. 9.5. The position of the wave along the $\xi$ axis is specified by fixing $v(0) = v(\xi_1) = \alpha$. As yet, $\xi_1$ is unknown, and will be determined when we construct the solution. Note that the place where $v = \alpha$ is the place that the dynamics change (since $\alpha$ is the point of discontinuity of $f$). Let I, II, and III denote, respectively, the regions $\xi < 0$, $0 < \xi < \xi_1$, and $\xi_1 < \xi$. In each region, the differential equation is linear and so can be solved analytically. The three regional solutions are then joined at $\xi = 0$ and $\xi = \xi_1$ by stipulating that $v$ and $w$ be continuous at the boundaries and that $v$ have a continuous derivative there. These constraints are sufficient to determine the solution unambiguously.

In regions I and III, $v < \alpha$, and so the differential equation is

$$\epsilon^2 v_{\xi\xi} + c\epsilon v_\xi - v - w = 0, \tag{9.54}$$

$$cw_\xi + v = 0. \tag{9.55}$$

Looking for solutions of the form $v = A \exp(\lambda\xi)$, $w = B \exp(\lambda\xi)$, we find that $A$ and $B$ must satisfy

$$\begin{pmatrix} \lambda^2\epsilon^2 + c\epsilon\lambda - 1 & -1 \\ 1 & c\lambda \end{pmatrix} \begin{pmatrix} A \\ B \end{pmatrix} = \begin{pmatrix} 0 \\ 0 \end{pmatrix}, \tag{9.56}$$

which has a nontrivial solution if and only if

$$\begin{vmatrix} \lambda^2\epsilon^2 + c\epsilon\lambda - 1 & -1 \\ 1 & c\lambda \end{vmatrix} = 0. \tag{9.57}$$

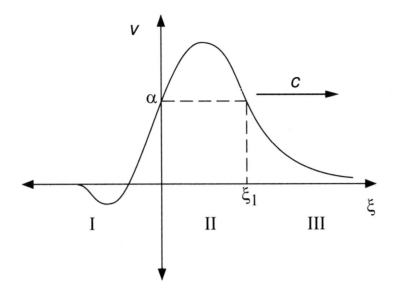

**Figure 9.5** Schematic diagram of the traveling pulse.

Hence, $\lambda$ must be a root of the characteristic polynomial

$$\epsilon^2 p(\lambda) = \epsilon^2 \lambda^3 + \epsilon c \lambda^2 - \lambda + 1/c = 0. \tag{9.58}$$

There is exactly one negative root, call it $\lambda_1$, and the real parts of the other two roots, $\lambda_2$ and $\lambda_3$, are positive.

In region II, the differential equation is

$$\epsilon^2 v_{\xi\xi} + c\epsilon v_\xi + 1 - v - w = 0, \tag{9.59}$$

$$cw_\xi + v = 0. \tag{9.60}$$

The inhomogeneous solution is $w = 1, v = 0$, and the homogeneous solution is a sum of exponentials of the form $e^{\lambda_i \xi}$.

Since we want the solution to approach zero in the limit $\xi \to \pm\infty$, the traveling pulse can be represented as the exponential $e^{\lambda_1 \xi}$ for large positive $\xi$, the sum of the two exponentials $e^{\lambda_2 \xi}$ and $e^{\lambda_3 \xi}$ for large negative $\xi$, and the sum of all three exponentials for the intermediate range of $\xi$ for which $v(\xi) > \alpha$. We take

$$w(\xi) = \begin{cases} Ae^{\lambda_1 \xi} & \text{for } \xi \geq \xi_1, \\[2mm] 1 + \displaystyle\sum_{i=1}^{3} B_i e^{\lambda_i \xi} & \text{for } 0 \leq \xi \leq \xi_1, \\[2mm] \displaystyle\sum_{i=2}^{3} C_i e^{\lambda_i \xi} & \text{for } \xi \leq 0, \end{cases} \tag{9.61}$$

with $v = -cw_\xi$. Now we require that $w(\xi), v(\xi)$, and $v_\xi(\xi)$ be continuous at $\xi = 0, \xi_1$, and that $v(0) = v(\xi_1) = \alpha$.

There are six unknown constants and two unknown parameters $c$ and $\xi_1$ that must be determined from the six continuity conditions and the two constraints. Following some calculation, we eliminate the coefficients $A, B_i,$ and $C_i$, leaving the two constraints

$$e^{\lambda_1 \xi_1} + \epsilon^2 p'(\lambda_1)\alpha - 1 = 0, \tag{9.62}$$

$$\frac{e^{-\lambda_2 \xi_1}}{p'(\lambda_2)} + \frac{e^{-\lambda_3 \xi_1}}{p'(\lambda_3)} + \frac{1}{p'(\lambda_1)} + \epsilon^2 \alpha = 0. \tag{9.63}$$

There are now two unknowns, $c$ and $\xi_1$, and two equations. In general, (9.62) could be solved for $\xi_1$, and (9.63) could then be used to determine $c$ for each fixed $\alpha$ and $\epsilon$. However, it is convenient to approach these equations in a slightly different manner, by treating $c$ as known and $\alpha$ as unknown, and finding $\alpha$ for any given $c$. So, we set $s = e^{\lambda_1 \xi_1}$, in which case (9.63) becomes

$$h(s) = 2 - s + \frac{p'(\lambda_1)}{p'(\lambda_2)} e^{-\lambda_2 \ln(s)/\lambda_1} + \frac{p'(\lambda_1)}{p'(\lambda_3)} e^{-\lambda_3 \ln(s)/\lambda_1} = 0, \tag{9.64}$$

where we have eliminated $\alpha$ using (9.62). We seek a solution of $h(s) = 0$ with $0 < s < 1$.

We begin by calculating that $h(0) = 2, h(1) = 0, h'(1) = 0,$ and $h''(1) = p'(\lambda_1)/\lambda_1^2 - 2$. The first of these relationships follows from the fact that the real parts of $\lambda_2$ and $\lambda_3$ are of different sign from $\lambda_1$, and therefore, in the limit as $s \to 0$, the exponential terms disappear as the real parts of the exponents approach $-\infty$. The second relationship, $h(1) = 0$, follows from the fact that $1/p'(\lambda_1) + 1/p'(\lambda_2) + 1/p'(\lambda_3) = 0$ (Exercise 8). The final two relationships are similar and are left as exercises (Exercises 8, 9).

If $h''(1) < 0$, then the value $s = 1$ is a local maximum of $h(s)$, so for $s$ slightly less than 1, $h(s) < 0$. Since $h(0) > 0$, a root of $h(s) = 0$ in the interval $0 < s < 1$ is assured.

When $\lambda_2$ and $\lambda_3$ are real, $h(s)$ can have at most one inflection point in the interval $0 < s < 1$. This follows because the equation $h''(s) = 0$ can be written in the form $e^{(\lambda_2 - \lambda_3)\xi_1} = c$, which can have at most one root. Thus, if $h''(1) < 0$, there is precisely one root, while if $h''(1) > 0$ there can be no roots. If the roots $\lambda_2$ and $\lambda_3$ are complex, uniqueness is not assured, although the condition $h''(1) < 0$ guarantees that there is at least one root.

Differentiating the defining polynomial (9.58) with respect to $\lambda$, we observe that the condition $h''(1) < 0$ is equivalent to requiring $\epsilon^2 \lambda_1^2 + 2c\epsilon\lambda_1 - 1 < 0$. Furthermore, from the defining characteristic polynomial, we know that $\epsilon^2 \lambda_1^2 - 1 = -c\epsilon\lambda_1 + \epsilon^2/(\lambda_1 c)$, and thus it follows that $h''(1) < 0$ if $\lambda_1 < -\frac{1}{c\sqrt{\epsilon}}$. Since the polynomial $p(\lambda)$ is increasing at $\lambda_1$, we are assured that $\lambda_1 < -\frac{1}{c\sqrt{\epsilon}}$ if $p(-\frac{1}{c\sqrt{\epsilon}}) > 0$, i.e., if

$$c^2 > \epsilon. \tag{9.65}$$

Thus, whenever $c > \sqrt{\epsilon}$, a root of $h(s) = 0$ with $0 < s < 1$ is guaranteed to exist.

Once $s$ is known, $\alpha$ can be found from the relationship (9.62) whereby

$$\alpha = \frac{1 - s}{\epsilon^2 p'(\lambda_1)}. \tag{9.66}$$

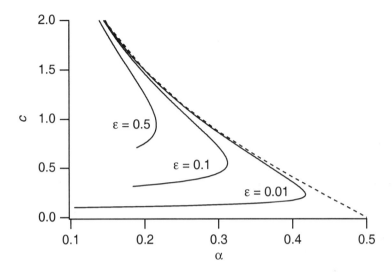

**Figure 9.6** Speed $c$ as a function of $\alpha$ for the traveling pulse solution of the piecewise linear FitzHugh–Nagumo system, shown for $\epsilon = 0.5, 0.1, 0.01$. The dashed curve shows the asymptotic limit as $\epsilon \to 0$, found by singular perturbation arguments.

In Fig. 9.6, we show the results of solving (9.64) numerically. Shown plotted is the speed $c$ against $\alpha$ for a sampling of values of $\epsilon$. The dashed curve is the asymptotic limit (9.12) for the curves in the limit $\epsilon \to 0$. The important feature to notice is that for each value of $\alpha$ and $\epsilon$ small enough there are two traveling pulses, while for large $\alpha$ there are no traveling pulses. In Fig. 9.7 is shown the fast traveling pulse, and in Fig. 9.8 is shown the slow traveling pulse, both for $\alpha = 0.1, \epsilon = 0.1$, and with $v(\xi)$ shown solid and $w(\xi)$ shown dashed.

Note that the amplitude of the slow pulse in Fig. 9.8 is substantially smaller than that of the fast pulse in Fig. 9.7. Generally speaking, the fast pulse is stable (Jones, 1984; Yanagida, 1985), and the slow pulse is unstable (Maginu, 1985). Also note that there is nothing in the construction of these wave solutions requiring $\epsilon$ to be small.

### Singular perturbation theory

The next way to extract information about the traveling pulse solution of (9.48)–(9.49) is to exploit the smallness of the parameter $\epsilon$ (Keener, 1980; for a different approach, see Rauch and Smoller, 1978). One reason we expect this to be fruitful is because of similarities with the flow for the FitzHugh–Nagumo equations without diffusion, shown in Fig. 4.17. By analogy, we expect the solution to stay close to the nullcline $f(v, w) = 0$ wherever possible, with rapid transitions between the two outer branches.

The details of this behavior follow from singular perturbation analysis. (This analysis was first given for a simplified FitzHugh–Nagumo system by Casten, Cohen, and Lagerstrom, 1975.) The first observation follows simply from setting $\epsilon$ to zero in (9.48).

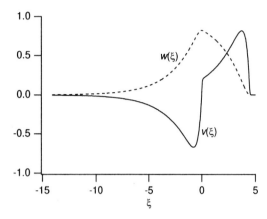

**Figure 9.7** Plots of $v(\xi)$ and $w(\xi)$ for the fast traveling pulse ($c = 2.66$) for the piecewise linear FitzHugh–Nagumo system with $\alpha = 0.1, \epsilon = 0.1$.

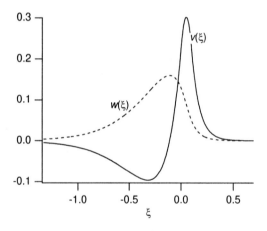

**Figure 9.8** Plots of $v(\xi)$ and $w(\xi)$ for the slow traveling pulse ($c = 0.34$) for the piecewise linear FitzHugh–Nagumo system with $\alpha = 0.1, \epsilon = 0.1$.

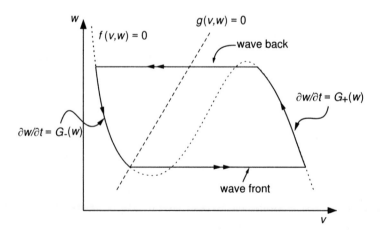

**Figure 9.9** Sketch of the phase portrait of the fast traveling solitary pulse for FitzHugh–Nagumo dynamics in the singular limit $\epsilon \to 0$.

Doing so, we obtain the *outer equations*

$$w_t = g(v, w), \qquad f(v, w) = 0. \tag{9.67}$$

Because the equation $f(v, w) = 0$ is assumed to have three solutions for $v$ as a function of $w$, and only two of these solutions, the upper and lower solution branches, are stable (cf. Fig. 4.15 and the discussion in Section 4.2), the outer equations (9.67) reduce to

$$\frac{\partial w}{\partial t} = G_\pm(w). \tag{9.68}$$

A region of space in which $v = V_+(w)$ is called an *excited region*, and a region in which $v = V_-(w)$ is called a *recovering region*. The outer equation is valid whenever diffusion is not large. However, we anticipate that there are regions of space (*interfaces*) where diffusion is large and in which (9.68) cannot be correct.

To find out what happens when diffusion is large we rescale space and time. Letting $y(t)$ denote the position of the wave front, we set $\tau = t$ and $\xi = \frac{x - y(t)}{\epsilon}$, after which the original system of equations (9.48)–(9.49) becomes

$$v_{\xi\xi} + y'(\tau)v_\xi + f(v, w) = \epsilon \frac{\partial v}{\partial \tau}, \tag{9.69}$$

$$-y'(\tau)w_\xi = \epsilon \left( g(v, w) - \frac{\partial w}{\partial \tau} \right). \tag{9.70}$$

Upon setting $\epsilon = 0$, we find the reduced *inner equations*

$$v_{\xi\xi} + y'(\tau)v_\xi + f(v, w) = 0, \tag{9.71}$$

$$y'(\tau)w_\xi = 0. \tag{9.72}$$

Even though the inner equations (9.71)–(9.72) are partial differential equations, the variable $\tau$ occurs only as a parameter, and so (9.71)–(9.72) can be solved as if they were ordinary differential equations. This is because the traveling wave is stationary in the moving coordinate system $\xi$, $\tau$. It follows that $w$ is independent of $\xi$ (but not necessarily $\tau$). Finally, since the inner equation is supposed to provide a transition layer between regions where outer dynamics hold, we require the *matching condition* that $f(v, w) \to 0$ as $\xi \to \pm\infty$. Note that here we use a more general $y(t)$ to locate the wave front, rather than $ct$ as before. In general, $y'(\tau)$ is the local wave velocity.

We recognize (9.71) as a bistable equation for which there are heteroclinic orbits. That is, for fixed $w$, if the equation $f(v, w) = 0$ has three roots, two of which are stable as solutions of the equation $dv/dt = f(v, w)$, then there is a number $c = c(w)$ for which the equation

$$v'' + c(w)v' + f(v, w) = 0 \tag{9.73}$$

has a heteroclinic orbit connecting the two stable roots of $f(v, w) = 0$. This heteroclinic orbit corresponds to a moving transition layer, traveling with speed $c$. It is crucial to note that since the roots of $f(v, w) = 0$ are functions of $w$, $c$ is also a function of $w$. To be specific, we define $c(w)$ to be the unique parameter value for which (9.73) has a solution with $v \to V_-(w)$ as $\xi \to \infty$, and $v \to V_+(w)$ as $\xi \to -\infty$. In the case that $c(w) > 0$, we describe this transition as an "upjump" moving to the right. If $c(w) < 0$, then the transition is a "downjump" moving to the left.

We are now able to describe a general picture of wave propagation. In most of space, outer dynamics (9.68) are satisfied. At any transition between the two types of outer dynamics, continuity of $w$ is maintained by a sharp transition in $v$ that travels at the speed $y'(t) = c(w)$ if $v = V_-(w)$ on the right and $v = V_+(w)$ on the left, or at speed $y'(t) = -c(w)$ if $v = V_+(w)$ on the right and $v = V_-(w)$ on the left, where $w$ is the value of the recovery variable in the interior of the transition layer. As a transition layer passes any particular point in space, there is a switch of outer dynamics from one to the other of the possible outer solution branches.

This singular perturbation description of wave propagation allows us to examine in more detail the specific case of a traveling pulse. The phase portrait for a solitary pulse is sketched in Fig. 9.9. A traveling pulse consists of a single excitation front followed by a single recovery back. We suppose that far to the right, the medium is at rest, and that a wave front of excitation has been initiated and is moving from left to right. Of course, for the medium to be at rest there must be a rest point of the dynamics on the lower branch, say $G_-(w_+) = 0$. Then, a wave that is moving from left to right has $v = V_-(w_+)$ on its right and $v = V_+(w_+)$ on its left, traveling at speed $y'(t) = c(w_+)$. Necessarily, it must be that $c(w_+) > 0$. Following the same procedure used to derive (9.8), one can show that

$$c(w) = \frac{\int_{V_-(w)}^{V_+(w)} f(v,w)\,dv}{\int_{-\infty}^{\infty} v_\xi^2\,d\xi}, \tag{9.74}$$

and thus $c(w_+) > 0$ only if

$$\int_{V_-(w_+)}^{V_+(w_+)} f(v,w_+)\,dv > 0. \tag{9.75}$$

If (9.75) fails to hold, then the medium is not sufficiently excitable to sustain a propagating pulse. It is important also to note that if $f(v,w)$ is of generalized FitzHugh–Nagumo form, then $c(w)$ has a unique zero in the interval $(W_*, W^*)$, where $W_*$ and $W^*$ are defined in Section 4.2.

Immediately to the left of the excitation front, the medium is excited and satisfies the outer dynamics on the upper branch $v = V_+(w)$. Because (by assumption) $G_+(w) > 0$, this can hold for at most a finite amount of time before the outer dynamics force another transition layer to appear. This second transition layer provides a transition between the excited region on the right and a recovering region on the left and travels with speed $y'(t) = -c(w)$, where $w$ is the value of the recovery variable in the transition layer. The minus sign here is because the second transition layer must be a "downjump." For this to be a steadily propagating traveling pulse, the speed of the upjump and the speed of the downjump must be identical. Thus, the value of $w$ at the downjump, say $w_-$, must be such that $c(w_-) = -c(w_+)$.

It may be that the equation $c(w_-) = -c(w_+)$ has no solution. In this case, the downjump must occur exactly at the knee, and then the wave is called a "phase wave," since the timing of the downjump is determined solely by the timing, or phase, of the outer dynamics, and not by any diffusive processes. That such a wave can travel at any speed greater than some minimal speed can be shown using standard arguments. The dynamics for phase waves are different from those for the bistable equation because the downjump must be a heteroclinic connection between a saddle point and a "saddle-node." That is, at the knee, two of the three steady solutions of the bistable equation are merged into one. The demonstration of the existence of traveling waves in this situation is similar to the case of Fisher's equation, where the nonlinearity $f(v, w)$ in (9.73) has two simple zeros, rather than three in the bistable case. In the phase wave problem, however, one of the zeros of $f(v, w)$ is not simple, but quadratic in nature, a canonical example of which is $f(v, w) = v^2(1 - v)$. We do not pursue this further except to say that such waves exist (see Exercise 14).

In summary, from singular perturbation theory we learn that the value of $w$ ahead of the traveling pulse is given by the steady-state value $w_+$, and the speed of the rising wave front is then determined from the bistable equation (9.73) with $w = w_+$. The wave front switches the value of $v$ from $v = V_-(w_+)$ (ahead of the wave) to $v = V_+(w_+)$ (behind the wave front). A wave back then occurs at $w = w_-$, where $w_-$ is determined from $c(w_-) = -c(w_+)$. The wave back switches the value of $v$ from $v = V_+(w_-)$ to $v = V_-(w_-)$. The duration of the excited phase of the traveling pulse is

$$T_e = \int_{w_+}^{w_-} \frac{dw}{G^+(w)}. \tag{9.76}$$

The duration of the absolute refractory period is

$$T_{\mathrm{ar}} = \int_{w_-}^{w_0} \frac{dw}{G_-(w)}, \tag{9.77}$$

where $w_0$ is that value of $w$ for which $c(w) = 0$ (Exercise 10). This approximate solution is said to be a singular solution, because derivatives of the solution become infinite (are singular) in the limit $\epsilon \to 0$.

## 9.4.2  The Hodgkin–Huxley Equations

The traveling pulse for the Hodgkin–Huxley equations must be computed numerically in one of two ways. The simplest way is to simulate the partial differential equation on a long one-dimensional spatial domain, or one can use the technique of shooting. In fact, shooting was used by Hodgkin and Huxley in their 1952 paper to demonstrate that the Hodgkin–Huxley equations support a traveling wave solution. Shooting is also the method by which a rigorous proof of the existence of traveling waves has been given (Hastings, 1975; Carpenter, 1977).

The shooting argument is as follows. We write the Hodgkin–Huxley equations in the form

$$\tau_m \frac{\partial v}{\partial t} = \lambda_m^2 \frac{\partial^2 v}{\partial x^2} + f(v, m, n, h), \tag{9.78}$$

$$\frac{dw}{dt} = \alpha_w(v)(1 - w) - \beta_w(v)w, \text{ for } w = n, m, \text{ and } h. \tag{9.79}$$

Now we look for solutions in $x, t$ that are functions of the translating variable $\xi = x/c + t$, and find the system of ordinary differential equations

$$\frac{\lambda_m^2}{c^2} \frac{d^2 v}{d\xi^2} + f(v, m, n, h) - \tau_m \frac{dv}{d\xi} = 0, \tag{9.80}$$

$$\frac{dw}{d\xi} = \alpha_w(v)(1 - w) - \beta_w(v)w, \text{ for } w = n, m, \text{ and } h. \tag{9.81}$$

Linearizing the system (9.80) and (9.81) about the resting solution at $v = 0$, one finds that there are four negative eigenvalues and one positive eigenvalue. A reasonable approximation to the unstable manifold is found by neglecting variations in $g_K$ and $g_{Na}$, from which

$$v(t) = v_0 e^{\mu t}, \tag{9.82}$$

where

$$\frac{\lambda_m^2}{c^2} \mu^2 - \tau_m \mu - 1 = 0$$

or

$$\mu = \frac{1}{2} \left( \tau_m \frac{c^2}{\lambda_m^2} + \frac{c}{\lambda_m} \sqrt{\tau_m^2 \frac{c^2}{\lambda_m^2} + 4} \right).$$

To implement shooting, one chooses a value of $c$, and initial data close to the rest point but on the unstable manifold (9.82), and then integrates numerically until (in all likelihood) the potential becomes very large. It could be that the potential becomes either large positive or large negative. In fact, once values of $c$ are found that do both, one uses bisection to home in on the homoclinic orbit that returns to the rest point in the limit $\xi \to \infty$.

For the Hodgkin–Huxley equations one finds a traveling pulse for $c = 3.24 \lambda_m$ ms$^{-1}$. Using typical values for squid axon (from Table 8.1, $\lambda_m = 0.65$ cm), we find $c = 21$ mm/ms, which is close to the value of 21.2 mm/ms found experimentally by Hodgkin and Huxley. Hodgkin and Huxley estimated the space constant for squid axon as $\lambda_m = 0.58$ cm, from which they calculated that $c = 18.8$ mm/ms. Their calculated speeds agreed very well with experimental data and thus their model, which was based only on measurements of ionic conductance, was used to predict accurately macroscopic behavior of the axon. It is rare that quantitative models can be applied so successfully. Propagation velocities for several types of excitable tissue are listed in Table 9.2.

---

**Table 9.2** Propagation velocities in nerve and muscle.

---

| Excitable Tissue | velocity (m/sec) |
|---|---|
| Myelinated nerve fibers | |
|   Large diameter (16–20 $\mu$m) | 100–120 |
|   Mid-diameter (10–12 $\mu$m) | 60–70 |
|   Small diameter (4–6 $\mu$m) | 30-50 |
| Nonmyelinated nerve fibers | |
|   Mid-diameter (3–5 $\mu$m) | 15-20 |
| Skeletal muscle fibers | 6 |
| Heart | |
|   Purkinje fibers | 1.0 |
|   Cardiac muscle | 0.5 |
|   Smooth muscle | 0.05 |

# 9.5 Periodic Wave Trains

Excitable systems are characterized by both excitability and refractoriness. That is, after the system has responded to a superthreshold stimulus with a large excursion from rest, there is a period of refractoriness during which no subsequent responses can be evoked, followed by a period of recovery during which excitability is gradually restored. Once excitability is restored, another wave of excitation can be evoked. However, the speed at which subsequent waves of excitation travel depends strongly on the time allowed for recovery of excitability following the last excitation wave. Generally (but not always), the longer the period of recovery, the faster the new wave of excitation can travel.

One might guess that a nerve axon supports, in addition to a traveling pulse, periodic wave trains of action potentials. With a periodic wave train, if recovery is a monotonic process, one expects propagation to be slower than for a traveling pulse, because subsequent action potentials occur before the medium is fully recovered, so that the sodium upstroke is slower than for a traveling pulse. The relationship between the speed and period is called the *dispersion curve*.

The dispersion curve for the Hodgkin–Huxley equations can be calculated numerically in one of two ways. The most direct method is to construct a ring, that is, a one-dimensional domain with periodic boundary conditions, initiate a pulse that travels in one direction on the ring, and solve the equations numerically until the solution becomes periodic in time. After recording the length of the ring and the speed of propa-

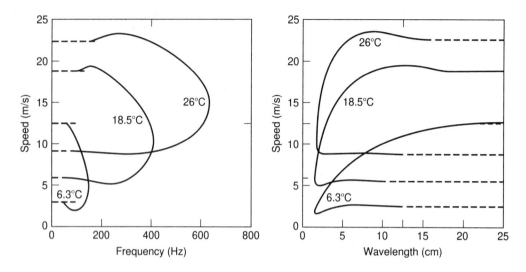

**Figure 9.10** Numerically computed dispersion curve (speed vs. frequency and speed vs. wavelength for various temperatures) for the Hodgkin–Huxley equations. (Miller and Rinzel, 1981, Figs. 1 and 2.)

gation, one could use the waveform as initial data for a ring of slightly different length, and do the calculation again. While this method is relatively easy, its principal disadvantage is that it requires the periodic solution to be stable. Dispersion curves often have sections whose periodic solutions are unstable, and this method cannot find those. Of course, only the stable solutions are physically realizable, so this disadvantage may not be so serious to the realist.

The second method is to look for periodic solutions of the equations in their traveling wave coordinates (9.80)–(9.81), using a numerical continuation method (an automatic continuation program such as AUTO recommends itself here). With this method, periodic solutions are found without reference to their stability, so that the entire dispersion curve can be calculated.

Dispersion curves for excitable systems have a typical shape, depicted in Figs. 9.10 and 9.11. Here we see a dispersion curve having two branches, one denoting fast waves, the other slow. The two branches meet at a knee or corner at the *absolute refractory period*, and for shorter periods no periodic solutions exist. The solutions on the fast branch are typical of action potentials and are usually (but not always) stable. The solutions on the slow branch are small amplitude oscillations and are unstable.

### 9.5.1 Piecewise Linear FitzHugh–Nagumo Equations

The dispersion curve in Fig. 9.11 was found for the FitzHugh–Nagumo system (9.48)–(9.49) with piecewise linear functions (9.52) and (9.53). The calculation is similar to that for the traveling pulse (Rinzel and Keller, 1973). Since this system is piecewise linear, we can express its solution as the sum of three exponentials,

$$w(\xi) = \sum_{i=1}^{3} A_i e^{\lambda_i \xi}, \tag{9.83}$$

on the interval $0 \le \xi < \xi_1$, and as

$$w(\xi) = 1 + \sum_{i=1}^{3} B_i e^{\lambda_i \xi} \tag{9.84}$$

on the interval $\xi_1 \le \xi \le \xi_2$, where $v = -cw_\xi$. We also assume that $v > \alpha$ on the interval $\xi_1 \le \xi \le \xi_2$. The numbers $\lambda_i, i = 1, 2, 3$, are roots of the characteristic polynomial (9.58).

We require that $w(\xi), v(\xi)$, and $v'(\xi)$ be continuous at $\xi = \xi_1$, and that $w(0) = w(\xi_2), v(0) = v(\xi_2)$, and $v'(0) = v'(\xi_2)$ for periodicity. Finally, we require that $v(0) = v(\xi_1) = \alpha$. This gives a total of eight equations in nine unknowns, $A_1, \ldots, A_3, B_1, \ldots, B_3$, $\xi_1, \xi_2$, and $c$. After some calculation (Exercise 11), we find two equations for the three unknowns $\xi_1, \xi_2$, and $c$ given by

$$\frac{e^{\lambda_1(P-\xi_1)} - 1}{p'(\lambda_1)(e^{\lambda_1 P} - 1)} + \frac{e^{\lambda_2(P-\xi_1)} - 1}{p'(\lambda_2)(e^{\lambda_2 P} - 1)} + \frac{e^{\lambda_3(P-\xi_1)} - 1}{p'(\lambda_3)(e^{\lambda_3 P} - 1)} + \epsilon^2 \alpha = 0, \tag{9.85}$$

$$\frac{e^{\lambda_1 P} - e^{\lambda_1 \xi_1}}{p'(\lambda_1)(e^{\lambda_1 P} - 1)} + \frac{e^{\lambda_2 P} - e^{\lambda_2 \xi_1}}{p'(\lambda_2)(e^{\lambda_2 P} - 1)} + \frac{e^{\lambda_3 P} - e^{\lambda_3 \xi_1}}{p'(\lambda_3)(e^{\lambda_3 P} - 1)} - \epsilon^2 \alpha = 0, \tag{9.86}$$

where $P = \xi_2/c$. It is important to note that since there are only two equations for the three unknowns, (9.85) and (9.86) define a family of periodic waves, parametrized by either the period or the wave speed. The relationship between the period and the speed of this wave family is the dispersion curve. In Fig. 9.11 are shown examples of the dispersion curve for a sampling of values of $\epsilon$ with $\alpha = 0.1$. Changing $\alpha$ has little qualitative effect on this plot. The dashed curve shows the limiting behavior of the upper branch (the fast waves) in the limit $\epsilon \to 0$. Of significance in this plot is the fact that there are fast and slow waves, and in the limit of large wavelength, the periodic waves approach the solitary traveling pulses represented by Fig. 9.6 (Exercise 11). In fact, periodic solutions look much like evenly spaced periodic repeats of (truncated) solitary pulses.

The dispersion curve for the piecewise linear FitzHugh–Nagumo system is typical of dispersion curves for excitable media, with a fast and slow branch meeting at a corner. In general, the location of the corner depends on the excitability of the medium (in this case, the parameter $\alpha$) and on the ratio of time scales $\epsilon$.

## 9.5.2 Singular Perturbation Theory

The fast branch of the dispersion curve can be found for a general FitzHugh–Nagumo system in the limit $\epsilon \to 0$ using singular perturbation theory. A periodic wave consists of an alternating series of upjumps and downjumps, separated by regions of outer dynamics. The phase portrait for a periodic wave train is sketched in Fig. 9.12. To be

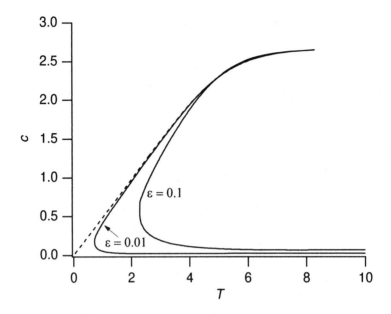

**Figure 9.11** Dispersion curves for the piecewise linear FitzHugh–Nagumo equations shown for $\epsilon = 0.1$ and 0.01. The dashed curve shows the singular perturbation approximation to the dispersion curve.

periodic, if $w_+$ is the value of the recovery variable in the upjump, traveling with speed $c(w_+)$, then the value of $w$ in the downjump must be $w_-$, where $c(w_+) = -c(w_-)$. The amount of time spent on the excited branch is

$$T_e = \int_{w_+}^{w_-} \frac{dw}{G_+(w)}, \tag{9.87}$$

and the amount of time spent on the recovery branch is

$$T_r = \int_{w_-}^{w_+} \frac{dw}{G_-(w)}. \tag{9.88}$$

The dispersion curve is then the relationship between speed $c(w_+)$ and period

$$T = T_e + T_r, \tag{9.89}$$

parametrized by $w_+$. This approximate dispersion curve (calculated numerically) is shown in Fig. 9.11 as a dashed curve.

The slow branch of the dispersion curve can also be found using perturbation methods, although in this case since the speed is small of order $\epsilon$, a regular perturbation expansion is appropriate. The details of this expansion are beyond the scope of this book, although the interested reader is referred to Dockery and Keener (1989). In general, the slow periodic solutions are unstable (Maginu, 1985) and therefore are of less

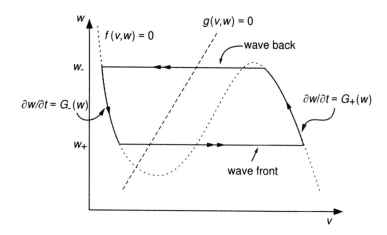

**Figure 9.12** Sketch of the phase portrait for the fast traveling periodic wave train for FitzHugh–Nagumo dynamics in the singular limit $\epsilon \to 0$.

physical interest than the fast solutions. Again, stability theory for the traveling wave solutions is beyond the scope of this book.

### 9.5.3 Kinematics

Not all waves are periodic. There can be wave trains with action potentials that are irregularly spaced and that travel with different velocities. A *kinematic theory* of wave propagation is one that attempts to follow the progress of individual action potentials without tracking the details of the structure of the pulse (Rinzel and Maginu, 1984). The simplest kinematic theory is to interpret the dispersion curve in a local way. That is, suppose we know the speed as a function of period for the stable periodic wave trains, $c = C(T)$. We suppose that the wave train consists of action potentials, and that the $n$th action potential reaches position $x$ at time $t_n(x)$. To keep track of time of arrival at position $x$ we note that

$$\frac{dt_n}{dx} = \frac{1}{c}. \tag{9.90}$$

We complete the description by taking $c = C(t_n(x) - t_{n-1}(x))$, realizing that $t_n(x) - t_{n-1}(x)$ is the instantaneous period of the wave train that is felt by the medium at position $x$.

A more sophisticated kinematic theory can be derived from the singular solution of the FitzHugh–Nagumo equations. For this derivation we assume that recovery is always via a phase wave, occurring with recovery value $W^*$. Suppose that the front of the $n$th action potential has speed $c(w_n)$, corresponding to the recovery value $w_n$ in the transition layer. Keeping track of the time until the next action potential, we find

$$t_{n+1}(x) - t_n(x) = T_e(x) + T_r(x) \tag{9.91}$$

$$= \int_{w_n}^{W^*} \frac{dw}{G_+(w)} + \int_{W^*}^{w_{n+1}} \frac{dw}{G_-(w)}. \tag{9.92}$$

Differentiating (9.92) with respect to $x$, we find the differential equation for $w_{n+1}(x)$:

$$\frac{1}{G_-(w_{n+1})} \frac{dw_{n+1}}{dx} = \frac{1}{G_+(w_n)} \frac{dw_n}{dx} + \frac{1}{c(w_{n+1})} - \frac{1}{c(w_n)}. \tag{9.93}$$

With this equation one can track the variable $w_{n+1}$ as a function of $x$ given $w_n(x)$ and from it reconstruct the speed and time of arrival of the $(n+1)$st action potential wave front.

While this formulation is useful for FitzHugh–Nagumo models, it can be given a more general usefulness as follows. Since there is a one-to-one relationship between the speed of a front and the value of the recovery variable $w$ in that front, we can represent these functions in terms of the speeds of the fronts as

$$t_{n+1}(x) - t_n(x) = A(c_n) + t_r(c_{n+1}), \tag{9.94}$$

where $A(c_n) = T_e$ is the *action potential duration* (APD) since the $n$th upstroke, and $t_r(c_{n+1}) = T_r$ is the recovery time preceding the $(n+1)$st upstroke. When we differentiate this conservation law with respect to $x$, we find a differential equation for the speed of the $(n+1)$st wave front as a function of the speed of the $n$th wave front, given by

$$t_r'(c_{n+1}) \frac{dc_{n+1}}{dx} = \frac{1}{c_{n+1}} - \frac{1}{c_n} - A'(c_n) \frac{dc_n}{dx}. \tag{9.95}$$

The advantage of this formulation is that the functions $A$ and $t_r$ may be known for other reasons, perhaps from experimental data. It is generally recognized that the action potential duration is functionally related to the speed of the previous action potential, and it is also reasonable that the speed of a subsequent action potential can be related functionally to the time of recovery since the end of the last action potential. Thus, the model (9.95) has applicability that goes beyond the FitzHugh–Nagumo context. For an example of how this idea has been used for the Beeler–Reuter dynamics, see Courtemanche et al. (1996).

## 9.6  EXERCISES

1.  Show that the $(U, W)$ phase-plane of the bistable equation, (9.6) and (9.7), has three steady states, two of which are saddle points. What is the nature of the third steady state? Show that the slope of the unstable manifold at the origin is given by the positive root of $\lambda^2 - c\lambda + f'(0) = 0$ and is always larger than $c$. What is the slope of the stable manifold at $(U, W) = (1, 0)$? Show that the slopes of both these manifolds are increasing functions of $c$.

2.  Use cable theory to find the speed of propagation for each of the axons listed in Table 8.1, assuming that the ionic currents are identical and the speed of propagation for the squid giant axon is 21 mm/ms.

3.  Find the space constant for a myelinated fiber.

4. Construct a traveling wave solution to the piecewise linear bistable equation, (9.1) and (9.3). Show that the wave travels with speed $\frac{1-2\alpha}{\sqrt{\alpha-\alpha^2}}$. (Construct the solution by using the techniques presented in Section 9.4.1.)

5. Find the shape of the traveling wave profile for (9.1) in the case that the function $f(V)$ is

$$
f(v) = \begin{cases} -v, & \text{for } v < \alpha/2, \\ v - \alpha, & \text{for } \alpha/2 < v < \dfrac{1+\alpha}{2}, \\ 1 - v, & \text{for } v > \dfrac{1+\alpha}{2}. \end{cases} \tag{9.96}
$$

6. Find the speed of traveling fronts for barnacle muscle fiber using the Morris–Lecar model (Chapter 4, (4.95)–(4.96)).
   Answer: 17 cm/s .

7. Write a program to find numerically the speed of propagation for the bistable equation. Use the program to determine the effect of sodium channel density (Table 9.1) on the speed of propagation in various axons, assuming that all other currents are the same as for the Hodgkin–Huxley model.

8. Show that $1/p'(\lambda_1) + 1/p'(\lambda_2) + 1/p'(\lambda_3) = 0$, where $p$ is defined by (9.58). Hence, show that $h(1) = 0$, where $h$ is defined by (9.64). Show also that $\lambda_1/p'(\lambda_1) + \lambda_2/p'(\lambda_2) + \lambda_3/p'(\lambda_3) = 0$ and thus $h'(1) = 0$. Finally, show that $h''(1) = p'(\lambda_1)/\lambda_1^2 - 2$.

9. The results of Exercise 8 can be generalized. Use contour integration in the complex plane to show that for an $n$th order polynomial $p(z) = z^n + \cdots$ with simple roots $z_k, k = 1, \ldots, n$,

$$
\sum_{k=1}^{n} \frac{z_k^j}{p'(z_k)} = 0, \tag{9.97}
$$

provided that $n > j + 1$. In addition, show that

$$
\sum_{k=1}^{n} \frac{z_k^{n-1}}{p'(z_k)} = 1. \tag{9.98}
$$

10. Show that the duration of the absolute refractory period of the traveling pulse for the generalized FitzHugh–Nagumo model is (approximately)

$$
T_{\text{ar}} = \int_{w_-}^{w_0} \frac{dw}{G_-(w)}, \tag{9.99}
$$

where $w_0$ is that value of $w$ for which $c(w) = 0$.

11. Derive (9.85) and (9.86). Show that in the limit as the period approaches infinity, these equations reduce to the equations for a solitary pulse (9.62) and (9.63).

12. Suppose a nearly singular ($\epsilon$ small) FitzHugh–Nagumo system has a stable periodic oscillatory solution when there is no diffusive coupling. How do the phase portraits for the spatially independent solutions and the periodic traveling waves differ? How are these differences reflected in the temporal behavior of the solutions?

13. (Fisher's equation.) Suppose $f(0) = f(1) = 0, f'(0) > 0, f'(1) < 0$, and $f(v) > 0$ for $0 < v < 1$.

    (a) Show that there are values of $c$ (for example $c = 0$) for which there are no trajectories with $v \geq 0$ connecting the two rest points $v = 0$ and $v = 1$.

(b)   Show that if there is a value of $c$ for which there is no heteroclinic connection with $v$ positive, then there is no such connecting trajectory for any smaller value of $c$.

(c)   Show that if there is a value of $c$ for which there is a heteroclinic connection with $v$ positive, there there is a similar connecting trajectory for every larger value of $c$.

(d)   Let $\mu$ be the smallest positive number for which $f(v) \leq \mu v$ for $0 \leq v \leq 1$. Show that a heteroclinic connection exists for all $c > \mu$.

14.  (Phase waves.) Suppose $f(0) = f(1) = f'(0) = 0, f'(1) < 0$ and $f(v) > 0$ for $0 < v < 1$. Show that all the statements of Exercise 13 hold.

(a)   Show that there are values of $c$ for which there are no trajectories connecting critical points for the equation $V'' + cV' + f(V) = 0$, with $f(0) = f'(0) = f(1) = 0$, and $f(v) > 0$ for $0 < v < 1$. Hint: What is the behavior of trajectories for $c = 0$?

(b)   Show that if a connecting trajectory exists for one value of $c < 0$, then it exists for all smaller (larger in absolute value) values of $c$. Hint: What happens to trajectories when $c$ is decreased (increased in absolute value) slightly? How do trajectories for different values of $c$ compare?

15.  Do traveling wave solutions exist for the equation $v_t = v_{xx} + f(v)$ with $v(-\infty, t) = 0$ and $v(+\infty, t) = 1$ where $f(v) = 0$ for $0 < v < q$, $f(v) > 0$ for $q < v < 1$, and $f(1) = 0$? If so, find the speed of propagation in the case that $f(v) = 1 - v$ for $q < v < 1$.

16.  Given a dispersion curve $c = C(T)$, use the simple kinetic theory

$$\frac{dt_n(x)}{dx} = \frac{1}{C(t_n(x) - t_{n-1}(x))} \tag{9.100}$$

to determine the stability of periodic waves on a ring of length $L$.
Hint: On a ring of length $L$, $t_n(x) = t_{n-1}(x - L)$. Suppose that $T^*C(T^*) = L$. Perform linear stability analysis for the solution $t_n(x) = \frac{x}{C(T^*)}$.

17.  Estimate the error in the calculated speed of propagation when Euler's method (forward differencing in time) with second-order centered differencing in space is used to approximate the solution of the bistable equation.
Hint: Use perturbation arguments to approximate the solution of the numerical problem with a traveling front solution of the bistable equation. (See Section 9.3.1.)

18.  Generalize the kinematic theory (9.91)–(9.93) to the case in which wave backs are not phase waves, by tracking both fronts and backs and the corresponding recovery value.

# CHAPTER 10

# Wave Propagation in Higher Dimensions

Not all cellular media can be viewed as one-dimensional cables, neither is all propagated activity one-dimensional. Tissues for which one-dimensional descriptions of cellular communication are inadequate include skeletal and cardiac tissue, the retina, and the cortex of the brain. To understand communication and signaling in these media requires more complicated mathematical analysis than for one-dimensional cables.

When beginning a study of two- or three-dimensional wave propagation, it is tempting simply to generalize the cable equation to higher dimensions by replacing first derivatives in space with spatial gradients, and second spatial derivatives with the Laplacian operator. Indeed, all the models in this chapter are of this type. However, be warned that this replacement may not always be justifiable.

Some cells, such as *Xenopus* oocytes (frog eggs), are sufficiently large so that waves of chemical activity can be sustained within a single cell. This is unusual, however, as most waves in normal physiological situations serve the purpose of communication between cells. Waves that coordinate activity within a single cell typically occur in reproducing cells for which the division process (*mitosis*) must be coordinated. For chemical waves in single cells, a reasonable first guess is that spatial coupling is by chemical diffusion. In that case, if the local chemical dynamics are described by the differential equation $u_t = kf$, then with spatial coupling the dynamics are represented by

$$u_t = \nabla \cdot (D\nabla u) + kf, \tag{10.1}$$

where $D$ is the (scalar) diffusion coefficient of the chemical species and $\nabla$ is the three-dimensional gradient operator. Here we have included the time constant $k$ (with units time$^{-1}$), so that $f$ has the same dimensional units as $u$.

For many cell types, intercellular communication is through gap junctions between immediate neighbors, so that diffusion is not spatially uniform. In this case, the first guess (or hope) is that the length constant of the phenomenon to be described is much larger than the typical cell size, and homogenization can be used to find an effective diffusion coefficient, $D_e$, as described in Chapter 7. Then (10.1) with the effective diffusion coefficient is a reasonable model. Note that there is no a priori reason to believe that cellular coupling is isotropic or that $D_e$ is a scalar quantity.

If this assumption, that the space constant of the signaling phenomenon is larger than the size of the cell, and hence that homogenization gives a valid approximation, is not justified, then we are stuck with the unpleasant business of studying communication between discretely coupled cells.

For electrically active cells, such as cardiac or muscle cells, the situation is complicated further by the fact that the membrane potential contains the signal, so that the intracellular and extracellular potentials must both be followed. We address this problem in Chapter 11, where we discuss waves in myocardial tissue. For now, suffice it to say that (10.1) is a reasonable qualitative model, but it is not certain that all the results derived from this equation are directly applicable to cellular media.

## 10.1  Propagating Fronts

### 10.1.1  Plane Waves

The simplest wave to look for in an excitable medium is a plane wave. Suppose that the canonical problem

$$U'' + c_0 U' + f(U) = 0 \tag{10.2}$$

(with dimensionless independent variable) is bistable and has a wave front solution $U(\xi)$ for some unique value of $c_0$, the value of which depends on $f$. The behavior of this solution was discussed in Chapter 9.

To find plane wave solutions of (10.1), we suppose that $u$ is a function of the single variable $\xi = \mathbf{n} \cdot \mathbf{x} - ct$, where $\mathbf{n}$ is a unit vector pointing in the forward direction of wave front propagation. In the traveling wave coordinate $\xi$, the time derivative $\frac{d}{dt}$ is replaced by $-c\frac{d}{d\xi}$, and the spatial gradient operator $\nabla$ is replaced by $\mathbf{n}\frac{d}{d\xi}$, so that the governing equation reduces to the ordinary differential equation

$$(\mathbf{n} \cdot \mathbf{Dn})u'' + cu' + kf(u) = 0. \tag{10.3}$$

We compare (10.3) with the canonical equation (10.2) and note that the solution of (10.3) can be found by a simple rescaling to be

$$u(\mathbf{x}, t) = U\left(\frac{\mathbf{n} \cdot \mathbf{x} - ct}{\Lambda(\mathbf{n})}\right), \tag{10.4}$$

where $c = c_0 k \Lambda(\mathbf{n})$ is the (directionally dependent) speed and $\Lambda(\mathbf{n}) = \sqrt{\frac{\mathbf{n} \cdot D \mathbf{n}}{k}}$ is the directionally dependent space constant.

## 10.1.2 Waves with Curvature

Wave fronts in two- or three-dimensional media are not expected to be plane waves. They are typically initiated at a specific location, and so might be circular in shape. Additionally, the medium may be structurally inhomogeneous or have a nonsimple geometry, all of which introduce curvature into the wave front.

It is known that curvature plays an important role in the propagation of a wave front in an excitable medium. A physical explanation makes this clear. Suppose that a circular wave front is moving inward, so that the circle is collapsing. Because different parts of the front are working to excite the same points, we expect the region directly in front of the wave front to be excited more quickly than if the wave were exactly planar. Similarly, an expanding circular wave front should move more slowly than a plane wave because the efforts of the wave to excite its neighbors are more spread out, and excitation is slower than for a plane wave.

While these curvature effects are well known in many contexts, we are interested here in a quantitative description of this effect. In this section we derive an equation for action-potential spread called the *eikonal-curvature equation*, the purpose of which is to show the contribution of curvature to wave-front velocity. Eikonal-curvature equations have been used in a number of biological contexts, including the study of wave-front propagation in the excitable Belousov–Zhabotinsky reagent (Foerster et al., 1989; Keener, 1986; Keener and Tyson, 1986; Tyson and Keener, 1988; Ohta et al., 1989), calcium waves in *Xenopus* oocytes (Lechleiter et al., 1991b; Sneyd and Atri, 1993; Jafri and Keizer, 1995) and in studies of myocardial tissue (Keener, 1991a; Colli-Franzone et al., 1990, 1993; also see Chapter 11). Eikonal-curvature equations have a long history in other scientific fields as well, including crystal growth (Burton et al., 1951) and flame front propagation (Frankel and Sivashinsky, 1987, 1988).

The derivation of the eikonal-curvature equation uses standard mathematical arguments of singular perturbation theory, which we summarize here. The key observation is that hidden inside (10.1) is the bistable equation (10.2), and the idea to be explored is that in some moving coordinate system that is yet to be determined, (10.1) is well approximated by the bistable equation (10.2).

Our goal is to rewrite equation (10.1) in terms of a moving coordinate system chosen so that it takes the form of (10.2). In three dimensions, we must have three spatial coordinates, one of which is locally orthogonal to the wave front, while the other two are coordinates describing the wave-front surface. By assumption, the function $u$ is approximately independent of the wave-front coordinates. We will introduce a scaling of the variables such that the derivatives with respect to the first variable are of most importance, and all other derivatives are less so. From this computation, we will learn that these restrictions on the coordinate system determine how it must move in order to maintain itself as a wave-front coordinate system, and this law of coordinate system motion will be the eikonal-curvature equation.

To begin, we introduce a general (as yet unknown) moving coordinate system

$$\mathbf{x} = \mathbf{X}(\xi, \tau), \qquad t = \tau. \tag{10.5}$$

According to the chain rule,

$$\frac{\partial}{\partial \xi_i} = \frac{\partial X_j}{\partial \xi_i} \frac{\partial}{\partial x_j}, \qquad \frac{\partial}{\partial \tau} = \frac{\partial}{\partial t} + \frac{\partial X_j}{\partial \tau} \frac{\partial}{\partial x_j}. \tag{10.6}$$

Here and in what follows, the summation convention will be followed (i.e., unless otherwise noted, repeated indices are summed from 1 to 3). It follows that

$$\frac{\partial}{\partial x_i} = \alpha_{ij} \frac{\partial}{\partial \xi_j}, \qquad \frac{\partial}{\partial t} = \frac{\partial}{\partial \tau} - \frac{\partial X_j}{\partial \tau} \alpha_{jk} \frac{\partial}{\partial \xi_k}, \tag{10.7}$$

where the matrix with entries $\alpha_{ij}$ is the inverse of the matrix with entries $\frac{\partial X_j}{\partial \xi_i}$ (the Jacobian of the coordinate transformation (10.5)).

We identify the variable $\xi_1$ as the coordinate normal to level surfaces of $u$, so $\xi_2$ and $\xi_3$ are the coordinates of the moving level surfaces. Then, we define the tangent vectors $\mathbf{r}_i = \frac{\partial X_i}{\partial \xi_i}, i = 1, 2, 3$, and the normal vectors $\mathbf{n}_i = \mathbf{r}_j \times \mathbf{r}_k$, where $i \neq j, k$, and $j < k$. Without loss of generality we take $\mathbf{r}_1 = \sigma(\mathbf{r}_2 \times \mathbf{r}_3)$, so that $\mathbf{r}_1$ is always normal to the level surfaces of $u$. Here $\sigma$ is an arbitrary (unspecified) scale factor. The vectors $\mathbf{r}_2$ and $\mathbf{r}_3$ are tangent to the moving level surface, although they are not necessarily orthogonal. While one can force the vectors $\mathbf{r}_2$ and $\mathbf{r}_3$ to be orthogonal, the actual construction of such a coordinate description on a moving surface is generally quite difficult. Furthermore, it is preferred to have an equation of motion that does not have additional restrictions, since the motion should be independent of the coordinate system by which the surface is described.

We can calculate the entries $\alpha_{ij}$ explicitly. It follows from Cramer's rule (Exercise 2) that

$$\alpha_{ij} = \frac{(\mathbf{n}_j)_i}{\mathbf{r}_j \cdot \mathbf{n}_j} (\text{no summation}), \tag{10.8}$$

where by $(\mathbf{n}_j)_i$ we mean the $i$th component of the $j$th normal vector $\mathbf{n}_j$.

Now we can write out the full change of variables. We calculate that (treating the coefficients $\alpha_{ij}$ as functions of $\mathbf{x}$)

$$\frac{\partial u}{\partial t} = \frac{\partial u}{\partial \tau} - \frac{\partial X_j}{\partial \tau} \alpha_{jk} \frac{\partial u}{\partial \xi_k}, \tag{10.9}$$

$$\nabla^2 u = \alpha_{ip} \alpha_{ik} \frac{\partial^2 u}{\partial \xi_p \partial \xi_k} + \frac{\partial \alpha_{ip}}{\partial x_i} \frac{\partial u}{\partial \xi_p}, \tag{10.10}$$

and rewrite (10.1) in terms of these new variables, finding (in the case that $D$ is a constant scalar) that

$$0 = D\alpha_{ip}\alpha_{iq}\frac{\partial^2 u}{\partial \xi_p \partial \xi_q} + D\frac{\partial \alpha_{ip}}{\partial x_i}\frac{\partial u}{\partial \xi_p} - \left(u_\tau - \frac{\partial X_j}{\partial \tau}\alpha_{jk}\frac{\partial u}{\partial \xi_k}\right) + kf(u). \tag{10.11}$$

There are two important assumptions that are now invoked, namely that the spatial scale of variation in $\xi_1$ is much shorter than the spatial scale for variations in the variables $\xi_2$ and $\xi_3$. We quantify this by supposing that there is a small parameter $\epsilon$ and that $\alpha_{j1} = O(1)$, while $\alpha_{jk} = O(\epsilon)$ for all $j$ and $k \neq 1$. In addition, we assume that to leading order in $\epsilon$, $u$ is independent of $\xi_2, \xi_3$, and $\tau$. Consequently, not all of the terms in (10.11) are of equal importance. If we take into account the $\epsilon$ dependence of $\alpha_{ij}$, then (10.11) simplifies to

$$D|\alpha|^2\frac{\partial^2 u}{\partial \xi_1^2} + \left(D\nabla \cdot \alpha + \frac{\partial \mathbf{X}}{\partial \tau} \cdot \alpha\right)\frac{\partial u}{\partial \xi_1} + kf(u) = O(\epsilon), \tag{10.12}$$

where $\alpha$ is the vector with components $\alpha_{j1}$, and hence, from (10.8), proportional to the normal vector $\mathbf{n}_1$. All of the terms on the left-hand side of (10.12) are large compared to $\epsilon$.

Here we see an equation that resembles the bistable equation (10.2). If the coefficients of (10.12) are constant, we can identify (10.12) with (10.2) by setting

$$\frac{\partial \mathbf{X}}{\partial \tau} \cdot \alpha + D\nabla \cdot \alpha = kc_0, \tag{10.13}$$

while requiring $D|\alpha|^2 = k$.

Equation (10.13) tells us how the coordinate system should move, and since $\alpha$ is proportional to $\mathbf{n}_1$, setting $D|\alpha|^2 = k$ determines the scale of the coordinate normal to the wave front, i.e., the thickness of the wave front. In reality, the coefficients of (10.12) are not constants, and these two requirements overdetermine the full coordinate transformation $\mathbf{X}(\xi, \tau)$. To overcome this difficulty, we assume that since the wave front and the coordinate system are slowly varying in space, we interpret (10.13) as determining the motion of only the midline of the coordinate system, at the location of the largest gradient of the front, rather than the entire coordinate system.

Equation (10.13) is the equation we seek that describes the motion of an action potential front, called the *eikonal-curvature equation*. However, for numerical simulations it is essentially useless. Numerical algorithms to simulate this equation reliably are extremely hard to construct. Instead, it is useful to introduce a function $S(x,t)$ that acts as an indicator function for the fronts (think of $S$ as determining the "shock" location). That is, if $S(x,t) > 0$, the medium is activated, while if $S(x,t) < 0$, the medium is in the resting state. Taking $\alpha$ to be in the direction of forward wave-front motion means that $\alpha = -\sqrt{\frac{k}{D}}\frac{\nabla S}{|\nabla S|}$, where we have used the fact that $|\alpha| = \sqrt{k/D}$. Since the zero level surface of $S(x,t)$ denotes the wave-front location, and thus $S$ is constant along the wave front, it follows that $0 = \nabla S \cdot X_t + S_t$, so that $X_t \cdot \alpha = \sqrt{\frac{k}{D}} \frac{S_L}{|\nabla S|}$. Hence,

$$S_t = |\nabla S| c_0 \sqrt{Dk} + D|\nabla S| \nabla \cdot \left( \frac{\nabla S}{|\nabla S|} \right). \tag{10.14}$$

The use of an indicator function $S(x,t)$ to determine the motion of an interface is called the *level set method* (Osher and Sethian, 1988), and it is both powerful and easy to implement.

Equation (10.14) is called the eikonal-curvature equation because of the physical interpretation of each of its terms. If we ignore the diffusive term, then we have the eikonal equation

$$\frac{\partial S}{\partial t} = |\nabla S| c_0 \sqrt{Dk}. \tag{10.15}$$

If **R** is a level surface of the function $S(x,t)$ and if **n** is the unit normal vector to that surface at some point, then (10.15) implies that the normal velocity of the surface **R**, denoted by $\mathbf{R}_t \cdot \mathbf{n}$, satisfies

$$\mathbf{R}_t \cdot \mathbf{n} = c_0 \sqrt{Dk}. \tag{10.16}$$

In other words, the front moves in the normal direction **n** with speed $c = c_0 \sqrt{Dk}$.

Equation (10.16) is the basis of a geometrical "Huygens" construction for front propagation, but the numerical integration of either (10.15) or (10.16) is fraught with difficulties. In particular, cusp singularities develop, and the indicator function $S(x,t)$ becomes ill-defined in finite time (usually very quickly). The second term of the right-hand side of (10.14) is a curvature correction, appropriately named because the term $\nabla \cdot \left( \frac{\nabla S}{|\nabla S|} \right)$ is twice the mean curvature (in three-dimensional space) or the curvature (in two-dimensional space) of the level surfaces of $S$ (see Exercise 3). In fact, the eikonal-curvature equation can be written as

$$\mathbf{R}_t \cdot \mathbf{n} = c_0 \sqrt{Dk} - D\kappa, \tag{10.17}$$

or

$$\tau \mathbf{R}_t \cdot \mathbf{n} = c_0 \Lambda - \Lambda^2 \kappa, \tag{10.18}$$

where $\kappa$ is the curvature (in two dimensions) or twice the mean curvature (in three dimensions) of the front, $\Lambda = \sqrt{D/k}$ is the space constant, and $\tau = 1/k$ is the time constant. Even though it usually represents only a small correction to the normal velocity of fronts, the curvature correction is important for physical and stability reasons, to prevent singularity formation. The sign of the curvature correction is such that a front with ripples is gradually smoothed into a plane wave.

Experiments on the Belousov–Zhabotinsky reagent have verified this relationship between speed and curvature of propagating fronts. For example, Foerster et al. (1988) measured the speed and curvature at different positions of a rotating spiral wave and at intersections of two spiral waves (thus obtaining curvatures of different signs) and found that the relationship between normal velocity and curvature was well approximated by a straight line with slope that was the diffusion coefficient of the rapidly reacting species.

## 10.2 Spatial Patterns and Spiral Waves

Now that we have some idea of how wave fronts propagate in an excitable medium, we next wish to determine the spatial patterns that may result. The most common pattern is created by, and spreads outward from, a single source. If the medium is sufficiently large so that more than one wave front can exist at the same time, then these are referred to as *target patterns*. Target patterns require a periodic source and so cannot exist in a homogeneous nonoscillatory medium.

A second type of spatial pattern is a spiral wave. Spiral waves do not require a periodic source for their existence, as they are typically self-sustained. Because they are self-sustained, spirals usually occur only in pathophysiological situations. That is, it is usually not a good thing for a system that relies on faithful propagation of a signal to be taken over by a self-sustained pattern. Thus, spirals on the heart are fatal, spirals in the cortex may lead to epileptic seizures, and spirals on the retina or visual cortex may cause hallucinations. The most famous example of spiral waves in an excitable medium is in the Belousov–Zhabotinsky reaction (Winfree, 1972, 1974), which we do not discuss here.

The mathematical discussion of spiral waves centers on the nature of periodic solutions of a system of differential equations with excitable dynamics spatially coupled by diffusion. A specific example is the FitzHugh–Nagumo equations with diffusive coupling in two spatial dimensions,

$$\epsilon \frac{\partial v}{\partial t} = \epsilon^2 \nabla^2 v + f(v, w), \qquad (10.19)$$

$$\frac{\partial w}{\partial t} = g(v, w). \qquad (10.20)$$

The leading-order singular perturbation analysis (i.e., with $\epsilon = 0$) suggests that the domain be separated into two, in which outer dynamics

$$\frac{\partial w}{\partial t} = G_\pm(w) \qquad (10.21)$$

hold (using the notation of Section 9.4.1). The region in which $\frac{\partial w}{\partial t} = G_+(w)$ is identified as the excited region, and the region in which $\frac{\partial w}{\partial t} = G_-(w)$ is called the recovering region. Separating these will be moving interfaces in which $v$ changes rapidly (with space and time constant $\epsilon$), so that diffusion is important, while $w$ remains essentially constant. At any point in space the solution should be periodic in time, so at large radii, where the wave fronts are nearly planar, the solution should lie on the dispersion curve.

The first guess as to how the interface should move is to assume that the interface is nearly planar and therefore has the same velocity as a plane wave, namely

$$\mathbf{R}_t = c(w)\mathbf{n}, \tag{10.22}$$

where $\mathbf{R}$ is the position vector for the interface, $\mathbf{n}$ is the unit normal vector of $\mathbf{R}$, and $c(w)$ is the plane-wave velocity as a function of $w$.

To see the implications of the eikonal equation, we suppose that the spiral interface is a curve $\mathbf{R}$ given by

$$X(r,t) = r\cos(\theta(r) - \omega t), \qquad Y(r,t) = r\sin(\theta(r) - \omega t). \tag{10.23}$$

Note that the interface is a curve parametrized by $r$, and so the tangent is $(X_r, Y_r)$. We then calculate that

$$\mathbf{R}_t = \begin{pmatrix} -\omega r \sin(\theta - \omega t) \\ \omega r \cos(\theta - \omega t) \end{pmatrix} \tag{10.24}$$

and

$$\sqrt{1 + r^2\theta'^2}\,\mathbf{n} = \begin{pmatrix} -\sin(\theta - \omega t) - r\theta'\cos(\theta - \omega t) \\ \cos(\theta - \omega t) - r\theta'\sin(\theta - \omega t) \end{pmatrix}, \tag{10.25}$$

so that the eikonal equation becomes

$$c(w)\sqrt{1 + r^2\theta'^2} = \omega r. \tag{10.26}$$

An integration then gives

$$\theta(r) = \rho(r) - \tan(\rho(r)), \qquad \rho(r) = \sqrt{\frac{r^2}{r_0^2} - 1}, \tag{10.27}$$

where $r_0 = c/\omega$, so that the interface is given by

$$X = r_0\cos(s) + r_0\rho(r)\sin(s), \qquad Y = r_0\sin(s) - r_0\rho(r)\cos(s), \tag{10.28}$$

where $s = \rho(r) - \omega t$. This interface is the involute of a circle of radius $r_0$. (The involute of a circle is the locus of points at the end of a string that is unwrapped from a circle.)

There are significant difficulties with this as a spiral solution, the most significant of which is that it exists only for $r \geq r_0$. The parameter $r_0$ is arbitrary, but positive, so that this spiral is rotating about some hole of finite size. The frequency of rotation is determined by requiring consistency with the dispersion curve. Note that the spiral has wavelength $2\pi r_0$, and period $\frac{2\pi}{\omega}$, and so $c = r_0\omega$. However, since the dispersion curve generally has a knee, and thus periodic waves do not exist for small enough wavelength, there is a lower bound on the radii for which this can be satisfied. Numerical studies of spirals suggest no such lower bound on the inner core radius and also suggest that there is a unique spiral frequency for a medium without a hole at the center. Unfortunately, the use of the eikonal equation gives no hint of the way a unique frequency is selected, so a different approach, using the eikonal-curvature equation, is required.

To apply the eikonal-curvature equation to find rotating spiral waves, we assume that the wave front is expressed in the form (10.23), so that the curvature is

$$\kappa = \frac{X'Y'' - Y'X''}{(X'^2 + Y'^2)^{3/2}} = \frac{\psi'}{(1 + \psi^2)^{3/2}} + \frac{\psi}{r(1 + \psi^2)^{1/2}}, \tag{10.29}$$

where $\psi = r\theta'(r)$ is called the shape function. Thus the eikonal-curvature equation (10.18) becomes

$$r\frac{d\psi}{dr} = (1 + \psi^2)\left[\frac{rc(w)}{\epsilon}(1 + \psi^2)^{1/2} - \frac{wr^2}{\epsilon} - \psi\right]. \tag{10.30}$$

If we suppose that $w$ is constant along the spiral front, then (10.30) can be solved numerically by "shooting" from $r = \infty$. A portrait of sample trajectories in the $(r, \psi)$ plane is shown in Fig. 10.1. The trajectories of (10.30) are "stiff," meaning that for large $r$, the trajectory $c(1 + \psi^2)^{1/2} - wr = 0$ is a strong attractor. This stiffness can be readily observed when (10.30) is written in terms of the variable $\phi = \frac{r\psi}{\sqrt{1+\psi^2}}$ as

$$\frac{\epsilon}{r}\frac{d\phi}{dr} = c - \omega\sqrt{r^2 - \phi^2}, \tag{10.31}$$

since $\epsilon$ multiplies the derivative term in (10.31).

Integrating from $r = \infty$, trajectories of (10.30) approach the origin by either blowing up or down near the origin $r = 0$. Since the origin is a saddle point, if parameters are chosen exactly right, the trajectory approaches $\psi = 0$. Thus, there is a unique relationship between $\omega$ and $c$ of the form $\omega/\epsilon = F(c/\epsilon)$ that yields trajectories that go all the way to the origin $r = 0, \psi = 0$.

Notice that the rescaling of variables $r \to \alpha r, c \to c/\alpha, \omega \to \omega/\alpha^2$ leaves (10.30) invariant, so that the relationship between $\omega$ and $c$ for which a trajectory approaches the saddle point at the origin must be of the form

$$\frac{\omega}{\epsilon\alpha^2} = F\left(\frac{c}{\epsilon\alpha}\right). \tag{10.32}$$

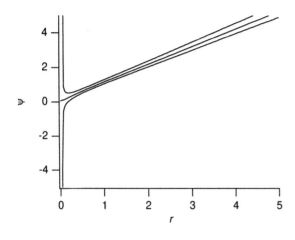

**Figure 10.1** Trajectories of (10.30) with $\omega/\epsilon = 2.8, 9m^*$, and 3.2, and $c/\epsilon = 3.0$.

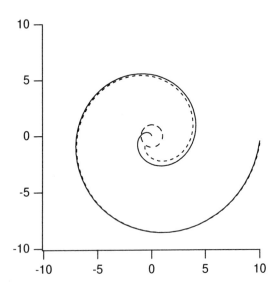

**Figure 10.2** Spiral arm corresponding to the trajectory of (10.30) that approaches the origin (shown solid), compared with the involute spiral (shown dashed) with the same parameter values. For this plot, $c/\epsilon = 3.0$.

It follows that $F(c/\alpha) = \frac{1}{\alpha^2} F(c)$ and thus $F(x) = m^* x^2$, for some constant $m^*$, so that

$$\omega = \frac{c^2 m^*}{\epsilon}. \tag{10.33}$$

Numerically, one determines that $m^* = 0.330958$.

An example of this spiral front is shown in Fig. 10.2, compared with the comparable involute spiral (10.28), shown dashed. For this figure $c/\epsilon = 3.0$.

At this point we have a family of possible spiral trajectories that have correct asymptotic (large $r$) behavior and approach $\psi = 0$ as $r \to 0$. This family is parametrized by the speed $c$. To determine which particular member of this family is the correct spiral front, we also require that the spirals be periodic waves; that is, they must satisfy the dispersion relationship. These two requirements uniquely determine the spiral properties. To see that this is so, in Fig. 10.3 are plotted the critical curve (10.33) and the approximate dispersion curve (9.89). For this plot we used piecewise linear dynamics, $f(v, w) = H(v - \alpha) - v - w, g(v, w) = v - \gamma w$ with $\alpha = 0.1, \gamma = 0, \epsilon = 0.05$.

## 10.2.1 More About Spirals

This discussion of higher-dimensional waves is merely the tip of the iceberg (or tip of the spiral), and there are many interesting unresolved questions.

While the spirals that are observed in physiological systems share certain qualitative similarities, their details are certainly different. The FitzHugh–Nagumo model discussed here shows only the qualitative behavior for generic excitable systems and so has little quantitative relevance. An analysis for the two-pool model of calcium wave propagation (Chapter 12) is similar in style, but much more difficult in detail because

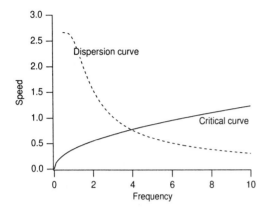

**Figure 10.3** Critical curve (10.33) and the approximate dispersion curve (9.89) using piecewise linear dynamics $f(v, w) = H(v - \alpha) - v - w$, $g(v, w) = v - \gamma w$ with $\alpha = 0.1$, $\gamma = 0$, $\epsilon = 0.05$.

the eikonal-curvature relationship is nonlinear and the dispersion curve is more difficult to obtain. Other physiological systems are likely governed by other dynamics. For example, a model for spreading cortical depression in the cortex has been proposed by Tuckwell and Miura (1978; Miura, 1981), and numerical simulations have shown rotating spirals. However, a detailed mathematical study of these equations has not been given.

This analytical calculation for the FitzHugh–Nagumo system is based on singular perturbation theory and therefore is not mathematically rigorous. In fact, as yet there is not a rigorous proof of the existence of spiral waves in an excitable medium. While the approximate solution presented here is known to be asymptotically valid, the structure of the core of the spiral is not correct. This problem has been addressed by Pelce and Sun (1991) and Keener (1992) for FitzHugh–Nagumo models with a single diffusing variable, and by Keener (1994) and Kessler and Kupferman (1996) for FitzHugh–Nagumo models with two diffusing variables, relevant for chemical reaction systems.

A second issue of concern is the stability of spirals. This also is a large topic, which is not addressed here. The interested reader should consult the work of Winfree (1991), Jahnke and Winfree (1991), Barkley (1994), Karma (1993, 1994), Panfilov and Hogeweg (1995), Kessler and Kupferman (1996).

Because the analytical study of excitable media is so difficult, simpler models have been sought, with the result that finite-state automata are quite popular. A finite-state automaton divides the state space into a few discrete values (for example $v = 0$ or 1), divides the spatial domain into discrete cells, and discretizes time into discrete steps. Then, rules are devised for how the states of the cells change in time. Finite-state automata are extremely easy to program and visualize. They give some useful insight into the behavior of excitable media, but they are also beguiling and can give "wrong" answers that are not easily detected. The literature on finite-state automata is vast (see, for instance, Moe et al., 1964; Smith and Cohen, 1984; and Gerhardt et al., 1990).

The obvious generalization of a spiral wave in a two-dimensional region to three dimensions is called a *scroll wave* (Winfree, 1973, 1991; Keener and Tyson, 1992). Scroll waves have been observed numerically (Jahnke et al., 1988; Lugosi and Winfree, 1988), in three-dimensional BZ reagent (Gomatam and Grindrod, 1987), and in cardiac tissue (Chen et al., 1988), although in experimental settings they are extremely difficult to visualize. In numerical simulations it is possible to initiate scroll waves with interesting topology, including closed scroll rings, knotted scrolls, or linked pairs of scroll rings.

The mathematical theory of scroll waves is also in its infancy. Attributes of the topology of closed scrolls were worked out by Winfree and Strogatz (1983a,b,c; 1984), and a general asymptotic theory for their evolution has been suggested (Keener, 1988b) and tested against numerical experiments on circular scroll rings and helical scrolls. There is not sufficient space here to discuss the theory of scroll waves. However, scroll waves are mentioned again briefly in later chapters on cardiac waves and rhythmicity.

## 10.3   EXERCISES

1.  What is the eikonal-curvature equation for (10.1) when the medium is anisotropic and $D$ is a symmetric matrix, slowly varying in space?
    Hint: Generalize (10.11) by calculating $\nabla \cdot (D\nabla u)$ using components of $D$ as $d_{ij}$, and use this to generalize (10.12).
    Answer: (10.14) becomes $S_t = \sqrt{\nabla S \cdot D \nabla S}\left(c_0 \sqrt{k} + \nabla \cdot \left(\frac{D\nabla S}{\sqrt{\nabla S \cdot D \nabla S}}\right)\right)$.

2.  Verify (10.8).
    Hint: Use Cramer's rule to find the inverse of a matrix with three column vectors, say $t_1, t_2$, and $t_3$. Use the fact that the determinant of such a matrix is $t_1 \cdot (t_2 \times t_3)$. Then apply this to the transpose of the Jacobian matrix $\frac{\partial x_j}{\partial \xi_i}$.

3.  Verify that in two spatial dimensions, $\nabla \cdot \left(\frac{\nabla S}{|\nabla S|}\right)$ is the curvature of the level surface of the function $S$.
    Hint: If $x = X(t), y = Y(t)$ is the parametric representation of a smooth level-surface curve, then $S(X(t), Y(t)) = 0$. Use this and derivatives of this expression with respect to $t$ to show that $\nabla \cdot \left(\frac{\nabla S}{|\nabla S|}\right) = \pm \frac{Y_t X_{tt} - Y_{tt} X_t}{(X_t^2 + Y_t^2)^{3/2}}$.

4.  The following are the rules for a simple finite automaton on a rectangular grid of points:

    (a)  The state space consists of three states, 0, 1, and 2, 0 meaning at rest, 1 meaning excited, and 2 meaning refractory.

    (b)  A point in state 1 goes to state 2 on the next time step. A point in state 2 goes to 0 on the next step.

    (c)  A point in state 0 remains in state 0 unless at least one of its nearest neighbors is in state 1, in which case it goes to state 1 on the next step.

    Write a computer program that implements these rules. What initial data must be supplied to initiate a spiral? Can you initiate a double spiral by supplying two stimuli at different times and different points?

5. (a) Numerically simulate spiral waves for the Pushchino model of Chapter 4, Exercise 13.

   (b) Numerically simulate spiral waves for the Pushchino model with

$$f(V) = \begin{cases} C_1 V & \text{when } V < V_1, \\ -C_2 V + a & \text{when } V_1 < V < V_2, \\ C_3(V - 1) & \text{when } V > V_2, \end{cases} \tag{10.34}$$

   and

$$\tau(V, w) = \begin{cases} \tau_1 & \text{when} \quad V_1 < V < V_2 \\ \tau_1 & \text{when} \quad V < V_1, w > w_1, \\ \tau_2 & \text{when} \quad V > V_2, \\ \tau_3 & \text{when} \quad V < V_1, w < w_1. \end{cases}$$

   Use the parameters $V_1 = 0.0026$, $V_2 = 0.837$, $w_1 = 1.8$, $C_1 = 20$, $C_2 = 3$, $C_3 = 15$, $a = 0.06$, $\tau_1 = 75$, $\tau_2 = 1.0$, $\tau_3 = 2.75$, and $k = 3$. What is the difference between these spirals and those for the previous model?
   Answer: There are no stable spirals for this model, but spirals continually form and break apart, giving a "chaotic" appearance.

# CHAPTER 11

# Cardiac Propagation

Cardiac cells perform two functions in that they are both excitable and contractile. They are excitable, enabling action potentials to propagate, and the action potential causes the cells to contract, thereby enabling the pumping of blood. The electrical activity of the heart is initiated in a collection of cells known as the *sinoatrial node* (SA node) located just below the superior vena cava on the right atrium. The cells in the SA node are autonomous oscillators. The action potential that is generated by the SA node is then propagated through the atria by the atrial cells.

The atria and ventricles are separated by a septum composed of nonexcitable cells, which normally acts as an insulator, or barrier to conduction, of action potentials. There is one pathway for the action potential to continue propagation and that is through another collection of cells, known as the *atrioventricular node* (AV node), located at the base of the atria.

Conduction through the AV node is quite slow, but when the action potential exits the AV node, it propagates through a specialized collection of fibers called the *bundle of HIS*, which is composed of Purkinje fibers. The Purkinje fiber network spreads via tree-like branching into the left and right *bundle branches* throughout the interior of the ventricles, ending on the *endocardial surface* of the ventricles. As action potentials emerge from the Purkinje fiber–muscle junctions, they activate the ventricular muscle and propagate through the ventricular wall outward to the epicardial surface. A schematic diagram of the cardiac conduction system is shown in Fig. 11.1.

It should be apparent from this introduction that in the heart there is one-dimensional wave propagation, for example, along a Purkinje fiber, and there is higher-dimensional propagation in the atrial and ventricular muscle.

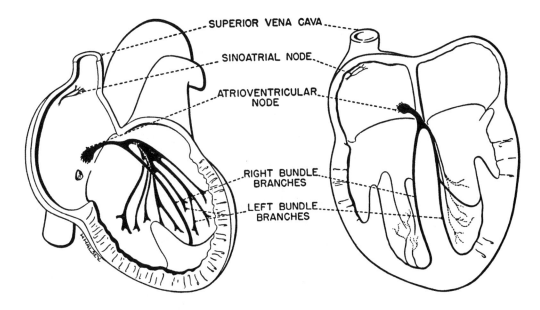

SUPERIOR VENA CAVA

SINOATRIAL NODE

ATRIOVENTRICULAR
NODE

RIGHT BUNDLE
BRANCHES

LEFT BUNDLE
BRANCHES

**Figure 11.1** Schematic diagram of the cardiac conduction system. (Rushmer, 1976, Fig. 3-9, p. 87).

# 11.1   Cardiac Fibers

## 11.1.1   Cellular Coupling

Myocardial cells are cable-like, roughly cylindrical, typically 100 $\mu$m long and 15 $\mu$m in diameter. They are packed together in a three-dimensional irregular brick-like packing, surrounded by extracellular medium (Fig. 11.2). Each cell has specialized contacts with its neighboring cells, mainly in end-to-end fashion, facilitated by a step-like surface that locks into neighboring cells. The opposing cell membranes form the *intercalated disk* structure. While the end-to-end cell membranes are typically separated by about 250 Å, there are places, called *junctions*, where the pre- and postjunctional membranes are fused together. The mechanical adhesion of cells is provided by adhering junctions in the intercalated disk, known as *desmosomes* or *tight junctions*. The electrical coupling of cells is provided by gap junctions (Chapter 7).

The intercellular channels provided by the gap junctions are around 20 Å in diameter and are characterized as "low-resistance" because the effective resistance is considerably less than what would result from two cell membranes butted together. However, compared to the intracellular cytoplasm, the gap junctions are of high resistance, simply because the cross-sectional area for electrical conduction through gap junctions is greatly reduced (about two percent of the total cross-sectional area).

To model a cardiac fiber, we consider a simple one-dimensional collection of cylindrical cells (with perimeter $p$) coupled in end-to-end fashion via gap junctions. From

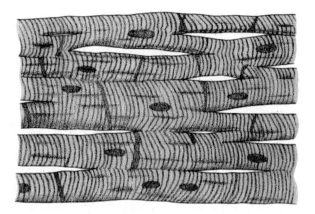

**Figure 11.2**  Cardiac cell structure. (Guyton and Hall, 1996, Fig. 9-2, p. 108.)

Chapter 8, we know that between gap junctions we have the conservation equation

$$p\left(C_m \frac{\partial V}{\partial t} + I_{\text{ion}}\right) = \frac{\partial}{\partial x}\left(\frac{A_i}{R_c}\frac{\partial V_i}{\partial x}\right) = -\frac{\partial}{\partial x}\left(\frac{A_e}{R_e}\frac{\partial V_e}{\partial x}\right). \tag{11.1}$$

where $R_c$ and $R_e$ are the resistivities of intracellular and extracellular space, respectively, and $A_i$ and $A_e$ are the average cross-sectional areas for intracellular and extracellular space per cell. At the ends of cells (each of length $L$), there is a possible jump in intracellular potential, but the current $-\frac{A_i}{R_c}\frac{\partial V_i}{\partial x}$ must be continuous, so that

$$\frac{[V_i]}{r_g} = \frac{A_i}{R_c}\frac{\partial V_i}{\partial x} \tag{11.2}$$

at the ends of cells, where $[V_i]$ is the jump in intracellular potential across the gap junctions. The parameter $r_g$ is the effective gap-junctional resistance, or inverse permeability. The extracellular potential and current are continuous, unaffected by the gap junctions.

The time constant for this fiber is the same as (8.16). The space constant, however, is affected by the gap-junctional resistance.

To find the space constant, we take $I_{\text{ion}} = -V/R_m$ and look for a geometrically decaying solution, with $V_i(x+L) = \mu V_i(x)$, $V_e(x+L) = \mu V_e(x)$ for some constant $\mu$. The constant $\mu$ relates to the space constant $\lambda_g$ through $\mu = e^{-L/\lambda_g}$.

The solution of this problem can be found analytically. For the $n$th cell, the solution is proportional to

$$\begin{pmatrix} V_i \\ V_e \end{pmatrix} = \mu^n \Phi(\mu, x), \tag{11.3}$$

where

$$\Phi(\mu, x) = \left\{ \left[ (\mu - \frac{1}{E})e^{\lambda x} + (\mu - E)e^{-\lambda x} \right] \begin{pmatrix} q_i \\ -q_e \end{pmatrix} \right.$$
$$\left. + 2q_e \frac{(\mu - \frac{1}{E})(\mu - E)}{\mu - 1} \begin{pmatrix} 1 \\ 1 \end{pmatrix} \right\} \tag{11.4}$$

for $0 < x < L$, where $E = e^{\lambda L}$, $\lambda L = \sqrt{q_e + q_i} = \sqrt{Q}$, and $q_i = \frac{L^2 S R_c}{v_i R_m}$, $q_e = \frac{L^2 S R_e}{v_e R_m}$, where $v_i = LA_i$ and $v_e = LA_e$ are the intracellular and extracellular volumes (per unit cell), respectively, $S = Lp$ is the cell surface area and $p = 2\sqrt{A_i \pi}$ is the perimeter. The number $\mu$ must be a root of the characteristic equation

$$r = 2\sqrt{Q} \frac{(\mu - \frac{1}{E})(\mu - E)}{\mu(E - \frac{1}{E})}, \tag{11.5}$$

where $r = \frac{S r_g}{R_m}$ is the effective nondimensional gap-junctional resistance.

The behavior of the extracellular and intracellular potentials is depicted in Fig. 11.3. As can be seen from this plot, the extracellular potential decays smoothly, but the intracellular potential decays with discrete jumps across the gap junctions.

It is not possible to measure the intracellular potential with the detail shown in Fig. 11.3, because cells are usually too small to invade with multiple intracellular electrodes without irreversibly damaging the cell membrane. In fact, the space constant is usually measured by fitting a decaying exponential to the intracellular or extracellular potential. The formula (11.5) can then be used to calculate the effective gap-junctional resistance.

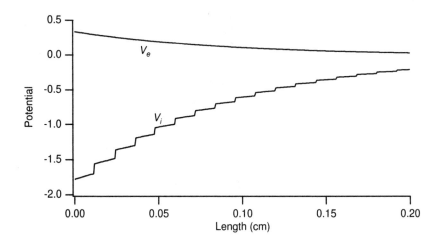

**Figure 11.3** Plot of intracellular and extracellular potentials as a function of space, with a constant subthreshold potential applied at a single point. For this plot cells had $L = 0.012$ cm, $A_i = 4.0 \times 10^{-6}$ cm$^2$, and a space constant $\lambda_g = 0.09$ cm. $R_m = 7000\,\Omega$cm$^2$, $R_c = 150\,\Omega$cm, $q_e = 0.5 q_i$, $q_i = 5.47 \times 10^{-3}$. The vertical scale on this plot is arbitrary.

If the gap-junctional resistance is small compared with the cytoplasmic resistance, then (11.5) has a simplified solution. If $r/Q$ is small, or if $L/\lambda_g$ is small, then the solution of (11.5) is given approximately by

$$\frac{L^2}{\lambda_g^2} = r + Q = r + q_i + q_e. \tag{11.6}$$

The formula (11.6) is used routinely in bioengineering and in linear circuit theory; it implies that resistance along the cable is additive. This is exactly the same answer that one would find for the space constant of a uniform continuous cable if the gap-junctional resistance were uniformly distributed throughout the cytoplasm. For most normal cells, the approximation (11.6) is valid. It is only when the gap-junctional resistance is excessively large, such as if there is *ischemia* or if the cells are treated with certain alcohols that block gap junctions, that this formula is substantially wrong.

The effective gap-junctional resistance $r_g$ can be calculated from (11.6). For example, using frog myocardial cells (which are longer than many other cells, $L = 131\,\mu\text{m}$, radius $= 7.5\,\mu\text{m}$), Chapman and Fry (1978) measured a space constant $\lambda_g = 0.328$ cm yielding $(\frac{L}{\lambda_g})^2 = 0.159$. From this they inferred (using $R_m = 1690\,\Omega\text{cm}^2$ and $q_e = 0$) that the effective cytoplasmic resistivity was $R_c = 588\,\Omega\text{cm}$. They were also able to measure the cytoplasmic resistivity directly, and they found $R_c = 282\,\Omega\text{cm}$ so that $q_i = 0.076$. Thus only 48% of the total resistivity of the cell was attributable to cytoplasmic resistance. The remaining 52% must be from gap-junctional resistance. We can calculate the effective gap-junctional resistance as $r_g = \frac{rR_m}{Lp} = 2.27\,\text{M}\Omega$ per cell.

The study of wave propagation in the cable equation (11.1) with jump conditions (11.2) is quite difficult, and results are limited. It is therefore useful to consider simplified models. The simplest of these is a continuous model in which gap junctions are ignored, or more precisely, in which the intracellular resistance is adjusted to incorporate gap-junctional resistance in a homogeneous (or averaged) sense. The second simplification is to take the opposite limit of small cytoplasmic resistance and to assume that cells are isopotential and all resistance is concentrated in the gap junctions. In this limit, conservation of current implies that

$$Lp\left(C_m\frac{\partial V_j}{\partial t} + I_{\text{ion}}\right) = d(V_{j+1} - 2V_j + V_{j-1}), \tag{11.7}$$

where $V_j$ is the (iso-) potential of the $j$th cell. We pick the coupling parameter $d$ such that

$$d\left(\mu - 2 + \frac{1}{\mu}\right) = \frac{Lp}{R_m}, \tag{11.8}$$

where $\mu = e^{-L/\lambda_g}$, so that the decay rate for a linear cable matches with the space constant of the medium.

## 11.1.2 Propagation Failure

One-dimensional propagation in cardiac fibers is expected to occur, since the equations are similar to those of Chapter 9. It is perhaps more interesting and of greater clinical significance to understand the causes of propagation failure. Here we mention two.

**Branching**

On leaving the AV node of the heart, the action potential enters the bundle of HIS. The bundle divides near the upper ventricular septum into right and left branches. The right bundle continues with little arborization toward the apex of the heart. The left bundle branch divides almost immediately into two major divisions: one anterior and superior, and the second posterior and inferior.

A bundle branch block occurs when the action potential fails to propagate through the entire branch. To understand something about the cause of bundle branch block, we consider a model of propagation in a one-dimensional fiber that divides into two. Wave fronts are governed by the cable equation

$$C_m R_m \frac{\partial V}{\partial t} = \frac{R_m}{p} \frac{\partial}{\partial x} \left( \frac{A}{R_c} \frac{\partial V}{\partial x} \right) + f(V). \tag{11.9}$$

We can also use this equation for propagation in a bundle of fibers by letting $p$ and $A$ be the total membrane perimeter and cross-sectional area, respectively, for the bundle.

Suppose there is a junction at which a cable splits into two. On both sides of this junction the cable equation (11.9) holds, but with different parameters $p$ and $A$, say $p_1, A_1$ for $x < 0$, and $p_2, A_2$ for $x > 0$. At the junction the potential $V$ and the axial current $\frac{A}{R_c} \frac{\partial V}{\partial x}$ must be continuous.

Using upper and lower solution techniques (Fife, 1979), one can demonstrate an important *comparison property* for the cable equation (11.9): If $V_1(x)$ and $V_2(x)$ are two functions that are ordered, with $V_1(x) \leq V_2(x)$, then the solutions of (11.9) with initial data $V_1(x)$ and $V_2(x)$, say $V_1(x, t)$ and $V_2(x, t)$ with $V_1(x, 0) = V_1(x)$ and $V_2(x, 0) = V_2(x)$, then $V_1(x, t) \leq V_2(x, t)$ for all time $t \geq 0$.

The importance of this theorem is that if we can establish the existence of a standing transitional profile, then traveling profiles of similar type are precluded. The standing wave is an upper bound for solutions and thereby prevents propagation (Pauwelussen, 1981).

Suppose the function $f(V)$ has three zeros, at $V = 0, \alpha$, and $1$. We look for a standing profile that connects $V = 0$ at $x = -\infty$ with $V = 1$ at $x = \infty$. The standing profile must satisfy the ordinary differential equation

$$\frac{R_m}{p_i} \left( \frac{A_i}{R_c} V_x \right)_x + f(V) = 0 \tag{11.10}$$

with $i = 1$ for $x < 0$ and with $i = 2$ for $x > 0$. Multiplying these equations by $V_x$ and integrating, we obtain

$$\frac{1}{2}\frac{R_m A_i}{R_c p_i}V_x^2 + F(V) = \begin{cases} 0 & \text{if } i = 1, \\ F(1) & \text{if } i = 2, \end{cases} \tag{11.11}$$

where $F(V) = \int_0^V f(u)du$. Sketches of these two curves are depicted in Fig 11.4 in the case $F(1) > 0$. A connecting trajectory exists if these two curves intersect at the same level of current. We express the profiles (11.11) in terms of the axial current $I = -\frac{A}{R_c}V_x$ and obtain

$$\frac{1}{2}\frac{R_m R_c}{A_i p_i}I^2 + F(V) = \begin{cases} 0 & \text{if } i = 1, \\ F(1) & \text{if } i = 2. \end{cases} \tag{11.12}$$

Intersections of these two curves occur if there is a solution of

$$F(V)\left(\frac{A_1 p_1}{A_2 p_2} - 1\right) = -F(1) \tag{11.13}$$

with $F(V) < 0$ in the range $0 < V < 1$. Since the minimum for $F(V)$ is at $V = \alpha$, there is a solution whenever

$$\frac{A_1 p_1}{A_2 p_2} \geq 1 - \frac{F(1)}{F(\alpha)}. \tag{11.14}$$

In the special case that $f(V)$ is the cubic polynomial $f(V) = V(V - 1)(\alpha - V)$, this condition becomes

$$\frac{A_1 p_1}{A_2 p_2} \geq 1 + \frac{1 - 2\alpha}{\alpha^3(2 - \alpha)}. \tag{11.15}$$

The interpretation is clear. If at a branch point of a fiber the product $pA$ increases by a sufficient amount, as specified by (11.14), then propagation through the branch point in the direction of increasing $pA$ is not possible. Of course, this criterion for propagation block depends importantly on the excitability of the fiber as expressed through the ratio $\frac{F(1)}{F(\alpha)}$, and propagation failure is more likely when the fiber is less excitable. Hence propagation block is time-dependent in that if inadequate recovery time is provided, or if the recovery mechanism is slower than normal, the likelihood of block at a branch point is increased.

### Gap-junctional coupling

We expect that gap-junctional resistance can have the similar effect of precluding propagation. To see how gap-junctional resistance affects the success or failure of propagation, we consider the idealized situation of cells of length $L$ coupled at their ends by gap junctions, as described by (11.2) and the cable equation (11.1) with piecewise linear ionic current

$$I_{\text{ion}}(V) = \frac{1}{R_m}\left[H(V - \alpha) - V\right]. \tag{11.16}$$

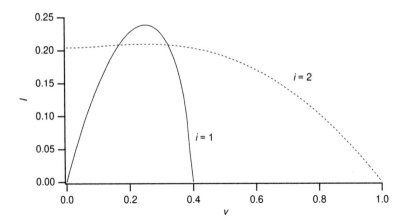

**Figure 11.4** The curves (11.12) with $i = 1$ and $\frac{R_m R_c}{2A_i p_i} = 0.04$ and with $i = 2$ and $\frac{R_m R_c}{2A_i p_i} = 1.0$, for $f(v) = v(v-1)(\alpha - v)$, $\alpha = 0.25$. An intersection of the solid curve with the dashed curve guarantees propagation failure.

This model recommends itself because it can be solved explicitly, even though it lacks quantitative reliability.

As before, we look for a standing solution on the assumption that the existence of a standing solution precludes the possibility of propagation.

To find a solution of the standing wave problem (Keener, 1991b), we use the solution (11.5) of the linear problem to find the solution for the $n$th cell to be

$$\begin{pmatrix} V_i \\ V_e \end{pmatrix}_n = A\mu^n \Phi(\mu, x) \tag{11.17}$$

for $n \geq 0$ and

$$\begin{pmatrix} V_i \\ V_e \end{pmatrix}_n = A\mu^{n+1} \Phi\left(\frac{1}{\mu}, x\right) + \begin{pmatrix} 1+C \\ C \end{pmatrix} \tag{11.18}$$

for $n < 0$, where $\mu < 1$ is a root of (11.5). Here $A$ and $C$ are as yet undetermined. However, this proposed solution has the feature that $V = V_i - V_e$ approaches 0 as $n \to \infty$, and it approaches 1 as $n \to -\infty$. Furthermore, the intracellular and extracellular currents are continuous at all junctions.

Now, to determine the coefficients $A$ and $C$, we require that $V_e$ be continuous at the junction between cell $n = -1$ (at $x = L$) and cell $n = 0$ (at $x = 0$) and that the junctional condition (11.2) be satisfied there as well.

A plot of this solution is shown in Fig. 11.5. The solution thus determined has $V < \frac{1}{2}$ for cell $n = 0$ at $x = 0$ and $V > \frac{1}{2}$ for cell $n = -1$ at $x = L$ whenever $r_g$ is positive. In particular,

$$V_0(0) = \frac{R_c \sqrt{Q}(E^2 - 2E\mu + 1)}{2\sqrt{Q}R_c(E^2 - 2E\mu + 1) + q_i r_g A_i(E^2 - 1)}. \tag{11.19}$$

However, to be a valid solution for the piecewise linear ionic current (11.16), it must be that $V_0(0) < \alpha$. This leads to the condition

$$r \geq \frac{1 - 2\alpha}{\alpha} \sqrt{Q} \left( \frac{E - 2\mu + \frac{1}{E}}{E - \frac{1}{E}} \right), \qquad (11.20)$$

where $r = \frac{Sr_g}{R_m}$.

The critical gap-junctional resistance is that value, say $r^*$, such that block occurs whenever $r \geq r^*$, and it is also that value of $r$ for which (11.20) is an equality. Then, using (11.5) we find a quadratic polynomial for $r^*$ as a function of $E$ and $\alpha$. In the limit of small $Q$, the positive root of this quadratic equation is

$$r^* = \frac{(1 - 2\alpha)^2}{\alpha(1 - \alpha)} + Q \left( \frac{1 - 2\alpha + 2\alpha^2}{2\alpha(1 - \alpha)} \right) + O(Q^{3/2}). \qquad (11.21)$$

In general, one can show that for $0 < \alpha < \frac{1}{2}$, $r^*$ is an increasing function of $Q$. When $Q = 0$, this reduces to the same as found in (9.47), with the coupling coefficient $D = \frac{1}{r}$.

## 11.2 Myocardial Tissue

### 11.2.1 The Bidomain Model

Coupling in cardiac tissue is complicated by the fact that the signal is the membrane potential, and this necessitates that the intracellular and extracellular spaces be continuously connected and intertwined, so that one can move continuously between any two points within one space without traversing through the opposite space. This is possible only in a three-dimensional domain.

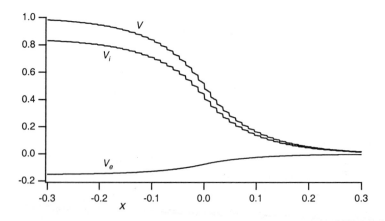

**Figure 11.5** Plot of the standing wave solution for cells of length $L$ coupled at their ends by resistive gap junctions. For this plot, cells have $L = 0.012$ cm, $A_i = 4.0 \times 10^{-6}$ cm$^2$, and a space constant $\lambda_g = 0.09$ cm. $R_m = 7000$ $\Omega$cm$^2$, $R_i = 150$ $\Omega$cm, $q_e = 0.5q_i$.

It is impossible to write and solve equations that take into account the fine-structure details of the geometry of these two interleaving spaces. However, the microstructure can be averaged (homogenized) to yield equations that describe the potentials in an averaged, or smoothed, sense, and these are adequate for most situations.

In this averaged sense, we view the tissue as a two-phase medium, as if every point in space is composed of a certain fraction of intracellular space and a fraction of extracellular space. Accordingly, at each point in space there are two electrical potentials $V_i$ and $V_e$, as well as two currents $i_i$ and $i_e$, with subscripts $i$ and $e$ denoting intracellular and extracellular space, respectively.

The relationship between current and potential is ohmic,

$$i_i = -\sigma_i \nabla V_i, \quad i_e = -\sigma_e \nabla V_e, \tag{11.22}$$

where $\sigma_i$ and $\sigma_e$ are conductivity tensors. The principal axes of the conductivity tensors are the same, owing to the cylindrical nature of the cells, but the conductivities in these directions are possibly different. At any point in space the total current is $i_t = i_i + i_e$, and unless there are extraneous current sources, the total current is conserved, so that $\nabla \cdot i_t = 0$, or

$$\nabla \cdot (\sigma_i \nabla V_i + \sigma_e \nabla V_e) = 0. \tag{11.23}$$

At every point in space there is a membrane potential

$$V = V_i - V_e. \tag{11.24}$$

The transmembrane current $i_T$ is the current that leaves the intracellular space to enter the extracellular space,

$$i_T = \nabla \cdot (\sigma_i \nabla V_i) = -\nabla \cdot (\sigma_e \nabla V_e). \tag{11.25}$$

For a biological membrane, the total transmembrane current is the sum of ionic and capacitive currents,

$$i_T = \chi \left( C_m \frac{\partial V}{\partial t} + I_{\text{ion}} \right) = \nabla \cdot (\sigma_i \nabla V_i). \tag{11.26}$$

Here $\chi$ is the membrane surface-to-volume ratio, needed to convert transmembrane current per unit area into transmembrane current per unit volume. In the typical scaling, $I_{\text{ion}} = -\frac{f(V)}{R_m}$.

Equation (11.26) shows how cardiac tissue is coupled, and it, together with (11.23), is called the *bidomain model*. The bidomain model was first proposed in the late 1970s by Tung and Geselowitz (Tung, 1978) and is now the generally accepted model for electrical behavior of cardiac tissue (Henriquez, 1993).

Boundary conditions for the bidomain model usually assume that there is no current across the boundary that enters directly into the intracellular space, whereas if there is an injected current, it enters the tissue through the extracellular domain.

## Derivation of the bidomain equations

The derivation of the bidomain model follows quickly from the homogenization of a cellular domain presented in Section 11.3 (see Neu and Krassowska, 1993, and Keener and Panfilov, 1996). The notation used here is introduced in Section 11.3.

We define the potentials in the intracellular and extracellular domains as $\phi_i$ and $\phi_e$, respectively. Then the membrane potential is the difference between the two potentials across the membrane boundary $\Gamma_m$ between the two domains:

$$\phi = (\phi_i - \phi_e)\,|_{\Gamma_m}. \tag{11.27}$$

At each point of the cell membrane the outward transmembrane current is given by

$$I_m = C_m \frac{d\phi}{dt} + \frac{1}{R_m} f_m(\phi), \tag{11.28}$$

where $C_m$ is the membrane capacitance and $f_m/R_m$ represents the transmembrane ionic current. The parameter $R_m$ is the membrane resistance.

It follows from homogenization (11.76), (11.77) that

$$\phi_i = V_i(x) + \epsilon W_i\left(\frac{x}{\epsilon}\right) \cdot T^{-1}\nabla V_i(x) + O(\epsilon^2 V_i), \tag{11.29}$$

$$\phi_e = V_e(x) + \epsilon W_e\left(\frac{x}{\epsilon}\right) \cdot T^{-1}\nabla V_e(x) + O(\epsilon^2 V_e), \tag{11.30}$$

and that $V_i(x)$ and $V_e(x)$ satisfy the averaged equations

$$\nabla \cdot (\sigma_i \nabla V_i) = -\nabla \cdot (\sigma_e \nabla V_e) = \frac{1}{v}\int_{\Gamma_m} I_m(x, \xi)\,dS_\xi, \tag{11.31}$$

where $I_m$ is the transmembrane current (positive outward). We calculate (using $\int_{\Gamma_m} W_i dS_\xi = \int_{\Gamma_m} W_e dS_\xi = 0$) that

$$\int_{\Gamma_m} I_m(x, \xi)\,dS_\xi = C_m S_m \frac{\partial V}{\partial t} + \int_{\Gamma_m} \frac{1}{R_m} F_m\left(V + \epsilon H(\xi, x)\right)dS_\xi, \tag{11.32}$$

where

$$H(\xi, x) = W_i(\xi) \cdot T^{-1}\nabla V_i(x) - W_e(\xi) \cdot T^{-1}\nabla V_e(x), \tag{11.33}$$

$$V = V_i - V_e. \tag{11.34}$$

It follows that

$$\frac{R_m}{\chi}\nabla \cdot (\sigma_i \nabla V_i) = -\frac{R_m}{\chi}\nabla \cdot (\sigma_e \nabla V_e) = C_m R_m \frac{\partial V}{\partial t} + \frac{1}{S_m}\int_{\Gamma_m} f_m\left(V + \epsilon H(\xi, x)\right)dS_\xi. \tag{11.35}$$

The parameter $\chi = \frac{S_m}{v}$ is the ratio of cell surface area per unit volume. In the limit $\epsilon = 0$, the equations (11.35) reduce to the standard bidomain model. With $\epsilon \neq 0$, this model can be used to study defibrillation (Chapter 14).

## Monodomain reduction

Equation (11.26) can be reduced to a monodomain equation for the membrane potential in one special case. Notice that

$$\nabla V_i = (\sigma_i + \sigma_e)^{-1}(\sigma_e \nabla V - i_t), \tag{11.36}$$

so that the balance of transmembrane currents becomes

$$\chi\left(C_m \frac{\partial V}{\partial t} + I_{\text{ion}}\right) = \nabla \cdot \left(\sigma_i(\sigma_i + \sigma_e)^{-1}\sigma_e \nabla V\right) - \nabla \cdot \sigma_i(\sigma_i + \sigma_e)^{-1}i_t. \tag{11.37}$$

Here we see that there is possibly a contribution to the transmembrane current from the divergence of the total current. We know that $\nabla \cdot i_t = 0$, so this source term is zero if the matrix $\sigma_i(\sigma_i + \sigma_e)^{-1}$ is proportional to a constant multiple of the identity matrix. In other words, if the two conductivity matrices $\sigma_i$ and $\sigma_e$ are proportional, $\sigma_i = \alpha\sigma_e$, with $\alpha$ a constant, then the source term disappears, and the bidomain model reduces to the monodomain model.

$$\chi\left(C_m \frac{\partial V}{\partial t} + I_{\text{ion}}\right) = \nabla \cdot (\sigma\nabla V), \tag{11.38}$$

where $\sigma = \sigma_i(\sigma_i + \sigma_e)^{-1}\sigma_e$. When $\sigma_i = \alpha\sigma_e$, the tissue is said to have equal anisotropy ratios. A one-dimensional model with constant conductivities can always be reduced to a monodomain problem.

## Plane waves

Cardiac tissue is strongly anisotropic, with wave speeds that differ substantially depending on their direction. For example, in human myocardium, propagation is about 0.5 m/s along fibers and about 0.17 m/s transverse to fibers. To see the relationship between the wave speed and the conductivity tensor we look for plane-wave solutions of the bidomain equations. Plane waves are functions of the single variable $\xi = \mathbf{n} \cdot \mathbf{x} - ct$, where $\mathbf{n}$ is a unit vector pointing in the direction of wave-front propagation. We assume that the ionic current is such that the canonical problem

$$u'' + c_0 u' + f(u) = 0 \tag{11.39}$$

has a wave-front solution $U(x)$ for some unique value of $c_0$, the value of which depends on $f$. The behavior of this solution was discussed in Chapter 9.

In terms of the traveling wave coordinate $\xi$, the bidomain equations reduce to the two ordinary differential equations

$$\frac{R_m}{\chi}\mathbf{n} \cdot \sigma_i \mathbf{n} V_i'' + cC_m R_m V' + f(V) = 0, \tag{11.40}$$

$$\mathbf{n} \cdot \sigma_i \mathbf{n} V_i'' + \mathbf{n} \cdot \sigma_e \mathbf{n} V_e'' = 0. \tag{11.41}$$

Using that $V = V_i - V_e$, we find that

$$V_i' = \frac{\mathbf{n} \cdot \sigma_e \mathbf{n}}{\mathbf{n} \cdot (\sigma_i + \sigma_e)\mathbf{n}} V', \tag{11.42}$$

$$V_e' = -\frac{\mathbf{n} \cdot \sigma_i \mathbf{n}}{\mathbf{n} \cdot (\sigma_i + \sigma_e)\mathbf{n}} V', \tag{11.43}$$

and

$$\frac{R_m}{\chi} \frac{(\mathbf{n} \cdot \sigma_i \mathbf{n})(\mathbf{n} \cdot \sigma_e \mathbf{n})}{\mathbf{n} \cdot (\sigma_i + \sigma_e)\mathbf{n}} V_i'' + cC_m R_m V' + f(V) = 0. \tag{11.44}$$

Now we compare (11.44) with (11.39) and find that the solutions are related through

$$V(\xi) = U\left(\frac{\xi}{\Lambda}\right), \tag{11.45}$$

where $\Lambda(\mathbf{n})^2 = \frac{R_m}{\chi} \frac{(\mathbf{n} \cdot \sigma_i \mathbf{n})(\mathbf{n} \cdot \sigma_e \mathbf{n})}{\mathbf{n} \cdot (\sigma_i + \sigma_e)\mathbf{n}}$ ($\Lambda(\mathbf{n})$ is the directionally dependent space constant), and the plane-wave velocity is

$$c = \frac{\Lambda(\mathbf{n})}{C_m R_m} c_0. \tag{11.46}$$

From this we learn that the speed of propagation depends importantly on direction $\mathbf{n}$, but the membrane potential profile is independent of direction except in its spatial scale $\Lambda$. This observation allows us to determine the coefficients of the conductivity tensors $\sigma_i$ and $\sigma_e$. This we do by observing from (11.43) that the total deflection of extracellular potential is dependent on direction. If we denote the total deflection of potentials during the upstroke by $\Delta V$ and $\Delta V_e$, then

$$r_d = \frac{\Delta V_{ed}}{\Delta V} = \left(\frac{\sigma_{id}}{\sigma_{id} + \sigma_{ed}}\right), \quad d = L, T \tag{11.47}$$

where the subscript $d$ denotes the longitudinal ($L$) or the transverse ($T$) fiber direction and $\sigma_{id} = \mathbf{n}_d \cdot \sigma_i \mathbf{n}_d$ with $\mathbf{n}_L$ a unit vector along the fiber axis and $\mathbf{n}_T$ a unit vector transverse to the fiber axis, and similarly for $\sigma_{ed}$. It follows that

$$\frac{\sigma_{ed}}{\sigma_{id}} = \frac{1 - r_d}{r_d}. \tag{11.48}$$

Measurements on dog myocardium (Roberts and Scher, 1982) find that $\Delta V_{eL} = 74 \pm 7$ mV, $\Delta V_{eT} = 43 \pm 6$ mV. With a typical membrane potential upstroke deflection of $\Delta V = 100$ mV, it follows that

$$\frac{\sigma_{eL}}{\sigma_{iL}} = 0.35, \quad \frac{\sigma_{eT}}{\sigma_{iT}} = 1.33, \tag{11.49}$$

implying that myocardial tissue has unequal anisotropy ratios.

## The eikonal-curvature relationship

Action potential wave fronts in the myocardium are not plane waves. To understand the effects of cardiac geometry and fiber orientation on action potential propagation requires numerical simulation of the bidomain equations. This task is made difficult for a number of reasons, including the fact that the action potential upstroke is sharp, requiring a fine spatial and temporal resolution. To overcome this restriction, it is useful to have a model of propagation that tracks the location of the action potential upstroke without following the fine details of the upstroke kinetics, thereby permitting simulation of the activation sequence on larger spatial domains with much less computational effort.

For physiologically realistic descriptions of the dynamics of ionic currents, the action potential front extends over only a few cell lengths, demanding a spatial grid size on the order of the cell size. Certainly, a whole-heart simulation using several grid points per cell is well beyond the memory capacity of most computers, and it is not apparent that the results would justify the effort.

In Section 10.1.2 we derived the eikonal-curvature equation for waves in a homogeneous excitable medium. Because cardiac tissue is an anisotropic bidomain, the derivation of the eikonal-curvature equation for cardiac tissue is a bit more tedious, but essentially the same.

The eikonal-curvature model for cardiac tissue was developed and used to study propagation in normal myocardium by a number of authors (Keener, 1986, 1991a; Tyson and Keener, 1988; Colli-Franzone et al., 1990, 1993; Colli-Franzone and Guerri, 1993). The eikonal-curvature equation describes the evolution of a wave-front surface in three-dimensional tissue. If $S(x, t) = 0$ denotes the location of an action potential wave front, then

$$C_m R_m S_t = \sigma c_0 + \sigma \cdot \left( D_{\text{eff}} \frac{\nabla S}{\sigma} \right), \tag{11.50}$$

where $D_{\text{eff}} = \beta_e^2 D_i + \beta_i^2 D_e$, $\beta_i = \frac{\nabla S \cdot D_i \nabla S}{\nabla S \cdot D \nabla S}$, $\beta_e = \frac{\nabla S \cdot D_e \nabla S}{\nabla S \cdot D \nabla S}$, $D = D_e + D_i$, $\sigma^2 = \beta_e \beta_i (\nabla s \cdot D \nabla S)$, $D_i = \frac{R_m \sigma_i}{\chi}$, and $D_e = \frac{R_m \sigma_e}{\chi}$. The derivation of this equation is only slightly more difficult than the derivation in Chapter 10 for a monodomain (see Keener and Panfilov, 1997).

Equation (11.50) is called the eikonal-curvature equation because of its physical interpretation. If we ignore the diffusive term (take $D_{\text{eff}} = 0$), then we have the eikonal equation

$$\frac{\partial S}{\partial t} = \frac{\sigma c_0}{C_m R_m}. \tag{11.51}$$

According to this equation, if $\mathbf{R}$ is a level surface of the function $s(x, t)$ and if $\mathbf{n}$ is the unit normal vector to that surface at some point, then the normal velocity of the surface $\mathbf{R}$, denoted by $\mathbf{R}_t \cdot \mathbf{n}$, satisfies

$$\mathbf{R}_t \cdot \mathbf{n} = \frac{\Lambda(\mathbf{n}) c_0}{C_m R_m}. \tag{11.52}$$

In other words, the front moves in the normal direction $\mathbf{n}$ with the plane wave velocity (11.46) for the bidomain equations.

To assign boundary conditions for (11.50), it is assumed that the tissue is insulated, and hence there is no current across the cardiac surface. It is further assumed that fibers are parallel to tissue boundaries, so that the action potential front is orthogonal to the boundary, and this implies that

$$\nabla S \cdot \mathbf{n} = 0 \qquad (11.53)$$

on the boundary of the tissue. The initial specification for $S(x, t)$ can be any smooth function that has $S(x, 0) = 0$ on the boundary of the initially stimulated region, is positive inside the stimulated region, and is negative outside the stimulated region.

## The activation sequence

The usefulness of the eikonal-curvature becomes apparent when one attempts to compute the *activation sequence*. The activation sequence is the spatial and temporal sequence in which the medium is activated by a wave initiated by the SA node.

Determination of the activation sequence is complicated by a number of features. First, as mentioned above, the medium is strongly anisotropic. Further, the fiber orientation in myocardial tissue varies through the thickness of the tissue, rotating approximately 120 degrees from epicardium to endocardium. Additionally, the geometry of the ventricles is complicated, and the initiation of action potentials occurs at numerous places on the endocardial surface at the termini of the Purkinje fiber network.

Without belaboring the details, an example of a computed action potential activation sequence using the eikonal-curvature equation for an anisotropic medium is shown in Fig. 11.6. Here is shown a sequence of wave-front surfaces at 20 ms intervals following stimulation on the top surface of a slab of tissue measuring 6 cm × 6 cm × 1 cm. The fiber orientation rotates continuously through 120 deg from top to bottom, and the velocity of propagation along fibers was taken to be three times faster than transverse to fibers. The most noticeable feature of these wave-front surfaces is the distortion from elliptical that occurs because of the rotational fiber orientation. Furthermore, there is a rotation of the elliptical axes, following the rotation of the fiber axes. However, the fastest propagation is not in the longitudinal fiber direction, as the ellipses rotate by only about 60 degrees. One can also determine (from the simulations) that normal wave-front velocity is always slower than the maximal plane-wave velocity, because of curvature slowing and fiber rotation.

Simulations of this type have also been done for whole heart with realistic geometry and fiber orientation incorporated into the conductivity tensors. An example of such is shown in Fig. 11.7, where a cutaway section of the intact dog ventricles is shown. Data for the geometry and fiber orientation from Hunter's laboratory (Nielsen et al., 1991) were incorporated into the conductivity tensors. The velocity of propagation along fibers was taken as 0.6 mm/s, and the ratio of longitudinal to transverse velocities was taken to be 2:1. Here, the stimulus was applied at the apex of the heart, although this is not the normal situation. Distortions of the propagating wave front from spherical are due primarily to the variable fiber orientation. For example, in this simulation, there is considerable propagation delay in the mid-myocardial wall of the left ventricle. A number of color pictures of this type can be found in Keener and Panfilov (1996).

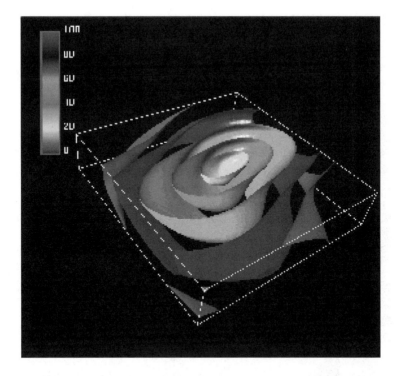

**Figure 11.6** Action potential wave fronts at 20 ms intervals in a slab of tissue with rotational anisotropy.

## 11.3 Appendix: The Homogenization of a Periodic Conductive Domain

The homogenization technique in Chapter 7 (see also Exercise 12 in Chapter 7) was used to find the effective diffusion coefficient for cells coupled via gap junctions. A similar technique can be used to determine effective electrical properties of cardiac cells.

The geometry of cardiac cells is slightly different from that described previously. We assume that an individual cardiac cell is some small periodic subunit $\Omega$ contained in a small rectangular box (Fig. 11.8). The rectangular box is divided into intracellular space $\Omega_i$ and extracellular space $\Omega_e$, separated by cell membrane $\Gamma_m$. The cells are connected to each other at the sides of the boxes through gap junctions, which are simply parts of the box wall that are contiguous with intracellular space. Thus the boundary of the cellular subunit, $\partial\Gamma$, is composed of two components, cell membrane $\Gamma_m$ and sides of the box $\Gamma_b$. Figure 11.8 is not intended to suggest a particular distribution of gap junctions in the cell membrane.

In either of the intracellular or extracellular spaces, currents are driven by a potential and satisfy Ohm's law $r_c i = -\nabla\phi$, where $r_c$ is the cytoplasmic resistance (a scalar).

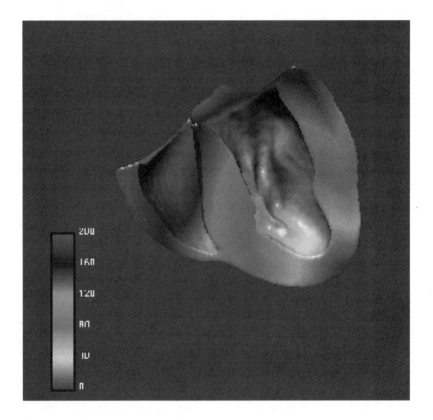

**Figure 11.7** The activation sequence in intact canine ventricles following stimulation at the apex.

On the interior of the region, current is conserved, so that

$$\nabla^2 \phi = 0.$$ (11.54)

Current enters the domain only across boundaries, as a transmembrane current, according to

$$\mathbf{n} \cdot \frac{1}{r_c} \nabla \phi = I_m$$ (11.55)

applied in the cell membrane, denoted by $\Gamma_m$, and where $\mathbf{n}$ is the outward unit normal to the membrane boundary.

Suppose that $x$ is the original Cartesian coordinate space. To allow for a variable fiber structure we assume that the orientation of the rectangular boxes is slowly varying (so that they are not exactly rectangular, but close enough), and that the axes of the rectangular cellular boxes form a natural "fiber" coordinate system. At each point in space the orientation of the rectangular box is determined by three orthogonal tangent vectors, forming the rows of a matrix $T(x)$. Then the fiber coordinate system is related

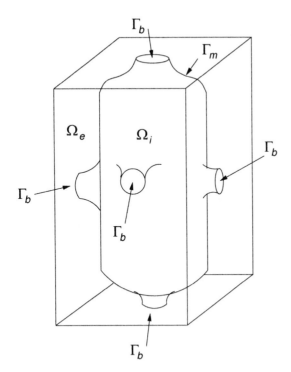

**Figure 11.8**   Sketch of idealized single cardiac cell.

to the original Cartesian coordinate system through

$$y = Y(x) = \int T(x)dx \tag{11.56}$$

and in the $y$-coordinate system the Laplacian operator is

$$\nabla^2 \phi = \nabla_y^2 \phi + \kappa \cdot \nabla_y \phi. \tag{11.57}$$

The vector $\kappa$ is the curvature vector, whose components are the mean curvatures of the coordinate level surfaces. If the components of the matrix $T$ are given by $t_{ij}$, then the coordinates of $\kappa$ are $\kappa_j = t_{ik} \frac{\partial t_{ij}}{\partial x_k}$.

Now we take into account that the boundary of the cells is varying rapidly on the scale of the fiber coordinate system, so we introduce the "fast" variable $\xi = \frac{y}{\epsilon}$, where $\epsilon$ is the small dimensionless parameter $\epsilon = \frac{l}{\Lambda}$, $l$ is the length of the cell, and $\Lambda$ is the natural length scale along fibers. We let $z = y$ be the slow variable and assume that $\kappa$ is a function solely of $z$, since variations of fiber direction are not noticeable at the cellular level.

Now we apply the homogenization technique introduced in Chapter 7. For this particular problem, this was first done by Neu and Krassowska (1993; see also Keener and Panfilov, 1997). Here we have a problem on two scales that we solve by making the usual "two-variable" assumption. That is, we treat $z$ and $\xi$ as independent variables,

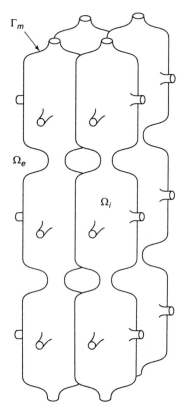

$\Gamma_m$

$\Omega_e$

$\Omega_i$

**Figure 11.9**  Periodic cell structure.

and following the chain rule, write

$$\nabla_z \rightarrow \nabla_z + \frac{1}{\epsilon}\nabla_\xi. \tag{11.58}$$

In terms of the two variables $z$ and $\xi$, we wish to solve

$$\frac{1}{\epsilon^2}\nabla_\xi^2\phi + \frac{2}{\epsilon}\nabla_\xi \cdot \nabla_z\phi + \nabla_z^2\phi + \frac{1}{\epsilon}\kappa \cdot \nabla_\xi\phi + \kappa \cdot \nabla_z\phi = 0, \tag{11.59}$$

subject to the boundary conditions

$$\frac{1}{r_c}\mathbf{n} \cdot \left(\frac{1}{\epsilon}\nabla_\xi\phi + \nabla_z\phi\right) = I_m(z,\xi) \tag{11.60}$$

on $\Gamma_m$ in $\xi$. We seek a solution that is periodic in the fast variable $\xi$.

The solution of this expanded partial differential equation is solved using the power series in $\epsilon$

$$\phi(z,\xi) = \Phi(z) + \epsilon\phi_1(z,\xi) + \epsilon^2\phi_2(z,\xi) + O(\epsilon^3), \tag{11.61}$$

where all functions are taken to be periodic in $\xi$. We create a hierarchy of equations to be solved by substituting the assumed solution form (11.61) into the governing equation

(11.59) and collecting like powers of $\epsilon$, with the result that

$$\nabla_\xi^2 \phi_1 = 0 \tag{11.62}$$

$$\nabla_\xi \cdot (\nabla_\xi \phi_2 + \nabla_z \phi_1) + (\nabla_z + \kappa) \cdot (\nabla_\xi \phi_1 + \nabla_z \Phi) = 0. \tag{11.63}$$

In a similar fashion we find a hierarchy of boundary conditions

$$\frac{1}{r_c} \mathbf{n}_\xi \cdot (\nabla_\xi \phi_1 + \nabla_z \Phi) = 0 \tag{11.64}$$

$$\frac{1}{r_c} \mathbf{n}_\xi \cdot (\nabla_\xi \phi_2 + \nabla_z \phi_1) = I_m(z, \xi) \tag{11.65}$$

applied on $\Gamma_m$, the membrane wall. Here, $\mathbf{n}_\xi = \epsilon \mathbf{n}$ is the unit outward normal vector normalized in units of $\xi$.

Now we solve this hierarchy of equations, one at a time. At this stage $\Phi(z)$ is not known. However, since $\Phi(z)$ is independent of $\xi$, the solution $\phi_1$ of (11.62) with boundary condition (11.64) is of the form

$$\phi_1(z, \xi) = W(\xi) \cdot \nabla_z \Phi(z) + \Phi_1(z). \tag{11.66}$$

Here, $W(\xi)$ is a fundamental solution vector, periodic in $\xi$ with zero surface average value $\int_{\Gamma_m} W(\xi) dS_\xi = 0$, and it satisfies the vector partial differential equation

$$\nabla_\xi^2 W(\xi) = 0 \tag{11.67}$$

subject to the boundary condition

$$\mathbf{n}_\xi \cdot (\nabla_\xi W(\xi) + I) = 0 \tag{11.68}$$

on $\Gamma_m$, the membrane wall. Here $I$ is the identity matrix. This fundamental problem separates into three independent problems for the three components of $W(\xi)$. Because the governing problem is linear, we can take $\Phi_1(z) = 0$ without loss of generality.

According to the divergence theorem, for any differentiable vector valued function $f(\xi)$,

$$\int_\Omega \nabla \cdot f dV_\xi = \int_{\partial\Omega} \mathbf{n}_\xi \cdot f \, dS_\xi. \tag{11.69}$$

Furthermore, if $f(\xi)$ is periodic in $\xi$, then

$$\int_{\Gamma_b} \mathbf{n}_\xi \cdot f dS_\xi = 0. \tag{11.70}$$

It follows that

$$\int_\Omega \nabla_\xi \cdot (\nabla_\xi \phi_2 + \nabla_z \phi_1) dV_\xi = \int_{\Gamma_m} \mathbf{n}_\xi \cdot (\nabla_\xi \phi_2 + \nabla_z \phi_1) dS_\xi. \tag{11.71}$$

Thus, a necessary condition that (11.63) have a solution is that

$$\int_\Omega \frac{1}{r_c} (\nabla_z + \kappa) \cdot (\nabla_\xi \phi_1 + \nabla_z \Phi) dV_\xi = -\int_{\Gamma_m} I_m(z, \xi) dS_\xi. \tag{11.72}$$

Substituting the solution (11.66) into the integral condition (11.72), we find the condition

$$(\nabla_z + \kappa) \cdot \left( \frac{1}{r_c} \int_\Omega (\nabla_\xi W(\xi) + I) dV_\xi \right) \nabla_z \Phi = - \int_{\Gamma_m} I_m(z, \xi) dS_\xi. \tag{11.73}$$

We identify the average conductivity tensor

$$\Sigma = \frac{1}{r_c v} \int_\Omega (\nabla_\xi W(\xi) + I) dV_\xi, \tag{11.74}$$

where $v$ is the volume of the rectangular box containing the cell. The quantity $\Sigma$ is the inverse of effective resistance per unit length, a measurable quantity. We then write (11.73) in terms of the original Cartesian coordinate variable $x$ as

$$\nabla \cdot (\sigma_{\text{eff}} \nabla \Phi) = -\frac{1}{v} \int_{\Gamma_m} I_m(z, \xi) dS_\xi, \tag{11.75}$$

where $\sigma_{\text{eff}}(x) = T \Sigma T^{-1}$.

In summary, we have found $\phi$ to be of the form

$$\phi = \Phi(z) + \epsilon W \left( \frac{z}{\epsilon} \right) \cdot T^{-1} \nabla_z \Phi(z) + O(\epsilon^2 \Phi), \tag{11.76}$$

where $\Phi$ satisfies the averaged Poisson equation

$$\nabla \cdot (\sigma_{\text{eff}} \nabla \Phi) = -\frac{1}{v} \int_{\Gamma_m} I_m(z, \xi) dS_\xi. \tag{11.77}$$

This homogenized solution is now readily used to derive the bidomain equations (Section 11.2.1).

## 11.4  EXERCISES

1. Compare the solution of (11.5) with the approximation (11.6). In particular, plot $\frac{\lambda_g^2}{L^2}(Q + r)$ as a function of $Q$ for different values of $\frac{L}{\lambda_g}$.

2. Use cable theory to estimate the effective coupling resistance for cardiac cells in the longitudinal and transverse directions. Assume that cells are 0.01 cm long, 0.00167 cm wide, $\chi = S/v = 2400$ cm$^{-1}$, $R_m = 7000$ $\Omega$-cm$^2$, $R_c = 150$ $\Omega$-cm, $R_e = 0$, with a longitudinal space constant of 0.09 cm and transverse space constant 0.03 cm (appropriate for canine crista terminalis). What difference do you observe with a transverse space constant of 0.016 cm (appropriate for sheep epicardium)?

3. Show that the choice of coupling (11.8) gives a decay rate for the discrete linear model (11.7) with $I_{\text{ion}} = -V/R_m$ that matches the space constant of the medium.

4. Using that the longitudinal and transverse cardiac action potential deflections are $\Delta V_{eL} = 74$ mV, $\Delta V_{eT} = 43$ mV, that the membrane potential has $\Delta V = 100$ mV (independent of direction), and that the axial speed of propagation in humans is 0.5 m/s in the longitudinal fiber direction and 0.17 m/s in the transverse direction, determine the coefficients of the conductivity tensors $\frac{\sigma_{eL}}{\sigma_{iL}}$, $\frac{\sigma_{eT}}{\sigma_{iL}}$, and $\frac{\sigma_{iT}}{\sigma_{iL}}$. What are these ratios in dog if the ratio of longitudinal to transverse speeds is 2:1?

# CHAPTER 12

# Calcium Waves

As we discussed in Chapter 5, the concentration of intracellular calcium is often observed to oscillate, with periods ranging from a few seconds to more than a minute. However, often these oscillations do not occur uniformly throughout the cell, but are organized into repetitive intracellular waves (Rooney and Thomas, 1993). In large cells such as *Xenopus* oocytes, the intracellular waves develop a high degree of spatial organization, forming concentric circles, plane waves, and multiple spirals (Lechleiter et al., 1991a, b; Lechleiter and Clapham, 1992). The speed of intracellular $Ca^{2+}$ waves is remarkably similar (5–20 $\mu ms^{-1}$) across a wide range of cell types (Jaffe, 1991). In intact livers, slices of hippocampal brain tissue, epithelial and glial cell cultures, and many other preparations, calcium waves have also been observed propagating from cell to cell. These intercellular waves are, in general, independent of extracellular $Ca^{2+}$, and can initiate intracellular oscillations in some cell types (Sanderson et al., 1990, 1994; Charles et al., 1991, 1992; Robb-Gaspers and Thomas, 1995).

Just as the exact function of $Ca^{2+}$ oscillations is not clear, neither is the function of $Ca^{2+}$ waves. It is widely believed that they enable communication from one side of a cell to another, or between cells, and can serve to synchronize a global, multicellular, response to a local stimulus. Nevertheless, the exact message carried by the wave is unknown in most cases. One good example of the function of a $Ca^{2+}$ wave is found in ciliated tracheal epithelial cells and is discussed in detail later in this chapter.

Although there is controversy about the exact mechanisms by which $Ca^{2+}$ waves propagate (and it is certainly true that the mechanisms differ from cell type to cell type), it is widely believed that in many cell types, intracellular $Ca^{2+}$ waves are driven by the diffusion of $Ca^{2+}$ between $Ca^{2+}$ release sites. According to this hypothesis, the $Ca^{2+}$ released from one $Ca^{2+}$-sensitive pool diffuses to neighboring pools and initiates further $Ca^{2+}$ release from there, via calcium-induced calcium release (Chapter 5). Repetition of this process can then generate an advancing wave front of high $Ca^{2+}$ concentration, i.e., a $Ca^{2+}$ wave. However, as we will see, this is not the only mechanism for the generation of $Ca^{2+}$ waves.

# 12.1   Waves in the Two-Pool Model

To show that $Ca^{2+}$ diffusion between release sites is a possible mechanism for $Ca^{2+}$ wave propagation, we consider first the two-pool model (Chapter 5). Inclusion of $Ca^{2+}$ diffusion into the (nondimensional) model equations gives

$$\epsilon u_t = \epsilon^2 u_{xx} + \epsilon(\mu - u) - \gamma f(u,v), \tag{12.1}$$

$$\epsilon v_t = f(u,v), \tag{12.2}$$

where the spatial variable, $x$, has been scaled by $\epsilon/\sqrt{D}$, $D$ being the diffusion coefficient of $Ca^{2+}$. Although there is some difficulty in defining a diffusion coefficient of $Ca^{2+}$ in the cytosol (because $Ca^{2+}$ is heavily buffered and may not obey a diffusion equation), we ignore these complications for now, treating them in a later section. Diffusion of $v$ is not included in (12.2), since it is assumed that $Ca^{2+}$ in the $Ca^{2+}$-sensitive store is not free to diffuse.

The two-pool model is capable of reproducing the waves observed in a variety of cell types (Dupont and Goldbeter, 1994). By varying the parameter values, the model can produce a wave with a sharp front and a sharp back, as is observed in cardiac cells, as well as the smoother rise followed by a homogeneous return to steady levels, as is observed in hepatocytes. Further, in two-dimensional space, the model can exhibit spiral waves similar to those observed in *Xenopus* oocytes (Girard et al., 1992).

Although the temporal behavior of the model can be understood by comparison with the FitzHugh–Nagumo model, the same is not true of the model when $Ca^{2+}$ diffusion is included. In particular, (12.1) and (12.2) cannot be put into the same form as the FitzHugh–Nagumo model, and therefore the previous theory of wave propagation does not necessarily apply.

## 12.1.1   A Piecewise Linear Model

Just as a piecewise linear simplification of the FitzHugh–Nagumo model allows much greater analytical understanding of the model behavior, so too does a piecewise linear simplification of the two-pool model. We construct a piecewise linear version of the two-pool model by replacing the nonlinear function $f(u,v)$ by the piecewise linear function $g(u,v)$ such that the curve $f(u,v) = 0$ is approximated by the curve $g(u,v) = 0$ (Sneyd et al., 1993). Graphs of these two curves are shown in Fig. 12.1.

If the steady state is shifted so that the steady state is always at $(0,0)$ for any $\mu$, the piecewise linear model equations are

$$\epsilon u_t = \epsilon^2 u_{xx} - \epsilon u - \gamma g(u,v), \tag{12.3}$$

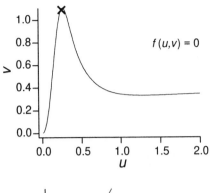

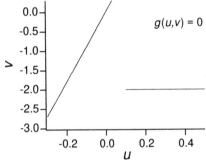

**Figure 12.1** Nullclines of the two-pool model (above) and its piecewise linear simplification (below). The x in the upper panel denotes the position of the steady state when $\mu = 0.4$. (Sneyd et al., 1993, Fig. 6.)

$$\epsilon v_t = g(u, v), \tag{12.4}$$

$$g(u, v) = \begin{cases} \beta_1 u - v, & u \leq a, \\ \beta_2 - v, & u > a, \end{cases} \tag{12.5}$$

where $a > 0$ and $\beta_2 < 0$ are functions of $\mu$, and $\beta_1 > 0$ is a constant.

To find a traveling wave solution, we follow a procedure similar to that described for the piecewise linear FitzHugh–Nagumo model, and thus the details are omitted. We convert to the traveling wave variable $\xi = x + st$, where $s$ is the wave speed, and look for solutions $u(\xi)$ of the form shown in Fig. 9.5. We divide the $\xi$ axis into three regions: region I, $\xi < 0$; region II, $0 < \xi < \xi_1$; and region III, $\xi_1 < \xi$. We then solve (12.3)–(12.5) in each region and determine the unknown constants by requiring continuity of $u$, $v$, and $u'$ at the boundaries, i.e., at $\xi = 0$ and $\xi = \xi_1$. Although the constraint equations cannot be solved analytically when $\epsilon > 0$, the solutions are easily computed numerically; a typical result is shown in Fig. 12.2. As $\epsilon \to 0$, the upjumps and downjumps in the wave become steeper, until at $\epsilon = 0$, they form shocks, or discontinuities, in the wave.

The wave speed is a decreasing function of $a$ (and thus an increasing function of $\mu$), and it can be shown (Exercise 2) that the traveling wave is the limit of a family of periodic plane waves as the period tends to infinity.

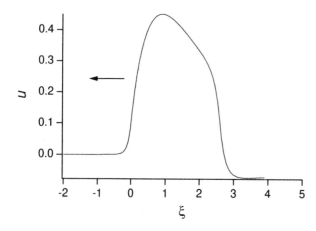

**Figure 12.2** Traveling wave of the piecewise linear model. (Sneyd et al., 1993, Fig. 10.)

## 12.1.2   Numerical Study of the Nonlinear Model

Isolated traveling waves and periodic plane waves in the nonlinear model can be found numerically, either by direct solution of the differential equations or by using a bifurcation tracking program such as AUTO (Doedel, 1986). Since the isolated traveling waves in the piecewise linear model arise as the limit of periodic plane waves as the period tends to infinity, it is convenient to begin the numerical study of the nonlinear model by looking for periodic plane waves.

Writing the model equations (12.1–12.2) in the traveling wave variable, $\xi = x + st$, gives

$$\epsilon^2 u'' - \epsilon s u' + \epsilon(\mu - u) - \gamma f(u,v) = 0, \tag{12.6}$$

$$\epsilon s v' - f(u,v) = 0. \tag{12.7}$$

Periodic plane waves correspond to periodic solutions of (12.7), and these arise (at least in this case) via Hopf bifurcations. The period $T$ of each periodic plane wave is a function both of $s$ and of $\mu$, and thus $T = T(s,\mu)$ defines a dispersion "surface" above the $(s,\mu)$ plane. A cross-section of the dispersion surface for fixed $\mu$ gives the traditional dispersion curve (i.e., period vs. wave speed), while a cross-section for fixed $s$ gives the period as a function of $\mu$. In Fig. 12.3 we plot the locus of Hopf bifurcation points (the Hopf curve) in the $(s,\mu)$ plane. For $s$ greater than some critical value (here, about 1), the two branches of the Hopf curve are connected by a branch of periodic orbits, while for $s$ less than the critical value, the branch of periodic orbits arising on the right branch of the Hopf curve terminates in a homoclinic bifurcation, i.e., an isolated traveling wave. The curve of homoclinic bifurcations is shown in Fig. 12.3 as the line labeled HC. For each value of $\mu$, the faster wave is stable. As can be seen from the figure, the speed of the isolated traveling wave increases as $\mu$ increases, as was predicted from the piecewise linear analysis.

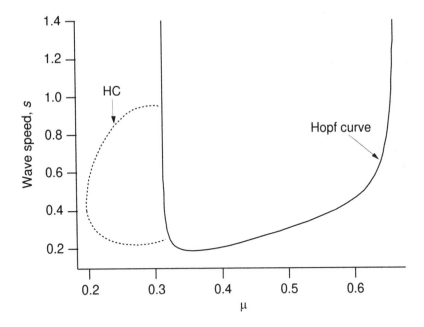

**Figure 12.3** The locus of Hopf bifurcations (Hopf curve) to periodic plane waves, and the curve of infinite period waves (i.e., isolated traveling pulses labeled HC) in the two-pool model.

### 12.1.3 The Speed–Curvature Equation

The two-pool model for calcium waves provides another example for which the speed–curvature relationship is different from that in (10.18) (Sneyd and Atri, 1993). The reason for this difference is that the fronts of calcium waves in the two-pool model are not governed by a bistable equation.

The equations governing the wave front are modifications of (12.1), (12.2), ignoring recovery, and are thus

$$u_t = \nabla^2 u - f(u, v), \tag{12.8}$$
$$v_t = f(u, v), \tag{12.9}$$

where space and time have been rescaled to remove the $\epsilon$ dependence. As in Chapter 10, we introduce a traveling coordinate and look for translation-invariant solutions in the wave-front coordinate. In the new coordinate system, the model equations reduce to

$$u_{\xi\xi} - (N + \kappa)u_\xi = f(u, v), \tag{12.10}$$
$$-Nv_\xi = -f(u, v), \tag{12.11}$$

where $N$ is the normal speed and $\kappa$ is the curvature of coordinate lines.

We now suppose that the ordinary differential equations

$$u_{\xi\xi} - s_1 u_\xi = f(u, v), \tag{12.12}$$

$$-s_2 v_\xi = -f(u,v), \tag{12.13}$$

have a heteroclinic trajectory whenever a special relationship between $s_1$ and $s_2$, denoted by $H(s_1, s_2) = 0$, is satisfied. In particular, plane waves occur at speed $s$, provided that $H(s,s) = 0$. We obtain a relationship between wave speed and curvature by setting $s_1 = N + \kappa, s_2 = N$, and requiring

$$H(N + \kappa, N) = 0. \tag{12.14}$$

We can find an explicit representation for the function $H$ in the special case of the piecewise linear model (12.5). Recalling that $a$ is the point of discontinuity of the function $g$, we let region I denote $\xi \leq 0, u \leq a$, region II denote $\xi \geq 0, u \geq a$, and look for a wave-front solution, i.e., a solution that rises from 0 to a steady level. Then,

$$u_I = a \exp(\lambda\xi), \tag{12.15}$$

$$V_i = \frac{\beta_1 a}{1 + s_2\lambda} \exp(\lambda\xi), \tag{12.16}$$

where $\lambda$ is the positive root of

$$s_2\lambda^2 + (1 - s_1 s_2)\lambda - (s_1 + s_2\beta_1) = 0. \tag{12.17}$$

In region II,

$$u_{II} = B - (B - a)\exp\left(-\frac{\xi}{s_2}\right), \tag{12.18}$$

$$V_{II} = \beta_2 + (B - a)\left(\frac{s_1}{s_2} + \frac{1}{s_2^2}\right)\exp\left(-\frac{\xi}{s_2}\right), \tag{12.19}$$

for some unknown constant $B$. Requiring $u_\xi$ and $v$ to be continuous at $\xi = 0$ then gives

$$\lambda a s_2 = B - a, \tag{12.20}$$

$$\frac{\beta_1 a}{1 + s_2\lambda} = \beta_2 + (B - a)\left(\frac{s_1}{s_2} + \frac{1}{s_2^2}\right). \tag{12.21}$$

We use (12.20) to eliminate $B - a$ from (12.21) to obtain

$$\frac{\beta_1 a}{1 + s_2\lambda} = \beta_2 + \lambda a s_2\left(\frac{s_1}{s_2} + \frac{1}{s_2^2}\right). \tag{12.22}$$

Equations (12.17), (12.22) constitute a relationship between $s_1$ and $s_2$ that must be satisfied for there to be a heteroclinic solution. As described above, replacing $s_1$ by $N + \kappa$ and $s_2$ by $N$ in these equations gives a formal relationship between the normal speed of a wave and its curvature. Numerical computations, as demonstrated in Fig. 12.4, show that waves in two dimensions satisfy this relationship.

## 12.2 Spiral Waves in *Xenopus*

In 1991, it was discovered by Lechleiter and Clapham and their coworkers that intracellular $Ca^{2+}$ waves in immature *Xenopus* oocytes showed remarkable spatiotemporal

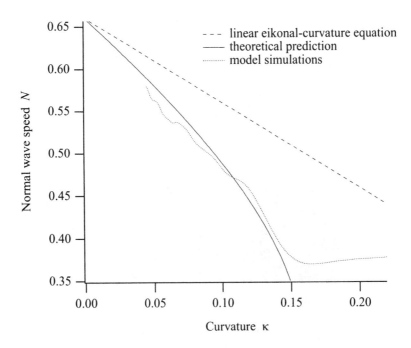

**Figure 12.4** Speed–curvature equation for the piecewise linear two-pool model. (Sneyd and Atri, 1993, Fig. 4.)

organization. By loading the oocytes with a $Ca^{2+}$-sensitive dye, releasing IP$_3$, and observing $Ca^{2+}$ release patterns with a confocal microscope, Lechleiter and Clapham (1992; Lechleiter et al., 1991b) observed spiral $Ca^{2+}$ waves *in vivo*. Typical experimental results are shown in Fig. 12.5A. The crucial feature of *Xenopus* oocytes that makes these observations possible is their large size. *Xenopus* oocytes can have a diameter larger than 600 $\mu$m, an order of magnitude greater than most other cells. In a small cell, a typical $Ca^{2+}$ wave (often with a width of close to 100 $\mu$m) cannot be observed in its entirety, and there is not enough room for a spiral to form. However, in a large cell it may be possible to observe both the wave front and the wave back, as well as spiral waves, and this has made the *Xenopus* oocyte one of the most important systems for the study of $Ca^{2+}$ waves. Of course, what is true for *Xenopus* oocytes is not necessarily true for other cells, and so one must be cautious about extrapolating to other cell types.

One common experimental procedure for initiating waves is to photorelease a bolus of IP$_3$ inside the cell and observe the subsequent $Ca^{2+}$ activity (Lechleiter and Clapham, 1992). After sufficient time, $Ca^{2+}$ wave activity disappears as IP$_3$ is degraded, but in the short term, the observed $Ca^{2+}$ activity is the result of $Ca^{2+}$ diffusion and IP$_3$ diffusion. Another technique is to release IP$_3$S$_3$, a nonhydrolyzable analogue of IP$_3$, which has a similar effect on IP$_3$ receptors but is not degraded by the cell. In this case, after sufficient time has passed, the IP$_3$S$_3$ is at a constant concentration in all parts of the cell. Wave

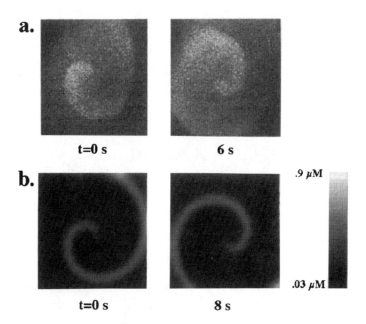

**Figure 12.5**  A: Spiral $Ca^{2+}$ wave in the *Xenopus* oocyte. The image size is $420 \times 420$ $\mu$m. The spiral has a wavelength of about 150 $\mu$m and a period of about 8 seconds. B: A spiral wave in the heuristic model, simulated on a domain of size $250 \times 250$ $\mu$m, with $[IP_3] = 95$ nM. (Atri et al., 1993, Fig. 11.)

activity can then be observed over a period of many minutes while the background level of $IP_3S_3$ remains constant, so that diffusion of $IP_3S_3$ has no further effect.

To model $Ca^{2+}$ waves in *Xenopus*, we choose to use the heuristic model (Chapter 5), as this model is based on experimental data from *Xenopus*. However, any of the other models would give qualitatively similar results. We extend (5.42) and (5.41) by the addition of $Ca^{2+}$ diffusion, and the diffusion and degradation of $IP_3$. Thus,

$$\frac{\partial p}{\partial t} = D_p \nabla^2 p - k_p p, \tag{12.23}$$

$$\frac{\partial c}{\partial t} = D_c \nabla^2 c + k_f \left( \mu_0 + \frac{\mu_1 p}{p + k_\mu} \right) \left( b + \frac{(1-b)c}{k_1 + c} \right) h - \frac{\gamma c}{k_\gamma + c} + \beta, \tag{12.24}$$

$$\tau_h \frac{dh}{dt} = \frac{k_2^2}{k_2^2 + c^2} - h. \tag{12.25}$$

$D_p$ and $D_c$ are the diffusion coefficients of $IP_3$ and $Ca^{2+}$, respectively.

Equations (12.23)–(12.25) were solved on a square domain with no-flux boundary conditions. The use of $IP_3$ was modeled by setting $k_p = 0.2$ s$^{-1}$, while the use of $IP_3S_3$ was modeled by setting $k_p = 0$. Standard parameters of the $Ca^{2+}$ wave model are given in Table 12.1. When a bolus of $IP_3S_3$ is released in the middle of the domain, it causes the release of a large amount of $Ca^{2+}$ at the site of the bolus. The $IP_3S_3$ then diffuses across the cell, releasing $Ca^{2+}$ in the process. Activation of $IP_3$ receptors by the released $Ca^{2+}$

**Table 12.1** Standard parameters of the model of $Ca^{2+}$ waves. One important point to note is the difference between $D_p$ and $D_c$. Although $IP_3$ is a much larger molecule than $Ca^{2+}$, it diffuses faster through the cell because it is not buffered (Allbritton et al., 1992).

| | | | | | |
|---|---|---|---|---|---|
| $b$ | = | 0.11 | $k_f$ | = | $8.1\ \mu Ms^{-1}$ |
| $\beta$ | = | $0.02\ \mu Ms^{-1}$ | $D_p$ | = | $300\ \mu m^2 s^{-1}$ |
| $\gamma$ | = | $2\ \mu Ms^{-1}$ | $D_c$ | = | $20\ \mu m^2 s^{-1}$ |
| $\tau_h$ | = | $2\ s$ | $k_p$ | = | $0 - 0.2\ s^{-1}$ |
| $k_1$ | = | $0.7\ \mu M$ | $k_\mu$ | = | $4\ \mu M$ |
| $k_2$ | = | $0.7\ \mu M$ | $\mu_0$ | = | 0.567 |
| $k_\gamma$ | = | $0.1\ \mu M$ | $\mu_1$ | = | 0.433 |

can lead to periodic $Ca^{2+}$ release from the stores, and the diffusion of $Ca^{2+}$ between $IP_3$ receptors serves to stabilize the waves, giving regular periodic traveling waves. These periodic waves are the spatial analogues of the oscillations seen in the temporal model, and arise from the underlying oscillatory kinetics. After sufficient time, $[IP_3S_3]$ is at the same level throughout the entire cell, a level that is dependent on how much was added in the original bolus. We saw before that if $[IP_3]$ is in the appropriate range, the model has a stable limit cycle. Thus, if $[IP_3S_3]$ is in this intermediate range over the entire cell, every part of the cell cytoplasm is in an oscillatory state. It follows from the standard theory of reaction–diffusion systems with oscillatory kinetics (see, for example, Kopell and Howard, 1973; Duffy et al., 1980; Neu, 1979; Murray, 1989) that periodic and spiral waves can exist for these values of $[IP_3S_3]$. When $IP_3$, rather than $IP_3S_3$, is released, the wave activity lasts for only a short time, which is consistent with the theoretical results. When the wave front is broken, a spiral wave of $Ca^{2+}$ often forms (Fig. 12.5B). Depending on the initial conditions, these spiral waves can be stable or unstable. In the unstable case, the branches of the spiral can intersect themselves and cause breakup of the spiral, in which case a region of complex patterning emerges in which there is no clear spatial structure (McKenzie and Sneyd, 1998).

## 12.3 Calcium Buffering

Calcium is heavily buffered in all cells, with about 99% of the available $Ca^{2+}$ bound to large proteins. The chemical reaction for calcium buffering can be represented by the reaction

$$P + Ca^{2+} \underset{k_-}{\overset{k_+}{\rightleftharpoons}} B, \tag{12.26}$$

where P is the buffering protein and B is buffered calcium. If we let $b$ denote the concentration of buffer with $Ca^{2+}$ bound, and $c$ the concentration of free $Ca^{2+}$, then a simple model of calcium buffering is

$$\frac{\partial c}{\partial t} = D_c \nabla^2 c + f(c) + k_- b - k_+ c(b_t - b), \tag{12.27}$$

$$\frac{\partial b}{\partial t} = D_b \nabla^2 b - k_- b + k_+ c(b_t - b), \tag{12.28}$$

where $k_-$ is the rate of $Ca^{2+}$ release from the buffer, $k_+$ is the rate of $Ca^{2+}$ uptake by the buffer, $b_t$ is the total buffer concentration, and $f(c)$ denotes all the other reactions involving free $Ca^{2+}$ (for example, release from the $IP_3$ receptors, reuptake by pumps, etc.).

## 12.3.1  Buffers with Fast Kinetics

If the buffer has fast kinetics, its effect on the intracellular $Ca^{2+}$ dynamics can be analyzed simply. For if $k_-$ and $k_+$ are large compared to the time constant of calcium reaction, then we take $b$ to be in the quasi-steady state

$$k_- b - k_+ c(b_t - b) = 0, \tag{12.29}$$

and so

$$b = \frac{b_t c}{K + c}, \tag{12.30}$$

where $K = k_-/k_+$. It follows that

$$\frac{\partial c}{\partial t} + \frac{\partial b}{\partial t} = (1 + \theta)\frac{\partial c}{\partial t}, \tag{12.31}$$

where

$$\theta = \frac{b_t K}{(K + c)^2}. \tag{12.32}$$

Combining (12.31) with (12.27) and (12.28), we obtain

$$\frac{\partial c}{\partial t} = \frac{1}{1 + \theta}\left(\nabla^2\left(D_c c + D_b b_t \frac{c}{K + c}\right) + f(c)\right) \tag{12.33}$$

$$= \frac{D_c + D_b \theta}{1 + \theta}\nabla^2 c - \frac{2 D_b \theta}{(K + c)(1 + \theta)}|\nabla c|^2 + \frac{f(c)}{1 + \theta}. \tag{12.34}$$

Note that we are assuming that $b_t$ is a constant, and doesn't vary in either space or time.

We see that nonlinear buffering changes the model significantly. In particular, $Ca^{2+}$ obeys a nonlinear diffusion–advection equation, where the advection is the result of $Ca^{2+}$ transport by a mobile buffer (Wagner and Keizer, 1994). Notice that the effective diffusion coefficient is a convex linear combination of the two diffusion coefficients $D_c$ and $D_b$, and so lies somewhere between the two. Since buffers are large molecules, the effective diffusion coefficient is decreased from that of unbuffered diffusion. Note also that if the buffer is not mobile, i.e., $D_b = 0$, then (12.34) reverts to a reaction–diffusion equation, with a reduced diffusion coefficient. Also, when $Ca^{2+}$ gradients are small, the nonlinear advective term can be ignored (Irving et al., 1990). Finally, the buffering also affects the qualitative nature of the nonlinear reaction term, $f(c)$, which is divided by $1 + \theta$, a function of $c$ (12.32). This may change many properties of the model, including oscillatory behavior and the nature of wave propagation.

To facilitate numerical simulations, the calcium and buffer concentrations are scaled by the approximate cytoplasmic calcium concentration at the peak of a wave, $c_0 = 1 \ \mu M$. Although this does not change the numerical values of any parameters, it is important to keep in mind that $c$ never gets much larger than $c_0$ and is usually considerably smaller. Typical parameter values are (Allbritton et al., 1992) $D_c = 300 \ \mu m^2 s^{-1}$, $D_b = 50 \ \mu m^2 s^{-1}$, $b_t = 150$, and $K = 10$, where $b_t$ and $K$ have both been scaled by $c_0$ and are thus dimensionless. In some cells, such as the *Xenopus* oocyte, $K$ is as large as 10 (Allbritton et al., 1992), but in other cells it can be considerably smaller.

Despite the complexity of (12.34) it retains the advantage of being a single equation. However, if the buffer kinetics are not fast relative to the $Ca^{2+}$ kinetics, the only way to proceed is with numerical simulations of the complete system, a procedure followed by a number of groups (Backx et al., 1989; Sala and Hernández-Cruz, 1990; Nowycky and Pinter, 1993).

### 12.3.2 The Existence of Buffered Waves

Since the presence of fast $Ca^{2+}$ buffers changes the nature of the $Ca^{2+}$ transport equation, it is of considerable interest to determine how $Ca^{2+}$ buffering affects the properties of waves. For instance, can the addition of a buffer eliminate wave activity? How much do buffers affect the speed of traveling waves? Does the addition of exogenous buffer, such as a fluorescent $Ca^{2+}$ dye, affect the existence or the speed of the $Ca^{2+}$ waves? First, we address the question of whether buffers can eliminate wave activity.

The form of (12.33) suggests the change of variables

$$w = D_c c + D_b b_t \frac{c}{K + c},$$
(12.35)

so that $w$ is a monotone increasing function of $c$, since

$$\frac{dw}{dc} = D_c + D_b \theta(c).$$
(12.36)

The unique inverse of this function is denoted by

$$c = \phi(w).$$
(12.37)

In terms of $w$, (12.33) becomes

$$\frac{\partial w}{\partial t} = \frac{D_c + D_b \Theta}{1 + \Theta} \left( \frac{\partial^2 w}{\partial x^2} + f(\phi(w)) \right),$$
(12.38)

where $\Theta = \frac{b_t K}{(K+\phi(w))^2}$.

Now, we assume that $f(c)$ is of bistable form, with three zeros, $C_1 < C_2 < C_3$, of which $C_1$ and $C_3$ are stable. It immediately follows that $f(\phi(w))$ has three zeros $W_1 < W_2 < W_3$, with $W_1$ and $W_3$ stable. The proof of existence of a traveling wave solution for (12.38) uses exactly the same arguments as that for the bistable equation presented in Chapter 9 (Sneyd et al., 1998). It follows that a traveling wave-front solution providing a transition from $W_1$ to $W_3$ exists if and only if

$$\int_{W_1}^{W_3} f(\phi(w))dw > 0. \tag{12.39}$$

Using (12.36), we write this condition in terms of $c$ as

$$\int_{C_1}^{C_3} f(c)(D_c + D_b\theta(c))dc > 0. \tag{12.40}$$

In general, this integral cannot be evaluated explicitly. However, for the simple case of cubic bistable kinetics $f(c) = c(1-c)(c-a)$, $0 < a < 1/2$, explicit evaluation of the integral (12.40) shows that traveling waves exist if and only if

$$a < a_c = \frac{1}{2}\frac{D_c - 12D_bb_tK[(3K^2+2K)\ln(\frac{K+1}{K}) - (3K+\frac{1}{2})]}{D_c + 12D_bb_tK[(K+\frac{1}{2})\ln(\frac{K+1}{K}) - 1]}. \tag{12.41}$$

One conclusion that can be drawn immediately from (12.40) is that a stationary buffer (i.e., one with $D_b = 0$) has no effect on the existence of traveling waves. For when $D_b = 0$, the condition (12.40) for the existence of the traveling wave reduces to

$$\int_0^1 f(c)dc > 0, \tag{12.42}$$

which is exactly the condition for the existence of a wave in the absence of a buffer. Hence a stationary buffer, no matter what its kinetic properties, cannot eliminate $Ca^{2+}$ waves. This conclusion is valid even if the calcium release is discrete (see Exercise 9).

Note that $a_c$ is a monotonically decreasing function of $D_bb_t/D_c$, and

$$a_c \to a_{c,min}(K) = -\frac{1}{2}\frac{[(3K^2+2K)\ln(\frac{K+1}{K}) - (3K+\frac{1}{2})]}{[(K+\frac{1}{2})\ln(\frac{K+1}{K}) - 1]} \quad \text{as } D_b \to \infty. \tag{12.43}$$

In Fig. 12.6 we give a plot of $a_{c,min}(K)$ against $K$. When $K$ is large, the minimum value of $a_c$ is close to 0.5, and thus wave existence is insensitive to $D_b$. However, when $K$ is small, $a_c$ can also become small as $D_b$ increases, and so in this case, a mobile buffer can easily stop a wave.

## 12.3.3   The Shape and Speed of Buffered Waves

The question of wave speed and shape is much more difficult and is not completely solved. It is not known exactly how general buffers affect the wave speed and shape in

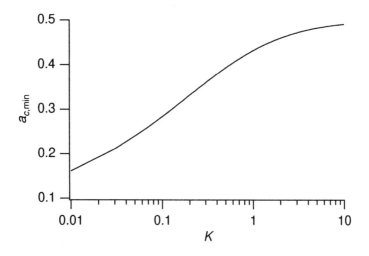

**Figure 12.6** Plot of $a_{c,\min}$ against $K$.

the FitzHugh–Nagumo model, far less in more general $Ca^{2+}$ wave models, and this is an area of current research. However, some results for simpler models may be obtained.

Let $\gamma = b_t/K$ and $\epsilon = 2/K$. When $\epsilon$ is small, then $\theta = \gamma(1 - \epsilon c) + O(\epsilon^2)$, and we can expand the differential equation (12.34) in powers of $\epsilon$, retaining only the leading-order terms. To lowest order in $\epsilon$, the buffered equation in one spatial dimension is

$$\frac{\partial c}{\partial t} = D_{\text{eff}} \frac{\partial^2 c}{\partial x^2} + \frac{f(c)}{1 + \gamma}, \tag{12.44}$$

where the effective diffusion coefficient $D_{\text{eff}} < D_c$ is given by

$$D_{\text{eff}} = \frac{D_c + D_b \gamma}{1 + \gamma}. \tag{12.45}$$

Rescaling time by $1 + \gamma$ and space by $\sqrt{D_c + D_b \gamma}$, we see that if the unbuffered system has a wave of speed $s\sqrt{D_c}$, the buffered system has a wave of speed

$$s \frac{\sqrt{D_c + \gamma D_b}}{1 + \gamma}. \tag{12.46}$$

In particular, for the buffered bistable equation (i.e., $f(c) = c(1 - c)(c - a)$) we get

$$s = \frac{(1 - 2a)}{1 + b_t/K} \sqrt{\frac{D_c + D_b b_t/K}{2}}. \tag{12.47}$$

Hence, the wave speed is an increasing function of $D_b$ for $K$ sufficiently large.

Recall from Chapter 9 that the heteroclinic orbit corresponding to the traveling wave solution of the bistable equation must leave in the direction of the unstable manifold at the origin and enter along the direction of the stable manifold at $c = 1$. It follows that the slope $\lambda_u$ of the heteroclinic orbit at the origin must be the positive root of

$$\lambda_u^2 - \left(\frac{s}{\phi_1(0)}\right)\lambda_u - \frac{\phi_3(0)a}{\phi_1(0)} = 0, \tag{12.48}$$

where

$$\phi_1(c) = \frac{D_c + D_b\theta(c)}{1 + \theta(c)}, \tag{12.49}$$

$$\phi_3(c) = \frac{1}{1 + \theta(c)}. \tag{12.50}$$

Similarly, if $\lambda_s$ is the slope of the heteroclinic orbit at $(1, 0)$, then $\lambda_s$ must be the negative root of

$$\lambda_s^2 - \left(\frac{s}{\phi_1(1)}\right)\lambda_s - \frac{\phi_3(1)(1-a)}{\phi_1(1)} = 0. \tag{12.51}$$

Although (12.47) was derived only for the case when $K$ is large, it turns out to be approximately correct for $K$ small also. The reason for this is not clear, but the fact can be used to derive an approximate expression for the wave front when $K$ is small. For if $s$ is given approximately by (12.47), one can construct approximate expressions for $\lambda_u$ and $\lambda_s$ in terms of the model parameters, and then use these expressions to construct an approximation to the wave front (Exercise 10).

However, what is more interesting is the shape of the wave. When $K$ is large, $\lambda_u \approx -\lambda_s$, and thus the rate at which the traveling wave rises is the same as the rate at which it saturates to its maximum value of 1. However, as $K$ decreases, this symmetry is lost, until, for example, when $K = 0.1$, $\lambda_u = 0.09$ and $\lambda_s = -0.99$. Thus, the buffered traveling wave has a noticeably different shape from that of the unbuffered traveling wave, as illustrated in Fig. 12.7. The buffered wave has a slow rise and a fast saturation, while the unbuffered wave is symmetrical around the midpoint. Since both $\lambda_u$ and $\lambda_s$ are easily measured experimentally, this provides a convenient way to determine the effects of buffers on the observed waves. In *Xenopus* oocytes, for example, experimental measurements show that the traveling waves are symmetrical about their midpoints. Thus, the theory predicts that the $Ca^{2+}$ buffers in *Xenopus* have a large effective $K$ value. This indeed turns out to be the case; experimental measurements in *Xenopus* give a value for $K$ of around 10, which is consistent with the theoretical prediction. As yet, the theory has not been applied to other cell types, nor has it been extended to more complex models.

## 12.4   Discrete Calcium Sources

Skeletal muscle and cardiac cells are invaded by *T-tubules* (T for transverse), which allow communication with the extracellular space (see Fig. 18.1 in Chapter 18). T-tubules have voltage-dependent calcium channels allowing the influx of calcium into the cell in response to an action potential. Directly opposite the calcium channels are

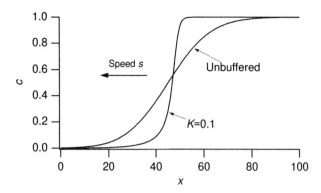

**Figure 12.7** Buffered and unbuffered traveling waves in the bistable equation. For the unbuffered wave, the diffusion coefficient was chosen such that the wave traveled with the same speed as in the buffered case.

the ryanodine receptors, through which additional calcium is released from *terminal* sarcoplasmic reticulum. Uptake of calcium into the SR is via calcium ATPases, which are uniformly distributed along the *junctional* SR. The physical arrangement of calcium release sites means that in these cell types it may not be appropriate to view the release of calcium as spatially uniform. In cardiac cells, calcium waves do not normally propagate without T-tubule stimulus. Thus, the discreteness of calcium release sites prevents the spontaneous propagation of a calcium wave, which would lead to spontaneous (uncontrolled) muscular contraction. One can also imagine situations in which the discreteness of release sites could have major negative consequences. For example, in hypertension, cardiac cells compensate for increased pressure by growing larger (hypertrophy). It is possible that in this hypertrophied state, the separation between T-tubules and ryanodine receptors is increased, leading to less effective coupling between action potentials and calcium release, and impaired contraction (Yue, 1997).

Here we examine the possibility of propagation of an intracellular calcium wave, propagating discretely, from one ryanodine receptor to the next, without the benefit of stimulus from an action potential. To explore this possibility, we assume that calcium is released from discrete release sites but is removed continuously, and so we consider the equation for calcium concentration

$$\frac{\partial c}{\partial t} = D_c \frac{\partial^2 c}{\partial x^2} - kc + L \sum_n \delta(x - nL) f(c). \tag{12.52}$$

The function $f$ is your favorite (bistable) description of calcium release, and $L$ is the spatial separation between release sites. Notice that the average spatial calcium release is normalized to be independent of $L$.

Waves of this type fail to propagate if there is a standing wave solution. Standing waves for (12.52) satisfy

$$0 = D_c \frac{\partial^2 c}{\partial x^2} - kc + L \sum_n \delta(x - nL) f(c). \tag{12.53}$$

On the intervals $nL < x < (n+1)L$, this becomes

$$0 = D_c \frac{\partial^2 c}{\partial x^2} - kc. \tag{12.54}$$

We find jump conditions at $x = nL$ by (formally) integrating from $nL^-$ to $nL^+$ to obtain

$$D_c c_x|_{nL^-}^{nL^+} + Lf(c_n) = 0, \tag{12.55}$$

where $c_n = c(nL)$.

Now we solve (12.54) to obtain

$$c(x) = (c_{n+1} - c_n \cosh \beta) \frac{\sinh\left(\frac{\beta x}{L}\right)}{\sinh \beta} + c_n \cosh \left(\frac{\beta x}{L}\right) \tag{12.56}$$

where $c_n = c(nL)$, $\beta^2 = \frac{kL^2}{D}$, so that

$$c_x(nL^+) = (c_{n+1} - c_n \cosh \beta) \frac{\beta}{L \sinh \beta}. \tag{12.57}$$

Similarly,

$$c_x(nL^-) = -(c_{n-1} - c_n \cosh \beta) \frac{\beta}{L \sinh \beta}. \tag{12.58}$$

It follows that (12.55) is the difference equation

$$\frac{k}{\beta \sinh \beta}(c_{n+1} - 2c_n \cosh \beta + c_{n-1}) + f(c_n) = 0, \tag{12.59}$$

which is a difference equation for $c_n$ that has standing wavelike solutions if $\beta$ is sufficiently large. For a different approach to this problem, see Keizer et al. (1998).

## 12.5   Intercellular Calcium Waves

In many cell types, a mechanical stimulus (for instance, poking a single cell with a micropipette) can initiate a wave of increased intracellular $Ca^{2+}$ that spreads from cell to cell to form an intercellular wave (Sanderson et al., 1994). Typical experimental results from airway epithelial cells are shown in Fig. 12.8. The epithelial cell culture forms a thin layer of cells, connected by gap junctions. When a cell in the middle of the culture is mechanically stimulated, the $Ca^{2+}$ in the stimulated cell increases quickly. After a time delay of a second or so, the neighbors of the stimulated cell also show an increase in $Ca^{2+}$, and this increase spreads sequentially through the culture. An intracellular wave moves across each cell, is delayed at the cell boundary, and then initiates an intracellular wave in the neighboring cell. The intercellular wave moves via the sequential propagation of intracellular waves. Of particular interest here is the fact that in the absence of extracellular $Ca^{2+}$, the stimulated cell shows no response, but an intercellular wave still spreads to other cells in the culture. It thus appears that a rise

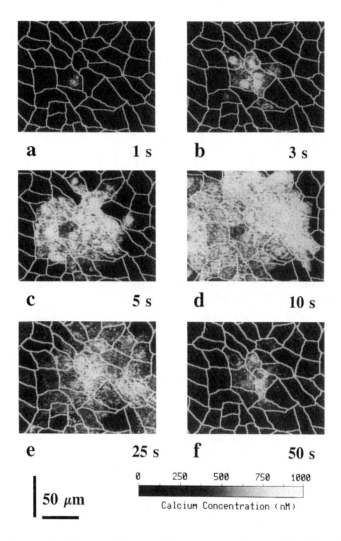

a    1 s    b    3 s

c    5 s    d    10 s

e    25 s    f    50 s

50 μm

0   250   500   750   1000
Calcium Concentration (nM)

**Figure 12.8** Mechanically stimulated intercellular wave in airway epithelial cells. The time after mechanical stimulation is given in seconds in the lower right corner of each panel. (Sneyd et al., 1995b, Fig. 4A.)

in $Ca^{2+}$ in the stimulated cell is not necessary for wave propagation. Neither is a rise in $Ca^{2+}$ sufficient to initiate an intercellular wave. For example, epithelial cells in culture sometimes exhibit spontaneous intracellular $Ca^{2+}$ oscillations, and these oscillations do not spread from cell to cell. Nevertheless, a mechanically stimulated intercellular wave does spread through cells that are spontaneously oscillating.

Little is known about the physiological importance of these intercellular $Ca^{2+}$ waves, although educated guesses can be made. Airway epithelial cells have cilia, whose function is to move mucus along the trachea. The rate at which cilia beat is closely re-

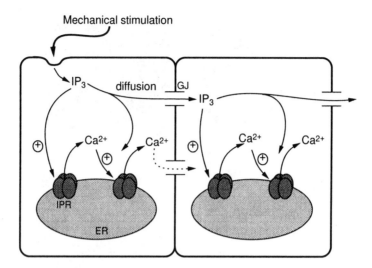

**Figure 12.9** Schematic diagram of the model of intercellular Ca$^{2+}$ waves. GJ: gap junction; ER: endoplasmic reticulum; IPR: IP$_3$ receptor.

lated to the concentration of intracellular Ca$^{2+}$ in the cell; as [Ca$^{2+}$] goes up, so does the ciliary beat frequency. Hence, the intercellular wave could be one way in which a group of ciliated cells coordinates a rise in beat frequency over the entire group. This would have the advantage that if one cell is mechanically stimulated by a foreign object in the airway, it could initiate faster ciliary beating in that cell and all its neighbors, which would serve to clear the airway more efficiently. However, despite the plausibility of this hypothesis, it remains conjectural. In nonciliated cells such as glial cells, or endothelial cells, the physiological purpose of the calcium wave is even less clear.

Sanderson and his colleagues (Boitano et al., 1992; Sanderson et al., 1994; Sneyd et al., 1994, 1995a,b) proposed a model of intercellular Ca$^{2+}$ waves (Fig. 12.9) based on this and other experimental evidence. They proposed that mechanical stimulation causes the production of large amounts of IP$_3$ in the stimulated cell, and this IP$_3$ moves through the culture by passive diffusion, moving from cell to cell through gap junctions. Since IP$_3$ releases Ca$^{2+}$ from the endoplasmic reticulum (as described in detail in Chapter 5), the diffusion of IP$_3$ from cell to cell results in a corresponding intercellular Ca$^{2+}$ wave. Since experimental results indicate that the movement of Ca$^{2+}$ between cells does not play a major role in wave propagation, the model assumes that intercellular movement of Ca$^{2+}$ is negligible. Relaxation of this assumption makes little difference to the model behavior, as it is the movement of IP$_3$ through gap junctions that determines the intercellular wave properties.

### Model equations

In the model, the epithelial cell culture is modeled as a square grid of square cells. It is assumed that IP$_3$ moves by passive diffusion and is degraded with saturable kinetics.

Thus, if $p$ denotes [IP$_3$], then

$$\frac{\partial p}{\partial t} = D_p \nabla^2 p - \frac{V_p p k_p}{k_p + p}. \tag{12.60}$$

When $p \ll k_p$, $p$ decays with time constant $1/V_p$. Ca$^{2+}$ is also assumed to move by passive diffusion, but it is released from the endoplasmic reticulum (ER) by IP$_3$ and pumped back into the ER by Ca$^{2+}$ ATPase pumps. The equations are

$$\frac{\partial c}{\partial t} = D_c \nabla^2 c + J_{\text{flux}} - J_{\text{pump}} + J_{\text{leak}}, \tag{12.61}$$

$$\tau_h \frac{dh}{dt} = \frac{k_2^2}{k_2^2 + c^2} - h, \tag{12.62}$$

$$J_{\text{flux}} = k_f \mu(p) h \left[ b + \frac{(1-b)c}{k_1 + c} \right], \tag{12.63}$$

$$J_{\text{pump}} = \frac{\gamma c^2}{k_\gamma^2 + c^2}, \tag{12.64}$$

$$J_{\text{leak}} = \beta, \tag{12.65}$$

$$\mu(p) = \frac{p^3}{k_\mu^3 + p^3}. \tag{12.66}$$

These are essentially the same as those of the heuristic model, described in detail in Chapter 5, and in equations (12.23)–(12.25). As before, $J_{\text{flux}}$ refers to the flux of Ca$^{2+}$ through the IP$_3$ receptors in the ER membrane and is a function of $p$, $c$, and a slow variable $h$, which denotes the proportion of IP$_3$ receptors that have not been inactivated by Ca$^{2+}$. Similarly, $J_{\text{pump}}$ denotes the removal of Ca$^{2+}$ from the cytoplasm by Ca$^{2+}$ ATPases in the ER membrane. $J_{\text{leak}}$ is an unspecified leak of Ca$^{2+}$ into the cytoplasm, either from outside the cell or from the ER. However, there are some minor differences between the heuristic model and the present one. For instance, to obtain better agreement with experimental evidence (Lytton et al., 1992), the Ca$^{2+}$ pump that removes Ca$^{2+}$ from the cytoplasm is assumed to have a Hill coefficient of 2, while the activation of the IP$_3$ receptor, described by the term $\mu(p)$ is assumed to have a Hill coefficient of 3. Values of the model parameters are given in Table 12.2.

Finally, the internal boundary conditions are given in terms of the flux of IP$_3$ from cell to cell. If cell $n$ has [IP$_3$] $= p_n$, it is assumed that the flux of IP$_3$ from cell $n$ to cell $n+1$ is given by $F(p_n - p_{n+1})$.

### Numerical results

Initially, a single cell was injected with IP$_3$, which was then allowed to diffuse from cell to cell, thereby generating an intercellular Ca$^{2+}$ wave. Figure 12.10 shows a density plot of a numerical solution of the model equations in two dimensions. An intercellular wave can be seen expanding across the grid of cells and then retreating as the IP$_3$ degrades. As expected for a process based on passive diffusion, the intracellular wave speed (i.e., the speed at which the intercellular wave moves across an individual cell)

**Table 12.2**  Parameters of the model of intercellular Ca$^{2+}$ waves.

| | | | | |
|---|---|---|---|---|
| $k_f$ | = | 3 $\mu$Ms$^{-1}$ | $b$ | = | 0.11 |
| $k_1$ | = | 0.7 $\mu$M | $k_2$ | = | 0.7 $\mu$M |
| $\tau_h$ | = | 0.2 s | $k_y$ | = | 0.27 $\mu$M |
| $\gamma$ | = | 1 $\mu$Ms$^{-1}$ | $\beta$ | = | 0.15 $\mu$Ms$^{-1}$ |
| $k_\mu$ | = | 0.01 $\mu$M | $V_p$ | = | 0.08 s$^{-1}$ |
| $k_p$ | = | 1 $\mu$M | $F$ | = | 2 $\mu$ms$^{-1}$ |
| $D_c$ | = | 20 $\mu$m$^2$s$^{-1}$ | $D_p$ | = | 300 $\mu$m$^2$s$^{-1}$ |

decreases with distance from the stimulated cell, and the arrival time and the intercellular delay increase exponentially with distance from the stimulated cell. For the values chosen for $F$, ranging from 1 to 8 $\mu$ms$^{-1}$, the model agrees well with experimental data from endothelial and glial cells (Demer et al., 1993; Charles et al., 1992). Although it is not possible to compare the model exclusively to data from epithelial cells, the agreement with data from other cell types indicates that the model provides a reasonable quantitative description of the intercellular wave. However, the most important model prediction is the value of $F$ needed to obtain such agreement. If $F$ is lower than about 1 $\mu$ms$^{-1}$, the intercellular wave moves too slowly to agree with experimental data. Since the actual value of $F$ is unknown, this prediction provides a way in which the underlying hypothesis (of passive diffusion of IP$_3$) may be tested.

## 12.6  EXERCISES

1. Construct the traveling wave solution of the piecewise linear version of the two-pool model as $\epsilon \to 0$. Show that the traveling wave in this case exhibits two shocks. Using the solution for $\epsilon = 0$ as a starting point, find the solution for small but nonzero $\epsilon$ and plot the wave solution. Compare to the solution when $\epsilon = 0$. Compare to the solution obtained by solving the differential equations numerically.

2. Construct a family of periodic plane-wave solutions to the piecewise linear two-pool model. Show that the traveling wave obtained in Section 12.1.1 is the limit of this family of periodic plane waves as the period tends to infinity.

3. Convert the model in Section 12.2 (i.e., the model for Ca$^{2+}$ waves based on the heuristic IP$_3$ receptor model) to traveling wave coordinates, find the points of Hopf bifurcation to periodic plane waves, and plot cross-sections of the dispersion surface. Are there any homoclinic bifurcations to traveling waves?

4. Construct the piecewise exponential solution to (12.8) and (12.9) in one space dimension. Show that the wave without recovery travels at the same speed as the wave with recovery and satisfies the same existence condition.

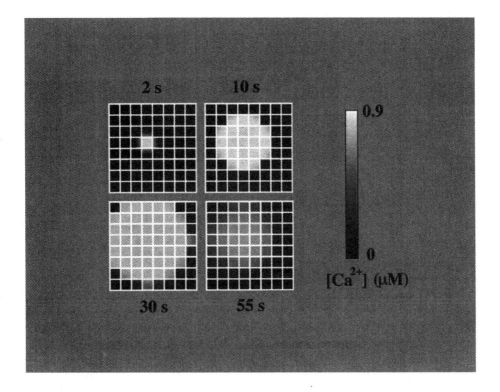

**Figure 12.10**  Density plot of a two-dimensional intercellular $Ca^{2+}$ wave, computed numerically from the intercellular $Ca^{2+}$ wave model. (Sneyd et al., 1995a, Fig. 4.)

5.  Show that (12.20) and (12.21) with $s_1 = N + k$ and $s_2 = N$ have two solutions for $N$ that appear in a tangent bifurcation as $\kappa$ is decreased.

6.  Generalize (12.34) to the case of multiple buffers, both mobile and immobile.

7.  Verify the details of the proof that (12.38) has a traveling wave solution that is a transition between $W_1$ and $W_3$ if and only if

$$\int_{W_1}^{W_3} f(\phi(w))dw > 0. \tag{12.67}$$

8.  By looking for solutions to (12.27) and (12.28) of the form $c = A \exp(\xi/\lambda)$, $b = B \exp(\xi/\lambda)$, where $\lambda$ is the space constant of the wave front and $\xi = x + st$, show that the speed of the wave $s$ is related to the space constant of the wave front by

$$s = \frac{D_c}{\lambda} + \lambda \left[ f'(0) - \frac{k_+ b_t(\lambda s - D_b)}{\lambda^2 k_- + \lambda s - D_b} \right]. \tag{12.68}$$

What is the equation in the limit $k_+, k_- \to \infty$, with $k_-/k_+ = K$? Hence show that for the generalized bistable equation,

$$\lambda s < D_{\text{eff}}. \tag{12.69}$$

9.  Suppose the source term $f(c)$ for the calcium buffering system in (12.27) is replaced by the discrete source term $L \sum_n \delta(x - nL)f(c)$. Look for standing-wave solutions of the resulting

fast buffering system and show that the equation for standing solutions can be reduced to the finite difference equation of the form

$$w_{n+1} - 2w_n + w_{n-1} + L^2 f(\phi(w_n)) = 0, \tag{12.70}$$

where $\phi(w)$ satisfies $w = D_c\phi - D_b b_t \frac{K}{K+\phi}$. What conclusions can you draw about the effect of buffering on propagation when calcium is released from discrete sites?

10. Use (12.48), (12.51), and (12.47) to construct an approximate expression for the traveling wave profile for the buffered bistable equation. Proceed by constructing a trajectory in the $c, c'$ phase-plane, connecting the origin to $(1, 0)$ such that it has slope $\lambda_u$ at the origin and slope $\lambda_s$ at $(1, 0)$. Then integrate to get

$$\frac{1}{\lambda_u} \ln(c) + \frac{1}{\lambda_s} \ln(1 - c) + \frac{\lambda_u + \lambda_s}{\lambda_u \lambda_s} \ln\left(c - \frac{\lambda_u}{\lambda_u + \lambda_s}\right) = \xi. \tag{12.71}$$

11. Repeat the analysis of 12.3.1, without assuming that $b_t$ is constant. Under what conditions may we assume that $b_t$ is constant?

# CHAPTER 13

# Regulation of Cell Function

The regulation of a cell's activity all takes place within its nucleus. Within the nucleus there are *nucleic acids*, which control the production of proteins necessary for cell function. The nucleic acids are large polymers of smaller molecular subunits called *nucleotides*, which themselves are composed of three basic molecular groups,

- a *nitrogenous base*, which is an organic ring containing nitrogen,
- a 5-carbon (pentose) sugar, either *ribose* or *deoxyribose*,
- an inorganic phosphate group.

Nucleotides may differ in the first two of these components, and consequently there are two specific types of nucleic acids, *deoxyribonucleic acid* (DNA) and *ribonucleic acid* (RNA).

There may be any one of five different nitrogenous bases present in the nucleotides: *adenine* (A), *cytosine* (C), *guanine* (G), *thymine* (T), and *uracil* (U). These are most often denoted by the letters A, C, G, T, and U, rather than by their full names.

The DNA molecule is a long double strand of nucleotide bases, which can be thought of as a twisted, or helical, ladder. The backbone (or sides of the ladder) is composed of alternating sugar and phosphate molecules, the sugar, deoxyribose, having one fewer oxygen atom than ribose. The "rungs" of the ladder are complementary pairs of nitrogenous bases, with G always paired with C, and A always paired with T. The bond between pairs is a weak hydrogen bond that is easily broken and restored during the replication process.

The ordering of the base pairs along the DNA molecule is called the genetic code, because it is this ordering of symbols from the four-letter alphabet of A, C, G, and T that controls all cellular biochemical functions. The nucleotide sequence is organized into code triplets, called *codons*, which code for amino acids as well as other signals, such as "start manufacture of a protein molecule" and "stop manufacture of a protein molecule." Segments of DNA that code for a particular product are called *genes*, of which there are about 100,000 in human DNA. Typically, a gene contains start and stop codons as well as the code for the gene product, and can include large segments of DNA whose function is unclear. One of the simplest known living organisms, *Mycoplasma genitalian*, has 470 genes and about 500,000 base pairs.

An RNA molecule is a single strand of nucleotides. It is different from DNA in that the sugar in the backbone is ribose, and the base U is substituted for T. Cells generally contain two to eight times as much RNA as DNA. There are three types of RNA, each of which plays a major role in cell physiology. For our purposes here, *messenger RNA* (mRNA) is the most important, as it carries the code for the manufacture of specific proteins. *Transfer RNA* (tRNA) acts as a carrier of one of the twenty amino acids that are to be incorporated into a protein molecule that is being produced. Finally, *ribosomal RNA* constitutes about 60% of the *ribosome*, a structure in the cellular cytoplasm on which proteins are manufactured.

The two primary functions that take place in the nucleus are the reproduction of DNA and the production of RNA. RNA is formed by a process called *transcription*, as follows. An enzyme called *RNA polymerase* (or, more precisely, a polymerase complex, as many other proteins are also needed) attaches to some starting site on the DNA, breaks the bonds between base pairs in that local region, and then makes a complementary copy of the nucleotide sequence for one of the DNA strands. As the RNA polymerase moves along the DNA strand, the RNA molecule is formed, and the DNA crossbridges reform. The process stops when the RNA polymerase reaches a transcriptional termination site and disengages from the DNA.

Proteins are manufactured employing all three RNA types. After a strand of mRNA that codes for some protein is formed in the nucleus, it is released to the cytoplasm. There it encounters ribosomes that "read" the mRNA much like a tape recording. As a particular codon is reached, it temporarily binds with the specific tRNA with the complementary codon carrying the corresponding amino acid. The amino acid is released from the tRNA and binds to the forming chain, leading to a protein with the sequence of amino acids coded for by the DNA.

Synthesis of a cellular biochemical product usually requires a series of reactions, each of which is catalyzed by a special enzyme. In prokaryotes, formation of the necessary enzymes is often controlled by a sequence of genes located in series on the DNA strand. This area of the DNA strand is called an *operon*, and the individual genes within the operon are called *structural genes*. At the beginning of the operon is a segment called a *promoter*, which is a series of nucleotides that has a specific affinity for RNA polymerase. The polymerase must bind with this promoter before it can begin traveling along the DNA strand to synthesize RNA. In addition, in the promoter region, there is an area called a *repressor operator*, where a regulatory repressor protein can bind, preventing the attachment of RNA polymerase, thereby blocking the transcription of the genes of the operon. Repressor protein generally exists in two allosteric forms, one that can bind with the repressor operator and thereby repress transcription, and one that does not bind. A substance that changes the repressor so that it breaks its bond with the operator is called an *activator*, or *inducer*.

## 13.1 The *lac* Operon

To illustrate how these reactions can all work together to regulate cell function, we give a simple description of the utilization of lactose by the bacterium *Escherichia coli (E. coli)*. Lactose is not generally available to *E. coli* as a food substrate, so the bacterium does not usually synthesize the enzymes necessary for its metabolic use, although they are available in very small quantities. However, there is an operon, called the *lac operon*, normally turned off, that codes for the three enzymes *β-galactoside permease*, *β-galactosidase*, and *β-thiogalactoside acetyl transferase*. If the bacterium is exposed to lactose, the permease mediates the transport of lactose into the cell, and *β*-galactosidase isomerizes lactose into *allolactose* (an allosteric isomer of lactose),

$$\text{lactose} + \text{E} \rightleftharpoons \text{allolactose} + \text{E}, \tag{13.1}$$

and also into simple hexose sugars, glucose and galactose,

$$\text{lactose} + \text{E} \rightleftharpoons \text{glucose} + \text{galactose} + \text{E}, \tag{13.2}$$

that can be metabolized for energy. The allolactose binds with a repressor molecule to keep it from repressing the production of mRNA. Thus, production of allolactose turns on the production of mRNA, which then leads to production of more enzyme, enabling the conversion of more lactose to allolactose. Hence we have the potential for an autocatalytic reaction. (The function of the transferase is not known.)

A mathematical model for this process is similar to a model of Griffith (1971) (see Exercise 2) and uses familiar principles. We suppose that the production of the enzyme is turned on by $m$ molecules of the product allolactose, denoted by P, according to

$$G + m\text{P} \underset{k_{-1}}{\overset{k_1}{\rightleftharpoons}} X,$$

where G is the inactive, or repressed, state of the gene, and X is the induced, or active, state of the gene.

Normally we would use the law of mass action to write an equation for the production of enzymes. Here, however, there is only one gene, and so we must invent a different argument. Notice that in a large population of genes, the percentage of genes in the active state at any time is given by the chemical equilibrium

$$p = \frac{[\text{P}]^m}{k_{\text{eq}}^m + [\text{P}]^m}, \tag{13.3}$$

where $k_{eq}^m = k_{-1}/k_1$. Thus, $p$ is the average production rate of a single "typical" gene, so that the average production of mRNA is described by the differential equation

$$\frac{dM}{dt} = M_0 + \frac{k_1[P]^m}{k_{eq}^m + [P]^m} - k_2 M, \tag{13.4}$$

where $M$ is the concentration of mRNA that codes for the enzymes. The term $M_0$ is added here because trace amounts of mRNA are produced simply because the binding of the repressor site is a stochastic process with random fluctuations, even in the absence of allolactose, allowing the production of small amounts of mRNA.

We next assume that the enzymes are produced at a rate linearly proportional to available mRNA and are degraded, so that the concentrations of enzymes permease (denoted $E_1$) and $\beta$-galactosidase (denoted $E_2$) are determined by

$$\frac{dE_1}{dt} = c_1 M - d_1 E_1, \tag{13.5}$$

$$\frac{dE_2}{dt} = c_2 M - d_2 E_2. \tag{13.6}$$

Since their codes are part of the same mRNA, the production rates of $E_1$ and $E_2$ are the same ($c_1 = c_2$). Lactose that is exterior to the cell, with concentration $S_0$, is brought into the cell to become the lactose substrate, with concentration $S$, at a Michaelis–Menten rate proportional to $E_1$, via

$$\frac{dS_0}{dt} = -\sigma_0 E_1 \frac{S_0}{k_0 + S_0}. \tag{13.7}$$

Once inside the cell, lactose substrate is converted to allolactose, and then allolactose is converted to glucose and galactose via enzymatic reaction with $\beta$-galactosidase, so

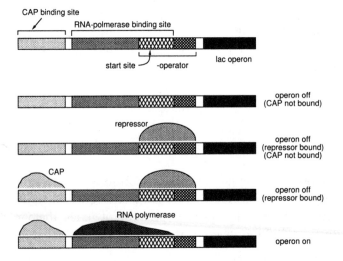

**Figure 13.1**  Control sites and control states for the *lac* operon.

that

$$\frac{dS}{dt} = \sigma_0 E_1 \frac{S_0}{k_0 + S_0} - \sigma_1 E_2 \frac{S}{k_s + S} \tag{13.8}$$

and

$$\frac{d[P]}{dt} = \sigma_1 E_2 \frac{S}{k_s + S} - \sigma_2 E_2 \frac{[P]}{k_p + [P]}. \tag{13.9}$$

This is a simplification of the full reaction mechanism, but it is adequate for our purposes here.

We expect this system of equations to have a biochemical switch to turn on the production of enzymes and then to switch it off when there is no more lactose to convert. To uncover this behavior we make a few simplifying assumptions. First, we take the mRNA to be in quasi-steady state, so that

$$M = \frac{k_1}{k_2}\left(\frac{[P]^m}{k_{eq}^m + [P]^m}\right) + \frac{M_0}{k_2}. \tag{13.10}$$

Second, the decay rates $d_1$ and $d_2$ include effects of degradation of enzyme and dilution because of increase in cell volume from cell growth. Typically, enzyme degradation is slow compared to cell growth, so it is reasonable to take $d_1 = d_2$, so that

$$\frac{dE_1}{dt} = \frac{c_1 M_0}{k_2} + \frac{c_1 k_1}{k_2}\left(\frac{[P]^m}{k_{eq}^m + [P]^m}\right) - d_1 E_1 \tag{13.11}$$

and $E_2 = E_1$. Next we assume that there is no delay in the conversion of the lactose substrate into allolactose. This allows us to eliminate (13.8) and replace (13.9) with

$$\frac{d[P]}{dt} = \sigma_0 E_1 \frac{S_0}{k_0 + S_0} - \sigma_2 E_1 \frac{[P]}{k_p + [P]}. \tag{13.12}$$

The three equations (13.7), (13.11), and (13.12) form a closed system. We introduce dimensionless variables $S_0 = k_0 s, [P] = k_p p, E_1 = e_0 e$, and $t = t_0 \tau$ to obtain

$$\frac{de}{d\tau} = m_0 + \frac{p^m}{\kappa^m + p^m} - \epsilon e, \tag{13.13}$$

$$\frac{dp}{d\tau} = \mu e\left(\frac{s}{s+1} - \lambda\frac{p}{p+1}\right), \tag{13.14}$$

$$\frac{ds}{d\tau} = -e\frac{s}{s+1}, \tag{13.15}$$

where $e_0^2 = c_1 k_0 k_1/(\sigma_0 k_2), t_0 = k_0/(e_0 \sigma_0), \lambda = \sigma_2/\sigma_0, \mu = k_0/k_p, \kappa = k/k_p, m_0 = M_0/k_1$, and $\epsilon = t_0 d_1$. This system of equations is relatively easy to understand. We assume that at time $\tau = 0$ an initial amount of lactose $s = s(0)$ is presented to a cell in which there are trace amounts of enzyme $e$. There are two possible responses. If the amount of lactose is too small, then the lactose is gradually depleted, although there is no increase in enzyme concentration. However, if the lactose dose is sufficiently large, then there is

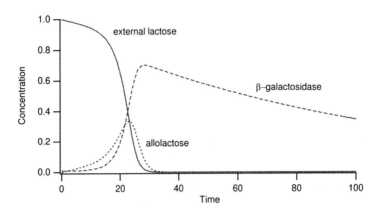

**Figure 13.2** Solution of the differential equations (13.13), (13.14), (13.15). Parameters are $\mu = \lambda = \kappa = 1$, $\epsilon = 0.01$, $m_0 = 0.001$, $m = 2$, and initial data were $s = 1.0$, $e = 0.01$, $p = 0.0$.

an autocatalytic response in enzyme production, as the *lac* operon is turned on and enzyme is produced. The production of enzyme shuts down when the lactose stimulus has been consumed, and the enzyme concentration then gradually declines (Fig. 13.2).

## 13.1.1 Glucose Oscillations

An additional feature of *E. coli* is that lactose is not utilized when there is adequate glucose. The mechanism for this control is as follows: preceding the promoter region of the *lac* operon where the RNA polymerase must bind to begin transcription, there is another region, called a CAP site (*catabolic gene activator protein*), which can be bound by a dimeric molecule CAP. CAP by itself has no influence on transcription unless it is bound to cyclic AMP (cAMP), but when CAP is bound to cAMP the complex can bind to the CAP site, thereby promoting the binding of RNA polymerase to the promoter region, allowing transcription.

The connection with glucose is that one of the catabolites of glucose (a product of its breakdown) lowers the amount of intracellular cAMP (by an unknown mechanism), thereby decreasing the amount of bound CAP, decreasing the activator activity of the CAP site. This is believed to be the mechanism underlying oscillatory usage of lactose, as follows. When the glucose concentration is low and lactose is available, CAP site activity promotes the production of allolactose and $\beta$-galactosidase, thereby turning on the production of glucose from lactose. However, as glucose concentrations rise, the CAP activity is turned off, so that lactose utilization ceases, at least until glucose levels fall. These oscillations of $\beta$-galactosidase activity with a period of approximately 50 minutes have been observed experimentally (Knorre, 1968). Experimental data for these oscillations are shown in Fig. 13.3.

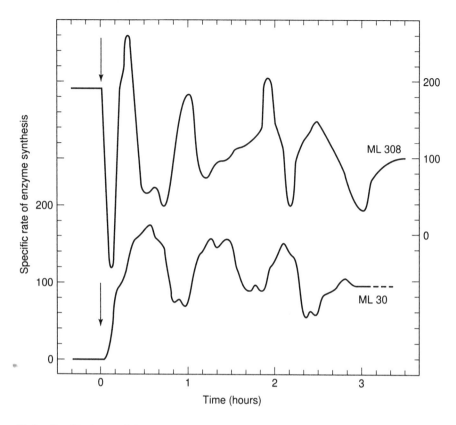

**Figure 13.3** Oscillations of the specific rate of $\beta$-galactosidase synthesis in *E. coli* strains ML30 and ML 308. (Knorre, 1968).

# 13.2 Cell Cycle Control

The *cell-division cycle* is that process by which a cell duplicates its contents and then divides in two. The adult human must manufacture many millions of new cells each second simply to maintain the status quo, and if all cell division is halted, the individual will die within a few days. On the other hand, abnormally rapid cell proliferation, i.e., *cancer*, can also be fatal, as rapidly proliferating cells interfere with the function of "normal" cells and organs. Control of the cell cycle involves, at a minimum, control of cell growth and replication of nuclear DNA in such a way that the size of the individual cells remains, on average, constant.

The cell cycle is traditionally divided into several distinct phases (shown schematically in Fig. 13.4), the most dramatic of which is mitosis or *M phase*. Mitosis is characterized by separation of previously duplicated nuclear material, nuclear division, and finally the actual cell division, called *cytokinesis*. In most cells the whole of *M* phase takes only about an hour, a small fraction of the total cycle time. The much longer period of time between one *M* phase and the next is called *interphase*. The por-

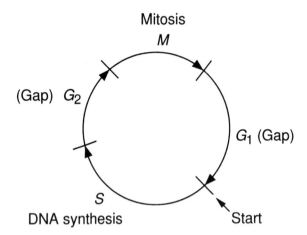

**Figure 13.4** Schematic diagram of the cell cycle.

tion of interphase following cytokinesis is called $G_1$ *phase* ($G$ = gap), during which cell growth occurs. When the cell is sufficiently large, DNA replication in the nucleus is initiated and continues during $S$ *phase* ($S$ = synthesis). Following $S$ phase is $G_2$ *phase*, providing a safety gap during which the cell is presumably preparing for $M$ phase, to ensure that DNA replication is complete before the cell plunges into mitosis.

There are actually two controlled growth processes. There is the chromosonal cycle, in which the genetic material is exactly duplicated and two nuclei are formed from one for every "turn" of the cycle. Accuracy is essential to this process, since each daughter nucleus must receive an exact replica of each chromosome. A less tightly controlled process, the cytoplasmic cycle, duplicates the cytoplasmic material, including all of the structures (mitochondria, organelles, sarcoplasmic reticulum, etc.). This growth is continuous during the $G_1$, $S$, and $G_2$ phases, pausing briefly only during mitosis.

In mature organisms these two processes operate in coordinated fashion, so that the ratio of cell mass to nuclear mass remains essentially constant. However, it is possible for these two to be uncoupled. For example, during *oogenesis*, a single cell (an *ovum*) grows in size without division. After fertilization, during *embryogenesis*, the egg undergoes twelve rapid synchronous mitotic divisions to form a ball consisting of 4096 cells, called the *blastula*.

There is strong evidence that these early embryonic divisions are controlled by a cytoplasmic biochemical limit cycle oscillator. For example, if fertilized (*Xenopus*) frog eggs are enucleated, they continue to exhibit periodic "twitches" or contractions, as if the cytoplasm continued to generate a signal in the absence of a nucleus. Enucleated sea urchin eggs go a step further by actually dividing a number of times before they notice that they contain no genetic material and consequently die.

The cell cycle has been studied most extensively for frogs and yeast. Frog eggs are useful because they are large and easily manipulated. Yeast cells are too small for these kinds of studies, but are suitable for cloning and identification of the involved genes and gene products. The budding yeast *Saccharomyces cerevisiae*, used by brewers and bakers, divides by first forming a bud that is initiated and grows steadily during $S$ and $G_2$ phases, and finally separates from its mother after mitosis.

## 13.2.1 The $G_1$ Checkpoint

The autonomous cell cycle oscillations seen in early embryos are unusual. Most cells proceed through the division cycle in fits and starts, pausing at "checkpoints" to ensure that all is ready for the next phase of the cycle. There are checkpoints at the end of the $G_1, G_2$, and $M$ phases of the cell cycle, although not all cells use all of these checkpoints. During early embryogenesis, however, the checkpoints are inoperable, and cells divide as rapidly as possible, driven by the underlying limit cycle oscillation.

The $G_1$ checkpoint is often called *Start*, because here the cell determines whether all systems are ready for $S$ phase and the duplication of DNA. Before Start, newly born cells are able to leave the mitotic cycle and differentiate (into nondividing cells with specialized function). However, after Start, they have passed the point of no return and are committed to another round of DNA synthesis and division.

As with all cellular processes, the cell cycle is regulated by genes and the proteins that they encode. There are two classes of proteins that form the basis of the cell-cycle control system. The first is the family of *cyclin-dependent protein kinases* (Cdk), which induce a variety of downstream events by phosphorylating selected proteins. The second family are the *cyclins*, so named because the first members to be identified are cyclically synthesized and degraded in each division cycle of the cell. Cyclins bind to Cdk molecules and control their ability to phosphorylate target proteins, but without cyclin, Cdk is inactive. In budding yeast cells there is only one major Cdk and nine cyclins, leading to a possibility of nine active cyclin–Cdk complexes. In mammals, the story is substantially more complicated, as there are (at last count) six Cdks and more than a dozen cyclins.

The critical chemicals for getting through the $G_1$ and $G_2$ checkpoints are known as *S-phase promoting factor* (SPF) and *M-phase promoting factor* (MPF), respectively. These are *heterodimers* because they consist of two essential subunits, a Cdk and a cyclin. A schematic diagram of the cell cycle is shown in Fig. 13.5.

The molecular events that constitute Start are most thoroughly understood for budding yeast. The major events triggered by Start are DNA synthesis and bud emergence. DNA synthesis appears to be triggered by a Cdk called Cdc28 (Cdc for cell-division-cycle; remember that in yeast there is only one major Cdk) in association with either cyclin Clb5 or cyclin Clb6. Bud emergence seems to depend on Cdk association with cyclin Cln1 or Cln2. These four cyclins are subject to rapid degradation, so their levels in the cell are controlled by their rates of transcription. Their transcription, in turn, is controlled by two transcription factors, SBF and MBF. As the cyclins accumulate and associate with Cdc28, they activate their own transcription factors, so their rate of accumulation increases. These positive feedback loops lead to autocatalytic production of the cyclins, ensuring that the cells pass Start decisively and irreversibly by an explosive activation of Start kinase. In addition, there is another activator, a complex of Cdc28

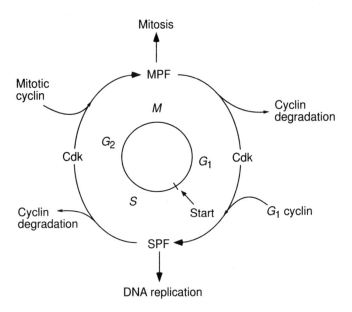

**Figure 13.5** Schematic diagram of the primary chemical reactions of the cell cycle.

with cyclin Cln3, which is thought to activate SBF and MBF in a cell-size-dependent fashion, although the mechanism of this size dependence is not known.

A simple mathematical model of this process (following Tyson et al., 1995) can reveal how the $G_1$ checkpoint works. We assume that there is one transcription factor, SBF, denoted by $S_a$ when active, and by $S_i$ when inactive. The transcription factor is rendered active by a Cln cyclin (the only cyclin in the model), denoted by N, and by a starter kinase (Cdc28-Cln3), denoted by A. The transcription factor is rendered inactive by another chemical, a phosphatase, denoted by E. The cyclin is produced via activation of SBF and degrades naturally according to

$$\xrightarrow{S_a} \text{N} \xrightarrow{k_2},$$   (13.16)

so that

$$\frac{d[\text{N}]}{dt} = \frac{k_1[S_a]}{k_s + [S_a]} - k_2[\text{N}].$$   (13.17)

The SBF is activated by both Cln and the starter kinase and is deactivated by the phosphatase via

$$S_i \underset{E}{\overset{N+A}{\rightleftarrows}} S_a,$$   (13.18)

leading to the equation

$$\frac{d[S_a]}{dt} = k_3([\text{A}] + [\text{N}]) \frac{[S_i]}{k_i + [S_i]} - k_4[\text{E}] \frac{[S_a]}{k_a + [S_a]}.$$   (13.19)

We assume that $[S_a] + [S_i] = C$, a constant. Then, in terms of the nondimensional variables $[S_a] = Cs, [N] = k_1/k_2 n, t = \tau/k_2$, we have

$$\frac{dn}{d\tau} = \frac{s}{\kappa_s + s} - n, \tag{13.20}$$

$$\frac{ds}{d\tau} = (\alpha + \lambda n)\frac{1 - s}{\kappa_i + 1 - s} - \mu\frac{s}{\kappa_a + s}, \tag{13.21}$$

where $\alpha = \frac{k_3[A]}{k_2 C}, \lambda = \frac{k_3 k_1}{k_2^2 C}, \mu = \frac{k_4[E]}{k_2 C}, \kappa_s = k_s/C, \kappa_i = k_i/C, \kappa_a = k_a/C$.

The behavior of this system is readily exposed by its phase portrait. The nullclines are the curves $\frac{dn}{d\tau} = 0$ and $\frac{ds}{d\tau} = 0$. It is apparent that for the $\frac{ds}{d\tau} = 0$ nullcline, $n$ is decreasing as a function of the control parameter $\alpha$, for fixed $s$. In Fig. 13.6 are shown examples of the nullclines, $\frac{dn}{d\tau} = 0$ a dashed curve, and $\frac{ds}{d\tau} = 0$ as solid curves, for two different values of $\alpha = 2.5$ and 3.8. Other parameter values are $\kappa_s = 1.0, \kappa_i = 0.1, \kappa_a = 0.001, \lambda = 37.0, \mu = 4.0$.

The behavior of this model for the $G_1$ checkpoint is now easily described. There are either one or three steady solutions. The steady solution with $s$ saturated (near 1) is a stable steady state that always exists, and corresponds to high levels of cyclin and activation of Start. For large values of the control parameter $\alpha$, corresponding to large cell size, this is the only steady solution. However, for small cell size, and hence, small values of the control parameter $\alpha$, there are three steady states, the smallest and largest being stable and the intermediate being an unstable saddle point.

We suppose that during $G_1$ phase with $\alpha$ small the system sits at the small steady-state solution. However, as the cell grows, the value of $\alpha$ increases (by an unknown mechanism), reaching a critical value at which the small steady-state solution disappears (through a saddle-node bifurcation), and the system quickly switches to the large steady state, at which production of cyclin is high, enabling the rapid production of S-phase promoting factor.

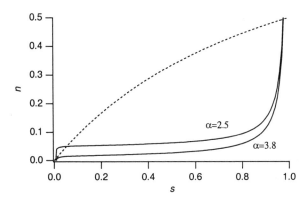

**Figure 13.6** Nullclines for the system of equations (13.20), (13.21), with the nullcline $\frac{dn}{d\tau} = 0$ for (13.20) shown as a dashed curve and nullclines $\frac{ds}{d\tau} = 0$ for (13.21) shown as solid curves, shown for $\alpha = 2.5$ and 3.8.

## 13.2.2   The $G_2$ Checkpoint

In budding yeast, the $G_2$ checkpoint works by a similar mechanism to the $G_1$ checkpoint. Here there is a cyclin Clb-kinase and its transcription factor Mcm1. The Clb cyclins are different from the Cln cyclins in that the Cln cyclins degrade naturally and rapidly, so that they must be continually synthesized, whereas the degradation of the Clb cyclins is mediated by other enzymes. We describe a bit of this regulatory pathway below. For this model, however, we assume that Cln-kinase (the dimer complex of a Cln and Cdc28 = MPF) acts enzymatically to inhibit the degradation of Clb. Furthermore, the Cln synthesis is inhibited by the presence of Clb.

A mathematical model combining both of these processes is as follows: We assume that there are two cyclins, a Cln (denoted by $N$) and a Clb (denoted by $B$). The production of Cln follows the reaction mechanism (13.16) and is governed by the differential equation (13.17), and the concentration of active SBF is governed by

$$\mathrm{S}_i \underset{E_1}{\overset{N+A_1}{\rightleftarrows}} \mathrm{S}_a, \tag{13.22}$$

from which we find an equation similar to (13.19), modified only to account for Clb inactivation,

$$\frac{d[\mathrm{S}_a]}{dt} = k_3([\mathrm{A}_1] + [\mathrm{N}])\frac{[\mathrm{S}_i]}{k_{s_i} + [\mathrm{S}_i]} - k_4([\mathrm{E}_1] + [\mathrm{B}])\frac{[\mathrm{S}_a]}{k_{s_a} + [\mathrm{S}_a]}. \tag{13.23}$$

The production of Clb (B) follows

$$\overset{R_a}{\longrightarrow} \mathrm{B} \overset{k_-(N)}{\longrightarrow} \tag{13.24}$$

and is governed by the equation

$$\frac{d[\mathrm{B}]}{dt} = \frac{k_5[\mathrm{R}_a]}{k_r + [\mathrm{R}_a]} - k_6[\mathrm{B}]\frac{k_n}{k_n + [\mathrm{N}]}. \tag{13.25}$$

Finally, the production of active Mcm1 ($R_a$, the transcription factor for the Clb) follows the mechanism

$$\mathrm{R}_i \underset{E_2}{\overset{B+A_2}{\rightleftarrows}} \mathrm{R}_a \tag{13.26}$$

and is governed by

$$\frac{d[\mathrm{R}_a]}{dt} = k_7([\mathrm{A}_2] + [\mathrm{B}])\frac{[\mathrm{R}_i]}{k_{r_i} + [\mathrm{R}_i]} - k_4[E_2]\frac{[\mathrm{R}_a]}{k_{r_a} + [\mathrm{R}_a]}. \tag{13.27}$$

Here we have also assumed that there is an additional enzyme $A_2$ that activates Mcm1, thereby activating the transcription of Clb.

The nondimensional version of this system of equations is found by taking $[\mathrm{S}_a] + [\mathrm{S}_i] = C_s$, $[\mathrm{R}_a] + [\mathrm{R}_i] = C_r$, $[\mathrm{R}_a] = C_r r$, $[\mathrm{S}_a] = C_s s$, $[\mathrm{N}] = \frac{k_1}{k_2}n$, $[\mathrm{B}] = \frac{k_5}{k_2}b$, from which it follows that

$$\frac{dn}{d\tau} = \frac{s}{\kappa_s + s} - n, \tag{13.28}$$

$$\frac{ds}{d\tau} = (\alpha + \lambda_n n)\frac{1 - s}{\kappa_{s_i} + 1 - s} - (\mu_1 + \lambda_b b)\frac{s}{\kappa_{s_a} + s}, \tag{13.29}$$

$$\frac{db}{d\tau} = \frac{r}{\kappa_r + r} - \mu_b \frac{b}{\kappa_n + n}, \tag{13.30}$$

$$\frac{dr}{dt} = (\beta + \lambda_r b)\frac{1 - r}{\kappa_{r_i} + 1 - r} - \mu_2 \frac{r}{\kappa_{r_a} + r}, \tag{13.31}$$

where $\alpha = \frac{k_3[A_1]}{k_2 C_s}$, $\beta = \frac{k_7[A_2]}{k_2 C_r}$, $\lambda_n = \frac{k_3 k_1}{k_2^2 C_s}$, $\lambda_b = \frac{k_4 k_5}{k_2^2 C_s}$, $\lambda_r = \frac{k_5 k_7}{k_2^2 C_r}$, $\mu_1 = \frac{k_4[E_1]}{k_2 C_s}$, $\mu_2 = \frac{k_4[E_2]}{k_2 C_r}$, $\mu_b = k_6 k_n/k_1$, $\kappa_n = k_2 k_n/k_1$, $\kappa_{s_a} = k_{s_a}/C_s$, $\kappa_{s_i} = k_{s_i}/C_s$, $\kappa_{r_i} = k_{r_i}/C_r$, $\kappa_{r_a} = k_{r_a}/C_r$.

We can understand something about the behavior of this system of four differential equations by looking at the nullclines in the $(n, b)$ plane. This is not a true phase portrait because the system is four-dimensional, but the steady-state behavior is correctly depicted. There are two curves that can be plotted: The curve on which both $\frac{dn}{d\tau} = \frac{ds}{d\tau} = 0$, and the curve on which both $\frac{db}{d\tau} = \frac{dr}{d\tau} = 0$. For example, the curve in the $(n, b)$ plane on which $\frac{dn}{d\tau} = \frac{ds}{d\tau} = 0$ can be found as a parametric curve with $s$ ranging between 0 and 1 as the parameter, and the curve on which $\frac{db}{d\tau} = \frac{dr}{d\tau} = 0$ can be found in a similar way as a parametric curve with $r$ as the parameter.

In Fig. 13.7 are shown two examples of each of these nullclines, the solid curves showing the $n$ nullcline and the dashed curve depicting the $b$ nullcline. The nullclines depend in monotonic fashion on the parameters $\alpha$ and $\beta$. That is, the $n$ nullcline moves up with increasing $\alpha$ and the $b$ nullcline moves leftward ($n$ values decrease for each fixed $b$) with increasing $\beta$.

Intersections of the nullclines locate steady-state solutions of the full system. Because the nullclines are both "s"-shaped, there are up to nine such steady-state values, depending on parameter values. In Fig. 13.7 there can be as many as five fixed points, depending on the values of $\alpha$ and $\beta$.

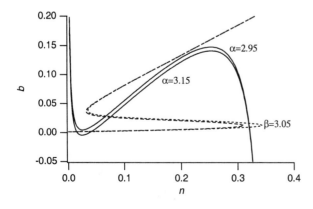

**Figure 13.7** Nullclines for the system of equations (13.28), (13.29), (13.30), (13.31), with the nullclines for $n$ (nondimensional Cln concentration) shown as solid curves and nullclines for $b$ (nondimensional Clb concentration) shown as dashed curves, shown for $\alpha = 2.95$ and 3.15, and $\beta = 3.05$ and 3.1. Other parameter values are $\kappa_s = 2.0$, $\lambda_n = 26.0$, $\mu_1 = 4.0$, $\kappa_{s_i} = 0.1$, $\kappa_{s_a} = 0.01$, $\lambda_b = 24.0$, $\lambda_r = 50.0$, $\mu_2 = 4.0$, $\mu_b = 0.33$, $\kappa_{r_i} = 0.1$, $\kappa_r = 17.0$, $\kappa_{r_a} = 0.01$, $\kappa_n = 0.08$.

The steady-state solutions of most interest are for small $n$ and $b$ corresponding to the $G_1$ checkpoint, and for small $b$ and large $n$ corresponding to the $G_2$ checkpoint. The first of these can be eliminated by increasing $\alpha$, and the second is eliminated by increasing $\beta$. Thus we can make up the following story by supposing that the size of $\alpha$ is dependent upon the size of the cell and that the size of $\beta$ is dependent upon the amount of DNA replication. A small cell is presumed to sit at the $G_1$ checkpoint until $\alpha$ increases enough to eliminate the rest point, whereupon the dynamics take $n$ and $b$ to the second checkpoint. There, the cell awaits sufficient increase in the parameter $\beta$ to eliminate the steady state and initiate mitosis.

It is noteworthy that if both of these checkpoints are eliminated by increasing both $\alpha$ and $\beta$ (the values $\alpha = 3.15$ and $\beta = 3.1$ suffice), then the dynamic motion is a periodic limit cycle. This limit cycle is shown in Fig. 13.8.

In budding yeast, cells do not divide with equal size, so that a mother and daughter cell can be identified. Because they are much larger, mother cells typically reenter mitosis almost immediately following division, while daughter cells must grow to adequate size, while sitting at the $G_1$ checkpoint, before initiating mitosis. In terms of this model, the mother cell has $\alpha$ large while the daughter cell has $\alpha$ small.

### 13.2.3   Control of M Phase

In fertilized *Xenopus* oocytes, cell division takes place without any cell growth, so the $G_1$ checkpoint is removed (or inoperable). The MPF that is critical for getting through the $G_2$ checkpoint is a dimer of a Cdk called Cdc2 and a mitotic cyclin. (The enzyme Cdc2 in *Xenopus* can be substituted for Cdc28 in budding yeast with no loss of function, although the two enzymes are slightly different.)

Cdc2 is one member of a class of enzymes, coded by *cell-division-cycle* (*cdc*) genes, that are required to get the cell past some specific point in the cell division cycle. In yeast, there are now about 70 known *cdc* genes, many of which have been cloned and sequenced. *cdc* deficient mutants, i.e., cells with mutated *cdc* genes, are characterized

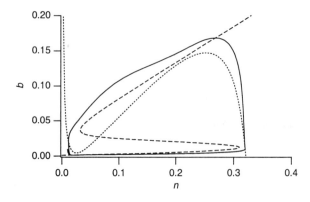

**Figure 13.8**  Limit cycle for the system of equations (13.28), (13.29), (13.30), (13.31), with the nullclines shown as dashed curves, with $\alpha = 3.15$ and $\beta = 3.1$. Other parameter values are as in Fig. 13.7.

by their inability to progress through the normal cell cycle, and since they continue to grow without dividing, by their large size.

A second type of gene, called *wee*, from the Old English meaning tiny, is so named because mutant cells divide at smaller size than normal. Thus wee mutants are presumed to be deficient in a gene product that normally inhibits passage through a checkpoint, so that they pass through unimpeded and hence are too "wee."

The *cdc2* gene encodes a cyclin-dependent protein kinase, Cdc2, which in combination with B-type cyclins (a Clb) forms MPF, which induces entry into *M* phase. The activity of the Cdc2-cyclin B dimer (MPF) is also controlled by phosphorylation at two sites, tyrosine-15 and threonine-167. (Tyrosine and threonine are two of the twenty amino acids that are strung together to form a protein molecule. The number 15 or 167 denotes the location on the protein sequence of Cdc2.) These two sites define four different phosphorylation states. MPF is active when it is phosphorylated at threonine-167 only. The other three phosphorylation states are inactive. Active MPF initiates a chain of reactions that controls mitotic events.

Movement between different phosphorylation states is mediated by other enzymes. The Wee1 enzyme, the gene product of the normal *wee* gene, inactivates MPF by adding a phosphate to the tyrosine-15 site. Cdc25 reverses this by dephosphorylating the tyrosine-15 site.

We can now describe a piece of the cell cycle regulating mitosis in oocytes. A schematic diagram of this regulation is shown in Fig. 13.9. Cyclin B is synthesized from amino acids and binds with free Cdc2 to form an inactive MPF dimer. The dimer is quickly phosphorylated on threonine-167 (by a protein kinase called CAK) and dephosphorylated at the same site by an unknown enzyme. Simultaneously, Wee1 can phosphorylate the dimer at the tyrosine-15 site, rendering it inactive, and Cdc25 can dephosphorylate the same site. Mitosis is initiated when a sufficient quantity of MPF is active.

All of this becomes dynamically interesting when we include the known feedback loops. Active MPF phosphorylates many proteins, including Cdc25 and Wee1. Phosphorylated Cdc25 is active, meaning it is able to dephosphorylate the MPF at tyrosine-15, and phosphorylated Wee1 is inactive, so that it is unable to phosphorylate MPF at

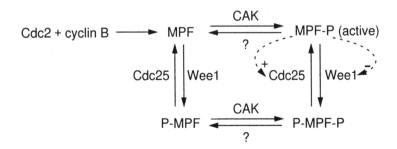

**Figure 13.9** Schematic diagram of the regulatory pathway of MPF.

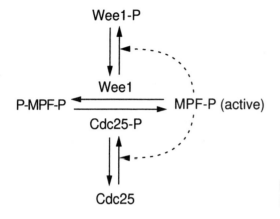

**Figure 13.10** Schematic diagram of the feedback control of Cdc25 and Wee1.

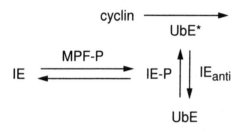

**Figure 13.11** Schematic diagram of the regulation of cyclin degradation.

tyrosine-15. Both of these actions by active MPF, the phosphorylation of Cdc25 and Wee1, act as positive feedback to allow autocatalytic production of active MPF. These feedbacks are depicted in Fig. 13.10.

There is also a negative feedback loop involving the degradation of cyclin. Cyclin is degraded after it is first "labeled" for destruction by a *ubiquitin-conjugating enzyme* (UbE). UbE is activated by active MPF, although with a substantial delay, suggesting that the activation of UbE is at the end of some chain reaction involving one or more intermediate enzymes, none of which have been identified. For this model we assume that there is one such intermediate enzyme, denoted by IE. These assumed reactions are depicted in Fig. 13.11.

We now have a complete verbal description of a model for the initiation of mitosis. In summary, as cyclin is produced, it combines with Cdc2 to form MPF. MPF is quickly phosphorylated to its active form. Active MPF turns on its own autocatalytic production by activating Cdc25 and inactivating Wee1. By activating UbE, which activates the destruction of cyclin, active MPF also turns on its own destruction, but with a delay, thus completing the cycle.

Of course, this verbal description is incomplete, because there are many other features of *M* phase control that have not been included. It also does not follow from verbal arguments alone that this model actually controls mitosis in a manner consistent with experimental observations. To check that this model is indeed sufficient to explain some features of the cell cycle, it is necessary to present it in quantitative form (Novak and Tyson, 1993).

The chemical species that must be tracked include the Cdc2 and cyclin monomers, the dimer MPF in its active and inactive states, as well as the four regulatory enzymes Wee1, Cdc25, IE, and UbE in their phosphorylated and unphosphorylated states.

First, cyclin (with concentration $y$) is produced at a steady rate and is degraded or combines with Cdc2 (with concentration $c$) to form the MPF dimer ($r$):

$$\frac{dy}{dt} = k_1[A] - k_2 y - k_3 Yc. \tag{13.32}$$

The MPF dimer can be in one of four phosphorylation states, with phosphate at tyrosine-15 ($s$), at threonine-167 (concentration $m$), at both sites (concentration $n$), or at none (concentration $r$). The movement among these states is regulated by the enzymes Wee1, Cdc25, CAK, and one unknown enzyme ("?"). Thus,

$$\frac{dr}{dt} = -(k_2 + k_{CAK} + k_{wee})r + k_3 yc + k_? n + k_{25} s, \tag{13.33}$$

$$\frac{ds}{dt} = -(k_2 + k_{CAK} + k_{25})s + k_? n + k_{wee} r, \tag{13.34}$$

$$\frac{dm}{dt} = -(k_2 + k_? + k_{wee})m + k_{CAK} r + k_{25} n, \tag{13.35}$$

$$\frac{dn}{dt} = -(k_2 + k_? + k_{25})n + k_{wee} m + k_{CAK} s. \tag{13.36}$$

Notice that in the above equations, cyclin degradation at rate $k_2$ is permitted for free cyclin as well as for cyclin that is combined with Cdc2. If cyclin degrades directly from a phosphorylated dimer, we assume that the phosphate is also immediately removed to form free Cdc2. Thus,

$$\frac{dc}{dt} = k_2(r + s + n + m) - k_3 cy. \tag{13.37}$$

These six equations would form a closed system were it not for feedback. Notice that the last equation (13.37) is redundant, since $m + r + s + n + c = $ constant. The feedback shows up in the nonlinear dependence of rate constants on the enzymes Cdc25, Wee1, IE, and UbE. This we express as

$$k_{25} = V'_{25}[\text{Cdc25}] + V''_{25}[\text{Cdc25\_P}], \tag{13.38}$$

$$k_{wee} = V'_{wee}[\text{Wee1\_P}] + V''_{wee}[\text{Wee1}], \tag{13.39}$$

$$k_2 = V'_2[\text{UbE}] + V''_2[\text{UbE}^*]. \tag{13.40}$$

In addition, the active states of Cdc25, Wee1, IE, and UbE are determined by Michaelis–Menten rate laws of the form

$$\frac{d[\text{Cdc25\_P}]}{dt} = \frac{k_a m[\text{Cdc25}]}{K_a + [\text{Cdc25}]} - \frac{k_b[\text{PPase}][\text{Cdc25\_P}]}{K_b + [\text{Cdc25\_P}]}, \tag{13.41}$$

$$\frac{d[\text{Wee1\_P}]}{dt} = \frac{k_e m[\text{Wee1}]}{K_e + [\text{Wee1}]} - \frac{k_f[\text{PPase}][\text{Wee1\_P}]}{K_f + [\text{Wee1\_P}]}, \tag{13.42}$$

$$\frac{d[\text{IE\_P}]}{dt} = \frac{k_g m[\text{IE}]}{K_g + [\text{IE}]} - \frac{k_h[\text{PPase}][\text{IE\_P}]}{K_h + [\text{IE\_P}]}, \tag{13.43}$$

$$\frac{d[\text{UbE}^*]}{dt} = \frac{k_c[\text{IE\_P}][\text{UbE}]}{K_c + [\text{UbE}]} - \frac{k_d[\text{IE}_{\text{anti}}][\text{UbE}^*]}{K_d + [\text{UbE}^*]}. \tag{13.44}$$

The quantities with _P attached correspond to phosphorylated forms of the enzyme quantity. The total amounts of each of these enzymes are assumed to be constant, so equations for the inactive forms are not necessary. PPase denotes a phosphatase that dephophorylates Cdc25_P.

This forms a complete model with nine differential equations having 8 Michaelis–Menten parameters and 18 rate constants. There are two ways to gain an understanding of the behavior of this system of differential equations: by numerical simulation of the full system of equations using reasonable parameter values, or by approximating the system to a smaller system of equations and studying the simpler system by analytical means.

The parameter values used by Novak and Tyson to simulate *Xenopus* oocyte extracts are shown in Tables 13.1 and 13.2.

While numerical simulation of these nine differential equations is not difficult, to gain an understanding of the basic behavior of the model it is convenient to make some simplifying assumptions. Suppose $k_{\text{CAK}}$ is large and $k_?$ is small, as experiments suggest. Then the phosphorylation of Cdc2 on threonine-167 occurs immediately after formation of the MPF dimer. This allows us to ignore the quantities $r$ and $s$. Next we assume that the activation and inactivation of the regulatory enzymes is rapid and is in quasi-steady state, depending on the level of active MPF. This leaves only three

**Table 13.1**  Michaelis–Menten constants for the cell-cycle model of Novak and Tyson (1993).

| | | | | | |
|---|---|---|---|---|---|
| $K_a/[\text{Cdc25}_{\text{total}}]$ | = | 0.1 | $K_b/[\text{Cdc25}_{\text{total}}]$ | = | 0.1 |
| $K_c/[\text{UbE}_{\text{total}}]$ | = | 0.01 | $K_d/[\text{UbE}_{\text{total}}]$ | = | 0.01 |
| $K_e/[\text{Wee1}_{\text{total}}]$ | = | 0.3 | $K_f/[\text{Wee1}_{\text{total}}]$ | = | 0.3 |
| $K_g/[\text{IE}_{\text{total}}]$ | = | 0.01 | $K_h/[\text{IE}_{\text{total}}]$ | = | 0.01 |

**Table 13.2**  Rate constants for the cell-cycle model of Novak and Tyson (1993).

| | | | | | |
|---|---|---|---|---|---|
| $k_1[A]/[\text{Cdc2}_{\text{total}}]$ | = | 0.01 | $k_3[\text{Cdc2}_{\text{total}}]$ | = | 1.0 |
| $V_2'[\text{UbE}_{\text{total}}]$ | = | 0.015 (0.03) | $V_2''[\text{UbE}_{\text{total}}]$ | = | 1.0 |
| $V_{25}'[\text{Cdc25}_{\text{total}}]$ | = | 0.1 | $V_{25}''[\text{Cdc25}_{\text{total}}]$ | = | 2.0 |
| $V_{\text{wee}}'[\text{Wee1}_{\text{total}}]$ | = | 0.1 | $V_{\text{wee}}''[\text{Wee1}_{\text{total}}]$ | = | 1.0 |
| $k_{\text{CAK}}$ | = | 0.25 | $k_?$ | = | 0.25 |
| $k_a[\text{Cdc2}_{\text{total}}]/[\text{Cdc25}_{\text{total}}]$ | = | 1.0 | $k_b[\text{PPase}]/[\text{Cdc25}_{\text{total}}]$ | = | 0.125 |
| $k_c[\text{IE}_{\text{total}}]/[\text{UbE}_{\text{total}}]$ | = | 0.1 | $k_d[\text{IE}_{\text{anti}}]/[\text{UbE}_{\text{total}}]$ | = | 0.095 |
| $k_e[\text{Cdc2}_{\text{total}}]/[\text{Wee1}_{\text{total}}]$ | = | 1.33 | $k_f[\text{PPase}]/[\text{Wee1}_{\text{total}}]$ | = | 0.1 |
| $k_g[\text{Cdc2}_{\text{total}}]/[\text{IE}_{\text{total}}]$ | = | 0.65 | $k_h[\text{PPase}]/[\text{IE}_{\text{total}}]$ | = | 0.087 |

equations for the three unknowns $y$ (free cyclin), $m$ (active MPF), and Cdc2 monomer $(q)$ as follows

$$\frac{dy}{dt} = k_1 - F_2 y - k_3 y q, \tag{13.45}$$

$$\frac{dm}{dt} = k_3 y q - F_2 m + F_{25} n - k_{\text{wee}} m, \tag{13.46}$$

$$\frac{dq}{dt} = -k_3 y q + F_2(m + n), \tag{13.47}$$

where $m + n + q = c$ is the total Cdc2. It follows that the total cyclin $l = y + m + n$ satisfies the differential equation

$$\frac{dl}{dt} = k_1 - F_2 l. \tag{13.48}$$

In addition, the rates $F_2$ and $F_{25}$ depend upon active MPF through

$$F_2 = k_2 + k_2' m^2, \tag{13.49}$$

$$F_{25} = k_{25} + k_{25}' m^2. \tag{13.50}$$

Any three of these four equations describes the behavior of the system. However, in the limit that $k_3$ is large compared to other rate constants, the system can be reduced to a two-variable system for which phase-plane analysis is applicable. If we set $v = k_3 y$, then

$$\frac{dv}{dt} + F_2 v = k_3(k_1 - qv), \tag{13.51}$$

so that $qv = k_1$ to leading order. If $k_1$ is small, then $y$ is small as well, so that (13.46) becomes

$$\frac{dm}{dt} = k_1 - F_2 m + F_{25}(l - m) - k_{\text{wee}} m. \tag{13.52}$$

The two equations (13.48) and (13.52) form a closed system that can be studied using the phase portrait in the $(l, m)$ plane. In this approximation, $q = c - l$. The nullclines are described by the equations

$$\frac{dl}{dt} = 0 : l = \frac{k_1}{F_2(m)} \tag{13.53}$$

and

$$\frac{dm}{dt} = 0 : l = \frac{k_{\text{wee}} m + F_2(m) m - k_1}{F_{25}(m)} + m, \tag{13.54}$$

which are shown plotted in Fig. 13.12.

For these parameter values, there is a unique unstable steady-state solution surrounded by a limit cycle oscillation. By adjusting parameters one can have a stable fixed point on the leftmost branch of the "n"-shaped curve corresponding to the $G_2$ checkpoint, or one can have a stable fixed point on the rightmost branch, yielding an

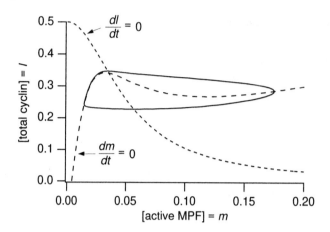

**Figure 13.12** Solution trajectory and nullclines for the system of equations (13.48), (13.52). Parameter values are $k_1 = 0.004$, $k_{wee} = 0.9$, $k_2 = 0.008$, $k_2' = 3.0$, $k_{25} = 0.03$, $k_{25}' = 50.0$.

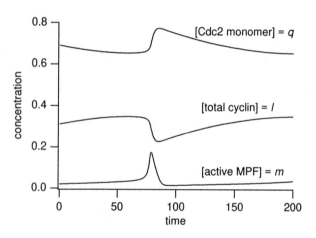

**Figure 13.13** Solution for the system of equations (13.46), (13.47), (13.48), plotted as a function of time. Parameter values are the same as in Fig. 13.12.

$M$ phase checkpoint. Here a possible control parameter is $k_{wee}$, since increasing $k_{wee}$ causes the $m$ nullcline to increase. Thus, increasing $k_{wee}$ creates a $G_2$ checkpoint on the leftmost branch of the "n"-shaped curve.

## 13.2.4  Conclusion

We are a long way from a complete understanding of the biochemistry of the cell cycle. Here we have seen a few of the main players, although there are major portions of the cell cycle (the $M$ phase checkpoint, for example) that are still a mystery. With the modern development of biochemistry, many similar stories relating to regulation of cell function are being unfolded. It is likely that in the coming years many of the details of this story will change and many new details will be added. However, the basic modeling process of turning a verbal description into a mathematical model and the consequent analysis of the model will certainly remain an important tool to aid our understanding of these complicated and extremely important processes. Furthermore, as the details of the stories become more complicated (as they are certain to do), mathematical analysis will become even more important in helping us understand how these processes work.

## 13.3 Exercises

1. The genes for the enzymes necessary for the production of the amino acid tryptophan are sequentially arranged in the DNA of *E. coli*, following a single operator site. If the operator site is occupied (bound by a repressor), then production of mRNA is blocked. However, the repressor protein binds to the operator site only when it is activated by binding with two molecules of tryptophan.

    Write a model for the production of tryptophan that incorporates this control mechanism. How would you describe the effect of this control mechanism (positive feedback, or negative feedback)?

2. Suppose that the production of an enzyme is turned on by $m$ molecules of the enzyme according to

$$\mathrm{G} + m\mathrm{P} \underset{k_-}{\overset{k_+}{\rightleftharpoons}} \mathrm{X},$$

    where G is the inactive state of the gene and X is the active state of the gene. Suppose that mRNA is produced when the gene is in the active state and the enzyme is produced by mRNA and is degraded at some linear rate. Find a system of differential equations governing the behavior of mRNA and enzyme. Give a phase portrait analysis of this system and show that it has a "switch-like" behavior.

3. Develop a mathematical model for the breakdown of lactose into glucose that incorporates the interplay between the two. Are oscillations in the uptake and utilization of lactose possible? How well does your model agree with the data shown in Fig. 13.3?

    Hint: Include an additional equation for glucose concentration and assume that glucose interferes with the production of mRNA. If this does not suffice, how might you modify the model to better reflect the biochemistry and to produce oscillations?

4. Goldbeter (1996) has developed and studied a "minimal" cascade model for the mitotic oscillator. The model assumes that cyclin B ($C$) is synthesized at a constant rate and activates Cdc25 kinase. The activated Cdc25 kinase in turn activates Cdc2 kinase ($M$), and the activated Cdc2 kinase is inactivated by the kinase Wee1. There is also a cyclin protease $X$ that is activated by Cdc2 kinase and inactivated by an additional phosphatase. The differential equations for this reaction scheme (shown in Fig. 13.14) are

$$\frac{dc}{dt} = v_i - v_d x \frac{c}{K_d + c} - k_d c, \tag{13.55}$$

$$\frac{dm}{dt} = V_1 \frac{1-m}{K_1 + 1 - m} - V_2 \frac{m}{K_2 + m}, \tag{13.56}$$

$$\frac{dx}{dt} = V_3 \frac{1-x}{K_3 + 1 - X} - V_4 \frac{x}{K_4 + X}, \tag{13.57}$$

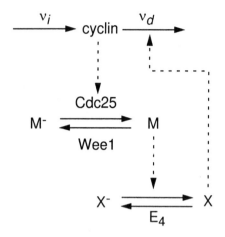

**Figure 13.14**  Diagram for the minimal cascade mitosis model of Goldbeter (1996).

**Table 13.3**  Parameter values for the Goldbeter minimal mitotic cycle model.

| $K_1$ | = | 0.1 | $V_{M1}$ | = | 0.5 min$^{-1}$ |
|---|---|---|---|---|---|
| $K_2$ | = | 0.1 | $V_2$ | = | 0.167 min$^{-1}$ |
| $K_3$ | = | 0.1 | $V_{M3}$ | = | 0.2 min$^{-1}$ |
| $K_4$ | = | 0.1 | $V_4$ | = | 0.1 min$^{-1}$ |
| $v_i$ | = | 0.023 $\mu$M/min | $v_d$ | = | 0.1 $\mu$M/min |
| $K_c$ | = | 0.3 $\mu$M | $K_d$ | = | 0.02 $\mu$M |
| $k_d$ | = | 3.33$\times$10$^{-3}$ min$^{-1}$ | | | |

and

$$V_1 = V_{M1} \frac{c}{K_c + c}, \qquad V_3 = V_{M3}m, \tag{13.58}$$

where $c$ denotes cyclin concentration, and $m$ and $x$ denote the fraction of active Cdc2 kinase and the fraction of active cyclin protease, respectively. The parameters $v_i$ and $v_d$ are the constant rate of cyclin synthesis and maximum rate of cyclin degradation by protease X, achieved at $x = 1$. $K_d$ and $K_c$ denote Michaelis constants for cyclin degradation and cyclin activation, while $k_d$ is the rate of nonspecific degradation of cyclin. The remaining parameters $V_i$ and $K_i, i = 1, \ldots, 4$, are the effective maximum rate and Michaelis constants for each of the four enzymes Cdc25, Wee1, Cdc2, and the protease phosphatase ($E_4$), respectively. Typical parameter values are shown in Table 13.3.
Numerically simulate this system of equations to show that there is a stable limit cycle oscillation for this model. What is the period of oscillation?

# PART II

# SYSTEMS PHYSIOLOGY

# CHAPTER 14

# Cardiac Rhythmicity

We have seen in previous chapters that cardiac cells are excitable, and some, such as sinoatrial (SA) nodal cells, are autonomous oscillators. We have also seen that these cells are coupled and can support propagated waves. It remains to understand how this activity is coordinated to produce a regular heartbeat, or how this regular activity may fail.

The study of cardiac electrical activity often takes on a quantitative flavor, in which one begins with one or more of the detailed cellular models discussed in Chapter 4 and builds a large-scale numerical simulation that incorporates a variety of features of the cardiac conduction system. In this chapter we take an entirely different approach, devoting our attention solely to qualitative behavior, using the simplest possible models that give insight into the important phenomena.

## 14.1 The Electrocardiogram

### 14.1.1 The Scalar ECG

One of the oldest and most important tools for evaluating the status of the heart and the cardiac conduction system is the *electrocardiogram* (ECG). It has been known since 1877, when the first ECG recording was made, that the action potential of the heart generates an electrical potential field that can be measured on the body surface. When an action potential is spreading through cardiac tissue, there is a wave front surface across which the membrane potential experiences a sharp increase. Along the same wave front, the extracellular potential experiences a sharp decrease. From a distance, this sharp decrease in potential looks like a Heaviside jump in potential. This rapid change in extracellular potential results from a current source (or sink) because ions are moving into or out of the extracellular space as transmembrane currents.

The body is a *volume conductor*, so when there is a current source somewhere in the body, such as during action potential spread, currents spread throughout the body. Although the corresponding voltage potential is quite weak, no larger than 4 mV, potential differences can be measured between any two points on the body using a sufficiently sensitive voltmeter.

Potential differences are observed whenever the current sources are sufficiently strong. There are three such events. When the action potential is spreading across the atria, there is a measurable signal, called the *P wave*. When the action potential is propagating through the wall of the ventricles, there is the largest of all deflections, called the *QRS complex*. Finally, the recovery of ventricular tissue is seen on the ECG as the *T wave*. The recovery of the atria is too weak to be detected on the ECG. Similarly, SA nodal firing, AV nodal conduction, and Purkinje network propagation are not detected on the normal body surface ECG because they do not involve sufficient muscle mass or generate enough extracellular current.

In Fig. 14.1 is shown a sketch of a typical single electrical ECG event, and a continuous recording is shown in Fig. 14.2a. In hospitals, ECG recordings are made routinely using oscilloscopes, or, if a permanent record is required, on a continuous roll of paper. The paper speed is standardized at 25 mm per second, with a vertical scale of 1 mV per cm, and the paper is marked with a lined grid of 1 mm and darkened lines with 0.5 cm spacing.

The most important use of the single-lead ECG is to detect abnormalities of rhythm. For example, a continuous oscillatory P wave pattern suggests *atrial flutter* (Fig. 14.2b) or *atrial fibrillation* (Fig. 14.2c). A rapid repetition of QRS complexes is *ventricular tachycardia* (Fig. 14.2d), and a highly irregular pattern of ventricular activation is called *ventricular fibrillation* (Fig. 14.2e). The normal appearance of P waves with a few skipped QRS complexes implicates a conduction failure in the vicinity of the AV node. Broadening of the QRS complex suggests that propagation is slower than normal, possibly because of conduction failure in the Purkinje network (Fig. 14.3). Spontaneously appearing extra deflections correspond to *extrasystoles*, arising from sources other than the SA or AV nodes.

## 14.1.2  The Vector ECG

There is much more information contained in the ECG than is available from a single lead. Some of this information can be extracted from the *vector electrocardiogram*. The mathematical basis for the vector ECG comes from an understanding of the nature of a volume conductor. The human body is an inhomogeneous volume conductor, meaning that it is composed of electrically conductive material. If we assume that biological tissue is ohmic, there is a linear relationship between current and potential

$$I = -\sigma \nabla \phi. \tag{14.1}$$

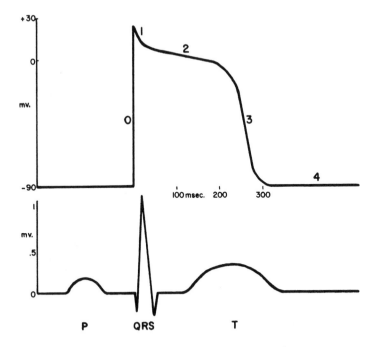

**Figure 14.1** Cellular transmembrane potential and electrocardiogram. The upper tracing represents the transmembrane potential of a single ventricular myocyte and the lower tracing shows the body surface potential during the same electrical event. The numbers on the upper tracing designate phases in the action potential cycle: 0: the upstroke, 1: the brief spike, 2: the plateau, 3: the rapid recovery, 4: resting potential. (Rushmer, 1976, Fig. 8–4, p. 286.)

The conductivity tensor $\sigma$ is inhomogeneous, because it is different for bone, lung, blood, etc., and it is anisotropic, because of muscle fiber striation, for example. Obviously, current is conserved, so that

$$\nabla \cdot I = -\nabla \cdot (\sigma \nabla \phi) = S, \tag{14.2}$$

where $S$ represents all current sources.

The most significant current source in the human body is the spreading action potential wave front in the heart. The spreading cardiac action potential is well approximated as a surface of current dipoles. The rapid increase in membrane potential (of about 100 mV) translates into an extracellular decrease of about 40 mV that extends spatially over a distance (the wave front thickness) of about 0.5 mm. If the exact location and strength of this dipole surface and the conductivity tensor for the entire body were known, then we could (in principle) solve the Poisson equation (14.2) to find the body surface potential at all times during the cardiac cycle. This problem is unsolved, and is known as the *forward problem of electrocardiography*.

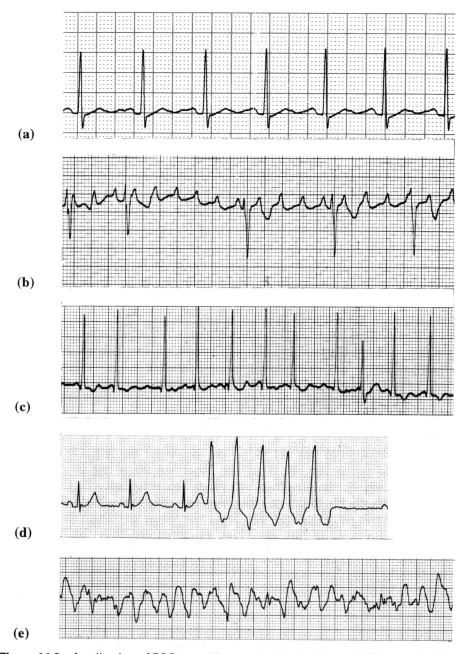

**(a)**

**(b)**

**(c)**

**(d)**

**(e)**

**Figure 14.2**   A collection of ECG recordings, including (a) Normal ECG recording (lead II) from a sedated 18-year-old male (JPK's son). (b) Atrial flutter showing rapid, periodic P waves, only some of which lead to QRS complexes. (Rushmer, 1976, Fig. 8-29, p. 316.) (c) Atrial fibrillation showing rapid, nonperiodic atrial activity and irregular QRS complexes. (Rushmer, 1976, Fig. 8-28, p. 315.) (d) (Monomorphic) ventricular tachycardia in which ventricular activity is rapid and regular (nearly periodic). (Davis, Holtz, and Davis, 1985, Fig. 17-24, p. 346.) (e) Ventricular fibrillation in which ventricular activity is rapid and irregular. (Rushmer, 1976, Fig. 8-30, p. 317.)

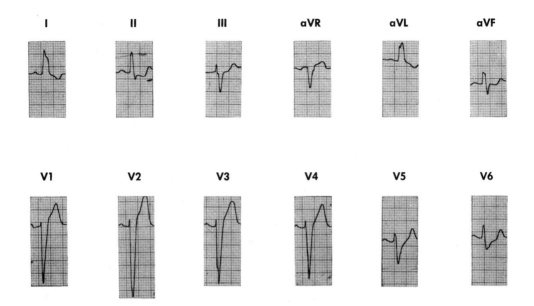

**Figure 14.3** ECG from the twelve standard leads, showing left bundle branch block (LBBB), diagnosed as such because of the lengthened QRS complex (0.12 ms), a splitting of the QRS complex in leads $V_1$ through $V_4$ into two signals, and a leftward deflection of the heart vector, indicated, for example, by the amplitude shift in lead $V_6$. (Rushmer, 1976, Fig 8-46, p. 338.)

What we would really like to know is the operator, say $T$, called a *transfer function*, that solves (14.2) and yields the *body surface potential* $\phi_B$, denoted by

$$\phi_B(t) = T \cdot S(t). \tag{14.3}$$

Even more useful, if the transfer function $T$ were known, one could determine the sources by inverting the forward problem

$$S(t) = T^{-1} \cdot \phi_B(t). \tag{14.4}$$

This problem, known as the *inverse problem of electrocardiography*, is even harder to solve than the forward problem, because it is a numerically unstable mathematical problem.

Since these problems are yet unsolved, we do well to make some simplifications. Our first simplification is to view the action potential upstroke surface as a single current dipole, known as the *heart dipole vector*. We define the heart dipole vector as

$$\mathbf{H}(t) = \int_V \mathbf{J} dV, \tag{14.5}$$

where $\mathbf{J}$ represents the dipole density at each point of the heart, and $V$ is the heart volume. The heart dipole vector is assumed to be located at a fixed point in space, changing only in orientation and strength as a function of time.

Next we assume that the volume conductor is homogeneous and infinite with unit conductance. Then, from standard potential theory (see Exercise 1), at any point $x$ in space,

$$\phi(x,t) = \frac{\mathbf{H}(t) \cdot x}{4\pi|x|^3},\tag{14.6}$$

where the dipole is assumed to be located at the origin. Thus, at each point on the body surface,

$$\phi_B(x,t) = l_x \cdot \mathbf{H}(t),\tag{14.7}$$

where $l_x$ is a vector, called the *lead vector*, associated with the electrode lead at position $x$. Of course, for a real person, the lead vector is not exactly $\frac{x}{4\pi|x|^3}$, and some other method must be used to determine $l_x$. However, (14.7) suggests that we can think of the body surface potential as the dot product of some vector $l_x$ with the heart vector $\mathbf{H}(t)$, and that $l_x$ has more or less the same orientation as a vector from the heart to the point on the body where the recording is made.

Since $\mathbf{H}$ is a three-dimensional vector, if we have three leads with linearly independent lead vectors, then three copies of (14.7) yields a matrix equation that can be inverted to find $\mathbf{H}(t)$ uniquely. In other words, if our goal is to determine $\mathbf{H}(t)$, then knowledge of the full transfer function is not necessary. In fact, additional measurements from other leads should give redundant information.

Of course, the information from additional leads is not truly redundant, but it is nearly so. Estimates are that a good three-lead system can account for 85% of the information concerning the nature of the dipole sources. Discrepancies occur because the sources are not exactly consolidated into a single dipole, or because the lead vectors are not known with great accuracy, and so on. However, for clinical purposes, the information gleaned from this simple approximation is remarkably useful and accurate.

The next simplification is to standardize the position of the body-surface recordings and to determine the associated lead vectors. Then, with experience, a clinician can recognize features of the heart vector by looking at recordings of the potential at the leads. Or sophisticated (and expensive) equipment can be built that inverts the lead vector matrix and displays the heart vector on a CRT display device.

Cardiologists have settled on 12 standard leads. The first three were established by Einthoven, the "father of electrocardiography" (1860–1927, inventor of the string galvanometer in 1905, 1924 Nobel Prize in physiology) and are still used today. These are the left arm (LA), the right arm (RA), and the left leg (LL). One cannot measure absolute potentials, but only potential differences. There are three ways to measure potential differences with these three leads, namely,

$$V_{\mathrm{I}} = \phi_{\mathrm{LA}} - \phi_{\mathrm{RA}},\tag{14.8}$$

$$V_{\mathrm{II}} = \phi_{\mathrm{LL}} - \phi_{\mathrm{RA}},\tag{14.9}$$

$$V_{\mathrm{III}} = \phi_{\mathrm{LL}} - \phi_{\mathrm{LA}},\tag{14.10}$$

and of course, since the potential drop around any closed loop is zero,

$$V_I + V_{III} = V_{II}. \tag{14.11}$$

With these three differences, there are three lead vectors associated with the orientation of the leads, and the potential difference is the amplitude of the projection of the heart vector **H** onto the corresponding lead vector. Thus, $L_j = l_j \cdot \mathbf{H}$, and $V_j = |L_j|$ for $j = $ I, II, III.

Einthoven hypothesized that the lead vectors associated with readings $V_I, V_{II}, V_{III}$ form an equilateral triangle in the vertical, frontal plane of the body, given by the unit vectors (ignoring an amplitude scale factor) $l_I = (1, 0, 0)$, and $l_{II} = (\frac{1}{2}, \frac{1}{2}\sqrt{3}, 0)$. The Einthoven triangle is shown in Fig. 14.4. Here the unit coordinate vector $(1, 0, 0)$ is horizontal from right arm to left arm, $(0, 1, 0)$ is vertical pointing downward, and the vector $(0, 0, 1)$ is the third coordinate in a right-handed system, pointing in the posterior direction, from the front to back of the chest. Associated with the frontal plane is a polar coordinate system, centered at the heart, with angle $\theta = 0$ along the $x$ axis, and $\theta = 90°$ vertically downward along the positive $y$ axis.

Of course, the lead vectors of Einthoven are not very accurate. Experiments to measure the lead vectors in a model of the human torso filled with electrolytes produced measured lead vectors $l_I = (0.923, -0.298, 0.241)$, and $l_{II} = (0.202, 0.972, -0.121)$ (Burger and van Milaan, 1948), which are not in the frontal plane. These lead vectors are known as the *Burger triangle*.

It is fairly easy to glean information about the direction of the heart vector by recognizing the information that is contained in (14.7). The vector ECG is actually a time-varying vector loop (shown in front, top, and side views in Fig. 14.6), and deducing time-dependent information is best done with an oscilloscope. However, one can estimate the mean direction of the vector by estimating the mean amplitude of a wave and then using (14.7) to estimate the mean heart vector. The mean (or time average) of

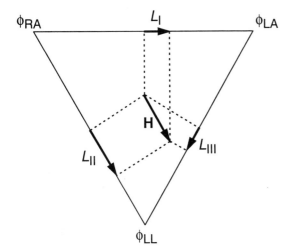

**Figure 14.4** The Einthoven triangle showing a typical heart vector **H** and associated lead vectors $L_I, L_{II}$ and $L_{III}$. Because the body is approximately planar, the lead vectors are assumed to be in the frontal plane.

the QRS complex is approximately proportional to the sum of the (positive) maximum and the (negative) minimum.

Since the lead voltage is a dot product of two vectors, a change in mean amplitude of a particular wave suggests either a change in amplitude of the heart vector or a change in direction of the heart vector. For example, the normal QRS and T wave mean dipoles are oriented about $45°$ below horizontal to the left (see Exercise 5). This is close to orthogonal to the lead vector $l_{III}$, and more or less aligned with lead vector $l_{II}$. Thus, on a normal ECG, we expect the mean amplitude of a QRS to be small in lead III, large in lead II, and intermediate to these two in lead I. Shifts in these relative amplitudes suggest a shift in the orientation of the heart dipole. For example, an increase in the relative amplitude of the potential difference at lead III and a decrease in amplitude at lead II suggests a shift of the heart vector to the right, away from the left, suggesting a malfunction of the conduction in the left heart.

Although two orthogonal lead vectors suffice to determine the orientation of the heart vector in the vertical plane, for ease of interpretation it is helpful to have more leads. For this reason, there are three additional leads on the frontal plane that are used clinically. To create these leads one connects two of the three Einthoven leads to a central point with 5000 $\Omega$ resistors to create a single terminal that is relatively indifferent to changes in potential and then takes the difference between this central potential and the remaining electrode of the Einthoven triangle. These measurements are denoted by aVR, aVL, or aVF, when the third unipolar lead is the right arm, the left arm, or the left foot, respectively. The initial "a" is used to denote an *augmented* unipolar limb lead.

For standard cardiographic interpretation the lead vectors for leads I, aVR$^-$, II, aVF, III, and aVL$^-$ are assumed to divide the frontal plane into equal $30°$ sectors. For example, $l_I$ is horizontal, $l_{aVR^-}$ is declined at $30°$, while $l_{aVF}$ is vertical, etc. The superscript for aVR$^-$ denotes the negative direction of the lead vector $l_{aVR}$ (Fig. 14.5).

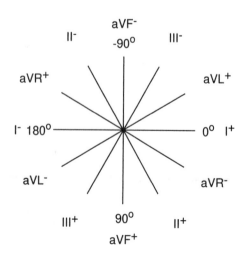

**Figure 14.5** The standard six leads for the electrocardiogram (and their negatives).

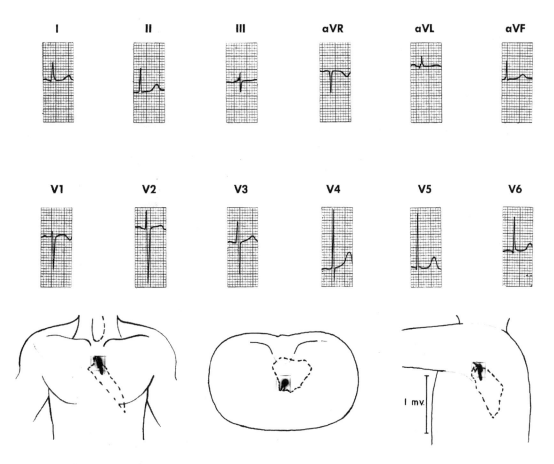

**Figure 14.6** Normal ECG and VCG recording from the standard twelve leads in a nine-year old girl. (Rushmer, 1976, Fig. 8-33, p. 320, originally from Guneroth, 1965.)

With these six leads, vector interpretation of the frontal plane orientation of the heart dipole is fast. One looks for the leads with the largest and smallest deflections, and surmises that the lead vector with largest mean amplitude is most parallel to the heart dipole, and the lead vector with the smallest mean deflection is nearly orthogonal to the heart dipole. Thus in the normal heart situations, readings at leads II and aVR should be the largest in mean amplitude, with positive deflection at lead II, and negative deflection at lead aVR, while the mean deflections from leads III and aVL should be the smallest, being the closest to orthogonal to the normal heart dipole (Fig. 14.6). Deviations from this suggest conduction abnormalities.

Six additional leads have been established to obtain the orientation of the heart dipole vector in a horizontal plane. For these leads, the three leads of Einthoven are connected with three 5000 Ω resistors to form a "zero reference," called the *central terminal of Wilson*. This is compared to a unipolar electrode reading taken from six different locations on the chest. These are denoted by $V_1, V_2, \ldots, V_6$ and are located on

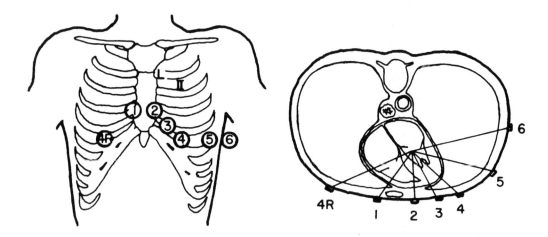

**Figure 14.7** Frontal and horizontal cross-sectional views of the thorax in relation to the V-lead positions of Wilson. (Rushmer, 1976, Fig. 8-10, p. 294.)

the right side of the sternum ($V_1$), the left side of the sternum ($V_2$) between the third and fourth ribs, and proceeding around the left chest following just below the fourth rib, ending on the side of the chest directly under the armpit ($V_6$) (Fig. 14.7).

While a detailed discussion of interpretation of a vector ECG is beyond the scope of this text, there are several features of cardiac conduction that are easy to recognize. Notice from Fig. 14.6 that the normal T wave and the normal QRS complex deflect in the same direction on leads I, II, and aVR (up on I and II, down on aVR). However, the QRS complex corresponds to the upstroke and the T wave to the downstroke of the action potential, so it must be that the activation (upstroke) and recovery (downstroke) wave fronts propagate in opposite directions. Said another way, the most recently activated tissue is the first to recover. The reason for the retrograde propagation of the wave of recovery is not fully understood. Second, an inverted wave (i.e., inverted from what is normal) implies that either the wave is propagating in the retrograde direction, or more typically with novice medical technicians, that the leads have been inadvertently reversed (see Exercise 4a).

The amplitude of the QRS complex reflects the amount of muscle mass involved in propagation. Thus, if the QRS amplitude is extraordinarily large, it suggests *ventricular hypertrophy*. If the ECG vector is leftward from normal, it suggests left ventricular hypertrophy (Fig. 14.8), while a rightward orientation suggests right ventricular hypertropy (Fig. 14.9). On the other hand if an amplitude decrease is accompanied by a rightward change in orientation, a diagnosis of *myocardial infarction* in the left ventricle is suggested, while a leftward orientation with decreased amplitude suggests a myocardial infarction of the right ventricle, as the heart vector is deflected away from the location of the infarction (see Exercises 6 and 7).

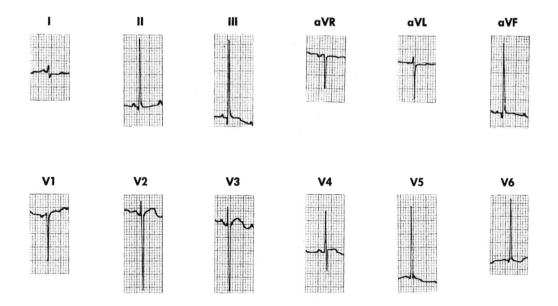

**Figure 14.8** Twelve-lead ECG recording for severe left ventricular hypertrophy, particularly noticeable in leads III and aVL. (Rushmer, 1976, Fig. 8-24, p. 331, originally from Guneroth, 1965.)

## 14.2 Pacemakers

### 14.2.1 Pacemaker Synchrony

The sinoatrial (SA) node is a clump of self-oscillatory cells located on the atrium near the superior vena cava. These cells fire regularly, initiating an action potential that propagates throughout the atrium, eventually terminating at the atrio-ventricular septum, or conducting into the AV node. SA nodal cells are not identical, but nevertheless fire at the same frequency. They are not synchronous in their firing (i.e., firing all at once), but they are *phase locked*, meaning that during each cycle, each cell fires once and there is a regular pattern to the firing of the cells. The variation of cellular properties in the SA node has two dominant features. There are gradual spatial gradients of the period of the oscillators and random deviations of individual cells from this average gradient.

Three questions concerning the SA pacemaker are of interest. First, since the individual cells all have different natural frequencies, what determines the frequency of the collective SA node? Second, what determines the details of the firing sequence, specifically, the location of the cell that fires earliest in the cycle and the subsequent firing order of the cells in the node? One might anticipate that the leader of the pack is the cell with the highest intrinsic frequency, but as we will see, this is not the case. Third, under what conditions does the SA node lose its ability to initiate the heartbeat (called *sinus node dysfunction*), and is it possible for other regions of (abnormal) oscillatory

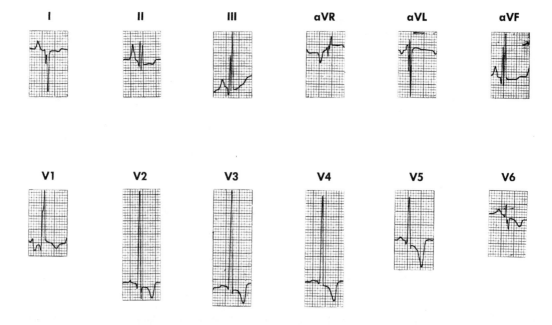

**Figure 14.9** Twelve-lead ECG recording for severe right ventricular hypertrophy, particularly noticeable in leads III, V2, and V3. (Rushmer, 1976, Fig. 8-23, p. 330, originally from Guneroth, 1965.)

cells to initiate a wave? Furthermore, since sinus node dysfunction is a potentially fatal condition, it is of clinical interest to understand how to treat this condition.

To address these questions, we suppose that the SA node is composed of self-oscillatory cells coupled together in a network, and that the action potential of the $i$th cell can be described by a vector of state variables $u_i$, which, in the absence of coupling, has dynamics

$$\frac{du_i}{dt} = F(u_i) + \epsilon G_i(u_i). \tag{14.12}$$

Here, the term $F(u)$ represents typical dynamics applicable for every cell, and $\epsilon G_i(u)$ represents the deviation of the dynamics for the $i$th cell from the average. The parameter $\epsilon$ is assumed to be a small positive number, indicating that the variation among cells is small. To specify $F$, one might use the YNI model or the FitzHugh–Nagumo model (as in section 4.3), adjusted to allow for autonomous oscillations.

Next we assume that the cells are isopotential and connected discretely through resistive gap-junction coupling and that the extracellular medium is isopotential. Then, when the cells are coupled, we obtain the system of equations

$$\frac{du_i}{dt} = F(u_i) + \epsilon G_i(u_i) + \epsilon D \sum_j d_{ij}(u_j - u_i). \tag{14.13}$$

Here $D$ is a diagonal matrix with entries of zero or one to indicate which of the state variables participate in the coupling. In neuromuscular media, only the intracellular and extracellular potentials participate in the coupling. However, because we have assumed that the extracellular potential is uniform, $D$ has only one nonzero entry, namely that one corresponding to the intracellular potential, which in this case is the same as the membrane potential.

The coefficients $d_{ij}$ are the coupling coefficients for the network of cells, where $d_{ij}$ is equal to the (positive) coupling strength (inversely proportional to the resistance) between cells $i$ and $j$. Of course, $d_{ij} = 0$ if cells $i$ and $j$ are not directly coupled, and coupling is symmetric, so that $d_{ij} = d_{ji}$. Without loss of generality, $d_{ii} = 0$.

A simple example for the coupling matrix comes from considering a one-dimensional chain of cells coupled by nearest-neighbor coupling, for which $d_{i,i+1} = d_{i,i-1} = d$, the coupling strength between cells, and all other coupling coefficients are zero. The general formulation (14.13) of the problem allows us to consider a wide variety of coupling networks, including anisotropically coupled rectangular grids and hexagonal grids. The parameter $\epsilon$ scales the coupling term to indicate that the coupling is weak, so that currents through gap junctions are small compared to transmembrane currents. The evidence for weak coupling is that the wave speed in the SA node is very slow, on the order of 2–5 cm/s, compared with 50 cm/s in myocardial tissue and 100 cm/s in Purkinje fiber. (See Exercise 12 for a possible explanation of why weak coupling is advantageous.)

Suppose the stable periodic solution of the equation $\frac{du}{dt} = F(u)$ is given by $U(t)$. Then, because $\epsilon$ is assumed to be small, one can use multiscale methods (see Section 14.5) to find that $u_i(t) = U(\omega(\epsilon)t + \delta\theta_i(t)) + O(\epsilon)$, where $\omega(\epsilon) = 1 + \epsilon\Omega_1 + O(\epsilon^2)$ and the phase shift $\delta\theta_i$ of each oscillator satisfies the equation

$$\frac{d}{dt}\delta\theta_i = \epsilon \left( \xi_i - \Omega_1 + \sum_{j \neq i} d_{ij}[h(\delta\theta_j - \delta\theta_i) - h(0)] \right). \tag{14.14}$$

The periodic coupling function $h$ and the numbers $\xi_i$ are specified in Section 14.5 and the scalar $\Omega_1$ is as yet undetermined. Notice that each oscillator has frequency $\omega(\epsilon)$ and that the phase is slowly varying by comparison with the underlying oscillation and represents only the variation from the typical oscillation.

While there are many interesting questions that could be addressed at this point, of greatest interest here is to determine the *firing sequence* in a collection of phase-locked oscillators. By firing sequence, we mean the order of firing of the individual cells. If the firing sequence of cells is spatially ordered, then the firing of cells appears as a spreading wave, although it is not a propagated wave, but a *phase wave*. (It is not propagated because it would remain for some time even if coupling were set to zero. The role of coupling is merely to coordinate, not to initiate, the wavelike behavior.)

To determine the approximate firing sequence, we suppose that the cells are phase-locked and that the steady-state phase differences are not too large. This is the case for normal SA nodal cells, since all of the SA nodal cells fire within a few milliseconds of each other during an oscillatory cycle of about one second duration. If the steady-state phase differences are small enough, we can replace $h(\delta\theta_j - \delta\theta_i) - h(0)$ in (14.14) by its local linearization $h'(0)(\delta\theta_j - \delta\theta_i)$. Then the steady states of (14.14) are determined as solutions of the linear system of equations

$$\sum_j d_{ij} h'(0)(\delta\theta_j - \delta\theta_i) = \Omega_1 - \xi_i. \tag{14.15}$$

We rewrite (14.15) in matrix notation by defining a matrix $A$ with entries $a_{ij} = d_{ij}$ if $i \neq j$ and $a_{ii} = -\sum_{j\neq i} d_{ij}$, and then (14.15) becomes

$$A\Phi = \frac{1}{h'(0)}(\vec{\Omega}_1 - \vec{\xi}), \tag{14.16}$$

where $\Phi$ is the vector with entries $\delta\theta_i$, $\vec{\Omega}_1$ is a vector with all entries $\Omega_1$, and the entries of $\vec{\xi}$ are the numbers $\xi_i$.

A few observations about the matrix $A$ are important. Notice that $A$ is symmetric and has a nontrivial null space, since $\sum_j a_{ij} = 0$. For consistency, we must choose $\Omega_1$ such that the sum of all rows of (14.16) is zero, so that

$$\Omega_1 = \frac{1}{N}\sum_i \xi_i. \tag{14.17}$$

Thus the bulk frequency of the SA node is determined as the average of the frequencies of the individual oscillators. This is a democratic process in which one cell equals one vote, regardless of coupling strength.

Next, since all the nonzero elements of $d_{ij}$ (and hence the off-diagonal elements of $A$) are positive and $A$ has zero row sums, all the nonzero eigenvalues of $A$ have negative real part. Furthermore, since $A$ is real, symmetric, and nonpositive definite, it has a complete set of $N$ mutually orthogonal, real eigenvectors, say $\{y_k\}$, with corresponding real eigenvalues $\lambda_k$. All of the eigenvalues $\lambda_k$ are negative or zero. If the matrix of coupling coefficients $d_{ij}$ is irreducible, then the constant eigenvector $y_1$ is the unique null vector of $A$, and $\lambda_k < 0$ for $k > 1$ (see Chapter 6, Exercise 6). The matrix of coupling coefficients is *irreducible* if all the cells are connected by some electrical path, so that there are no electrically isolated clumps of cells. Suppose also that the eigenvectors are ordered by increasing amplitude of the eigenvalue. The solution of (14.16) is readily expressed in terms of the eigenvectors and eigenvalues of $A$ as

$$\Phi = -\frac{1}{h'(0)}\sum_{k\neq 1}\langle\vec{\xi}, y_k\rangle\frac{y_k}{\lambda_k}. \tag{14.18}$$

The scalar $\Omega_1$ drops out of this expression because the eigenvector $y_1$ is the constant vector and $\langle y_k, y_1\rangle = 0$ for all $k \neq 1$.

The firing sequence is now determined from $\Phi$. That is, if $\delta\theta_k$ is the largest element of $\Phi$, then the phase of the $k$th cell is the most advanced and therefore the first to fire, and so on in decreasing order. It remains to gain some understanding of the relationship between the natural frequencies $\vec{\xi}$ and the firing sequence $\Phi$.

The general principle of how the firing sequence is determined from the natural frequencies $\vec{\xi}$ is apparent from (14.18). The firing sequence is a superposition of the eigenvectors $\{y_k\}$ with amplitudes $\lambda_k^{-1}\langle\vec{\xi},y_k\rangle$. Thus, eigenvector components of $\vec{\xi}$ that are most influential on the firing sequence are those components for which $\lambda_k^{-1}\langle\vec{\xi},y_k\rangle$ is largest in amplitude. The expression (14.18) is a filter that suppresses, or filters out, certain components of $\vec{\xi}$. It follows that a single cell with high natural frequency compared to its coupled neighbors is not necessarily able to lead the firing sequence.

The expression (14.18) for the firing sequence does not give much geometrical insight. Furthermore, it is usually not a good idea to solve matrix problems such as (14.16) using eigenfunction expansions, since direct numerical methods are much faster and easier. To illustrate how (14.18) works, we consider, as an example, a two-dimensional grid of cells coupled by nearest-neighbor coupling. The natural frequencies of the cells are randomly distributed, with the fastest cells concentrated near the center of the grid. The distribution of frequencies found by direct numerical solution is depicted in Fig. 14.10, with darker locations representing the slowest intrinsic frequencies. The firing sequence for this collection of cells is shown in Fig. 14.11, where cells with advanced (or largest) phase fire earliest in the firing sequence. The initiation of the firing sequence is at the site of a group of fast, but not necessarily the fastest, oscillators. Notice that the phase is smoothed, giving the appearance of wave-like motion moving from the location of largest phase to smallest phase, even though these are phase waves rather than propagated waves. In this figure, the scale of the phase variable is arbitrary.

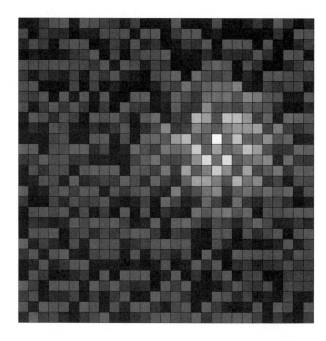

**Figure 14.10** Natural frequencies $\xi_i$ for a collection of oscillatory cells.

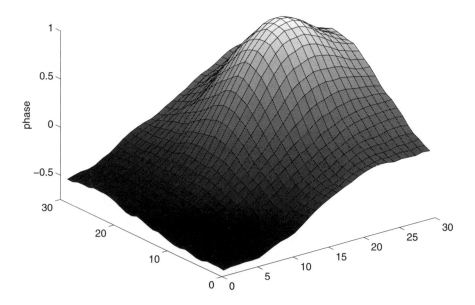

**Figure 14.11**  Phase for oscillators in a collection of coupled cells with nearest-neighbor coupling and natural frequencies as depicted in Fig. 14.10.

## 14.2.2   Critical Size of a Pacemaker

The SA node is a small clump of self-oscillatory cells in a sea of excitable (but nonoscillatory) cells whose function is to initiate the cardiac action potential. SA nodal cells have no contractile function and therefore no contractile machinery. Thus, when viewed in terms of contractile efficiency, SA nodal cells are a detriment to contraction and a waste of important cardiac wall space. On the other hand, the SA node cannot be made too small because presumably, it would not be able to generate the current necessary to entrain the rest of the heart successfully. Thus it is important to have some measure of the critical size of the SA node.

An *ectopic focus* is a collection of cells other than the SA node or AV node that are normally not oscillatory but that for some reason (for example, increased extracellular potassium) become self-oscillatory and manage to entrain the surrounding tissue into a rapid beat. In some situations, particularly in people with scar tissue resulting from a previous heart attack, the appearance of an ectopic focus may be life-threatening.

We model the behavior of a clump of oscillatory cells in an otherwise nonoscillatory medium in a simple way, using FitzHugh–Nagumo dynamics:

$$\frac{\partial v}{\partial t} = \nabla^2 v + f(v) - w, \tag{14.19}$$

$$\frac{\partial w}{\partial t} = \epsilon(v - \gamma w - \alpha(r/\sigma)), \tag{14.20}$$

where $v$ represents the membrane potential and $w$ the recovery variable for our excitable medium. The function $f(v)$ is of typical "cubic" shape (cf. Chapter 4). The function $\alpha(r)$ is chosen to specify the intrinsic cell behavior as a function of the radial variable $r$. The number $\sigma$ is a scale factor that measures the size of the oscillatory region. We take $\epsilon$ to be a small positive number and require $\gamma > 0, f'(v)\gamma < 1$ for all $v$. This requirement on $\gamma$ guarantees that the steady-state solution of (14.19)–(14.20) is unique. If the domain is bounded, typical boundary conditions are Neumann (no-flux) conditions. Notice that space has been scaled to have unit space constant. We assume radial symmetry for the SA node as well as for the entire spatial domain.

When there is no spatial coupling, there are two possible types of behavior, exemplified by the phase portraits in Figs. 4.17 and 4.19. In these examples, the system has a unique steady-state solution that is globally stable (Fig. 4.17) or has an unstable steady-state solution surrounded by a stable periodic orbit (Fig. 4.19), depending on the location of the intercept of the two nullclines.

The transition from a stable to an unstable steady state is a subcritical Hopf bifurcation. The Hopf bifurcation is readily found from standard linear analysis. Suppose $v^*$ is the equilibrium value for $v$ ($v^*$ is a function of $\alpha$). Then the characteristic equation for (14.19)–(14.20) (with no diffusion) is

$$f'(v^*) = \lambda + \frac{\epsilon}{\lambda + \epsilon\gamma}, \tag{14.21}$$

where $\lambda$ is an eigenvalue of the linearized system. There is a Hopf bifurcation (i.e., $\lambda$ is purely imaginary) when

$$f'(v^*) = \epsilon\gamma, \tag{14.22}$$

provided that $\epsilon\gamma^2 < 1$. If $f'(v^*) > \epsilon\gamma$, the steady-state solution is an unstable spiral point, whereas if $f'(v^*) < \epsilon\gamma$, the steady-state solution is linearly stable. If $\epsilon$ is small, most of the intermediate (increasing) branch of the curve $f(v)$ is unstable, with the Hopf bifurcation occurring close to the minimal and maximal points. Thus, there is a range of values of $\alpha$, which we denote by $\alpha_* < \alpha < \alpha^*$, for which the steady solution is unstable.

We wish to model the physical situation in which a small collection of cells (like the SA node or an ectopic focus) is intrinsically oscillatory, while all other surrounding cells are excitable, but not oscillatory. To model this, we assume that $\alpha(r)$ is such that the steady solution is unstable for small $r$, but stable and excitable for large $r$, so that $\lim_{r\to\infty} \alpha(r) = a < \alpha_*$ and $\lim_{r\to\infty} f'(v^*(r)) < \epsilon\gamma$. As an example, we might have the bell-shaped curve

$$\alpha(r) = a + (b - a)\exp\left(-\frac{r^2}{R^2}\right), \tag{14.23}$$

$$R^2 = \log\left(\frac{b - a}{\alpha_* - a}\right), \tag{14.24}$$

with $a < \alpha_* < b < \alpha^*$. The scale factor $R$ was chosen such that $\alpha(1) = \alpha_*$, so that cells with $r < 1$ are self-oscillatory and the cells outside unit radius are nonoscillatory.

Another way to specify $\alpha(r)$ is simply as the piecewise constant function

$$\alpha(r) = \begin{cases} b, & \text{for } 0 < r < 1, \\ a, & \text{for } r > 1, \end{cases} \tag{14.25}$$

with $a < \alpha_* < b < \alpha^*$. The specification (14.25) is particularly useful when used in combination with the piecewise linear function

$$f(v) = \begin{cases} -v, & \text{for } v < \dfrac{1}{4}, \\ v - \dfrac{1}{2}, & \text{for } \dfrac{1}{4} < v < \dfrac{3}{4}, \\ 1 - v, & \text{for } v > \dfrac{3}{4}, \end{cases} \tag{14.26}$$

since then all the calculations that follow can be done explicitly (see Exercises 11 and 12).

There are two parameters whose influence we wish to understand and that we expect to be most significant, namely, $a$, the asymptotic value of $\alpha(r)$ as $r \to \infty$, and $\sigma$, which determines the size of the oscillatory region. Note that as $a$ decreases, the cells become less excitable and the wavespeed of fronts decreases. We expect the behavior to be insensitive to variations in $b$, although this should be verified as well.

With a nonuniform $\alpha(r)$, the uncoupled medium has a region of cells with unstable steady states and a region with stable steady states. With diffusive coupling, the steady state is smoothed and satisfies the elliptic equation

$$\nabla^2 v + F(v,r) = 0, \tag{14.27}$$

$$F(v,r) = f(v) - w, \tag{14.28}$$

$$w = \frac{1}{\gamma}\left(v - \alpha\left(\frac{r}{\sigma}\right)\right). \tag{14.29}$$

For each $r$, the function $F(v,r)$ is a monotone decreasing function of $v$ having a unique zero, say $v = v^*(r)$, $F(v^*(r),r) = 0$. It follows that there is a unique, stable solution of (14.29), denoted by $v_0(r), w_0(r)$. In fact, this unique solution is readily found numerically as the unique steady solution of the nonlinear parabolic equation

$$\frac{\partial y}{\partial t} = \nabla^2 y + F(y,r). \tag{14.30}$$

This steady-state solution is shown in Fig. 14.12. Here are shown three different steady-state solutions of (14.19)–(14.20); the uncoupled solution (the steady states for the uncoupled medium, i.e., with no diffusion), the solution for a symmetric one-dimensional medium, and the solution for a spherically symmetric three-dimensional medium. The three-dimensional solution with spherical symmetry is not much harder to find than the one-dimensional solution, because the change of variables $y = Y/r$ transforms (14.30) in three spatial dimensions into

$$\frac{\partial Y}{\partial t} = \frac{\partial^2 Y}{\partial r^2} + rF\left(\frac{Y}{r}, r\right). \tag{14.31}$$

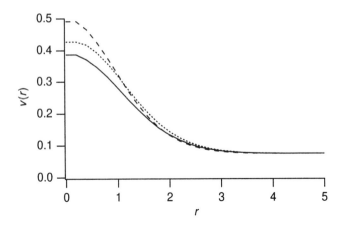

**Figure 14.12** Three steady-state solutions with $f(v) = 10.0v(v-1)(0.5-v)$, $\gamma = 0.1$, and $\alpha(r)$ given by (14.23), with $a = 0.104$, $b = 0.5$, $\sigma = 2.25$. The short dashed curve shows the uncoupled solution, the long dashed curve shows the solution for a symmetric one-dimensional medium, and the solid curve shows the solution for a spherically symmetric three-dimensional medium.

Solutions of this partial differential equation are regular at the origin if we require $Y = 0$ at $r = 0$. Diffusion obviously smooths the steady-state solution in the oscillatory region.

The issue of collective oscillation is determined by the stability of the diffusively smoothed steady state as a solution of the partial differential equation system (14.19)–(14.20). To study the stability of the steady state, we look for a solution of (14.19)–(14.20) of the form $v(r) = v_0(r) + V(r)e^{\lambda t}$, $w = w_0(r) + W(r)e^{\lambda t}$ and linearize. We obtain the linear system

$$\lambda V = \nabla^2 V + f'(v_0(r))V - W, \tag{14.32}$$

$$\lambda W = \epsilon(V - \gamma W). \tag{14.33}$$

Because of the special form of this linear system, it can be simplified to a single equation, namely,

$$\nabla^2 V + f'(v_0(r))V = \mu V, \tag{14.34}$$

where $\mu = \lambda + \frac{\epsilon}{\lambda + \epsilon\gamma}$. Equation (14.34) has a particularly nice form, being a *Schrödinger equation*. In quantum physics, the function $-f'(v_0(r))$ is the potential-energy function, and the eigenvalues $\mu$ are the energy levels of bound states. In the present context, we are interested in determining the sign of the real part of $\lambda$ through $\mu = \lambda + \frac{\epsilon}{\lambda + \epsilon\gamma}$. Notice that the relationship between $\mu$ and $\lambda$ here is of exactly the same form as the characteristic equation for individual cells (14.21). This leads to a nice interpretation for the Schrödinger equation (14.34). Because it is a self-adjoint equation, the eigenvalues $\mu$ of (14.34) are real. Therefore, there is a Hopf bifurcation for the medium whenever $\mu = \epsilon\gamma$. The entire collection of coupled cells is stable when the largest eigenvalue

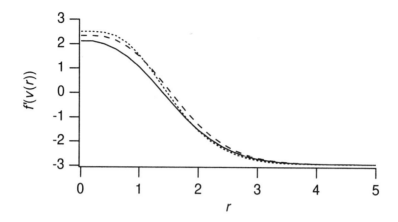

**Figure 14.13**  The potential function $f'(v(r))$ for three steady profiles $v(r)$. The short dashed curve corresponds to the uncoupled solution, the long dashed curve to the symmetric one-dimensional medium, and the solid curve to a spherically symmetric three-dimensional medium.

satisfies $\mu < \epsilon\gamma$ and unstable if the largest eigenvalue has $\mu > \epsilon\gamma$.

In Fig. 14.13 is shown the potential function $f'(v(r))$ for the three steady profiles of Fig. 14.12. The largest eigenvalue of (14.34) represents an average over space of the influence of $f'(v_0(r))$ on the stability of the steady state. When this value is larger than $\epsilon\gamma$ (the critical slope of $f(v)$ at which Hopf bifurcations of the uncoupled system occur), then the entire medium loses stability to a Hopf bifurcation and gives rise to an oscillatory solution. The condition $\mu > \epsilon\gamma$ is therefore the condition that determines whether a region of oscillatory cells is a source of oscillation. If $\mu < \epsilon\gamma$, the oscillatory cells are masked by the rest of the medium.

Some observations about the size of the eigenvalues $\mu$ are immediate. Because $\lim_{r\to\infty} v_0(r) = \lim_{r\to\infty} v^*(r)$, it follows that $f'(v_0(r)) < \epsilon\gamma$ for large $r$. For there to be a bounded solution of (14.34) that is exponentially decaying at $\pm\infty$, there must be a region of sinusoidal behavior in which $\mu < f'(v_0(r))$. Thus, the largest eigenvalue of (14.34) is guaranteed to be smaller than the maximum of $f'(v_0(r))$. Therefore, if $v_0(r) < \alpha_*$ (so that $f'(v_0(r)) < \epsilon\gamma$ for all $r$), there are no oscillatory cells, and the steady solution is stable. Furthermore, since the largest eigenvalue is strictly smaller than the maximum of $f'(v_0(r))$ and it varies continuously with changes in $v_0(r)$, there are profiles $\alpha(r)$ having a nontrivial collection of oscillatory cells that is too small to render the medium unstable. That is, there is a critical mass of oscillatory cells necessary to cause the medium to oscillate. Below this critical mass, the steady state is stable, and the oscillation of the oscillatory cells is quenched.

Suppose $f'(v)$ is a monotone increasing function of $v$ in some range $v < v^+$, and suppose that $\alpha(r)$ is restricted so that $v_0(r) < v^+$ for all $r$. Suppose further that $\alpha(r)$ is a monotone increasing function of its asymptotic value $a$ and a monotone decreasing function of $r$. Then the steady-state solution $v_0(r)$ is an increasing function (for each point $r$) of both $a$ and $\sigma$. Therefore, the function $f'(v_0(r))$ is an increasing function of $a$ and $\sigma$ for all values of $r$, from which it follows—using standard comparison arguments for eigenfunctions (Keener, 1988, or Courant and Hilbert, 1953)—that $\mu(a,\sigma)$, the largest eigenvalue of (14.34), is an increasing function of both $a$ and $\sigma$. As a result, if $\alpha(r)$ is restricted so that $v_0(r) < v^+$ for all $r$, there is a monotone decreasing function of $\sigma$, denoted by $\sigma = \Sigma(a)$, along which the largest eigenvalue $\mu(a,\sigma)$ of (14.34) is precisely $\epsilon\gamma$.

This summary statement shows that to build the SA node, one must have a sufficiently large region of oscillatory tissue, and that the critical mass requirement increases if the tissue becomes less excitable or if the coupling becomes stronger. Strong coupling inhibits oscillations, because increasing coupling increases the space constant, and $\sigma$ was measured in space constant units. Therefore, an increase of the space constant increases the critical size requirement of the oscillatory region. In Fig. 14.14 is shown the critical Hopf curve $\sigma = \Sigma(a)$ for a one-dimensional domain and for a three-dimensional domain (taking $\epsilon = 0$), both found numerically.

Having established that there is a critical size for a self-oscillatory region above which oscillations occur and below which oscillations are prevented, we would like to examine the behavior of the oscillations. Two types of oscillatory behavior are possible. If the far field $r \to \infty$ is sufficiently excitable, then the oscillations of the oscillatory region excite periodic waves that propagate throughout the medium, as depicted in Fig. 14.15. On the other hand, it may be that there are oscillations that fail to propagate throughout the entire medium, as depicted in Fig. 14.16. In Fig. 14.15, the oscillatory region successfully drives oscillatory waves that propagate throughout the entire medium. Here, $a = 0.2$, $\sigma = 3.0$. In Fig. 14.16, the oscillatory region is incapable of driving periodic waves into the nonoscillatory region. For this figure, $a = 0.0$, and $\sigma = 3.0$, so that the medium at infinity does not support front propagation.

The issue of whether or not the entire medium is entrained to the central oscillator is decided by the relationship between the period of the oscillator and the dispersion curve for the far medium. Roughly speaking, if the period of the central oscillator is large enough compared to the absolute refractory period of the far medium (the knee

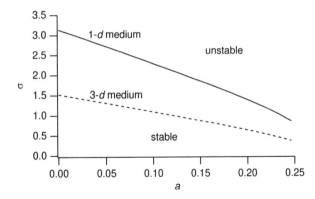

**Figure 14.14** The critical curve $\sigma = \Sigma(a)$ along which there is a Hopf bifurcation for the system (14.19)–(14.20), shown solid for a one-dimensional and dashed for a three-dimensional medium.

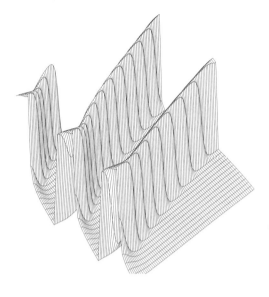

**Figure 14.15** Waves generated by an oscillatory core that propagate into the nonoscillatory region. The nonlinearity here is the same as in Fig. 14.12 with $\epsilon = 0.1$, $\gamma = 0.1$, $a = 0.2$, $b = 0.5$, and $\sigma = 3.0$.

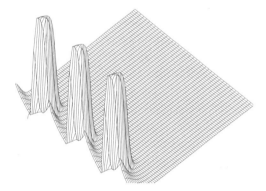

**Figure 14.16** Waves generated by an oscillatory core that fail to propagate into the nonoscillatory region. The nonlinearity here is the same as in Fig. 14.12 with $\epsilon = 0.1$, $\gamma = 0.1$, $a = 0$, $b = 0.5$, and $\sigma = 3.0$.

of the dispersion curve), then waves can be expected to propagate into the far field in one-to-one entrainment. On the other hand, if the frequency of the oscillation is below the knee of the dispersion curve, we expect partial or total block of propagation. Block of propagation occurs as the excitability of the far field, parametrized by $a$, decreases.

We can summarize how the oscillations of the medium depend on coupling strength. For a medium with fixed asymptotic excitability, if the size of the oscillatory region is large enough, there is oscillatory behavior. However, this critical mass is an increasing function of coupling strength. With sufficiently large coupling, the oscillations of any finite clump of oscillatory cells (in an infinite domain of nonoscillatory cells) are quenched. If coupling is decreased, the critical mass for oscillation decreases. Thus, any clump of oscillatory cells oscillates if coupling is weak enough. However, if coupling is too weak, then effects of discrete coupling may become important, and the oscillatory clump of cells may lose its ability to entrain the entire medium. It follows that if the medium is sufficiently excitable, there is a range of coupling strengths, bounded above and below, in which a mass of oscillatory cells entrains the medium. If the coupling is too large, the oscillations are suppressed, while if the coupling is too weak, the oscillations are localized and cannot drive oscillations in the medium far away from the oscillatory source. On the other hand, if the far region is not sufficiently excitable, then one of these two mechanisms suppresses entrainment for all coupling strengths.

# 14.3 Cardiac Arrhythmias

Cardiac arrhythmias are disruptions of the normal cardiac electrical cycle. They are generally of two types. There are temporal disruptions, which occur when cells act out of sequence, either by firing autonomously or by refusing to respond to a stimulus from other cells, as in AV nodal block or a bundle branch block. A collection of cells that fires autonomously is called an ectopic focus. Generally speaking, these arrhythmias cause little disruption to the ability of the heart muscle to pump blood, and so if they do not initiate some other kind of arrhythmia, are generally not life-threatening.

The second class of arrhythmias are those that are reentrant in nature and can occur only because of the spatial distribution of cardiac tissue. If they occur in the ventricles, reentrant arrhythmias are of serious concern and life-threatening, as the ability of the heart to pump blood is greatly diminished. Reentrant arrhythmias on the atria are less dangerous, since the pumping activity of the atrial muscle is not necessary to normal function with minimal physical activity.

A classic example of a reentrant rhythm is Wolff–Parkinson–White (WPW) syndrome, in which an action potential circulates continuously between the atria and the ventricles through a loop, exiting the atria through the AV node and reentering the atria through an *accessory pathway* (or vice versa). Since conduction through the AV node is quite slow compared to other propagation, an accessory pathway that circumvents the AV node usually reveals itself on the ECG by an early, broad deflection of the QRS complex (Fig. 14.17). This deflection is broadened because it depicts propagation through myocardial tissue, which is slow compared to normal propagation through the Purkinje network. WPW syndrome is life-threatening if not detected and treated, because it allows for the possibility of rapid reentrant rhythms. However, WPW syndrome is usually curable, as cardiac surgeons can use localized radio frequency waves to burn and permanently obliterate the accessory pathway, restoring a normal single pathway conduction and a normal ECG.

## 14.3.1 Atrioventricular Node

In the normal heart, the only pathway for an action potential to travel to the ventricles is through the AV node. As noted above, propagation through the AV node is quite slow compared to propagation in other cardiac cells. This slowed conduction is primarily due to a decreased density of sodium channels, which yields a decreased upstroke

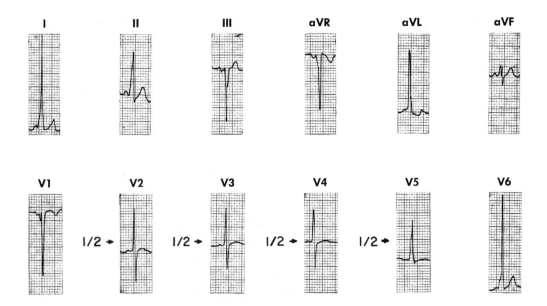

**Figure 14.17** Twelve-lead ECG recording of Wolff–Parkinson–White syndrome, identifiable by the shortened P–Q interval (because the AV delay is circumvented) and the slowed QRS upstroke, particularly noticeable in leads II, aVR, and V6. (Rushmer, 1976, Fig. 8-47, p. 339; originally from Guneroth, 1965).

velocity. With a decrease of sodium channel density there is also an increase in the likelihood of conduction failure.

Propagation failure in the AV node leads to skipped QRS complexes on the ECG, or, more prosaically, skipped heartbeats. A skipped heartbeat once in a while is not particularly dangerous, but it is certainly noticeable. During *diastole* (the period of ventricular relaxation during the heartbeat cycle), the ventricles fill with blood. Following an abnormally long diastolic period, the heart becomes enlarged, and when the next compression (*systole*) occurs, *Starling's law* (i.e., that compression is stronger when the heart is more distended initially, cf. Chapter 15) takes control, and compression is noticeably more vigorous, giving the subject a solid thump in the chest.

AV nodal conduction abnormalities are sorted into three classes. They are all readily visible from ECG recordings by looking at the time interval between the P wave and the QRS complex, i.e., the *P–R interval*. Type I AV nodal block shows itself as an increase in the P–R interval as the SA pacing rate increases. Type III AV nodal block corresponds to no AV nodal conduction whatever and total absence of a QRS complex.

Type II AV nodal block is phenomenologically the most interesting. In the simplest type, there is one QRS complex for every two P waves, a 2:1 pattern. A more complicated pattern is as follows: on the ECG (Fig. 14.18), P waves remain periodic, although the P–R interval is observed to increase gradually until one QRS complex is skipped. Following the skipped beat, the next P–R interval is quite short, but then the

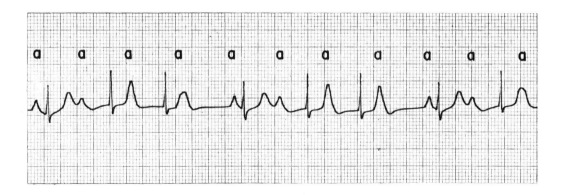

**Figure 14.18** ECG recording of a Wenckebach pattern in which every fourth or fifth atrial beat is not conducted. (Rushmer, 1976, Fig. 8-24, p. 313.)

P–R lengthening begins again, leading to another skipped beat, and so on. A pattern with $n$ P waves to $n - 1$ QRS complexes is called an $n$-to-$(n - 1)$ *Wenckebach pattern*, after the German cardiologist Wenckebach (1904).

A simple mathematical description of AV nodal signal processing can be given as follows: we view the AV node as a collection of cells that fire when they are excited, which happens if their potential reaches a threshold, $\theta(t)$. Immediately after firing, the cells become refractory but then gradually recover. Effectively, at firing, the threshold increases dramatically but then decreases back to its steady-state value as recovery proceeds. This model ignores the fact that the AV node is self-oscillatory and will fire without stimulus with a low frequency of 30–40 per minute. The self-oscillatory nature of the AV node becomes evident only in cases of SA nodal failure or at very low SA nodal firing rates. Thus, the model discussed here is valid at high stimulus rates (appropriate for AV nodal block) but not at low stimulus rates.

Input to the AV node comes from the action potential propagating through the atria from the SA node. The AV node experiences a periodic, time-varying potential, say $\phi(t)$. Firing occurs if the input signal reaches the threshold. Therefore, at the $n$th firing time, denoted by $t_n$,

$$\phi(t_n) = \theta(t_n). \tag{14.35}$$

Subsequent to firing, the threshold evolves according to

$$\theta(t) = \theta_0 + [\theta(t_n^+) - \theta_0]e^{-\gamma(t-t_n)}, \qquad t > t_n. \tag{14.36}$$

Note that $\theta \to \theta_0$ as $t \to \infty$, and thus $\theta_0$ denotes the base value of the threshold. Further, $\theta = \theta(t_n^+)$ at $t = t_n$, and thus $\theta(t_n^+) - \theta(t_n^-)$ denotes the jump in the threshold caused by the firing of an action potential. To complete the model we must specify $\theta(t_n^+)$. The important feature of $\theta(t_n^+)$ is that it must have some memory, that is, depend in some way on $\theta(t_n^-)$. Therefore, we take

$$\theta(t_n^+) = \theta(t_n^-) + \Delta\theta. \tag{14.37}$$

The simple choice used here is to take $\Delta\theta$ a constant. However, consideration of the threshold in FitzHugh–Nagumo models suggests that (in a more general model) $\Delta\theta$ could also be some decreasing function of $\theta(t_n^-)$, i.e., $\Delta\theta = \Delta\theta(\theta(t_n^-)) = \Delta\theta(\phi(t_n))$, since $\phi(t_n) = \theta(t_n^-)$.

Now we can find the next firing time as the smallest solution of the transcendental equation

$$\phi(t_{n+1}) = \theta_0 + [\theta(t_n^+) - \theta_0]e^{-\gamma(t_{n+1}-t_n)}. \tag{14.38}$$

Equation (14.38) can be rearranged into an equation of the form

$$F(t_{n+1}) = F(t_n) + \Delta\theta e^{\gamma t_n} = G(t_n), \tag{14.39}$$

where

$$F(t) = (\phi(t) - \theta_0)e^{\gamma t}. \tag{14.40}$$

Plots of typical functions $F(t)$ and $G(t)$ are shown in Fig. 14.19. Here we have taken $\phi(t) - \theta_0 = \sin^4(\pi t)$. The dashed lines in this figure follow a few iterates of the map.

The key observation is that the map $t_n \mapsto t_{n+1}$ as defined by (14.39) is the lift of a circle map. Before proceeding with this example, we give a brief introduction to the theory of circle maps.

The first application of circle maps to the behavior of neurons was given by Knight (1972). More detailed discussions of maps and chaos and the like with application to a wide array of biological problems can be found in Glass and Mackey (1988), Glass and Kaplan (1995), and Strogatz (1994).

## Circle maps

A circle map is a map of the circle to itself, $f : S^1 \to S^1$, but it is often easier to describe a circle map in terms of its *lift* $F : R \to R$, where $F$ is a monotone increasing function and $F(x+1) = F(x) + 1$. The two functions $f$ and $F$ are related by

$$f(x) \equiv F(x \bmod 1) \bmod 1. \tag{14.41}$$

(For convenience we normalize the circumference of the circle to be of length 1, rather than $2\pi$.)

The primary challenge from a circle map is to determine when the behavior is periodic and to understand the possible nonperiodic behaviors. The simplest periodic behavior is a period 1 solution, say a point $x_0$ for which $F(x_0) = x_0 + 1$. This orbit is also said to have *rotation number* one because it rotates around the circle once on each iterate. This is also described as 1:1 phase locking between input and output. A more complicated periodic orbit would be a point $x_0$ and its iterates $x_i$ with the property that $x_n = x_0 + m$. In other words, the iterates rotate around the circle $m$ times in $n$ iterates. The rotation number is $m/n$, and there is $m : n$ phase locking, with $m$ output cycles for every $n$ input cycles.

The key fact to understand is that the asymptotic behavior of a circle map is characterized by its *rotation number*, $\rho$, defined by

$$\rho = \lim_{n \to \infty} \frac{F^n(x)}{n}. \tag{14.42}$$

$F^n(x)$ is the $n$th iterate of the point $x$,

$$F^n(x) = F(F^{n-1}(x)), \tag{14.43}$$

where $F^0(x) = x$ and $F^1(x) = F(x)$.

If $F$ is a continuous function, the rotation number has the following properties:

1. $\rho$ exists and is independent of $x$.
2. $\rho$ is rational if and only if there are periodic points.
3. If $\rho$ is irrational, then the map $F$ is equivalent to a rigid rotation by the amount $\rho$.
4. If there is a continuous family of maps $F_\lambda$, then $\rho(\lambda)$ is a continuous function of $\lambda$. Furthermore, if $F_\lambda$ is a monotone increasing function of $\lambda$, then $\rho$ is a nondecreasing function of $\lambda$.
5. Generically, if $\rho(\lambda)$ is rational at some value of $\lambda_0$, it is constant on an open interval containing $\lambda_0$.

Here is what this means in practical terms. Since $\rho$ exists, independent of $x$, all orbits have the same asymptotic behavior, orbiting the circle at the same rate. If $\rho$ is rational, the asymptotic behavior is periodic, whereas if $\rho$ is irrational, the motion is equivalent to a rigid rotation. In this case, the behavior is aperiodic, but not complicated, or "chaotic." There are no other types of behavior for a continuous circle map.

The last two features of $\rho$ make the behavior of the orbits so unusual, being a function that is continuous, monotone nondecreasing (if $F_\lambda$ is an increasing function of $\lambda$), yet locally constant at all the rational levels. Such a function is called the *Devil's staircase*. Notice that if $\rho$ is rational on an open interval of parameter space, then phase locking is robust.

The reason for this robustness is that a periodic point with $\rho = p/q$ corresponds to a root of the equation $F^q(x) - x = p$, and roots of equations are generally, but not always, robust, or transversal (i.e., the derivative of $F^q(x) - x$ at a root is nonzero). If a root is transversal, then arbitrarily small perturbations to the equation do not destroy the root, and it persists for a range of parameter values. However, the existence of a periodic point is no guarantee that it is robust. For example, the simple shift $F(x) = x + \lambda$ has periodic points whenever $\lambda$ is rational, but these periodic points are never isolated or robust.

A detailed exposition on continuous circle maps and proofs of the above statements can be found in Coddington and Levinson (1984, chapter 17).

Now we attempt to apply this theory of circle maps to (14.39). Notice that this is indeed the lift of a circle map, since if $t_n$ and $t_{n+1}$ satisfy (14.39), then so do $t_n + T$ and $t_{n+1} + T$. To find a circle map, we let $k_n$ be the largest integer less than $t_n/T$ and define

$\psi_n = (t_n - k_n T)/T$. In these variables the map (14.39) can be written as

$$f(\psi_{n+1}) = (f(\psi_n) + \Delta\theta e^{\gamma T \psi_n})e^{\gamma T \Delta k_n}, \qquad (14.44)$$

where

$$f(\psi) = (\Phi(\psi) - \theta_0)e^{\gamma T \psi}, \qquad \Phi(\psi) = \phi(T\psi), \qquad (14.45)$$

and $\Delta k_n = k_{n+1} - k_n$.

We can make a few observations about the map $\psi_n \mapsto \psi_{n+1}$. First, and most disconcerting, the map is not continuous. In fact, it is apparent that there are values of $t$ on the unit interval that can never be firing times. For $t$ to be permitted as a firing time it must be the first point at which $F(t)$ reaches the level $G(t_n)$, i.e., the first time that the threshold is reached. At such a point, $F'(t) > 0$. Since there are regions for which $F'(t) < 0$, which can therefore never be firing times, this is a map of the unit interval *into*, but not onto, itself. However, the map $t_n \mapsto t_{n+1}$ is order preserving, since $G(t)$ is increasing whenever $F(t)$ is increasing.

Since the entire unit interval is not covered by the map, it is only necessary to examine the map on its range. Examples of the map $\psi_n \mapsto \psi_{n+1}$ are shown in Figs. 14.20–14.23. Here we have plotted the map only on the attracting range of the unit interval. These show important and typical features, namely that the map consists of either one or two continuous, monotone increasing branches. The first branch, with values above the one-to-one curve, corresponds to firing in response to the subsequent input (with $k_{n+1} = k_n + 1$), and the second, with values below the one-to-one curve, corresponds to firing after skipping one beat (with $k_{n+1} = k_n + 2$). The skipped beat occurs because when the stimulating pulse arrives, it is subthreshold and so does not evoke a response.

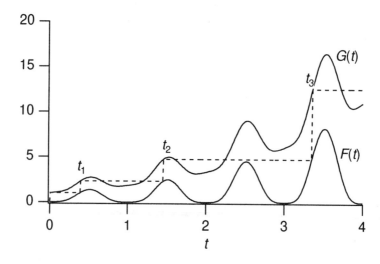

**Figure 14.19**  Plot of the functions $F(t)$ and $G(t)$ with $\Delta\theta = 1.0$, $\gamma = 0.6$.

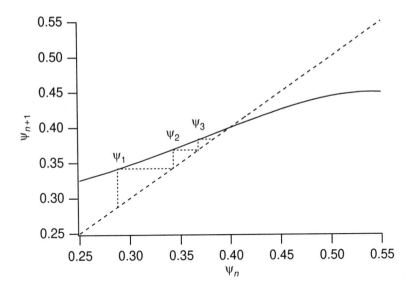

**Figure 14.20**  Plot of the map $\psi_n \mapsto \psi_{n+1}$ with $\Delta\theta = 1.0$, $\gamma T = 0.8$.

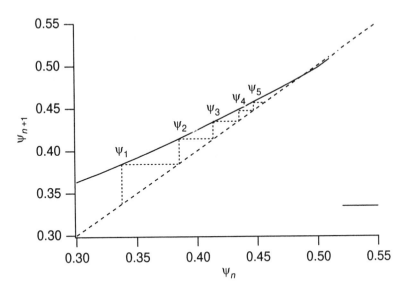

**Figure 14.21**  Plot of the map $\psi_n \mapsto \psi_{n+1}$ with $\Delta\theta = 1.0$, $\gamma T = 0.695$.

The sequence of figures in Figs. 14.20–14.23 is arranged according to decreasing values of $\gamma T$. Note that as $\gamma$ decreases, the rate of recovery from inhibition decreases. For $\gamma T$ sufficiently large, there is a unique fixed point, corresponding to firing in 1:1 response to the input signal. This makes intuitive sense, for when $\gamma$ is large, the recovery from inhibition is fast, and thus the AV node can be driven at the frequency of the

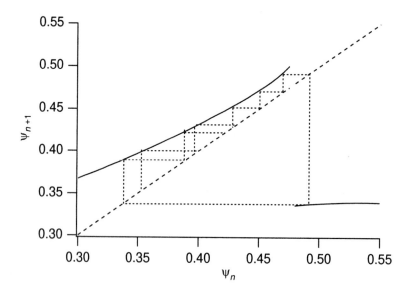

**Figure 14.22** Plot of the map $\psi_n \mapsto \psi_{n+1}$ with $\Delta\theta = 1.0$, $\gamma T = 0.67$.

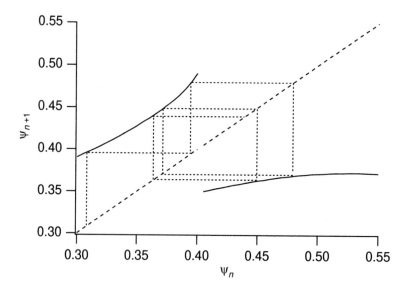

**Figure 14.23** Plot of the map $\psi_n \mapsto \psi_{n+1}$ with $\Delta\theta = 1.0$, $\gamma T = 0.55$.

SA node. For large $\gamma T$ the map is relatively insensitive to changes in parameters. As $\gamma T$ decreases, the first branch of the map increases and the value of the fixed point increases, corresponding to a somewhat delayed firing. Furthermore, because the slope of the map in the vicinity of the fixed point is close to 1, the fixed point is sensitive to changes in parameter values (depicted in Fig. 14.21), corresponding to type I AV block.

As the parameter $\gamma T$ is decreased further, the fixed point is lost and a second branch to the map appears (as in Fig. 14.22). Iterations show that subsequent firings become later and later in the input cycle until one beat is skipped, followed by a firing that is relatively early in the input cycle. For this region of parameter space, the map replicates the Wenckebach pattern.

Finally, as $\gamma T$ decreases further, the second branch "slides over" to the left and eventually intersects the one-to-one line, yielding a fixed point. This fixed point corresponds to a periodic pattern of one skipped beat for each successful firing, a two-to-one pattern, and replicates type II AV block.

The behavior of the map in the region with no fixed point can be described by the rotation number. For maps of the type (14.39) the rotation number can be defined, analogously to our earlier definition, by

$$\rho = \lim_{n \to \infty} \frac{t_n}{nT}. \tag{14.46}$$

The following features of the rotation number $\rho$ can be verified (Keener, 1980a, 1981):

1. $\rho$ exists and is independent of initial data.
2. $\rho$ is a monotone decreasing function of $\gamma T$.
3. $\rho$ attains *every* rational level between 0 and 1 on an open interval of parameter space.

For continuous circle maps, it is not certain that every rational level is attained on an open interval of parameter space.

The main consequence of this result is that between 1:1 phase locking and 2:1 AV block, for every rational number there is an open interval of $\gamma T$ on which the rotation with that rational number is attained.

## 14.3.2   Reentrant Arrhythmias

### Point stimuli

Reentrant arrhythmias usually form spontaneously, although they can also be intentionally initiated. The way in which they are intentionally initiated probably has little to do with how they form spontaneously. However, intentional initiation gives important insight into the nature of excitable media. It is well known that reentrant arrhythmias can be initiated intentionally by the correct application of point stimuli. This procedure has been described beautifully by Winfree (1987), with many gorgeous color plates, so here we content ourselves with a shorter, less colorful, verbal description of the process.

When a current is injected at some point to resting cardiac tissue, cells in the vicinity of the stimulus are depolarized. If the stimulus is of sufficient amplitude and duration, the cells closest to the stimulating electrode may receive a superthreshold stimulus and become excited. Cells further away from the stimulus site receive a subthreshold stimulus, so they return to rest when the stimulus ends. At the border between subthreshold and superthreshold stimulus, a wave front is formed.

Once a transition front is formed, the nonlinear dynamics and curvature determine whether it moves forward or backward. That is, if the undisturbed medium is sufficiently excitable, and the initially excited domain is sufficiently large, the wave front moves outward into the unexcited region. If, however, the unaffected medium is not excitable, but partially refractory, or the excited domain is too small, the wave front recedes and collapses.

If the stimulated medium is initially uniform, these two are the only possible responses to a stimulus. However, if the state of the medium in the vicinity of the stimulating electrode is not uniform, then there is a third possible response. Suppose, for example, that there is a gradual gradient of recovery so that a portion of the stimulated region is excitable, capable of supporting wave fronts (with positive wave speed) and the remaining portion of the stimulated region cannot support wave fronts, but only wave backs (i.e., fronts with negative speed). Then, the result of the stimulus is to produce both wave fronts and wave backs.

With a mixture of wave fronts and wave backs, a portion of the wave surface will expand, and a portion will retract. Allowed to continue in this way, a circular (two-dimensional) domain evolves into a double-armed spiral, and a spherical (three-dimensional) domain evolves into a scroll. If the domain is sufficiently large, these become self-sustained reentrant patterns.

In resting tissue with no pacemaker activity, two stimuli are required to initiate a reentrant pattern. The first is required to set up a spatial gradient of recovery. Then, if the timing and location of the second is within the appropriate range, a single action potential that propagates in the backward, but not forward, direction can be initiated. This window of time and space is called the *vulnerable window* or *vulnerable period*. If the tissue mass is large enough or if there is a sufficiently long closed one-dimensional path, the retrograde propagation initiates a self-sustained reentrant pattern of activation.

## Sudden cardiac death

Death following a heart attack probably occurs via a much different mechanism. A heart attack occurs when there is a sudden occlusion of a coronary artery, stopping the flow of blood to a portion of the ventricular wall. Following this occlusion, cells become anoxic, and cell metabolism changes. There is a subsequent change in the internal osmotic pressure, followed by swelling of the cell. To prevent swelling, stretch-activated potassium channels release large quantities of potassium into the extracellular space, possibly rendering the cell self-oscillatory, but certainly changing the cell's resting potential. Gradually, gap junctions fail, and cells become electrically decoupled. Eventually, the cells die (a myocardial infarction), and form nonfunctioning scar tissue.

It is during the period of potassium extrusion preceding complete electrical decoupling that a reentrant arrhythmia is most likely to form. While the details are not known, there are several ingredients associated with their formation that are certain. First, there must be a region where propagation is blocked. Clearly, this is not sufficient, because in general, propagation simply goes around the blocked region and continues merrily on its way. However, if the region of block is on a one-dimensional conducting pathway and there is an alternative route by which an action potential can reach the blocked region from the opposite side, then a reentrant pattern is formed if the returning action potential successfully propagates through the blocked region in the retrograde direction. Such a region is called a region of *one-way block*, and we know that such regions can exist, for example, at points of fiber arborization (Section 11.1.2). One-dimensional paths with one-way block may be created by infarctions. Following occlusion of a coronary artery, tissue is highly inhomogeneous, and all sorts of strange conductive arrangements are possible, indeed likely.

A simple model shows why the initiation of a reentrant pattern is so dramatic. Suppose that there are cells located next to the exit from a one-dimensional path with one-way block (Fig. 14.24), and suppose that these cells are normally stimulated by some external pacemaker, with period $T$. We define the instantaneous frequency of stimulus as $\Delta T_{n+1} = t_{n+1} - t_n$, where $t_n$ is the $n$th firing. Now we take a simple kinematic description of propagation in the one-way path and suppose that the speed of propagation in the path is a function of the instantaneous period, $c = c(\Delta T)$. (Typically, $c$ is an increasing function of $\Delta T$.) Then the travel time around the one-way loop is $\frac{L}{c(\Delta T)}$. In cardiac tissue, the speed of an action potential is on the order of 0.5 m/s, so that travel time around the loop is much shorter than the period of external stimulus. Thus the wave on the loop typically returns to the stimulus site long before the next external stimulus arrives (i.e., we assume that $L/c < T$). If this travel time is larger than the absolute refractory period $T_r$ of the cells but smaller than $T$, the period of the external stimulus, then it stimulates the cells and initiates another wave around the loop. Thus,

$$\Delta T_{n+1} = t_{n+1} - t_n = \frac{L}{c(\Delta T_n)}, \tag{14.47}$$

provided that $T > \frac{L}{c(\Delta T_n)} > T_r$.

On the other hand, if this travel time is smaller than $T_r$, the stimulus is not successful, and the cells must await the next external stimulus before they fire, so that

$$\Delta T_{n+1} = t_{n+1} - t_n = T \tag{14.48}$$

if $\frac{L}{c(\Delta T_n)} < T_r$.

With this information, we can construct the map $\Delta T_n \mapsto \Delta T_{n+1}$ (shown in Fig. 14.25). There are obviously two branches for this map (shown as solid curves). Of interest are the fixed points of this map, corresponding to a periodic pattern of stimulus. The fixed point on the upper branch corresponds to the normal stimulus pattern from the external source, whereas the fixed point on the lower branch corresponds to a high-frequency, reentrant, pattern. The key feature of this map is that there is hysteresis between the two fixed points. In a "normal" situation (Fig. 14.25A), with $L$ small and $T$ large, the period of stimulus is fixed at $T$. However, as $L$ increases or as $T$ decreases, rendering $L > T_r c(\delta T_n)$, there is a "snap" onto the smaller-period fixed point, corresponding to initiation of a reentrant pattern (Fig. 14.25B). The pernicious nature

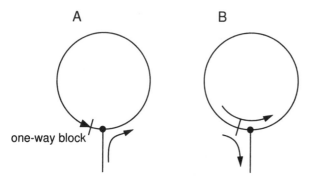

**Figure 14.24** Diagram of a conducting path with one-way block, preventing conduction from right to left. A: Conduction of a stimulus around the loop until it encounters refractoriness and fails to propagate further. B: Conduction of a reentrant pattern circulating continuously around the loop and exiting via the entry pathway on every circuit.

of the reentrant pattern is demonstrated by the fact that increasing the period back to previous levels does not restore the low-frequency pattern—the iterates of the map stay fixed at the lower fixed point, even though there are two possible fixed points. This is because the circulating pattern acts as a retrograde source of high-frequency stimulus on the original stimulus site, thereby masking its periodic activity.

Note that there are a number of ways that this reentrant pattern might be initiated. First, following a heart attack, a growing infarcted region may lead to a gradual increase in $L$, initiating the reentrant pattern while keeping $T$ fixed. On the other hand, an infarcted region may exist but remain static ($L$ fixed), and the reentrant pattern is initiated following a decrease in $T$, for example, during strenuous exercise. Thus, a static one-way loop acts like a "period bomb" (rather than a time bomb), ready to go off whenever the period is sufficiently low.

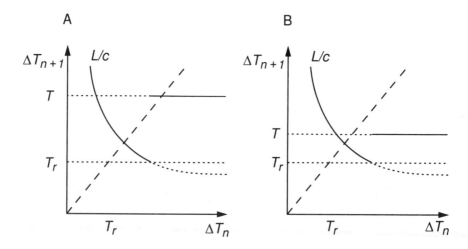

**Figure 14.25**  Next-interval map for a one-way conducting loop in two cases. A: With $T$ large, so that two stable steady solutions exist. B: With $T$ small, so that the only steady solution corresponds to reentry.

It is possible to initiate a reentrant arrhythmia in any healthy human heart using the two-stimulus protocol. However, the reason that we do not all succumb to reentrant arrhythmias (although many of us will) is that the spontaneous generation of reentrant patterns seems to require a substantial mass of damaged tissue (in this model, sufficiently large, but not too large, $L$). The most dangerous time for onset of a reentrant arrhythmia following a heart attack is when the infarcted area is in this critical size domain. It is perhaps not surprising that reentrant patterns also occur with high likelihood during reperfusion after an occlusion has been removed and blood flow restored to a region of tissue damage (although a full explanation of this requires a much more detailed model and analysis than that given here).

### Tachycardia and fibrillation

The two primary reentrant arrhythmias are *tachycardia* and *fibrillation*. Both of these can occur on the atria (*atrial tachycardia* and *atrial fibrillation*) or on the ventricles (*ventricular tachycardia* and *ventricular fibrillation*). When they occur on the atria, they are not life-threatening because there is little disruption of blood flow. However, when they occur on the ventricles, they are life-threatening. Ventricular fibrillation is fatal if it is not terminated quickly. Symptoms of ventricular tachycardia include dizziness or fainting, and sometimes rapid "palpitations."

Tachycardia is often classified as being either *monomorphic* or *polymorphic*, depending on the assumed morphology of the activation pattern. Monomorphic tachycardia is identified as having a simple periodic ECG, while polymorphic tachycardia is usually quasiperiodic, apparently the superposition of more than one periodic oscillation. A typical example of a polymorphic tachycardia is called *torsades de pointes*, and appears on the ECG as a rapid oscillation with slowly varying amplitude (Fig. 14.26). A vectorgram interpretation suggests a periodically rotating mean heart vector.

The simplest reentrant pattern is one for which the path of travel is a one-dimensional path. These were first studied by Mines (1914) when he intentionally cut a ring of tissue from around the superior vena cava and managed to initiate waves that traveled in only one direction. More complicated monomorphic tachycardias correspond to single spirals on the atrial surface (known as *atrial flutter*) or single scroll waves in the ventricular muscle. A three-dimensional view of a (numerically computed) monomorphic V-tach is shown in Fig. 14.27.

Stable monomorphic ventricular tachycardia is rare, as most reentrant tachycardias become unstable and degenerate into fibrillation. Thus, the clinical occurrence of stable monomorphic V-tach is considered an anomaly rather than the typical case.

Fibrillation is believed to correspond to the presence of many reentrant patterns moving throughout the ventricles in continuous, perhaps erratic, fashion, leading to an uncoordinated pattern of ventricular contraction and relaxation. A surface view of a (numerically computed) fibrillatory pattern is shown in Fig. 14.28.

The likely reason that monomorphic V-tach is rare is because there are a number of potential instabilities, although the mechanism of the instability has not been decisively determined. Some possibilities are discussed by a number of authors (Courtemanche

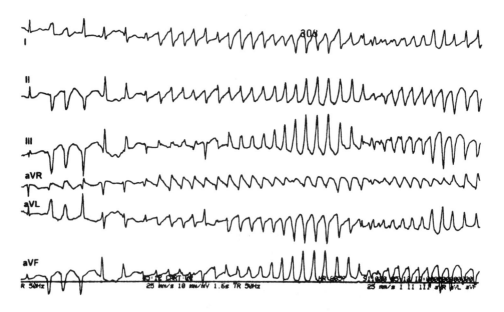

**Figure 14.26** A six-lead ECG recording of *torsades de pointes*. (Zipes and Jalife, 1995, Fig. 79-1, p. 886.)

and Winfree, 1991; Karma, 1993, 1994; Panfilov and Holden, 1990; Panfilov and Hogeweg, 1995; Bar and Eiswirth, 1993; Courtemanche et al., 1993). Suffice it to say, whatever the form and evolution of a reentrant pattern, all are dangerous, so we now devote our attention to the important problem of how to get rid of a reentrant pattern, whether stable or erratic.

## 14.4   Defibrillation

Nearly everyone in the United States knows something about defibrillation. They have all seen television shows where the paramedic places paddles on the chest of a man who has unexpectedly collapsed, yells "Clear!" and then a jolt of electricity shakes the body of the victim. Mysteriously, the victim revives. Since they were first made in 1947, defibrillators have saved many lives, and the recent development of implantable defibrillators will no doubt extend the lives of many people who in a previous era would have died from their first heart attack. The goal of a defibrillator is clear. Since during fibrillation different regions of tissue are in different phases of electrical activity, some excited, some refractory, some partially recovered, the purpose of defibrillation is to give an electrical impulse that stimulates the entire heart, so that the electrical activity is once again coordinated and will return to rest as a whole to await the next normal SA nodal stimulus. Said another way, the purpose is to reset the phase of each cardiac cell so that all cells are in phase.

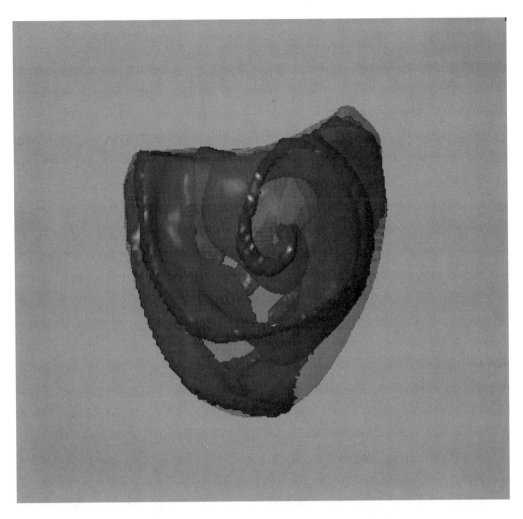

**Figure 14.27** Numerically computed scroll wave in ventricular muscle. (Panfilov and Keener, 1995, p. 685, Fig. 3a.)

While it is known that defibrillation works (and is accomplished thousands of times daily around the world), the dilemma we face is that simple mathematical models fail to explain how this can happen, and indeed, seem to suggest that defibrillation shocks cannot achieve their goal.

To understand the dilemma, consider the numerical calculation shown in Fig. 14.29. Here is shown the result of applying a stimulus to the ends of a bidomain cable. The stimulus on the left is depolarizing and on the right is hyperpolarizing. On the left, a right-moving wave is initiated almost immediately, and on the right a left-moving wave is initiated via anode break excitation (see Chapter 4, Exercises 7 and 12 ). The dilemma, however, is that local stimuli can only have local effects.

**Figure 14.28** Surface view of fibrillatory reentrant activity in the ventricles (computed by A. Panfilov).

A similar conclusion is drawn from Fig. 14.30. Here is shown a periodic traveling wave on a one-dimensional bidomain cable, traveling from left to right. (If the left and right ends were connected, this wave would circulate around the ring indefinitely.) What cannot be seen in this figure is that a large stimulus was applied at the ends of the cable between the first and second traces, simulating defibrillation. What can be seen from this figure is that the stimulus has essentially no effect on the traveling wave. This stimulus has no chance of defibrillating the cable, since the effects of the stimulus are local only.

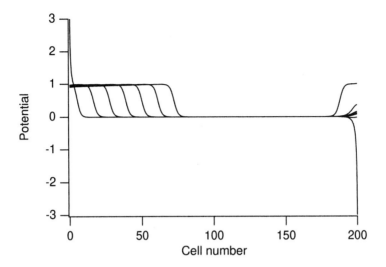

**Figure 14.29** Response of a uniform cable to a stimulus of duration $t = 0.2$ applied at the ends of the cable. Traces shown start at time $t = 0.1$ and with equal time steps $\Delta t = 0.2$.

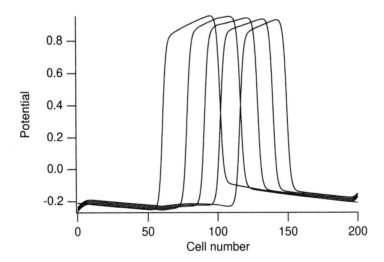

**Figure 14.30** A traveling wave in a uniform cable following application of a stimulus at the ends of the cable at a time between the first and second traces.

The question, then, is how can defibrillation work, if only those regions close to the stimulating source are excited by the stimulus. The likely answer is that the medium into which the stimulus is applied has small inhomogeneities that are not accounted for in a uniform cable model.

To see the effect these spatial inhomogeneities might have, we consider what happens when a brief but large current is applied to an inhomogeneous one-dimensional strand of cardiac tissue. The inhomogeneity comes from the fact that much of the intracellular resistance is concentrated in the gap junctions. We take the bidomain model equations (11.1), but now assume that the conductivity $\sigma_i = A_i/R_i$ is continuous but nonconstant to reflect the occurrence of gap junctions. We take $\sigma_e = A_e/R_e$ to be constant. Then, the monodomain reduction for a one-dimensional cable gives from (11.37)

$$p\left(C_m \frac{\partial V}{\partial t} - \frac{f(V)}{R_m}\right) = \frac{\partial}{\partial x}\left(\frac{\sigma_i \sigma_e}{\sigma_i + \sigma_e} \frac{\partial V}{\partial x}\right) + \frac{\partial}{\partial x}\left(\frac{\sigma_i}{\sigma_i + \sigma_e}\right) I(t). \qquad (14.49)$$

The important observation is that if $\sigma_i$ is nonconstant, the new term in (14.49) acts as a source term everywhere throughout the medium, and it is this source term that we wish to exploit.

A physical explanation of this current source term is as follows. A current that is applied at one end of the fiber will flow to the opposite end of the fiber following the path of least resistance by dividing itself between two paths, the intracellular and extracellular paths. If both paths are homogeneous, the current is quickly divided into two parts, where it stays until it is forced back into the extracellular space at the far end. However, the intercellular inhomogeneities of resistance act much like speed bumps on a two-lane thoroughfare. As traffic nears the speed bump in one lane, it will merge into the clear lane to avoid slowing down, but once past the speed bump, it will again split back into a two-lane flow. It is the merging of traffic between the two lanes that is the analogy of a transmembrane current and that makes it possible to depolarize and hyperpolarize individual cells at their ends.

It is useful to introduce dimensionless variables $\tau = t/\tau_m, y = x/\lambda_m$, where $\tau_m = C_m R_m, \lambda_m^2 = \sigma_e D R_m/p$, and we then obtain

$$V_\tau - f(V) = \left(\frac{d}{D} V_y\right)_y - J(\tau)\left(\frac{d}{D}\right)_y, \qquad (14.50)$$

where $d = \sigma_i/(\sigma_i + \sigma_e)$, $D^{-1}$ is the average value of $d^{-1}$, and $J(\tau) = D R_m I(\tau)/(p\lambda_m)$. In addition, we have boundary conditions $V_y = -J(\tau)$ at $y = 0$ and at $y = Y = L/\lambda_m$.

There are potentially many different spatial scales for resistive inhomogeneities. However, here we want to focus on the effect of resistive inhomogeneities on the spatial scale of individual cells, and so, we suppose that $d/D$ is a periodic function of $y$, with period $\epsilon = l/\lambda_m$, where $l$ is the cell length. Specifically, we take

$$d = d\left(\frac{y}{\epsilon}\right), \qquad (14.51)$$

with $d(y)$ a function of period 1. Typically, $\epsilon$ is a small number, on the order of 0.1.

Now we are able to use homogenization arguments to separate (14.50) into two equations that describe the behaviors on different spatial scales (Exercise 14). The result of a standard multiscale calculation gives

$$V(z, \tau, \eta) = u_0(z, \tau) + \epsilon J(\tau) W(\eta) - \epsilon W(\eta) \frac{\partial u_0(z, \tau)}{\partial z}, \tag{14.52}$$

where $\eta = z/\epsilon$,

$$\frac{dW}{d\eta} = 1 - \frac{D}{d(\eta)}, \tag{14.53}$$

and

$$\frac{\partial u_0}{\partial \tau} - \overline{f\,(V(z, \tau, \eta))} = \frac{\partial^2 u_0}{\partial z^2}, \tag{14.54}$$

where

$$\overline{f\,(V(z, \tau, \eta))} = \frac{\epsilon}{Y} \int_0^{Y/\epsilon} f\,(V(z, \tau, \eta))\, d\eta. \tag{14.55}$$

The interpretation of (14.54) is significant. While a current stimulus is being applied, the response at the cellular level has an effect that is communicated to the tissue on a macroscopic scale through the nonlinearity of the ionic currents. (If the ionic current $f(V)$ were linear, the applied stimulus would have no global effect, since it would have zero average.)

To get some insight into the dynamics while the stimulus is applied, we take the simple model for gap-junctional resistance

$$r_i = r_c + \frac{r_g}{l} \delta(\eta) \tag{14.56}$$

on the interval $0 \le \eta < 1$, and periodically extended from there. Here, $r_c$ is the intracellular cytoplasmic resistance per unit length, and $r_g$ is the gap-junctional resistance per cell. The function $\delta(\eta)$ is any positive function with small support and area one unit that represents the spatial distribution of the gap-junctional resistance, for example, the Dirac delta function.

In the specific case that $\delta(\eta)$ is the Dirac delta function, we calculate that

$$W'(\eta) = R_g(1 - \delta(\eta)), \tag{14.57}$$

where $R_g = \frac{r_g}{r_g + l(r_c + r_e)}$ is the fraction of the total resistance per unit length that is concentrated into the gap junctions. Here we used the fact that $r_i = 1/\sigma_i$. Then, $W(\eta)$ is given by

$$W(\eta) = R_g\left(\eta - \frac{1}{2}\right), 0 \le \eta < 1, \tag{14.58}$$

and $W(\eta + 1) = W(\eta)$.

The function $W(\eta)$ is a *sawtooth function*, and according to (14.52), when a stimulus is applied, the membrane potential is the sum of two components, the sawtooth function $W(\eta)$ on the spatial scale of cells and $u(y, \tau)$ on the macroscopic scale of the

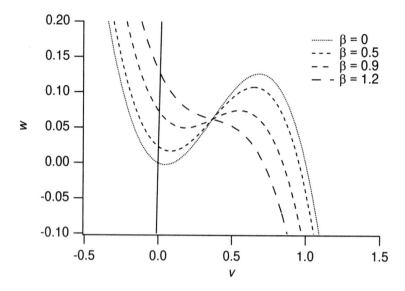

**Figure 14.31** Nullclines for FitzHugh–Nagumo dynamics, $w = v(1 - v)(\alpha - v) + \frac{\beta^2}{12}(1 + \alpha - 3v)$, $w = v/\gamma$, modified to include the effects of a current stimulus, for several values of $\beta$.

tissue. The effect of the small-scale oscillatory behavior on the larger scale problem is found by averaging, whereby

$$\overline{F}(u, \beta) = \overline{f(V(z, \tau, \eta))} = \int_{-\frac{1}{2}}^{\frac{1}{2}} f(u + \beta\eta) \, d\eta \tag{14.59}$$

and $\beta = R_g \epsilon(J(t) - \frac{\partial u}{\partial z})$. In other words, the effect of the current stimulus is to modify the ionic current through local averaging over the cell. The details of the structure of $W(\eta)$ are not important because they are felt only in an average sense. For the cubic model $f(V) = V(V - 1)(\alpha - V)$, $\overline{F}$ can be calculated explicitly to be

$$\overline{F}(V, \beta) = f(V) + \frac{\beta^2}{12}(1 + \alpha - 3V). \tag{14.60}$$

A plot of this function for different values of $\beta$ is shown in Fig. 14.31.

Before examining this model for its ability to explain defibrillation, we discuss the simpler problem of direct activation of resting tissue.

## 14.4.1   The Direct Stimulus Threshold

Direct activation (or field stimulation) occurs if all or essentially all of the tissue is activated simultaneously without the aid of a propagated wave front. According to the model (14.54), it should be possible to stimulate cardiac tissue directly with brief stimuli of sufficiently large amplitude.

A numerical simulation demonstrating how direct stimulation can be accomplished for the bistable equation is shown in Fig. 14.32. In this simulation a one-dimensional array of 200 cells was discretized with five grid points per cell, and a brief, large current was injected at the left end and removed at the right end of the cable. In Fig. 14.29 is shown the response to the stimulus when the cable is uniform. Shown here is the membrane potential, beginning at time $t = 0.1$, and at later times with equal time steps $\Delta t = 0.2$. The stimulus duration was $t = 0.2$, so its effects are seen as a depolarization on the left and hyperpolarization on the right in the first trace. As noted above, a wave is initiated from the left from superthreshold depolarization, and a wave from the right is initiated by anode break excitation.

The same stimulus protocol produces a substantially different result if the cable has nonuniform resistance. In Fig. 14.32 is shown the response of the discretized cable with high resistance at every fifth node, at times $t = 0.15, 0.25$, and $0.35$, with a stimulus duration of $0.2$. The first curve, at time $t = 0.15$, is blurred because the details of the membrane potential cannot be resolved on this scale. However, the overall effect of the rapid spatial oscillation is to stimulate the cable directly, as seen from the subsequent traces.

To analyze this situation, note that since direct activation occurs without the benefit of propagation, it is sufficient to ignore diffusion and the boundary conditions and simply examine the behavior of the averaged ordinary differential equation

$$\frac{dV}{d\tau} = \overline{F}(V, \beta). \tag{14.61}$$

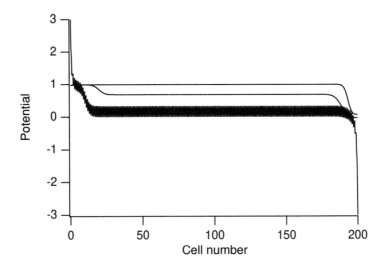

**Figure 14.32** Response of a nonuniform cable with regularly spaced high-resistance nodes to a stimulus of duration $t = 0.2$ applied at the ends of the cable. The traces show the response at time $t = 0.15$, during the stimulus, and at times $t = 0.25, 0.35$ after the stimulus has terminated.

For any resting excitable system, it is reasonable to assume that $f(V) < 0$ for $0 < V < \theta$, where $\theta$ is the threshold that must be exceeded to stimulate an action potential. To directly stimulate a medium that is initially at rest with a constant stimulus, one must apply the stimulus until $V > \theta$. The minimal time to accomplish this is given by the strength–duration relationship,

$$T = \int_0^\theta \frac{dV}{\overline{F}(V, \beta)}. \tag{14.62}$$

Clearly, this expression is meaningful only if $\beta$ is sufficiently large that $\overline{F}(V, \beta) > 0$ on the interval $0 < V < \theta$. In other words, there is a minimal stimulus level (a threshold) below which the medium cannot be directly stimulated.

## 14.4.2 The Defibrillation Threshold

While its threshold cannot be calculated in the same way as for direct stimulus, the mechanism of defibrillation can be understood from simple phase-plane arguments. To study defibrillation, we must include the dynamics of recovery in our model equations. Thus, for simplicity and to be specific we take FitzHugh–Nagumo dynamics

$$I_{\text{ion}}(v, w) = -f(v) + w, \tag{14.63}$$
$$w_\tau = g(v, w), \tag{14.64}$$

with $f(v) = v(v - 1)(\alpha - v)$, and assume that the parameters are chosen such that reentrant waves are persistent. This could mean that there is a stable spiral solution, or it could mean that the spiral solution is unstable but some nonperiodic reentrant motion is persistent. Either way, we want to show that there is a threshold for the stimulating current above which reentrant waves are terminated.

The mechanism of defibrillation is easiest to understand for a periodic wave on a one-dimensional ring, but the idea is similar for higher-dimensional reentrant patterns. For a one-dimensional ring, the phase-portrait projection of a rotating periodic wave is a closed loop. From singular perturbation theory, we know that this loop clings to the leftmost and rightmost branches of the nullcline $w = f(v)$ and has two rapid transitions connecting these branches, and these correspond to wave fronts and wave backs (recall Fig. 9.12).

According to our model, the effect of a stimulus is to temporarily change the $v$ nullclines and thereby to change the shape of the closed loop. After the stimulus has ended, the distorted closed loop will either go back to a closed loop, or it will collapse to a single point on the phase portrait and return to the rest point. If the latter occurs, the medium has been "defibrillated."

Clearly, if $\beta$ is small and the periodic oscillation is robust, then the slight perturbation is insufficient to destroy it. On the other hand, if $\beta$ is large enough, then the change is substantial and collapse may result.

There are two ways that this collapse can occur. First, and easiest to understand, if the nullcline for nonzero $\beta$ is a monotone curve (as in Fig. 14.31 with $\beta = 1.2$), then the open loop collapses rapidly to a double cover of the single curve $w = \overline{F}(v, \beta)$, from where it further collapses to a single point in phase space. For the specific cubic model, this occurs if $\beta^2 > \frac{4}{3}(1 - \alpha + \alpha^2)$.

The $v$ nullcline need not be monotone to effect a collapse of the periodic loop. In fact, as $\beta$ increases, the negative-resistance region of the nullcline becomes smaller, and the periodic loop changes shape into a loop with small "thickness" (i.e., with little separation between the front and the back) and with fronts and backs that move with nearly zero speed. Indeed, if the distorted front is at a large enough $w$ "level" (in sense of singular perturbation theory), it cannot propagate at all and stalls, leading to a collapse of the wave. Another way to explain this is to say that with large enough $\beta$, the "excitable gap" between the refractory tail and the excitation front is excited, pushing the wave front forward (in space) as far as possible into the refractory region ahead of it, thereby causing it to stall, and eventually to collapse.

This scenario can be seen in Fig. 14.33, where the results of numerical simulations for a one-dimensional nonuniform cable are shown. Earlier, in Fig. 14.30, we showed a wave, propagating to the right, at equal time steps of $\Delta t = 0.55$. To simulate a reentrant arrhythmia, this wave was chosen so that were this cable a closed loop, the wave would circulate around the loop indefinitely without change of shape. It is not apparent from this previous figure that a stimulus of duration $t = 0.3$ was applied at the ends of the cable between the first and second traces, because in a uniform cable a stimulus at the boundary has little effect on the interior of the medium.

In Fig. 14.33 is shown exactly the same sequence of events for a nonuniform cable. This time, however, the applied stimulus (between the first and second traces) induces a rapidly oscillating membrane potential on the spatial scale of cells (not shown), which has the average effect (because of nonlinearity) of "pushing" the action potential forward as far as possible. This new front cannot propagate forward because it has been pushed into its refractory tail and has stalled. In fact, the direction of propagation reverses, and the action potential collapses as the front and back move toward each other.

To illustrate further this mechanism of defibrillation in a two-dimensional domain, numerical simulations were performed using a standard two-variable model of an excitable medium and using the full bidomain model derived in Section 11.3 (Keener and Panfilov, 1996). Parameters for the excitable dynamics were chosen such that spirals are not stable, but exhibit breakup and develop into "chaotic" reentrant patterns (Panfilov and Hogeweg, 1995), thereby giving a reasonable model of cardiac fibrillation (see Chapter 10, Exercise 5).

Some time after initiating a reentrant wave pattern, a constant stimulus (of duration about 2.5 times the duration of the action potential upstroke) was applied uniformly to the sides of the rectangular domain. Because the stimulus was applied uniformly along the sides of the domain, the stimulus parameter $\beta$ was constant throughout the medium.

In Fig. 14.34a is shown an irregular reentrant pattern just before the stimulus is applied. In this picture, the darkest regions are excited tissue, white denotes recovered

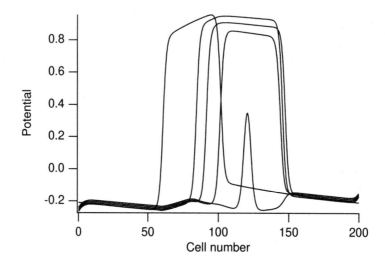

**Figure 14.33** A traveling wave in a nonuniform cable following application of a stimulus at the boundary.

and excitable tissue, and grey is refractory tissue. Following a stimulus (with $\beta = 0.86$), the excited region expanded to include essentially all of the recovered tissue, as the excitable gap was eliminated by a polarizing stimulus. Shortly thereafter (Fig. 14.34c, $t = 12$) the activation collapsed, leaving behind only recovered or refractory tissue, which shortly thereafter returned to uniform rest. The extensive patterning seen in this last figure shows a mixture of refractory and recovered tissue, but since it contains no excited tissue (except a small patch at the lower left corner that is propagating out of the domain), it cannot become reexcited, but must return to rest. (The similar fates of recovered and refractory regions can be seen in Fig. 14.35b, where the patterning has nearly disappeared.)

Defibrillation is unsuccessful with a smaller stimulus $\beta = 0.84$. The pattern at time $t = 12$ is shown in Fig. 14.35a and is similar to Fig. 14.34c, which was successful. Here, however, after the stimulus and subsequent collapse of much of the excitation, one small excited spot remains at the upper left-hand corner of the medium, which eventually evolves into a double spiral pattern (Fig. 14.35b,c), reestablishing a reentrant arrhythmia.

## 14.5  Appendix: The Phase Equations

Because coupled oscillators arise so frequently in mathematical biology and physiology, there is an extensive literature devoted to their study. In this book, coupled oscillators play a role in the sinoatrial node and in the digestive system. An important model for

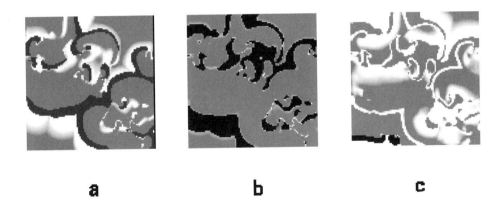

**Figure 14.34** Successful defibrillation of a two-dimensional region.

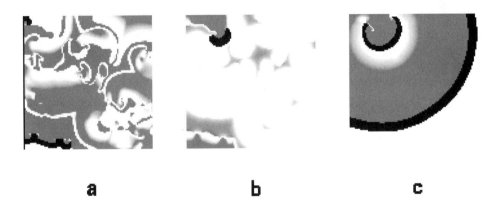

**Figure 14.35** Unsuccessful defibrillation of a two-dimensional region.

the study of coupled oscillators has been the *phase equation*, an equation describing the evolution of the phases of a loosely coupled collection of similar oscillators.

In this appendix we give a derivation of the phase equation. This derivation is similar to that of Neu (1979), using perturbation techniques and a multiscale analysis. A more technical derivation is given by Ermentrout and Kopell (1984). The attempt to describe populations of oscillators in terms of phases was first made by Winfree (1967), and in the context of reaction–diffusion systems, by Ortoleva and Ross (1973, 1974). This was improved by Kuramoto (Kuramoto and Tsuzuki, 1976; Kuramoto and Yamada, 1976). Neu's derivation is also discussed by Murray (1989).

To set the stage, consider the system of equations

$$\frac{dx}{dt} = \Lambda(r)x - \omega(r)y, \tag{14.65}$$

$$\frac{dy}{dt} = \omega(r)x + \Lambda(r)y, \tag{14.66}$$

where $r^2 = x^2 + y^2$. Systems of this form are called *lambda–omega* ($\Lambda$–$\omega$) systems, and are special because by changing to polar coordinates, $x = r\cos\theta$, $y = r\sin\theta$, (14.65) and (14.66) can be written as

$$\frac{dr}{dt} = r\Lambda(r), \tag{14.67}$$

$$\frac{d\theta}{dt} = \omega(r). \tag{14.68}$$

This system has a stable limit cycle at any radius $r > 0$ for which $\Lambda(r) = 0$, $\Lambda'(r) < 0$. The periodic solution travels around this circle with angular velocity $\omega(r)$. Starting from any given initial conditions, the solution of (14.65) and (14.66) will eventually settle onto a regular oscillation with fixed amplitude and period $\frac{\omega(r)}{2\pi}$. Hence, in the limit as $t \to \infty$, the system is described completely by its angular velocity around a circle.

Now suppose that we have two similar systems, one with a limit cycle of amplitude $R_1$ and angular velocity $\omega_1$, and the other with a limit cycle of amplitude $R_2$ and angular velocity $\omega_2$. If there is no coupling between the systems, each will oscillate at its own frequency, unaffected by the other, and in the four-dimensional phase space the solutions will approach the torus $r_1 = R_1$, $r_2 = R_2$, moving around the torus with angular velocities $\theta'_1 = \omega_1$, $\theta'_2 = \omega_2$. Since all solutions eventually end up winding around the torus, and since any solution that starts on the torus cannot leave it, the torus is called an *attracting invariant torus*. The flow on the torus can be described entirely in terms of the rates of change of $\theta_1$ and $\theta_2$. In this case $(\theta_1 - \theta_2)' = \omega_1 - \omega_2$, and so the phase difference increases at a constant rate. Thus, analogously to the one-dimensional system discussed above, in the limit as $t \to \infty$, the original system of four-differential equations can be reduced to a two-dimensional system describing the flow on a two-dimensional torus.

If our two similar systems are now loosely coupled, so that each oscillator has only a small effect on the other, it is reasonable to expect (and indeed it can be proved; see Rand and Holmes (1980) for a nice discussion of this, and Hirsch et al. (1977) for a proof) that the invariant torus persists, changing its shape and position by only a small amount. In this case, the longtime solutions for $r_1$ and $r_2$ remain essentially unchanged, with $r_1 = R_1 + O(\epsilon)$ and $r_2 = R_2 + O(\epsilon)$, where $\epsilon \ll 1$ is the strength of the coupling. However, the flow *on* the torus could have a drastically different nature, as the phase difference need no longer simply increase at a constant rate. Hence, although the structure of the torus is preserved, the properties of the flow on the torus are not.

In general, the flow on the torus is described by

$$\frac{d\theta_i}{dt} = \omega_i + f_i(r_1, r_2, \theta_1, \theta_2, \epsilon), \qquad i = 1, 2, \tag{14.69}$$

but since $r_1 = R_1 + O(\epsilon)$ and $r_2 = R_2 + O(\epsilon)$, to lowest order in $\epsilon$ this simplifies to

$$\frac{d\theta_i}{dt} = \omega_i + h_i(\theta_1, \theta_2, \epsilon), \qquad i = 1, 2. \tag{14.70}$$

In general, $R_1$ and $R_2$ appear in (14.70), the so-called *phase equation*, but the independent variables $r_1$ and $r_2$ do not. It follows that the full four-dimensional system that describes the two coupled oscillators can be understood in terms of a simpler system describing the flow on a two-dimensional invariant torus.

To derive the equations describing this flow on a torus, we assume that we have a coupled oscillator system that can be written in the form

$$\frac{du_i}{dt} = F(u_i) + \epsilon G_i(u_i) + \epsilon \sum_{j=1}^{N} a_{ij} H(u_j). \tag{14.71}$$

Here, $u_i$ is the vector of state variables for the $i$th oscillator, the coefficients $a_{ij}$ represent the coupling strength, and the function $H(u)$ determines the effect of coupling. For simplicity, we assume that $H$ is independent of $i$ and $j$.

To get the special case of SA nodal coupling in (14.13), we take $a_{ij} = d_{ij}$ for $i \neq j$, $a_{ii} = -\sum_{j \neq i} d_{ij}$, and $H(u) = Du$. We take this general form of $H$ to allow for synaptic as well as diffusive coupling.

Next, we assume that when $\epsilon = 0$ we have a periodic solution, i.e., that the equation

$$\frac{du}{dt} = F(u) \tag{14.72}$$

has a stable periodic solution, $U(t)$, scaled to have period one. Note that because of the functions $G_i$, we are not assuming that each oscillator is identical. Thus, the natural frequency of each oscillator is close to, but not exactly, one.

The model system (14.71) is a classic problem to which the *method of averaging* or the *multiscale method* can be applied. Specifically, since $\epsilon$ is small, we expect the behavior of (14.71) to be dominated by the periodic solution $U(t)$ of the unperturbed problem and that deviations from this behavior occur on a much slower time scale. To accommodate two different time scales, we introduce two time-like variables, $\sigma = \omega(\epsilon)t$ and $\tau = \epsilon t$, as fast and slow times, respectively. Here, $\omega$ is a function, as yet unknown, of order 1. Treating $\sigma$ and $\tau$ as independent variables, we find from the chain rule that

$$\frac{d}{dt} = \omega(\epsilon)\frac{\partial}{\partial \sigma} + \epsilon \frac{\partial}{\partial \tau}, \tag{14.73}$$

and accordingly, (14.71) becomes

$$\omega(\epsilon)\frac{\partial u_i}{\partial \sigma} + \epsilon \frac{\partial u_i}{\partial \tau} = F(u_i) + \epsilon G_i(u_i) + \epsilon \sum_{j=1}^{N} a_{ij} H(u_j). \tag{14.74}$$

Next we suppose that $u_i$ and $\omega(\epsilon)$ have power series expansions in $\epsilon$, given by

$$u_i = u_i^0 + \epsilon u_i^1 + \cdots, \qquad \omega(\epsilon) = 1 + \epsilon \Omega_1 + \cdots \tag{14.75}$$

Note that the first term in the expansion for $\omega$ is the frequency of the unperturbed solution, $U$. Expanding (14.74) in powers of $\epsilon$ and gathering terms of like order, we find a hierarchy of equations, beginning with

$$\frac{\partial u_i^0}{\partial \sigma} = F(u_i^0),\tag{14.76}$$

$$\frac{\partial u_i^1}{\partial \sigma} - F_u(u_i^0)u_i^1 = G_i(u_i^0) + \sum_{j=1}^{N} a_{ij}H(u_j^0) - \Omega_1\frac{\partial u_i^0}{\partial \sigma} - \frac{\partial u_i^0}{\partial \tau}.\tag{14.77}$$

Equation (14.76) is easy to solve by taking

$$u_i^0 = U(\sigma + \delta\theta_i(\tau)).\tag{14.78}$$

The phase shift $\delta\theta_i(\tau)$ allows each cell to have different phase shift behavior, and it is yet to be determined.

Next, observe that $\frac{d}{d\sigma}(\frac{dU}{d\sigma} - F(U)) = \frac{\partial U'}{\partial \sigma} - F_u(u_i^0)U' = 0$, so that the operator $LU = \frac{\partial U}{\partial \sigma} - F_u(u_i^0)U$ has a null space spanned by $U'(\sigma + \delta\theta_i(\tau))$. The null space is one-dimensional because the periodic solution is assumed to be stable. It follows that the adjoint operator $L^*y = -\frac{\partial y}{\partial \sigma} - F_u(u_i^0)^Ty$ has a one-dimensional null space spanned by some periodic function $y = Y(\sigma + \delta\theta_i(\tau))$. (It is a consequence of Floquet theory that the Floquet multipliers of the operator $L$, say $\mu_i$, and the Floquet multipliers of $L^*$, say $\mu_i^*$, are multiplicative inverses, $\mu_i\mu_i^* = 1$ for all $i$. Since a periodic solution has Floquet multiplier 1 and there is only one periodic solution for $L$, there is also precisely one periodic solution for the adjoint operator $L^*$. See Exercise 18.) Without loss of generality we scale $Y$ so that $\int_0^1 U'(\sigma) \cdot Y(\sigma)d\sigma = 1$. Therefore, for there to be a periodic solution of (14.77), the right-hand side of (14.77) must be orthogonal to the null space of the adjoint operator $L^*$. This requirement translates into the system of differential equations for the phase shifts

$$\frac{d}{d\tau}\delta\theta_i = \xi_i - \Omega_1 + \sum_j a_{ij}h(\delta\theta_j - \delta\theta_i),\tag{14.79}$$

where

$$\xi_i = \int_0^1 Y(\sigma)G_i(U(\sigma))d\sigma,\tag{14.80}$$

$$h(\phi) = \int_0^1 Y(\sigma)H(U(\sigma + \phi))d\sigma.\tag{14.81}$$

The numbers $\xi_i$ are important because they determine the approximate natural (i.e., uncoupled) frequency of the $i$th cell. This follows from the fact that when $a_{ij} = 0$, a simple integration gives $\delta\theta_i = \epsilon t(\xi_i - \Omega_1)$, and thus

$$u_i^0 = U(\omega(\epsilon)t + \delta\theta_i) = U((1 + \epsilon\xi_i)t).\tag{14.82}$$

Hence, the uncoupled frequency of the $i$th cell is $2\pi(1 + \epsilon\xi_i)$. Therefore, $\xi_i$ can (presumably) be measured or estimated without knowing the function $G_i(u)$. The function $h(\phi)$ can be determined numerically (see, for example, Exercise 17.)

The function $h(\phi)$ has an important physical interpretation, being a *phase resetting function*. That is, $h(\phi)$ shows the effect of one oscillator on another when the two have phases that differ by $\phi$. For example, in the case of two identical oscillators, the two phase shifts are governed by

$$\frac{d}{d\tau}\delta\theta_1 = -\Omega_1 + a_{12}h(\delta\theta_2 - \delta\theta_1) + a_{11}h(0), \tag{14.83}$$

$$\frac{d}{d\tau}\delta\theta_2 = -\Omega_1 + a_{21}h(\delta\theta_1 - \delta\theta_2) + a_{22}h(0). \tag{14.84}$$

Thus, when $a_{12}h(\delta\theta_2 - \delta\theta_1) > 0$, the phase of oscillator 1 is advanced, while if $a_{12}h(\delta\theta_2 - \delta\theta_1) < 0$, the phase of oscillator 1 is retarded. Furthermore, it is not necessarily the case that $h(0) = 0$, so that identical oscillators with identical phases may nonetheless exert a nontrivial influence on each other.

The system of equations (14.79) can be written in terms of phase differences by defining $\Phi$ as the vector of consecutive phase differences and defining $\overline{\delta\theta}$ as the average of the phase shifts, $\overline{\delta\theta} = \frac{1}{N}\sum_{i=1}^{N}\delta\theta_i$. In terms of these variables the system (14.79) can be written in the form

$$\frac{d\Phi}{d\tau} = \Delta + C(\Phi), \tag{14.85}$$

where $\Delta$ is the vector of consecutive differences of $\xi_i$. This is a closed system of $N - 1$ equations. *Phase locking* is defined as the situation in which there is a stable steady solution of (14.85), a state in which the phase differences of the oscillators do not change (see Chapter 21).

## 14.6 Exercises

1. (a) The fundamental solution of Poisson's equation in free space with a unit source at the origin ($\nabla^2\phi = -\delta(x)$) is $\phi(x) = \frac{1}{4\pi|x|}$.
   Find the solution of Poisson's equation with a source at the origin and a sink of equal strength at $x = x_1$, and let $|x_1| \to 0$. What must you assume about the strength of the source and the sink in order to obtain a nonzero limiting potential?

   (b) Find the solution of Poisson's equation with a dipole source by solving the problem $\nabla^2\phi = \frac{1}{\epsilon}(\delta(x - \epsilon v) - \delta(x))$ with $|v| = 1$, and then taking the limit $\epsilon \to 0$.

2. Determine the heart rate for the ECG recording in Fig. 14.2a. The subject was sedated at the time this recording was made. What are the effects of the sedation?

3. Identify the different deflections in the ECG recording shown in Fig. 14.36. What can you surmise about the nature of propagation for the extra QRS complex (called an *extrasystole*). Because of the apparent periodic coupling between the normal QRS and the extrasystole, this rhythm is called a *ventricular bigeminy*.

4. (a) Suggest a diagnosis for the ECG recording in Fig. 14.37.
   Hint: What does the inverted P-wave suggest? What is the heart rate?

   (b) What possible mechanisms can account for the failure of the SA node to generate the heartbeat?

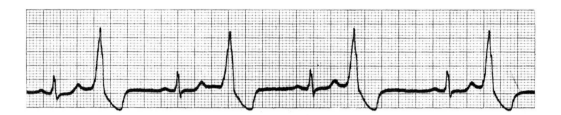

**Figure 14.36** ECG recording of a ventricular extrasystole for exercise 3. (Rushmer, 1976, Fig 8-25, p. 314, originally from Guneroth, 1965.)

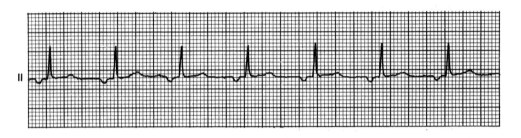

**Figure 14.37** ECG for Exercise 4a. Can you determine the nature of this abnormality? (Goldberger and Goldberger, 1994, p. 45.)

5. Estimate the mean deflection of the QRS complex in each of the six standard leads (I, II, III, aVR, aVL, aVF) in Fig. 14.6 and then estimate the mean heart vector for the normal heartbeat.

6. Improve your skill at reading ECGs by finding the mean heart vector for the QRS complexes in Figs. 14.3, 14.8, and 14.9. Why is hypertrophy the diagnosis for Figs. 14.8 and 14.9? In what direction are the heart vectors deflected in these figures?

7. Find the mean heart vector for the ECG recording shown in Fig. 14.38. What mechanism can you suggest that accounts for this vector? Hint: Notice that the amplitude is substantially smaller than normal, suggesting loss of tissue mass (infarction), and determine the location of this loss by determining the deflection of the heart vector from normal. Is the deflection toward or away from the location of tissue loss?

8. Consider the following simple model of a forced periodic oscillator, called the *Poincaré oscillator* (also called a *radial isochron clock* or a *snap back oscillator*; Guevara and Glass, 1982; Hoppensteadt and Keener, 1982; Keener and Glass, 1984; Glass and Kaplan, 1995). A point is moving counterclockwise around a circle of radius 1. At some point of its phase, the point is moved horizontally by an amount $A$ and then allowed to instantly "snap back" to radius 1 moving along a radial line toward the origin (see Fig. 14.39).

   (a) Determine the *phase resetting curve* for this process. That is, given the phase $\theta$ before resetting, find the phase $\phi$ after resetting the clock. Plot $\phi$ as a function of $\theta$ for several values of $A$.

   (b) Show that for $A < 1$ the phase resetting function is a *type 1* map, satisfying $\phi(\theta + 2\pi) = \phi(\theta) + 2\pi$.

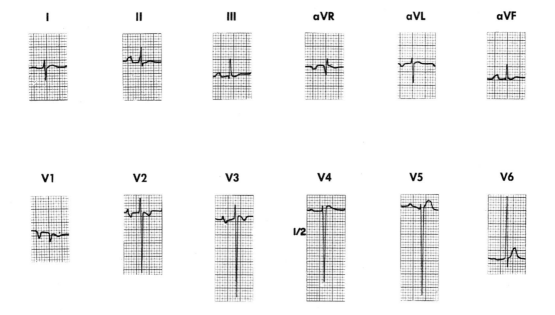

I  II  III  aVR  aVL  aVF

V1  V2  V3  V4  V5  V6

**Figure 14.38** ECG for Exercise 7. Can you determine the nature of this abnormality? (Rushmer, 1976, Fig. 8-51, p. 343.)

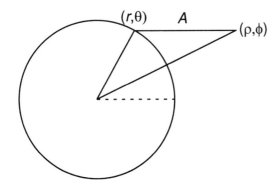

$(r,\theta)$ $A$ $(\rho,\phi)$

**Figure 14.39** Diagram of phase resetting for the Poincaré oscillator (Exercise 8).

(c) Show that for $A > 1$, the phase resetting function is a *type 0* map, for which $\phi(\theta + 2\pi) = \phi(\theta)$.

(d) Show that there is a *phase singularity*, that is, values of $A$ and $\theta$ for which a new phase is not defined (Winfree, 1980).

(e) Construct a map $\theta_n \mapsto \theta_{n+1}$ by resetting the clock every $T$ units of phase, so that

$$\theta_{n+1} = \phi(\theta_n) + T. \tag{14.86}$$

Show that for $A < 1$, this is a circle map. Determine the values of $A$ and $T$ for which there is one-to-one phase locking.

9.  A simple model of the response of an excitable cell to a periodic stimulus is provided by the *integrate-and-fire* model (Knight, 1972; Keener, Hoppensteadt and Rinzel, 1981). In this model the membrane potential $v(t)$ is assumed to behave linearly unless the threshold is reached, whereupon the membrane "fires" and is immediately reset to zero. The equations describing the evolution of the membrane potential are

$$\frac{dv}{dt} = -\gamma v + S(t), \tag{14.87}$$

and $v(t^+) = 0$ if $v(t) = v_T$, the threshold. We take the periodic stimulus to be a simple sinusoidal function, $S(t) = S_0 + S_m \sin(\omega t)$.

Let $T_n$ be the time of the $n$th firing. Formulate the problem of determining $T_{n+1}$ from $T_n$ as a circle map. For what parameter values is this a continuous circle map and for what parameter values is it a discontinuous circle map? (Guevara and Glass, 1982.)

10. Suppose the membrane potential rises at a constant rate $\lambda$ until it reaches a threshold $\theta(t)$, at which time the potential is reset to zero. Suppose that the threshold is the simple sinusoidal function $\theta(t) = \theta_0 + \theta_m \sin(\omega t)$. Let $T_n$ be the time of the $n$th firing. Formulate the problem of determining $T_{n+1}$ from $T_n$ as a circle map. For what parameter values is this a continuous circle map and for what parameter values is it a discontinuous circle map? (Glass and Mackey, 1979.)

11. Suppose a constant-current stimulus $I$ is added to a cable with FitzHugh–Nagumo dynamics and piecewise linear function $f$ as in (14.26).

(a)  Find the steady-state solution as a function of input current $I$.

(b)  Examine the stability of this steady-state solution. Show that the eigenvalues are eigenvalues of a Schrödinger equation with a square well potential. Find the critical Hopf bifurcation curve. Show that for $\epsilon$ sufficiently small, the solution is stable if $I$ is small or large, but there is an intermediate range of $I$ for which the solution is unstable (Rinzel and Keener, 1983).

Hint: Because the function $f(v)$ is piecewise linear, the potential for the Schrödinger equation is a square well potential. Solve the resulting transcendental equations numerically.

12. (a)  Carry out the calculations of Section 14.2.2 for a one-dimensional piecewise linear model (14.26) with $\alpha(r)$ specified by (14.25), $b = \frac{1}{2}$. Determine the critical stability curve.

(b)  Generalize this calculation by supposing that the oscillatory cells have coupling coefficient $D$. What is the effect of the coupling coefficient of the oscillatory cells on the critical stability curve? Show that oscillatory behavior is more likely with weak coupling.

13. A heart attack corresponds to the sudden occlusion of a coronary artery and is rapidly followed by anoxia and the increase of extracellular potassium in the region of decreased blood flow. The increase in extracellular potassium can lead to spontaneous oscillations, but anoxia also leads to cell death and loss of excitability (as the resting potential goes to zero). This gradient of resting potential between normal and damaged tissue leads to a current between the damaged cells and surrounding normal tissue, called a *current of injury*, and the border zone between damaged cells and normal cells may become self-oscillatory and may drive oscillations in the surrounding medium.

Devise a model of FitzHugh–Nagumo type that shows the qualitative effect of increased potassium and anoxia in regions of decreased blood flow. Following the analysis of Section (14.2.2), determine conditions under which this region becomes an ectopic focus.

14. Use homogenization to separate (14.50) into two equations on different spatial scales. Show that $V \approx u_0(z, \tau) + \epsilon W(\eta)\left(J(\tau) - \frac{\partial u_0(z, \tau)}{\partial z}\right)$, where $W'(\eta) = 1 - \frac{D}{d(\eta)}$ and $u_0$ satisfies the averaged equation (14.54). Show that this answer is valid even if $J(\tau) = O(\frac{1}{\epsilon})$.

15. Suppose that the nonlinear function $f(V)$ for an excitable medium is well represented by

$$f(V) = V\left(\frac{V}{\theta} - 1\right), \tag{14.88}$$

at least in the vicinity of the rest point and the threshold. Find the relationship between minimal time and stimulus strength to directly stimulate the medium.
Hint: Evaluate 14.62.

16. Find the relationship between minimal time and stimulus strength to directly stimulate the medium (as in the previous problem) for the Beeler–Reuter model of myocardial tissue (Section 4.3). To do this, set $m = m_\infty(V)$ and set all other dynamic variables to their steady-state values and then evaluate (14.62).

17. Suppose a collection of FitzHugh–Nagumo oscillators described by

$$\delta\frac{dv}{dt} = f(v) - w, \tag{14.89}$$

$$\frac{dw}{dt} = v - \alpha, \tag{14.90}$$

with $f(v) = v(v - 1)(\alpha - v)$ with parameter values $\delta = 0.05$, $\alpha = 0.4$, is coupled through the variable $v$. Calculate (numerically) the coupling function

$$h(\phi) = \int_0^P y_1(\sigma)v(\sigma + \phi)d\sigma, \tag{14.91}$$

where $y_1$ is the first component of the periodic adjoint solution.

18. (a) Consider a linear system of differential equations $\frac{dy}{dt} = A(t)y$ and the corresponding adjoint system $\frac{dv}{dt} = -A^T(t)v$, where $A(t)$ is periodic with period $P$. Let $Y(t)$ and $V(t)$ be matrix solutions of these equations and suppose that $V^T(0)Y(0) = I$. Show that $V^T(P)Y(P) = I$.

(b) The eigenvalues of the matrix $Y(P)Y^{-1}(0)$ are called the Floquet multipliers for the system $\frac{dy}{dt} = A(t)y$. What does the fact that $V^T(0)Y(0) = V^T(P)Y(P)$ imply about the Floquet multipliers for $\frac{dy}{dt} = A(t)y$ and its adjoint system?

# CHAPTER 15

# The Circulatory System

The circulatory system forms a closed loop for the flow of blood that carries oxygen from the lungs to the tissues of the body and carries carbon dioxide from the tissues back to the lungs (Figs. 15.1 and 15.2). Because it is a closed loop system, there are two pumps to overcome the resistance and maintain a constant flow. The left heart receives oxygen-rich blood from the lungs and pumps this blood into the *systemic arteries*. The systemic arteries form a tree of progressively smaller vessels, beginning with the aorta, branching to the small arteries, then to the arterioles, and finally to the capillaries. The exchange of gases takes place in the capillaries. Leaving the systemic capillaries, the blood enters the *systemic veins*, through which it flows in vessels of progressively increasing size toward the right heart. The systemic veins consist of venules, small veins, and the venae cavae. The right heart pumps blood into the *pulmonary arteries*, which form a tree that distributes the blood to the lungs. The smallest branches of this tree are the pulmonary capillaries, where carbon dioxide leaves and oxygen enters the blood. Leaving the pulmonary capillaries, the oxygenated blood is collected by the pulmonary veins, through which it flows back to the left heart. It takes about a minute for a red blood cell to complete this circuit.

While there is an apparent structural symmetry between the pulmonary and systemic circulations, there are significant quantitative differences in pressure and blood volume. Nevertheless, the output of the right and left sides of the heart must always balance, even though the cardiac output, or total amount of blood pumped by the heart, varies widely in response to the metabolic needs of the body. One of the goals of this chapter is to understand how the cardiac output is determined and regulated in response to the metabolic needs of the body. Questions of this nature have been studied for many years, and many books have been written on the subject (see, for example,

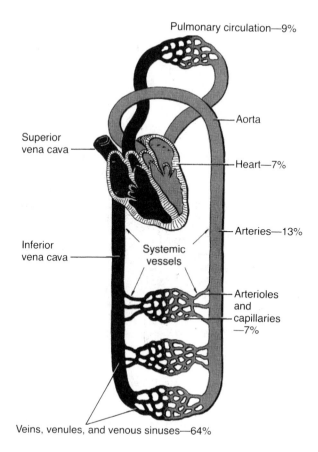

**Figure 15.1** Schematic diagram of the circulatory system, showing the systemic and pulmonary circulations, the chambers of the heart, and the distribution of blood volume throughout the system. (Guyton and Hall, 1996, Fig. 14-1, p. 162.)

Guyton, 1963, or Reeve and Guyton, 1967). Here, we consider only the simplest models for the control of cardiac output.

Each beat of the heart sends a pulse of blood through the arteries, and the form of this arterial pulse changes as it moves away from the heart. An interesting problem is to understand these changes and their clinical significance in terms of the properties of the blood and the arterial walls. Again, this problem has been studied in great detail, and we present here a brief look at the earliest and simplest models of the arterial pulse.

## 15.1 Blood Flow

The term *blood pressure* refers to the force per unit area that the blood exerts on the walls of blood vessels. Blood pressure varies both in time and distance along the circulatory system. *Systolic pressure* is the highest surge of pressure in an artery, and results

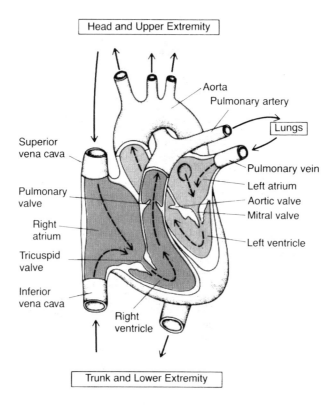

**Figure 15.2**  Schematic diagram of the heart as a pump. (Guyton and Hall, 1996, Fig. 9-1, p. 108.)

from the ejection of blood by the ventricles during ventricular contraction, or *systole*. *Diastolic pressure* is the lowest pressure reached during ventricular relaxation and filling, called *diastole*. In the aorta of a normal human, systolic pressure is about 120 mm Hg and diastolic pressure is about 80 mm Hg.

If we ignore the effects of gravity (which we do throughout this book), then we may asume that blood flows in response to pressure gradients. The simplest way to characterize a blood vessel is as a *resistance vessel* in which the radius is constant and the flow is linearly proportional to the pressure drop. In a linear resistance vessel, the flow, $Q$, is related to pressure by the ohmic relationship

$$Q = \frac{\Delta P}{R},\tag{15.1}$$

where $\Delta P$ is the pressure drop and $R$ is the resistance. The relationship between resistance and radius of the vessel is dramatic. To understand this dependence suppose that a viscous fluid moves slowly and steadily through a cylindrical vessel of fixed radius. The velocity of the fluid is described by a vector $\mathbf{u}$ that has axial, radial, and angular components. Because the fluid is incompressible and fluid is conserved, it must be that

$$\nabla \cdot \mathbf{u} = 0. \tag{15.2}$$

Furthermore, because momentum is conserved, the Navier–Stokes equations (Segel, 1977) hold:

$$\rho(\mathbf{u}_t + \mathbf{u} \cdot \nabla \mathbf{u}) = -\nabla P + \mu \nabla^2 \mathbf{u}, \tag{15.3}$$

where $\rho$ is the constant fluid density, $P$ is the fluid pressure, and $\mu$ is the fluid viscosity. (A brief derivation of the conservation and momentum equations in the case of zero viscosity is given in Chapter 23, Section 23.2.1.) If we assume that the flow is steady and that the nonlinear terms are small compared to viscosity (in an appropriate nondimensional scaling), then

$$\mu \nabla^2 \mathbf{u} = \nabla P. \tag{15.4}$$

This simplification of the Navier–Stokes equation is called the Stokes equation.

The applicability of the Stokes equation to blood flow is suspect for several reasons. The viscosity contribution (the Laplacian) in the Stokes equation is derived from an assumed constitutive law relating stresses and strains in the fluid (Segel, 1977) that is known not to hold in fluids containing long polymers or other complicated chemical structures, including red blood cells. Furthermore, in the capillaries, the large size of the red blood cells compared to the typical diameter of a capillary suggests that a continuum description is not appropriate. However, we do not concern ourselves with these issues here and accept the Stokes equation description as adequate.

We look for a solution of the Stokes equation whose only nonzero component is the axial component. We define coordinates on the cylinder in the usual fashion, letting $x$ denote distance along the cylinder in the axial direction and letting $r$ denote the radial direction. The angular direction, with coordinate $\theta$, does not enter into this analysis.

First observe that with only axial flow, the incompressibility condition (15.2) implies that $\frac{\partial u}{\partial x} = 0$, where $u$ is the axial component of the velocity vector. Thus, $u$ is independent of $x$. Then, with a steady flow, (15.4) reduces to the ordinary differential equation

$$\mu \frac{1}{r} \frac{d}{dr} \left( r \frac{d}{dr} u \right) = \frac{dP}{dx} = P_x, \tag{15.5}$$

where $P_x$ is the axial pressure gradient along the vessel. Note also that $P_x$ must be constant, independent of $r$ and $x$. Because of viscosity, the velocity must be zero at the wall of the cylindrical vessel, $r = r_0$.

It is easy to calculate that

$$u(r) = -\frac{P_x}{4\mu}(r_0^2 - r^2), \tag{15.6}$$

from which it follows that the total flux through the vessel is (*Poiseuille's law*)

$$Q = 2\pi \int_0^{r_0} u(r)r\,dr = -\frac{\pi P_x}{8\mu}r_0^4. \tag{15.7}$$

This illustrates that the total flow of blood through a vessel is directly proportional to the fourth power of its radius, so that the radius of the vessel is by far the most important factor in determining the rate of flow through a vessel. In terms of the cross-sectional area $A_0$ of the vessel, the flux through the cylinder is

$$Q = -\frac{P_x}{8\pi\mu}A_0^2, \tag{15.8}$$

while the average fluid velocity over a cross-section of the cylinder is given by

$$v = \frac{Q}{A_0} = -\frac{P_x}{8\pi\mu}A_0. \tag{15.9}$$

Note that a positive flow is defined to be in the increasing $x$ direction, and thus a negative pressure gradient $P_x$ drives a positive flow $Q$. Important controls of the circulatory system are vasodilators and vasoconstrictors, which, as their names suggest, dilate or constrict vessels and thereby adjust the vessel resistance by adjusting the radius.

   If there are many parallel vessels of the same radius, then the total flux is the sum of the fluxes through the individual vessels. If there are $N$ vessels, each of cross-sectional area $A_0$ and with total cross-sectional area $A = NA_0$, the total flux through the system is

$$Q = -\frac{P_x}{8\pi\mu}A_0(NA_0) = -\frac{P_x}{8\pi\mu}A_0A, \tag{15.10}$$

and the corresponding average velocity is

$$v = \frac{Q}{A}. \tag{15.11}$$

   Now, for there to be no stagnation in any portion of the systemic or pulmonary vessels, the total flux $Q$ must be the same constant everywhere, implying that $P_xA_0A$ must be constant. Thus, for a vessel with constant cross-sectional area, the pressure drop must be linear in distance. Furthermore, the pressure drop per unit length must be greatest in that part of the circulatory system for which $A_0A$ is smallest.

   In Table 15.1 are shown the total cross-sectional areas and pressures at entry to different components of the vascular system. The largest pressure drop occurs in the arterioles, and the pressure in the small veins and venae cavae is so low that these are often collapsed. The pressure at the capillaries must be low to keep them from bursting, since they have very thin walls. The numbers in Table 15.1 suggest that the pressure drop per unit length is greatest in the arterioles, about a factor of three times greater than in the capillaries, even though the capillaries are substantially smaller than the arterioles.

   When the diameter of a vessel decreases, the velocity must increase if the flux is to remain the same. In the circulatory system, however, a decrease in vessel diameter is accompanied by an increase in total cross-sectional area (i.e., an increase in the total number of vessels), so that the velocity at the capillaries is small, even though the capillaries have very small radius. In fact, according to (15.11), the velocity in a

**Table 15.1** Diameter, total cross-sectional area, mean blood pressure at entrance, and mean fluid velocity of blood vessels.

| Vessel | $D$ (cm) | $A$ (cm$^2$) | $P$(mm Hg) | $v$ (cm/s) |
|---|---|---|---|---|
| Aorta | 2.5 | 2.5 | 100 | 33 |
| Small arteries | 0.5 | 20 | 100 | 30 |
| Arterioles | $3 \times 10^{-3}$ | 40 | 85 | 15 |
| Capillaries | $6 \times 10^{-4}$ | 2500 | 30 | 0.03 |
| Venules | $2 \times 10^{-3}$ | 250 | 10 | 0.5 |
| Small veins | 0.5 | 80 | | 2 |
| Venae cavae | 3.0 | 8 | 2 | 20 |

vessel is independent of the radius of the individual vessel but depends solely on the total cross-sectional area of the collection of similar vessels. Once again, from Table 15.1 we see that the velocity drops continuously from aorta to arteries to arterioles to capillaries and then rises from capillaries to venules to veins to venae cavae. The velocity at the vena cava is about half that at the aorta.

## 15.2 Compliance

Because blood vessels are elastic, there is a relationship between distending pressure and volume. Suppose we have an elastic vessel of volume $V$, with a uniform internal pressure $P$. The simplest assumption one can make is that $V$ is linearly related to $P$, and thus

$$V = V_0 + CP, \tag{15.12}$$

for some constant $C$, called the *compliance* of the vessel, where $V_0$ is the volume of the vessel at zero pressure. Although this linear relationship is not always accurate, it is good enough for the simple models that we use here.

The compliance of the venous compartment is about 24 times as great as the compliance of the arterial system, because the veins are both larger and weaker than the arteries. It follows that large amounts of blood can be stored in the veins with only slight changes in venous pressure, so that the veins are often called storage areas. The blood vessels in the lungs are also much more compliant than the systemic arteries.

It is possible for veins and arteries to collapse and for blood flow through the vessel to cease; i.e., the radius becomes zero if the pressure is sufficiently negative. Negative pressures are possible if one takes into account that there is a fluid pressure in the body exterior to the vessels, and $P$ actually refers to the drop in pressure across the vessel wall. The flow of whole blood is stopped at a nonzero radius, primarily because of the nonzero diameter of red blood cells. Thus, when the arterial pressure falls below about 20 mm Hg, the flow of whole blood is blocked, whereas blockage of plasma in arterioles occurs between 5 and 10 mm Hg.

Equation (15.12) is applicable when the vessel has the same internal pressure throughout. When $P$ is not uniform, we model the compliance of the vessel by relating the cross-sectional area to the pressure. Again, the simplest assumption to make is that the relationship is linear, and thus

$$A = A_0 + cP, \tag{15.13}$$

for some constant $c$. Note that $c$ is the compliance per unit length, since in a cylindrical vessel of length $L$ and uniform internal pressure, $V = AL$, so that $C = cL$. However, here we refer to both $C$ and $c$ as compliance.

For a given flow, the pressure drop in a compliance vessel is different from the pressure drop in a resistance vessel. Further, in a compliance vessel, the flow is not a linear function of the pressure drop. We know from (15.8) that the flux through a vessel is proportional to the product of the pressure gradient and the square of the area. Thus, for a compliance vessel,

$$8\pi\mu Q = -P_x A^2(P), \tag{15.14}$$

where $A(P)$ is the relationship between cross-sectional area and pressure for the chosen vessel. In steady state, the flux must be the same everywhere, so that

$$x = -\frac{1}{8\pi\mu Q} \int_{P_0}^{P(x)} A^2(P)dP \tag{15.15}$$

determines the pressure as a function of distance $x$. If the cross-sectional area of the vessel is given by (15.13), then the flux through a vessel of length $L$ is related to the input pressure $P_0$ and output pressure $P_1$ by

$$RQ = \frac{1}{3\gamma}(1 + \gamma P)^3\big|_{P_1}^{P_0} \tag{15.16}$$

$$= (P_0 - P_1)\left(1 + \gamma(P_0 + P_1) + \frac{\gamma^2}{3}(P_0^2 + P_0P_1 + P_1^2)\right), \tag{15.17}$$

where $R = 8\pi\mu L/A_0^2$ and $\gamma = c/A_0$. In the limit of zero compliance, this reduces to the linear ohmic law (15.1). Note that $1/\gamma$ has units of pressure, while $R$ has units of pressure/flow. Thus, $R$ can be interpreted as the flow resistance, as in (15.1).

Since $Q$ is an increasing function of $\gamma$, it follows that a given flow can be driven by a smaller pressure drop in a compliance vessel than in a noncompliance vessel. This relationship is viewed graphically in Fig. 15.3, where we plot the scaled flux $RQ$ as a function of pressure drop $\Delta P = P_0 - P_1$ for fixed $\gamma$ and $P_0$. Clearly, the higher the compliance, the smaller the pressure drop required to drive a given fluid flux. This explains, for example, why the pressure drop in the veins can be much less than in the arteries.

We also calculate the volume of blood contained in a vessel with input pressure $P_0$ and output pressure $P_1$ to be

$$V = \int_0^L A(x)\,dx = \int_{P_0}^{P_1} A(P)x'(P)\,dP = -\frac{1}{8\pi\mu Q}\int_{P_0}^{P_1} A^3(P)dP. \tag{15.18}$$

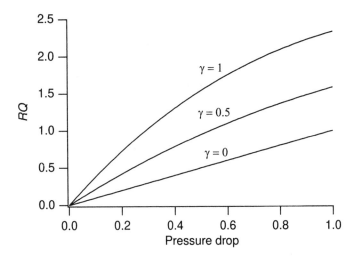

**Figure 15.3** Scaled flow $RQ$ (with units of pressure) as a function of pressure drop $\Delta P = P_0 - P_1$ for different values of compliance $\gamma$. For all curves $P_0 = 1.0$.

For a linear compliance vessel (15.13) this is

$$\frac{V}{V_0} = \frac{3}{4}\left(\frac{(1+\gamma P)^4 |_{P_1}^{P_0}}{(1+\gamma P)^3 |_{P_1}^{P_0}}\right) \tag{15.19}$$

$$= 1 + \frac{\gamma}{2}(P_0 + P_1) + \frac{\gamma^2}{6}(P_0 - P_1)^2 + O(\gamma^3), \tag{15.20}$$

where $V_0$ is the volume of the vessel at zero pressure. It is left as an exercise (Exercise 2) to show that when there is no pressure drop across the vessel, so that $P_0 = P_1$,

$$V = V_0 + V_0\gamma P_0 = V_0 + \frac{V_0}{A_0}cP, \tag{15.21}$$

which is the same as (15.12).

# 15.3 The Microcirculation and Filtration

The purpose of the circulatory system is to provide nutrients to and remove waste products from the cellular interstitium. To do so requires continuous filtration of the interstitium. This filtration is accomplished primarily at the level of capillaries, as fluid moves out of the capillaries at the arteriole end and back into the capillaries at the venous end.

The efflux or influx of fluid from or into the capillaries is determined by the local pressure differences across the capillary wall. In normal situations, the pressure drop through the capillaries is substantial, about 25 mm Hg.

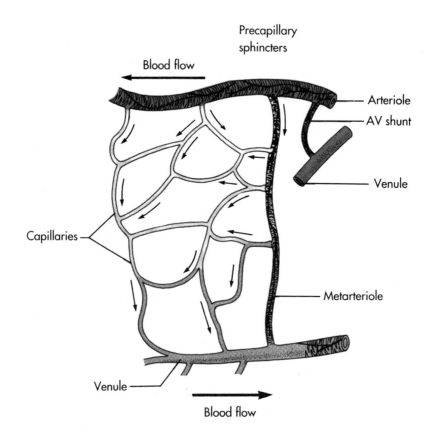

**Figure 15.4**  Diagram of the capillary microcirculation. (Berne and Levy, 1993, Fig. 28-1, p. 466.)

A schematic diagram of the capillary network is shown in Fig. 15.4. To get some understanding of how filtration works and why a capillary pressure drop is necessary, we use a simple one-dimensional model of a capillary. We suppose that there is an influx $Q_i$ at $x = 0$ that must be the same as the efflux at $x = L$, where $L$ is the length of the capillary. At each point $x$ along the capillary, there is blood flow $q$. The (hydrostatic) pressure $P_c$ at each point along the capillary is determined by

$$\frac{dP_c}{dx} = -\rho q, \tag{15.22}$$

where $\rho$ is the coefficient of capillary resistance. Flow into or out of the capillary (across the porous capillary wall) is determined by the difference between total internal pressure and total external (interstitial) pressure. The interstitial hydrostatic pressure $P_i$ is typically about $-3$ mm Hg, and the interstitial fluid colloidal osmotic pressure $\pi_i$ is about 8 mm Hg, while the plasma colloidal osmotic pressure $\pi_c$ averages about 28 mm Hg (although it necessarily varies along the length of the capillary; see Exercise 11).

Thus, the flow into the capillary is determined by

$$\frac{dq}{dx} = K_f(-P_c + P_i + \pi_c - \pi_i), \tag{15.23}$$

where $K_f$ is the capillary filtration rate.

We assume that only $P_c$ varies along the length of the capillary. Clearly, when $P_c$ is high, fluid is forced out of the capillary, and when $P_c$ drops sufficiently low, fluid is reabsorbed into the capillary.

Equations (15.22) and (15.23) form a linear system of equations, which is readily solved. We find that

$$P_c(x) = P_i + \pi_c - \pi_i - \frac{\Delta P_c}{2} \frac{\sinh \beta(x - L/2)}{\sinh \frac{\beta L}{2}}, \tag{15.24}$$

where $\Delta P_c$ is the total pressure drop across the capillary and $\beta^2 = \rho K_f$. It also follows that

$$q(x) = Q_i \frac{\cosh \beta(x - L/2)}{\cosh \frac{\beta L}{2}}, \tag{15.25}$$

where

$$Q_i = \frac{\Delta P_c}{R} \frac{\beta L/2}{\tanh \frac{\beta L}{2}} \tag{15.26}$$

is the total influx at $x = 0$ and $R = \rho L$ is the total resistance of the capillary. Notice the similarity of this formula with (15.1), describing the relationship between flow and pressure drop in a nonleaky vessel. Apparently, leakiness in the vessel ($\beta \neq 0$) has the effect of decreasing the overall resistance of the capillary flow.

Notice that $q(x)$ is minimal, and therefore the interstitial flow is maximal, at $x = L/2$. We define the filtration rate $Q_f$ to be the maximal flux through the interstitium, in this case,

$$Q_f = Q_i - q(L/2). \tag{15.27}$$

It follows that

$$\frac{Q_f}{Q_i} = 1 - \operatorname{sech} \frac{\beta L}{2}. \tag{15.28}$$

The filtration rate depends on the single dimensionless parameter $\beta L = \sqrt{\rho K_f} L$. Thus filtration is enhanced in vessels that are "leaky" (large $K_f$) and of small radius, since vessel resistance $\rho$ is inversely proportional to $r^4$.

## 15.4 Cardiac Output

During a heartbeat cycle, the pressure and volume of the heart change in a highly specific way, shown in Fig. 15.5. Notice that the pressure–volume loops are of rectangular

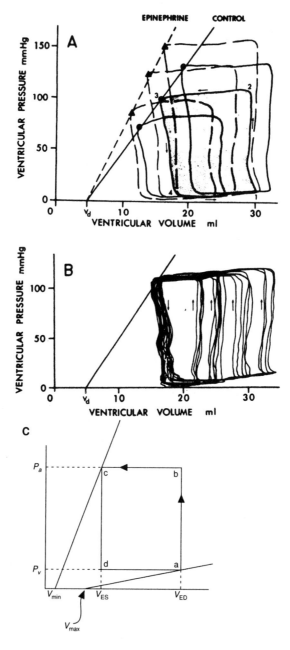

**Figure 15.5** Experimental data of the pressure–volume relationship during the heartbeat cycle in the denervated left ventricle of the dog. (Sagawa et al., 1978, Fig. 11.4.) A: Three beats from different end-diastolic volumes and against different arterial pressures are shown in solid lines, with the broken lines representing the same beat cycles in the presence of epinephrine, which enhances the contraction. B: Pressure–volume loops of the same ventricle for four different end-diastolic volumes, but against the same arterial pressure. C: Schematic diagram of the pressure–volume loop (adapted from Hoppensteadt and Peskin, 1992, Fig. 5.5). a: inflow valve closes, b: outflow valve opens, c: outflow valve closes, d: inflow valve opens.

shape, and thus the pressure and volume change at different places in the cycle. For example, on the right ascending side of the loop, the pressure is increasing while volume is constant, which corresponds to the ventricle contracting while the outflow valve is closed. At the top right-hand corner the outflow valve opens, and the volume decreases as the blood is pumped out at a constant pressure. The constant pressure at the top of the loop corresponds to the arterial pressure, while the constant pressure along the bottom of the loop corresponds to the venous pressure. Notice also that the top left corners of the loops lie on the same straight line. Thus, the ventricular volume at the end of a contraction (the end-systolic volume $V_{ES}$) is a linear function of the arterial blood pressure. Further, in panel B of the same figure, the top left corner of the loop (the end-systolic pressure and volume) is constant, independent of the total volume of blood pumped by the ventricle, suggesting that it depends solely on the arterial pressure.

This suggests that

$$V_{ES} = V_{\min} + C_s P_a, \tag{15.29}$$

where $V_{\min}$ is the intercept on the $V$ axis, and $C_s$ is the slope of the line connecting the three labeled points in Fig. 15.5A. The key observation is that $C_s$ and $V_{\min}$ are independent of the arterial pressure and the end-diastolic volume, and are thus intrinsic properties of the ventricle. $C_s$ is, in fact, the compliance of the ventricle at the end of systole. By connecting the lower right corners of the pressure–volume loops, as shown in Fig. 15.5C, we reach a similar conclusion for the end-diastolic volume.

These observations can be summarized in a simple model, in which we determine the total amount of blood pumped by the ventricle, i.e., the *cardiac output*, but ignore the time-dependent changes in volume and pressure over the beat cycle. We view the heart as a compliance vessel whose basal volume and compliance change with time. Thus,

$$V = V_0(t) + C(t)P. \tag{15.30}$$

During diastole, when the heart is filling, the heart is relaxed and compliant, so that $V_0$ and $C$ are large. During this time, the aortic valve is closed, preventing backflow, so that the pressure is essentially the same as the venous pressure. During systole, the heart is contracting and much less compliant, so that $C$ and $V_0$ are decreased compared to diastolic values. At this time, the mitral valve is closed, preventing backflow into the veins, so that the pressure in the heart is the same as the arterial pressure. Accordingly, the minimal volume (end systolic) is given by (15.29), and the maximal volume $V_{ED}$ (end diastolic) is

$$V_{ED} = V_{\max} + C_d P_v, \tag{15.31}$$

where $P_a$ and $P_v$ are the arterial and venous pressures, respectively, and $C_s$ and $C_d$ are the compliances of the heart during systole and diastole, respectively. This implies that the *stroke volume* is

$$V_{\text{stroke}} = V_{\max} - V_{\min} + C_d P_v - C_s P_a, \tag{15.32}$$

and the total cardiac output is

$$Q = FV_{\text{stroke}}, \tag{15.33}$$

where $F$ is the heart rate in beats per unit time.

This expression of cardiac output has some features that agree with reality. For example, if $C_s$ is small compared to $C_d$, then the cardiac output depends primarily on the venous pressure, or on the rate of venous return. This phenomenon, that cardiac output increases with increasing filling, is commonly referred to as *Starling's law*. While there is some decrease in output due to arterial loading, this effect is not nearly as significant as the increase in output resulting from an increase in venous pressure.

According to this formula, cardiac output is a linear function of venous pressure. This is not a terrible approximation in normal physiological ranges, although a more accurate formula would show saturation as a function of venous pressure. That is, cardiac output approaches a constant as a function of venous pressure for venous pressure above 10 mm Hg. Cardiac output also saturates at high frequencies because of inadequate fluid filling.

## 15.5 Circulation

### 15.5.1 A Simple Circulatory System

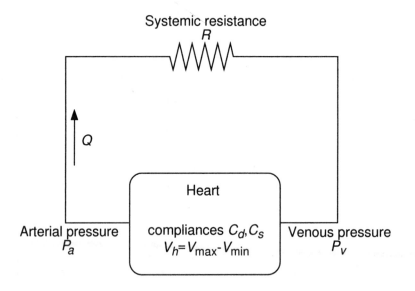

**Figure 15.6** Schematic diagram of the simplest circulation model, with a single-chambered heart and a single loop.

To illustrate how all the above pieces fit together to give a model of the circulatory system, consider a simple circulatory system with one loop and a single-chambered heart (Fig. 15.6). To begin with, suppose we have only a heart and a resistive closed loop. For the resistive closed loop, we suppose that the total flux is related to the pressure drop through

$$Q = (P_a - P_v)/R, \tag{15.34}$$

so that in steady state the flux through the loop must match the cardiac output, yielding

$$Q = F(V_h + C_d P_v - C_s P_a) = (P_a - P_v)/R, \tag{15.35}$$

where $V_h = V_{max} - V_{min}$. Equation (15.35) gives a relationship between arterial and venous pressure that must be maintained in a steady-state condition. Unfortunately, these are not uniquely determined by this equation. The reason the solution is not completely determined is primarily because we have not allowed the circulation loop to be a compliance vessel. If we allow the loop to be a compliance vessel, then there is an additional relationship between pressure and total volume that must be satisfied.

To see how this works for a simple system, suppose that the circulatory loop consists of a compliance vessel with cross-sectional area given by (15.13). It follows from (15.16) that

$$Q = \frac{1}{3R\gamma}\{(1 + \gamma P_a)^3 - (1 + \gamma P_v)^3\}, \tag{15.36}$$

and the total volume of the vessel is given by

$$\frac{V}{V_0} = \frac{3}{4}\left\{\frac{(1 + \gamma P_a)^4 - (1 + \gamma P_v)^4}{(1 + \gamma P_a)^3 - (1 + \gamma P_v)^3}\right\}. \tag{15.37}$$

These two equations, together with

$$Q = F(V_h + C_d P_v - C_s P_a), \tag{15.38}$$

give a system of three equations in terms of the four unknowns $Q, P_a, P_v$, and $V$. (Of course, it is also possible to regard $P_a$ and $P_v$ as known, i.e., measured, quantities, and then view $\gamma$ and $R$ as unknowns. This would determine the resistance and compliance corresponding to a given pressure difference.) This is too many unknowns for the number of equations, and so we must find another equation before the solution is uniquely determined. The final equation comes from conservation of blood. Because blood is assumed to be incompressible, and because the heart chambers are assumed to have a fixed volume (as cardiac output is expressed in terms of the average output), it follows that $V$ must be constant. The system is then completely determined. However, because it is nonlinear, a closed-form solution is not apparent, and the easiest way to obtain a solution is to solve the equations numerically.

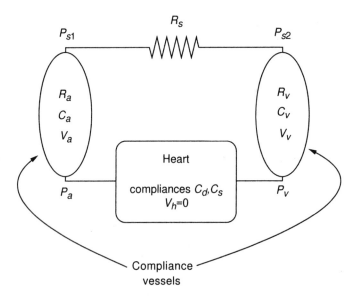

**Figure 15.7** Schematic diagram of the two-compartment model of the circulation. The heart and pulmonary system are combined into a single vessel, and the systemic capillaries are modeled as a resistance vessel. The larger arteries and veins are modeled as compliance vessels.

## 15.5.2 A Simple Linear Circulatory System

It is difficult to extend the analysis of the previous section to more realistic models because of the complexity of the resulting nonlinear equations. However, much can be learned using linear approximations to the governing equations. The simplest linear model is due to Guyton (1963), in which the circulatory system is represented as a closed loop with two compliance vessels and one pure resistance vessel (Fig. 15.7). The large arteries and veins are each treated as compliance vessels with linearized flow equations, and the systemic capillaries are treated as a resistance vessel with no compliance.

The equations describing this model can be conveniently divided into two groups: those describing the arterial system and those describing the venous system.

**Arterial system:** Cardiac output is described in terms of the compliance of the ventricle, and so

$$Q = F(C_d P_v - C_s P_a), \tag{15.39}$$

where, for simplicity, we have assumed that $V_h = 0$, i.e., that $V_{max} = V_{min}$. The larger arteries are modeled as a compliance vessel, and thus, from (15.17),

$$Q = \frac{P_a - P_{s1}}{R_a}, \tag{15.40}$$

where we have ignored the nonlinear terms. This approximation is accurate if the compliance is small. The parameters are defined in Fig. 15.7. For example, $P_a$ is the arterial pressure, while $P_{s1}$ is the blood pressure at the (somewhat arbitrary) border between the larger arteries and the arterial capillaries, and $R_a$ is the resistance of the larger arteries. Note that although this expression for the flux looks as though the arteries are treated as resistance vessels, this is only because the nonlinear terms are omitted. The compliance of the arteries appears in the relationship between the pressure and the volume of the arteries,

$$V_a = \frac{\gamma V_0}{2}(P_a + P_{s1}) = \frac{C_a}{2}(P_a + P_{s1}), \tag{15.41}$$

where $C_a = c_a V_0/A_0$ is the compliance of the systemic arteries. Note that we have assumed that the volume of the systemic arteries is zero at zero pressure.

**Venous system:** We have three similar equations for the venous system, except that the equation for the cardiac output (15.39) is replaced by an equation describing the flow through the capillaries,

$$Q = \frac{P_{s1} - P_{s2}}{R_s}. \tag{15.42}$$

The remaining two equations describe the flow through the veins,

$$Q = \frac{P_{s2} - P_v}{R_v}, \tag{15.43}$$

and the volume of blood in the veins,

$$V_v = \frac{C_v}{2}(P_v + P_{s2}). \tag{15.44}$$

At this point, we have a system of six equations in seven unknowns (four pressures, two volumes, and $Q$). The final equation comes from conservation of volume, according to which

$$V_a + V_v = V_0, \tag{15.45}$$

where $V_0$ is a given constant.

These seven equations, being linear, can be solved for the unknowns. However, before doing so, we make two further simplifications. First, we assume that the systolic compliance, $C_s$, is nearly zero, and second, that the pressure drops across the larger vessels are small, so that $R_a$ and $R_v$ are quite small, with the result that $P_a = P_{s1}$ and $P_v = P_{s2}$, to a good approximation. This removes two of the variables, leaving us with a system of five equations:

$$Q = FC_d P_v, \tag{15.46}$$

$$Q = \frac{P_a - P_v}{R_s}, \tag{15.47}$$

$$V_a = C_a P_a, \tag{15.48}$$

$$V_v = C_v P_v, \tag{15.49}$$

$$V_a + V_v = V_0. \tag{15.50}$$

The solution of this system is easily found to be

$$P_a = \frac{(1 + FC_dR_s)V_0}{C_v + (1 + FC_dR_s)C_a}, \tag{15.51}$$

$$P_v = \frac{V_0}{C_v + (1 + FC_dR_s)C_a}, \tag{15.52}$$

$$Q = \frac{FC_dV_0}{C_v + (1 + FC_dR_s)C_a}. \tag{15.53}$$

A number of qualitative features of the circulation can be seen from this solution.

1. As the heart rate increases, the arterial pressure $P_a$ increases to a maximum of $V_0/C_a$, but as the heart rate falls, the arterial pressure decreases to a minimum of $V_0/(C_v + C_a)$.
2. Conversely, as the heart rate falls, the venous pressure $P_v$ increases to a maximum of $V_0/(C_v + C_a)$. Hence, in heart failure, the arterial pressure falls and the venous pressure rises, until they are equal. With no pressure drop, there is no flow.
3. An increase in the systemic resistance, $R_s$, leads to a decrease in the cardiac output, an increase in the arterial pressure, and a decrease in the venous pressure.
4. Since $V_a = C_aP_a$ and $V_v = C_vP_v$, an increase in systemic resistance is accompanied by a shift in the blood volume from the venous system to the arterial system, i.e., $V_v$ decreases, and $V_a$ increases.

In reality, systemic resistance varies widely (decreasing, for example, during exercise), but the cardiac output compensates for this variation, keeping the arterial pressure relatively constant. Thus, the above model, which includes no control of cardiac output, needs to be modified to agree with experimental data. Later in this chapter we consider some simple models for regulation of the circulation. Before doing so, however, we consider a more complex model of the circulation, incorporating more compartments.

## 15.5.3 A Multicompartment Circulatory System

To construct a more detailed linear model of the circulatory system, we assume that the systemic and pulmonary loops each consist of two compliance vessels, the arterial and venous systems, connected by the capillaries, a pure resistance. Further, we assume that the heart has two chambers, the left and right hearts. A schematic diagram of the model is given in Fig. 15.8. We must write equations for the flow through each of these compartments and keep track of the total volume of blood contained in the system. Unfortunately, the notation can be difficult to follow. We let subscripts $a$, $v$, $s$, and $p$ denote, respectively, arterial, venous, systemic, and pulmonary. So, for example, $P_{sa}$ is the pressure at the entrance to the systemic arteries, and $C_{sa}$ is the compliance of the systemic arteries. Also, subscripts $r, l, d$, and $s$ denote, respectively, right, left, diastolic

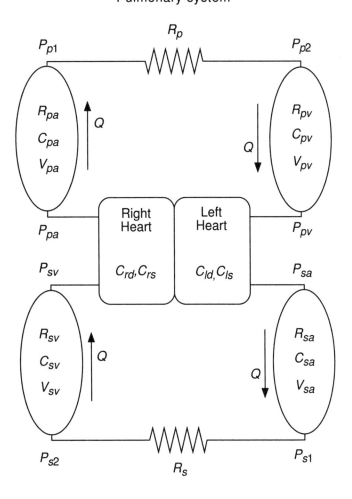

Figure 15.8 Schematic diagram of the multicompartment model of the circulation.

and systolic. Thus, $C_{ld}$ denotes the diastolic compliance of the left heart. Finally, $P_{s1}$ denotes the pressure at the ill-defined border between the systemic arteries and the systemic capillaries, with similar definitions for $P_{s2}$, $P_{p1}$, and $P_{p2}$ as indicated in Fig. 15.8.

As before, we write down the governing equations in groups.

**Systemic arteries:**

$$Q = \frac{P_{sa} - P_{s1}}{R_{sa}}, \tag{15.54}$$

$$Q = F(C_{ld}P_{pv} - C_{ls}P_{sa}),  \tag{15.55}$$

$$V_{sa} = V_0^s + \frac{C_{sa}}{2}(P_{sa} + P_{s1}).  \tag{15.56}$$

Note that here we have assumed that the volume of the systemic arteries at zero pressure is $V_0^s$, not zero, as was assumed in the previous model.

**Systemic veins:**

$$Q = \frac{P_{s2} - P_{sv}}{R_{sv}},  \tag{15.57}$$

$$Q = \frac{P_{s1} - P_{s2}}{R_s},  \tag{15.58}$$

$$V_{sv} = \frac{C_{sv}}{2}(P_{sv} + P_{s2}).  \tag{15.59}$$

For the venous system, it is reasonable to take the basal volume as zero, because if the blood pressure falls to zero, these vessels collapse. In the arterial system, however, such an approximation is not realistic.

**Pulmonary arteries:**

$$Q = \frac{P_{pa} - P_{p1}}{R_{pa}},  \tag{15.60}$$

$$Q = F(C_{rd}P_{sv} - C_{rs}P_{pa}),  \tag{15.61}$$

$$V_{pa} = V_0^p + \frac{C_{pa}}{2}(P_{pa} + P_{p1}).  \tag{15.62}$$

**Pulmonary veins:**

$$Q = \frac{P_{p2} - P_{pv}}{R_{pv}},  \tag{15.63}$$

$$Q = \frac{P_{p1} - P_{p2}}{R_p},  \tag{15.64}$$

$$V_{pv} = \frac{C_{pv}}{2}(P_{pv} + P_{p2}).  \tag{15.65}$$

At this stage we have 12 equations for 13 unknowns (8 pressures, 4 volumes, and $Q$). The final equation, as before, comes from the conservation of blood volume, whereby

$$V_{sa} + V_{sv} + V_{pa} + V_{pv} = V_0.  \tag{15.66}$$

The capillary and heart volumes need not be included in this equation because they are assumed to be fixed.

This system of equations can be treated in a number of ways. First, using a symbolic manipulation package such as Maple or Mathematica, it is not difficult to find the solution directly. A second approach is to make a number of simplifying assumptions, as was done in the simpler model discussed above. For example, if we assume that there is no pressure drop over the arteries or veins (both pulmonary and systemic), then we find that $P_{sa} = P_{s1}$, $P_{sv} = P_{s2}$, and similarly for the pulmonary equations. This removes

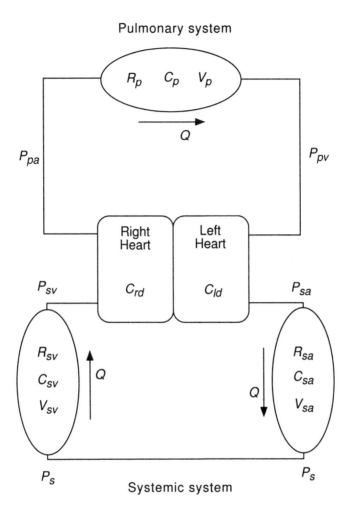

**Figure 15.9** Schematic diagram of the simplified three-compartment model, obtained from the one in Fig. 15.8 by letting $R_s$ and $R_p$ tend to zero and combining all the pulmonary vessels into one.

four variables and four equations, giving a system of nine equations for the remaining nine unknowns. This variation of the model has been discussed by Hoppensteadt and Peskin (1992), and its further study is left as an exercise (Exercise 3).

A second approximation of the full system is to omit the pure systemic resistance and combine all the pulmonary vessels into a single compliance vessel. This results in a model consisting of only three compliance vessels (Fig. 15.9). It is left as an exercise (Exercise 4) to show that this approximation results from letting $R_s$ and $R_p$ approach zero and by setting $C_{pa} = C_{pv}$ and $R_{pa} = R_{pv}$. For convenience, we write $C_p$ in place of $C_{pa}$, and $R_p$ in place of $R_{pa}$. We also assume that the systolic compliances are negligible.

The equations governing this simplified three compartment model are as follows.

**Systemic arteries:**

$$Q = \frac{P_{sa} - P_s}{R_{sa}}, \tag{15.67}$$

$$Q = FC_{ld}P_{pv}, \tag{15.68}$$

$$V_{sa} = V_0^s + \frac{C_{sa}}{2}(P_{sa} + P_s). \tag{15.69}$$

**Systemic veins:**

$$Q = \frac{P_s - P_{sv}}{R_{sv}}, \tag{15.70}$$

$$V_{sv} = \frac{C_{sv}}{2}(P_{sv} + P_s). \tag{15.71}$$

**Pulmonary system:**

$$Q = \frac{P_{pa} - P_{pv}}{R_p}, \tag{15.72}$$

$$Q = FC_{rd}P_{sv}, \tag{15.73}$$

$$V_p = \frac{C_p}{2}(P_{pa} + P_{pv}). \tag{15.74}$$

Note that we have assumed that the basal pulmonary volume is zero.

**Conservation of volume:**

$$V_{sa} + V_{sv} + V_p = V_0. \tag{15.75}$$

Because this is a linear system, one can solve (using symbolic manipulation) to obtain

$$P_{sa} = Q\left(\frac{1}{FC_{rd}} + R_{sa} + R_{sv}\right), \tag{15.76}$$

$$P_s = Q\left(\frac{1}{FC_{rd}} + R_{sv}\right), \tag{15.77}$$

$$P_{sv} = \frac{Q}{FC_{rd}}, \tag{15.78}$$

$$P_{pa} = Q\left(\frac{1}{FC_{ld}} + R_p\right), \tag{15.79}$$

$$P_{pv} = \frac{Q}{FC_{ld}}, \tag{15.80}$$

$$Q = \frac{V_e}{\alpha + \frac{C_p}{FC_{ld}} + \frac{C_{sv} + C_{sa}}{FC_{rd}}}, \tag{15.81}$$

where $\alpha = R_{sv}(C_{sa} + C_{sv}/2) + R_{sa}C_{sa}/2 + C_pR_p/2$, and $V_e = V - V_0^s - V_0^p$ is the excess volume beyond that which is necessary to fill the system at zero pressure. Clearly, the cardiac output saturates for large $F$, and

$$\lim_{F \to \infty} Q = Q_\infty = \frac{V_e}{\alpha} = \frac{2V_e}{R_{sv}(2C_{sa} + C_{sv}) + R_{sa}C_{sa} + C_p R_p}. \qquad (15.82)$$

We can also see that, as in the simpler model, systemic arterial pressure is an increasing function of the heart rate, and systemic venous pressure is a decreasing function of the heart rate.

Cardiac output depends linearly on excess blood volume $V - V_0^s - V_0^p$. In trauma, if there is substantial blood loss, the cardiac output drops rapidly. If there is no compensatory control (increased heart rate or change of resistance and compliance), loss of 15–20% of blood volume over a period of half an hour is fatal. If reflexes are intact, loss of 30–40% in half an hour is fatal. Notice that in this model (taking $V = 5$ liters, $V_0^p + V_0^s = 1.2$ liters) a 20% loss of blood with no compensatory control leads to a 26% loss of cardiac output.

There are ten physical parameters in this system of nine equations, giving nine relationships between parameters if the solution is known from data. We take as typical volumes $V_{sa} = 1$, $V_{sv} = 3.5$, $V_p = 0.5$ (liters), and typical pressures $P_{sa} = 100$, $P_s = 30$, $P_{sv} = 2$, $P_{pa} = 15$, $P_{pv} = 5$ (mm Hg). Total cardiac output is about 5.6 liters/min, with a heart rate of 80 beats per minute and a stroke volume of 0.07 liter. Using these, we find estimates for parameters of $C_{sv} = 0.22$, $C_p = 0.05$ liter/mm Hg, $R_{sa} = 12.5$, $R_{sv} = 5.0$, $R_p = 1.78$ mm Hg/(liters/min), and $C_{ld} = 0.014$, $C_{rd} = 0.35$ (liter/mm Hg)/stroke.

To estimate $V_0^s$ and $C_{sa}$, we note that at the heart the arterial pressure is varying on a beat-to-beat basis, as is the volume of blood in the arteries. During one beat, the blood ejected from the heart must be accommodated by the compliance of the arteries. Thus,

$$\Delta V = C_{sa} \Delta P, \qquad (15.83)$$

where $\Delta V$ is the stroke volume, about 0.07 liter, and $\Delta P$ is the difference between systolic and diastolic pressure, about 40 mm Hg. Once $C_{sa}$ is known ($C_{sa} = 0.0018$ liter/mm Hg), the resting volume $V_0^s$ can be determined using the given values of volume and pressure in (15.69) as $V_0^s = 0.94$ liter.

These numbers correlate with some known features of the adult circulatory system. For example, the venous system is about 24 times more compliant than the arterial system. The ratio found here is $C_{sv}/C_{sa} = 122$, which is too high, but this does not create significant errors in interpretation. Total resistance of the systemic circulation is larger than the pulmonary resistance, and the compliance of the left heart is less than that of the right heart, simply because the left ventricular wall is much thicker than the right.

To gain some understanding of the dependence of the solution on parameters, we calculate the sensitivity $\sigma_{yx}$, which is the sensitivity of dependence of the dependent variable $y$ upon changes in the independent variable $x$. Thus, $\sigma$ is the proportional

**Table 15.2**  Sensitivities for the three compartment circulatory system (expressed as percentages).

|          | Normal | $C_{sa}$ 0.0018 | $C_{sv}$ 0.22 | $C_p$ 0.05 | $R_{sa}$ 12.5 | $R_{sv}$ 5.0 | $R_p$ 1.78 | $C_{ld}$ 0.014 | $C_{rd}$ 0.035 | $F$ 80 |
|----------|--------|------|-------|-------|------|-------|------|-------|-------|-------|
| $P_{sa}$ | 100    | −5.4 | −82.8 | −11.8 | 67.1 | −46.8 | −5.9 | 5.9   | 8.5   | 14.4  |
| $P_s$    | 30     | −5.4 | −82.8 | −11.8 | −2.9 | 18.5  | −5.9 | 5.9   | 3.8   | 9.8   |
| $P_{sv}$ | 2      | −5.4 | −82.8 | −11.8 | −2.9 | −74.8 | 6.1  | 5.9   | −89.5 | 84.6  |
| $P_{pa}$ | 15     | −5.4 | −82.8 | −11.8 | −2.9 | −74.8 | 60.7 | −27.5 | 10.5  | −17.0 |
| $P_{pv}$ | 5      | −5.4 | −82.8 | −11.8 | −2.9 | −74.8 | −5.9 | −94.1 | 10.5  | −84.6 |
| $V_{sa}$ | 1.0    | 21.6 | −18.9 | −2.7  | 11.6 | −7.2  | −1.3 | 1.3   | 1.7   | 3.0   |
| $V_{sv}$ | 3.5    | −5.4 | 17.2  | −11.8 | −2.9 | 12.7  | −5.9 | 5.9   | −2.0  | 3.9   |
| $V_p$    | 0.5    | −5.4 | −82.8 | 88.2  | −2.9 | −74.8 | 44.0 | −44.2 | 10.5  | −33.6 |
| $Q$      | 5.6    | −5.5 | −82.8 | −11.8 | −2.9 | −74.8 | −5.9 | 5.9   | 10.5  | 16.4  |
| $Q_\infty$ | 6.7  | −6.2 | −86.7 | −7.1  | −3.5 | −89.5 | −7.0 | 0.0   | 0.0   | 0     |

change in $y$ for a given proportional change in $x$, and so

$$\sigma_{yx} = \frac{\Delta y/y}{\Delta x/x} = \frac{x}{y}\frac{\partial y}{\partial x},\tag{15.84}$$

in the limit as $\Delta x$ goes to zero. In Table 15.2 are shown the sensitivities, expressed as percentages, for the three-compartment loop using normal parameter values, as shown in the table. For example, the first number in the first column, −5.4, is the sensitivity of the systemic arterial pressure $P_{sa}$ to changes in the systemic arterial compliance $C_{sa}$.

From this table we infer some interesting features of the human circulatory system. First, the system is relatively insensitive to changes in the arterial compliance $C_{sa}$. In fact, compliance of the arterial system is insignificant compared to compliance of the other compartments and to a first approximation can be ignored. On the other hand, the system is strongly sensitive to changes in $C_{sv}$, the venous compliance. Similarly, the solution is relatively insensitive to changes in arterial resistance, $R_{sa}$, but is relatively sensitive to changes in venous resistance, $R_{sv}$. Much of the regulation of the cardiac systems occurs through changes in the compliance and resistance of the venous system, and this result demonstrates the efficacy of that choice.

The most common cause of coronary occlusion is *atherosclerosis*, hardening of the arteries. (About half of all deaths in the United States and Europe are the result of atherosclerosis and two thirds of those are the result of a thrombosis (clotting) of one or more coronary arteries.) This occurs when excess cholesterol and fats are deposited in the arteries. These deposits are invaded by fibrous tissue and frequently become calcified, resulting in atherosclerotic plaques and stiffened arterial walls that can be neither constricted nor dilated. While systemic compliance is not extremely important in this model, systemic resistance is significant, and increases in systemic resistance produce increases in arterial pressure.

A person with higher than normal mean arterial pressure is said to have hypertension, or *high blood pressure*. Life expectancy is shortened substantially when mean arterial pressure is 50 percent or more above normal. The lethal effects of hypertension are

1. Increased cardiac workload, leading to congestive heart disease or coronary heart disease, often leading to a fatal heart attack;
2. rupture of a major blood vessel in the brain (a *stroke*), resulting in paralysis, dementia, blindness, or multiple other brain disorders;
3. multiple hemorrhages in the kidneys, leading to renal destruction and eventual kidney failure and death.

One other parameter that has an important effect is the diastolic compliance of the left heart, $C_{ld}$. As expected, if this compliance decreases, there is a reduction in systemic arterial pressure and a reduction in cardiac output. There is also a noticeable increase in pulmonary blood volume, $V_p$. Thus, left heart failure, which corresponds to a weakening of the left ventricular muscles and hence decreased cardiac efficiency and decreased compliance, results in excess fluid and fluid congestion in the lungs, known as *pulmonary edema*. Notice that in this model, failure of the left or right heart does not influence the maximal cardiac output $Q_\infty$, although it certainly requires a higher heart rate to effect the same output. Notice, also, that with left or right heart failure, systemic volume changes little, so that one does not expect peripheral edema.

## 15.6  Cardiac Regulation

The circulatory system is equipped with a complex system for the control of blood flow. There are three major types of control mechanisms:

1. Local control of blood flow in the individual tissue, determined mainly by the tissue's need for blood perfusion.
2. Neural control, by which the overall vesicular resistance and cardiac activity are controlled.
3. Humoral control, in which substances dissolved in the blood, such as hormones, ions, or other chemicals, cause changes in flow properties.

### 15.6.1  Autoregulation

*Autoregulation* is a local mechanism that makes flow through a tissue responsive to local oxygen demand but relatively insensitive to arterial pressure. In tissue for which the delivery of oxygen is of central importance (for example, the brain or the heart) the local blood flow is controlled to be slightly higher than required, but no higher.

In dead organs, an increase in arterial pressure produces a linear increase in blood flow, suggestive of a linear-resistance vessel. However, in normally functioning tissue, the arterial pressure can be changed over a large range with little effect on the blood flow (Fig. 15.10). For example, in muscle, with an arterial pressure between 75 mm Hg and 175 mm Hg the blood flow remains within ±10-15% of normal.

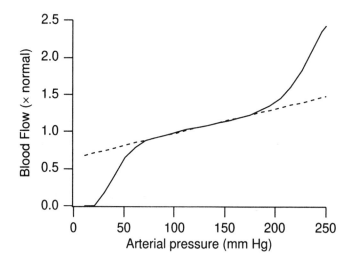

**Figure 15.10**   Blood flow as a function of arterial pressure if pressure is raised over a period of a few minutes. (Data (solid curve) from Guyton and Hall, 1996, Fig. 17-4, p. 203.) Dashed curve shows the result from the model (15.88).

The flow through an artery is known to be responsive to the need for oxygen. For example, an eightfold increase in metabolism produces a fourfold increase in blood flow (as shown in Fig. 15.11B). Similarly, if oxygen content falls because of anemia, high altitude, or carbon monoxide poisoning, the blood flow increases to compensate. For example, a reduction to 25% of normal oxygen saturation produces a threefold increase in blood flow, not quite enough to compensate fully for the loss (see Fig. 15.11A).

Although the mechanism for autoregulation is not completely understood, it is most likely that resistance of tissue is responsive to biochemical measures of how hard it is working, such as concentrations of $H^+$, $CO_2$, $O_2$, and lactic acid. The arterioles are highly muscular, and their diameters can change manyfold. The metarterioles (terminal arterioles) are encircled by smooth muscle fibers at intermittent points and are also used to regulate flow.

The arterial blood has the same composition for all tissues of the body so cannot be used as a local control mechanism. This is problematic because it is the arterioles whose resistance is regulated. Here we assume that the resistance of arterioles is a function of the concentration of oxygen in the venous blood. This is possible, since arteries and veins tend to run side by side, and venous concentrations may regulate arterial resistance by release and diffusion of regulatory substances, called *vasodilators*.

An example of how this may work is provided by cardiac tissue. If cardiac activity increases and the utilization of oxygen exceeds the supply, ATP is degraded, increasing the concentration of *adenosine*. Adenosine is a vasodilator, which leaks out of the cells into the venous flow to cause local dilation of coronary arteries.

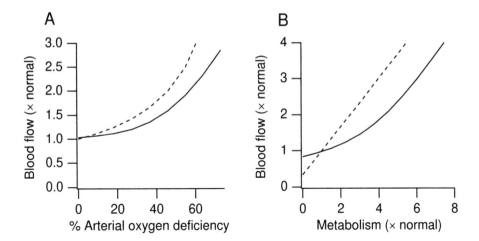

**Figure 15.11** A: Blood flow as a function of the percentage of arterial oxygen deficiency, keeping arterial pressure and metabolic rate fixed. (Data (solid curve) from Guyton and Hall (1996), Fig. 17-2, p 200.) Dashed curve shows the result from the model. B: Blood flow as a function of metabolism, keeping arterial pressure and arterial oxygen content fixed at normal levels. (Data (solid curve) from Guyton and Hall, 1996, Fig. 17-1, p. 200.) Dashed curve shows the result from the model.

A simple model for autoregulation is as follows (Huntsman et al., 1978; Hoppensteadt and Peskin, 1992). We keep track of oxygen consumption and blood flow via

$$([O_2]_a - [O_2]_v)Q = M, \tag{15.85}$$

$$P_a - P_v = RQ, \tag{15.86}$$

where $[O_2]_a$ and $[O_2]_v$ are the arterial and venous oxygen concentrations, respectively, $M$ is the metabolic rate (oxygen consumption per unit time), $P_a$ and $P_v$ are the arterial and venous pressures driving the flow $Q$ through tissue with total resistance $R$. In this model we treat $[O_2]_a$ as a given constant and $[O_2]_v$ as variable. Now we assume that there is some linear relationship between arterial resistance and venous oxygen content, say

$$R = R_0 \left(1 + A[O_2]_v\right). \tag{15.87}$$

The assumption of linearity is reasonable in restricted ranges of oxygen content. Here, the parameter $A$ denotes the sensitivity of resistance to oxygen; if $A = 0$ the resistance is unregulated.

We can solve these equations for the flow rate $Q$ to get

$$Q = \frac{1}{1 + A[O_2]_a} \left(MA + \frac{P_a}{R_0}\right), \tag{15.88}$$

where we have assumed that $P_v = 0$. This is a linear relationship between flow rate and arterial pressure, which, when $A = 0$, reproduces the unregulated situation. However, with $A > 0$, the sensitivity of the flow to changes in arterial pressure varies with arterial oxygen content. Furthermore, this expression shows linear dependence of blood flow on metabolism.

We can estimate the parameters $A$ and $R_0$ using the data from Fig. 15.10. In the range of pressures between 75 and 175 mm Hg, the curve is well represented by the straight line

$$\frac{Q}{Q^*} = \frac{1}{3} + \frac{2}{3}\frac{P_a}{P^*}, \tag{15.89}$$

where $Q^*$ and $P^*$ are the normal values of flow and pressure ($Q^* = 5.6$ liters/min., $P^* = 100$ mm Hg). (Remark: Any straight line through $P^*$ and $Q^*$ must be of the form $\frac{Q}{Q^*} = \alpha + (1 - \alpha)\frac{P_a}{P^*}$.) Comparing this to the regulated curve (15.88) at normal values,

$$Q = \frac{1}{1 + A[O_2]_a^*}\left(M^*A + \frac{P_a}{R_0}\right), \tag{15.90}$$

where $[O_2]_a^*$ and $M^*$ are normal values of arterial oxygen and metabolism, respectively ($M^* = Q^*([O_2]_a^* - [O_2]_v^*)$, $[O_2]_a^* = 104$ mm Hg, $[O_2]_v^* = 40$ mm Hg), we find that

$$A = \frac{2Q^*}{3M^* - 2Q^*[O_2]_a^*}, \tag{15.91}$$

$$R_0 = P^*\frac{2Q^*(M^* - [O_2]_a^*)}{Q^*M^*}, \tag{15.92}$$

so that

$$\frac{Q}{Q^*} = \frac{2\frac{M}{M^*} + \frac{P}{P^*}}{3 + \frac{13}{4}\left(\frac{[O_2]_a}{[O_2]_a^*} - 1\right)}, \tag{15.93}$$

using typical values for $\frac{Q^*[O_2]_a^*}{M^*} = \frac{13}{8}$.

In Fig. 15.10 are shown the data for blood flow as a function of arterial pressure, compared with the model (15.93). The good agreement over the linear range is the result of fitting.

The relationship (15.88) reproduces two other features of autoregulation that are qualitatively correct. It predicts that the flow rate increases as the arterial oxygen content decreases, and increases linearly with metabolic rate. In Fig. 15.11A is shown the blood flow plotted as a function of arterial oxygen deficiency. Here, the solid curve is taken from data, and the dashed curve is from (15.93). Similarly, in Fig. 15.11B is shown the blood flow plotted as a function of metabolic rate. As before, the solid curve is from data, and the dashed curve is from (15.93). Clearly, the model gives reasonable qualitative agreement for blood flow as a function of arterial oxygen content, and for blood flow as a function of metabolism.

## 15.6.2 The Baroreceptor Loop

The *baroreceptor loop* is a global feedback control mechanism using the nervous system to adjust the heart rate, the venous resistance, and thereby the venous pressure in order to maintain the arterial pressure at a given level, with the ultimate goal of regulating the cardiac output.

The need to regulate cardiac output is apparent. During exercise, when the demand for oxygen goes up, cardiac output normally rises at a linear rate, with slope about 5 (since 5 liters of blood are required to supply 1 liter of oxygen). In normal situations, the cardiac output and heart rate are roughly proportional, indicating that the stroke volume remains essentially constant. However, if heart rate is artificially driven up with a pacemaker, with no increase in oxygen consumption, then the cardiac output remains virtually the same, indicating a decrease in stroke volume. Similarly, in exercise with a fixed heart rate (set by a pacemaker), total cardiac output increases to meet the demand.

The primary nervous mechanism for the control of cardiac output is the *baroreceptor reflex*. This reflex is initiated by stretch receptors, called *baroreceptors* or *pressoreceptors*, located in the walls of the *carotid sinus* and *aortic arch*, large arteries of the systemic circulation. A rise in arterial pressure is detected and causes a signal to be sent to the central nervous system from which feedback signals are sent through the autonomic nervous system to the circulatory system, thereby enabling the regulation of arterial pressure. For example, the baroreceptor reflex occurs when a person stands up after having been lying down. Immediately upon standing, the arterial pressure in the head and upper body falls, with dizziness or loss of consciousness a distinct possibility. The falling pressure at the baroreceptors elicits an immediate reflex, resulting in a strong sympathetic discharge throughout the entire body, thereby minimizing the decrease in blood pressure in the head. This observation suggests that the larger dinosaurs required a well-tuned baroreceptor reflex in order not to faint every time they raised their heads.

The most important part of the autonomic nervous system for regulation of the circulation is the *sympathetic nervous system*, which innervates almost all the blood vessels, with the exception of the capillaries. The primary effects of sympathetic nervous stimulation are

1. the contraction of small arteries and arterioles (by stimulation of the surrounding smooth muscle) to increase blood flow resistance and thereby decrease blood flow in the tissues;
2. constriction of the veins, thereby decreasing the amount of blood in the peripheral circulation;
3. the stimulation of the heart muscle, thereby increasing both the heart rate and stroke volume.

The *parasympathetic* or *vagus* nerves have the opposite effect on the heart, namely to decrease the heart rate and decrease the strength of contractility. Strong sympathetic stimulation can increase the heart rate in adult humans to 180–200 beats per minute, and even as high as 220 beats per minute in young adults. Strong parasympathetic stimulation can lower the heart rate to 20–40 beats per minute and can decrease the strength of contraction by 20 to 30 percent.

The effect of the baroreceptors is to increase sympathetic stimulation and decrease parasympathetic stimulation when there is a drop in arterial pressure. This increase in sympathetic activity, in turn, increases the heart rate, the systemic resistance, and the cardiac compliances, and decreases the venous compliance. Notice from Table 15.2 that in the unregulated circulation these changes effect an increase of arterial pressure. Thus, the overall effect of the baroreceptor loop is to maintain the arterial pressure at a desired level.

The sympathetic nervous system is also stimulated by the brain vasomotor center, where an increase in carbon dioxide in the brain acts to cause a widespread vasoconstriction throughout the body. A sympathetic response is also stimulated by fright or anger, and is called the *alarm reaction*.

The sympathetic nervous system acts by three mechanisms. First, it stimulates the contraction of vessels by the release of vasoconstrictors, primarily *norepinephrine*. Simultaneously, the adrenal medullae are stimulated to release epinephrine (adrenaline) and norepinephrine into the circulating blood. These two hormones are carried in the bloodstream to all parts of the body, where they act directly on blood vessels, usually to cause contraction. (Some tissues respond to epinephrine by dilation rather than constriction.) The action of secreted norepinephrine lasts about 30 minutes. Finally, sympathetic nervous activity acts to increase heart rate and heart contractility.

To include the baroreceptor loop and the sympathetic nervous system in our circulation model, we suppose that the level of sympathetic stimulation is given by $S$, and that $S$ is related to the deviation of the arterial pressure $P_{sa}$ through a simple linear relationship

$$S = S^* + \beta(P_{sa}^* - P_{sa}), \tag{15.94}$$

so that as arterial pressure decreases, sympathetic stimulation increases. Here, $S^*$ and $P_{sa}^*$ are "normal" values. In animal experiments, blocking all sympathetic activity leads to a drop of arterial pressure from 100 to 50 mm Hg, indicating a continuous basal level of sympathetic firing at normal pressure ($S^* \neq 0$), called *sympathetic tone*, known to be about one impulse per second.

Next, we assume that heart rate $F$, arterial resistance $R_{sa}$, and cardiac compliances $C_{ld}$ and $C_{rd}$ are (unspecified) increasing functions of $S$, while the venous compliance $C_{sv}$ is a decreasing function of $S$. We account for the metabolic need of the tissue through

$$([O_2]_a - [O_2]_v)Q = M, \tag{15.95}$$

and suppose that the metabolic need is communicated to the tissue through autoregulation via

$$R_{sa} = R(S) + A[O_2]_{sv}. \tag{15.96}$$

This representation is slightly different from that in (15.87), but it has the same interpretation.

Combined with the balance equations (15.67)–(15.75), we have a closed system of equations, which can be solved to find the cardiac output $Q$ as a function of metabolic need $M$. The solution is complicated, and so we leave the details to the interested reader.

However, it is useful to view the solution of these equations in a slightly different way. We suppose that the effect of baroreceptor feedback is to hold the arterial pressure fixed at some target level and to adjust other parameters such as arterial resistance, venous compliance, and heart rate so that this target pressure is maintained. Then we can view $P_{sa}$ as a parameter of the model and let the heart rate, say, be an unknown. Thus we solve the governing equations, not for the pressures and cardiac output as functions of the heart rate, as with the unregulated flow, but for heart rate and cardiac output as functions of arterial pressure and metabolism.

We then obtain

$$F = \frac{\left(\frac{1}{C_{rd}}(2C_{sv} + C_{sa}) + \frac{2}{C_{ld}}C_p\right)(AM + P_{sa}) + \frac{1}{C_{rd}}(P_{sa}C_{sa} - 2V_e)}{(2V_e - P_{sa}C_{sa})(R_{sa} + R_{sv}) - (AM + P_{sa})(C_{sv}R_{sv} + C_pR_p + C_{sa}R_{sv})}, \quad (15.97)$$

$$Q = \frac{\left(C_{ld}(2C_{sv} + C_{sa}) + 2C_{rd}C_p\right)(AM + P_{sa}) + C_{ld}(P_{sa}C_{sa} - 2V_e)}{C_{ld}\left(R_{sa}(2C_{sv} + C_{sa}) + R_{sv}C_{sv} - C_pR_p\right) + 2C_pC_{rd}(R_{sa} + R_{sv})}. \quad (15.98)$$

Although these formulae are somewhat complicated and obscure, here we see a number of features for the controlled circulation that are markedly different from those for the uncontrolled circulation. Most obvious is that heart rate and cardiac output respond to changes in metabolic need $M$. In fact, the cardiac output can be increased by increasing the arterial pressure or decreasing the systemic resistances.

These formulae also show some difficulties that the control system faces. Notice that there are parameter ranges for which either the numerator or denominator are negative. These are parameter values for which the solution is not valid, or, said another way, that are outside the range of physical possibility. Thus, for example, certain target pressures $P_{sa}$ cannot be maintained if $V_e$ is either too large or too small. Similarly (and not surprisingly), there are some large values of metabolism and pressure $(AM + P_{sa})$ that are impossible to maintain.

If the heart rate cannot be controlled by the baroreceptor loop, as for instance when there is an implanted pacemaker, then $F$ must be viewed as a parameter of the model rather than an unknown. Instead, some other variables, such as cardiac compliance, are the unknowns. If we suppose that the cardiac compliances always maintain the same ratio, then it is easy to see that an increase in heart rate leads to an exactly compensating decrease in compliance and stroke volume, so that the same total output is maintained.

## 15.7  Fetal Circulation

Because the fetus receives all of its oxygen through the umbilical cord and the placenta, the lungs of the fetus are not used for gas exchange. Instead, the lungs are collapsed and have high resistance to blood flow: only 12% of the blood flow is through the lungs. This situation is reversed at birth when the newborn takes its first breath, expanding the lungs, and when the umbilical cord constricts.

Necessitated by the high resistance of the pulmonary circulation, the fetal circulatory system has a connection between the pulmonary artery and the aorta, called the *ductus arteriosus*, that shunts blood from the outflow of the right heart directly into the systemic arteries. After birth, the ductus gradually closes.

The ventricular chambers of the developing fetal heart are nearly equal in size. It is only after birth that the load on the left ventricle increases, necessitating additional growth of the left ventricular wall to accommodate an increased demand. To equalize the output of the two hearts, there is a small opening in the interatrial septum, called the *foramen ovale*. On the left side of the septum there is a small flap of tissue that allows flow from the right atrium to the left but prevents the reverse from occurring. In the fetus, this flap is open, but at birth it closes for reasons that will become clear below.

To model the fetal circulatory system, we use the same three-compartment model as above with additional connections allowed by the ductus arteriosus and the foramen ovale (Fig. 15.12). Since there is no longer a single loop, we must keep track of the flows in each compartment. These flows are governed by the following equations:

**Systemic arteries:**

$$Q_s = \frac{P_{sa} - P_s}{R_{sa}}. \tag{15.99}$$

**Systemic veins:**

$$Q_s = \frac{P_s - P_{sv}}{R_{sv}}. \tag{15.100}$$

**Pulmonary system:**

$$Q_p = \frac{P_{pa} - P_{pv}}{R_p}. \tag{15.101}$$

**Left heart:**

$$Q_l = F(C_{ld}P_{pv} - C_{ls}P_{sa}). \tag{15.102}$$

**Right heart:**

$$Q_r = F(C_{rd}P_{sv} - C_{rs}P_{pa}). \tag{15.103}$$

The equations for the volumes are unchanged from before.

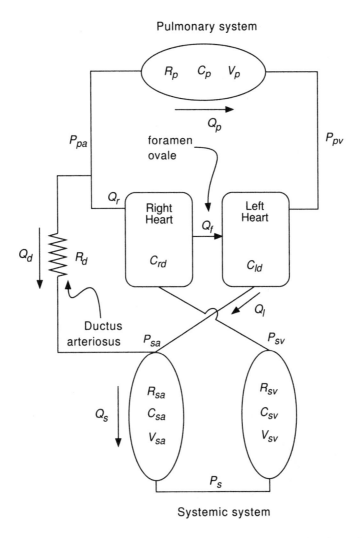

**Figure 15.12** Schematic diagram of the fetal circulation. The model is based on the three-compartment model in Fig. 15.9, with additional connections to model the ductus arteriosus and the foramen ovale.

Since fluid is conserved, flow into any junction must equal the flow out. There are four junctions, and thus we have the four conservation laws

$$Q_l + Q_d = Q_s, \tag{15.104}$$

$$Q_r = Q_d + Q_p, \tag{15.105}$$

$$Q_s = Q_r + Q_f, \tag{15.106}$$

$$Q_p + Q_f = Q_l, \tag{15.107}$$

where $l$ = left heart, $r$ = right heart, $s$ = systemic, $p$ = pulmonary, $f$ = foramen ovale, $d$ = ductus arteriosus. Notice that there are only three independent relationships here, as the first three imply the fourth, for the simple reason that total fluid is conserved. As a result, any three flows can be determined as functions of the remaining three.

A quick count now shows us that there are 14 variables (5 pressures, 3 volumes, and 6 flows), but only 12 equations so far, including the three equations for the volumes and the equation for the conservation of volume, which are not shown explicitly here. Thus, to characterize the system completely, we need two further equations. These are the equation for the ductus,

$$Q_d = \frac{P_{pa} - P_{sa}}{R_d}, \tag{15.108}$$

and the equation for the foramen, modeled as an ideal valve,

$$P_{sv} = P_{pv}, \quad \text{if } Q_f > 0, \tag{15.109}$$

$$Q_f = 0, \quad \text{if } P_{sv} < P_{pv}. \tag{15.110}$$

Here, there is no resistance to flow in the forward direction if the valve is open, and if the valve is closed, there is no flow in the backward direction.

There are two possible solutions. First, for the "foramen open" solution, we set $P_{sv} = P_{pv}$ and then solve the governing system of equations. This yields a valid solution for all parameter values for which $Q_f > 0$. On the other hand, if we take $Q_f = 0$ (the "foramen closed" solution) and determine all the pressures, we have a valid solution for all parameter values for which $P_{sv} < P_{pv}$. For any set of parameter values there should be one, but only one, solution set.

We begin by looking for the "foramen open" solution. For simplicity, we again suppose that the systolic compliances are negligible, i.e., that $C_{rs} = C_{ls} = 0$. We obtain

$$Q_f = Q_r \frac{R_d(C_{ld} - C_{rd}) + R_p C_{ld} - C_{rd}(R_{sa} + R_{sv})}{C_{rd}(R_{sa} + R_{sv} + R_p + R_d)}, \tag{15.111}$$

$$Q_d = Q_r \frac{C_{ld}(R_{sv} + R_{sa}) - R_p C_{rd}}{C_{rd}(R_{sv} + R_{sa} + R_d + R_p)}, \tag{15.112}$$

$$Q_s = Q_r \frac{R_p(C_{rd} + C_{ld}) + R_d C_{ld}}{C_{rd}(R_{sa} + R_{sv} + R_p + R_d)}, \tag{15.113}$$

$$Q_p = Q_r \frac{(R_{sa} + R_{sv})(C_{rd} + C_{ld}) + R_d C_{rd}}{C_{rd}(R_{sa} + R_{sv} + R_p + R_d)}, \tag{15.114}$$

$$Q_l = Q_r \frac{C_{ld}}{C_{rd}}. \tag{15.115}$$

In the developing fetus, the left and right hearts are nearly the same. If $C_{rd} = C_{ld} = C_d$, the outputs from the left and right hearts are the same, and $Q_l = Q_r = Q$. With this simplification we find that

$$Q_f = Q_d = Q \frac{R_p - R_{sa} - R_{sv}}{R_{sa} + R_{sv} + R_p + R_d}, \tag{15.116}$$

$$Q_s = Q \frac{2R_p + R_d}{R_{sa} + R_{sv} + R_p + R_d}, \tag{15.117}$$

$$Q_p = Q \frac{2R_{sa} + 2R_{sv} + R_d}{R_{sa} + R_{sv} + R_p + R_d}. \tag{15.118}$$

As long as the pulmonary resistance is larger than the total systemic resistance, the flow $Q_f$ is positive, as required by our initial assumption. Thus, by adjusting the pulmonary resistance $R_p$, the foramen and the ductus allow blood to be shunted from the lungs to the systemic circulation. Notice that in the extreme case of $R_p = \infty$, we have $Q_f = Q_d = Q, Q_p = 0, Q_s = 2Q$. In other words, if $R_p = \infty$, there is no pulmonary flow, the flow returning from the systemic circulation is equally divided between the left and right hearts for pumping, and the blood pumped by the right heart is shunted from the lungs to the systemic arteries.

At birth, the lungs fill with air and expand, dramatically decreasing the resistance of blood flow in the lungs. Simultaneously, the umbilical cord constricts, dramatically increasing the total systemic resistance $R_{sa} + R_{sv}$. When this happens, the flow through the foramen reverses, and closes the foramen. To find the flow solution in this "foramen closed" situation, we take $Q_f = 0$ and drop the restriction that $P_{sv} = P_{pv}$. It follows immediately from (15.106) and (15.107) that $Q_p = Q_l, Q_s = Q_r$, and $Q_d = Q_r - Q_l$. Thus, we see that the flow through the ductus is used to balance the outputs of the left and right sides of the heart.

Furthermore,

$$P_{pv} = \frac{Q_r}{FC_{ld}}, \tag{15.119}$$

$$P_{sv} = \frac{Q_l}{FC_{rd}}, \tag{15.120}$$

$$\frac{P_{sa}}{P_{pa}} = \left( \frac{\frac{1}{FC_{rd}} + R_{sv} + R_{sa}}{\frac{1}{FC_{rd}} + R_{sa} + R_{sv} + R_d} \right) \left( \frac{\frac{1}{FC_{ld}} + R_p + R_d}{\frac{1}{FC_{ld}} + R_p} \right), \tag{15.121}$$

and

$$\frac{P_{pv}}{P_{sv}} = \frac{FC_{rd}(R_{sa} + R_{sv} + R_d) + 1}{FC_{ld}(R_p + R_d) + 1}, \tag{15.122}$$

which is greater than one (as required) as long as $R_p < R_{sa} + R_{sv}$. Thus, remarkably, as soon as the first breath is drawn and the pulmonary resistance drops, the pulmonary venous pressure exceeds the systemic venous pressure, keeping the foramen closed, allowing the skin flap to gradually grow over and seal tightly. In addition,

$$Q_d = Q_r \frac{(\frac{1}{FC_{ld}} + R_p) - (\frac{1}{FC_{ld}} + R_{sa} + R_{sv})}{\frac{1}{FC_{ld}} + R_d + R_p}. \tag{15.123}$$

Thus, immediately after birth, when the left and right heart compliances are the same, the flow through the ductus reverses direction, so that the left heart output exceeds the right heart output, with

$$\frac{Q_l}{Q_r} = \frac{\frac{1}{FC_{rd}} + R_{sa} + R_{sv} + R_d}{\frac{1}{FC_{ld}} + R_p + R_d}. \tag{15.124}$$

For reasons that are not completely understood (probably because of the increased concentration of oxygen in the blood), the ductus gradually closes, so that $R_d$ grows, eventually to $\infty$. As it does so, the arterial pressure $P_{sa}$ increases, causing the left ventricle to thicken gradually, decreasing its compliance. The end result (taking $R_d \to \infty$) is the solution of the single-loop system found in the previous section, although with parameter values that are not yet the "adult" values.

## 15.7.1  Pathophysiology of the Circulatory System

Occasionally, the heart or its associated blood vessels are malformed during fetal life, leaving the newborn infant with a defect called a *congenital anomaly*. There are three major types of congenital abnormalities:

1. A blockage, or *stenosis*, of the blood flow at some part of the heart or a major vessel.
2. An abnormality that allows blood to flow directly from the left heart or aorta to the right heart or pulmonary artery, bypassing the systemic circulation.
3. An abnormality that allows blood to flow from the right heart or pulmonary artery to the left heart or aorta, thereby bypassing the lungs.

### Patent ductus arteriosus (PDA)

While the ductus arteriosus constricts to a small size shortly after birth, it is several months before flow is completely occluded. In about 1 out of 5500 babies, the ductus never closes, a condition known as *patent ductus arteriosus*. In a child with a patent ductus, there is a substantial backflow from the left heart into the lungs, so that the blood is well oxygenated, but there is decreased cardiac reserve and respiratory reserve, because insufficient blood is supplied to the systemic arteries. As the child grows and systemic pressure increases, the backflow through the ductus also increases, sometimes causing the diameter of the ductus to increase, thereby worsening the condition. Symptoms of patent ductus include fainting or dizziness during exercise, and there is usually hypertrophy of the left heart.

It can happen that the lungs respond to the excess pulmonary flow by increasing pulmonary resistance, thereby, according to (15.123), reversing the flow in the ductus, shunting blood from the right heart to the aorta, carrying deoxygenated blood directly into the systemic arteries.

### Closed foramen ovale in utero

In this situation the circulation is like the circulation after birth, except that the pulmonary resistance exceeds the systemic resistance, $R_p > R_{sa} + R_{sv}$. According to

(15.124), the output of the left heart is low compared to the output of the right heart, so that development of the left heart is impaired and the right heart is overdeveloped at birth.

### Atrial septal defect (ASD)

If the foramen does not close properly at birth, there remains a hole in the septum between the left and right atria, allowing oxygenated blood to leak from the left heart to the right heart. Assuming that the ductus closes successfully, so that $Q_d = 0$, it follows that

$$Q_p = Q_r = FC_{rd}P_{sv}, \tag{15.125}$$

$$Q_s = Q_l = FC_{ld}P_{pv}, \tag{15.126}$$

$$Q_f = Q_s - Q_r = Q_l - Q_p, \tag{15.127}$$

so that

$$\frac{Q_p}{Q_s} = \frac{Q_r}{Q_l} = \frac{C_{rd}}{C_{ld}}. \tag{15.128}$$

If the left heart has smaller compliance than the right heart, as would be true in an adult, the pulmonary flow exceeds the systemic flow.

### ASD and PDA

The configuration here is the same as with the fetal circulation, except that there is no valve to prohibit flow from the left to right atrium. The solution is the "foramen open" solution, for which

$$Q_s - Q_p = Q_f + Q_d = Q_s \left( 1 - \frac{R_dC_{rd} + (R_{sa} + R_{sv})(C_{ld} + C_{rd})}{R_dC_{ld} + R_p(C_{ld} + C_{rd})} \right), \tag{15.129}$$

which is negative for typical parameter values. This shows that it is possible to reduce the shunted flow and equalize the pulmonary and systemic flows by *banding* or surgically constricting the pulmonary artery, thus increasing $R_p$. The banding procedure works, however, only if $R_d \neq \infty$, that is, only if there is flow through the ductus. Banding has no effect in ASD when the ductus is closed, because in the limit $R_d \to \infty$, the flow through the foramen is

$$Q_f = Q_s \left( 1 - \frac{C_{rd}}{C_{ld}} \right), \tag{15.130}$$

independent of $R_p$.

## 15.8   The Arterial Pulse

The above analysis treats the circulation as if the various pressures in the blood vessels were constant over time. Of course, since the heart pumps blood in a pulsatile manner, this is not the case. Each beat of the heart forms a pressure wave that travels along the arteries, changing shape as it moves away from the heart. Typical experimental data, taken from a dog artery, are shown in Fig. 15.13. It is evident that closer to the heart the pressure pulse is wider and does not have a distinct second wave, and the velocity and pressure waves have different forms. However, as the pulse moves away from the heart, the pressure wave becomes steeper, a second wave develops following the first, and the velocity profile becomes similar to the pressure profile. Since variations in the form of the arterial pulse are often used as clinical indicators (for example, the second wave is usually absent in patients with diabetes or atherosclerosis), it is important to gain an understanding of the physical mechanisms underlying the shape of the pulse in normal physiology. Models to explain the shape of the arterial pulse range from simple linear ones to complex models incorporating the tapering of the arterial walls and its branching structure (Pedley, 1980; Lighthill, 1975; Peskin, 1976), and the modern literature on models of the arterial pulse is vast. Here, we restrict our attention to only the simplest models.

## 15.8.1 The Conservation Laws

Consider flow in a blood vessel with cross-sectional area $A(x, t)$. For simplicity we assume that the flow is a *plug flow*, with velocity that is a scalar quantity $u$ and is a function of axial distance along the vessel only. Poiseuille flow becomes plug flow in the limit of zero viscosity, and so in the following analysis we omit consideration of viscous forces. The volume of the vessel of length $L$ is $\int_0^L A(x, t)\, dx$, and thus conservation

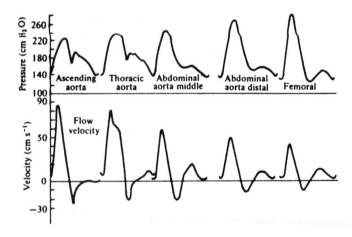

**Figure 15.13** The form of the arterial pulse measured in the arteries of a dog. The top panel shows the pressure waveform, and the bottom panel shows the velocity. (Pedley, 1980, Fig. 1.14, taken from McDonald, 1974.)

of mass requires that

$$\frac{\partial}{\partial t}\left(\int_0^L A(x,t)\,dx\right) = u(0)A(0,t) - u(L)A(L,t). \qquad (15.131)$$

Taking the partial derivative of (15.131) with respect to $L$ and replacing $L$ by $x$ gives

$$A_t + (Au)_x = 0. \qquad (15.132)$$

According to Newton's law, the rate of change of momentum is equal to the total force exerted. Thus, if $P$ is defined to be the excess pressure generated by the heart (i.e., the difference between the actual pressure and the resting pressure), then conservation of momentum demands that

$$\frac{\partial}{\partial t}\left(\rho\int_0^L A(x,t)u(x,t)\,dx\right) = \rho A(x,t)u^2(x,t)\big|_{x=L}^0 + P(x,t)A(x,t)\big|_{x=L}^0. \qquad (15.133)$$

Note that $\rho A(0,t)u(0,t)$ is the rate at which mass enters the vessel across the surface $x=0$, so that $\rho A(0,t)u^2(0,t)$ is the rate at which momentum enters the vessel across this surface. Differentiating (15.133) with respect to $L$ and replacing $L$ by $x$, we find that

$$\rho\left((Au)_t + (Au^2)_x\right) = -(PA)_x. \qquad (15.134)$$

This second equation can be simplified by expanding the derivatives and using (15.132) to get

$$\rho(u_t + uu_x) = -P_x \qquad (15.135)$$

as the equation for the conservation of momentum.

For simplicity we assume that the vessel is a linear compliance vessel with

$$A(P) = A_0 + cP. \qquad (15.136)$$

In the analysis that follows, one can use a more general relationship between area and pressure, but the basic conclusions remain unchanged. With this expression for the cross-sectional area, the conservation equation becomes

$$c(P_t + uP_x) + A(P)u_x = 0. \qquad (15.137)$$

### 15.8.2  The Windkessel Model

One of the earliest models of the heart, dating back to the past century (Frank, 1899; translated by Sagawa et al., 1990), is the *windkessel* model, from the German word meaning an air chamber, or bellows. (The name originally arose because of the similarities between the mechanical conditions in the arterial system and the operation of the *windkessel*, or bellows, of a nineteenth-century fire engine.)

The *windkessel* model is obtained from (15.135) and (15.137) by letting $\rho \to 0$, in which case $P_x = 0$, so that the pressure is a function only of time. Thus, we model the greater arteries as a compliance vessel, extending from $x=0$ to $x=L$, with a

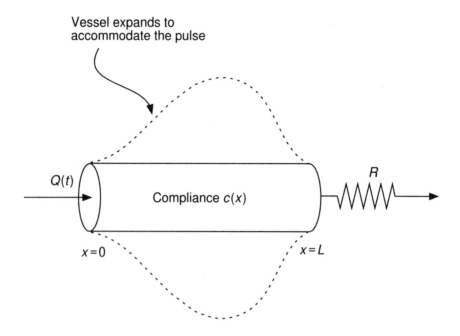

**Figure 15.14** Schematic diagram of the *windkessel* model.

pressure and volume varying over time; inflow at $x = 0$ is from the heart, and outflow at $x = L$ is into the peripheral arterial system (Fig. 15.14). However, although the pressure is uniform inside the vessel, the compliance is not, so that the cross-sectional area of the vessel varies with $x$, the distance along the vessel. In fact, we assume that $c(0) = c(L) = 0$, but that the compliance $c(x)$ is nonzero inside the vessel. Finally, we assume that the outflow to the peripheral system is through a resistance $R$.

From (15.137) it follows that

$$u_x(x,t) = \frac{-c(x)P_t}{A_0 + c(x)P},$$ (15.138)

and thus, integrating from $x = 0$ to $x = L$, we get

$$A_0 u(0,t) = \theta(P)P_t + A_0 u(L,t),$$ (15.139)

where

$$\theta(P) = \int_0^L \frac{A_0 c(x)}{A_0 + c(x)P} \, dx.$$ (15.140)

We now note that $A_0 u(0,t)$ is the flow into the vessel from the heart, and thus we write it as $Q(t)$. Further, $A_0 u(L,t)$ is the outflow from the vessel into the peripheral vessels, which we can write as $P/R$ (since the compliance at $x = L$ is zero, and the resistance is $R$). Hence, we have the differential equation for $P$,

$$Q(t) = \theta(P)P_t + \frac{P}{R}.$$ (15.141)

From this differential equation we learn that when the heart ejects blood, $Q(t)$ increases quickly, leading to a corresponding increase in $P$ (i.e., the vessel fills up and expands). When the flow stops, $Q$ becomes zero, and the pressure decreases to zero according to $P_t = -P/(\theta(P)R)$. Thus, the major arteries act as a bellows, inflating to accommodate the blood from the heart and then contracting to pump the blood through the periphery.

### 15.8.3 A Small-Amplitude Pressure Wave

If we assume that $p$ are $u$ are small, so that all nonlinear terms in (15.135) and (15.137) can be ignored, we then obtain the linear system

$$\rho u_t + P_x = 0,$$ (15.142)
$$cP_t + A_0 u_x = 0.$$ (15.143)

By cross-differentiation, we can eliminate $u$ and find a single equation for $P$, namely

$$P_{tt} = \frac{A_0}{c\rho}P_{xx},$$ (15.144)

which is well known as the *wave equation*.

Solutions of the wave equation include traveling wave solutions, which are functions whose shape is invariant but that move at the velocity $s = \sqrt{\frac{A_0}{c\rho}}$. For arteries, this velocity is on the order of 4 m/s, as can be verified by comparing the arrival times of the pressure pulse at the carotid artery in the neck and at the posterior tibial artery at the ankle.

The general solution of the wave equation (15.144) can be written in the form

$$P(x,t) = f(t - x/s) + g(t + x/s),$$ (15.145)

where $f$ and $g$ are arbitrary functions. Note that $f(t - x/s)$ denotes a wave, with profile $f(x)$, traveling from left to right, while $g(t + x/s)$ denotes a wave traveling in the opposite direction. It follows that the general solution for $u$ is

$$u = \frac{1}{\rho s}[f(t - x/s) - g(t + x/s)].$$ (15.146)

### 15.8.4 Shock Waves in the Aorta

Although the linear wave equation can be used to gain an understanding of many features of the arterial pulse, such as reflected waves and waves in an arterial network (Lighthill, 1975), there are experimental indications that nonlinear effects are also important (Anliker et al., 1971a,b). One particular nonlinear effect that we investigate here is the steepening of the wave front as it moves away from the heart. If the wave front becomes too steep, the top of the front overtakes the bottom, and a shock, or discontinuity, forms, a solution typical of hyperbolic equations. Of course, physiologically, a true shock is not possible, as blood viscosity and the elastic properties of the arterial wall preclude the formation of a discontinuous solution. Nevertheless, it might be possible to generate very steep pressure gradients within the aorta.

Under normal conditions, no such shocks develop. However, in conditions where the aorta does not function properly, allowing considerable backflow into the heart, the heart compensates by an increase in the ejection volume, thus generating pressure waves that are steeper and stronger than those observed normally. Furthermore, the *pistol-shot* phenomenon, a loud cracking sound heard through a stethoscope placed at the radial or femoral artery, often occurs in patients with aortic insufficiency. It has been postulated that the pistol-shot is the result of the formation of a shock wave within the artery, a shock wave that is possible because of the increased amplitude of the pressure pulse.

To model this phenomenon, recall that the governing equations are

$$c(P_t + uP_x) + A(P)u_x = 0, \tag{15.147}$$

$$\rho(u_t + uu_x) + P_x = 0, \tag{15.148}$$

which can be written in the form

$$w_t + Bw_x = 0, \tag{15.149}$$

where

$$w = \begin{pmatrix} u \\ P \end{pmatrix} \tag{15.150}$$

and

$$B = \begin{pmatrix} u & \dfrac{1}{\rho} \\ \dfrac{A(P)}{c} & u \end{pmatrix}. \tag{15.151}$$

Using the method of characteristics (Whitham 1974; Pedley, 1980; Peskin, 1976), we can determine some qualitative features of the solution. Roughly speaking, a characteristic is a curve $C$ in the $(x, t)$ plane along which information about the solution propagates. For example, the equation $u_t + cu_x = 0$ has solutions of the form $u(x, t) = U(x - ct)$, so that information about the solution propagates along curves $x - ct = $ constant in the $(x, t)$ plane. Similarly, characteristics for the wave equation (15.144) are curves of the form $t \pm x/s = $ constant, because it is along these curves that information about the solution travels.

To find characteristic curves, we look for curves in $x, t$ along which the original partial differential equation behaves like an ordinary differential equation. Suppose a characteristic curve $C$ is defined by

$$x = x(\lambda), \quad t = \lambda. \tag{15.152}$$

Derivatives of functions $w(x, t)$ along this curve are given by

$$\frac{dw}{d\lambda} = w_t + w_x \frac{dx}{d\lambda}. \tag{15.153}$$

Notice that with $dx/d\lambda = c$, the partial differential equation $u_t + cu_x = 0$ reduces to the simple ordinary differential equation $u_\lambda = 0$. Thus, curves with $dx/dt = c$ are characteristic curves for this simple equation.

To reduce the system (15.149) to characteristic form, we try to find appropriate linear combinations of the equations that transform the system to an ordinary differential equation. Thus, suppose the matrix $B$ has a left eigenvector $\xi^T$ with corresponding eigenvalue $s$, so that $\xi^T B = s\xi^T$. If we multiply (15.149) by $\xi^T$, we find that with the identification $dx/d\lambda = s$,

$$0 = \xi^T(w_t + Bw_x) = \xi^T(w_t + sw_x) = \xi^T w_\lambda. \tag{15.154}$$

In other words, along the curve $dx/dt = s$, the original system of equations reduces to the simple ordinary differential equation $\xi^T w_\lambda = 0$.

It is an easy matter to determine that the eigenvalues of $B$ are

$$s = u \pm K(P), \tag{15.155}$$

where

$$K(P) = \sqrt{\frac{A(P)}{\rho c}}, \tag{15.156}$$

with corresponding left eigenvector

$$\xi^T = (\xi_1, \xi_2) = (\rho K(P), \pm 1). \tag{15.157}$$

It follows from $\xi^T w_\lambda = 0$ that

$$u_\lambda \pm \frac{1}{\rho K(P)} P_\lambda = 0 \tag{15.158}$$

along the characteristic curve $dx/d\lambda = u \pm K(P)$, which we denote by $C_\pm$. Now, notice that

$$\frac{d}{dP} K(P) = \frac{1}{2} \frac{1}{\rho K(P)}, \tag{15.159}$$

so that

$$\frac{d}{d\lambda}(u \pm 2K(P)) = 0. \tag{15.160}$$

In other words, $u + 2K(P)$ is conserved (remains constant) along $C_+$, the characteristic curve with slope $dx/dt = u + K(P)$, and $u - 2K(P)$ is conserved along $C_-$, the characteristic curve with slope $dx/dt = u - K(P)$.

Now, to see how this reduction allows us to solve a specific problem, consider the region $x \geq 0$, $t \geq 0$ with $u(0, t) > 0$ specified. For example, $u(0, t)$ could be the velocity

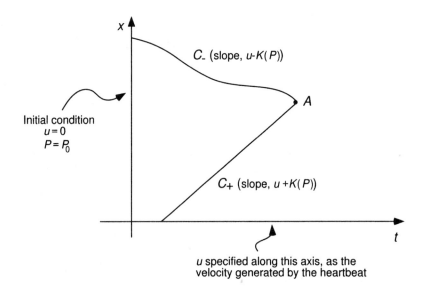

**Figure 15.15**  Diagram of the characteristics of the arterial pulse equations in the $(t, x)$ plane.

pulse generated by a single heartbeat. We suppose that initially, $u(x, 0) = 0, P(x, 0) = P_0$ for all $x \geq 0$, where $P_0$ is the diastolic pressure.

Pick any point $A$ in the region $x \geq 0, t \geq 0$ (Fig. 15.15). There are two characteristics passing through $A$, one, $C_+$, with positive slope $u + K(P)$ and one, $C_-$, with negative slope $u - K(P)$. (Here and in the following we assume that $u$ is small enough so that $C_-$ always has negative slope.) Following $C_-$ up and to the left, we see that it intersects the vertical axis, where $u = 0$ and $P = P_0$ (because of the specified initial data). Since the quantity $u - 2K(P)$ is conserved on $C_-$, it must be that $u - 2K(P) = -2K(P_0)$ at the point $A$. Since $A$ is arbitrary, it follows that $u = 2K(P) - 2K(P_0)$ everywhere in the first quadrant. Thus, we know that $u + 2K(P) = 4K(P) - 2K(P_0)$ is constant along $C_+$. Hence, $K(P)$ is conserved along $C_+$, as are both $P$ and $u$, so that $C_+$ is a straight line. The slope of $C_+$ is the value of $u + K(P)$ at the intersection of $C_+$ with the horizontal axis.

To be specific, suppose $u(0, t)$ first increases and then decreases as a function of $t$, as shown in Fig. 15.16A. (In this figure $u(0, t)$ is shown as piecewise linear, but this is simply for ease of illustration.) To be consistent, since $u = 2K(P) - 2K(P_0)$ everywhere in the first quadrant, $K(P(0, t)) = K(P_0) + \frac{1}{2}u(0, t)$, so that the slope of the $C_+$ characteristics is $s(t) = \frac{3}{2}u(0, t) + K(P_0)$, which also increases and then decreases as a function of $t$. With increasing slopes, the characteristics converge, resulting in a steepening of the wave front. If characteristics meet, the solution is not uniquely defined by this method, and shocks develop.

The place a shock first develops can be found by determining the points of intersection of the characteristics. Suppose we have two $C_+$ characteristics, one emanating from the $t$-axis at $t = t_1$, described by $x = s(t_1)(t - t_1)$, and the other emanating at $t = t_2$,

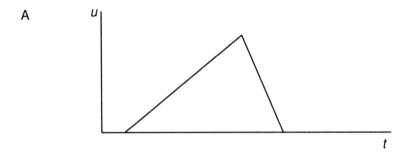

A

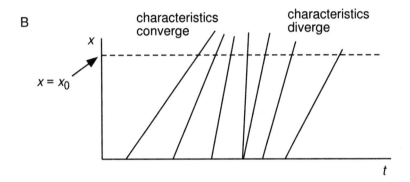

B

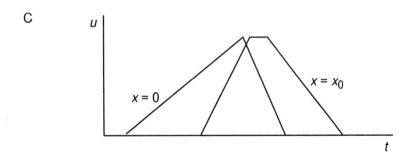

C

**Figure 15.16** A: Sketch of $u(0, t)$. B: Characteristics generated by $u(0, t)$ in the previous figure. C: Plots of $u(x, t)$ for $x = 0$ and $x = x_0$, obtained by taking cross-sections for a fixed $x$ (as indicated by the dotted line in B).

described by $x = s(t_2)(t - t_2)$. They intersect at any point $(t_i, x_i)$, where

$$t_i = \frac{s(t_2)t_2 - s(t_1)t_1}{s(t_2) - s(t_1)}, \tag{15.161}$$

$$x_i = \frac{s(t_2)s(t_1)(t_2 - t_1)}{s(t_2) - s(t_1)}. \tag{15.162}$$

In the limit $t_2 \to t_1$,

$$t_i = \frac{s(t)}{s'(t)} + t, \tag{15.163}$$

$$x_i = \frac{s^2(t)}{s'(t)}, \tag{15.164}$$

which defines parametrically the envelope of intersection points as a function of $t$, the time of origin of one of the characteristics.

The important point to note is that the first point of shock formation is where $x_i$ is smallest. In other words, for data with $s'(t)$ large, the shock develops quickly, and close to $x = 0$. Thus, generally speaking, the steeper the pulse generated by the heart, the sooner and closer a shock will form. This may explain why the pistol-shot occurs in patients with aortic insufficiency but not in other patients. Using numerical simulations of the model equations, Anliker et al. (1971a,b) have shown that under conditions of aortic insufficiency, a steep pressure gradient can develop within 40 cm of the heart, well within the physiological range.

It is also noteworthy that the slope $s$ depends on diastolic pressure $P_0$ through $s = \frac{3}{2}u + K(P_0)$. Thus, a decrease in $K(P_0)$, caused either by a decrease of $P_0$ or a decrease of the function $A(P)$, leads to a decrease of the first location of shock formation $x_i$.

Notice also that if $s'(t) < 0$, so that $u(0,t)$ is decreasing, no shock can form for positive $x$. This can also be seen from Fig. 15.16, since, if $u(0,t)$ is a decreasing function of $t$, the characteristics fan out and do not intersect for positive $x$.

## 15.9 Exercises

1. Equation (15.7) was derived assuming that the radius of the vessel and pressure drop along the vessel were constant, but then it was used in (15.14) as if the pressure were variable. Under what conditions is this a valid approximation?

2. Show that

$$\lim_{x \to y} \frac{x^4 - y^4}{x^3 - y^3} = \frac{4}{3}y, \tag{15.165}$$

   so that (15.21) follows.

3. Simplify the six-compartment model of the circulation by assuming that there are no pressure drops over the arterial and venous systems (either systemic or pulmonary), and thus, for example, $P_{sa} = P_{s1}$. Solve the resultant equations and compare with the behavior of the three-compartment model presented in the text. How do the parameter values change? Are the sensitivities altered? (Calculate the sensitivities using a symbolic manipulation program.)

4. Show that (15.67)–(15.75) can be derived from (15.54)–(15.66) by letting $R_s$ and $R_p$ approach zero and by letting $C_{pa} = C_{pv}$ and $R_{pa} = R_{pv}$.

5. In the three-compartment circulatory model it was assumed that the base volume of the pulmonary circulation $V_0^p$ was zero. Using the fact that the systolic pressure in the pulmonary artery is about 22 mm Hg and the diastolic pressure is about 7 mm Hg, determine

the typical value for the volume of the pulmonary circulation. How does this change to the model affect the results?

6. Find the pressures as a function of cardiac output assuming $V_l \neq 0, V_r \neq 0$, where $V_l$ is the basal volume of the left heart, and similarly for $V_r$ (cf. equation (15.35)). Show that

$$Q_\infty = \frac{2C_p V_l/C_{ld} + 2(C_{sv} + C_{sa})V_r/C_{rd} + 2(V - V_0^p - V_0^s)}{C_p R_p + C_{sv} R_{sv} + C_{sa}(R_{sa} + 2R_{sv})}. \tag{15.166}$$

7. Explore the behavior of the autoregulation model with $R = R_0 (1 + a[O_2]_v)/(1 + b[O_2]_v)$.

8. What symptoms in the circulation would you predict from anemia?
   Hint: Anemia refers simply to an insufficient quantity of red blood cells, which results in decreased resistance and oxygen-carrying capacity of the blood.

9. Devise a simple single loop model for the fetal circulation that treats the systemic flow and the placental flow as parallel flows. How should the flow be split between system and placenta to support the highest metabolic rate?

10. In the model for autoregulation $P_a$ and $P_v$ are given. More realistically, they would be determined, in part at least, by $R_s$. Construct a more detailed model for autoregulation, including the effects of $R_s$ on the pressures, and show how the arterial and venous pressures, and the cardiac output, depend on $M$ and $A$.

11. Modify the model for capillary filtration by allowing the plasma osmotic pressure to vary along the capillary distance, by setting $\pi_c = RTc_c \frac{Q_i}{q}$, where $c_c$ is the local concentration of osmolites. If incoming pressure is unchanged from the first model, what is the effect of osmotic pressure on total filtration? What changes must be made to the incoming pressure to maintain the same total filtration?
    Hint: Study the phase portrait for this system of equations.

12. Derive a simplified *windkessel* model by starting with a single vessel with volume $V(t) = V_0 + CP(t)$. Assume that the flow leaves through a resistance $R$ and that there is an inflow (from the heart) of $Q(t)$. Derive the differential equation for $P$ and compare it to (15.141).

13. Frank (1899) described a method whereby the flux of blood out of the heart could be estimated from a knowledge of the pressure pulse, even when the arterial resistance is unknown. Starting with (15.141), assume that during the second part of the arterial pulse, $Q(t) \equiv 0$. Write down equations for the first and second parts of the pulse, eliminate $R$, and find an expression for $Q$. Give a graphical interpretation of the expression for $Q$.

14. In the model for the arterial pulse, set $u(0,t) = at$ for some constant $a$, and determine the curve in the $(t, x)$ plane along which characteristics form an envelope. Determine the first value of $x$ at which a shock forms.

# CHAPTER 16

# Blood

Blood is composed of two major ingredients, the liquid *blood plasma* and several types of cellular elements suspended within the plasma. The cellular elements constitute approximately 40% of the total blood volume and are grouped into three major categories: *erythrocytes* (red blood cells), *leukocytes* (white blood cells), and *thrombocytes* (platelets).

## 16.1 Blood Plasma

The blood plasma is 89–95% water, with a variety of dissolved substances. The dissolved substances with small molecular weight include bicarbonate, chloride, phosphorus, sodium, calcium, potassium, magnesium, urea, and glucose. There are also large protein molecules including *albumin* and $\alpha$-, $\beta$-, and $\gamma$-*globulins*. Of these proteins, albumin has the highest molar concentration in plasma and this makes the greatest contribution to the plasma osmotic pressure.

In addition, gases, such as carbon dioxide and oxygen, are dissolved in the blood plasma. For an ideal gas, the pressure, volume, and temperature are related by the ideal gas law,

$$PV = nkT, \tag{16.1}$$

where $P$ is the pressure, $V$ is the volume, $n$ is the number of gas molecules, $k$ is Boltzmann's constant, and $T$ is temperature in Kelvin. Since concentration is $c = n/V$, for an ideal gas

$$P = ckT. \tag{16.2}$$

This representation of concentration is in units of molecules per volume, and while this seems natural, it is not the usual way that concentrations are represented. To express concentration in terms of moles per unit volume, we multiply and divide (16.2) by Avogadro's number $N_A$ to obtain

$$P = CRT, \tag{16.3}$$

where $C = c/N_A$ and $R = kN_A$ is the universal gas constant.

Air is a mixture of different gases, with 78% nitrogen and 21% oxygen. Each of these gases contributes to the total pressure of the mixture via its *partial pressure*. The partial pressure of gas $i$, $P_i$, is defined by

$$P_i = x_i P, \tag{16.4}$$

where $x_i$ is the mole fraction of gas $i$, and $P$ is the total pressure of the gas mixture. Thus, by definition, the total pressure of a gas mixture is the sum of the partial pressures of each individual gas in the mixture. In an ideal mixture the partial pressure of gas $i$ is equal to the pressure that gas $i$ would exert if it alone were present.

When a gas with partial pressure $P_i$ comes into contact with a liquid, some of the gas will dissolve in the liquid. When a steady state is reached, the amount of gas dissolved in the liquid is a function of the partial pressure of the gas above the liquid. If the concentration of the dissolved gas is low enough, thus forming an ideally dilute solution, then $P_i$ is related to the concentration of gas $i$ by

$$c_i = \sigma_i P_i, \tag{16.5}$$

where $c_i$ is the concentration of gas $i$, and $\sigma_i$ is called the *solubility*. In general $\sigma_i$ is a function of the temperature and the total pressure above the liquid. In Table 16.1 are shown the solubilities of important respiratory gases in blood, where it can be seen, for example, that the solubility of carbon dioxide in blood is about 20 times larger than that of oxygen.

**Table 16.1** Solubility of respiratory gases in blood plasma.

| Substance | $\sigma$ (Molar/mm Hg) |
|---|---|
| $O_2$ | $1.4 \times 10^{-6}$ |
| $CO_2$ | $3.3 \times 10^{-5}$ |
| $CO$ | $1.2 \times 10^{-6}$ |
| $N_2$ | $7 \times 10^{-7}$ |
| He | $4.8 \times 10^{-7}$ |

## 16.2  Erythrocytes

Erythrocytes (red blood cells) are small biconcave discs measuring about 8 $\mu$m in diameter. They are flexible, allowing them to change shape and to pass without breaking through blood vessels with diameters as small as 3 $\mu$m. Their function is the transport of oxygen from the lungs to the rest of the body, and they accomplish this with the help of a large protein molecule called *hemoglobin*, which binds oxygen in the lungs, later releasing it in tissue. Hemoglobin is the principal protein constituent of mature erythrocytes. A similar protein, *myoglobin*, is used to store and transport oxygen within muscle; mammals that dive deeply, such as whales and seals, have skeletal muscle that is especially rich in myoglobin.

### 16.2.1  Myoglobin and Hemoglobin

The binding of oxygen with myoglobin and hemoglobin serves as an excellent example of relatively simple chemical reactions that are of fundamental importance in blood physiology. We get some understanding of this process by examining the experimentally determined saturation function for hemoglobin and myoglobin as a function of the partial pressure of oxygen. For myoglobin the saturation curve is much like a standard Michaelis–Menten saturation function, while for hemoglobin it is S-shaped (Fig. 16.1). From these curves we see that when the partial pressure of oxygen is at 100 mm Hg (about what it is in the lungs), hemoglobin is 97% saturated. This amount is affected only slightly by small changes in oxygen partial pressure, because at this level

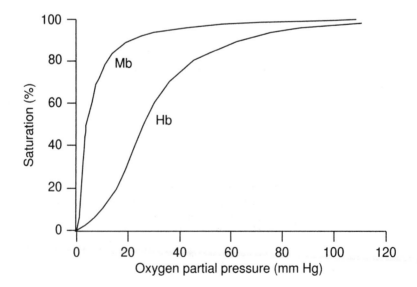

**Figure 16.1**  Uptake of oxygen by myoglobin and hemoglobin. (Rubinow, 1975, Fig. 2.13, p. 82, taken from Changeux, 1965.)

the saturation curve is relatively flat. In veins or tissue, however, where the partial pressure of oxygen is about 40 mm Hg, the saturation is about 75%. Furthermore, because this is on a steep portion of the saturation curve, if the metabolic demand for oxygen should decrease the oxygen pressure to, say, 20 mm Hg, then hemoglobin gives up its oxygen readily, reducing its saturation to about 35%. At this value of oxygen partial pressure the saturation of myoglobin is at 90%. Thus, if the tissue is muscle, the oxygen will be transferred from hemoglobin to myoglobin. These curves illustrate that the affinity of myoglobin for oxygen is greater than that of hemoglobin.

These saturation curves are of fundamental importance to blood chemistry, so it is of interest to understand why the saturation curves of the two are as they are. We can derive models of these saturation curves from the underlying chemistry (see Section 1.2.4 on cooperativity). Myoglobin consists of a polypeptide chain and a disc-shaped molecular ring called a *heme group*, which is the active center of myoglobin. At the center of the heme group is an iron atom, which can bind with oxygen, forming oxymyoglobin. Hemoglobin consists of four such polypeptide chains (called *globin*) and four heme groups, allowing the binding of four oxygen molecules. When bound with oxygen, the iron atoms in hemoglobin and myoglobin give them their red color. Myoglobin content accounts for the difference in color between red meat such as beef, and white meat such as chicken.

A simple reaction scheme describing the binding of oxygen with myoglobin is

$$O_2 + Mb \underset{k_-}{\overset{k_+}{\rightleftharpoons}} MbO_2.$$

The dynamics for this scheme are described by (1.4), and at equilibrium $k_+[Mb][O_2] = k_-[MbO_2]$, so that the percentage of occupied sites is

$$Y = \frac{[MbO_2]}{[Mb] + [MbO_2]} = \frac{[O_2]}{K + [O_2]}, \tag{16.6}$$

where $K = k_-/k_+$.

To compare the function (16.6) with the saturation curve for myoglobin in Fig. 16.1, we must relate the oxygen concentration to the oxygen partial pressure via $[O_2] = \sigma P_{O_2}$. Then, (16.6) becomes

$$Y = \frac{P_{O_2}}{K/\sigma + P_{O_2}} = \frac{P_{O_2}}{K_P + P_{O_2}}, \tag{16.7}$$

and we get a good fit of the myoglobin uptake curve in Fig. 16.1 with $K_P = 2.6$ mm Hg. Notice that the equilibrium constant $K_P$ is in units of pressure rather than concentration, as is more typical. The equilibrium constant $K$ is related to $K_P$ through $K = \sigma K_P$. For the myoglobin saturation curve $K = 3.7$ $\mu$M; however, because it is typical to describe concentrations of dissolved gases in units of pressure, it is also typical to write the equilibrium constant $K$ in these units as $K = 2.6\sigma$ mm Hg. A comparison of the curve (16.7) with the data is shown in Fig. 16.2.

The primary reason that the saturation curve for hemoglobin is significantly different from that for myoglobin is that it has four oxygen binding sites instead of one.

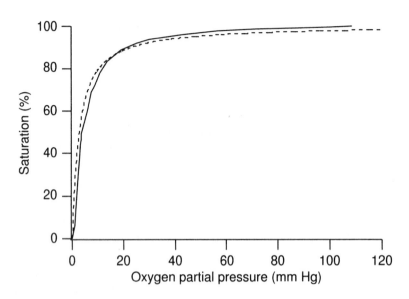

**Figure 16.2** Comparison of myoglobin saturation curve (solid) with the curve (16.7) (dashed) with $K_P = 2.6$ mm Hg.

A simple kinetic scheme for the formation of oxyhemoglobin is

$$4O_2 + Hb \underset{k_-}{\overset{k_+}{\rightleftharpoons}} Hb(O_2)_4,$$

with the corresponding differential equation

$$\frac{d[Hb]}{dt} = k_-[Hb(O_2)_4] - k_+[Hb][O_2]^4. \tag{16.8}$$

At steady state, the percentage of available hemoglobin sites that are bound to oxygen is

$$Y = \frac{[Hb(O_2)_4]}{[Hb(O_2)_4] + [Hb]} = \frac{[O_2]^4}{[O_2]^4 + K^4}, \tag{16.9}$$

where $K^4 = k_-/k_+$. We use the half-saturation level from the hemoglobin uptake curve in Fig. 16.1 to estimate $K$ as $K = 26\sigma$ mm Hg.

While (16.9) (shown as a short dashed curve in Fig. 16.3) reproduces some features of the uptake curve that are qualitatively correct, it is not quantitatively accurate. In fact, one can achieve a much better fit of the data with the Hill equation

$$Y = \frac{[O_2]^n}{[O_2]^n + K^n}, \tag{16.10}$$

with $n = 2.5$ and $K = 26\sigma$ mm Hg. However, there is no adequate theoretical basis for such a model.

A better model keeps track of the elementary reactions involved in the binding process, and is given by

$$O_2 + H_{j-1} \underset{k_{-j}}{\overset{k_j}{\rightleftharpoons}} H_j, \qquad j = 1, 2, 3, 4,$$

where $H_j = Hb(O_2)_j$. The steady state for this reaction is attained at

$$[H_j] = \frac{k_{+j}}{k_{-j}}[H_{j-1}][O_2] = \frac{[H_{j-1}][O_2]}{K_j}, \tag{16.11}$$

and the saturation function is

$$Y = \frac{\sum_{j=0}^{4} j H_j}{4 \sum_{j=0}^{4} H_j}. \tag{16.12}$$

Substituting (16.11) into (16.12) we obtain the saturation function

$$Y = \frac{\sum_{j=0}^{4} j \alpha_j [O_2]^j}{4 \sum_{j=0}^{4} \alpha_j [O_2]^j}, \tag{16.13}$$

where $\alpha_j = \prod_{i=1}^{j} K_i^{-1}$, $K_j = k_{-j}/k_{+j}$, $\alpha_0 = 1$.

One can fit the saturation function (16.13) to the hemoglobin uptake curve shown in Fig. 16.1, with the result $K_1 = 45.9, K_2 = 23.9, K_3 = 243.1, K_4 = 1.52\sigma$ mm Hg (Roughton et al., 1972). The striking feature of these numbers is that $K_4$ is much smaller than $K_1, K_2$, or $K_3$, indicating that there is apparently a greatly enhanced affinity of oxygen for hemoglobin if three oxygen molecules are already bound to it. Hemoglobin prefers to be "filled up" with oxygen. The mechanism for this positive cooperativity is not completely understood. (If the binding sites were independent, then $K_1$ would be the smallest and $K_4$ would be the largest equilibrium constant; see Exercise 2 and Section 1.2.4.) The most widely known model of hemoglobin cooperativity is that of Monod, Wyman, and Changeux (1965). Models of this type were discussed in Section 1.2.4, and so construction of an MWC model of hemoglobin is left as an exercise (Exercise 4).

Notice that the affinity of oxygen for myoglobin is greater than for any of the binding sites of hemoglobin. In Fig. 16.3 is shown a comparison between the data and the approximate curves (16.9) and (16.13). The Hill equation fit (16.10) is not shown because it is nearly identical to (16.13).

## 16.2.2 Hemoglobin Saturation Shifts

There are a number of factors that affect the binding of oxygen to hemoglobin, the most important of which is the hydrogen ion, which is an allosteric inhibitor of oxygen binding (Chapter 1). As we will see, the interactions between oxygen concentration and carbon dioxide concentration (which indirectly changes the hydrogen ion concentration) are important for transport of both oxygen and carbon dioxide.

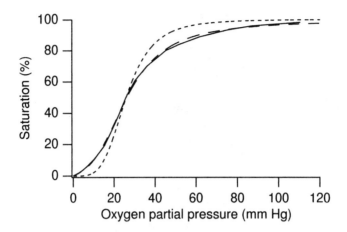

**Figure 16.3** Comparison of hemoglobin saturation curve (solid) with the curves (16.9) (short dashed) and (16.13) (long dashed).

Carbon monoxide combines with hemoglobin at the same binding site as oxygen (and is a competitive inhibitor), but with an affinity more than 200 times greater. Therefore the carbon monoxide saturation curve is almost identical to the oxygen saturation curve, except that the abcissa is scaled by a factor of about 200. At a carbon monoxide partial pressure of 0.5 mm Hg, and in the absence of oxygen, hemoglobin is 97% saturated with carbon monoxide. If oxygen is present at atmospheric concentrations, then it takes a carbon monoxide partial pressure of only 0.7 mm Hg (about 0.1 percent) to cause oxygen starvation in the tissues (Chapter 17).

*Fetal hemoglobin*, a different type of hemoglobin found in the fetus, has a considerable leftward shift for its oxygen saturation curve. This allows fetal blood to carry as much as 30% more oxygen at low oxygen partial pressures than can adult hemoglobin. This is important since the oxygen partial pressure in the fetus is always low. The left-shift of the fetal hemoglobin saturation curve is also important for the transfer of oxygen from mother to fetus.

Because it is important in the next section, we construct a simple model of the allosteric inhibition by hydrogen ions of oxygen binding to hemoglobin. As illustrated in Fig. 16.4, we assume that the hemoglobin molecule can exist in four different states: with $H^+$ bound (concentration Z), with $O_2$ bound (concentration Y), with neither bound (concentration X), or with both bound (concentration W). This is, of course, an extreme simplification, as it ignores the cooperative nature of oxygen binding as discussed in the previous section and in Chapter 1, but nevertheless the results are qualitatively correct.

Assuming that each reaction is at equilibrium we find

$$O^4 X = K_1 Y, \tag{16.14}$$

$$hX = K_2 Z, \tag{16.15}$$

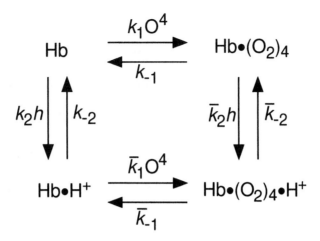

**Figure 16.4** Binding diagram for the allosteric binding of hydrogen ions and oxygen to hemoglobin. We assume a single hydrogen ion binding site, and a simplified mechanism for oxygen binding. Hb denotes hemoglobin.

$$O^4 Z = \bar{K}_1 W, \tag{16.16}$$

$$X + Y + Z + W = T_{\text{Hb}}, \tag{16.17}$$

where $h$ denotes $[H^+]$, $O$ denotes $[O_2]$, $T_{\text{Hb}}$ denotes the total concentration of hemoglobin, $K_1 = k_{-1}/k_1$ and similarly for $K_2, \bar{K}_1$ and $\bar{K}_2$ (which is used below). Solving these four equations we find

$$Y + W = \frac{O^4 T_{\text{Hb}}}{\phi(h) + O^4}, \tag{16.18}$$

where

$$\phi(h) = \frac{K_1 \bar{K}_1 (K_2 + h)}{K_2 \bar{K}_1 + h K_1}. \tag{16.19}$$

Note that we are interested in $Y + W$ as a function of $O$, since $Y + W$ is the total concentration of hemoglobin with oxygen bound, and thus plotting $Y + W$ as a function of $O$ gives the oxygen saturation curve.

It is easily seen from (16.18) that $h$ does not change the maximal saturation, although it shifts the mid-point of the curve. Since hydrogen ions are an allosteric inhibitor of oxygen binding, we assume that $\bar{K}_1 > K_1$. Note that, in this case, $\phi(h)$ is an increasing function of $h$, and thus increasing $h$ shifts the saturation curve to the right, as required.

Before we can discuss the importance of the allosteric effect of hydrogen ions on oxygen binding it is necessary to discuss the mechanism of carbon dioxide transport.

## 16.2.3  Carbon Dioxide Transport

Just as oxygen is taken up in the lungs and transported to the tissues, so carbon dioxide must be transported from the tissues to the lungs for removal from the body. In the blood, $CO_2$ is transported in three main forms. In venous blood a significant amount (about 6%) is present as dissolved $CO_2$. A slightly greater amount (about 7%) is bound to the globin part of hemoglobin as carbamino compounds, but most $CO_2$ (87 %) is present in the form of bicarbonate ions.

In the tissues $CO_2$ diffuses down its concentration gradient into the plasma and into the red blood cells. In both plasma and red blood cells it combines with water to form carbonic acid ($H_2CO_3$), which then dissociates quickly into hydrogen ions and bicarbonate ions. Thus,

$$CO_2 + H_2O \underset{r_{-1}}{\overset{r_1}{\rightleftharpoons}} H_2CO_3 \underset{r_{-2}}{\overset{r_2}{\rightleftharpoons}} H^+ + HCO_3^-. \tag{16.20}$$

This reaction proceeds slowly in the plasma but much more rapidly in the red blood cells because of the presence there of the enzyme *carbonic anhydrase*, which increases the speed of $CO_2$ hydration by more than a thousand times. The $H^+$ formed by the dissociation of carbonic acid binds to the globin part of hemoglobin, and the bicarbonate ion diffuses into the plasma in exchange for $Cl^-$.

In the lungs the reaction is reversed, as $CO_2$ diffuses down its concentration gradient to be excreted in the alveolar air and then the expired air. It is important to emphasize that the direction of the carbonic anhydrase reaction (16.20) is determined by the local concentration of $CO_2$. In the tissues, $[CO_2]$ is high, which drives reaction (16.20) from left to right, thus storing $CO_2$ in the blood. In the lungs, $[CO_2]$ is low, driving the reaction from right to left, thus removing $CO_2$ from the blood. Of course, carbonic anhydrase speeds up the reaction in both directions; without this increase in speed not enough $CO_2$ can be stored in the blood to remove it from the body fast enough.

The importance of the allosteric effect of $H^+$ on oxygen binding to hemoglobin is now apparent. In the tissues, because of the high local $CO_2$ concentration, the hydration of $CO_2$ causes an increase in the local concentration of $H^+$ (i.e., the blood pH falls slightly, from about 7.4 to about 7.35), which in turn results in a decreased affinity of hemoglobin for oxygen, thus increasing oxygen release to the tissues. In the lungs, the reverse occurs; the low local $CO_2$ concentration causes a decrease in $H^+$ concentration which results in an increase in hemoglobin oxygen affinity, and thus increased oxygen uptake. This effect of $CO_2$ concentration on oxygen transport (mediated by the carbonic anhydrase reaction and hydrogen ions), is known as the *Bohr effect*.

It is interesting to note that, from the principle of detailed balance applied to the reaction scheme shown in Fig. 16.4 (i.e., from consistency of the four equilibrium equations) it must be that

$$\frac{K_1}{\bar{K}_1} = \frac{K_2}{\bar{K}_2}. \tag{16.21}$$

It thus follows that, if $\bar{K}_1 > K_1$ we must also have $\bar{K}_2 > K_2$. In other words, if $H^+$ is an allosteric inhibitor of oxygen binding, then oxygen must also be an allosteric inhibitor of $H^+$ binding. Hence, as $CO_2$ influences oxygen transport, so too oxygen affects $CO_2$ transport. In the tissues, where $[O_2]$ is low, binding of $H^+$ to hemoglobin is enhanced. This lowers the local $H^+$ concentration, thus driving the carbonic anhydrase reaction from left to right, and increasing $CO_2$ storage. The reverse occurs at the lungs. The enhancement of $CO_2$ transport by low levels of oxygen is called the *Haldane effect*. Note that, according to the principle of detailed balance (at least in our simple model) the Bohr effect implies the Haldane effect, and vice versa.

To construct a mathematical model of $CO_2$ transport, we assume that the dissociation of carbonic acid is fast, and thus

$$[H_2CO_3] = R_2[H^+][HCO_3^-], \tag{16.22}$$

where $R_2 = r_{-2}/r_2$. At steady state,

$$[CO_2] = R_1 R_2 [H^+][HCO_3^-], \tag{16.23}$$

where $R_1 = r_{-1}/r_1$.

Carbon dioxide enters this system from the tissues and leaves at the lungs. When it does so, bicarbonate is produced or removed. However, since the carbonic anhydrase reaction produces exactly one hydrogen ion for each bicarbonate ion it produces, and since these hydrogen ions must either be free, or bound to hemoglobin, it follows that

$$[HCO_3^-] = h + Z + W - T_0, \tag{16.24}$$

where $T_0 = h_0 + Z_0 + W_0 - [HCO_3^-]_0$ is some measured reference level. In reality each hemoglobin molecule can bind many hydrogen ions, and so the conservation equation should be

$$[HCO_3^-] = h + n(Z + W) - T_0, \tag{16.25}$$

where $n$ can be as large as 10 or 20, and $T_0 = h_0 + n(Z_0 + W_0) - [HCO_3^-]_0$. The number $n$ is important, because without it (if $n = 0$), the pH fluctuates widely with changes in bicarbonate (Exercise 5a), whereas in normal blood, practically all the $H^+$ produced by the carbonic anhydrase reaction is absorbed by the hemoglobin. This demonstrates the extreme importance of hemoglobin as a hydrogen ion buffer. Note that, to be consistent, the factor $n$ should also be included in the model for hemoglobin. However, as this would greatly increase the complexity of the binding model without adding anything fundamentally new, we cheat slightly by including $n$ in the bicarbonate conservation equation, but not in the binding diagram. A more accurate model gives the same qualitative result.

In arterial and venous blood, the oxygen and carbon dioxide concentrations are known. Their precise values are set by the rate of gas exchange in the lungs and the tissues, and depend to some extent on the properties of the carbonic anhydrase reaction, among other things. Thus, to be strictly correct, we should not treat them as constants, but solve for them as part of a more complicated model. We omit these complications here and treat $O$ and $[CO_2]$ as known constants. Our goal is to find the other unknowns $(X, Y, Z, W$ and $h)$ as functions of the gas concentrations.

We now have five equations

$$O^4 X = K_1 Y, \tag{16.26}$$

$$h X = K_2 Z, \tag{16.27}$$

$$O^4 Z = \bar{K}_1 W, \tag{16.28}$$

$$X + Y + Z + W = T_{\text{Hb}}, \tag{16.29}$$

$$[CO_2] = R_1 R_2 h[h + n(Z + W) - T_0], \tag{16.30}$$

to solve for the five unknowns. It is an easy matter to solve (16.26)-(16.29) for $X, Y, Z$, and $W$ in terms of $O^4, h$ and parameters, and substitute these into (16.30). This yields a single equation for $[CO_2]$ as a function of $O^4$ and $h$. This equation can be readily solved numerically for $h$ as a function of $[CO_2]$ and $O^4$. Solution of this equation is left for the exercises (Exercise 5).

### 16.2.4   Red Blood Cell Production

One cubic millimeter of blood contains 4.2–5.4 million red blood cells. These cells have an average lifetime of 120 days and are estimated to travel through about 700 miles of blood vessels during their life span. Because of aging and rupturing, red blood cells must be constantly replaced. On average, the body must produce $3 \times 10^9$ new erythrocytes for each kilogram of body weight every day.

Red blood cells are produced by the bone marrow. In a child before the age of 5, blood cells are produced in the marrow of essentially all the bones. However, with age, the marrow of the long bones becomes quite fatty and so produces no more blood cells after about age 20. In the adult, most red blood cells are produced in the marrow of membranous bones, such as the vertabrae, sternum, ribs, and ilia.

In the bone marrow there are cells, called *pluripotential hemopoietic stem cells*, from which all of the cells in the circulating blood are derived. As these cells grow and reproduce, a portion of them remains exactly like the original pluripotential cells, maintaining a more or less constant supply of these cells. The larger portion of the reproduced stem cells differentiates to form other cells, called *committed stem cells*. The committed stem cells produce colonies of specific types of blood cells, called *formed elements*, including erythrocytes, granulocytes, monocytes, and megakaryocytes.

Growth and reproduction of stem cells are controlled by multiple proteins called *growth inducers*, and differentiation is controlled by another set of proteins, the *differentiation inducers*. Formation of growth inducers and differentiation inducers is itself controlled by factors outside the bone marrow such as, in the case of red blood cells, low oxygen concentration for an extended period of time.

The principal factor stimulating red blood cell production is the hormone *erythropoietin*. About 90% of the erythropoietin is secreted by renal tubular epithelial cells when blood is unable to deliver sufficient oxygen. The remainder is produced by other tissues (mostly the liver). When both kidneys are removed or destroyed by renal disease, the person invariably becomes anemic because of insufficient production of erythropoietin.

The role of erythropoietin in bone marrow is twofold. First, it stimulates the production of pre-erythrocytes, called *proerythroblasts*, and it also controls the speed at which the developing cells pass through the different stages. Normal production of red blood cells from stem cells takes 5–7 days, with no appearance of new cells before 5 days, even at high levels of erythropoietin. At high erythropoietin levels the rate of red blood cell production (number per unit time) can be as much as ten times normal, even though the maturation rate of an individual red blood cell varies much less. Details of the regulatory system governing red blood cells can be found in Williams (1990).

In most people the production of red blood cells is relatively constant. However, there are pathological conditions that exhibit oscillatory behavior. The most common oscillatory hematological disease is called *periodic hematopoiesis* (PH) (Milton and Mackey, 1989). In humans, PH is a disease characterized by 17–28 day periodic oscillations in numbers of all the circulating formed elements of blood. All grey collies have PH. In a related disease known as periodic chronic myelogenous leukemia (CML), the number of neutrophils (see Section 16.3) varies from approximately normal levels to barely detectable numbers. Data showing oscillations of white blood cell count in a twelve-year-old girl with periodic CML are shown in Fig. 16.5.

Periodic erythropoiesis can be induced in rabbits by the injection of an incompatible red cell isoantibody (Orr et al., 1968) or in mice by the administration of a single

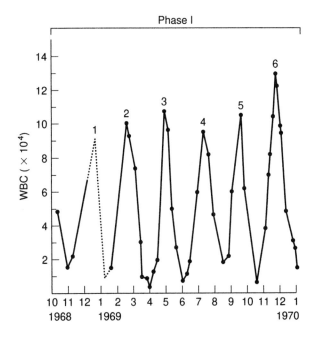

**Figure 16.5** White blood cell count as a function of time for a twelve-year-old girl with periodic CML. (Gatti et al., 1973, Fig. 1.)

dose of marrow-seeking radioisotope $^{89}$ SR (Gurney et al., 1981). In patients with CML, periodic oscillations of granulocytes may be induced by chemotherapy.

We can model the blood production cycle as follows (Belair et al., 1995). We let $n(x, t)$ be the number of red blood cells at time $t$ that are $x$ units old. That is, they were released into the bloodstream at time $t - x$. We suppose that as they age, a certain percentage of them die, but at some age $X$ all remaining cells die. The conservation law for the number of cells is

$$\frac{\partial n}{\partial t} + \frac{\partial n}{\partial x} = -\beta n, \qquad (16.31)$$

where $\beta$ is the death rate. In general, we expect the death rate to be a function of age, so that $\beta = \beta(x)$. However, for this model we take the death rate to be independent of age. At any given time the total number of red blood cells in circulation is

$$N(t) = \int_0^X n(x, t) dx. \qquad (16.32)$$

Now we suppose that the production of red blood cells is controlled by $N$, and that once a cohort of cells is formed in the bone marrow, they will emerge into the bloodstream as mature cells some fixed time $d$ later, about 5 days. Here we are ignoring the fact that at high levels of erythropoietin (low oxygen) cells mature a bit more rapidly. Thus,

$$n(0, t) = F(N(t - d)), \qquad (16.33)$$

where $F$ is some nonlinear production function that is monotone decreasing in its argument. The function $F$ is related to the rate of secretion of erythropoietin in response to the red blood cell population size.

The steady-state solution for this model is easy to determine. We set $\partial n / \partial t = 0$ and find that

$$n(x) = \begin{cases} n(0) e^{-\beta x}, & x < X, \\ 0, & x > X, \end{cases} \qquad (16.34)$$

where $n(0)$ is yet to be determined. If we define $N_0$ to be the total steady-state number of blood cells, then

$$N_0 = \int_0^X n(x) \, dx, \qquad (16.35)$$

from which it follows that

$$N_0 = \int_0^X n(0) e^{-\beta x} \, dx = \frac{n(0)}{\beta} (1 - e^{-\beta X}), \qquad (16.36)$$

and thus

$$F(N_0) = \frac{\beta N_0}{1 - e^{-\beta X}}. \qquad (16.37)$$

Since $F(N_0)$ is a monotone decreasing function of $N_0$, (16.37) is guaranteed to have a unique solution. In fact, the solution is a monotone decreasing function of the

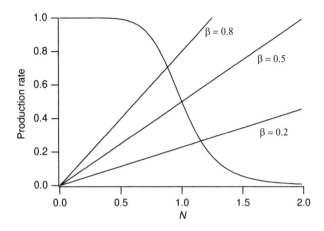

**Figure 16.6** Plot of left- and right-hand sides of (16.37) for three different values of $\beta$ and for $F(N) = \frac{1}{1+N^7}$ and $X = 10$.

parameter $\beta$, indicating that at higher death rates, the cell population drops while the production of cells increases. An illustration of these facts is provided by the graph in Fig. 16.6, where the two curves $F(N)$ and $\frac{\beta N}{1-e^{-\beta X}}$ are plotted as functions of $N$. Here the function $F(N)$ is taken to be $F(N) = \frac{A}{1+N^7}$, as suggested by data from autoimmune-induced hemolytic anemia in rabbits (Belair et al., 1995).

The next interesting question to ask is whether this steady solution is stable or unstable. We linearize the governing equation about the steady state $N_0$ and find that the deviation of $n(x,t)$ from steady state, denoted by $\delta n$, satisfies (16.31). To find the linearized initial condition, we use (16.33) and write

$$n(0,t) = F(N_0) + \delta n(0,t) = F(N_0 + \delta N(t-d)) \approx F(N_0) + F'(N_0)\delta N(t-d), \qquad (16.38)$$

so that

$$\delta n(0,t) = F'(N_0)\delta N(t-d). \qquad (16.39)$$

We seek the characteristic equation, found by setting $\delta N(t) = e^{\lambda t}$, from which it follows that

$$F'(N_0)e^{-\lambda d}\frac{1 - e^{-(\lambda+\beta)X}}{\lambda + \beta} = 1. \qquad (16.40)$$

The roots of this equation determine the stability of the linearized solution. If all the roots have negative real part, then the solution is stable, whereas if there are roots with positive real part, the steady solution is unstable. For the remainder of this discussion, we take $\beta = 0$. This implies that all cells die at exactly age $X$. A different simplification, taking $X \to \infty$, leads to a delay differential equation that is discussed in Chapter 17 (see also Exercise 9).

In the limit $\beta \to 0$, the characteristic equation (16.40) is

$$F'(N_0)(e^{-\lambda d} - e^{-\lambda(d+X)}) = \lambda. \qquad (16.41)$$

Since $F'(N_0) < 0$, there are no positive real roots. (The root at $\lambda = 0$ is spurious.)

There is possibly one negative real root; all other roots are complex. It follows that even if the steady solution is stable, the return to steady state will be oscillatory rather than monotone. Thus, following rapid disruptions of blood cell population, such as traumatic blood loss or transfusion, or a vacation at a high-altitude ski resort, the red blood cell population will oscillate about its steady state.

The only possible way to have a root with positive real part is if it is complex. Furthermore, a transition from stable to unstable can occur only if a complex root changes the sign of its real part, leading to a Hopf bifurcation. If a Hopf bifurcation occurs, it does so with $\lambda = i\omega$. We substitute $\lambda = i\omega$ into (16.41) and separate this into its real and imaginary parts to obtain

$$F'(N_0)(\cos(\omega d) - \cos(\omega(d + X))) = 0, \tag{16.42}$$

$$F'(N_0)(\sin(\omega d) - \sin(\omega(d + X))) = -\omega. \tag{16.43}$$

It follows from (16.42), because of symmetry, that $\omega(2d + X) = 2n\pi$, for any positive integer $n$. It could also be that $\omega X = 2n\pi$, but this fails to work in (16.43). With $\omega(2d + X) = 2n\pi$, (16.43) becomes

$$2dF'(N_0)\sin(\omega d) = -\omega d, \tag{16.44}$$

or

$$2dF'(N_0) = -\frac{2n\pi}{2 + \frac{X}{d}} \frac{1}{\sin\left(\frac{2n\pi}{2+\frac{X}{d}}\right)}. \tag{16.45}$$

Finally, we use that $F(N_0) = N_0/X$ to write

$$\frac{N_0 F'(N_0)}{F(N_0)} = -\frac{1}{2} \frac{X}{d} \frac{2n\pi}{2 + \frac{X}{d}} \frac{1}{\sin(\frac{2n\pi}{2+\frac{X}{d}})}. \tag{16.46}$$

For each integer $n$, this equation defines a relationship between $N_0$ and $X/d$ at which there is a Hopf bifurcation. If we take $F$ to be of the special form

$$F(x) = \frac{A}{1 + x^p}, \tag{16.47}$$

then we can use (16.37) (in the limit $\beta \to 0$) and (16.46) to find an analytic relationship between $dA(= dF(0))$ and $X/d$ at which Hopf bifurcations occur (see Exercise 7). Shown in Fig. 16.7 is this curve for $n = 1$ and $p = 7$. The case $n = 1$ is the only curve of interest, since it is the first instability. That is, if a curve for $n > 1$ is crossed, the steady solution is already unstable from the $n = 1$ instability.

The implications of this calculation are interesting. If the nondimensional parameters $X/d$ and $dF(0)$ are such that they lie above the curve in Fig. 16.7, then the steady solution is unstable, and a periodic or oscillatory solution is likely (but not rigorously proven). On the other hand, if these parameters lie below or to the far right of this curve, the steady solution is stable.

From this we learn that there are three mechanisms by which cell production can be destabilized, and these are by changing the maximal production rate $F(0)$, the expected

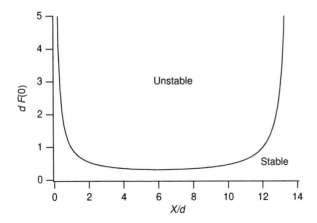

**Figure 16.7** Critical stability curve (Hopf bifurcation curve) for cell growth.

lifetime $X$, or the production delay $d$. If $X/d$ is sufficiently large ($> \approx 14$), the system cannot be destabilized. However, if $X/d$ is small enough, increasing $F(0)$ is destabilizing. Increasing $d$ is also destabilizing. If $F(0)$ and $X$ are held fixed, then changing $d$ moves $y = dF(0)$ and $x = X/d$ along the hyperbola $yx$ = constant. Every such hyperbola has exactly one intersection with the critical stability curve shown in Fig. 16.7. Thus, decreasing $d$ is stabilizing, as it increases $X/d$, moving it out of and away from the unstable region.

For normal humans, with $d = 5$ days and $X = 120$ days, there is no instability, since $X/d = 24$. However, any mechanism that substantially shortens $X$ can have a destabilizing effect and can result in oscillatory production of blood cells (Exercise 8). Near the bifurcation, the period of oscillation is $T = \frac{2\pi}{\omega}$, where $\omega(2d + X) = 2\pi$, so that

$$T = 2d + X. \tag{16.48}$$

Thus, for example, a disorder that halves the normal lifetime of red blood cells to $X = 60$ days should result in oscillatory cell production with period on the order of 70 days, the approximate period of the oscillation depicted in Fig. 16.5.

## 16.3 Leukocytes

The *leukocytes* (white blood cells) are the mobile units of the body's immune system. There are six types of white blood cells normally found in the blood. These are the *neutrophils, eosinophils, basophils, monocytes, lymphocytes,* and *plasma cells*. The neutrophils, eosinophils, and basophils are called *granulocytes,* or in clinical terminology, *polymorphonuclear* (PMN) cells, because they have a granular appearance and have multiple nuclei. The normal adult human has about 7000 white blood cells per microliter of blood, approximately 62% of which are neutrophils and 30% of which are lymphocytes. The granulocytes and monocytes protect the body against invading organisms mainly by ingesting them, a process called *phagocytosis*.

## 16.3.1   Leukocyte Chemotaxis

Leukocytes crawl about in tissue by putting out pseudopodal extensions by which they adhere to the fibrous matrix of the tissue. In uniform chemical concentrations of chemical stimulus, their motion is that of a persistent random walk. At random times they undergo random changes in direction. The *persistence time*, the average time between changes of direction, is on the order of a few minutes, and the speed of migration is on the order of 2–20 $\mu$m/min.

One important question is how leukocytes are able to find their bacterial targets. The answer is that they move preferentially in the direction of increasing chemoattractant gradients. Exactly how this is accomplished, how this should be modeled, and how well the model represents this behavior is the topic of this section.

Here we derive a simple model for directed motion in a one-dimensional medium (Tranquillo and Lauffenberger, 1987). We assume that the population of cells, $c$, can be subdivided into two subpopulations, $c = n^+ + n^-$, where superscripts $+$ and $-$ denote right-moving and left-moving cells, respectively. If $v^+$ is the velocity of right-moving cells, and $v^-$ is the velocity of left-moving cells, then the flux of cells is given by

$$J_c = v^+ n^+ - v^- n^-. \tag{16.49}$$

We expect that the cell velocity should be a function only of local conditions, so that $v^+ = v^- = v$. In general, $v$ will be a function of $x$ and $t$. Now we write conservation equations for the directional cell species,

$$\frac{\partial n^+}{\partial t} = -\frac{\partial (v n^+)}{\partial x} + p^- n^- - p^+ n^+, \tag{16.50}$$

$$\frac{\partial n^-}{\partial t} = \frac{\partial (v n^-)}{\partial x} + p^+ n^+ - p^- n^-, \tag{16.51}$$

where $p^+$ is the probability per unit time that a right-moving cell changes direction to become a left-moving cell, and $p^-$ is the probability that a left-moving cell becomes a right-moving cell. These probabilities are also known as *turning rates*.

An equation governing the cell flux $J_c$ is found by differentiating (16.49) and using (16.50) and (16.51), yielding

$$\frac{\partial J_c}{\partial t} - \frac{J_c}{v}\frac{\partial v}{\partial t} = -J_c(p^+ + p^-) - v\frac{\partial (vc)}{\partial x} - vc(p^+ - p^-). \tag{16.52}$$

The steady-state flux is found by setting all time derivatives equal to zero, from which we find that

$$J_c = -v^2 T_p \frac{\partial c}{\partial x} + v(p^- - p^+)T_p c - T_p v \frac{\partial v}{\partial x} c, \tag{16.53}$$

where $T_p^{-1} = p^+ + p^-$.

Now we define phenomenological population migration parameters $\mu = T_p v^2$ as the *random motility coefficient* and $V_c = T_p v(p^- - p^+)$ as the *chemotactic velocity*. Then the equilibrium flux is

$$J_c = -\mu \frac{\partial c}{\partial x} + V_c c - T_p v \frac{\partial v}{\partial x} c. \tag{16.54}$$

Finally, the total cell density is governed by the equation

$$\frac{\partial c}{\partial t} = -\frac{\partial J_c}{\partial x}. \tag{16.55}$$

The movement of cells is governed by three terms in (16.54). The first term, $-\mu \frac{\partial c}{\partial x}$, represents purely random movement of cells, since it gives a diffusive term in (16.55). The second and third terms allow for directed cell movement, since they are proportional to $c$. The directed motion from the second term is due to a difference in the directional change probabilities, while the directed motion in the third term is due to variation in cell speed with spatial position. The second term is called *chemotaxis*, and the third term is *chemokinesis*.

The next problem is to determine the coefficients of these movement terms and in so doing to understand more about the sensory capabilities of the cells. It is known that cell speed can vary with stimulus concentration, yielding a chemokinetic effect, and changes in the direction of movements can be biased toward attractant concentration gradients, a chemotactic response. These responses are mediated by cell surface receptors for attractant molecules that can measure the attractant concentration and its spatial gradient.

There is no a priori theory for the dependence of cell speed on attractant concentration, so it must be measured experimentally. For example, with the tripeptide attractant formyl-norleucyl-leucyl-phenylalanine (FNLLP), the data show that leukocyte velocity is a linearly increasing function of the logarithm of concentration over the range of concentrations $10^{-9}$ M to $10^{-6}$ M, with velocity about 2–5 $\mu$m/min (Zigmond et al., 1981).

Leukocytes determine the presence of an attractant when it binds to receptors on the leukocyte cell surface. When there is a spatial gradient of the attractant, there is also a spatial gradient in the concentration of bound receptors. The side of the cell that experiences a higher concentration of attractant will have a higher concentration of occupied receptors. It has been found experimentally that the fraction of leukocytes that move toward higher attractant concentrations is dependent on this gradient in receptor occupancy. The simplest reasonable expression (Zigmond, 1977) is

$$f = \frac{1}{2} \left( 1 + \frac{\chi_0 \frac{\partial N_b}{\partial x}}{1 + \chi_0 \frac{\partial N_b}{\partial x}} \right), \tag{16.56}$$

where $f$ is the fraction of cells moving toward higher concentrations, $\chi_0$ is the chemotactic sensitivity, and $N_b$ is the number of bound cell receptors. Notice that $N_b$ is a function of $a$, the concentration of chemoattractant, and $a$ is a function of $x$, so the spatial gradient of $N_b$ is given by $\frac{\partial N_b(a)}{\partial x} = \frac{dN_b}{da}\frac{\partial a}{\partial x}$. For small gradients,

$$f \approx \frac{1}{2}\left(1 + \chi_0\frac{dN_b}{da}\frac{\partial a}{\partial x}\right), \tag{16.57}$$

while for large gradients, $f \approx 1$. Thus in small gradients, the fraction of cells moving toward higher concentrations is linearly proportional to the gradient, and this fraction approaches 1 as the gradient increases. From data for rabbit leukocytes responding to the peptide attractant formyl-methionyl-methionyl-methionine (FMMM) it is estimated that $\chi_0 = 2 \times 10^{-5}$ cm/receptor.

In a uniform steady state (for which $\frac{\partial n}{\partial x} = 0$), $n^+p^+ = n^-p^-$, so that

$$f = \frac{n^+}{n^+ + n^-} = \left(1 + \frac{p^+}{p^-}\right)^{-1}. \tag{16.58}$$

Since $T_p = (p^- + p^+)^{-1}$, we find the chemotactic velocity to be

$$V_c = (2f - 1)v = v\frac{\chi_0\frac{dN_b}{da}\frac{\partial a}{\partial x}}{1 + \chi_0\frac{dN_b}{da}\frac{\partial a}{\partial x}}. \tag{16.59}$$

For a single homogeneous population of cell receptors, the number of bound receptors is related to the concentration of attractant through a Michaelis–Menten relationship

$$N_b = \frac{N_T a}{K_d + a}, \tag{16.60}$$

where $K_d$ is the receptor dissociation constant and $N_T$ is the total number of cell receptors.

If the function $v = v(a)$ is known, we have a complete model for the flux of cells due to an attractant concentration. In the special case that cell velocity is independent of attractant concentration, and the attractant concentration and gradient are small, this reduces to a well-known model for chemotaxis (Keller and Segel, 1971),

$$J_c = -\mu\frac{\partial c}{\partial x} + \chi c\frac{\partial a}{\partial x}, \tag{16.61}$$

where $\chi = v\chi_0 N_b'(a)$.

## 16.3.2  The Inflammatory Response

Leukocytes respond to a bacterial invasion by moving up a gradient of some chemical attractant produced by the bacteria and then ingesting the bacterium when it is encountered. Here we present a one-dimensional model (Alt and Lauffenberger, 1987) to show if and when the leukocytes successfully defend against a bacterial invasion.

There are three concentrations that must be determined. These are the bacterial, attractant, and leukocyte concentrations, denoted by $b, a$, and $c$, respectively. The governing equations for these concentrations follow from the following assumptions concerning their behavior:

1. Bacteria diffuse, reproduce, and are destroyed when they come in contact with leukocytes:

$$\frac{\partial b}{\partial t} = \mu_b \frac{\partial^2 b}{\partial y^2} + (k_g - k_d c)b. \tag{16.62}$$

2. The chemoattractant is produced by bacterial metabolism and diffuses:

$$\frac{\partial a}{\partial t} = D\frac{\partial^2 a}{\partial y^2} + k_p b. \tag{16.63}$$

3. The leukocytes are chemotactically attracted to the attractant, and they die as they digest the bacteria, so that

$$\frac{\partial c}{\partial t} = -\frac{\partial J_c}{\partial y} - (g_0 + g_1 b)c. \tag{16.64}$$

For this model we assume that the leukocyte flux is given by (16.61), although more general descriptions are readily incorporated.

To specify boundary conditions we assume that $y = 0$ is the skin surface and that a blood-transporting capillary or venule lies at distance $y = L$ from the skin surface. We assume that the bacteria cannot leave the tissue domain, although the attractant may diffuse into the bloodstream. Leukocytes enter the tissue from the bloodstream at a rate proportional to the circulating leukocyte density $c_b$. When chemotactic attractant is present, the emigration rate increases, because leukocytes that would normally flow in the bloodstream tend to adhere to the vessel wall (*margination*) and then migrate into the interstitium. These considerations lead to the boundary conditions

$$\frac{\partial b}{\partial y} = 0 \text{ at } y = 0 \text{ and } y = L, \tag{16.65}$$

$$\frac{\partial a}{\partial y} = \begin{cases} 0 & \text{at } y = 0, \\ -h_a a & \text{at } y = L, \end{cases} \tag{16.66}$$

$$J_c = \begin{cases} 0 & \text{at } y = 0, \\ -(h_0 + h_1 a)(c_b - c) & \text{at } y = L. \end{cases} \tag{16.67}$$

The governing equations are made dimensionless by setting $x = y/L, \tau = k_g t, u = c/c_b, v = b/b_0$, and $w = a/a_0$. We find that

$$\frac{\partial v}{\partial \tau} = \rho_v \frac{\partial^2 v}{\partial x^2} + (1 - \xi u)v, \tag{16.68}$$

$$\frac{\partial w}{\partial \tau} = \rho_w \left( \frac{\partial^2 w}{\partial x^2} + v \right), \tag{16.69}$$

$$\frac{\partial u}{\partial \tau} = \rho_u \left( \frac{\partial^2 u}{\partial x^2} - \alpha \frac{\partial}{\partial x} \left( u\frac{\partial w}{\partial x} \right) \right) - \gamma_0 (1 + v)u, \tag{16.70}$$

where $a_0 = L^2 k_p b_0/D, b_0 = g_0/g_1, \alpha = \chi a_0/\mu, \rho_v = \frac{\mu_b}{k_g L^2}, \rho_u = \frac{\mu}{k_g L^2}, \rho_w = \frac{D}{k_g L^2}, \xi = k_d c_b/k_g, \gamma_0 = g_0/k_g$.

In nondimensional form the boundary conditions become

$$\frac{\partial v}{\partial x} = 0 \text{ at } x = 0 \text{ and } x = 1, \tag{16.71}$$

$$\frac{\partial w}{\partial x} = \begin{cases} 0 & \text{at } x = 0, \\ -\sigma w & \text{at } x = 1, \end{cases} \tag{16.72}$$

$$\rho_u \left( \frac{\partial u}{\partial x} - \alpha u \frac{\partial w}{\partial x} \right) = \begin{cases} 0 & \text{at } x = 0, \\ \gamma_0 (\beta_0 + \beta_1 w)(1 - u) & \text{at } x = 1, \end{cases} \tag{16.73}$$

where $\sigma = h_a L/D, \beta_0 = \frac{h_0}{g_0 L}, \beta_1 = \frac{h_1 a_0}{g_0 L}$.

There is at least one steady-state solution for this system of equations. It is the *elimination state*, in which $v = w = 0$ and

$$u(x) = \frac{1}{A} \cosh \left( \sqrt{\frac{\gamma_0}{\rho_u}} x \right), \tag{16.74}$$

where $A = \cosh \left( \sqrt{\frac{\gamma_0}{\rho_u}} \right) + \frac{\rho_u}{\gamma_0 \beta_0} \sqrt{\frac{\gamma_0}{\rho_u}} \sinh \left( \sqrt{\frac{\gamma_0}{\rho_u}} \right)$. In this state, all bacteria are eliminated, and the leukocyte density is independent of any bacterial properties. This should represent the normal state for healthy tissue. If $\gamma_0/\rho_u$ is small, then this steady distribution of leukocytes is nearly constant, at level $(1 + \frac{1}{\beta_0})^{-1}$.

Bacterial diffusion is generally much smaller than the diffusion of leukocytes or of chemoattractant. Typical numbers are $D = 10^{-6}$ cm²/s, $\mu = 10^{-7}$ cm²/s, $\mu_b < 10^{-8}$ cm²/s, $k_g = 0.5$ h⁻¹, and $L = 100$ μm. With these numbers, $\rho_u$ and $\rho_w$ are relatively large, while $\rho_v$ is small. This leads us to consider an approximation in which bacterial diffusion is ignored, while attractant and leukocyte diffusion are viewed as fast. In this approximation, airborne bacteria can attach to the surface, but they do not move much on the time scale of leukocyte and chemoattractant motion.

Our first approximation is to ignore bacterial diffusion (take $\rho_v = 0$) and then to assume that a bacterial invasion occurs at the skin surface $x = 0$. This is a reasonable assumption for periodontal, peritoneal, and epidermal infections, which are highly localized, slowly moving infections. Then, since we neglect bacterial diffusion, we specify the bacterial distribution by

$$v(x, \tau) = V(\tau)\delta(x), \tag{16.75}$$

where $\delta(x)$ is the Dirac delta function. The governing equation for $V(\tau)$ is

$$\frac{\partial V}{\partial \tau} = (1 - \xi u(0, \tau))V. \tag{16.76}$$

Since $v = 0$ for $x > 0$, the equations for $w$ and $u$ simplify slightly to

$$\frac{\partial w}{\partial \tau} = \rho_w \frac{\partial^2 w}{\partial x^2}, \tag{16.77}$$

$$\frac{\partial u}{\partial \tau} = \rho_u \left( \frac{\partial^2 u}{\partial x^2} - \alpha \frac{\partial}{\partial x} \left( u \frac{\partial w}{\partial x} \right) \right) - \gamma_0 u, \tag{16.78}$$

while the effect of the bacterial concentration at the origin is reflected in the boundary conditions at $x = 0$ (found by integrating (16.69) and (16.70) "across" the origin),

$$\frac{\partial w}{\partial x} = -V, \tag{16.79}$$

$$\rho_u \left( \frac{\partial u}{\partial x} - \alpha u \frac{\partial w}{\partial x} \right) = \gamma_0 V u. \tag{16.80}$$

An identity that will be important below is found by integrating (16.78) with respect to $x$ to obtain

$$\gamma_0^{-1} \frac{dU}{dt} = -U - Vu(0, \tau) + (\beta_0 + \beta_1 w(1, \tau))(1 - u(1, \tau)), \tag{16.81}$$

where $U(\tau) = \int_0^1 u(x, \tau) dx$ is the total leukocyte population within the tissue.

Our second approximation is to assume that the chemoattractant diffusion is sufficiently large, so that the chemoattractant is in quasi-steady state,

$$\frac{\partial^2 w}{\partial x^2} = 0. \tag{16.82}$$

This implies that $w(x)$ is a linear function of $x$ with gradient

$$\frac{\partial w}{\partial x} = -V. \tag{16.83}$$

Finally, we assume that $\rho_u$ is large (taking $\rho_u \to \infty$), so that the leukocyte density is also in quasi-steady state with $J_c = 0$, that is,

$$\frac{\partial u}{\partial x} + \alpha V u = 0. \tag{16.84}$$

We can solve this equation and find the leukocyte spatial distribution to be

$$u(x, \tau) = U(\tau) F(\alpha V) e^{-\alpha V x}, \tag{16.85}$$

where $F(z) = \frac{z}{1 - e^{-z}}$ is determined by requiring $U(\tau) = \int_0^1 u(x, \tau) dx$.

Now we are able to determine $u(0, \tau), u(1, \tau)$ from (16.85) and $w(1, \tau)$ from (16.72) and (16.83), which we substitute into the equation for total leukocyte mass (16.81) to obtain

$$\gamma_0^{-1} \frac{dU}{d\tau} = (\beta_0 + \beta V) \left( 1 - UF(\alpha V) e^{-\alpha V} \right) - (VF(\alpha V) + 1)U, \tag{16.86}$$

where $\beta = \beta_1 / \sigma$. Similarly, from (16.76) and (16.85), we find the equation governing $V$ to be

$$\frac{\partial V}{\partial \tau} = V \left( 1 - \xi U F(\alpha V) \right). \tag{16.87}$$

## Phase-plane analysis

The system of equations (16.86)–(16.87) is a two-variable system of ordinary differential equations that can be studied using standard phase-plane methods. In this analysis we focus on the influence of two parameters: $\beta$, which characterizes the enhanced leukocyte emigration from the bloodstream, and $\alpha$, which measures the chemotactic response of the leukocytes to the attractant.

One steady-state solution that always exists is $U = (1 + \frac{1}{\beta_0})^{-1}, V = 0$. This represents the elimination state in which there are no bacteria present. Any other steady solutions that exist with $V > 0$ are compromised states in which the bacteria are allowed to persist in the tissue.

We assume that the system is at steady state at time $\tau = 0$ with $U(0) = U_0 = (1 + \frac{1}{\beta_0})^{-1}$ when a bacterial challenge with $V(0) = V_0 > 0$ is presented. We begin the analysis with simple cases for which $\alpha = 0$.

**Case I:** $\alpha = 0, \beta = 0$.
In this case the system reduces to

$$\gamma_0^{-1} \frac{dU}{d\tau} = \beta_0 - (\beta_0 + 1)U - VU, \tag{16.88}$$

$$\frac{\partial V}{\partial \tau} = V(1 - \xi U). \tag{16.89}$$

There are three nullclines: $\frac{dV}{d\tau} = 0$ on the vertical line $U = \frac{1}{\xi}$ and on the horizontal line $V = 0$, and $\frac{dU}{d\tau} = 0$ on the hyperbola $V = \frac{\beta_0 - (\beta_0 + 1)U}{U}$.

Two types of behavior are possible. If $\xi U_0 < 1$, there are no steady states in the positive first quadrant. The only steady state is at $U = U_0, V = 0$. For $U \leq U_0, \frac{dV}{d\tau} > 0$, so that $U$ decreases and $V$ increases without bound. The bacterial challenge cannot be met. This situation is depicted in Fig. 16.8. In this and all the following phase portraits, the nullcline for $\frac{dV}{d\tau} = 0$ is shown as a short dashed curve, and the nullcline for $\frac{dU}{d\tau} = 0$ is shown as a long dashed curve. The solid curve shows a typical trajectory starting from initial data $U = U_0, V = V_0$.

If $\xi U_0 > 1$, there is a nontrivial steady state in the first quadrant, which is a saddle point. This means that there is a value $V^*$ for which a trajectory starting at $U = U_0, V =$

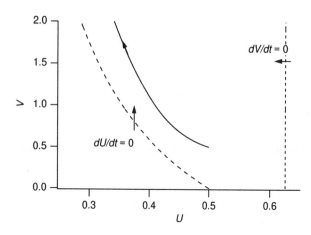

**Figure 16.8** Phase portrait for the system (16.86)–(16.87) with "small" $\xi = 1.6$, "small" $\beta = 0.1$, $\alpha = 0$. Other parameters are $\beta_0 = 1.0$, $\gamma_0 = 0.2$, so that $U_0 = 0.5$.

$V^*$ is on the stable manifold of this steady state and divides the line $U = U_0$ into two types of behavior. If $V < V^*$ initially, the trajectory evolves toward the elimination state, while if $V > V^*$ initially, the trajectory is unbounded. Thus, for large enough $\xi$ and small enough initial bacterial population, the challenge can be withstood, but for a larger initial bacterial challenge, the bacterial population wins the competition. The number $V^*$ is a monotone increasing function of $\xi$, and $\lim_{\xi \to \infty} V^* = \infty$. This follows because to the right of $U = \frac{1}{\xi}$ the stable manifold is an increasing curve as a function of $U$, so that $V^*$ lies above the the value of $V$ at the saddle point. However, as a function of $\xi$, the steady-state value of $V$ is monotone increasing as $\xi$ increases, approaching $\infty$ in the limit $\xi \to \infty$, so $V^* \to \infty$ as well.

The phase portrait for this situation is depicted in Fig. 16.9. In this situation the bacterial challenge is met only if $\xi$ is large enough and $V_0$ is small enough, so that the leukocytes are effective killers, although with $\alpha = \beta = 0$ they are not good hunters. Note that $\xi = k_d c_b / k_g$, where $k_d$ is the rate at which leukocytes kill bacteria, $k_g$ is the growth rate of the bacteria, and $c_b$ is the leukocyte density in the blood. Hence, large $\xi$ means that leukocytes are effective killers, since they kill bacteria at a rate exceeding the growth rate of the bacteria.

**Case II:** $\alpha = 0, \beta > 0$.
Here, the leukocytes can respond to the bacterial challenge by enhanced emigration from the bloodstream, but they cannot localize preferentially within the tissue. The system of equations becomes

$$\gamma_0^{-1} \frac{dU}{d\tau} = (\beta_0 + \beta V)(1 - U) - (V + 1)U, \tag{16.90}$$

$$\frac{\partial V}{\partial \tau} = V(1 - \xi U). \tag{16.91}$$

The nullclines for $\frac{dV}{d\tau}$ are unchanged from above. The nullcline $\frac{dU}{d\tau} = 0$ is the hyperbola $V = \frac{\beta_0 - (\beta_0 + 1)U}{(\beta + 1)U - \beta}$. For small $\beta$, with $\frac{\beta}{\beta + 1} < U_0$, the behavior of the system changes only slightly from Case I. These phase portraits are as depicted in Figs. 16.8 and 16.9.

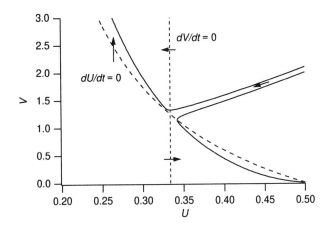

**Figure 16.9** Phase portrait for the system (16.86)–(16.87) with "large" $\xi = 3.0$, "small" $\beta = 0.1$, $\alpha = 0$. Other parameters are $\beta_0 = 1.0$, $\gamma_0 = 0.2$, so that $U_0 = 0.5$.

If $\xi U_0 < 1$, the bacterial population grows without bound, whereas if $\xi U_0 > 1$, the bacterial population can be eliminated if $V < V^*$ initially. The value $V^*$ is a monotone increasing function of $\beta$. Thus, with $\beta$ small, the leukocytes have an enhanced ability to eliminate a bacterial population. In fact, if $\xi\beta > \beta+1$ (phase portrait not shown), then $V^* = \infty$, so that a bacterial invasion of any size can be eliminated. Notice that in this case, the bacterial invasion is controlled because the leukocytes are effective killers and they effectively deploy troops to withstand the invasion. There is still no mechanism making them effective hunters.

In all of the above cases, the leukocyte population decreases initially, and if the bacterial population is controllable, the leukocyte population eventually rebounds back to normal. If $\beta$ is large enough, with $\frac{\beta}{\beta+1} > U_0$, then the response to a bacterial invasion is with an initial increase in leukocyte population. If $\xi\beta < \beta+1$, then the bacterial population is unbounded; the invasion cannot be withstood.

If $\xi\beta > \beta+1$ and $\xi U_0 < 1$, there is a nontrivial steady state in the positive first quadrant that is a stable attractor. All trajectories starting at $U = U_0$ go to this stable steady-state solution with $U > U_0$. Since $V > 0$ for this steady solution, the bacterial population is controlled but not eliminated. This situation is depicted in Fig. 16.10.

Finally, if $\xi U_0 > 1$, the leukocyte population initially increases and then decreases back to normal as the bacterial population is eliminated. This situation is depicted in Fig. 16.11.

The above information is summarized in Fig. 16.12, where four regions with differing behaviors are shown, plotted in the $(1/\beta, \xi)$ parameter space. The four regions are bounded by the curves $\xi = 1/U_0$ and $\xi = 1 + 1/\beta$ and are identified by the asymptotic state for $V$, $\lim_{\tau\to\infty} V(\tau)$. For $\xi > 1/U_0$ and $\xi > 1+1/\beta$, the bacteria are always eliminated. For $\xi > 1/U_0$ and $\xi < 1+1/\beta$, there are two possibilities, either elimination or unbounded bacterial growth, depending on the initial size of the bacterial population. For $\xi < 1/U_0$ and $\xi > 1+1/\beta$, the bacteria survive but are controlled at population size $V_p$, and finally, for $\xi < 1/U_0$ and $\xi < 1+1/\beta$, the bacterial population cannot be controlled but becomes infinite.

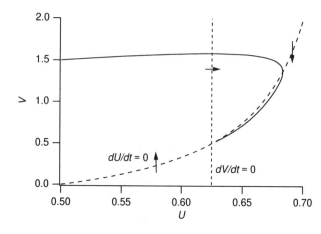

**Figure 16.10** Phase portrait for the system (16.86)–(16.87) with "small" $\xi = 1.6$, "large" $\beta = 3.0$, $\alpha = 0$. Other parameters are $\beta_0 = 1.0$, $\gamma_0 = 0.2$, so that $U_0 = 0.5$.

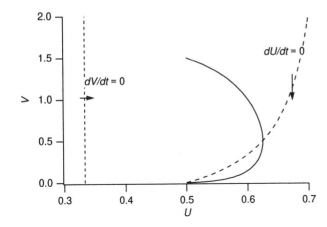

**Figure 16.11** Phase portrait for the system (16.86)–(16.87) with "large" $\xi = 3.0$, "large" $\beta = 3.0$, $\alpha = 0$. Other parameters are $\beta_0 = 1.0$, $\gamma_0 = 0.2$, so that $U_0 = 0.5$.

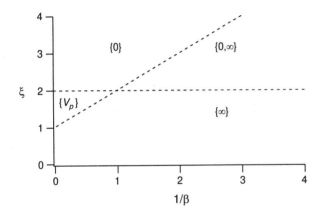

**Figure 16.12** Parameter space for the system (16.86)–(16.87) with $\alpha = 0$.

**Case III: $\alpha > 0$, $\beta > 0$.**

The primary goal of this model is to determine the effect of the chemotaxis coefficient on the performance of the leukocytes in warding off a bacterial invasion. We have seen so far that with $\alpha = 0$ there are three possible responses to an invasion. The bacteria may become unbounded, they may be controlled at a nonzero steady state, or they may be eliminated, depending on the sizes of the parameters $\xi$ and $\beta$. With $\alpha \neq 0$, we expect control and elimination to be enhanced, if only because the bacterial growth rate is a decreasing function of $\alpha$.

The effect of $\alpha \neq 0$ is seen first of all in the nullclines. The nullclines are the curves

$$\frac{dV}{d\tau} = 0 : U = \frac{1}{\xi F(\alpha V)}, \tag{16.92}$$

and

$$\frac{dU}{d\tau} = 0 : U = \frac{\beta_0 + \beta V}{(\beta_0 + \beta V)F(\alpha V)e^{-\alpha V} + VF(\alpha V) + 1}. \tag{16.93}$$

Both of these are decreasing functions of $\alpha$, and both asymptote to $U = 0$ as $V \to \infty$. A steady state occurs whenever there is an intersection of these two curves. This condition we write as

$$\frac{1}{\xi} = \frac{\alpha V(\beta_0 + \beta V)}{(1 - e^{-\alpha V}) + \alpha V^2 + \alpha V e^{-\alpha V}(\beta_0 + \beta V)} = G(V). \qquad (16.94)$$

One can easily see that $G(0) = U_0$ and that $\lim_{V \to \infty} G(V) = \beta$. This implies that there is an even number of roots if

$$\left(\frac{1}{\xi} - U_0\right)\left(\frac{1}{\xi} - \beta\right) > 0, \qquad (16.95)$$

and an odd number of roots otherwise. An odd number of roots implies that there is at least one steady-state solution in the first quadrant; with an even number there could be no steady states. This leads to four different possible outcomes separated by the curves $\xi = \frac{1}{U_0}$ and $\xi = \frac{1}{\beta}$. These are

1. $\xi < \frac{1}{U_0}, \xi < \frac{1}{\beta}$. There can be zero or two steady states. If there are no steady states, then $V$ becomes infinite. If there are two steady states, one of them is stable and the trajectories for sufficiently small initial bacterial populations approach the stable steady state, where they persist.

   We can find the boundary between these two cases by looking for a double root of (16.94). We do this by solving (16.94) and the equation $G'(V) = 0$ simultaneously. This gives a curve in the $(\beta, \xi)$ parameter plane parametrized by $V$, as follows: For each $V$, $\beta$ is a root of the quadratic equation

$$\alpha^2 V^4 \beta^2 - V(-2\alpha^2 V^2 \beta_0 + \alpha V - 2e^{\alpha V} + 2)\beta$$

$$+ \beta_0(\alpha^2 V^2 \beta_0 - \alpha V - 1 + e^{\alpha V}(1 - V^2\alpha)) = 0, \qquad (16.96)$$

   and then $\xi$ is given by (16.94) for each $V, \beta$. It is an easy matter to determine this curve numerically. The curve is plotted in Fig. 16.13 as a solid curve, shown for the three values of $\alpha = 0.5, 0.75$, and $1.0$.

   Below this curve in the $(\frac{1}{\beta}, \xi)$ parameter space, there are no steady-state solutions. The phase portrait for this case is similar to that of Fig. 16.8 and is left as an exercise (see Exercise 13). For all trajectories starting at $U = U_0$, $V(\tau) \to \infty$.

   Above the "double root" curve there are two steady solutions, one of which is stable. In this situation, some trajectories lead to persistent bacterial populations, while others (with larger initial values) become infinite. This phase portrait has similarities with Fig. 16.9 and is left as an exercise (see Exercise 13).

2. $\xi < \frac{1}{U_0}, \xi > \frac{1}{\beta}$. Here there is one stable steady state, which is a global attractor. All trajectories evolve to this steady state, so that the bacterial population is controlled, but it is not eliminated. The phase portrait for this case is quite similar to the previous case, except that there is only one nontrivial steady state, and no saddle point, so there is no separatrix, and all trajectories approach the persistent state. It should be noted that with $\xi < \frac{1}{U_0}$, the bacterial population can never be eliminated. However, with $\alpha > 0$, the population is more readily controlled than with $\alpha = 0$.

3. $\xi > \frac{1}{U_0}, \xi < \frac{1}{\beta}$. There is a single steady state in the first quadrant, which is a saddle point and which therefore divides the initial data into two types, those that are eliminated and those that become unbounded. The phase portrait for this case is similar to Fig. 16.9 and is left as an exercise (see Exercise 13).

4. $\xi > \frac{1}{U_0}, \xi > \frac{1}{\beta}$. Here there are no steady-state solutions in the positive quadrant, in which case the bacterial population is always eliminated. Here the effect of chemotaxis can be seen in the transient behavior of the leukocyte population. If the initial bacterial population is small, the leukocyte population initially increases before it decreases back to its equilibrium. If the initial bacterial population is large, then the leukocyte population initially decreases, then increases, and then finally decreases back to steady state, having eliminated the bacterial population. The phase portrait for this case has similarities with Fig. 16.11 and is left as an exercise (see Exercise 13).

In summary, to control a bacterial invasion, the leukocytes must be sufficiently lethal to the bacteria ($\xi$ sufficiently large). They must also be able to recruit new troops, and it is advantageous that they move chemotactically, since they are more effective if $\alpha > 0$. This result is not surprising. However, the significance of this approximate analysis is that the model behaves as we want it to behave, suggesting that it is a reasonable model, worthy of more detailed study and development.

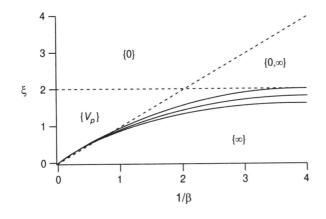

**Figure 16.13** Parameter space for the system (16.86)–(16.87) with $\beta_0 = 1.0$, $\alpha > 0$.

# 16.4   Clotting

## 16.4.1   The Clotting Cascade

The need for a clotting system is obvious. In any higher organism with a circulatory system, the loss of the transporters of vital metabolites and waste products has disastrous, perhaps fatal, consequences. However, the occurrence of clots in an otherwise normal circulatory system is also potentially disastrous, as it prevents a flow that is equally important to survival.

The clotting system must be fast reacting, and yet localized. Since all the ingredients for clotting are carried in the blood, there must be some control that prevents propagation. As we know from earlier chapters, a highly excitable system of diffusing species has the possibility, indeed the strong likelihood, of supporting traveling waves. For the clotting system, a propagating front would be as disastrous as failure of a clot to form. Thus, the dilemma we face is to understand the mechanisms underlying a highly excitable system of reacting and diffusing chemicals that does *not* support wave propagation.

In fact, there are more than 50 substances in blood and tissue that play a role in the clotting process. Crucial to the process is the enzyme *thrombin*. Thrombin acts enzymatically on *fibrinogen*, converting it to *fibrin*, which then forms the meshwork of the clot. However, this is not all, as thrombin is an extremely active enzyme, with many other regulatory roles.

Thrombin is formed when prothrombin, which is carried in the blood, is converted by an enzyme called *prothrombin activator*. Prothrombin activator is formed as the end result of two different enzymatic cascades, which are, however, closely linked. The fastest, called the *extrinsic pathway*, is initiated following tissue trauma. The second pathway, called the *intrinsic pathway*, is initiated following trauma to blood or contact of blood with collagen, or any negatively charged surface, and is not dependent on tissue trauma. However, this second pathway is much slower than the extrinsic pathway. Classic *hemophilia*, a tendency to bleed that occurs in 1 in every 10,000 males in the United States, results from a deficiency of one of the important enzymes in the intrinsic pathway.

Here we describe only the extrinsic pathway. Thirteen of the important factors in the clotting cascade are denoted using Roman numerals as factors I through XIII, although for historical reasons, they also have other names. Here we retain the Roman numeral notation. Of those that have active and inactive states, the active state is denoted by appending the letter "a" to its name. Thus, for example, Xa is the active form of factor X.

The extrinsic pathway can be described as follows: tissue trauma causes the release of a combination of agents called, collectively, *factor III* (or tissue thromboplastin, or tissue factor). Factor III consists primarily of certain phospholipids from the membranes of the damaged tissues, and acts enzymatically to activate factor VII, converting it to factor VIIa. Factor VIIa then acts enzymatically to activate factor X to Xa, which

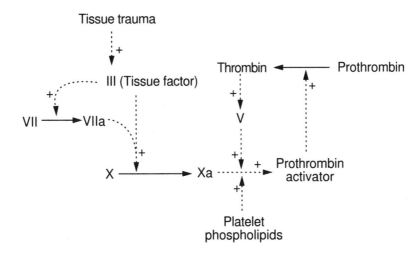

**Figure 16.14** Schematic diagram of the extrinsic pathway for blood clotting.

then combines with factor V and other phospholipids to form prothrombin activator. As mentioned above, prothrombin activator converts prothrombin to thrombin, from which the clot eventually forms.

The speed of the extrinsic pathway is increased by various positive feedback mechanisms. First, thrombin activates factor V, thus increasing the rate of formation of prothrombin activator. Second, thrombin is one of the substances that activates platelets (described below) to make them sticky and highly reactive. In their reactive form, platelets are a source of the phospholipids with which factor X combines to produce prothrombin activator. Finally, factor Xa is known to activate molecules of factor VII bound to factor III. The clotting cascade is summarized in Fig. 16.14.

Thrombin is also degraded, so that its activity is not permanent. One of the ways that thrombin is degraded is by binding to the fibrin network, so that thrombin is eventually degraded by the result of its own activity, a negative feedback loop. There are also anticoagulants that act to inactivate thrombin. Primary among these are antithrombin III and heparin. Heparin by itself has little or no anticoagulant effect, but in complex with antithrombin III, the effectiveness of antithrombin III is increased a hundred- to a thousandfold. The concentration of heparin is normally quite slight, although it is produced in large quantities by *mast cells*, located in the connective tissue surrounding capillaries. They are especially abundant in tissue surrounding the capillaries of the lung and of the liver. This is important, because these organs receive many clots that form in the slowly moving venous blood and must be removed. Heparin is widely used in medical practice to prevent intravascular clotting.

Using techniques that have been amply illustrated elsewhere in this book, it is possible to write differential equations describing the dynamics of thrombin activation and clot formation (Exercise 14). Models of certain aspects of this clotting network have been developed and studied by Nesheim et al. (1984, 1992), Willems et al. (1991), and

Jones and Mann (1994). Other models that are less specific but attempt to capture the excitability of the system include those of Jesty et al. (1993) and Beltrami and Jesty (1995).

The clotting system has the feature that wherever activated factor X binds with phospholipid, there also will occur thrombin and clotting. Thus, diffusion and transport of factor Xa could cause clots to spread in uncontrolled fashion. Similarly, if factor III (tissue factor) were to spread via diffusion and transport, one can imagine the havoc it would wreak on the circulatory system.

Clearly, it is dangerous to have the ingredients for an explosive (excitable) system floating around and diffusing in the blood. There must be some mechanism by which the rapid spread of clots is prevented. Indeed, there are several. First, the walls of blood vessels are quite smooth, preventing the excitable ingredients in blood from becoming activated. Second, when there is tissue trauma, factor III is exposed, but not released. That is, factor III remains bound to the membrane surface and turns tissue into a catalytic surface; it is not free to diffuse. Similarly, when platelets are activated, the surfaces of the platelets become catalytic reactors that carry the catalyst phospholipids. Thus, prothrombin activator is a membrane-bound catalyst: the clotting cascade is activated only in the vicinity of this activated catalytic surface. In other words, the clotting reaction is largely regulated by phospholipid surfaces (Fogelson and Kuharsky, 1998). For this reason, it is necessary to discuss the activity of platelets.

## 16.4.2  Platelets

Platelets are minute round or oval discs 2 to 4 micrometers in diameter. They are formed in the bone marrow from *megakaryocytes*, which are large cells in the bone marrow that fragment into platelets. There are normally between 150,000 and 300,000 platelets per microliter of blood, constituting only a small percentage of the volume ($\approx$ 0.3 percent by volume).

A platelet is an active structure with a half-life of 8 to 12 days. Since platelets are cell fragments with no nucleus, they cannot reproduce. A platelet normally circulates with the blood in a dormant, or inactivated state, in which it does not adhere to other platelets or to the blood vessel wall. However, when platelets come in contact with a damaged vascular surface or sufficient chemical triggers, they become activated and change their characteristics drastically, as follows:

1. The platelet's surface membrane is altered so that the platelet becomes sticky, capable of adhering to other activated platelets or the vessel wall.
2. The platelet secretes chemicals, including large amounts of ADP and thrombaxane $A_2$, which are capable of activating other platelets.
3. The platelets change from rigid discoidal to highly deformable, extending long, thin appendages called *pseudopodia*.

An important requirement for controlled clotting is that the circulating blood must be able to build a catalytic bed in the vicinity of the injury, even though there is fluid flow. The aggregation of platelets is the means by which a catalytic reactor bed is built, and so is an important part of the process by which the flow of blood from a damaged vessel is halted.

A mathematical model for the aggregation of platelets and the formation of platelet plugs has been formulated and studied by Fogelson (1992). The model is a continuum model that assumes that there is a concentration of activated and nonactivated platelets, denoted by $\phi_a$ and $\phi_n$, respectively. Platelets are immersed in blood and are neutrally buoyant, moving with the local fluid velocity $\mathbf{u}$. There is some chemical concentration $c$, say of ADP, that is released by platelets when they are activated and that has the effect of stimulating nonactivated cells. Activated cells are sticky and form aggregates when they come into contact with each other.

One can readily write conservation equations for the density of inactivated and activated platelets. However, because this is an exercise in fluid and continuum mechanics, which is beyond the scope of this text, we do not reproduce these here. Numerical simulation of these equations then demonstrates how platelet aggregates can form in the vicinity of tissue trauma.

In Figs. 16.15 and 16.16 are shown two snapshots of a (two-dimensional) fluid flow past an activated obstacle and a segment of damaged vessel wall. The figures show an aggregate of activated platelets growing from the obstacle and damaged vessel wall in a fluid flow moving from left to right, which, one surmises, gradually causes the occlusion of the flow.

The question remains (and is not addressed by these simulations) of why the platelet system does not exhibit traveling fronts of aggregation. The putative answer is that smooth (undamaged) vascular walls are nonsticky and that they contain inhibitors of

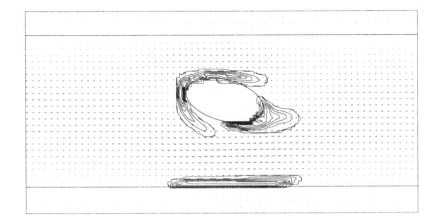

**Figure 16.15** Growth of a platelet aggregate in a fluid flow. Fluid is flowing from left to right with velocity vectors shown, (part 1). Contours in the aggregates depict the density of platelet "stickiness." (With permission of A. Fogelson).

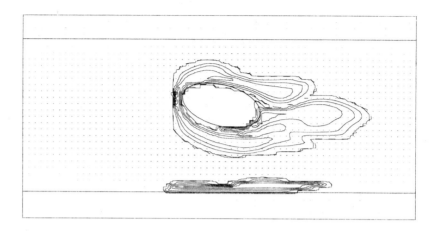

**Figure 16.16**  Growth of a platelet aggregate in a fluid flow, (part 2). Notice the increased aggregate size and decreased fluid velocity, even though the total pressure drop is unchanged from part 1. (With permission of A. Fogelson.)

ADP, the primary factor in the activation of platelets, and inhibitors of thrombin, and these prevent the uncontrolled spread of activated platelets.

A complete model of clotting would combine platelet aggregation with the surface-catalyzed production of thrombin via the pathway described above. This complete model would take into account the controlled construction of a reactive catalytic surface, the control of the reaction by the catalytic surface, and the ultimate construction of a clot that shuts off the blood flow.

While it is feasible to write down a reasonable model of the entire clotting process (since the biochemistry is reasonably well known), the understanding of such a model is far from complete, primarily because of the complicated interaction with the fluid flow. In fact, this is a subject of active research.

## 16.5  EXERCISES

1.  What is the volume (per mole) of an ideal gas at room temperature ($27°$ C) and 1 atm pressure? What is its volume at body temperature ($98°$ F)?

2.  Suppose that a carrier (like hemoglobin) of a molecule (like oxygen) has $n$ independent binding sites, with individual binding and unbinding rates $k_+$ and $k_-$. Let $c_j$ denote the concentrations of the state with $j$ molecules bound. Assume concentrations are in steady state.

    (a)  Show that $c_j = \binom{n}{j} x^j c_0$, where $\binom{n}{j} = \frac{n!}{j!(n-j)!}$ is the *binomial coefficient* where $x = s_0/K$, $K = k_-/k_+$, and $s_0$ is the concentration of the carrier molecule.
         Hint: Keep track of the total number of binding sites.

    (b)  Find the saturation function in the case that $n = 4$.

**Table 16.2**  Approximate numerical data for the hemoglobin saturation curve.

| $P_{O_2}$ (mm Hg) | percent saturation |
|---|---|
| 3.08 | 2.21 |
| 4.61 | 3.59 |
| 6.77 | 6.08 |
| 10.15 | 10.50 |
| 12.31 | 14.09 |
| 15.38 | 19.34 |
| 18.77 | 28.45 |
| 22.77 | 40.33 |
| 25.85 | 50.0 |
| 30.15 | 60.50 |
| 36.00 | 69.89 |
| 45.23 | 80.11 |
| 51.69 | 83.98 |
| 61.85 | 88.95 |
| 75.38 | 93.37 |
| 87.08 | 95.86 |
| 110.5 | 98.07 |

(c) Show that the four equilibrium constants $K_1, K_2, K_3, K_4$ defined in (16.11) are given by $(K_1, K_2, K_3, K_4) = K(\frac{1}{4}, \frac{2}{3}, \frac{3}{2}, 4)$.

(d) Estimate $K$ to give a good fit of this model to the hemoglobin saturation curve. How does this curve compare with the curve (16.13)?

(e) Determine whether the hemoglobin binding sites are independent. How close are the equilibrium constants here to those found in the text?

3. Approximate numerical data for the hemoglobin saturation curve are found in Table 16.2. Fit these data to a curve of the form (16.13).

Hint: Suppose we have data points $\{x_i, y_i\}, i = 1, \ldots, n$, that we wish to fit to some function $y = f(x)$, and that the function $f$ depends on parameters $\{\alpha_i\}, i = 1, \ldots, m$. A fit of the data is achieved when the parameters are picked such that the function

$$F = \sum_{j=1}^{n} (f(x_j) - y_j)^2 \tag{16.97}$$

is minimized. To find this fit, start with reasonable estimates for the parameters and then allow them to change dynamically (as a function of a time-like variable $t$) according to

$$\frac{d\alpha_k}{dt} = -\sum_{j=1}^{n} f(x_j) \frac{\partial f(x_j)}{\partial \alpha_k}. \tag{16.98}$$

With this choice,

$$\frac{dF}{dt} = \sum_{k=1}^{m} \sum_{j=1}^{n} f(x_j) \frac{\partial f(x_j)}{\partial \alpha_k} \frac{d\alpha_k}{dt} \leq 0, \tag{16.99}$$

so that $F$ is a decreasing function of $t$. A fit is found when numerical integration reaches a steady-state solution of (16.98).

4. Construct a Monod–Wyman–Changeux model (Section 1.2.4) for oxygen binding to hemoglobin and determine the saturation function. Fit to the experimental data given in Table 16.2 and compare to the fit of (16.13).

5. (a) If a 25 mM solution of sodium bicarbonate is equilibrated with carbon dioxide at 40 mm Hg partial pressure, the pH is found to be 7.4. If the partial pressure of carbon dioxide is increased until the pH is 6.0, what is the bicarbonate concentration? What is the carbon dioxide partial pressure at this pH? What is the difference if this experiment is carried out in whole blood instead of sodium bicarbonate solution?

   (b) Pick reasonable values for $K_1$, $K_2$ and $\bar{K}_1$ and find the oxygen saturation curve as a function of $h$. (Hint: the parameters must be chosen so that hemoglobin acts as a hydrogen ion buffer at physiological concentrations. Try $K_1^{1/4} = 26\,\sigma$ mm Hg. Alternately, pick $K_1$, $K_2$ and $\bar{K}_1$ so that $(\phi(h))^{1/4} = 26\,\sigma$ mm Hg at pH $= 7.4$.)

   (c) Solve (16.26) - (16.30) numerically and plot $X$, $Y$, $Z$, $W$ and $h$ as functions of $O$ and $[CO_2]$. How much $CO_2$ is transported from the tissues to the lungs? Remove the Bohr and Haldane effects by setting $K_1 = \bar{K}_1$. How does this change the amount of oxygen and carbon dioxide transported? Typical parameter values are: $P_{CO_2}$ in arterial blood, 39 mm Hg ; $P_{CO_2}$ in venous blood, 46 mm Hg; $P_{O_2}$ in arterial blood, 100 mm Hg; $P_{O_2}$ in venous blood, 40 mm Hg; $R_1R_2 = 10^{6.1}\mathrm{M}^{-1}$; $T_{Hb} = 3$ mM; $n = 10$, $[HCO_3^-] = 25$ mM and pH = 7.4 in arterial blood.

6. Develop a detailed model of oxygen and carbon monoxide binding with hemoglobin. How can the fact that CO has 210 times the affinity for binding be used to estimate the equilibrium coefficients?

7. Find an analytic relationship for the critical stability curve (Section 16.2.4), relating $dF(0)$ to $X/d$ as follows: Use that $F(N) = \frac{F(0)}{1+N^7}$ to solve (16.46) for $N_0$ as a function of $X/d$ and then determine $dF(0)$ using that $F(N_0) = N_0/X$.

8. A deficiency of vitamin $B_{12}$, or folic acid, is known to cause the production of immature red blood cells with a shortened lifetime of one-half to one-third of normal. What effect does this deficiency have on the population of red blood cells?

9. Suppose $X \to \infty$ in the red blood cell production model. Show that

$$\frac{dN}{dt} = F(N(t-d)) - \beta N. \qquad (16.100)$$

   (a) Find the stability characteristics for the steady-state solution of this equation.

   (b) Show that the period of oscillation $T = 2\pi/\omega$ at a Hopf bifurcation point is bounded between $2d$ and $4d$.

   Hint: Differentiate the equation $N(t) = \int_0^\infty n(x,t)dx$ with respect to $t$ and use the partial differential equation $n_t + n_x = -\beta n$ and the initial condition $n(x,t) = F(N(t-d))$ to eliminate $n(x,t)$.

10. The maturation rate of red blood cells in bone marrow varies as a function of erythropoietin levels. Suppose that $x$ denotes the maturity (rather than chronological age) of a red blood cell. Suppose further that cells are initially formed at maturity $x = -d$, are released into the bloodstream at maturity $x = 0$, age at the normal chronological rate, and die at age $x = X$. Suppose further that the rate of maturation $G$ is a decreasing function of the total circulating red blood cell count $N$ and that the rate of cell production at maturity $x = -d$ is $F(N)$.

(a) Replace the condition (16.33) with an evolution equation of the form (16.31) to account for maturities $x$ in the range $-d < x < 0$.

(b) Perform a stability analysis for this modified model. Does the variability of $G$ make the solution more or less likely to become unstable via a Hopf bifurcation?

11. Suppose $X$ is finite and $\beta = 0$ in the red blood cell production model. Show that the evolution of $N$ is described by the delay differential equation

$$\frac{dN}{dt} = F(N(t-d)) - F(N(t-d-X)). \tag{16.101}$$

12. Numerically simulate (16.31) with boundary data (16.33) with parameters chosen from the stable region and from the unstable region.

13. Sketch the phase portraits for the equations (16.86)–(16.87) in Case III ($\alpha > 0, \beta > 0$) as follows:

(a) $\xi < \frac{1}{U_0}, \xi < \frac{1}{\beta}$. (For example,

   i. $\xi = 1.0, \beta = 0.5, \alpha = 0.5, \beta_0 = 1.0, \gamma_0 = 0.2$, and

   ii. $\xi = 1.5, \beta = 0.5, \alpha = 1.3, \beta_0 = 1.0, \gamma_0 = 0.2$.)

(b) $\xi < \frac{1}{U_0}, \xi > \frac{1}{\beta}$ (For example, $\xi = 1.8, \beta = 0.6, \alpha = 0.5, \beta_0 = 1.0, \gamma_0 = 0.2$.)

(c) $\xi > \frac{1}{U_0}, \xi < \frac{1}{\beta}$. (For example, $\xi = 2.2, \beta = 0.3, \alpha = 0.5, \beta_0 = 1.0, \gamma_0 = 0.2$.)

(d) $\xi > \frac{1}{U_0}, \xi > \frac{1}{\beta}$. (For example, $\xi = 2.2, \beta = 2.0, \alpha = 0.5, \beta_0 = 1.0, \gamma_0 = 0.2$.)

Locate each of these cases in Fig. 16.13.

14. (a) Write a system of differential equations describing the clotting reaction.

(b) Suppose that the concentration of activated platelets is given by $\phi_a(x, t)$. Write a system of diffusion–transport–reaction equations for the clotting cascade that takes into account that phospholipid is bound to the membrane of activated platelets.

# CHAPTER 17

# Respiration

The respiratory system is responsible for gas transfer between the tissues and the outside air. Carbon dioxide that is produced by metabolism in the tissues must be moved by the blood to the lungs, where it is lost to the outside air, and oxygen that is supplied to the tissues must be extracted from the outside air by the lungs.

The nose, mouth, pharynx, larynx, trachea, broncheal trees, lung air sacs and respiratory muscles are the structures that make up the respiratory system (Fig. 17.1). The nasal cavities are specialized for warming and moistening inspired air and for filtering the air to remove large particles. The larynx, or "voice box," contains the vocal folds that vibrate as air passes between them to produce sounds. Below the larynx the respiratory system divides into *airways* and *alveoli*. The airways consist of a series of branching tubes that become smaller in diameter and shorter in length as they extend deeper into the lung tissue. They terminate after about 23 levels of branches in blind sacs, the alveoli. The *terminal bronchioles* represent the deepest point of the bronchial tree to which inspired air can penetrate by flowing along a pressure gradient. Beyond the terminal bronchioles, simple diffusion along concentration gradients is primarily responsible for the movement of gases.

Alveoli are thin-walled air sacs that provide the surface across which gases are exchanged (Fig. 17.2). Each lung contains about 300 million alveoli with a combined surface area of about 70–85 square meters. The alveoli are surrounded by respiratory membrane that serve to bring air and blood into close contact with a large surface area. In the lung capillaries, from 70 to 140 ml of blood is spread over the surface area of the lungs.

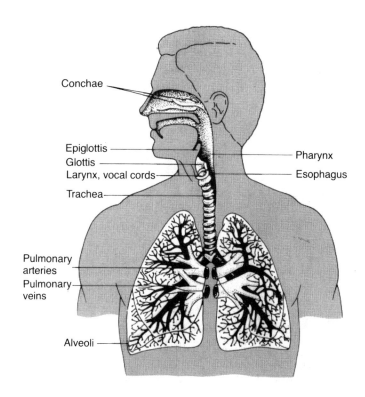

**Figure 17.1**  Diagram of the respiratory passages. (Guyton and Hall, 1996, Fig. 37-9, p. 486.)

## 17.1   Capillary–Alveoli Gas Exchange

### 17.1.1   Diffusion Across an Interface

In Chapter 16 we discussed how the partial pressure of a gas is defined as the mole fraction of the gas multiplied by the total pressure. If a gas with partial pressure $P_s$ is in contact with a liquid, the steady-state concentration $U$ of gas in the liquid is given by

$$U = \sigma P_s, \tag{17.1}$$

where $\sigma$ is the solubility of the gas in the liquid. Because of this, we can define the partial pressure of a dissolved gas with concentration $U$ to be $U/\sigma$.

Now suppose that a gas with partial pressure $P_g$ is brought into contact with a liquid within which that same gas is dissolved with concentration $U$, and thus partial pressure $U/\sigma$. If $U/\sigma$ is not equal to $P_g$, then there will be a net flow of gas across the interface. The simplest model (but not necessarily the most accurate) assumes that the flow is linearly proportional to the difference in partial pressures across the interface,

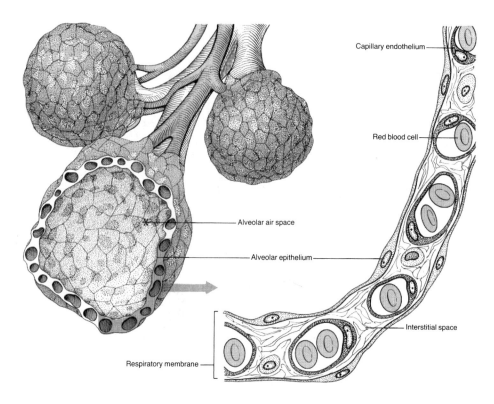

**Figure 17.2**   The alveoli, or air sacs, of the lung are covered by an extensive network of capillaries that form a thin layer of blood for the exchange of gases. (Davis, Holtz, and Davis, 1985, Fig. 19-4, p. 391.)

and thus

$$q = D_s \left( P_g - \frac{U_i}{\sigma_i} \right), \tag{17.2}$$

where $q$ is the net flux per unit area of the gas (positive when gas is flowing from the gaseous phase to the dissolved phase), and $D_s$ is the *surface diffusion constant*.

## 17.1.2   Capillary–Alveolar Transport

To understand something about the transport of a gas across the capillary wall into the alveolar space, we begin with the simplest possible model. We suppose that a gas such as oxygen or carbon dioxide is dissolved in blood at some concentration $U$ uniformly across the cross-section of the capillary. The blood is flowing along a capillary that is bounded by alveolar air space. The partial pressure of the gas in the alveolar space, $P_g$, is taken to be constant.

Consider a segment of the capillary, of length $L$, with constant cross-sectional area $A$ and perimeter $p$. The total amount of the dissolved gas contained in the capillary at any time is $A \int_0^L U(x,t)dx$. Since mass is conserved, we have

$$\frac{d}{dt}\left(A \int_0^L U(x,t)\,dx\right) = v(0)AU(0,t) - v(L)AU(L,t) + p\int_0^L q(x,t)\,dt, \qquad (17.3)$$

where $v(x)$ is the velocity of the fluid in the capillary, and $q$ is the flux (positive inward, with units of moles per time per unit area) of gas along the boundary of the capillary. This assumes that diffusion along the length of the capillary is negligible compared to diffusion across the capillary wall. Differentiating (17.3) with respect to $L$ and replacing $L$ by $x$ gives the conservation law

$$U_t + (vU)_x = \frac{pq}{A}. \qquad (17.4)$$

Finally, if we assume that the flow velocity $v$ is constant along the capillary, then using (17.2), we obtain

$$U_t + vU_x = \frac{pD_s}{A}\left(P_g - \frac{U}{\sigma}\right) = D_m(\sigma P_g - U), \qquad (17.5)$$

where $D_m = \chi D_s/\sigma$, and $\chi = p/A$ is the surface-to-volume ratio. Notice that $D_m$ has units of $(\text{time})^{-1}$, so it is the inverse of a time constant, the *membrane exchange rate*.

In steady state (independent of time), the conservation law (17.5) reduces to the first-order, linear ordinary differential equation

$$v\frac{dU}{dx} = D_m(\sigma P_g - U). \qquad (17.6)$$

Note that, as one would expect intuitively, the rate of change of $U$ at the steady state is inversely proportional to the fluid velocity. Now we suppose that the concentration $U$ at the inflow $x = 0$ is fixed at $U_0$ (at partial pressure $P_0 = U_0/\sigma$). In steady state, the concentration at each position $x$ is given by the exponentially decaying function

$$U(x) = \sigma P_g + (U_0 - \sigma P_g)e^{-D_m x/v}. \qquad (17.7)$$

If the exposed section of the capillary has length $L$, the total flux of gas across the wall is $Q = p\int_0^L q\,dx = vA[U(L) - U_0]$, which is

$$Q = vA\sigma(P_g - P_0)(1 - e^{-D_m L/v}). \qquad (17.8)$$

Plotted in Fig. 17.3 is the nondimensional flux

$$\bar{Q} = \frac{Q}{D_m LA\sigma(P_g - P_0)} = \frac{v}{D_m L}(1 - e^{\frac{-D_m L}{v}}). \qquad (17.9)$$

Note that

$$Q \to vA\sigma(P_g - P_0) \qquad (17.10)$$

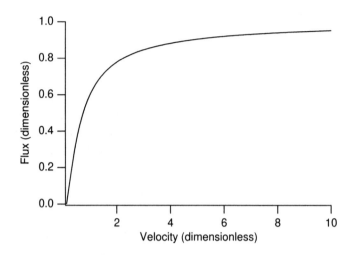

**Figure 17.3**   Dimensionless transmural flux $\bar{Q}$ as a function of dimensionless flow velocity $\frac{v}{D_m L}$ from (17.9).

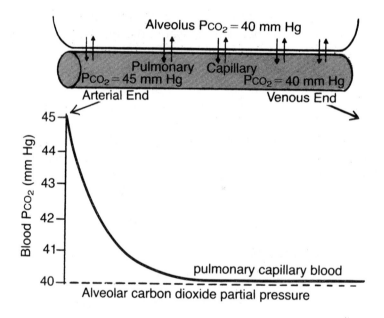

**Figure 17.4**   Loss of carbon dioxide from the pulmonary capillary blood into the alveolus. (The curve in this figure was constructed from data in Milhorn and Pulley, *Biophys. J.* 8:337, 1968. Figure from Guyton and Hall, 1996, Fig. 40-6, p. 515.)

in the limit $D_m L/v \to \infty$. Thus, an infinitely long capillary has only a finite total flux, as the dissolved gas concentration approaches the alveolar concentration along the length of the capillary.

Data on the diffusion of carbon dioxide from the pulmonary blood into the alveolus (Fig. 17.4) suggest that carbon dioxide is lost into the alveolus at an exponential rate, consistent with (17.7). Furthermore, because the solubility of carbon dioxide in water is quite high, the difference between the partial pressure for the entering blood and the alveolar air is small, about 5 mm Hg.

In contrast, the solubility of oxygen in blood is small (about 20 times smaller than carbon dioxide, see Table 16.1), and although the difference in partial pressures is larger, this is not adequate to account for the balance of oxygen inflow and carbon dioxide outflow. That is, if (17.10) is relevant, then a decrease in $\sigma$ by a factor of 20 requires a corresponding increase by a factor of 20 for the partial pressure differences to maintain similar transport. Thus, if this is the correct mechanism for carbon dioxide and oxygen transport, the difference $P_0 - P_g$ for oxygen should be about twenty times larger than for carbon dioxide. Since $104 - 40 \neq 20(45 - 40)$ (using typical numbers from Figs. 17.4 and 17.5), there is reason to doubt this model.

Second, the data in Fig. 17.5 suggest that the uptake of oxygen by the capillary blood is not exponential with distance, but nearly linear for the first third of the distance, where it becomes fully saturated. We consider a model of this below. First, however, we discuss the effects of blood chemistry on gas exchange, which was ignored in the above model.

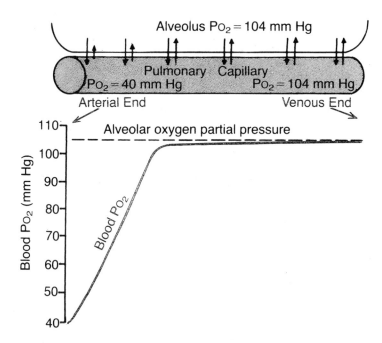

**Figure 17.5** Uptake of oxygen by the pulmonary capillary blood. (The curve in this figure was constructed from data in Milhorn and Pulley, *Biophys. J.* 8:337, 1968. Figure from Guyton and Hall, 1996, Fig. 40-1, p. 514.)

## 17.1.3  Carbon Dioxide Removal

Blood chemistry plays a significant role in facilitating the transport of gases between blood and alveoli. To understand something of this facilitation, we first consider a simple model for carbon dioxide transport that takes the carbon dioxide–bicarbonate chemistry into account. We assume that carbon dioxide is converted to bicarbonate via the reaction

$$CO_2 + H_2O \underset{k_{-1}}{\overset{k_1}{\rightleftharpoons}} HCO_3^- + H^+.$$

This is the carbonic anhydrase reaction discussed in Section 16.2.3. For convenience we ignore here the intermediary $H_2CO_3$. Since the dissociation of $H_2CO_3$ into $HCO_3^-$ and $H^+$ is fast, this makes no difference to the model.

Now we write conservation equations for the two chemical species $CO_2$ and $HCO_3^-$ (in steady state, and ignoring diffusion within the capillary) as

$$v\frac{dU}{dx} = D_{CO_2}(\sigma_{CO_2}P_{CO_2} - U) + k_{-1}[H^+]V - k_1U, \tag{17.11}$$

$$v\frac{dV}{dx} = k_1U - k_{-1}[H^+]V, \tag{17.12}$$

where $U = [CO_2]$, $V = [HCO_3^-]$. Notice that $D_{CO_2}$ is a rate constant, similar to $D_m$ above.

Although this is a linear problem and it can be solved exactly, it is illustrative to use an approximate, singular perturbation technique, as this technique will prove useful in the next section. First notice that we can add (17.11) and (17.12) to obtain

$$v\frac{d}{dx}(U + V) = D_{CO_2}(\sigma_{CO_2}P_{CO_2} - U). \tag{17.13}$$

Now we assume that $V$ equilibrates rapidly, so that it can be taken to be in quasi-steady state. Accordingly, we set $V = K_cU$, where $K_c = \frac{k_1}{k_{-1}[H^+]}$. It follows that, assuming that $[H^+]$ is constant,

$$v(1 + K_c)\frac{dU}{dx} = D_{CO_2}(\sigma_{CO_2}P_{CO_2} - U). \tag{17.14}$$

This equation is identical in form to (17.6). If we take the inlet conditions to be $U = U_0 = \sigma_{CO_2}P_0$ and $V = V_0 = K_cU_0$, then the total flux $Q$ is

$$Q = v(1 + K_c)\sigma_{CO_2}(P_0 - P_{CO_2})(1 - e^{-D_{CO_2}L/(v(1+K_c))}), \tag{17.15}$$

and in the limit as $D_{CO_2}L/v \to \infty$,

$$Q \to v(1 + K_c)\sigma_{CO_2}(P_0 - P_{CO_2}), \tag{17.16}$$

which is a factor of $1 + K_c$ larger than in (17.10). The only difference between this flux (17.15) and the original (17.8) is that the velocity $v$ has been multiplied by the factor $1 + K_c$. In other words, the conversion of carbon dioxide to bicarbonate via the carbonic anhydrase reaction effectively increases the flow rate by the factor $1 + K_c$.

The equilibrium constant for the bicarbonate–carbon dioxide reaction is given by $\log_{10}(\frac{k_1}{k_{-1}}) = -6.1$. Thus (since $pH = -\log_{10}[H^+]$ with $[H^+]$ in moles per liter), at $pH = 7.4$, we have $K_c = 20$, and the improvement in carbon dioxide transport because of the carbonic anhydrase reaction is substantial.

In words, the improvement in total flux arises because the conversion of bicarbonate to carbon dioxide continually replenishes the carbon dioxide that is lost to the alveolar air. Thus, the carbon dioxide concentration in the capillary does not fall so quickly, leading to an increase in the total flux.

## 17.1.4 Oxygen Uptake

The chemistry for the absorption of oxygen by hemoglobin has a similar, but nonlinear, effect. We take a simple model for the chemistry of hemoglobin (discussed in Section 16.2.1), namely

$$\text{Hb} + 4\text{O}_2 \underset{k_{-2}}{\overset{k_2}{\rightleftharpoons}} \text{Hb}(\text{O}_2)_4.$$

Of course, there are more detailed models of hemoglobin chemistry, but the qualitative behavior is affected little by these details. We write the conservation equations as

$$v\frac{dW}{dx} = D_{O_2}(\sigma_{O_2}P_{O_2} - W) + 4k_{-2}Y - 4k_2ZW^4, \tag{17.17}$$

$$v\frac{dY}{dx} = k_2ZW^4 - k_{-2}Y, \tag{17.18}$$

$$v\frac{dZ}{dx} = k_{-2}Y - k_2ZW^4, \tag{17.19}$$

where $W = [O_2]$, $Y = [\text{Hb}(O_2)_4]$, $Z = [\text{Hb}]$, and $D_{O_2}$ is the oxygen exchange rate constant. The last of these equations is superfluous, since total hemoglobin is conserved, and so we take $Z + Y = Z_0$. Notice further that (17.17) and (17.18) can be added to obtain

$$v\frac{d}{dx}(W + 4Y) = D_{O_2}(\sigma_{O_2}P_{O_2} - W). \tag{17.20}$$

We expect oxygen uptake by hemoglobin to be fast compared to the transmural exchange, so take $Y$ to be in quasi-steady state, setting

$$Y = Z_0\frac{W^4}{K_{O_2}^4 + W^4}, \tag{17.21}$$

where $K_{O_2}^4 = k_{-2}/k_2$. On substitution into (17.20) we find

$$v\frac{d}{dx}\left(W + 4Z_0\frac{W^4}{K_{O_2}^4 + W^4}\right) = D_{O_2}(\sigma_{O_2}P_{O_2} - W). \tag{17.22}$$

More generally, if $f(W)$ is the oxygen saturation curve for hemoglobin, then

$$v\frac{d}{dx}(W + 4Z_0f(W)) = D_{O_2}(\sigma_{O_2}P_{O_2} - W).\tag{17.23}$$

This equation is a nonlinear first-order ordinary differential equation, which, being separable, can be solved exactly. The solution is given implicitly by

$$\int_{W_1}^{W_2}\frac{W + 4Z_0f(W)}{\sigma_{O_2}P_{O_2} - W}dW = \frac{D_{O_2}L}{v}.\tag{17.24}$$

However, this exact solution does not provide much insight. It is more useful to compare (17.23) with (17.14), in which the flux of carbon dioxide was facilitated by the factor $K_c$. Here, there is facilitation of oxygen flux by the factor $1 + 4Z_0f'(W)$. Clearly, the two ways to exploit this facilitation are to have a high concentration of hemoglobin and to use a saturation curve with a steep slope in the range of operating values.

The total flux of oxygen is given by

$$Q = A\int_0^L D_{O_2}(\sigma_{O_2}P_{O_2} - W)dx\tag{17.25}$$

$$= Av\int_0^L \frac{d}{dx}(W + 4Z_0f(W))dx\tag{17.26}$$

$$= Av(W + 4Z_0f(W))|_{W_0}^{W_1}.\tag{17.27}$$

For oxygen, this enhancement is substantial. For normal blood at 100 mm Hg oxygen partial pressure, hemoglobin is 97% saturated, and the hemoglobin of 100 ml of blood carries 19.4 ml of oxygen. By contrast, the same 100 ml of blood contains only 0.3 ml of dissolved (unbound) oxygen. When this hemoglobin is chemically pure, it can combine with a total of 20 ml of oxygen. This implies that (using that 1 mole of a dissolved gas fills 24.6 liters at room temperature) $Z_0 = 2.0$ mM.

Incoming blood has a partial pressure of 40 mm Hg, at which hemoglobin is about 75% saturated and alveolar air is at 104 mm Hg. A reasonable fit of the oxygen saturation curve is given by the function $f(W) = \frac{W^4}{K_{O_2}^4 + W^4}$ with $K_{O_2} = 30\sigma$ mm Hg. A slightly better fit is obtained with the Hill equation $f(W) = \frac{W^n}{K_{O_2}^n + W^n}$, $n = 2.5$, $K_{O_2} = 26\sigma$ mm Hg. Either way, the oxygen flux is increased by hemoglobin by a factor of about 14.

Finally, using (17.24) we can find the rate of oxygen uptake as a function of length along the capillary. In Fig. 17.6 is shown the partial pressure of oxygen, plotted as a function of the dimensionless distance $D_{O_2}x/v$ along the capillary. The significant observation is that oxygen partial pressure rises steeply and nearly linearly, until it saturates, comparing well with the experimental data shown in Fig. 17.5.

## 17.1.5  Carbon Monoxide Poisoning

Carbon monoxide poisoning occurs because carbon monoxide competes with oxygen for hemoglobin binding sites. The goal of this section is to see how this competition

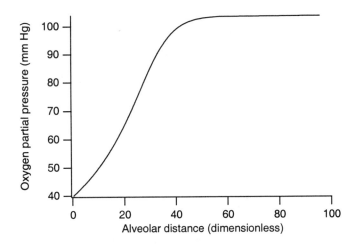

**Figure 17.6** Oxygen partial pressure as a function of nondimensional distance $\frac{D_{O_2}x}{v}$.

for hemoglobin affects oxygen exchange and how carbon monoxide can be eliminated from the blood.

To model this problem we assume that the chemistry for carbon monoxide binding with hemoglobin is the same as for oxygen, except that the affinity of carbon monoxide for hemoglobin is much larger (about 200 times) than the affinity of oxygen for hemoglobin. Thus,

$$Hb + 4CO \underset{k_{-2}}{\overset{k_2}{\rightleftharpoons}} Hb(CO)_4.$$

The conservation equations for carbon monoxide gaseous exchange in the alveolus are similar in form to those for oxygen, being (ignoring diffusion of the dissolved gases)

$$v\frac{dU}{dx} = D_{CO}(\sigma_{CO}P_{CO} - U) + 4k_{-3}S - 4k_3ZU^4, \tag{17.28}$$

$$v\frac{dS}{dx} = k_3ZU^4 - k_{-3}S, \tag{17.29}$$

where $U = [CO]$, $S = [Hb(CO)_4]$, $Z = [Hb]$, and $D_{CO}$ is the carbon monoxide exchange rate constant. The balance of oxygen is governed by (17.17) and (17.18). Conservation of hemoglobin implies that $Z + Y + S = Z_0$.

As before, (17.20) holds, as does

$$v\frac{d}{dx}(U + 4S) = D_{CO}(\sigma_{CO}P_{CO} - U). \tag{17.30}$$

Now we assume that both carbon monoxide and oxygen are in quasi-steady state, so that

$$K_{CO}^4 S = ZU^4, \qquad K_{O_2}^4 Y = ZW^4, \tag{17.31}$$

where $K_{CO}^4 = k_{-3}/k_3$, $K_{O_2}^4 = k_{-2}/k_2$. It is convenient to introduce scaled variables $w$ and $u$ with $w = K_{O_2}^{-1}W$, $u = K_{CO}^{-1}U$. It follows from $Z + Y + S = Z_0$ that

$$S = Z_0 \frac{u^4}{1 + w^4 + u^4}, \tag{17.32}$$

$$Y = Z_0 \frac{w^4}{1 + w^4 + u^4}, \tag{17.33}$$

so that

$$v \frac{d}{dx}\left(w + 4z_0 \frac{w^4}{1 + w^4 + u^4}\right) = D_{O_2}(w^* - w), \tag{17.34}$$

$$v \frac{d}{dx}\left(u + 4\beta z_0 \frac{u^4}{1 + w^4 + u^4}\right) = D_{CO}(u^* - u), \tag{17.35}$$

where $z_0 = Z_0/K_{O_2}$, $\beta = K_{CO}/K_{O_2}$, $w^* = K_{O_2}^{-1}\sigma_{O_2}P_{O_2}$, $u^* = K_{CO}^{-1}\sigma_{CO}P_{CO}$.

While we cannot solve this system of differential equations explicitly, the difficulty can be readily seen. Because $\beta$ is large (on the order of 200), the total carbon monoxide concentration changes as a function of $x$ slowly, much more slowly than does the total oxygen concentration. Thus, $w$ increases quickly to $w^*$, releasing some carbon monoxide as it does so, while $u + 4\beta z_0 \frac{u^4}{1 + w^4 + u^4}$ remains essentially fixed. As a result, in the length of the alveolus, oxygen is recharged, but very little carbon monoxide is eliminated.

The lethality of carbon monoxide can be seen from a simple steady-state analysis. Suppose that $u = u^*$ is at steady state with the environment, so that no carbon monoxide is gained or lost in the alveoli. The concentration of oxygen in the blood is proportional to $w + 4z_0 \frac{w^4}{1 + w^4 + u^4}$, so that the rate of oxygen transport is proportional to

$$M = w^* + 4z_0 \frac{(w^*)^4}{1 + (w^*)^4 + u^4} - w_0 - 4z_0 \frac{w_0^4}{1 + w_0^4 + u^4},$$

where $w_0$ is the alveolar input level and $w^*$ is the output level from the alveolus. When there is no carbon monoxide present (i.e., when $u = 0$), the input and output levels are 40 and 104 mm Hg, respectively, so that (with $K_{O_2} = 30\sigma$ mm Hg) $w_0 = 40/30 = 1.333$, $w^* = 104/30 = 3.47$. Thus, with normal metabolism, the required flow rate has $M = 53$, where we have set $z_0 = 52$. When carbon monoxide is present, this same flow rate must be maintained (as the need of the tissues for oxygen remains unchanged), but now the presence of $u$ in the denominator changes things. Keeping $M$ fixed, we see that if $u$ is greater than 4.64, then the incoming blood has $w_0 < 0$, so that the tissue is in oxygen debt. With $\beta = 200$, $u = 4.64$ is equivalent to a carbon monoxide partial pressure of 0.7 mm Hg, a mere 0.1% by volume. In other words, an ambient concentration of 0.1% carbon monoxide leads to certain death because of oxygen depletion.

Since $\beta_{z_0}$ is so large (on the order of $10^4$), we can approximate the dynamics of carbon monoxide by

$$4\beta_{z_0}v \frac{d}{dx}\left(\frac{u^4}{1+(w^*)^4+u^4}\right) = D_{CO}(u^* - u). \tag{17.36}$$

If we set

$$F = \frac{u^4}{1+(w^*)^4+u^4}, \tag{17.37}$$

we find that

$$\frac{dF}{dx} = -\frac{D_{CO}}{4\beta_{z_0}v}(1+(w^*)^4)^{1/4}\left(\frac{F}{1-F}\right)^{1/4}, \tag{17.38}$$

where we have taken $u^* = 0$, assuming that the victim is placed in a carbon-monoxide-free environment. Clearly, the rate of carbon monoxide elimination is proportional to $(1+(w^*)^4)^{1/4}$, which for large $w^*$ is linear in $w^*$. Thus, the rate of carbon monoxide elimination can be increased by placing the victim in an environment of high oxygen.

In hospitals it is typical to place a carbon monoxide poisoning victim in an environment of oxygen at 2–2.5 atm. At 2 atmospheres (1 atm = 760 mm Hg), $w^* = (2 \times 760/30) = 50.7$, compared to $w^* = 3.5$ at normal oxygen levels, giving an increase in the rate of carbon monoxide elimination of about 14.

## 17.2 Ventilation and Perfusion

Gaseous exchange is mediated by the combination of ventilation of the alveoli with inspired air and the perfusion of the capillaries with blood. It is the balance of these two that determines the gas content of the lungs and of the recharged blood.

To see how this balance is maintained, suppose that $\dot{V}$ is the volume flow rate of air that participates in the exchange of the alveolar content. Not all inspired air participates in this exchange, because some inspired air never reaches the terminal bronchioles. The parts of the lung that are ventilated but do not participate in gaseous exchange are called the *anatomical dead space*. In normal breathing, the total amount of inspired air is about 500 ml per breath (men 630 ml; women 390 ml). Of this, 150 ml is anatomical dead space, so only 350 ml participates in alveolar gaseous exchange. With 15 breaths per minute, $\dot{V}$ is about 5250 ml/min.

Now suppose that $Q$ is the volume flow rate of blood into and out of the alveolar capillaries. Cardiac output is about 70 ml per beat, so at 72 beats per minute, $Q$ is about 5000 ml/min. The ratio $\dot{V}/Q$ is called the *ventilation–perfusion ratio*, and it is the most important determinant of lung–blood gas content.

If $c_i$ and $c_a$ are the concentrations of a gas in the inspired air and in the alveolar air, respectively, then the flow of the gas is

$$\dot{V}(c_i - c_a). \tag{17.39}$$

Similarly, the flow of the gas into the blood is given by

$$Q(c_L - c_0), \tag{17.40}$$

where $c_0$ and $c_L$ are the input and output capillary gas concentrations. The fact that these two must be in balance leads to the equation

$$\frac{\dot{V}}{Q} = \frac{(c_L - c_0)}{(c_i - c_a)}. \tag{17.41}$$

From the previous section we learned that the two most important respiratory gases, carbon dioxide and oxygen, are equilibrated when they leave the alveolus in the capillaries. In other words, the partial pressures of carbon dioxide and oxygen in the alveolus and in the blood leaving the pulmonary capillary are the same. Of course, this is not true at high perfusion rates, but it is a satisfactory assumption at normal physiological flow rates.

Because carbon dioxide is quickly converted to bicarbonate, the total blood carbon dioxide (i.e., both free and converted) is given by

$$[CO_2] = \sigma_{CO_2}(1 + K_c)P_{CO_2}. \tag{17.42}$$

This implies that for carbon dioxide, the ventilation–perfusion ratio must satisfy

$$\frac{\dot{V}}{Q} = \sigma_{CO_2}RT(1 + K_c)\frac{(P_0 - P_a)}{P_a}, \tag{17.43}$$

where $P_0$ and $P_a$ are the inflow and alveolar carbon dioxide partial pressures. Note that we have taken the carbon dioxide partial pressure in the inspired air to be zero, we have assumed that $P_L = P_a$, and we have used the ideal gas law to express the atmospheric carbon dioxide concentrations in terms of pressures.

For oxygen, the relationship between partial pressure and total blood oxygen (both free and bound to hemoglobin) is determined from the hemoglobin saturation function $f(W)$ as

$$[O_2] = W + 4Z_0 f(W), \tag{17.44}$$

where $Z_0$ is the total hemoglobin concentration, as in the previous section.

In these terms the ventilation–perfusion ratio must be

$$\frac{\dot{V}}{Q} = \frac{RT}{(P_i - P_a)}\{W_a - W_0 + 4Z_0[f(W_a) - f(W_0)]\}, \tag{17.45}$$

where the subscripts $a, i$, and $0$ have the same interpretations as above. Here, as with carbon dioxide, we assume that the partial pressure of oxygen in the alveolar air is the same as the partial pressure in the blood leaving the alveolus, so that $W_a = \sigma_{O_2}P_a$.

A plot of the alveolar partial pressures of carbon dioxide and oxygen as a function of ventilation–perfusion ratio is shown in Fig. 17.7. This figure was determined as follows. First, using $W_a$ as a parameter and keeping $W_0 = 40\sigma_{O_2}$ mm Hg fixed, the ventilation–perfusion curve for oxygen was found using (17.45). For this curve we used

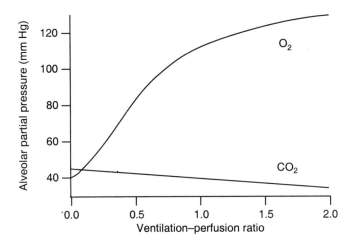

**Figure 17.7**  Alveolar partial pressure as a function of ventilation–perfusion ratio.

$f(W) = \frac{W^4}{K_{O_2}^4 + w^4}$ with $K_{O_2} = 30\sigma$ mm Hg, and $Z_0 = 2.2$ mM ($RT = 1.7 \times 10^4$ mm Hg/M). Then, we used (17.43) to find the carbon dioxide partial pressure as a function of ventilation–perfusion. For this plot, $P_{CO_2} = 45$ mm Hg, and we chose $K_c = 12$ because it gives a reasonable fit of the available data.

From this figure we see that the alveolar oxygen partial pressure is an increasing function of $\dot{V}/Q$, while the alveolar carbon dioxide partial pressure is a decreasing function thereof. In normal situations, the ventilation–perfusion ratio is about 1. An increase in this ratio is called *hyperventilation*, and a decrease is called *hypoventilation*. During hyperventilation, there is rapid removal of carbon dioxide, and the partial pressure of carbon dioxide in the arterial blood drops below the normal level of 40 mm Hg. This results in less carbon dioxide available for carbonic acid formation, and consequently blood pH rises above the normal level, resulting in *respiratory alkalosis*. In hyperventilation there is no substantial change in oxygen concentration because the hemoglobin is fully saturated.

The opposite situation, in which the ventilation–perfusion ratio drops, increases carbon dioxide content and decreases oxygen content of the arterial blood. The increase of carbon dioxide increases carbonic acid formation and decreases blood pH, a condition referred to as *respiratory acidosis*. To compensate for these changes, the blood gas concentration stimulates the carotid and aortic chemoreceptors to increase the rate of ventilation.

In Fig. 17.8 is shown the volume fraction of gaseous exchange as a function of ventilation–perfusion ratio. (Volume fraction of a gas is the fraction of gas in a given volume, found as the ratio of partial pressure to total pressure.) Typical partial pressures of the respiratory gases are shown in Table 17.1.

The oxygen that is taken in by the blood is consumed by metabolic processes to produce carbon dioxide. However, the amount of carbon dioxide produced is generally

**Table 17.1** Partial pressures (in mm Hg) of respiratory gases as they enter and leave the lungs.

| Substance | Atmospheric air | Humidified air | Alveolar air | Expired air |
|-----------|-----------------|----------------|--------------|-------------|
| $N_2$     | 597.9           | 563.5          | 569.0        | 566.0       |
| $O_2$     | 159.0           | 149.3          | 104.0        | 120.0       |
| $CO_2$    | 0.3             | 0.3            | 40.0         | 27.0        |
| $H_2O$    | 3.7             | 47.0           | 47.0         | 47.0        |

less than the amount of oxygen consumed. The *respiratory exchange rate R* is the ratio of carbon dioxide output to oxygen uptake, and is rarely more than one. When a person is using carbohydrates for body metabolism, $R$ is 1.0 because one molecule of carbon dioxide is formed for every molecule of oxygen consumed. On the other hand, when oxygen reacts with fats, a large share of the oxygen combines with hydrogen to form water instead of carbon dioxide. In this mode, $R$ falls to as low as 0.7. For a normal person with a normal diet, $R = 0.825$ is considered normal.

Since the respiratory exchange rate is just the ratio of the two curves shown in Fig. 17.8, one can use that figure to determine the ventilation–perfusion ratio as a function of the respiratory exchange rate, which is, in turn, determined by the metabolism.

For these figures, the inflow carbon dioxide and oxygen partial pressures were fixed at 45 and 40 mm Hg, respectively. If, however, the metabolic rate and the type of metabolism are taken into account, the inflow partial pressures are determined by those rates and are not fixed. For example, during strenuous exercise, the partial pressure of oxygen in the tissue can drop to as low as 15 mm Hg. However, the general result is the same, namely that alveolar carbon dioxide partial pressure decreases with increasing $\dot{V}/Q$ and alveolar oxygen partial pressure increases.

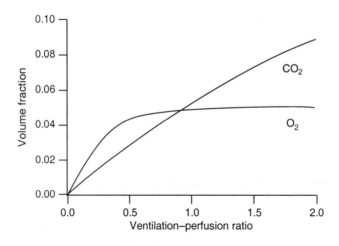

**Figure 17.8** Volume fraction of gaseous exchange as a function of ventilation–perfusion ratio.

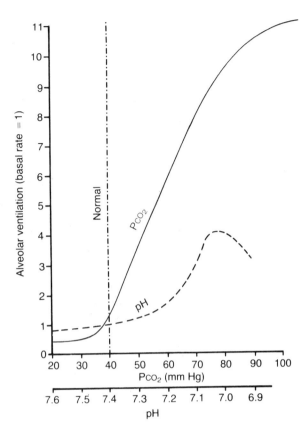

**Figure 17.9** Effects of increased arterial $P_{CO_2}$ and decreased arterial pH on the alveolar ventilation rate. (Guyton and Hall, 1996, Fig. 41-3, p. 528.)

## 17.3 Regulation of Ventilation

While the exchange of gases takes place in the lungs, the control of the rate of ventilation is accomplished in the brain. There, in the respiratory center, is located a chemosensitive area that is sensitive to the concentrations of chemicals in the blood, primarily carbon dioxide. Changes in blood $P_{CO_2}$ are detected, and this leads to changes in the rate of breathing by activating or inhibiting the inspiratory neurons (described in the next section). In Fig. 17.9 is shown the effect of carbon dioxide on ventilation rate.

To construct a model for this control, we let $x$ denote the partial pressure of carbon dioxide in the blood. Carbon dioxide is produced at rate $\lambda$ by metabolism and eliminated by ventilation at the lungs. Thus,

$$\frac{dx}{dt} = \lambda - \alpha x \dot{V}, \tag{17.46}$$

where $\dot{V}$ is the ventilation rate, and we assume that the transport of carbon dioxide through the lungs is linearly proportional to the concentration of carbon dioxide and the ventilation rate.

Now we take the ventilation to be of the form shown in Fig. 17.9, for example, the Hill equation

$$\dot{V}(x) = V_m \frac{x^n}{\theta^n + x^n}. \tag{17.47}$$

Furthermore, we recognize that there is a substantial delay between ventilation of the blood and the measurement of $P_{CO_2}$ at the respiratory center in the brain because the transport of blood from the lungs back to the heart and then to the brain takes time. Thus, our complete model becomes (Glass and Mackey, 1988)

$$\frac{dx}{dt} = \lambda - \alpha x \dot{V}(x(t - \tau)). \tag{17.48}$$

Typical physical parameter values for the model are given in Table 17.2.

Before proceeding further with the analysis of this equation, it is worthwhile to introduce dimensionless variables and parameters. We set $x = \theta y, t = \frac{s}{\alpha V_m}, \tau = \frac{\sigma}{\alpha V_m}$, and $\lambda = \theta \alpha V_m \beta$ and obtain

$$\frac{dy}{ds} = \beta - yF(y(s - \sigma)), \tag{17.49}$$

where $F$ is a sigmoidal function, monotone increasing with a maximum of 1 as $y \to \infty$.

Because the function $yF(y)$ is monotone increasing in $y$, there is a unique steady-state solution for (17.49). Furthermore, the steady-state solution is a monotone increasing function of the parameter $\beta$, indicating that blood $P_{CO_2}$ and ventilation increase as a function of steady metabolism. However, the dynamical (non-steady-state) situation may be quite different.

To understand more about the dynamic behavior of this system of equations, we perform a linear stability analysis. We suppose that the steady state is $y = y^*$, and set $y = y^* + Y$, substitute into (17.49), and assume that $Y$ is small enough so that only linear terms of the local Taylor series are necessary. The resulting linearized equation for $Y$ is

$$\frac{dY(s)}{ds} = -F(y^*)Y(s) - y^*F'(y^*)Y(s - \sigma). \tag{17.50}$$

---

**Table 17.2**  Physical parameters for the Mackey–Glass model of respiratory control.

| | | |
|---|---|---|
| $\lambda$ | = | 6 mm Hg/min |
| $V_m$ | = | 80 liter/min |
| $\tau$ | = | 0.25 min |

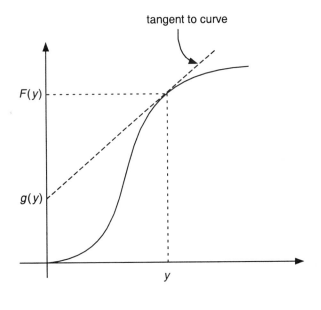

**Figure 17.10** Sketch of the construction of the function $g(y) = F(y) - yF'(y)$.

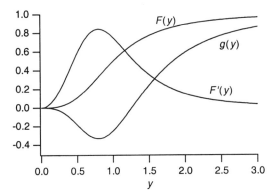

**Figure 17.11** Plots of $F(y) = \frac{y^n}{1+y^n}$, $F'(y)$ and $g(y)$, with $n = 3$.

Solutions of exponential form $Y = Y_0 e^{\mu s}$ exist provided that $\mu$ satisfies the characteristic equation

$$\mu + F(y^*) + y^* F'(y^*) e^{-\mu \sigma} = 0. \tag{17.51}$$

Since there is a monotone relationship between $y^*$ and $\beta$, it is convenient to view $y^*$ as an independent parameter.

The function

$$g(y) = F(y) - yF'(y) \tag{17.52}$$

is important to the analysis that follows and has a nice geometrical interpretation. This function is constructed by drawing a straight line from the point $(y, F(y))$ to $y = 0$ with slope $F'(y)$, as illustrated in Fig. 17.10. The three functions $F(y)$, $F'(y)$, and $g(y)$ are shown in Fig. 17.11, in the case $F(y) = \frac{y^3}{1+y^3}$.

We wish to understand the behavior of the roots of the characteristic equation (17.51). First, we observe that if $g(y^*)$ is positive, then all roots of the characteristic equation (17.51) have negative real part, so that the steady solution is stable. This follows, because if the real part of $\mu$ is positive, then $|\mu + F(y^*)| > |\mu + y^*F'(y^*)| > |y^*F'(y^*)e^{-\mu\sigma}|$. Note that we are assuming that $F, F'$, and $y$ are all positive.

The only real roots of (17.51) are negative. Thus, the only way the real part of a root can change sign is if it is complex, a Hopf bifurcation. To see whether Hopf bifurcations occur, we set $\mu = i\omega$. If this is a root of (17.51), then of necessity, $|i\omega + F(y^*)| = |y^*F'(y^*)|$, and thus $|F(y^*) + i\omega|^2 = [F(y^*)]^2 + \omega^2 = [y^*F'(y^*)]^2$. In this case, it follows that $y^*F'(y^*) > F(y^*)$, which implies that $g(y^*) < 0$. We split (17.51) into real and imaginary parts, obtaining

$$F(y^*) + y^*F'(y^*) \cos \omega\sigma = 0, \tag{17.53}$$

$$\omega - y^*F'(y^*) \sin \omega\sigma = 0. \tag{17.54}$$

It follows that $\omega = \sqrt{(y^*F'(y^*))^2 - (F(y^*))^2}$ (provided that $g(y^*) < 0$) and that

$$\tan \omega\sigma = -\frac{\omega}{F(y^*)}. \tag{17.55}$$

The smallest root of this equation is on the interval $\frac{\pi}{2} < \omega\sigma < \pi$, and for this root,

$$\sigma = \frac{1}{\omega}\left[\pi + \tan^{-1}\left(-\frac{\omega}{F(y^*)}\right)\right]. \tag{17.56}$$

We can view this information as follows. For a given $y^*$, we have the frequency $\omega$ and the critical delay $\sigma$ at which a Hopf bifurcation occurs. If the delay is smaller than this critical delay, then the steady solution is stable, while if the delay is larger, then the steady solution is unstable and an oscillatory solution is likely.

Plots of $\omega$ and $\sigma$ are shown in Fig. 17.12. Steady solutions having $\sigma$ greater than the critical value of delay (17.56) are unstable. In this case, numerical simulations show that there is a stable periodic solution of the governing equations, shown in Fig. 17.13. Here is shown the dimensionless concentration $y$ (shown solid) and the dimensionless

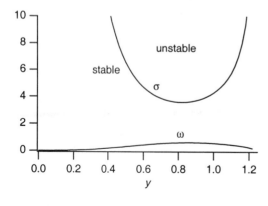

**Figure 17.12** Plots of $\omega$ and $\sigma$ at Hopf bifurcation points.

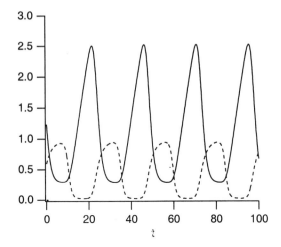

**Figure 17.13** Oscillatory solution of the Mackey–Glass equation (17.49) with parameters $y^* = 0.8, \sigma = 10.0$. Solution (carbon dioxide content) is shown as solid curve, ventilation rate is shown as dashed curve.

ventilation rate $F(y_\sigma)$ (shown dashed) as a function of time, with parameter values $\beta = 0.8, \sigma = 10.0$.

An episode of periodic fluctuation of ventilation, depicted by this periodic solution, is called *Cheyne–Stokes breathing*. In this condition, the person breathes deeply for a short interval and then breathes slightly or not at all for an additional interval, then repeats the cycle, with a period of 40 to 60 seconds. Notice that Cheyne–Stokes breathing can be caused by an increased delay in the transport of blood to the brain or an increase in the negative feedback gain (the slope of $F$). The first type (delayed transport) is likely to occur in patients with chronic heart failure, and the second type (increased gain) occurs mainly in patients with brain damage, and is often a signal of impending death.

## 17.4 The Respiratory Center

Breathing is controlled by a neural central pattern generator called the *respiratory center*. The respiratory center is composed of three major groups of neurons located at the base of the brain. The *dorsal respiratory group*, located in the dorsal portion of the medulla, mainly causes inspiration; the *ventral respiratory group* can cause either inspiration or expiration, depending upon which neurons in the group are stimulated; and the *pneumotaxic center*, located above the medulla in the superior portion of the pons, helps control the rate and pattern of breathing.

The basic rhythm of respiration is generated mainly by the dorsal group, by emitting repetitive bursts of inspiratory action potentials. While the basic cause of these bursts is unknown, in primitive animals, neural networks have been found in which one set of neurons stimulates a second set, which in turn inhibits the first set, leading to periodic bursting activity that lasts throughout the lifetime of the animal.

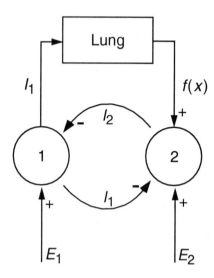

**Figure 17.14** A mutual inhibition network for the control of respiration.

The inspiratory signal is said to be a *ramp signal*, as it begins slowly and increases steadily for about 2 seconds, whereupon it abruptly ceases for the next 3 seconds before a new cycle begins. During normal quiet breathing, the ventral respiratory group is almost totally inactive. Expiration results primarily from elastic recoil of the lungs and thoracic cage.

In addition to neural mechanisms operating entirely within the brain, reflex signals from the periphery also help control respiration. Located in the walls of the bronchi and bronchioles throughout the lungs are stretch receptors that transmit signals to the dorsal respiratory group. Thus, when the lungs become overly inflated, the stretch receptors activate a feedback response that switches off the inspiratory ramp and stops further inspiration. This reflex is called the *Hering–Breuer inflation reflex*.

The real mechanism for the generation of the respiratory pattern is not known. However, a speculative, qualitative model for a neural network that can control breathing can be built using two neurons, or clumps of neurons, that inhibit each other (von Euler, 1980; Wyman, 1977), as illustrated in Fig. 17.14. (An alternate model is suggested in Exercise 12.) We suppose that there are two neurons with time-dependent outputs (their firing rates) $I_1$ and $I_2$ governed by

$$\tau_1 \frac{dI_1}{dt} + I_1 = F_1, \tag{17.57}$$

$$\tau_2 \frac{dI_2}{dt} + I_2 = F_2, \tag{17.58}$$

where $F_1$ and $F_2$ are related to the firing rates of inhibitory and excitatory inputs. For simplicity we assume that the arrangement is symmetric, so that the time constants of the neuronal output are the same, $\tau_1 = \tau_2 = \tau$. We further assume that the neurons have steady excitatory inputs, $E_1$ and $E_2$, respectively, and that they are cross-inhibited, so that the output from neuron 1 inhibits neuron 2, and vice versa. Thus we take $F_1 = F(E_1 - I_2)$ and $F_2 = F(E_2 - I_1)$. The function $F(x)$ is zero for $x < 0$ (so that the input and output are never negative), and a positive, increasing function of $x$ for $x > 0$. Thus, we have the system of differential equations

$$\tau\frac{dI_1}{dt} + I_1 = F(E_1 - I_2), \tag{17.59}$$

$$\tau\frac{dI_2}{dt} + I_2 = F(E_2 - I_1). \tag{17.60}$$

At this point there is no feedback from the lungs.

Equations (17.59) and 17.60) are easily studied using phase-plane analysis. There are three different possible phase portraits depending on the relative sizes of $E_1$ and $E_2$, two of which are shown in Figs. 17.15 and 17.16. In what follows we assume that $F' > 1$ for all positive arguments, although this restriction can be weakened somewhat. If $E_2$ is much larger than $E_1$, so that $E_1 < F(E_2)$ and $E_2 > F(E_1)$, then, as shown in Fig. 17.15, there is a unique stable fixed point at $I_2 = F(E_1), I_1 = 0$, in which neuron 2 is firing and neuron 1 is quiescent. If $E_1$ is much larger than $E_2$, then the reverse is true, namely, there is a unique stable fixed point at $I_1 = F(E_2), I_2 = 0$, with neuron 1 firing and neuron 2 quiescent. There is an intermediate range of parameter values when $E_1$ and $E_2$ are similar in size, $E_1 < F(E_2)$ and $E_2 < F(E_1)$, shown in Fig. 17.16, for which there are three steady states, the two on the axes, and one in the interior of the positive quadrant. The third (interior) steady state is a saddle point, and is therefore unstable.

This neural network exhibits hysteresis. Suppose we slowly modulate the parameter $E_1$. If it is initially small (compared to $E_2$, which is fixed at some positive level), then neuron 2 fires steadily and inhibits neuron 1. As $E_1$ is increased, this situation remains unchanged, even when $E_1$ and $E_2$ are of similar size, when two stable steady solutions exist. However, when $E_1$ becomes sufficiently large, the steady-state solution at $I_1 = F(E_2), I_2 = 0$ suddenly disappears, and the variables $I_1, I_2$ move to the opposite steady state at $I_2 = F(E_1), I_1 = 0$. Now if $E_1$ is decreased, when $E_1$ is small enough there

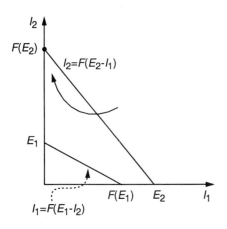

**Figure 17.15** Phase portrait for mutual inhibition network with $E_1 < F(E_2)$ and $E_2 > F(E_1)$.

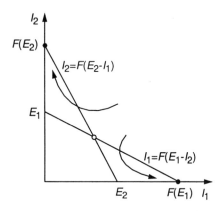

**Figure 17.16** Phase portrait for mutual inhibition network with $E_1 < F(E_2)$ and $E_2 < F(E_1)$.

is a reverse transition back to the steady state at $I_1 = F(E_2), I_2 = 0$, completing the hysteresis loop.

To use this hysteresis to control breathing, we model the diaphragm as a damped mass–spring system driven by $I_1$, the (firing rate) output from neuron 1, the inspiratory neuron:

$$m\frac{d^2x}{dt^2} + \mu\frac{dx}{dt} + kx = I_1. \tag{17.61}$$

We model the effect of the stretch receptors by a function $f(x)$ that is a monotone increasing function of diaphragm displacement $x$. The stretch receptors are assumed to excite only neuron 2, so that the output variables are governed by

$$\tau\frac{dI_1}{dt} + I_1 = F(E_1 - I_2), \tag{17.62}$$

$$\tau\frac{dI_2}{dt} + I_2 = F(E_2 - I_1 + f(x)). \tag{17.63}$$

We could allow stretch receptors to inhibit neuron 1 as well.

With this model, oscillation of the diaphragm is assured if the time constant $\tau$ is sufficiently small. The stretch receptors act to modulate the excitatory inputs, so that as the lung expands, they excite neuron 2. With $E_2 + f(x)$ sufficiently large, neuron 1, the inspiratory neuron, is switched off. With no inspiratory input, the lung relaxes, returning $f(x)$ toward zero and decreasing the excitation to neuron 2. This removes the inhibition to neuron 1 and allows it to fire once again. Thus, if parameters are adjusted properly, the hysteresis loop is exploited, and the inspiration–expiration cycle is established. The oscillations are robust and easily established.

This oscillation can be externally controlled. For example, by increasing $E_2$, the cycle can be stopped after expiration, whereas by increasing $E_1$ the inhibition of the stretch receptors can be overridden and inspiration lengthened (as in, take a deep breath). Decreasing $E_1$ shortens the inspiration time and can stop breathing altogether.

In Fig. 17.17 is shown a plot of the two inhibitory variables $I_1$ and $I_2$ (shown dashed) plotted as functions of time. Parameter values for this simulation were $\tau = 1.0, m =$

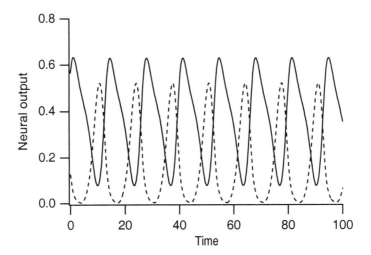

**Figure 17.17** Neural output variables $I_1$ and $I_2$ (dashed) shown as functions of time.

$0.5, \mu = 5.0, k = 1.0, E_1 = 0.5, E_2 = 0.3$. The function $F$ was specified as $F(x) = \frac{2x^2}{0.2+x}$ for positive $x$ and zero otherwise, and the stretch response curve was taken to be $f(x) = x^3/(1 + x^3)$.

## 17.5 Exercises

1. Give a "proper" mathematical derivation of (17.14) by introducing appropriate dimensionless parameters. What dimensionless parameter must be small for this approximation to be valid?
   Answer: $\epsilon = \frac{D_{CO_2}}{k_{-1}[H^+]}$.

2. (a) Develop a model of carbon dioxide and oxygen transport that includes the oxyhemoglobin buffering reaction and the effect of free hydrogen ions on the concentration of bicarbonate. Does the inclusion of proton exchange improve or hinder the rate at which oxygen and carbon dioxide are transported into or out of the blood?

   (b) Estimate the overall effect of this exchange by assuming that the pH of pulmonary venous blood is about 0.04 lower than that of arterial blood.

3. Develop a model of $CO_2$ transport that accounts for its competitive binding with Hb. What is the effect of this binding on total carbon dioxide and oxygen flux?

4. Construct a simple model for the total oxygen and total carbon monoxide in the blood. Assume that the circulatory system is a well-mixed container and that oxygen is removed by metabolism, while oxygen is added and carbon monoxide eliminated during transport through the lungs. Use the models of Section 17.1.5 to determine reasonable transfer rate functions. Estimate the parameters of the model and use numerical computations to determine the half-clearance times for elimination of carbon monoxide at different oxygen levels. How well does your model fit the experimental data shown in Table 17.3?

**Table 17.3** Experimental half-clearance times for elimination of carbon monoxide from the blood (Pace et al., 1950; also see Exercise 4).

| $O_2$ in atm | Half-clearance time (min) |
|---|---|
| 0.21 | 249 |
| 1.0 | 47 |
| 2.5 | 22 |

**Table 17.4** Alveolar gas concentration and oxygen saturation at different altitudes. The last column shows the alveolar $P_{O_2}$ when breathing pure oxygen at atmospheric pressure. At this pressure, $O_2$ saturation is 100%. (Guyton and Hall, 1996, Table 43-1, p. 550.)

| Altitude (ft) | Baròmetric Pressure (mm Hg) | $P_{O_2}$ in air (mm Hg) | Alveolar $P_{O_2}$ (in air) (mm Hg) | $O_2$ Saturation (in air) (%) | Alveolar $P_{O_2}$ (in oxygen) (mm Hg) |
|---|---|---|---|---|---|
| 0 | 760 | 159 | 104 | 97 | 673 |
| 10,000 | 523 | 110 | 67 | 90 | 436 |
| 20,000 | 349 | 73 | 40 | 73 | 262 |
| 30,000 | 226 | 47 | 18 | 24 | 139 |

5. Suppose the respiratory exchange rate is fixed. Show that there is a linear relationship between the alveolar carbon dioxide and oxygen partial pressures.

6. (a) Assume that regulatory mechanisms maintain the arterial oxygen partial pressure at 40 mm Hg and the ventilation–perfusion ratio at 1. Find the alveolar $P_{O_2}$ and the oxygen saturation, leaving the alveolus as a function of atmospheric $P_{O_2}$.

   (b) Data are shown in Table 17.4 for breathing normal air or breathing pure oxygen. What assumption from part 6a is apparently wrong? From the data, determine the arterial oxygen partial pressure.

7. Devise a different model in which metabolism and ventilation are held fixed. How do the alveolar $P_{O_2}$ and $O_2$ saturation vary as a function of atmospheric pressure?

8. Using data from Table 17.4, estimate the altitude at which incoming alveolar blood has zero $P_{O_2}$, at normal metabolism.

9. Determine the red blood cell count (concentration of hemoglobin) that is necessary to maintain constant venous oxygen partial pressure as a function of altitude at fixed metabolism.

10. Find the rate of carbon monoxide clearance as a function of external $P_{O_2}$, with fixed metabolism and ventilation.

11. (a) Determine the structure of stable steady solutions of equations (17.59–17.60) in the $(E_1, E_2)$ parameter plane using $F(x) = \frac{2x^2}{0.2+x}$ for positive $x$ and zero otherwise.

    (b) Numerically simulate the system of equations (17.61–17.63) using the parameters in the text. Plot the function $E_2 + f(x)$ as a function of time in the above parameter plane to see how hysteresis is exploited by this system.

12. Consider the following as a possible model for the respiratory center. Two neural FitzHugh–Nagumo oscillators have inhibitory synaptic inputs, so that

$$\frac{dv_i}{dt} = f(v_i, w_i) - s_i g_s (v_i - v_\theta), \tag{17.64}$$

$$\tau_v \frac{dw_i}{dt} = w_\infty(v) - w_i, \tag{17.65}$$

for $i = 1, 2$. The synaptic input $s_i$ is some neurotransmitter that is released when the opposite neuron fires:

$$\frac{ds_i}{dt} = \alpha_s(1 - s_i)x_j F(v_j) - \beta_s s_i, \quad j \neq i, \tag{17.66}$$

and the amplitude of the release $x_j$ decreases gradually when the neuron is firing, via

$$\frac{dx_i}{dt} = \alpha_x(1 - x_i) - \beta_x F(v_i)x_i. \tag{17.67}$$

(a) Simulate this neural network with $f(v, w) = 1.35v(1 - \frac{1}{3}v^2) - w$, $w_\infty(v) = \tanh(5v)$, $F(v) = \frac{1}{2}(1 + \tanh 10v)$, and with parameters $\tau_v = 5$, $v_\theta = -2$, $\alpha_s = 0.025$, $\beta_s = 0.002$, $\alpha_x = 0.001$, $\beta_x = 0.01$, $g_s = 0.19$.

(b) Give an approximate analysis of the fast and slow phase portraits for these equations to explain how the network works.

(c) How does this bursting oscillator compare with those discussed in Chapter 6?

(d) What features of this model make it a good model for the control of the respiratory system and what features are not so good?

# CHAPTER 18

# Muscle

Muscle cells resemble nerve cells in their ability to conduct action potentials along their membrane surfaces. In addition, however, muscle cells have the ability to translate the electrical signal into a mechanical contraction, which enables the muscle cell to perform work. There are three types of muscle cells, namely skeletal muscle, which moves the bones of the skeleton at the joints; cardiac muscle, whose contraction enables the heart to pump blood; and smooth muscle, which is located in the walls of blood vessels and contractile visceral organs. Skeletal and cardiac muscle cells have a banded appearance under a microscope, with alternating light and dark bands, and thus they are called *striated muscle*. They have similar (though not identical) contractile mechanisms. Smooth muscle, on the other hand, is not striated, and its physiology is considerably different from the other two types of muscle. Because of the tremendous diversity of smooth muscle physiology, in this chapter we discuss only the contractile mechanisms of striated muscle.

Single skeletal muscle cells are elongated cylindrical cells with several nuclei. Each cell contains numerous cylindrical structures, called *myofibrils*, surrounded by the membranous channels of the sarcoplasmic reticulum (Fig. 18.1). Myofibrils are the functional units of skeletal muscle, containing protein filaments that make up the contractile unit. Each myofibril is segmented into numerous individual contractile units called *sarcomeres*, each about 2.5 $\mu$m long. The sarcomere, illustrated schematically in Fig. 18.2, is made up primarily of two types of parallel filaments, designated as thin and thick filaments. Viewed end on, six thin filaments are positioned around each central thick filament in a hexagonal arrangement. Viewed along its length, there are regions where thin or thick filaments are overlapping or nonoverlapping. At the end of the sarcomere is a region, called the *Z-line*, where the line filaments are anchored. Thin filaments extend from the Z-lines at each end toward the center, where they overlap

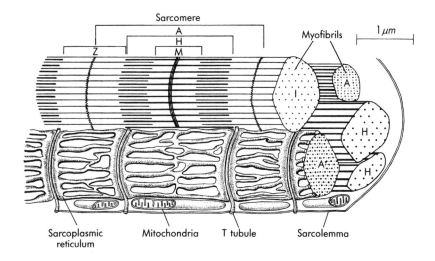

**Figure 18.1** Schematic diagram of a skeletal muscle cell. (Berne and Levy, 1993, p. 283, Fig. 17-2.)

with thick filaments. The regions where there is no overlap, containing only thin filaments, are called *I-bands*, and the regions containing myosin (thick) filaments (with some overlap with thin filaments) are called *A-bands*. The central region of the sarcomere, containing only thick filaments, is called the *H-zone*. During contraction, both the H-zone and the I-bands shorten as the overlap between thin and thick filaments increases.

Muscle contraction is initiated by an action potential transmitted across a synapse from a neuron. This action potential spreads rapidly across the muscle membrane, spreading into the interior of the cell along invaginations of the cell membrane called *T-tubules*. T-tubules form a network in the cell interior, near the junction of the A- and I-bands, and increase the surface area over which the action potential can spread. They enable the action potential to reach quickly into the cell interior. Voltage-gated $Ca^{2+}$ channels are opened by the action potential, and $Ca^{2+}$ enters the cell, initiating the release of further $Ca^{2+}$ from the sarcoplasmic reticulum (Chapter 5). The resulting high intracellular $Ca^{2+}$ concentration causes a change in the myofilament structure that allows the thick filaments to bind and pull on the thin filaments, resulting in muscle contraction.

Excellent reviews of muscle physiology, and the development of models for muscle are given by White and Thorson (1975) and Huxley (1980).

## 18.1 Crossbridge Theory

Thick filaments contain the protein myosin, which is made up of a polypeptide chain with a globular head. These heads constitute the *crossbridges* that interact with the thin

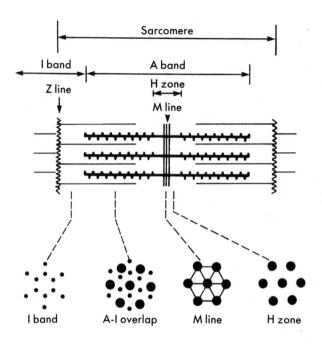

**Figure 18.2** Longitudinal section (top panel) and cross-section (lower panels) of a sarcomere showing its organization into bands. (Berne and Levy, 1993, p. 283, Fig. 17-3.)

filaments to form bonds that act in ratchet-like fashion to pull on the thin filaments. In addition, the myosin heads have the ability to dephosphorylate ATP as an energy source.

Thin filaments contain the three proteins actin, tropomyosin, and troponin. Each actin monomer is approximately spherical, with a radius of about 5.5 nm, and they aggregate into a double-stranded helix, with a complete twist about every 14 monomers. Because the coil is double-stranded, this structure repeats every 7 monomers, or about every 38 nm. Tropomyosin, a rod-shaped protein, forms the backbone of the double-stranded coil. The troponin consists of a number of smaller polypeptides, which include a binding site for calcium as well as a portion that blocks the crossbridge binding sites on the actin helix. When calcium is bound, the confirmation of the troponin–tropomyosin complex is altered just enough to expose the crossbridge binding sites. In Fig. 18.3 we show a scale drawing of the probable way in which the actin, tropomyosin, and myosin proteins fit together.

Contraction takes place when the crossbridges bind and generate a force causing the thin filaments to slide along the thick filaments. A schematic diagram of the cross-bridge reaction cycle is given in Fig. 18.4, with the accompanying physical arrangement shown in Fig. 18.5. Before binding and contraction, ATP is bound to the crossbridge heads of the myosin (M), and the concentration of calcium is low. When the calcium concentration increases, calcium ions bind to the troponin–tropomyosin complex, ex-

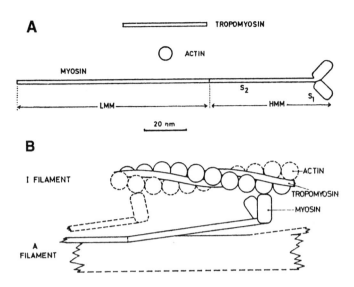

**Figure 18.3** A: Scale drawing of actin, myosin, and tropomyosin proteins. B: Scale drawing of the thick and thin filaments (labeled the A and I filaments here), showing the probable way in which the actin, myosin, and tropomyosin proteins fit together. Troponin, which is bound to tropomyosin, is not included in the diagram. (White and Thorson, 1975, Fig. 9, parts A and B (i).)

posing the crossbridge binding sites on the actin filament (A). Where possible, a weak bond between actin and myosin is formed. Release of the phosphate changes the weak bond to a strong bond and changes the preferred configuration of the crossbridge from nearly perpendicular to a bent (foreshortened) position. While the crossbridge is in anything but this energetically preferred, bent state, there is an applied force that acts to pull the thin filament along the thick filament. The movement of the crossbridge to its newly preferred configuration is called the *power stroke*. Almost immediately upon reaching the preferred bent configuration, the crossbridge releases its ADP and binds another ATP molecule, causing dissociation from the actin binding site and return to its initial perpendicular and unbound position. ATP is then dephosphorylated, yielding ADP, phosphate, and the stored mechanical energy for the next cycle. Thus, during muscle contraction, each crossbridge cycles through sequential binding and unbinding to the actin filament.

As we will see in the following sections, to construct quantitative models of crossbridge binding it is necessary to know how many actin binding sites are available to a single crossbridge. One possibility is that the crossbridge must be precisely oriented to the actin binding site, and thus, in each turn of the helix, only one binding site is available to each crossbridge. In other words, from the point of view of the crossbridge, the binding sites have an effective separation of about 38 nm. Because of the physical constraints on each crossbridge, this means that at any time, there is only a single

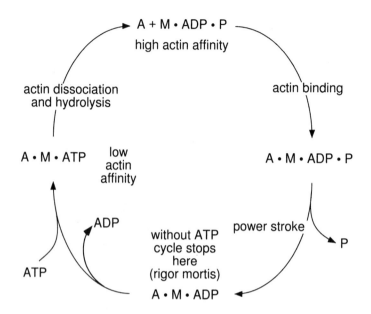

**Figure 18.4**  Major reaction steps in the crossbridge cycle. M denotes myosin, and A denotes actin.

binding site available to each crossbridge. This is the assumption behind the Huxley model, which we consider in detail below.

However, from the distribution of actin binding sites and crossbridges shown in Fig. 18.3, it is plausible that this assumption is not correct. Perhaps, depending on the flexibility of the actin filament, each crossbridge has a number of potential binding sites. In our discussion we concentrate on models for the two extreme cases: first, where each crossbridge has only a single available binding site, and second, where each crossbridge has a continuous array of available binding sites. Intermediate models, in which the crossbridge has a small number of discrete binding sites available, are considerably more complex and are mentioned only briefly.

Because of the sarcomere structure, the tension a muscle develops depends on the muscle length. In Fig. 18.6 we show a curve of isometric tension as a function of sarcomere length. By isometric tension, we mean the tension developed by a muscle when it is held at a fixed length and repeatedly stimulated (i.e., with a high-frequency periodic stimulus). Under these conditions the muscle goes into *tetanus*, a state, caused by saturating concentrations of calcium in the sarcoplasm, in which the muscle is continually attempting to contract. Note that the muscle cannot actually contract, because it is held at constant length, although it must go through the chemistry cycle of the power stroke, since the development of tension requires that energy be consumed.

At short lengths, overlap of the thin filaments causes a drop in tension, but as this overlap decreases (as the length increases) the tension rises. However, when the length is large, there is less overlap between the thick and thin filaments, so fewer crossbridges

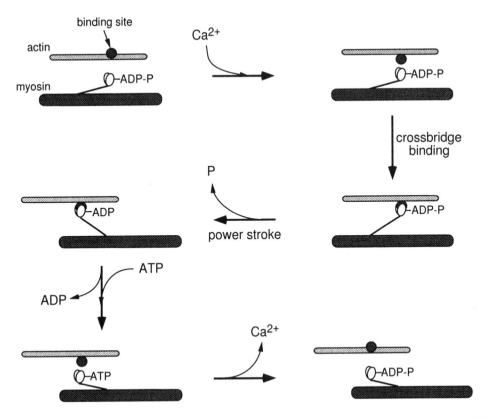

**Figure 18.5**  Position of crossbridge components during the major steps in the crossbridge cycle.

bind, and less tension develops. When there is no overlap between the thick and thin filaments, the muscle is unable to develop any tension.

Skeletal muscle tends to operate at lengths that correspond to the peak of the isometric length–tension curve, and thus in many experimental setups the tension the muscle develops does not depend significantly on the muscle length. However, the same is not true for cardiac muscle, which considerably complicates theoretical studies of this muscle type. For these reasons we restrict our attention to models based on data from skeletal muscle. Peskin (1975) presents a detailed description of some theoretical models of cardiac muscle.

## 18.2   The Force–Velocity Relationship: The Hill Model

One of the earliest models for a muscle is due to A.V. Hill (1938) and was constructed before the details of the sarcomere anatomy were known. Hill observed that when a muscle contracts against a constant load (an *isotonic* contraction), the relationship

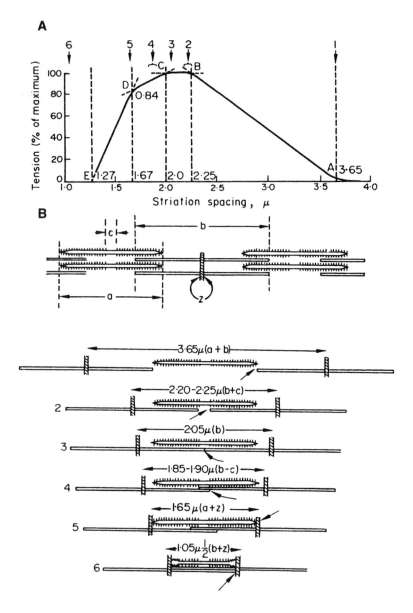

**Figure 18.6** A: Isometric tension as a function of the length of the sarcomere. B: schematic diagrams of the arrangement of the thick and thin filaments for the six different places indicated in panel A. (Gordon et al., 1966, reproduced in White and Thorson, 1975, Fig. 14.)

between the constant rate of shortening $v$ and the load $p$ is well described by the *force–velocity* equation

$$(p + a)v = b(p_0 - p), \qquad (18.1)$$

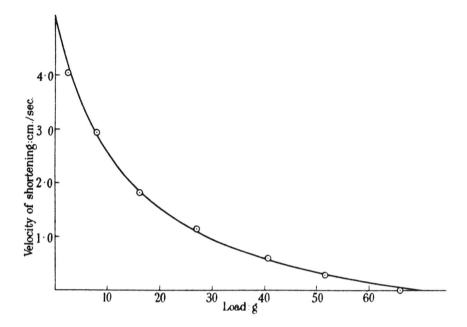

**Figure 18.7** The relationship between the load on a muscle and the velocity of contraction (Hill, 1938; Fig. 12). The symbols are the data points, while the smooth curve is calculated from (18.1) using the parameter values $a = 357$ grams (of weight) per square centimeter of muscle fiber (g-wt/cm$^2$), $a/p_0 = 0.22$, $b = 0.27$ muscle lengths per second.

where $a$ and $b$ are constants that are determined by fitting to experimental data in a way that we discuss presently. A typical force–velocity curve is plotted in Fig. 18.7. When $v = 0$, then $p = p_0$, and thus $p_0$ represents the force generated by the muscle when the length is held fixed; i.e., $p_0$ is the *isometric* force. As we discussed above, the tension generated by a skeletal muscle in isometric tetanus is approximately independent of length, and thus $p_0$ is approximately independent of length also. When $p = 0, v = bp_0/a$, which is the maximum speed at which a muscle is able to shorten.

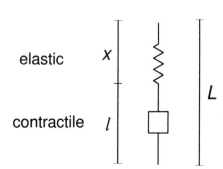

elastic

contractile

**Figure 18.8** Schematic diagram of Hill's two-element model for skeletal muscle. The muscle is assumed to consist of an elastic element in series with a contractile element with a given force–velocity relationship.

In an attempt to explain these observations, we model a muscle fiber as a contractile element with the given force–velocity relationship, in series with an elastic element (Fig. 18.8). In some versions of the model a parallel elastic element is included (see Exercise 1), but as it plays no essential role in the following discussion, it is omitted here. As shown in Fig. 18.8, we let $l$ denote the length of the contractile element, we let $x$ denote the length of the elastic element, and we let $L = l + x$ denote the total length of the fiber. Then, letting $v$ denote the velocity of contraction of the contractile element, we have

$$v = -\frac{dl}{dt},$$ (18.2)

where, by assumption, $v$ is related to the load on the muscle by the force–velocity equation (18.1). To derive a differential equation for the time dependence of $p$, we note that because the elastic element is in series with the contractile element, the two experience the same force. We assume that the load on the elastic element is a function of its length $p = P(x)$ and then use the chain rule and the force–velocity equation to obtain

$$\frac{dp}{dt} = \frac{dP}{dx}\frac{dx}{dt}$$

$$= \frac{dP}{dx}\left[\frac{dL}{dt} - \frac{dl}{dt}\right]$$

$$= \frac{dP}{dx}\left[\frac{dL}{dt} + v\right]$$

$$= \frac{dP}{dx}\left[\frac{dL}{dt} + \frac{b(p_0 - p)}{p + a}\right].$$ (18.3)

It remains to determine $dP/dx$.

Hill made the simplest possible assumption, that the elastic element is linear, and thus

$$P = \alpha(x - x_0),$$ (18.4)

where $x_0$ is its resting length. Thus, $dP/dx = \alpha$, and the differential equation for $p$ is

$$\frac{dp}{dt} = \alpha\left[\frac{dL}{dt} + \frac{b(p_0 - p)}{p + a}\right].$$ (18.5)

## 18.2.1  Fitting Data

Suppose a muscle in tetanus is held at a fixed tension until it reaches its isometric length, and then the tension is suddenly decreased and held fixed at a lower value. A typical result is shown in Fig. 18.9A, where we plot the muscle length against time. As soon as the tension is reduced, the muscle length decreases (plotted in the vertical direction) as the elastic element contracts. After a transition period during which the length exhibits small oscillations, the muscle decreases in length at a constant rate. Plotting the rate of decrease against the constant applied tension gives one point on the force–velocity curve. More specifically, if the tension is jumped from $p_0$ to $p_1$, the muscle contracts at the constant rate $v$, where

$$(p_1 + a)v = b(p_0 - p_1). \tag{18.6}$$

Repeating the experiment for tension jumps of different magnitudes (shown in Fig. 18.9B) one finds a series of points on the force–velocity curve, through which one can fit the force–velocity equation to obtain values for $a$, $b$, and $p_0$. Note that this procedure is valid only if $p_0$ does not change during the course of the experiment. In other words, as the muscle shortens with constant velocity, it must be that $p_0$ remains unchanged. As we discussed above, this is an acceptable assumption for skeletal muscle operating near the peak of the length–tension curve.

Similarly, the characteristics of the elastic element can be determined from the initial jump in length. If one extrapolates the line of constant speed back to the time of the tension jump (line $xyz$ in Fig. 18.9A), then the distance $0z$ is the change in length of the elastic element. This relies on the assumption that the force–velocity properties of the muscle change instantaneously with the change in tension.

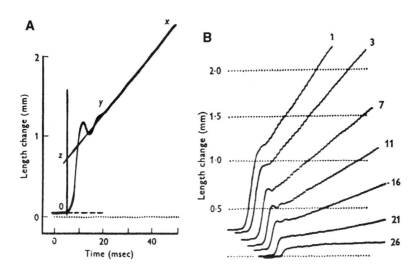

**Figure 18.9** A: Plot of length change against time after a step decrease in tension. The length decreases in a sudden jump, and then, after an initial oscillatory phase, decreases at constant velocity. (Jewell and Wilkie, 1958; reproduced in White and Thorson, 1975, Fig. 5.) B: Length change as a function of time from a series of tension step experiments. The baseline of each trace has been shifted for clarity, and each dot on the horizontal axis denotes 1 ms. For each step, the value to the right of the curve denotes the final value of the tension in grams of weight (g-wt). (Jewell and Wilkie, 1958; reproduced in White and Thorson, 1975, Fig. 6.)

## 18.2.2   Some Solutions of the Hill Model

### Isometric tetanus solution

If a muscle at rest is put into tetanus by repeated stimulation, the isometric tension builds up over a period of time. Because the tension is measured isometrically, the length of the muscle does not change, and thus $dL/dt = 0$. Hence, the differential equation for the tension is

$$\frac{dp}{dt} = \alpha \left[ \frac{b(p_0 - p)}{p + a} \right]. \tag{18.7}$$

This is a separable equation, so after separation, we integrate from 0 to $t$ and use the initial condition $p(0) = 0$ to obtain

$$-p - (p_0 + a)\log\left(\frac{p_0 - p}{p_0}\right) = \alpha bt, \tag{18.8}$$

which describes the time course of the change in tension. As $t \to \infty$, $p \to p_0$, as expected.

### Release at constant velocity

Suppose a muscle, held originally at its isometric tension $p_0$, is allowed to contract with constant velocity $u$. It seems reasonable that the muscle tension should decrease until it reaches the value $p_u$ determined from the force–velocity curve for a velocity $u$. The differential equation for $p$ is

$$\frac{dp}{dt} = \alpha \left[ -u + \frac{b(p_0 - p)}{p + a} \right], \tag{18.9}$$

with initial condition $p(0) = p_0$. As before, we assume that $p_0$ does not change during the course of the contraction. The solution is

$$p_0 - p + (p_u + a)\log\left(\frac{p_0 - p_u}{p - p_u}\right) = \alpha t(b + u), \tag{18.10}$$

where $p_u$ is defined by $(p_u + a)u = b(p_0 - p_u)$. As $t \to \infty$, $p \to p_u$, and thus our intuitive reasoning is confirmed.

### Response to a jump in length

Possibly the most interesting solution is the response to a step decrease in length, as this solution has been used to show that the Hill model does not provide an accurate description of all aspects of muscle behavior (Jewell and Wilkie, 1958).

First, Jewell and Wilkie determined the parameters of the Hill model by the series of experiments described above (Fig. 18.9). They then used the Hill model to predict the response of the muscle to a step decrease in muscle length. Suppose that a muscle,

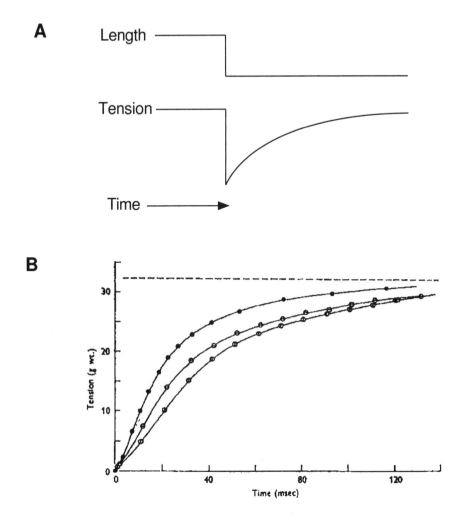

**Figure 18.10** A: Schematic diagram of the response to a step decrease in length. B: Comparison of the Hill model to the data of Jewell and Wilkie (1958). The closed circles are computed from the Hill model, while the open circles are data points from two slightly different experimental procedures. (Jewel and Wilkie, 1958; reproduced in White and Thorson, 1975, Fig. 7.)

originally held at its isometric tension $p_0$, is suddenly decreased in length. One expects the muscle tension to suddenly decrease, but then slowly increase back to $p_0$. This is because the isometric tension is independent of length but should take some time to develop at the new length. A typical solution is sketched schematically in Fig. 18.10A.

More precisely, suppose that the length of the muscle as a function of time is given by

$$L(t) = L_1 + L_0 - L_0 H(t), \tag{18.11}$$

where $H(t)$ is the usual Heaviside function, and where $L_1$ and $L_0$ are constants, $L_0$ being the magnitude of the length step. Thus

$$\frac{dL}{dt} = -L_0\delta(t), \tag{18.12}$$

where $\delta(t)$ denotes the Dirac delta function. Substituting this expression into the differential equation for $p$ gives

$$\frac{dp}{dt} = \alpha\left[-L_0\delta(t) + \frac{b(p_0 - p)}{p + a}\right], \tag{18.13}$$

$$p(0^-) = p_0. \tag{18.14}$$

If we integrate (18.13) (formally) from $t = -\epsilon$ to $t = \epsilon$ and then let $\epsilon \to 0$, we get

$$p(0^+) - p(0^-) = -\alpha L_0, \tag{18.15}$$

and thus the delta function causes a jump of $-\alpha L_0$ in $p$ at the origin. Hence, (18.13) and (18.14) can be written as the initial value problem

$$\frac{dp}{dt} = \alpha\left[\frac{b(p_0 - p)}{p + a}\right], \qquad t > 0, \tag{18.16}$$

$$p(0) = p_0 - \alpha L_0. \tag{18.17}$$

Since we have reduced the problem to the isometric tetanus problem studied earlier (although with a different initial condition), the solution is easily calculated.

When the solution is predicted from the Hill model in this way, it does not agree with experimental observations on the tension recovery following a step decrease in length. In fact, the tension recovers less quickly than is predicted by the model, as illustrated in Fig. 18.10B. Here, the model computations (shown as closed circles), consistently lie above the data points (shown as open circles). These observations, made possible by the improvements in experimental technique in the 20 years after Hill's model was first proposed, forced the conclusion that the Hill model has serious defects. In particular, the assumption that the force–velocity relationship (18.1) is satisfied immediately after a change in tension is a probable major source of error. At the same time that Hill's model was shown to have problems, much more was being discovered about the structure of the sarcomere. This motivated the construction of a completely different type of model, based on the kinetics of the crossbridges rather than on heuristic elastic and contractile elements. The first model of this new type was due to Huxley (1957), and has been the basis for the majority of subsequent models of muscle behavior.

## 18.3   A Simple Crossbridge Model: The Huxley Model

To formulate a mathematical model describing crossbridge interactions in a sarcomere, we suppose that a crossbridge can bind to an actin binding site at position $x$, where $x$ measures the distance along the thin filament to a binding site from the crossbridge, and $x = 0$ corresponds to the position in which the bound crossbridge exerts no force during the power stroke on the thin filament (Fig. 18.11). Crossbridges can be bound to a binding site with $x > 0$, in which case they exert a contractile force, or they can be bound to a site with $x < 0$, in which case they exert a force that opposes contraction. A crossbridge bound to a binding site at $x$ is said to have displacement $x$. In his original model Huxley assumed that the actin binding sites were sufficiently far apart that each crossbridge could be associated with a unique binding site. If we make this assumption, each crossbridge, whether bound or not, can be associated with a unique value of $x$. Let $\rho$ denote the number of crossbridges (either bound or unbound) with displacement $x$. For simplicity we assume that $\rho$ is independent of $x$ and $t$. Hence the distribution of bound crossbridges changes with time, but the distribution of all crossbridges does not change. We then define $n(x,t)$ to be the fraction of crossbridges with displacement $x$ that are bound. Note that it is not correct to let $\rho$ be a constant independent of $x$, as this implies that there are crossbridges with unrealistically large displacements. More accurately, one should assume that there is some constant $x_0$ such that $\rho(x)$ is a constant on the interval $-x_0 < x < x_0$ and is zero everywhere else. This eliminates any crossbridges with displacements that are unphysiologically large. However, as we will see, one can achieve a similar effect by choosing the model functions appropriately.

Next, we drastically simplify the reaction mechanism, and assume that a crossbridge can be in one of two states, namely either unbound (U), or strongly bound (B) and thereby generating a force. We suppose further that the binding and unbinding of crossbridges is described by the simple reaction scheme

$$U \underset{g(x)}{\overset{f(x)}{\rightleftarrows}} B,$$

where the rate constants are functions of the displacement $x$.

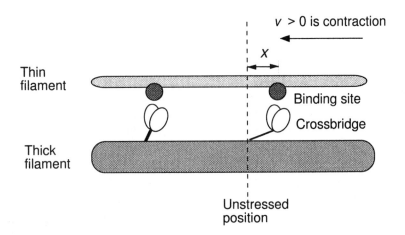

**Figure 18.11** Schematic diagram of the Huxley crossbridge model.

The conservation law for the fraction of bound crossbridges is

$$\frac{\partial n}{\partial t} - v(t)\frac{\partial n}{\partial x} = (1 - n)f(x) - ng(x), \tag{18.18}$$

where $v(t)$ is the velocity of the actin filament relative to the myosin filament. For notational consistency, we assume that $v > 0$ denotes muscle contraction.

Every time a crossbridge is bound, one ATP molecule is dephosphorylated, so the rate of energy release, $\phi$, for this process is given by

$$\phi = \rho\epsilon \int_{-\infty}^{\infty} (1 - n(x,t))f(x)dx, \tag{18.19}$$

where $\epsilon$ is the chemical energy released by one crossbridge cycle. Since $n$ is, in general, a function of the contraction velocity, $\phi$ is also. We also suppose that a bound crossbridge is like a spring, generating a restoring force $r(x)$ related to its displacement. Hence, the total force exerted by the muscle is

$$p = \rho \int_{-\infty}^{\infty} r(x)n(x,t)\,dx. \tag{18.20}$$

To find the force–velocity relationship for muscle, we assume that the fiber moves with constant velocity, so that $\partial n/\partial t = 0$. Then, the steady distribution $n(x)$ is the solution of the first-order differential equation

$$-v\frac{dn}{dx} = (1 - n)f(x) - ng(x). \tag{18.21}$$

The solution of this differential equation is easily understood. The function $n(x)$ "tracks" the quasi-steady-state solution $\frac{f(x)}{f(x)+g(x)}$ at a rate that is inversely proportional to $v$. Thus, if $v$ is small, $n(x)$ is well approximated by the quasi-steady-state solution, whereas if $v$ is large, $n(x)$ changes slowly as a function of $x$. From this we make two observations. First, the force is largest at small velocities. In fact, at zero velocity, the isometric force is

$$p_0 = \rho \int_{-\infty}^{\infty} r(x)\frac{f(x)}{f(x) + g(x)}dx. \tag{18.22}$$

Second, at large velocities, the distribution $n(x)$ has small amplitude, and so the force is small. The force decreases because the amount of time during which a crossbridge is close to a binding site is small, and so binding is less likely, with the result that a smaller fraction of crossbridges exerts a contractile force. Another factor is that at higher velocities a greater number of crossbridges are carried into the $x < 0$ region before they can dissociate, hence generating a force opposing contraction. It is intuitively reasonable that at some maximum velocity, the force generated by the crossbridges with $x < 0$ exactly balances the force generated by those with $x > 0$, at which point no tension is generated by the muscle, and the maximum velocity of shortening is attained. We have already seen that this occurs in the Hill force–velocity curve. Crossbridge theory provides an elegant explanation of this phenomenon.

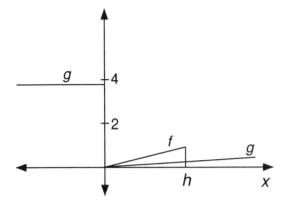

**Figure 18.12** The attachment and detachment functions, $f$ and $g$, in the Huxley model

To obtain quantitative formulas, one must make some reasonable guesses for the functions $f(x)$ and $g(x)$, and then calculate $n(x)$ and $p$ numerically or analytically. Although numerical solutions can always be obtained, there are several choices of $f(x)$ and $g(x)$ for which analytical solutions are possible. The functions that Huxley chose are illustrated in Fig. 18.12 and have the form

$$f(x) = \begin{cases} 0, & x < 0, \\ f_1 x/h, & 0 < x < h, \\ 0, & x > h, \end{cases} \tag{18.23}$$

$$g(x) = \begin{cases} g_2, & x < 0, \\ g_1 x/h, & x > 0. \end{cases} \tag{18.24}$$

In this model, the rate of crossbridge dissociation, $g$, is low when the crossbridge exerts a contractile force, but when $x$ is negative, the crossbridge opposes contraction, and $g$ increases. Similarly, crossbridges do not attach at a negative $x$ ($f = 0$ when $x < 0$), and as $x$ increases, the rate of crossbridge attachment increases as well. This ensures that crossbridge attachment contributes an overall contractile force. At some value $h$, the rate of crossbridge attachment falls to zero, as it is assumed that crossbridges cannot bind to a binding site that is too far away.

The corresponding solution for $n(x)$ is

$$n(x) = \begin{cases} \dfrac{f_1}{f_1 + g_1} \left[ 1 - \exp(-\phi/v) \right] \exp(xg_2/v), & x < 0, \\[3mm] \dfrac{f_1}{f_1 + g_1} \left\{ 1 - \exp\left[ \left( \dfrac{x^2}{h^2} - 1 \right) \dfrac{\phi}{v} \right] \right\}, & 0 < x < h, \\[3mm] 0, & x > h, \end{cases} \tag{18.25}$$

where $\phi = (f_1 + g_1)h/2$. This steady solution is plotted in Fig. 18.13 for four values of $v$. Note the unphysiological implication of this solution, that $n > 0$ for all $x < 0$. However, only a negligible number of crossbridges are bound at unphysiological displacements, so that these may be ignored in the model.

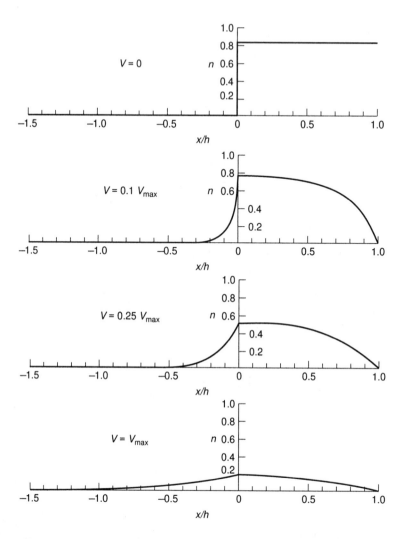

**Figure 18.13** Steady-state distributions of $n$ in the Huxley model, for different values of $v$. (Huxley, 1957, Fig. 7.) In this figure and the next, the parameter values were chosen by trial and error by Huxley so as to obtain a good fit with experimental data. The values are $g_1/(g_1 + f_1) = 3/16$, $g_2/(f_1 + g_1) = 3.919$, $f_1 + g_1 = 1/2$. Since $x$ and $v$ can be scaled by $h$, without loss of generality we may take $h = 1$.

Assuming that the crossbridge acts like a linear spring, so that $r(x) = kx$ for some constant $k$, the force generated by the muscle (defined by (18.20)) can be calculated as a function of the velocity of contraction, and the result compared to the Hill force–velocity equation (18.1). The force–velocity equation calculated from the Huxley model is

$$p = \frac{\rho k f_1}{f_1 + g_1} \frac{h^2}{2} \left\{ 1 - \frac{v}{\phi}(1 - e^{-\phi/v}) \left( 1 + \frac{1}{2} \left( \frac{f_1 + g_1}{g_2} \right)^2 \frac{v}{\phi} \right) \right\}, \qquad (18.26)$$

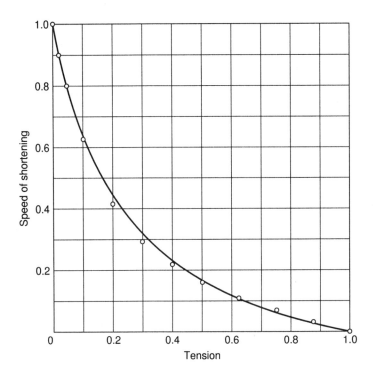

**Figure 18.14** The force–velocity curve of the Huxley model (solid curve) compared to Hill's data (open circles). (Huxley, 1957, Fig. 8.) For simplicity, $p$ has been scaled so that $p(0) = 1$. This determines the value used for $\rho k$. Further, the parameters have been scaled so that $v_{max} = 1$.

which for appropriate choice of parameters gives an excellent fit to the force–velocity curve, as illustrated in Fig. 18.14. Huxley chose the model parameters by a process of trial and error so that the rate of energy production also agreed with experimental data (see the caption for Fig. 18.13). We discuss these experimental data in more detail below when we consider a slightly different crossbridge model. We will also show how, in a slightly simpler model, the process of parameter estimation can be made more systematic.

## 18.3.1 Isotonic Responses

Thus far we have shown how the Huxley model can explain the Hill force–velocity curve using crossbridge dynamics. However, for the model to give an acceptable explanation of muscle dynamics, there is a great deal of additional experimental data with which it should agree. In particular, the model should explain the response of a muscle, first, to a step change in tension (isotonic response) and second, to a step change in length (isometric response). After all, the Hill model was rejected as a satisfactory explanation because of its inability to explain all such data.

It is instructive to consider how one calculates the response of the Huxley model to a step change in tension, as the procedure is not obvious. Suppose a muscle exerts its isometric tension, $p_0$, at some length $L$. Then, the steady-state crossbridge distribution is

$$n_s(x) = \frac{f(x)}{f(x) + g(x)}. \tag{18.27}$$

Now suppose the tension on the muscle is reduced to $p_1 < p_0$ so suddenly that no crossbridges are able to associate or dissociate during the reduction. In a typical experiment of Civan and Podolsky (1966), a muscle fiber of length 15,000 $\mu$m was subjected to a change in tension that changed the fiber length by less than 50 $\mu$m, a relative length change of 1/300. Hence, a typical sarcomere of length 2.5 $\mu$m changed in length by less than 10 nm, and so the length of each crossbridge was changed by less than 10 nm. A crossbridge is able to absorb such length changes without dissociating from the binding site. The extension of each crossbridge decreases by an unknown amount $\Delta L$, and so the crossbridge distribution suddenly changes to $n_s(x + \Delta L)$ and is no longer at steady state. The change in length is found by constraining the new tension to be $p_1$, and hence $\Delta L$ satisfies

$$p_1 = \int_{-\infty}^{\infty} k(x) n_s(x + \Delta L) \, dx. \tag{18.28}$$

Although (18.28) cannot in general be solved analytically, $\Delta L$ can be determined numerically, since it is easy to determine $p_1$ as a function of $\Delta L$.

Following the sudden change in tension, the crossbridge population is not at steady state, so it must change according to the differential equation (18.18) with initial condition $n(x, 0) = n_s(x + \Delta L)$ and subject to the constraint that the tension is constant at $p = p_1$. However, during this evolution, $v$ is not constant, as there is some transient behavior before the muscle reaches its steady contraction velocity (cf. Fig. 18.9). However, we can determine an expression for $v(t)$ in terms of $n(x, t)$ that guarantees that the tension remains constant at $p_1$.

Since $p_1$ is constant, it must be that $\frac{\partial p_1}{\partial t} = 0$, or

$$0 = \int_{-\infty}^{\infty} k(x) n_t(x, t) dx = \int_{-\infty}^{\infty} k(x) \left( v(t) \frac{\partial n}{\partial x} + (1 - n) f(x) - n g(x) \right) dx. \tag{18.29}$$

We solve this for $v(t)$ to get

$$-v(t) = \frac{\int_{-\infty}^{\infty} k(x) \left( (1 - n) f(x) - n g(x) \right) dx}{\int_{-\infty}^{\infty} k(x) \frac{\partial n}{\partial x} dx}. \tag{18.30}$$

Thus, for the tension to remain constant, the partial differential equation (18.18) must have the contraction velocity specified by (18.30).

Using a slightly different approach, Podolsky et al. (1969; Civan and Podolsky, 1966) showed that the Huxley model does not agree with experimental data in its response to a step change in tension. We saw in Fig. 18.9 that immediately after the tension reduction the muscle length changes also, and after an initial oscillatory period, the muscle contracts with a constant velocity. However, the Huxley model does not show any oscillatory behavior, the approach to constant velocity being monotonic.

Motivated by this discrepancy, Podolsky and Nolan (1972, 1973) and Podolsky et al. (1969) altered the form of the functions $f$ and $g$ to obtain the required oscillatory responses in the Huxley model. Of course, in Huxley's original model no physiological justification was given for the functions $f$ and $g$, and modification of these functions is therefore an obvious place to start when fiddling with the model to fit the data. Julian (1969) also showed that the Huxley model can be adjusted to give the correct responses to a step change in length. The details of these analyses do not concern us greatly; the main point is that Huxley's crossbridge model has enough flexibility to explain a wide array of experimental data.

## 18.3.2 Other Choices for Rate Functions

The simple choices for rate functions made by Huxley yield interesting analytical results. However, there are numerous other ways that the rate functions might be chosen. Suppose, for simplicity, that the rate functions $f(x)$ and $g(x)$ have nonoverlapping compact support. Where it is nonzero, we take $f(x)$ to be a constant, $f(x) = \alpha/\epsilon$, on a small interval near the maximum displacement $h$, say $h - \epsilon \leq x \leq h$, so that actin and myosin bind rapidly in a small interval near $h$. We expect that $\alpha$ depends on the local calcium concentration. On the support of $f(x)$, $n(x) = 1 - \exp[\alpha(x - h)/(\epsilon v)]$.

The role of $g(x)$ is to break crossbridge bonds. A simple way to accomplish this is to assume that all bonds break at exactly $x = \delta < 0$, in which case

$$n(x) = \begin{cases} 1 - e^{\alpha(x-h)/\epsilon v}, & h - \epsilon \leq x \leq h, \\ (1 - e^{-\alpha/v}), & \delta \leq x \leq h - \epsilon, \\ 0, & \text{elsewhere.} \end{cases} \tag{18.31}$$

It is left as an exercise (exercise 6) to show that in the limit as $\epsilon \to 0$ and $\delta \to 0$ the force–velocity curve for this model with a linear restoring force does not produce zero force at some positive velocity, a feature that appears in the Hill force–velocity curve.

A second option is to suppose that bonds break when $x < 0$, and thus to take $g(x) = \kappa/(\delta - x)$ on $\delta < x < 0$. Note that the rate of bond breakage is infinite at $x = \delta$, and thus all crossbridges are disassociated at distance $x = \delta$. Then

$$n(x) = \begin{cases} 1 - e^{\alpha(x-h)/\epsilon v}, & h - \epsilon \leq x \leq h, \\ 1 - e^{-\alpha/v}, & 0 \leq x \leq h - \epsilon, \\ (1 - e^{-\alpha/v})\left(1 - \dfrac{x}{\delta}\right)^{-\kappa/v}, & \delta \leq x < 0, \\ 0, & \text{elsewhere,} \end{cases} \tag{18.32}$$

in which case $n(x)$ is a continuous function of $x$ with compact support.

To find a force–velocity curve that produces zero force at some positive velocity, we note that to have zero force, the force generated by crossbridges with $x < 0$ must exactly balance the force generated by crossbridges with $x > 0$. There are two modifications that accomplish this balance of forces. The first is to shift the population $n(x)$ so that for large $v$, the density for $x > 0$ is increased, and the second is to allow those bonds with $x > 0$ to exert a larger force than those with $x < 0$. Notice from (18.21) that as $v$ increases, $n(x)$ responds less quickly to $f(x)$ and $g(x)$ (as a function of increasing $x$). Thus, if the support of $f(x)$ and/or $g(x)$ is not zero, the distribution $n(x)$ shifts to the left as $v$ increases, as illustrated in Fig. 18.13.

Yet another way to determine the functions $f$ and $g$ is to estimate the energy of the bond as a function of position, and then from Eyring rate theory to determine the rates of reaction of binding and unbinding. This is the approach followed, for example, by Pate (1997). In fact, now that the biochemistry of the crossbridge reactions is known, fairly sophisticated models of this type are possible (Marland, 1998).

## 18.4   Determination of the Rate Functions

So far we have seen that an ad hoc approach to the determination of the functions $f$, $g$, and $r$ can generate models that agree in varying degrees with experimental data. Obviously, it is desirable to find some way in which these functions can be determined more systematically, for example, to guarantee the correct form of the force–velocity curve. One way that this can be accomplished is by using a slightly different model for the crossbridge dynamics (Lacker and Peskin, 1986; Peskin, 1975, 1976).

### 18.4.1   A Continuous Binding Site Model

Recall that the Huxley model was based on the assumption that the actin binding sites are sufficiently separated so that each crossbridge can be associated with a unique binding site. Thus, even when a crossbridge is unbound, the distance of the crossbridge to the nearest binding site is defined. We now make the opposite assumption, that the actin binding sites are not discrete sites but are continuously distributed, so that myosin can bind anywhere along a thin filament. By analogy, one can think of the thin filament as flypaper, to which the myosin heads stick wherever they touch down. In this case, the variable $x$ denotes the distance between the crossbridge anchor and the binding position, and the crossbridge distribution is described by a function $n(x,t)$ such that $\int_a^b n(x,t)\,dx$ is the fraction of crossbridges (at time $t$) that are attached with distance to the binding site $x$ in the range $[a,b]$. Note that an unbound crossbridge cannot be associated with a value of $x$, as $x$ is meaningful only for a bound crossbridge. The total fraction of bound crossbridges is

$$N = \int_{-\infty}^{\infty} n(x,t)\,dx < 1, \qquad (18.33)$$

and the total fraction of unbound crossbridges is $1 - N$.

To derive the differential equation for $n$, we consider the conservation of cross-bridges with $x$ in the interval $[a,b]$. Let $P$ denote the pool of crossbridges that are bound with $x \in [a,b]$. If the muscle is contracting at velocity $v > 0$, crossbridges move out of $P$ at the rate $vn(a,t)$ and move into $P$ at the rate $vn(b,t)$. Further, if $f$ is defined such that $\int_a^b f(x)dx$ is the rate at which new crossbridges are formed with $x \in [a,b]$, and if $g(x)$ denotes the rate at which crossbridges with displacement $x$ detach, then the rate of change of crossbridges is

$$\frac{d}{dt}\int_a^b n(s,t)\,ds = v[n(b,t) - n(a,t)] + (1-N)\int_a^b f(s)\,ds - \int_a^b g(s)n(s,t)\,ds. \quad (18.34)$$

Replacing $b$ by $x$ and differentiating with respect to $x$, we obtain

$$\frac{\partial n}{\partial t} - v(t)\frac{\partial n}{\partial x} = (1-N)f(x) - ng(x). \quad (18.35)$$

Note that in the derivation of (18.35) we assume that the rate of crossbridge attachment is proportional to the fraction of unattached crossbridges, $1 - N$.

The equations for the continuous binding site model are similar to those of the Huxley model, the differences being, first, that the rate of crossbridge attachment is given by $(1 - N)f$ in the continuous binding site model and $(1 - n)f$ in the Huxley model, and, second, that $n$ and $f$ have different units in the two models. In the Huxley model $n$ is dimensionless, while in the continuous binding site model $n$ has dimension of length$^{-1}$. Similarly, $f$ has dimension of time$^{-1}$ in the Huxley model and dimension of length$^{-1}$ time$^{-1}$ in the continuous binding site model.

In the following discussion we restrict our attention to a simplified version of the continuous binding site model in which all crossbridges attach at some preferred displacement, say, $x = h$. In this case $f(x) = F\delta(x - h)$, where $F$ is the rate of crossbridge attachment. For this choice of $f$ it is most convenient to rewrite the differential equation to incorporate crossbridge attachment as a boundary condition. We do this by integrating (18.35) from $h - \epsilon$ to $h + \epsilon$ and letting $\epsilon \to 0$. The jump in $n$ at $x = h$ is then given by $F(1 - N)/v$, and so (18.35) can be written as

$$\frac{\partial n}{\partial t} - v(t)\frac{\partial n}{\partial x} = -ng(x), \qquad x < h, \quad (18.36)$$

$$n(h,t) = \frac{F(1-N)}{v}. \quad (18.37)$$

Although in general, $N$ and $v$ are functions of $t$, we consider only those cases in which they are constant. However, $N$ and $n$ are also functions of $v$, and we sometimes write $N(v)$ and $n(x,t;v)$ to emphasize this dependence.

## 18.4.2 A General Binding Site Model

Both the continuous binding site model and the Huxley model can be derived as limiting cases of a more general model (Peskin, 1975). Suppose that on the thin filament there are a discrete number of actin binding sites, with a regular spacing of $\Delta x$ (as illustrated

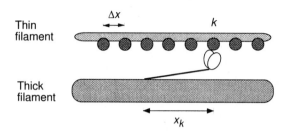

**Figure 18.15** Schematic diagram of a crossbridge model with discrete binding sites. The actin binding sites are separated by a distance $\Delta x$, and $x_k$ denotes the distance of the $k$th binding site from the unstressed position of the crossbridge.

in Fig. 18.15). We denote the horizontal distance from the crossbridge (on the thick filament) to the $k$th binding site by $x_k$. Finally, let $n_k(t)$ denote the probability that the crossbridge is attached to site $k$ at time $t$. Then, if $f(x_k)$ and $g(x_k)$ are the rates at which crossbridges attach and detach respectively from the $k$th site, we have

$$\frac{dn_k(t)}{dt} = f(x_k)\left[1 - \sum_i n_i(t)\right] - g(x_k)n_k(t). \qquad (18.38)$$

Note that the rate of crossbridge attachment is proportional to the fraction of unattached crossbridges, $1 - \sum_i n_i(t)$.

If we now assume that the $n_k(t)$ are samples of a smooth function, so that

$$n_k(t) = n(x_k(t), t), \qquad (18.39)$$

it follows that

$$\frac{dn_k}{dt} = -v\frac{\partial n}{\partial x_k} + \frac{\partial n}{\partial t}, \qquad (18.40)$$

where, for consistency with our assumption that $v$ is positive for a contracting muscle, we have defined $v = -dx_k/dt$. Substituting (18.40) into (18.38) gives

$$-v\frac{\partial n}{\partial x_k} + \frac{\partial n}{\partial t} = f(x_k)\left[1 - \sum_{i=-\infty}^{\infty} n(x_k + i\Delta x, t)\right] - g(x_k)n(x_k, t). \qquad (18.41)$$

Since this holds for any binding site, the subscript $k$ may be omitted, and thus

$$-v\frac{\partial n}{\partial x} + \frac{\partial n}{\partial t} = f(x)\left[1 - \sum_{i=-\infty}^{\infty} n(x + i\Delta x, t)\right] - g(x)n. \qquad (18.42)$$

By taking two different limits of (18.42) we obtain the continuous binding site and Huxley models. Suppose first that the binding sites are so widely spaced that at any given time only one is within reach of any crossbridge. This is modeled by assuming that $n(x,t) = 0$ if $|x| > \Delta x/2$. In this case, the only nonzero term in the sum in (18.42) is the term corresponding to $i = 0$. Thus (18.42) becomes

$$-v\frac{\partial n}{\partial x} + \frac{\partial n}{\partial t} = f(x)[1 - n(x,t)] - g(x)n(x,t), \qquad (18.43)$$

which is the Huxley model.

If, however, we make the assumption that $\Delta x$ is small and if we let $n = \hat{n}\Delta x, f = \hat{f}\Delta x$, we then get

$$-v\frac{\partial \hat{n}}{\partial x} + \frac{\partial \hat{n}}{\partial t} = \hat{f}(x)\left[1 - \sum_{i=-\infty}^{\infty} \hat{n}(x + i\Delta x, t)\Delta x\right] - g(x)\hat{n}. \qquad (18.44)$$

In the limit as $\Delta x \to 0$ the sum becomes a Riemann integral, so that

$$-v\frac{\partial \hat{n}}{\partial x} + \frac{\partial \hat{n}}{\partial t} = \hat{f}(x)\left[1 - \int_{-\infty}^{\infty} \hat{n}(s, t)\,ds\right] - g(x)\hat{n}, \qquad (18.45)$$

which is the continuous binding site model.

### 18.4.3  The Inverse Problem

The continuous binding site model (18.36) and (18.37) can be used to determine $F$, $g(x)$, and $r(x)$ directly from experimental data. (Recall that $r(x)$ is the restoring force generated by a crossbridge with extension $x$.) The steady-state solution of (18.36) and (18.37) can be written as

$$n(x; v) = \begin{cases} \dfrac{F[1 - N(v)]}{v}\exp\left(\displaystyle\int_x^h \frac{-g(s)}{v}\,ds\right), & x < h, \\[4mm] 0, & x > h. \end{cases} \qquad (18.46)$$

Integrating (18.46) from $-\infty$ to $\infty$, we obtain

$$N(v) = \int_{-\infty}^{\infty} n(x; v)\,dx = \frac{F[1 - N(v)]}{v}I(v), \qquad (18.47)$$

where

$$I(v) = \int_{-\infty}^{h} \exp\left(\int_x^h \frac{-g(s)}{v}\,ds\right)dx. \qquad (18.48)$$

Thus, we can solve for $N(v)$ as

$$N(v) = \frac{FI(v)}{FI(v) + v}. \qquad (18.49)$$

Substituting (18.49) into (18.46) we obtain an explicit solution for $n$,

$$n(x; v) = \frac{F}{FI(v) + v}\exp\left(\int_x^h \frac{-g(s)}{v}\,ds\right). \qquad (18.50)$$

Since the average force produced by a crossbridge is

$$p(v) = \int_{-\infty}^{\infty} r(x)n(x; v)\,dx, \qquad (18.51)$$

it follows that if $F$, $g$, and $r$ are known, then (18.50) can be used to derive an explicit expression for the force–velocity curve. This is the direct problem that we considered in the context of the Huxley model. Here we want to solve the inverse problem of determining $F$, $g$, and $r$ from our knowledge of $p$. However, we need additional information to do this.

The energy flux during constant contraction can be measured experimentally; in general it is a function of $v$. If we assume that the energy flux $\phi(v)$ is proportional to the rate at which crossbridges go through the cycle of binding and unbinding to the actin filament, then $\phi$ is proportional to the crossbridge turnover rate,

$$\phi(v) = \rho \epsilon F(1 - N), \tag{18.52}$$

where $\rho$ is the total number of crossbridges and $\epsilon$ is the energy released during each crossbridge cycle. If the fraction of attached crossbridges during isometric tetanus is known, then $F$ can be calculated from

$$\phi_0 = \rho \epsilon F(1 - N_0), \tag{18.53}$$

where $\phi_0 = \phi(0)$ and $N_0 = N(0)$. Next, $I(v)$ can be calculated from $\phi(v)$ by substituting (18.52) into (18.49), which gives

$$I(v) = v \left( \frac{F\rho\epsilon - \phi(v)}{F\phi(v)} \right) = \frac{\rho \epsilon v [\phi_0 - (1 - N_0)\phi]}{\phi\phi_0}. \tag{18.54}$$

Hence, from experimental knowledge of $N_0$ and $\phi(v)$ we can calculate explicit expressions for $F$ and $I(v)$.

To find $g$ from $I(v)$, we define the transformation

$$y(x) = \int_x^h g(s)\, ds. \tag{18.55}$$

Since $g$ is positive, $y$ is a monotonic function of $x$ and has an inverse that can be calculated explicitly. Differentiating (18.55) with respect to $x$, we obtain

$$\frac{dy}{dx} = -g(x), \tag{18.56}$$

from which it follows that

$$x(y) = h - \int_0^y \frac{ds}{\bar{g}(s)}, \tag{18.57}$$

where $\bar{g}$ is defined by $g(x) = g(x(y)) = \bar{g}(y)$, and where we have used the condition $y(h) = 0$, so that $x(0) = h$.

Using these definitions, and also defining $\sigma = 1/v$, we get

$$I(1/\sigma) = \int_{-\infty}^h e^{-\sigma y}\, dx, \tag{18.58}$$

$$= \int_\infty^0 -e^{-\sigma y} \frac{dy}{g(x)}, \tag{18.59}$$

$$= \int_0^\infty \frac{1}{\bar{g}(y)} e^{-\sigma y}\, dy. \tag{18.60}$$

The function $I(1/\sigma)$ is the Laplace transform of $1/\bar{g}(y)$, and so $1/\bar{g}(y)$ is obtained as the inverse Laplace transform of $I(1/\sigma)$. Furthermore, $g$ can be obtained as a function of $x$, since $x$ is defined as a function of $y$ by (18.57). Thus, for given $y$ we can calculate both $\bar{g}(y)$ and $x(y)$. Since $\bar{g}(y) = g(x(y))$, we thus have a parametric representation for $g(x)$.

An explicit formula for $\bar{g}(y)$ can be obtained by using the inversion formula for Laplace transforms. Thus,

$$\frac{1}{\bar{g}(y)} = \frac{1}{2\pi i} \int_{c-i\infty}^{c+i\infty} I(1/\sigma) e^{\sigma y} \, d\sigma, \tag{18.61}$$

where $c > 0$ is arbitrary.

In a similar way, $r(x)$ can be obtained from the force–velocity curve $p(v)$. It is left as an exercise to show that

$$\frac{\epsilon p(1/\sigma)}{\sigma \phi(1/\sigma)} = \int_0^\infty \frac{\bar{r}(y)}{\bar{g}(y)} e^{-\sigma y} \, dy, \tag{18.62}$$

where $\bar{r}(y) = r(x(y))$. Hence

$$\bar{r}(y) = \frac{\epsilon \bar{g}(y)}{2\pi i} \int_{c-i\infty}^{c+i\infty} \frac{p(1/\sigma)}{\sigma \phi(1/\sigma)} e^{\sigma y} \, d\sigma. \tag{18.63}$$

## A specific example

The above analysis can be used to calculate $F$, $g$, and $r$ to fit the Hill force–velocity curve and energy flux data (also observed by Hill). First, note that the force–velocity equation (18.1) can be written in the form

$$p(v) = \frac{bp_0 - av}{v + b}. \tag{18.64}$$

Second, Hill (1938) observed that at constant rate of contraction, the heat flux $\dot{q}$ generated by a contracting muscle is linear, given by

$$\dot{q} = av + \phi_0, \tag{18.65}$$

where the constant $a$ is the same as in the force–velocity equation, and where $\phi_0$ is the energy flux at zero velocity. The energy flux is the sum of two terms: the heat flux and the power used by the muscle. The power of a muscle contracting at speed $v$ is $pv$ (force times velocity), and thus the energy flux, $\phi(v)$, is given by

$$\phi(v) = \dot{q} + pv = \phi_0 + \frac{bv(a + p_0)}{v + b}. \tag{18.66}$$

Substituting the expression for $\phi$ into (18.54) and using (18.61), we find that

$$\frac{1}{\bar{g}(y)} = \frac{\rho \epsilon}{\phi_0} \frac{1}{2\pi i} \int_{c-i\infty}^{c+i\infty} \left[ \frac{\sigma_+ + N_0(\sigma - \sigma_*)}{\sigma(\sigma - \sigma_*)} \right] e^{\sigma y} \, d\sigma \tag{18.67}$$

and

$$\bar{r}(y) = \frac{\epsilon \bar{g}(y)}{2\pi i} \int_{c-i\infty}^{c+i\infty} \left[ \frac{p_0 \sigma - a/b}{\sigma(\sigma - \sigma_*)} \right] e^{\sigma y} \, d\sigma, \tag{18.68}$$

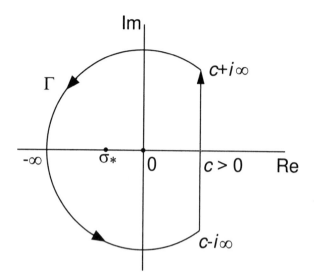

**Figure 18.16** Contour for the evaluation of the path integral in the continuous binding site model. (Adapted from Lacker and Peskin, 1986, Fig. 6.)

where $\sigma_+ = -(a + p_0)/\phi_0$ and $\sigma_* = -1/b + \sigma_+$. These integrals can be evaluated using the contour $\Gamma$ shown in Fig. 18.16.

From the residue theorem, we know that the integral around $\Gamma$ is the sum of the residues inside the contour. Further, it is not difficult to see that the integral over the semicircular part of the contour goes to zero as the radius of the semicircle becomes infinite. Hence

$$2\pi i \sum \text{residues} = \int_\Gamma = \int_{c-i\infty}^{c+i\infty}. \tag{18.69}$$

Both of the integrals (18.67) and (18.68) have two simple poles inside $\Gamma$, one at $\sigma = 0$, the other at $\sigma = \sigma_*$. For the integral (18.67),

$$\text{the pole at } \sigma = 0 \text{ has residue } N_0 - \sigma_+/\sigma_*; \tag{18.70}$$

$$\text{the pole at } \sigma = \sigma_* \text{ has residue } \sigma_+ e^{\sigma_* y}/\sigma_*; \tag{18.71}$$

while for the integral (18.68),

$$\text{the pole at } \sigma = 0 \text{ has residue } \frac{a}{b\sigma_*}; \tag{18.72}$$

$$\text{the pole at } \sigma = \sigma_* \text{ has residue } \left(p_0 - \frac{a}{b\sigma_*}\right) e^{\sigma_* y}. \tag{18.73}$$

Adding these residues for each integral gives, finally,

$$\frac{1}{\bar{g}(y)} = \frac{\rho\epsilon}{\phi_0}\left[N_0 + \frac{\sigma_+}{\sigma_*}(e^{\sigma_* y} - 1)\right], \tag{18.74}$$

$$\bar{r}(y) = \frac{\epsilon\bar{g}(y)}{\phi_0}\left[\frac{a}{b\sigma_*} + \left(p_0 - \frac{a}{b\sigma_*}\right)e^{\sigma_* y}\right]. \tag{18.75}$$

To calculate $x(y)$, since

$$x(y) = h - \int_0^y \frac{ds}{\bar{g}(s)}, \tag{18.76}$$

it follows that

$$x(y) = h - \frac{\rho \epsilon}{\phi_0} \left[ \left( N_0 - \frac{\sigma_+}{\sigma_*} \right) y + \frac{\sigma_+}{\sigma_*^2} (e^{\sigma_* y} - 1) \right]. \tag{18.77}$$

This gives a parametric definition of $g(x)$ and $r(x)$.

Finally, we note one important feature of this model. Each crossbridge exerts zero force at some value of $y = y_0$ such that $\bar{r}(y_0) = 0$. Solving for $y_0$ gives

$$y_0 = \frac{1}{\sigma_*} \ln \left( \frac{a}{a - p_0 b \sigma_*} \right). \tag{18.78}$$

Hence, if we wish $x = 0$ to correspond to the equilibrium state of the crossbridge (i.e., when it exerts no force) as it was in the Huxley model, we cannot set $h$ arbitrarily. In fact, $h$ must be chosen such that $x(y_0) = 0$, and thus

$$h = \frac{\rho \epsilon}{\phi_0} \left[ \left( N_0 - \frac{\sigma_+}{\sigma_*} \right) y_0 + \frac{\sigma_+}{\sigma_*^2} (e^{\sigma_* y_0} - 1) \right]. \tag{18.79}$$

Plots of $g$ and $r$ are shown in Fig. 18.17. From these curves we note that as the displacement of the crossbridge becomes more negative, its probability of detachment increases, but the force it exerts decreases. This allows a high isometric force without a corresponding reduction in the maximum contraction velocity; crossbridges initially exert a large force, but tend not to be carried into the region where they oppose contraction.

## 18.5 The Discrete Distribution of Binding Sites

The Huxley model assumes that at any one time, each crossbridge has only a single actin binding site available for binding, while the continuous binding site model assumes the opposite, that crossbridges can bind anywhere. However, the real situation is probably something in between these two extremes. Depending on the flexibility of the actin filament, it is probable that each crossbridge has a selection of more than one binding site, but it is unlikely that the binding sites are effectively continuous (cf. Fig. 18.3). T.L. Hill (1974, 1975) has constructed a detailed series of models that treat, with varying degrees of accuracy, the intermediate case when the actin binding sites are distributed discretely but more than one is within reach of a crossbridge at any time. Detailed consideration of models of this type is left for the exercises (Exercises 9 and 10).

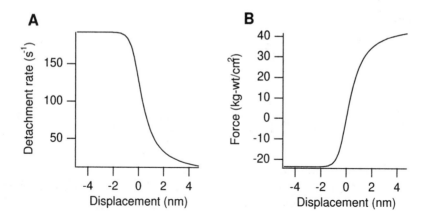

**Figure 18.17** Crossbridge detachment rate (A) and force (B) in the continuous binding site model, calculated from (18.74), (18.75), and (18.77), using the parameter values $p_0 = 3$ kg-wt/cm$^2$, $a/p_0 = 0.25$, $b = 0.325$ muscle lengths per second, $\phi_0 = ab$, $F = 125$ s$^{-1}$, $N_0 = 0.9$, $\phi_0/\rho\epsilon = F(1 - N_0) = 12.5$ s$^{-1}$. If we require that the crossbridge exert no force at $x = 0$, then all crossbridges must attach at $h = 4.78$ nm, assuming that the length of a half-sarcomere is 1.1 $\mu$m.

## 18.6   High Time-Resolution Data

All the models discussed so far treat crossbridge binding as a relatively simple phenomenon; either crossbridges are bound or they are not, and there is no consideration of the possibility that each crossbridge might have a number of different bound states. As we have seen, such assumptions do a good job of explaining muscle behavior on the time scale of tens of milliseconds. However, as the development of new experimental techniques allowed the measurement of muscle length and tension on much shorter time scales, the initial models were improved to take this high time-resolution data into account. One of the first models to do so was that of Huxley and Simmons (1971). The Huxley–Simmons model is quite different from models discussed above, giving a detailed description of how the force exerted by an *attached* crossbridge can vary with time over a short period, but it does not take into account the kinetics of crossbridge binding and unbinding to the thin filament.

### 18.6.1   High Time-Resolution Experiments

As we have already seen (Fig. 18.10B), when muscle length is decreased, the tension immediately decreases, and then, over a time period of 100 milliseconds or so, recovers to its original level. When this tension recovery is measured at a higher time resolution, two components of the recovery become apparent (Fig. 18.18). The initial drop in tension (which occurs simultaneously with the change in length) is followed by a rapid, partial recovery, followed in turn by a much slower complete recovery to the original

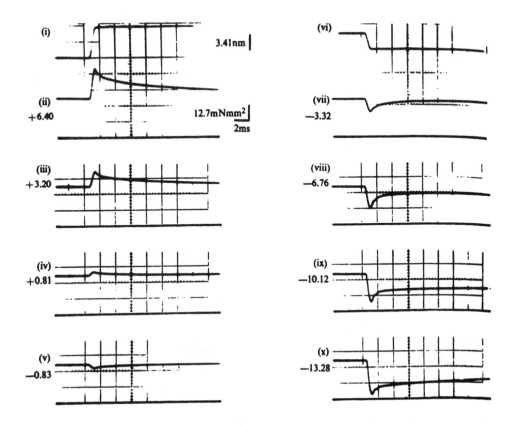

**Figure 18.18** Changes in tension after a sudden stretch (ii–iv) or a sudden shortening (v, vii–x). Traces (i) and (vi) show the time course of the length change for traces (ii) and (vii), respectively. The number to the left of each record denotes the amount of the length change (in nm) per half-sarcomere. Note the high time-resolution of the measurements. (Huxley and Simmons, 1971, Fig. 2.)

tension. The slower recovery process is described by the other models discussed in this chapter. Typical experimental results are shown in Fig. 18.19. In this figure, $T_1$ denotes the value of the tension after the initial drop, and $T_2$ denotes the value of the tension after the initial rapid recovery. For length increases ($y > 0$) $T_1$ is a linear function of the change in length, while for length decreases ($y < 0$), the decrease in tension is less than might be expected from the linear relation. It is likely that because the length step is not instantaneous but takes about a millisecond to complete, when a larger length decrease is applied, the rapid recovery process has already begun to take effect by the time the length decrease has been completed. If this is true (and it certainly appears plausible from the curves shown in Fig. 18.18), $T_1$ would be consistently overestimated for larger, negative, $y$. From the linearity of the curve for $y > 0$, it is reasonable to suppose that the relationship between $T_1$ and $y$ is linear over the entire range of $y$, as denoted by the dashed line in the figure.

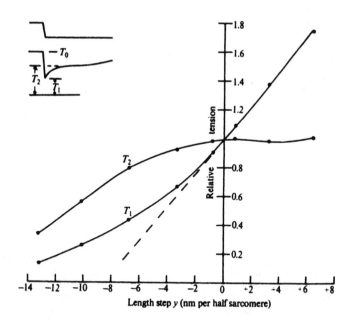

**Figure 18.19** Curves of $T_1$ and $T_2$ as functions of the length step. As depicted in the inset, $T_1$ is the minimal tension reached during the step, while $T_2$ is the value of the tension reached after the quick recovery phase. The upper trace of the inset depicts the time course of the length change. (Huxley and Simmons, 1971, Fig. 3.)

In contrast, $T_2$ is clearly a nonlinear function of $y$. For small length changes, the rapid process restores the tension to its original level, but for steps of larger length, the rapid process results in only partial recovery. The time course of the rapid recovery has a dominant rate constant, $r$, which is a decreasing function of $y$,

$$r = \frac{r_0}{2} \left( 1 + e^{-\alpha y} \right), \qquad (18.80)$$

where $r_0 = 0.4$ and $\alpha = 0.5$ are determined by fitting to experimental data.

## 18.6.2 The Model Equations

To model and give a possible explanation of the above results, we assume that a crossbridge consists of two parts: an elastic arm connected to a rotating head that can bind to the actin filament in two different configurations. As illustrated in Fig. 18.20, the head of the crossbridge contains three combining sites, $M_1, M_2, M_3$, each of which has the ability to bind to a corresponding site, $A_1, A_2, A_3$, on the actin filament. (To avoid confusion with previous terminology, the Ms and As are called combining sites, rather than binding sites.) The affinity between the combining sites is greatest for $M_3 A_3$, and smallest for $M_1 A_1$. As the head of the crossbridge rotates in the direction of increasing $\theta$, it moves through the sequence of binding configurations, $M_1 A_1$ only, $M_1 A_1$ and $M_2 A_2$, $M_2 A_2$ only, $M_2 A_2$ and $M_3 A_3$, $M_3 A_3$ only. During this progression the crossbridge arm is

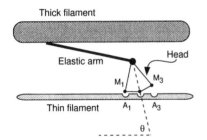

Thick filament

Elastic arm

Head

$M_1$    $M_3$

Thin filament    $A_1$    $A_3$

$\theta$

**Figure 18.20** Schematic diagram of the Huxley-Simmons crossbridge model. (Adapted from Huxley and Simmons, 1971, Fig. 5).

extended, and is thus under tension. The two stable configurations of the crossbridge are those in which two consecutive combining sites are attached simultaneously. Because the binding affinity is greater for $M_3A_3$, the energetically most favorable position for the crossbridge head is for $M_2$ and $M_3$ to be bound to $A_2$ and $A_3$ simultaneously.

An intuitive explanation of the experimental results is as follows. At steady state there is a balance between the tension on the crossbridge arm and the force exerted by the head of the crossbridge. The crossbridge head, in trying to rotate to a position of lower energy, places the elastic crossbridge arm under tension, and so the crossbridge arm, in turn, exerts a contractile force on the muscle. When the muscle is held at a constant length, the sum of all the crossbridge contributions gives the isometric force. If the length of the muscle is suddenly reduced, the tension on the crossbridge arm is suddenly reduced also, and this causes the instantaneous drop in tension seen experimentally. However, over the next few milliseconds, the reduced tension on the arm allows the crossbridge head to rotate to an energetically more favorable position, thus restoring the tension on the arm, and consequently restoring the muscle tension. Hence, the instantaneous drop in tension results from the fact that the elastic crossbridge arm responds instantaneously to a change in length, while the time course of the tension recovery is governed by how fast the crossbridge head rotates, which is, in turn, governed by the kinetics of attachment and detachment of the combining sites.

It is important to note that this interpretation of the experimental evidence assumes that during isometric tetanus, all crossbridges have a positive displacement, i.e., $x > 0$. Otherwise, if some crossbridges had $x < 0$, a shortening of the muscle fiber would *increase* the force exerted by these crossbridges, in conflict with the above interpretation. However, in the models discussed so far, this happens to be the case. For example, in the Huxley model the isometric tetanus solution is

$$n(x) = \frac{f(x)}{f(x) + g(x)}, \tag{18.81}$$

which is zero when $f(x) = 0$. Since $f$ is nonzero only when $x > 0$, it follows that all crossbridges have a positive displacement during isometric tetanus. Although this is the case for the Huxley model, this is not necessarily true in all experimental situations or in all models.

The model also neglects the effects of those few crossbridges that have such small displacements that a small decrease in length serves to shift them to negative displacements. However, the quantitative effects of this neglect are almost certainly small.

To express the model mathematically, we construct a potential-energy diagram for the crossbridge. Recall that the crossbridge head has two stable configurations, one when $M_1A_1$ and $M_2A_2$ bonds exist simultaneously, which we denote as position one, and the other when $M_2A_2$ and $M_3A_3$ bonds exist simultaneously which we denote as position two. Because these configurations are stable, the potential energy of the crossbridge head reaches a local minimum at positions one and two, and since position two is energetically favored over position one, it has a lower potential energy (see Fig. 18.21A). However, as the head rotates from position one to position two, the crossbridge arm is extended, which increases the total potential energy of the crossbridge. Thus, adding

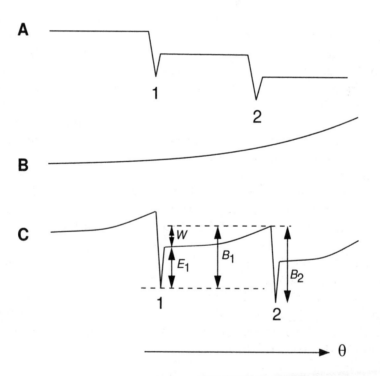

**Figure 18.21** A: Potential energy of the crossbridge head. As the head rotates, with increasing $\theta$ the combining sites bind consecutively. The potential energy decreases overall from left to right, as it is assumed that $M_3$ and $A_3$ have the highest affinity. The two local minima correspond to stable configurations when two consecutive combining sites are bound simultaneously; these are called position one and position two. B: Potential energy due to the elastic energy of the crossbridge arm. C: Total potential energy of the crossbridge, showing the notation used in the model.

the potential energy of the crossbridge arm (Fig. 18.21B) to the potential energy of the crossbridge head, we get the total potential energy of the crossbridge (Fig. 18.21C).

Now let $n_1$ and $n_2 = 1 - n_1$ denote the fraction of crossbridges in positions one and two, respectively, and let $y$ denote the displacement of the thick filament relative to the thin filament, such that $y = 0$ corresponds to the steady state before any length change is applied (i.e., the isometric case). Also, let $y_1$ and $y_2$ denote the lengths of the crossbridge arm when the head is at positions one and two respectively, let $y_0 = (y_1 + y_2)/2$ (the midway position), and let $h = y_2 - y_1$. Finally, let $F_1$ and $F_2$ denote the tension in the crossbridge arm when the head is in positions one and two, respectively. Then,

$$F_1 = K(y + y_0 - h/2), \qquad F_2 = K(y + y_0 + h/2), \tag{18.82}$$

where $K$ is the stiffness of the crossbridge arm, assumed to follow Hooke's law. Hence, the tension, $\phi$, on the crossbridge arm is given by

$$\phi = n_1 F_1 + n_2 F_2 = K(y + y_0 - h/2 + hn_2). \tag{18.83}$$

As the crossbridge head moves from position one to position two, the extending crossbridge arm does work, exerting an average force of approximately $(F_1 + F_2)/2$ over a distance $h$. Thus the work, $W$, is given by

$$W = h\frac{F_1 + F_2}{2} = Kh(y + y_0). \tag{18.84}$$

Now suppose that a crossbridge moves from position one to position two at rate $k_+$ and moves in the opposite direction at rate $k_-$. As with barrier models for the ionic current through a membrane channel (Chapter 3), we assume that each of these rates is an exponential function of the height of the potential-energy barrier that the crossbridge must cross in order to jump from one combining configuration to the other. With this assumption, $k_+ = \exp(\frac{-B_1}{kT})$ and $k_- = \exp(\frac{-B_2}{kT})$, where $T$ is the absolute temperature and $k$ is Boltzmann's constant. Note that $kT$ has units of joules, and so the jumps in the potential energy diagram must also be expressed in joules. Then,

$$\frac{k_+}{k_-} = \exp\left(\frac{B_2 - B_1}{kT}\right) \tag{18.85}$$

$$= \exp\left(\frac{B_2 - E_1 - W}{kT}\right) \tag{18.86}$$

$$= \exp\left(\frac{B_2 - E_1 - Kh(y + y_0)}{kT}\right). \tag{18.87}$$

We can now eliminate $E_1$ and $B_2$ by assuming that during isometric tetanus, $n_1 = n_2$. In this case $y = 0$ by definition (since $y$ is defined as the deviation from the isometric case), and $k_+ = k_-$, and so

$$B_2 - E_1 = Khy_0. \tag{18.88}$$

Substituting back into (18.87), we get

$$\frac{k_+}{k_-} = \exp\left(\frac{-Khy}{kT}\right). \tag{18.89}$$

When the length of the muscle is changed, the crossbridges redistribute themselves among the two configurations according to the differential equation

$$\frac{dn_2}{dt} = k_+ n_1 - k_- n_2 = k_+ - rn_2, \tag{18.90}$$

where

$$r = k_+ + k_- = k_-\left[1 + \exp\left(\frac{-yKh}{kT}\right)\right]. \tag{18.91}$$

Since $r$ is the time constant for the redistribution of crossbridges, it follows that $r$ is also the time constant for the development of the tension $T_2$ at the end of the quick recovery. Equation (18.91) agrees with (18.80) if $Kh = \alpha kT$, and thus the model has the correct time course for tension development. At steady state,

$$n_2 = \frac{k_+}{k_+ + k_-} = \frac{1}{2}\left[1 + \tanh\left(\frac{\alpha y}{2}\right)\right], \tag{18.92}$$

where we have used the fact that $k_+/k_- = \exp(-\alpha y)$. Hence, the steady-state tension (which corresponds to $T_2$) is given by

$$\phi = K(y + y_0 - h/2 + hn_2) = \frac{\alpha kT}{h}\left[y_0 + y - \frac{h}{2}\tanh\left(\frac{\alpha y}{2}\right)\right]. \tag{18.93}$$

A plot of $\phi$ is given in Fig. 18.22, from which it is seen that the model gives an excellent qualitative description of the experimental data.

In summary, although the Huxley–Simmons model does not take later events, such as crossbridge binding and unbinding, into account and is not intended to describe

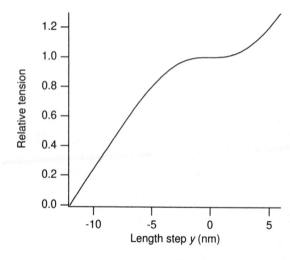

**Figure 18.22** Plot of relative steady tension $\phi$ (18.93) in the Huxley–Simmons model. Parameter values: $\alpha = 0.5\,\text{nm}$, $y_0 = 8\,\text{nm}$, $h = 8\,\text{nm}$. Because we plot relative tension, the value of $kT$ has no effect on the shape of the curve.

the full tension recovery in the manner of the Huxley model, it nevertheless provides an excellent qualitative description of the initial phase of tension recovery following a step change in length.

## 18.7 EXERCISES

1. Derive a differential equation for the load in a three-element Hill model with an elastic element in series with a contractile element (as shown in Fig. 18.8) in parallel with an additional elastic element of length $L$.
   Remark: The total load in the whole unit is divided between the two parallel subunits.

2. When a muscle cell dies, its ATP is depleted, with the result that the power stroke stalls and calcium cannot be withdrawn using the $Ca^{2+}$ ATPase. How does this explain rigor mortis?

3. Derive the force–velocity relationship (18.26) for the Huxley model.

4. Calculate the response of the Huxley model to a step change in length. In other words, calculate how $n$ changes as a function of $t$ and $x$, and hence calculate how the tension changes as a function of time.

5. Show that when

$$f(x) = \begin{cases} f_{max}e^{(x-h)/\lambda}, & x < h, \\ 0, & x > h, \end{cases} \tag{18.94}$$

$$g(x) = g_{max}\left(1 - e^{(x-h)/\lambda}\right), \tag{18.95}$$

$$r(x) = r_{max}\frac{e^{(x-h)/h} - \alpha}{1 - \alpha}, \tag{18.96}$$

the Huxley model reproduces the Hill force–velocity curve exactly. Note that the behavior of $g$ and $r$ for $x > h$ are irrelevant, as no crossbridges are ever bound there. Unfortunately, although this choice for $f$ and $g$ gives good agreement with the force–velocity curve, it gives poor agreement with the energy flux data.

6. Assuming that the crossbridges act as linear springs, calculate the force–velocity curve for (18.31) and show that in the limit as $\epsilon \to 0$ and $\delta \to 0$ it does not produce zero force for some positive velocity. Give an intuitive explanation for this.

7. Use the exact solution for the continuous binding site model to calculate the maximal shortening velocity $v_{max}$ and the isometric tension. Show that $v_{max}$ scales with $g$; i.e., if $g$ is multiplied by a constant factor, so is $v_{max}$. Give an intuitive explanation of this result. How does $g$ affect the isometric force? If $g$ is constant, is it possible to design a muscle that has both a high $v_{max}$ and a high isometric force? How could one design a muscle that has both a high $v_{max}$ and a high isometric force?

8. Show that if $g$ is a constant in the continuous binding site model, the force–velocity curve is the Laplace transform of $r(x)$.

9. This exercise and the next are based on the discrete binding site models of T.L. Hill (1974,1975). Suppose that each crossbridge is within reach of no more than two binding sites at one time, and that adjacent crossbridges do not "see" the same two binding sites. Suppose also that adjacent binding sites are separated by a distance $\Delta x$, and let $x$ denote the distance of the crossbridge from one of the binding sites, binding site 0, say. Define $x$ such that if the crossbridge is bound to site 0 and has $x = 0$, it exerts no force. Also, let

$n_i(x, t), i = 0, -1$, denote the fraction of crossbridges with displacement $x$ that are bound to binding site $i$.

(a) Show that the conservation equations are

$$-v\frac{dn_0}{dx} = f(x)[1 - n_0(x) - n_{-1}(x)] - g(x)n_0(x), \tag{18.97}$$

$$-v\frac{dn_{-1}}{dx} = f(x - \Delta x)[1 - n_0(x) - n_{-1}(x)] - g(x - \Delta x)n_{-1}(x), \tag{18.98}$$

where as usual, $v$ denotes the steady contraction velocity.

(b) Derive expressions for the isometric distributions of $n_0$ and $n_{-1}$. Compute the isometric force. Show that if the Huxley model is modified to include two binding sites, the isometric force is increased.

(c) Compute the force–velocity curve. Hint: For each $v$ solve the differential equations numerically, using the boundary conditions $n_0(h) = 0$, $n_{-1}(h + \Delta x) = 0$, then substitute the result into the expression for the force and integrate numerically.

(d) Modify the model to include slippage of the crossbridge from one binding site to another. Show that in the limit as slippage becomes very fast, the two differential equations (18.97) and (18.98) reduce to a single equation.

10. Consider the general binding site model (18.42) that incorporates the discrete distribution of binding sites. Why is this equation much harder to integrate than the models we have discussed previously? The isometric solution is considerably easier to calculate. Let $n_u(x) = 1 - \sum_{-\infty}^{\infty} n(x + i\Delta x)$. Show that

$$1 - n_u(x) = \sum_{-\infty}^{\infty} \frac{f(x + i\Delta x)}{g(x + i\Delta x)}. \tag{18.99}$$

Hence calculate the isometric solution $n(x)$. Details of this model, and others, are given by T.L. Hill (1974, 1975).

11. A muscle fiber must be able to produce a force even at negative velocities. For example, if you slowly lower a brick onto a table, your bicep is extending while simultaneously resisting the freefall of the brick. How might the models presented here be modified to allow this possibility? Hint: There must be some mechanism to break crossbridge bonds that are extended, i.e., have $x > 0$.

# CHAPTER 19

# Hormone Physiology

Hormones control a vast array of bodily functions, including sexual reproduction and sexual development, whole-body metabolism, blood glucose levels, plasma calcium concentration, and growth. Hormones are produced in, and released from, diverse places, including the hypothalamus and pituitary, the adrenal gland, the thyroid gland, the testes and ovaries, and the pancreas, and they act on target cells that are often at a considerable physical distance from the site of production. Since they are carried in the bloodstream, hormones are capable of a diffuse whole-body effect, as well as a localized effect, depending on the distance between the production site and the site of action. In many ways the endocrine system is similar to the nervous system, in that it is an intercellular signaling system in which cells communicate via cellular secretions. Hormones are, in a sense, neurotransmitters that are capable of acting on target cells throughout the body, or conversely, neurotransmitters can be thought of as hormones with a localized action.

There are a number of basic types of hormones. Some, such as epinephrine and norepinephrine, originate from the amino acid tyrosine. Other, water-soluble, hormones are derived from proteins or peptides, while the *steroid* hormones are derived from cholesterol and are thus lipid-soluble. The diversity of the chemical composition of hormones results in a corresponding diversity of mechanisms of hormone action.

Steroid hormones, being lipid-soluble, diffuse across the cell membrane and bind to receptors located in the cell cytoplasm. The resultant conformational change in the receptor leads to activation of specific portions of DNA, thus initiating the transcription of RNA, eventually (possibly hours or days later) resulting in the production of specific proteins that modify cell behavior. An example of one such hormone is aldosterone, whose effect on epithelial cells is to enhance the production of ion channel proteins, rendering the cell more permeable to sodium.

Other hormones, such as acetylcholine, act by binding to receptors located on the cell-surface membrane and causing a conformational change that results in the opening or closing of ionic channels.

Another important mechanism of hormone action is through second messengers, of which there are several examples in this book. Many hormone receptors are linked to G-proteins; binding of a hormone to the receptor results in the activation of the G-protein, and the triggering of a cascade of enzymatic reactions. For example, in the adenylate cyclase cascade, a wide variety of hormones (including adrenocorticotropin, luteinizing hormone, and vasopressin) cause activation of the G-protein, which in turn activates the membrane-bound enzyme adenylate cyclase. This activation results in an increase in the intracellular concentration of cAMP, and the consequent activation of a number of enzymes, with eventual effects on cell behavior; the specific effects depend on the cell type and the type of hormonal stimulus. In Chapters 5 and 12 we described the result of another signaling cascade, the phosphoinositide cascade, in which activation of cell-surface receptors leads to the activation of phospholipase C, the cleavage of phosphotidyl inositol 4,5-bisphosphate, and the resultant production of inositol 1,4,5-trisphosphate ($IP_3$) and diacylglycerol. As we saw, $IP_3$ releases $Ca^{2+}$ from internal stores, and this can lead to intracellular $Ca^{2+}$ oscillations and traveling waves.

Hormones can also act by directly converting the receptors into activated enzymes. For example, when insulin binds to a membrane receptor, the portion of the binding protein that protrudes into the cell interior becomes an activated kinase, which then promotes the phosphorylation of several substances inside the cell. The phosphorylation of proteins in the cell leads to a variety of other effects, including the enhanced uptake of glucose.

Much hormonal activity is characterized by oscillatory behavior, with the period of oscillation ranging from milliseconds ($\beta$-cell spiking) to minutes (insulin secretion) to hours ($\beta$-endorphin). In Table 19.1 are shown examples of pulsatile secretion of various hormones in man. The pulsatility of normal hormonal activity is not completely understood, but has significant implications for the treatment of hormonal abnormalities with drug therapies.

Despite the analogy with neural transmission, there is a significant difference between the endocrine and nervous systems that has important ramifications for mathematical modeling. Not only is the endocrine system extremely complicated, but the data that are presently obtainable are less susceptible to quantitative analysis than, say, voltage measurements in neurons. Further, the distance between the sites of hormone production and action, and the complexities inherent in the mode of transport, make it extraordinarily difficult to construct quantitative models of hormonal control. For these reasons, models in endocrinology are less mechanistic than many of the models presented elsewhere in this book, and thus, in some ways, are less realistic.

**Table 19.1**   Examples of pulsatile secretion of hormones in man (Brabant et al., 1992.)
Different values correspond to different primary sources.

| Hormone | Pulses/Day |
|---|---|
| Growth hormone | 9–16, 29 |
| Prolactin | 4–9, 7–22 |
| Thyroid-stimulating hormone | 6–12, 13 |
| Adrenocorticotropic hormone | 15, 54 |
| Luteinizing hormone | 7–15, 90–121 |
| Follicle-stimulating hormone | 4–16, 19 |
| $\beta$-Endorphin | 13 |
| Melatonin | 18–24, 12–20 |
| Vasopressin | 12–18 |
| Renin | 6, 8–12 |
| Parathyroid hormone | 24–139, 23 |
| Insulin | 108–144, 120 |
| Pancreatic polypeptide | 96 |
| Somatostatin | 72 |
| Glucagon | 103, 144 |
| Estradiol | 8–19 |
| Progesterone | 6– 6 |
| Testosterone | 8–12, 13 |
| Aldosterone | 6, 9–12 |
| Cortisol | 15, 39 |

# 19.1   Ovulation in Mammals

At birth, the human ovary contains approximately 2 million ovarian *follicles*, which consist of germ cells, or oocytes, surrounded by a cluster of endocrine cells that provide an isolated and protected environment for the oocyte. In the first stage of follicle development, occurring mostly before puberty, and taking anywhere from 13 to 50 years, the cells surrounding the oocyte (the granulosa cells) divide and form several layers around the oocyte, forming the so-called secondary follicle. Subsequent to puberty, these secondary follicles form a reserve pool from which follicles are recruited to begin the second stage of development. In this second stage, follicles increase to a final size of up to 20 mm before they rupture and release the oocyte to be fertilized. The release of the oocyte is called *ovulation*. Although many follicles begin this second developmental stage, few reach full maturity and ovulate, as the rest atrophy and die. In fact, the number of oocytes reaching full maturity is carefully controlled, so that litter sizes are generally restricted to within a relatively narrow range, and different species have different typical litter sizes. For example, to quote some interesting, if not particularly useful, facts from Asdell (1946), both the dugong and llama have a typical litter size of 1, the crestless Himalayan porcupine typically gives birth to two offspring, while the dingo produces, on average, 3. Different breeds of pigs have litter sizes ranging from 6 to 11.

There must therefore be a complex process that, despite the continuous recruitment of secondary follicles into the second developmental stage, allows precise regulation of the number remaining at ovulation. Further, the temporal periodicity of ovulation is tightly controlled, with ovulation occurring at regularly spaced time intervals.

In addition to questions related to the nature of the control of ovulation, there is the question of efficiency. It appears inefficient to regulate the final number of mature follicles by initiating the growth of many and killing off most of them. One might speculate that it would be more reasonable to initiate growth in only the required number and ensure that they all progress through to ovulation.

Normal ovulation involves growth in both ovaries. However, since removal of one ovary does not change the total number of eggs released during ovulation, the control mechanism is not a local one, but a global one, known to operate through the circulatory system. Maturation of follicles is stimulated by *gonadotropin*, which is released from the pituitary gland. Gonadotropin consists of two different hormones called *follicle-stimulating hormone* (FSH) and *luteinizing hormone* (LH). However, follicles themselves secrete *estradiol*, which stimulates the production of gonadotropin, forming a feedback control loop for the control of follicle maturation (Fig. 19.1).

## 19.1.1  The Control of Ovulation

One of the most elegant models of hormonal control is due to Lacker (1981; Lacker and Peskin, 1981; Akin and Lacker, 1984) and describes a possible mechanism by which mammals control the number of eggs released at ovulation. In the model it is assumed that each follicle interacts with other follicles only through the hormone concentrations in the bloodstream. As follicles mature they become more sensitive to gonadotropin, and their secretion of estradiol increases. The model of this feedback control loop is considerably oversimplified, since it does not incorporate a detailed mechanistic description of how estradiol production depends on gonadotropin or vice versa. However, it provides a phenomenological description of how a global interaction mechanism can be organized to give precise control over the final number of eggs reaching maturity.

The three basic assumptions of the model are that

1. The rate at which follicles secrete estradiol is a marker of follicle maturity.
2. The concentration of estradiol in the blood controls the release of FSH and LH from the pituitary.
3. The concentrations of FSH and LH control the rate of follicle maturation, and at any given instant, the response of each follicle to FSH and LH is a function of the follicle's maturity.

To express these assumptions mathematically, we define the following variables and parameters:

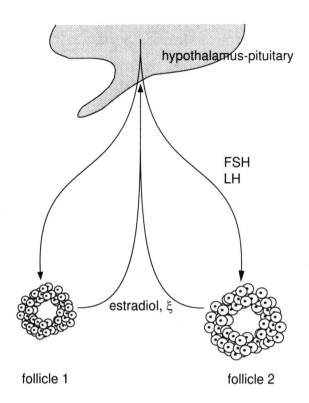

**Figure 19.1** Schematic diagram of the Lacker model for ovulation in mammals. (Adapted from Lacker, 1981, Fig. 1.)

$\xi$     concentration of estradiol,
$\gamma$     rate of clearance of estradiol from the blood,
$V$     plasma volume,
$s_i$     rate of secretion of estradiol from the $i$th follicle,
$N$     number of interacting follicles.

Here, all concentrations refer to serum concentrations (i.e., concentrations in the blood). Then, the rate of change of the total estradiol concentration is given by

$$V\frac{d\xi}{dt} = \sum_{i=1}^{N} s_i(t) - \gamma\xi. \tag{19.1}$$

Assuming that the rates of addition and removal of estradiol are much faster than the rate of follicle maturation, we take $\xi$ to be at pseudo-steady state, and thus

$$\xi = \frac{1}{\gamma}\sum_{i=1}^{N} s_i(t) = \sum_{i=1}^{N} \xi_i(t), \tag{19.2}$$

where $\xi_i(t) = s_i(t)/\gamma$ is the contribution that the $i$th follicle makes to $\xi$. In general, $d\xi_i/dt$ is a function of both $\xi_i$ and $\xi$, but does not depend directly on any other $\xi_j, j \neq i$. This is because we assume that local follicle–follicle interactions are not an important feature of the control mechanism, but that follicles interact only via the total estradiol concentration. Hence, the most general form of the model equations is

$$\frac{d\xi_i}{dt} = f(\xi_i, \xi), \qquad i = 1, \ldots, N. \tag{19.3}$$

The function $f$ is called the *maturation function*. Note that the concentrations of FSH and LH do not appear explicitly, as their effect on $\xi_i$ is modelled indirectly by assuming that $\frac{d\xi_i}{dt}$ depends on $\xi$.

Here we discuss one particular form of the maturation function. This form is not based on experimental evidence but is chosen such that the model behaves correctly. Specifically, we take

$$\frac{d\xi_i}{dt} = f(\xi_i, \xi) = \xi_i \phi(\xi_i, \xi), \qquad i = 1, \ldots, N, \tag{19.4}$$

where

$$\phi(\xi_i, \xi) = 1 - (\xi - M_1 \xi_i)(\xi - M_2 \xi_i). \tag{19.5}$$

The constants $M_1$ and $M_2$ are parameters that are the same for every follicle, so that each follicle obeys the same developmental rules. As a function of $\xi_i$, for fixed $\xi$, $\phi$ is an inverted parabola with a maximum at

$$\xi_{i,\max} = \frac{\xi}{2}\left(\frac{1}{M_1} + \frac{1}{M_2}\right). \tag{19.6}$$

If $\xi_i$ is large or small, the growth of $\xi_i$ is negative, and so this growth rate is fastest for those follicles with maturity within a narrow range, depending on the total estradiol concentration. Thus, with a given initial distribution of follicle maturities, those with $\xi_i$ close to $\xi_{i,\max}$ grow at the expense of the others. Further, since the growth rate $f$ is proportional to $\xi_i$, the selective growth of the $i$th follicle leads to an autocatalytic increase in $\xi_i$.

## Numerical solutions

Before we study the behavior of the model analytically, it is helpful to see some typical numerical solutions. The numerical solution of (19.4)–(19.5), with $M_1 = 3.85$, and $M_2 = 15.15$, starting with a group of 10 follicles with initial maturities randomly distributed between 0 and 0.1, shows that the maturity of four or five follicles goes to infinity in finite time, while the other follicles die (Fig. 19.2). Since ovulation is triggered by high, fast-rising estradiol levels, solutions that become infinite in finite time are interpreted as ovulatory solutions. Not only do a similar number of follicles ovulate in each run, they also ovulate at the same time. Hence, ovulatory solutions for $\xi_i$ and $\xi_j$, say, are ones in which $\xi_i(t)$ and $\xi_j(t) \to \infty$ as $t \to T < \infty$, with $\xi_i/\xi_j \to 1$. These numerical solutions show that the model has the correct qualitative behavior. However, analytic methods give a deeper understanding of how this control is accomplished.

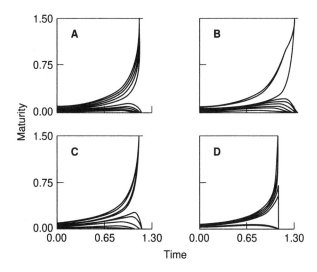

**Figure 19.2** Typical numerical solutions of the ovulation model. Each numerical simulation was started with a group of 10 follicles with maturities randomly distributed between 0 and 0.1. Parameter values are $M_1 = 3.85$, $M_2 = 15.15$. In panels A and D five follicles ovulate (their maturity blows up in finite time), while in panels B and C only four ovulate. All other follicles atrophy and die. (Lacker, 1981, Fig. 7.)

## Symmetric solutions

Much of the behavior of the ovulation model can be understood by considering symmetric solutions, in which $M$ of the follicles have the same maturity, while all others have zero maturity. Thus, $\xi_i = \xi/M, i = 1, \ldots, M$, and $\xi_i = 0, i = M + 1, \ldots, N$, in which case the model simplifies to

$$\frac{d\xi}{dt} = \xi + \mu\xi^3, \tag{19.7}$$

where $\mu = -(1 - M_1/M)(1 - M_2/M)$. The solution is given implicitly by

$$\frac{\xi}{\xi_0}\sqrt{\frac{1 + \mu\xi_0^2}{1 + \mu\xi^2}} = e^t, \tag{19.8}$$

where $\xi_0 = \xi(0)$ is the initial value. When $\mu > 0$, $t \to \log\left(\sqrt{(1 + \mu\xi_0^2)/(\mu\xi_0^2)}\right)$ as $\xi \to \infty$, while when $\mu < 0$, $t$ blows up to infinity as $\xi \to \sqrt{-1/\mu}$ (Fig. 19.3). Thus, when $\mu > 0$, $\xi$ becomes infinite in finite time, while when $\mu < 0$, $\xi$ goes to the steady state $\sqrt{-1/\mu}$ as $t \to \infty$. The former solution corresponds to an ovulatory solution, and the time of ovulation, $T$, is

$$T = \log\left(\sqrt{\frac{1 + \mu\xi_0^2}{\mu\xi_0^2}}\right). \tag{19.9}$$

It follows that if $M$ is between $M_1$ and $M_2$, all $M$ follicles progress to ovulation at time $T$ and the other follicles are suppressed, but if $M$ is outside this range, all $M$ follicles go

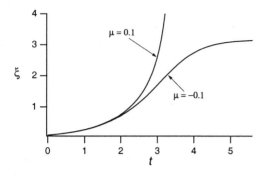

**Figure 19.3** Symmetric solutions of the ovulation model for two values of $\mu$. The initial condition was set arbitrarily at $\xi_0 = 0.1$. When $\mu > 0$, the solution blows up in finite time, while when $\mu < 0$, the solution approaches a steady state as $t \to \infty$.

to the steady (nonovulatory) state

$$\xi_M = \frac{1}{M}\sqrt{\frac{-1}{\mu}} = \frac{1}{\sqrt{(M - M_1)(M - M_2)}}. \tag{19.10}$$

Hence, in the symmetric case, ovulation numbers must be between $M_1$ and $M_2$.

### Solutions in phase space

To understand these symmetric solutions more fully, and to understand how they relate to the behavior of nonsymmetric solutions, it is helpful to consider the trajectories in the $N$-dimensional phase space defined by $\xi_i, i = 1, \ldots, N$. Each symmetric solution lies on a line of symmetry $l_M$ of the $M$-dimensional coordinate hyperplane. This is illustrated in Fig. 19.4 for the case $N = 3$: the $l_1$ lines are the $\xi_1, \xi_2$, and $\xi_3$ axes, the $l_2$ lines lie in the 2-dimensional coordinate planes, and the $l_3$ line makes a 45 degree angle with the $\xi_1, \xi_2$ plane. Note that not all the $l_1$ and $l_2$ lines of symmetry are included in the diagram. When $M$ is between $M_1$ and $M_2$, $l_M$ contains no critical point, and any trajectory starting on $l_M$ goes to infinity along $l_M$, reaching infinity in finite time $T$. However, when $M$ is outside the range of $M_1$ and $M_2$, $l_M$ contains a critical point, $P_M$, and solutions that start on $l_M$ stay on $l_M$, approaching $P_M$ as $t \to \infty$. In Fig. 19.4, $M_1 = 1.9$ and $M_2 = 2.9$, and so the only possible ovulation number is 2. Thus, each $l_1$ contains a critical point, $P_1$, that prevents the ovulation of single follicles, and similarly for $l_3$. The $l_2$ lines are the only lines of symmetry not containing a critical point.

The relationship between the symmetric solutions and the general solutions is most easily seen by analyzing the stability of the critical points $P_M$. Linearizing the model equations (19.3) around $P_M$ gives the linear system (after rearranging the variables)

$$\frac{d\tilde{P}}{dt} = A\tilde{P}, \tag{19.11}$$

where $\tilde{P}$ is a small perturbation from $P_M$ and

$$
A = \left. \begin{pmatrix}
\overbrace{\begin{array}{ccc|ccc}
a_1 + b_1 & & b_1 & & & \\
& \cdot & & & b_1 & \\
& \cdot & & & \vdots & \\
& \cdot & & & & \\
b_1 & & a_1 + b_1 & & & \\
\hline
& & & a_2 & & 0 \\
& 0 & & & \cdot & \\
& & & & & \cdot \\
& & & 0 & & a_2
\end{array}}^{M}
\end{pmatrix} \right\} M.
\qquad (19.12)
$$

The components of $A$ are the partial derivatives of the model equations evaluated at $P_M$, and so

$$
a_1 = \xi_M \frac{\partial \phi}{\partial \xi_i}\bigg|_{(\xi_M, M\xi_M)}, \qquad a_2 = \phi(0, M\xi_M), \qquad b_1 = \xi_M \frac{\partial \phi}{\partial \xi}\bigg|_{(\xi_M, M\xi_M)}. \qquad (19.13)
$$

The stability of $P_M$ is determined by the eigenvalues of $A$, which, because of the block structure of $A$, are the eigenvalues of the two diagonal block matrices of $A$. That is, if we write

$$
A = \begin{pmatrix} A_1 & B_1 \\ 0 & A_2 \end{pmatrix}, \qquad (19.14)
$$

then the eigenvalues of $A$ are the eigenvalues of $A_1$ and $A_2$. Hence, $A$ has an eigenvalue $\lambda_{\text{out}} = a_2$ of multiplicity $N - M$ (from $A_2$), an eigenvalue $\lambda_s = a_1 + M b_1 = -2$ of multiplicity 1, and an eigenvalue $\lambda_{\text{in}} = a_1$ of multiplicity $M - 1$, the latter two coming from $A_1$.

In the following discussion we let $Z = (\delta\xi_1, \ldots, \delta\xi_N)$ denote an eigenvector at $P_M$, and use subscripts to denote the different eigenvectors.

**Perturbations along** $l_M$**.** Corresponding to the simple eigenvalue $\lambda_s$ is the eigenvector $Z_s$ whose components satisfy $\delta\xi_i = 1, i = 1, \ldots, M, \delta\xi_i = 0, i = M + 1, \ldots, N$. Hence, $Z_s$ is in the direction of $l_M$. Since $\lambda_s < 0$, it follows that $l_M$ is on the stable manifold of $P_M$. Since symmetry is preserved along $l_M$, any solution that starts on $l_M$ goes to $P_M$ as $t \to \infty$.

**Perturbations orthogonal to** $l_M$ **in the coordinate hyperplane.** Corresponding to the eigenvalue $\lambda_{\text{in}}$ are the eigenvectors $Z_1, \ldots, Z_{M-1}$ whose components satisfy $\sum_{i=1}^{M} \delta\xi_i = 0, \delta\xi_i = 0, i = M + 1, \ldots, N$. $Z_1$ to $Z_{M-1}$ are independent vectors that lie in the coordinate hyperplane (since all have their last $M - N$ components equal

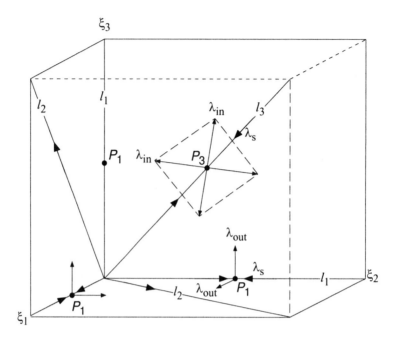

**Figure 19.4**   Phase space for a system of 3 interacting follicles ($N = 3$) with $M_1$ and $M_2$ chosen such that only two follicles ovulate. $\lambda_{out}$ denotes eigenvalues with eigenvectors that point out of the coordinate hyperplane, while $\lambda_{in}$ denotes eigenvalues with eigenvectors that are in the coordinate hyperplane. $\lambda_s$ denotes the eigenvalue with eigenvector in the direction of the line of symmetry.

to 0). Since they are also orthogonal to $l_M$, they span the orthogonal complement of $l_M$ in the coordinate hyperplane.

**Perturbations orthogonal to $l_M$ and the coordinate hyperplane.** Corresponding to the eigenvalue $\lambda_{out}$ are the eigenvectors $Z_{M+1}, \ldots, Z_{N-1}$ whose components satisfy $\sum_{i=M+1}^{N} \delta\xi_i = 0, \delta\xi_i = 0, i = 1, \ldots, M$. Finally, there is also the eigenvector $Z_N$ with components $\delta\xi_i = (M - N)b_1, i = 1, \ldots, M$ and $\delta\xi_i = (a_1 - a_2) + Mb_1, i = M+1, \ldots, N$. All the eigenvectors corresponding to $\lambda_{out}$ are orthogonal to both $l_M$ and the coordinate hyperplane and span the orthogonal complement of the coordinate hyperplane.

These eigenvectors are illustrated in Fig. 19.4. At the critical point $P_1$, situated on the $\xi_2$ axis, there are two independent eigenvectors corresponding to $\lambda_{out}$, and these are both orthogonal to $l_1$, the $\xi_2$ axis. Note that as the coordinate hyperplane is a line in this case, there are no eigenvectors corresponding to $\lambda_{in}$. At $P_3$ the converse is true. Here there are no eigenvectors corresponding to $\lambda_{out}$, as the coordinate hyperplane is the entire space. In three dimensions, the only critical point that could have eigenvectors corresponding to all the eigenvalues $\lambda_s, \lambda_{in}$, and $\lambda_{out}$ would be $P_2$. However, for these parameter values $P_2$ does not exist, as 2 lies between $M_1$ and $M_2$.

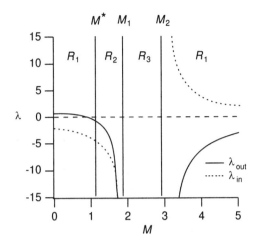

**Figure 19.5** The eigenvalues of $A$ as functions of $M$, calculated with the parameter values $M_1 = 1.9$, $M_2 = 2.9$. In the regions labeled $R_1$ the steady state $P_M$ is a saddle point; in the region labeled $R_2$, $P_M$ is stable; and in the region labeled $R_3$, there are no symmetric critical points, i.e., $P_M$ does not exist for those values of $M$. Note that only integer values for $M$ have any physical meaning. As described in the text, the eigenvalues $\lambda_{\text{out}}$ and $\lambda_{\text{in}}$ correspond, respectively, to eigenvectors pointing out and in of the symmetric hyperplane.

It remains to determine the stability of each critical point $P_M$. This is easily done by direct computation of the eigenvalues, which gives

$$\lambda_s = a_1 + Mb_1 = -2, \tag{19.15}$$

$$\lambda_{\text{in}} = a_1 = \frac{(M_1 + M_2)M - 2M_1M_2}{(M - M_1)(M - M_2)}, \tag{19.16}$$

$$\lambda_{\text{out}} = a_2 = -\frac{(M_1 + M_2)M - M_1M_2}{(M - M_1)(M - M_2)}. \tag{19.17}$$

Plots of $\lambda_{\text{in}}$ and $\lambda_{\text{out}}$ as functions of $M$ are shown in Fig. 19.5. Between $M_1$ and $M_2$, $P_M$ does not exist, but for $M > M_2$ and for $M < M^* = M_1M_2/(M_1 + M_2)$, $\lambda_{\text{in}}$ and $\lambda_{\text{out}}$ are of opposite signs. For $M^* < M < M_1$, the eigenvalues are both negative. It follows that if there are integers in the interval $(M^*, M_1)$, then there are stable critical points $P_M$, with $M^* < M < M_1$. All other symmetric critical points are unstable.

Finally, we note that there are critical points other than the symmetric ones discussed so far, but they are all unstable.

In summary, when there are no integers in the interval $(M^*, M_1)$ all the critical points are unstable and all the symmetric critical points are saddle points. In fact, from any starting point, all solutions approach infinity along one of the symmetric trajectories, $l_M$, where $M_1 < M < M_2$. These trajectories become infinite in finite time and are interpreted as ovulatory solutions. However, if there are integers in the interval $(M^*, M_1)$, there are corresponding stable critical points. Any solution that starts in the domain of attraction of one of these stable critical points, $P_{M_s}$ say, approaches $P_{M_s}$ as time increases, and the system becomes "stuck" there. No follicles ovulate, but $M_s$ follicles remain fixed at an intermediate maturity.

### Stability of $l_M$

Although one might expect to observe ovulation numbers anywhere in the range $M_1$ to $M_2$, numerical simulations show that only some of these actually occur. This is

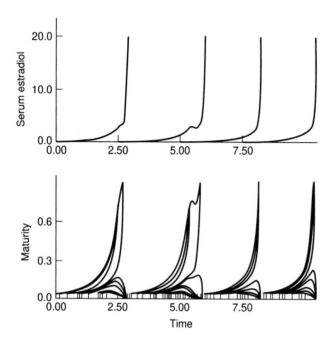

**Figure 19.6** Typical solutions when follicles begin to develop at random times, generated by a Poisson process. Each small tick on the horizontal axis marks the initiation of development in a single follicle. Although the parameter values are $M_1 = 3.85$, $M_2 = 15.15$, and thus one might expect to observe ovulation numbers ranging from 4 to 15, only the ovulation numbers 4 and 5 are observed. (Lacker and Peskin, 1981, Fig. 11.)

illustrated in Fig. 19.6, where $M_1 = 3.85$ and $M_2 = 15.15$. In the previous numerical simulations (Fig. 19.2) we started with a fixed number of follicles with random initial maturities normally distributed; in Fig. 19.6, however, follicles mature at random times (generated by a Poisson process), so that the simulation more accurately reflects the physiological situation.

Despite the random entry of follicles into the maturing pool, ovulation occurs at regular intervals, and the ovulation number varies little. Hence, the model generates periodic behavior from stochastic input. Furthermore, although we might expect to see ovulation numbers anywhere in the range 4 to 15, only the ovulation numbers 4 and 5 are observed. An explanation of this observation is found by examining the stability of the symmetric ovulatory solutions. This is done by transforming to a new coordinate system in which the ovulatory solutions, which become infinite in finite time, are transformed into finite critical points. The stability of these finite critical points can then be analyzed using standard linear stability methods.

We begin by noting that the initial ordering of a solution can never change. That is, if $\xi_i$ starts above $\xi_j$, it remains above $\xi_j$ for all time. This is true because

$$\frac{d}{dt}(\xi_i - \xi_j) = \xi_i\phi(\xi_i, \xi) - \xi_j\phi(\xi_j, \xi) = h(\xi_i, \xi_j, \xi)(\xi_i - \xi_j) \tag{19.18}$$

for some function $h$, and as long as $\xi_i$ and $\xi_j$ are bounded, so also is $h(\xi_i, \xi_j, \xi)$. Clearly,

$$\ln(\xi_i(t) - \xi_j(t)) = \ln(\xi_i(0) - \xi_j(0)) + \int_0^t h(\xi_i, \xi_j, \xi)dt. \tag{19.19}$$

If the right-hand side of this expression is bounded, so also is the left-hand side, so that $\xi_i(t) \neq \xi_j(t)$.

Since the original ordering of the maturities is preserved, we arrange the $N$ follicles in order of maturity, with $\xi_1$ denoting the follicle with the greatest maturity, and define a new time scale by

$$\tau(t) = \int_0^t \xi_1^2(s)\,ds. \tag{19.20}$$

As $t \to T$ (recall that $T$ is the finite time of ovulation), $\tau(t) \to \infty$. For as $\xi_1$ gets large, $d\xi_1/dt \approx \xi_1^3$, and hence $\xi_1^2$ behaves like $1/(T-t)$ as $t \to T$. Furthermore, $\xi_1^2$ is positive, and so $\tau$ is an increasing function of $t$ that is therefore invertible. We use the inverse to define new variables

$$\gamma_i(\tau) = \frac{\xi_i(t(\tau))}{\xi_1(t(\tau))}, \tag{19.21}$$

$$\Gamma(\tau) = \frac{\xi(t(\tau))}{\xi_1(t(\tau))}. \tag{19.22}$$

In terms of these new variables the model equations (19.4)–(19.5) become

$$\frac{d\gamma_i}{d\tau} = \gamma_i \Phi(\gamma_i, \Gamma), \qquad i = 1, \ldots, N, \tag{19.23}$$

$$\Gamma = \sum_{j=1}^{N} \gamma_j, \tag{19.24}$$

$$\Phi(\gamma_i, \Gamma) = (1 - \gamma_i)[M_1 M_2 (1 + \gamma_i) - \Gamma(M_1 + M_2)]. \tag{19.25}$$

Note that $\gamma_1(\tau) \equiv 1$, and $0 \leq \gamma_i(\tau) \leq 1$ for each $i$.

All ovulatory and anovulatory solutions correspond to critical points of (19.23)–(19.25) of the form

$$\gamma_i = \begin{cases} 1, & i = 1, \ldots, M, \\ 0, & i = M+1, \ldots, N. \end{cases} \tag{19.26}$$

Although ovulatory and anovulatory solutions now look the same, they can be distinguished by determining whether the original variable $\xi$ is finite. If so, the critical point corresponds to an anovulatory solution.

Equations (19.23)–(19.25) have only two distinct eigenvalues,

$$\lambda_1 = (M_1 + M_2)M - 2M_1 M_2, \tag{19.27}$$

$$\lambda_2 = -(M_1 + M_2)M + M_1 M_2, \tag{19.28}$$

which are plotted in Fig. 19.7. If $M$ lies between $M^*$ and $2M^*$, then both $\lambda_1$ and $\lambda_2$ are negative, so that the critical point is stable. Otherwise, the critical point is unstable. It follows that only ovulation numbers between $M_1$ and $2M^*$ are stable, and are therefore observable.

In the numerical simulations shown in Fig. 19.6, $M_1 = 3.85$ and $M_2 = 15.15$, in which case $2M^* = 6.14$. This is consistent with the numerical simulations in which

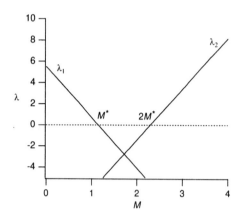

**Figure 19.7**  The eigenvalues $\lambda_1$ and $\lambda_2$ of the ovulation model in the transformed variables $\gamma_i$ and $\tau$.

only ovulation numbers 4 and 5 were observed. One possible reason why ovulation number 6 is not observed is that it lies close to the stability boundary. Thus, its domain of attraction is probably relatively small, and therefore the probability that a random process finds this domain of attraction is small.

### The effect of population size

With this model we can suggest an answer to the question of efficiency, namely, why do so many follicles begin the maturation process, only to atrophy and die? The answer appears to be that the mean time to ovulation is controlled more precisely by a large population than by a small one. This is illustrated in Fig. 19.8. For this figure, the model was simulated for 80 cycles, and the distribution of ovulation numbers and times was plotted for three different population sizes. Each population had the parameter values $M_1 = 6.1, M_2 = 5000$, and thus the expected ovulation numbers lie in the range 7 to 12. As the population size increases, the mean ovulation number decreases, but the shape of the distribution does not change a great deal. However, although the mean ovulation time (shown here centered at 0) does not change as the population size is increased, the distribution sharpens dramatically, and the range of observed ovulation times is dramatically reduced. Thus, while the majority of follicles atrophy and die, they have an important, although not immediately obvious, function: helping to regulate the timing of ovulation. This provides a possible explanation of why women near menopause (i.e., with fewer available oocytes) typically experience menstrual irregularities.

## 19.1.2    Other Models of Ovulation

Although this model of ovulation is one of the simplest and most elegant, other, more complex, models have been constructed. For example, Schwarz (1969) and Bogumil et al. (1972) have proposed models that incorporate large numbers of parameters and are based more directly on experimental data.

A recent model of a different type is due to Faddy and Gosden (1995). They construct a compartmental model of follicle dynamics over the lifetime of an individual female

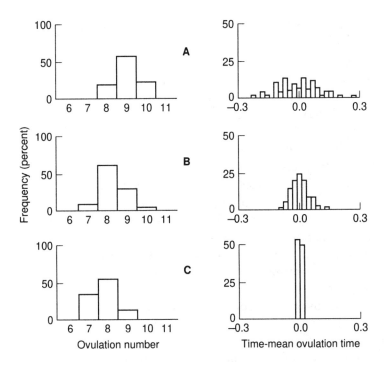

**Figure 19.8** The effect of the population size on the distribution of ovulation numbers. In panels A, B, and C, there are, respectively, 10, 100, and 1000 follicles interacting. Parameter values are $M_1 = 6.1$, $M_2 = 5000$. As the follicle population gets larger, the mean ovulation time decreases only slightly, while the standard deviation of the distribution of ovulation times decreases. Thus, larger populations allow more precise control over the ovulation number. (Lacker, 1981, Fig. 13.)

and fit their model to experimental data to obtain follicle growth and death rates as functions of the individual's age. Although this compartmental model does not provide insight into the mechanisms underlying periodic ovulation and a constant ovulation number, it provides an understanding of follicle dynamics over a larger time span.

## 19.2 Pulsatile Secretion of Luteinizing Hormone

Luteinizing hormone and follicle-stimulating hormone, known collectively as gonadotropin, have a monthly cycle (in humans) related to ovulation, and also vary periodically on a time scale of hours. Although the precise function of these hourly variations is unclear, they occur in both males and females, and are crucial to development and maturation in both sexes. Gonadotropin is produced by the pituitary gonadotrophs in response to gonadotropin-releasing hormone (GnRH), sometimes called luteinizing-hormone-releasing hormone, which is itself produced in the hypothalamus. Periodic variations in gonadotropin secretion are therefore the result of periodic variations in

GnRH secretion. In fact, if GnRH secretion is constant rather than pulsatile, the secretion of gonadotropin is greatly reduced, and thus the pulsatility of GnRH secretion has an important regulatory function (Knobil, 1981).

This observation has been used as the basis for clinical treatments of certain reproductive disorders. In women suffering from abnormal GnRH secretion, the pulsatile administration of GnRH can, in some cases, restore normal ovulation and fertility. However, the frequency of the pulse must be controlled carefully. Wildt et al. (1981) have shown that the secretion of gonadotropin in rhesus monkeys is approximately maximized by the administration of GnRH pulses with a frequency of one per hour. If the frequency of the GnRH pulse is increased to 2 per hour, gonadotropin secretion is inhibited. Conversely, if the frequency is decreased to one pulse every three hours, the rate of secretion of follicle-stimulating hormone (FSH) increases, while the rate of secretion of luteinizing hormone (LH) decreases.

An example of pulsatile secretion of LH and testosterone in males is shown in Fig. 19.9. Although the testosterone secretion is not obviously oscillatory, the fluctuations in LH secretion clearly are.

Similar mechanisms to those controlling ovulation are apparently at work here. In males, gonadotropin stimulates the production of testosterone from the testes, while in females it stimulates the production of estradiol from the ovaries. In the above model for ovulation, we saw that estradiol can stimulate further production of gonadotropin, forming a positive feedback loop. However, estradiol can have both positive and negative feedback effects on the production of gonadotropin. In models of pulsatile testosterone and gonadotropin secretion, it appears that negative feedback from estradiol and testosterone to gonadotropin production is the important mechanism.

An early model for LH levels in the rat is that of Shotkin (1974a,b), although this model did not consider oscillatory aspects. There have been a number of models of oscillatory GnRH release, starting with the work of Smith (1980, 1983), and of Cartwright and Husain (1986). These early models consisted of three compartments representing the hypothalamus, the pituitary, and the reproductive gland, or gonad, and the oscillations arose, either from a fixed time delay in the response of the gonads to the concentration of LH, or by feedback from LH and testosterone to the hypothalamus. However, there is little experimental evidence to support the assumptions of these models, and so they are essentially phenomenological. A more recent model of this type is due to Liu and Deng (1991), and was the first to make a serious attempt to determine model parameters by fitting to experimental data.

# 19.3  Pulsatile Insulin Secretion

Hormones secreted from cells in the pancreas are responsible for the control of glucose, amino acids, and other molecules that are necessary for metabolism. The pancreas contains a large number of secretory cells, grouped into about one million *islets of Langerhans* consisting of approximately 2,500 cells each. There are three principal

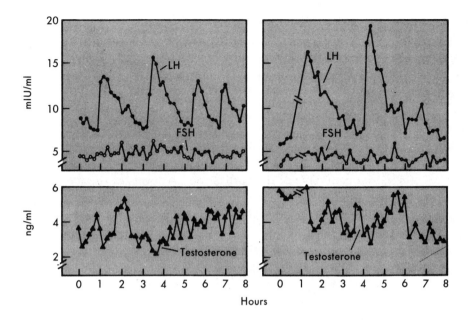

**Figure 19.9** Pulsatile secretion of LH, FSH, and testosterone in men. (Berne and Levy, 1993, Fig. 48-15, p. 912.)

secretory cell types: the $\alpha$-cells secrete glucagon, the $\beta$-cells secrete insulin, and the $\delta$-cells secrete somatostatin. Glucagon and insulin have complementary actions. A high concentration of glucose in the bloodstream (corresponding to an overabundance of nutrients) stimulates the production of insulin, which in turn induces storage of excess nutrient and decreases the rate at which nutrients are mobilized from storage areas such as adipose tissue or the liver. Insulin acts principally on three tissues: striated muscle (including the heart), liver, and adipose tissue. All the actions of insulin apparently stem from its interaction with a specific receptor in the plasma membrane of insulin-sensitive cells. How this interaction leads to the many actions of insulin on the cell is not fully understood. In striated muscle and adipose tissue, one important action of insulin is to stimulate the transport of glucose into the cell by a specific carrier (or carriers) in the plasma membrane. It appears to do this by recruiting glucose carriers to the plasma membrane from intracellular sites where they are inactive. Insulin thus increases the $V_{max}$ of transport, often as much as 10- to 20-fold. When glucose enters the cell it is rapidly phosphorylated and metabolized.

In the case of the liver, insulin does not increase the rate of transport of glucose into the cell (although it increases the net uptake of glucose). In the liver, insulin acts on a number of intracellular enzymes to increase glucose storage and decrease mobilization of glucose stores. The details of how insulin does this are far from clear.

Glucagon raises the concentration of glucose in the bloodstream. It acts mainly but not entirely on the liver, where it stimulates glycogen breakdown and the formation of glucose from noncarbohydrate precursors such as lactate, glycerol, and amino acids. Glucagon released in the islets stimulates the beta cells in the vicinity to secrete insulin.

Insulin secretion oscillates on a number of different time scales, ranging from tens of seconds to more than 100 minutes. The fast oscillations are caused (at least in part) by bursting electrical activity described in Chapter 6. During each burst of action potentials the cytosolic $Ca^{2+}$ concentration rises as $Ca^{2+}$ flows in through voltage-gated $Ca^{2+}$ channels, and this rise in $Ca^{2+}$ stimulates insulin secretion. Oscillations with a much larger period of around 100 minutes are also observed, and are called *ultradian oscillations* (Fig. 19.10). Finally, oscillations with intermediate frequencies of around 10 minutes or so also occur. One of the earliest observations of these oscillations was made in the rhesus monkey by Goodner et al. (1977), and some of their results are reproduced in Fig. 19.11. Glucagon and insulin oscillate out of phase, while insulin and glucose are in phase, with the increase of glucose leading the increase of insulin by an average of about one minute. Oscillations with intermediate frequency are also observed in isolated rat islets (Bergstrom et al., 1989; Berman et al., 1993), although, as can be seen from Fig. 19.12, spectral analysis is usually necessary to determine the principal underlying frequency. Once the underlying trend has been removed, a spectral decomposition of the data shows a frequency peak at about 0.07 $\text{min}^{-1}$, corresponding to a period of 14.5 minutes.

### Insulin units

Historically, a unit (U) of insulin was defined to be that amount of insulin (in cubic centimeters) that lowers the percentage of blood sugar in a normal rabbit to 0.045 in 2 to 6 hours. The crudity of such a unit was the natural result of the fact that it was not possible to purify insulin until relatively recently, and thus a bioassay was the only way of determining the amount. An excellent discussion of historical insulin units is given by Lacy (1967).

Later, mouse units became more convenient, and a unit was defined to be the amount of insulin required to produce convulsions in half the mice under standard conditions. A mouse unit is about 1/600 of a rabbit unit. Fortunately, insulin extracted from most animals has equivalent activity in rabbits, mice, and men, although the guinea pig and capybara are exceptions to this rule. Various modifications were made to the conditions of these assays, but with the advent of reasonably pure preparations of insulin the unit was redefined as 1/24 milligrams.

## 19.3.1  Ultradian Oscillations

Ultradian insulin oscillations have a number of observable features. First, oscillations occur during constant intravenous glucose infusion and are not dependent on periodic nutrient absorption from the gut. However, damped oscillations occur after a single stimulus such as a meal. Second, glucose and insulin concentrations are highly correlated, with the glucose peak occurring about 10–20 minutes earlier than that of insulin. Third, the amplitude of the oscillations is an increasing function of glucose concentra-

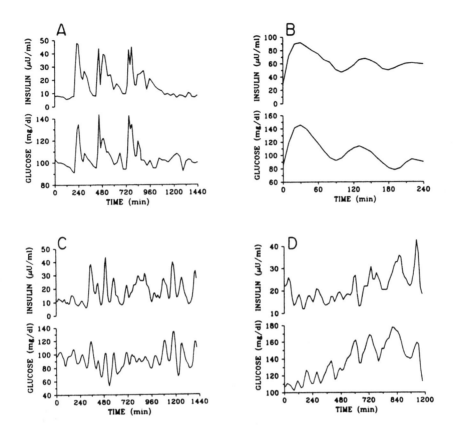

**Figure 19.10** Oscillations of insulin and glucose. A: During the ingestion of 3 meals. B: During oral glucose. C: During continuous nutrition. D: During constant glucose infusion. Oscillations with a period of around 120 minutes occur even during constant stimulation (i.e., constant glucose infusion), and occur in a damped manner after a single stimulus such as ingestion of a meal. (Sturis et al., 1991, Fig. 1.)

tion, while the frequency is not; and fourth, the oscillations do not appear to depend on glucagon.

Although there are many possible mechanisms that are consistent with the above observations, they can all be explained by a relatively simple model (Sturis et al., 1991) in which the oscillations are produced by interactions between glucose and insulin.

A schematic diagram of the model is shown in Fig. 19.13. There are three pools in the model, representing remote insulin storage in the interstitial fluid, insulin in the blood, and blood glucose. As we will see, two insulin pools are necessary, which is, by itself, an interesting model prediction. There are two delays, one explicit and the other implicit. Although plasma insulin regulates glucose production, it does so only after a delay of about 36 minutes. This delay is incorporated explicitly as a three-stage linear filter. An additional implicit delay arises because glucose utilization is regulated by the

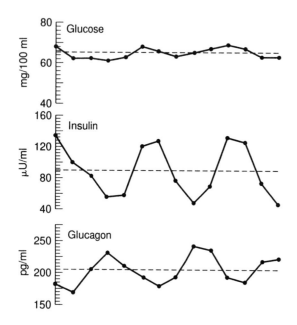

**Figure 19.11**  Intermediate frequency oscillations of glucose, insulin, and glucagon in monkeys. (Goodner et al., 1977, Fig. 1A.)

remote (interstitial) insulin, and not by the plasma insulin, while glucose has a direct effect (through insulin secretion from the pancreas) on plasma insulin levels.

We let $x, y$, in units of mU, denote the amounts of plasma insulin and remote insulin, respectively, and we let $z$, in units of mg, denote the total amount of glucose. Then the model equations follow from the following assumptions:

1.  Plasma insulin is produced at a rate $f_1(z)$ that is dependent on plasma glucose. The insulin exchange with the remote pool is a linear function of the concentration difference between the pools $x/V_1 - y/V_2$ with rate constant $E$, where $V_1$ is the plasma volume and $V_2$ is the interstitial volume. In addition, there is linear removal of insulin from the plasma by the kidneys and the liver, with rate constant $1/t_1$. Thus,

$$\frac{dx}{dt} = f_1(z) - \left(\frac{x}{V_1} - \frac{y}{V_2}\right) E - \frac{x}{t_1}. \tag{19.29}$$

Note that this equation and the two that follow are written in terms of total amounts of insulin and glucose, rather than concentrations. Formulations using concentrations or total quantities are equivalent, provided that the blood and interstitial volumes remain constant, which we assume.

2.  Remote insulin accumulates via exchange with the plasma pool and is degraded in muscle and adipose tissue at rate $1/t_2$:

$$\frac{dy}{dt} = \left(\frac{x}{V_1} - \frac{y}{V_2}\right) E - \frac{y}{t_2}. \tag{19.30}$$

3.  Plasma glucose is produced at a rate $f_5$ that is dependent on plasma insulin, but only indirectly, as $f_5$ is a function of $h_3$, the output of a three-stage linear filter. The

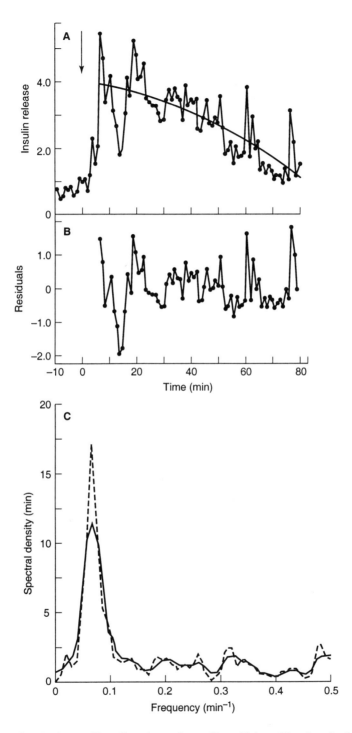

**Figure 19.12** A: Oscillations of insulin release in perifused islets. The data indicate a slow time scale decreasing trend (the smooth line) upon which are superimposed faster time scale oscillations. B: When the slow decrease is removed from the data, the residuals exhibit oscillations around 0. C: Spectral analysis of the residuals shows a frequency peak at about 0.07 min$^{-1}$, corresponding to oscillations with a period of 14.5 minutes. The dashed and continuous lines correspond to two different filters used in the spectral analysis. (Bergstrom et al., 1989. Figs. 1A, C, and 3.)

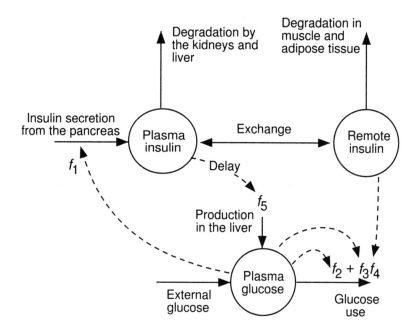

**Figure 19.13** Schematic diagram of the model of ultradian insulin oscillations.

input to the filter is $x$, so glucose production is regulated by plasma insulin but delayed by the filter. There is input $I$ from the addition of glucose from outside the system, by eating a meal, say. Finally, glucose is removed from the plasma by two processes. Thus,

$$\frac{dz}{dt} = f_5(h_3) + I - f_2(z) - f_3(z)f_4(y). \tag{19.31}$$

Glucose utilization is described by two terms: $f_2(z)$ describes utilization of glucose that is independent of insulin, as occurs, for instance, in the brain, and is an increasing function that saturates quickly. The second removal term, $f_3(z)f_4(y)$, describes insulin-dependent utilization of glucose. Both $f_3$ and $f_4$ are increasing functions, with $f_3$ linear and $f_4$ sigmoidal.

4. The three-stage linear filter sastifies the system of differential equations

$$t_3\frac{dh_1}{dt} = (x - h_1), \tag{19.32}$$

$$t_3\frac{dh_2}{dt} = (h_1 - h_2), \tag{19.33}$$

$$t_3\frac{dh_3}{dt} = (h_2 - h_3). \tag{19.34}$$

The specific functional forms used for $f_1, \ldots, f_5$ are

$$f_1(z) = \frac{209}{1 + \exp(-z/(300V_3) + 6.6)},$$ (19.35)

$$f_2(z) = 72\left[1 - \exp\left(\frac{-z}{144V_3}\right)\right],$$ (19.36)

$$f_3(z) = \frac{0.01z}{V_3},$$ (19.37)

$$f_4(y) = \frac{90}{1 + \exp\left(-1.772 \log\left\{y\left[\frac{1}{V_2} + \frac{1}{E t_2}\right]\right\} + 7.76\right)} + 4,$$ (19.38)

$$f_5(h_3) = \frac{180}{1 + \exp(0.29 h_3/V_1 - 7.5)},$$ (19.39)

and these are graphed in Fig. 19.14. The remaining model parameters are given in the caption to Fig. 19.15.

Numerical solution of the model equations shows that a constant infusion of glucose causes oscillations in insulin and glucose. As $I$ increases, the oscillation pe-

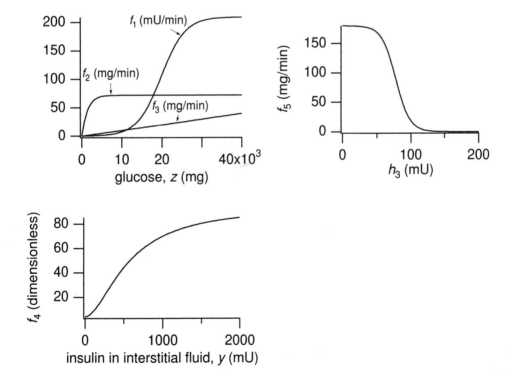

**Figure 19.14** Graphs of $f_1, \ldots, f_5$ in the model for ultradian insulin oscillations. The exact forms of these functions are not physiologically significant, but are chosen to give the correct qualitative behavior.

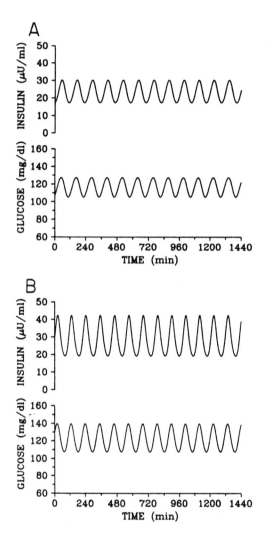

**Figure 19.15**  Ultradian insulin oscillations in the model. The glucose infusion rates are A: $I = 108$ mg/min, and B: $I = 216$ mg/min. Note that insulin and glucose are expressed in units of concentration. An amount is easily converted to a concentration by dividing by the volume of the appropriate space. Parameter values are $V_1 = 3$ liters, $t_1 = 6$ min, $V_2 = 11$ liters, $t_2 = 100$ min, $V_3 = 10$ liters, $t_3 = 12$ min, $E = 0.2$ liter/min. (Sturis et al., 1991, Fig. 5.)

riod remains practically unchanged, but the amplitude increases (Fig. 19.15), in good qualitative agreement with experimental data. However, it is interesting that these oscillations disappear if the compartment of remote insulin is removed from the model. This indicates that the division of insulin into two functionally separate stores could play an important role in the dynamic control of insulin levels. Another prediction of the model is that the oscillations are dependent on the delay in the regulation of glucose

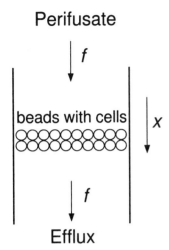

Perifusate

$f$

beads with cells

$x$

$f$

Efflux

**Figure 19.16** Schematic diagram of a typical perifusion system.

production. If the delay caused by the three-stage filter is either too large or too small, the oscillations disappear.

## 19.3.2 Insulin Oscillations with Intermediate Frequency

Insulin also oscillates with a period of about 10–20 minutes (Figs. 19.11 and 19.12). Because these oscillations occur in islets and the isolated pancreas, it appears that unlike the ultradian oscillations, they are caused by a mechanism intrinsic to the pancreatic islets.

These oscillations also occur in the experimental perifusion system depicted in Fig. 19.16. A thin layer of insulin-secreting $\beta$-cells is sandwiched between beads and exposed to the flow of a solution, the *perifusate*, with flow rate $f$. By collecting the solution exiting the bottom of the perifusion system, one can determine how the rate of insulin release of the cells in the bed depends on the composition and flow rate of the influx solution.

After a step increase in perifusate glucose concentration (to synchronize the cells), regular oscillations in the rate of insulin release are seen (Fig. 19.12). The insulin oscillations are influenced by both the flow rate and glucose concentration of the perifusate. This suggests that secretions from the cells into the fluid affect the rate of secretion. If the oscillatory mechanism were confined to the interior of the cells, differing flow rates (with the same glucose concentrations) should not alter the properties of the oscillations.

The release of insulin from a $\beta$-cell depends on the uptake of glucose by the cell. Glucose is taken up into cells by a family of glucose transporters, called GLUT-type transporters. They operate by mechanisms discussed in Chapter 2, and come in five subtypes distinguished by their affinities, capacities, and kinetic properties; the particular subtypes of concern to us here are the GLUT-1 and GLUT-2 transporters. The

GLUT-1 transporters are assumed to be activated by insulin, providing a positive feedback. Furthermore, in the presence of insulin, GLUT-1 carrier protein is recruited into the cell membrane from the cytoplasm. There is also evidence that insulin promotes the synthesis of GLUT-1 protein, thereby adding to the positive feedback effect. It is also proposed that insulin inhibits the uptake of glucose by GLUT-2 transporters, producing negative feedback.

## Model equations

A model for the release of insulin from $\beta$-cells using these assumptions was constructed by Maki and Keizer (1995a,b). If the volume flow rate, $f$, is large compared to the volume of the cells in the bed, $V_{bed}$, it is reasonable to approximate the bed as a continuously stirred mixture, and thus ignore spatial dependencies in the bed (see Exercise 4). We let $G$ and $I$ denote the concentrations of glucose and insulin, respectively, in the efflux solution, and let the subscript 0 denote the concentrations in the influx solution. Also, we let $k_0 = f/V_{bed}$. Then, the rate at which glucose flows out of the bed is $k_0G$, and the rate at which it flows into the bed is $k_0G_0$. Hence, we have the conservation laws for glucose and insulin,

$$\frac{dG}{dt} = -R_1 - R_2 - k_0(G - G_0), \tag{19.40}$$

$$\frac{dI}{dt} = R_s - k_0(I - I_0), \tag{19.41}$$

where $R_1$ is the rate of glucose uptake by GLUT-1 receptors, $R_2$ is the rate of glucose uptake by GLUT-2 receptors, and $R_s$ is the rate of insulin secretion by the cells in the bed.

When the flow rate is large enough and when the concentration of insulin in the influx is small enough, both $G$ and $I$ may be replaced by their pseudo-steady states. For if $k_0G$ is large compared to $R_1$ and $R_2$, and $k_0I/R_s$ is order 1, a simple asymptotic argument shows that

$$G = G_0, \tag{19.42}$$

$$I = I_0 + \frac{R_s}{k_0}. \tag{19.43}$$

Inside the cell, glucose is metabolized at the rate $R_m$. Thus, if $G_i$ denotes the interior concentration of glucose, then

$$\frac{dG_i}{dt} = R_1 + R_2 - R_m. \tag{19.44}$$

To complete the model equations we introduce a variable $J$, an inhibition variable (similar to $h$ or $n$ in the Hodgkin–Huxley context), which measures the extent to which insulin inhibits its own release. $J$ does not correspond directly to a measured physiological process, but is a phenomenological representation of a slow negative feedback process. The variable $J$ obeys the differential equation

$$\tau \frac{dJ}{dt} = J_\infty - J,$$   (19.45)

where

$$J_\infty = \frac{K_{\text{inh}}}{K_{\text{inh}} + I}.$$   (19.46)

Note that $J_\infty$ decreases as the concentration of insulin increases, and thus an increase in insulin leads to a decrease in $J$, with a time delay related to the time constant $\tau$.

In summary, the model equations are

$$G = G_0,$$   (19.47)

$$I = I_0 + \frac{R_s}{k_0},$$   (19.48)

$$\frac{dG_i}{dt} = R_1 + R_2 - R_m,$$   (19.49)

$$\tau \frac{dJ}{dt} = J_\infty - J.$$   (19.50)

To complete the model description, it remains to discuss the functional forms of the various rate terms. First, the rate of glucose metabolism is assumed to be an increasing function of glucose concentration. Thus,

$$R_m = \frac{V_m G_i}{K_m + G_i},$$   (19.51)

for some constants $V_m$ and $K_m$. In a similar way, the rates of the GLUT-1 and GLUT-2 transporters are assumed to be simple increasing functions of the external concentration of glucose $G_0$. $R_1$ is assumed to be an increasing function of $I$, which models the recruitment of GLUT-1 transporters by insulin and results in positive feedback,

$$R_1 = \left( \frac{V_1 G_0}{K_1 + G_0} \right) \left( \frac{I}{K_i + I} \right).$$   (19.52)

$R_2$ is assumed to be an increasing function of $J$, and thus a decreasing function of $I$, at least at steady state,

$$R_2 = J^m \frac{V_2 G_0}{K_2 + G_0} - L_g G_i.$$   (19.53)

A leak term $L_g G_i$ describes the leak of glucose out of the cell and is appended to $R_2$.

Finally, the rate of insulin secretion, $R_s$, is described by an empirical function determined by fitting to experimental data. By combining data on how $R_m$ depends on $G_0$ with data on how $R_s$ depends on $G_0$, one can determine the relationship between $R_s$ and $R_m$. We are then able to express $R_s$ in terms of $R_m$ and hence in terms of $G_i$. By doing so we circumvent the inconvenient fact that although the rate of insulin secretion depends in some way on internal glucose concentrations, this relationship has not

---

**Table 19.2**  Standard parameter values of the model for intermediate insulin oscillations.

---

| Fixed by experiment | $V_m$ | 0.24 mM/min |
|---|---|---|
| | $K_m$ | 9.8 mM |
| | $V_s$ | 0.034 mM/min |
| | $K_s$ | 0.13 mM/min |
| | $V_1$ | 34.7 mM/min |
| | $K_1$ | 1.4 mM |
| | $V_2$ | 32 mM/min |
| | $K_2$ | 17 mM |
| Experimentally variable | $k_0$ | 550/min |
| | $I_0$ | 0 mM |
| | $G_0$ | 8–19 mM |
| Adjustable | $K_{inh}$ | $1 \times 10^{-7}$ mM |
| | $K_i$ | $6 \times 10^{-8}$ mM |
| | $\tau$ | 20 min |
| | $L_g$ | 20/min |

been measured directly. The result is

$$R_s = \frac{V_s(R_m^4 + L)}{R_m^4 + K_s^4 + L} J^n. \tag{19.54}$$

The factor $J^n$ does not follow from the experimental data but is included here so that insulin exerts a direct negative feedback effect on the rate of insulin secretion.

Most of the model parameters can be determined from experimental data, and are summarized in Table 19.2. The adjustable parameters are $K_{inh}$, $K_i$, $\tau$, and $L_g$. With the exception of $L_g$, these are parameters associated with the various types of insulin feedback. Since this is the part of the model for which there is the least direct evidence, it is not surprising that these parameters cannot be determined directly from experimental data.

By choosing different values for $m$ and $n$, it is possible to vary the type of negative feedback. If $m = 0$ and $n = 1$, insulin directly inhibits the rate of insulin secretion, whereas if $m = 2$ and $n = 0$, insulin decreases the rate of glucose uptake by GLUT-2 receptors. Since this reduces the concentration of glucose inside the cell, it indirectly decreases the rate of insulin secretion.

In the direct inhibition model ($m = 0, n = 1$), as the concentration of glucose in the influx solution, $G_0$, is increased, the steady-state concentration of insulin in the efflux solution increases. This corresponds to an increase in the rate of insulin secretion. When $G_0$ is large enough, oscillations with a period of about 16 minutes arise via a Hopf bifurcation, and as $G_0$ increases further, the amplitude of the oscillations decreases until they disappear via another Hopf bifurcation. Although the period of the oscillations agrees well with experimental data, and oscillations occur at approximately the correct glucose concentrations, the decrease in amplitude with increasing $G_0$ is opposite to what is observed experimentally.

In the indirect inhibition model the opposite effect is seen. As before, the oscillations appear and disappear at Hopf bifurcations, but here the amplitude of the oscillations increases as $G_0$ increases, in better agreement with experiment. It thus appears that of the two hypotheses, indirect inhibition is the more plausible.

# 19.4 Adaptation of Hormone Receptors

It remains to answer the question of why hormone secretion is pulsatile in the first place. As with many oscillatory physiological systems, there is no completely satisfactory answer to this question. However, one plausible hypothesis has been proposed by Li and Goldbeter (1989). Based on a model of a hormone receptor first constructed by Segel, Goldbeter, and their coworkers (Segel et al., 1986; Knox et al., 1986), Li and Goldbeter constructed a model of a hormone receptor that responds best to stimuli of a certain frequency, thus providing a possible reason for the importance of pulsatility.

Closely linked to this hypothesis is the phenomenon of receptor adaptation. Often, the response to a constant hormone stimulus is much smaller than the response to a time-varying stimulus. In the extreme case, the receptor responds to a time-varying input, but has no response to a steady input, regardless of the input magnitude, a phenomenon called *exact adaptation*. We have seen a number of examples of adaptation in this book; for example, the models of the $IP_3$ receptor discussed in Chapter 5 show adaptation in their response to a step-function increase in $Ca^{2+}$ concentration; i.e., their response is an initial peak in the $Ca^{2+}$ release, followed by a decrease to a lower plateau as the receptor is slowly inactivated by $Ca^{2+}$. Similarly, in Chapter 22 we will see how biochemical feedback in photoreceptors can result in a system that displays remarkably precise adaptational properties, as embodied in Weber's law. Because of the importance of adaptation in physiological systems, it is interesting to study how adaptation arises in a simple receptor model.

The key assumption is that the hormone receptor can exist in two different conformational states, R and D, and each conformational state can have hormone bound or unbound (Fig. 19.17). For simplicity we assume that the active form of the receptor has hormone bound to the receptor in state R. The addition of hormone to the receptor system causes a change in the proportion of each receptor state, but the total receptor concentration stays fixed.

Letting $r, x, y, d$ denote $[R]/R_T, [RH]/R_T, [DH]/R_T$ and $[D]/R_T$ respectively, where $R_T$ is the total receptor concentration, we find the following equations for the receptor system:

$$\frac{dr}{dt} = -[k_1 + k_r H(t)]r + k_{-r}x + k_{-1}d, \tag{19.55}$$

$$\frac{dx}{dt} = k_r H(t)r - (k_2 + k_{-r})x + k_{-2}y, \tag{19.56}$$

$$\frac{dy}{dt} = k_2 x - (k_{-2} + k_{-d})y + k_d H(t)d. \tag{19.57}$$

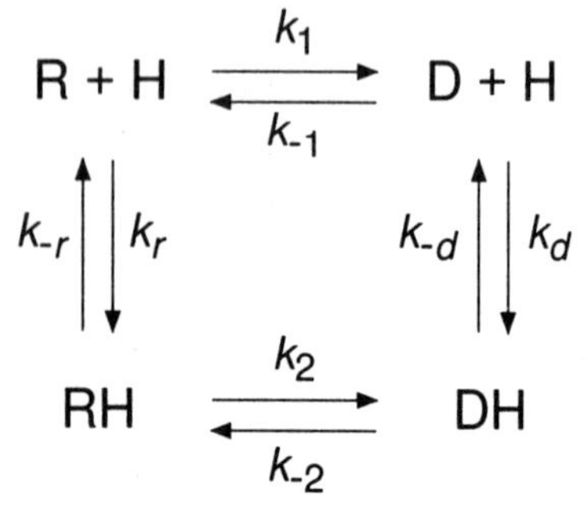

**Figure 19.17** Schematic diagram of a model of a hormone receptor.

Because of the conservation condition $r + x + y + d = 1$ there are only three independent variables, so only three equations are needed. The function $H(t)$ denotes the hormone concentration as a function of time, and is assumed known.

Each state of the receptor is assumed to have an intrinsic activity, and the total activity of the receptor is given by the sum over all the receptor states, weighted by the intrinsic activity of the state. Thus, if we let $A$ denote the total activity of the receptor, we have

$$A = a_1 r + a_2 x + a_3 y + a_4 d, \tag{19.58}$$

for some constants $a_1, \ldots, a_4$. Here, $a_1$ is the intrinsic activity of the receptor in state R, and similarly for the other $a$'s.

For simplicity, we assume that the binding of the ligand is essentially instantaneous, and thus

$$x = \frac{H(t)r}{K_r}, \tag{19.59}$$

$$y = \frac{H(t)d}{K_d}, \tag{19.60}$$

where $K_r = k_{-r}/k_r$ and $K_d = k_{-d}/k_d$. In this case, we have a single differential equation for the receptor,

$$\frac{1}{k_1}\frac{dr}{dt} = \frac{K_1 K_d}{H + K_d} - r\left(1 + \frac{K_1 K_d (K_r + H)}{(H + K_d)K_r}\right), \tag{19.61}$$

where $K_1 = k_{-1}/k_1$. The steady states are given by

$$r_0 = \frac{1}{\frac{K_d + H}{K_1 K_d} + \frac{K_r + H}{K_r}}, \tag{19.62}$$

$$x_0 = \frac{H r_0}{K_r}, \tag{19.63}$$

$$y_0 = \frac{H\left(1 - \frac{K_r + H}{K_r}r_0\right)}{K_d + H}. \tag{19.64}$$

If for simplicity we assume that $a_4 = 0$, so that the state $d$ is completely inactive, the steady-state activity of the receptor is

$$A = a_1 r_0 + a_2 x_0 + a_3 y_0 \tag{19.65}$$

$$= \frac{a_1 K_1 K_d K_r + H(a_2 K_1 K_d + a_3 K_r)}{K_r K_d (K_1 + 1) + H K_1 K_d K_r (K_r + K_1 K_d)}. \tag{19.66}$$

In general, this is a saturating curve as a function of $H$. However, exact adaptation occurs if $A$ is independent of $H$, in which case

$$A = A|_{H=0} = \lim_{H \to \infty} A, \tag{19.67}$$

so that

$$\frac{K_1 a_1}{1 + K_1} = \frac{a_3 K_r + K_1 K_d a_2}{K_r + K_1 K_d}. \tag{19.68}$$

Note that since the right-hand side of (19.68) is the weighted average of $a_2$ and $a_3$, exact adaptation is possible only when $a_1$ is greater than the smaller of $a_2$ and $a_3$. In general, $a_2$ is larger than $a_3$ (as the RH form of the receptor has a greater intrinsic activity than its inactivated form DH), and thus $a_1 > a_3$ is required for exact adaptation. In other words, the intrinsic activity of the unbound receptor (R form) must be higher than the intrinsic activity of the inactivated receptor, even when the hormone is bound.

In response to a step increase in hormone concentration, the receptor state is first quickly converted to the RH form, which has a high activity, and thus the overall activity initially increases. However, over a longer time period, the RH form gradually converts to the DH form, which has a lower activity than the R (unbound) form. This receptor inactivation decreases the activity back to the basal level. Thus, exact adaptation arises from a process of fast activation and slow inactivation, a mechanism that has appeared in many forms throughout this book.

## 19.5 Exercises

1. By taking partial derivatives of $f(\xi, \xi_i)$ with respect to $\xi_i$ confirm that the model (19.3), when linearized about $P_M$, takes the form given in (19.11)–(19.13). Calculate the eigenvalues and eigenvectors of the matrix $A$.

2. This exercise works through the derivation of a Lyapunov function for the Lacker model (Akin and Lacker, 1984). Define $\delta(\xi) = \xi^2$, $\rho(\xi) = \xi^{-2} - 1$, and $\phi(p_i) = p_i(M_1 + M_2 - M_1 M_2 p_i)$, where $p_i = \xi_i/\xi$. Show that

$$\frac{d\xi_i}{dt} = \delta(\xi)\xi_i[\rho(\xi) + \phi(p_i)], \tag{19.69}$$

$$\frac{d\xi}{dt} = \delta(\xi)\xi[\rho(\xi) + \bar{\phi}], \tag{19.70}$$

$$\frac{dp_i}{dt} = \delta(\xi)p_i[\phi(p_i) - \bar{\phi}], \tag{19.71}$$

where

$$\bar{\phi} = \sum_{i=1}^{n} p_i \phi(p_i). \tag{19.72}$$

Also, define a new time scale $\tau$ by

$$\frac{d\tau}{dt} = \delta(\xi), \tag{19.73}$$

and show that

$$\frac{dp_i}{d\tau} = p_i[\phi(p_i) - \bar{\phi}]. \tag{19.74}$$

Finally, show that

$$V(p_1, \ldots, p_n) = \sum_{i=1}^{n} \int_0^{p_i} \phi(s) \, ds \tag{19.75}$$

is a Lyapunov function for the model by showing that

$$\frac{dV}{d\tau} = \sum_{i=1}^{n} p_i[\xi(p_i) - \bar{\phi}]^2 \geq 0. \tag{19.76}$$

Hint: derive and use the fact that $\sum p_i(\xi(p_i) - \bar{\phi}) = 0$.

3.  Since $\gamma_1 \equiv 1$, (19.23)–(19.25) provide no information about $\xi_1$. Use the original variables to show that $\bar{\xi}_1(\tau) = \xi(t(\tau))$ satisfies the differential equation

$$\frac{1}{2}\frac{d}{d\tau}\bar{\xi}_1^2 = 1 - \bar{\xi}_1^2(\Gamma - M_1)(\Gamma - M_2). \tag{19.77}$$

Find $t$ as a function of $\tau$. Describe how the original variables $\xi_i(t)$ may be obtained once the $\gamma_i(\tau)$ have been obtained by numerical solution of (19.23)–(19.25). Why is it preferable to solve the model in the transformed variables $\gamma_i$ rather than the original variables $\xi_i$?

4.  Show that the conservation equation for a species in the perifusion column of Fig. 19.16 is

$$\frac{\partial \rho}{\partial t} + v\frac{\partial \rho}{\partial x} = r, \tag{19.78}$$

where $\rho$ is the concentration of the species, $v$ is the velocity of the flow, and $r$ is the rate of change due to reactions in the column. Integrate over the cell volume, $V_{bed}$, to get

$$\frac{\partial \bar{\rho}}{\partial t} = \bar{r} - k_0(\bar{\rho} - \rho_0), \tag{19.79}$$

where $\bar{r}$ and $\bar{\rho}$ are the average values of $r$ and $\rho$ in the cell layer, $\rho_0$ is the inflow value of $\rho$, and $k_0 = vA_{bed}/V_{bed} = f/V_{bed}$. Hint: use the approximation that the outflow value of $\rho$ is approximately equal to the average value of $\rho$ in the cell layer.

5.  Use phase-plane analysis to discuss the behavior of the insulin secretion model (19.47)–(19.50).

6.  For the model of a hormone receptor assuming fast ligand binding (Section 19.4), calculate the response to a step function, and then to a stimulus of the form $H(t) = 1$ for $0 < t < t_0$, $H(t) = 0$ otherwise (call this stimulus a step pulse). Calculate the response to a series of step pulses, and calculate the width of the pulse and the time between pulses so as to obtain the greatest average activity. Li and Goldbeter (1989) give the details.

7.  (From Loeb and Strickland, 1987.) Many cells respond maximally to a hormone concentration that is much too low to saturate the hormone receptors. This can be explained by

assuming that the response is dependent on a secondary mediator. Suppose that the hormone, H, combines reversibly with its receptor, $R^o$, to form the complex $HR^o$. Suppose that the secondary mediator M is formed at a rate proportional to $[HR^o]$ and is degraded with linear kinetics. Finally, suppose that M combines reversibly with its own receptor, R, to form MR, and that the cellular response is linearly proportional to [MR]. What is the fractional receptor occupancy as a function of [H]? Show that the fractional response (as a function of [H]) has the same shape as the fractional receptor occupancy curve, shifted to the left by a constant factor (when plotted against log[H]). Give a biological interpretation.

# CHAPTER 20

# Renal Physiology

The kidneys perform two major functions. First, they excrete most of the end products of bodily metabolism, and second, they control the concentrations of most of the constituents of the body fluids. The main goal of this chapter is to gain some understanding of the process by which the urine is formed and waste products removed from the bloodstream. The control of the constituents of the body fluids is discussed only secondarily.

The primary operating unit of the kidney is called a *nephron*, of which there are about a million in each kidney (Figs. 20.1 and 20.2). Each nephron is capable of forming urine by itself. The entrance of blood into the nephron is by the *afferent arteriole*, located in the renal cortex, and the tubules of the nephron and the associated peritubular capillaries extend deep into the renal medulla. The principal functional units of the nephron are the *glomerulus*, through which fluid is filtered from the blood; the *juxtaglomerular apparatus*, by which glomerular flow is controlled; and the *long tubule*, in which the filtered fluid is converted into urine.

## 20.1 The Glomerulus

The first stage of urine formation is the production of a filtrate of the blood plasma. The glomerulus, the primary filter, is a network of up to 50 parallel branching and anastomosing (rejoining) capillaries covered by epithelial cells and encased by *Bowman's capsule*. Blood enters the glomerulus by way of the afferent arteriole and leaves through the *efferent arteriole*. Pressure of the blood in the glomerulus causes the fluid to filter into Bowman's capsule, carrying with the filtrate all the dissolved substances of small molecular weight. The glomerular membrane is almost completely impermeable

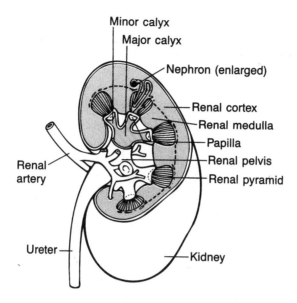

**Figure 20.1** The kidney. (Guyton and Hall, 1996, Fig. 26-2, p. 317.)

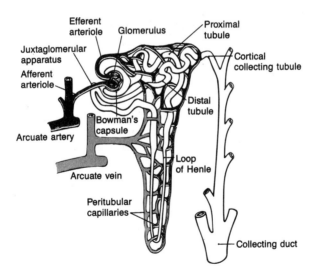

**Figure 20.2** The nephron. (Guyton and Hall, 1996, Fig. 26-3, p. 318.)

to all plasma proteins, the smallest of which is albumin (molecular weight 69,000). As a result, the glomerular filtrate is identical to plasma except that it contains no significant amount of protein.

The quantity of filtrate formed each minute is called the *glomerular filtration rate*, and in a normal person averages about 125 ml/min. The filtration fraction is the fraction

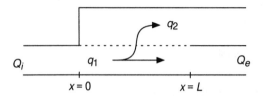

**Figure 20.3**  Schematic diagram of the glomerular filtration.

of renal plasma flow that becomes glomerular filtrate and is typically about 20 percent. Over 99 percent of the filtrate is reabsorbed in the tubules, with the remaining small portion passing into the urine.

There are three pressures that affect the rate of glomerular filtration. These are the pressure inside the glomerular capillaries that promote filtration, the pressure inside Bowman's capsule that opposes filtration, and the colloidal osmotic pressure (cf. Chapter 2, Section 2.7) of the plasma proteins inside the capillaries that opposes filtration.

A mathematical model of the glomerular filter can be described simply as follows. We assume that the glomerular capillaries comprise a one-dimensional tube with flow $q_1$ and that the surrounding Bowman's capsule is also effectively a one-dimensional tube with flow $q_2$ (Fig. 20.3). Since the flow across the glomerular capillaries is proportional to the pressure difference across the capillary wall, the rate of change of the flow in the capillary is

$$\frac{dq_1}{dx} = K_f(P_2 - P_1 + \pi_c), \tag{20.1}$$

where $P_1$ and $P_2$ are the hydrostatic fluid pressures in tubes 1 and 2, respectively, $\pi_c$ is the osmotic pressure of suspended proteins and formed elements of blood, and $K_f$ is the capillary filtration rate. The osmotic pressure of the suspended proteins is given by

$$\pi_c = RTc, \tag{20.2}$$

where $c$, the concentration expressed in moles per liter, is a function of $x$, since the suspension becomes more concentrated as it moves through the glomerulus. Since the large proteins bypass the filter, we have the conservation equation

$$c_i Q_i = c q_1, \tag{20.3}$$

where $c_i$ is the input concentration and $Q_i$ is the input flux. It follows that

$$\pi_c = \pi_i \frac{Q_i}{q_1}, \tag{20.4}$$

where $\pi_i = RTc_i$ is the input osmotic pressure. Since the hydrostatic pressure drop in the glomerulus is small compared to the pressure drop in the efferent and afferent arterioles, we take $P_1$ and $P_2$ to be constants.

Equation (20.1) along with (20.4) gives a first-order differential equation for $q_1$, which is easily solved. Setting $q_1(L) = Q_e$ we find that

$$\frac{Q_e}{Q_i} + \alpha \ln \left( \frac{\frac{Q_e}{Q_i} - \alpha}{1 - \alpha} \right) = 1 - K_f L \frac{\pi_i}{\alpha Q_i}, \tag{20.5}$$

where $Q_e$ is the efflux through the efferent arterioles, $L$ is the length of the filter, and $\alpha = \pi_i/(P_1 - P_2)$.

Finally, we assume that the pressures and flow rates are controlled by the input and output arterioles, via

$$P_a - P_1 = R_a Q_i, \tag{20.6}$$

$$P_1 - P_e = R_e Q_e, \tag{20.7}$$

and that the flow out of the glomerulus into the proximal tubule is governed by

$$P_2 - P_d = R_d(Q_i - Q_e), \tag{20.8}$$

where $P_a$, $P_e$, and $P_d$ are the afferent arteriole, efferent arteriole, and descending tubule pressures, respectively, and $R_a$, $R_e$, and $R_d$ are the resistances of the afferent and efferent arterioles and proximal tubule, respectively. Typical values are $P_1 = 60, P_2 = 18, P_a = 100, P_e = 18, P_d = 14 - 18, \pi_i = 25$ mm Hg, with $Q_i = 650, Q_d = Q_i - Q_e = 125$ ml/min.

The flow rates and pressures vary as functions of the arterial pressure. To understand something of this variation, in Fig. 20.4 is shown the renal blood flow rate $Q_i$ and the glomerular filtration flow rate as functions of the arterial pressure. It is no surprise that both of these are increasing functions of arterial pressure $P_a$.

The strategy for numerically computing this curve is as follows: with resistances $R_a$ and $R_e$ and pressures $P_e$, $P_d$, and $\pi_i$ specified and fixed at typical levels, we pick a value for glomerular filtrate $Q_d = Q_i - Q_e$. For this value, we solve (20.5) (using a simple bisection algorithm) to find both $Q_i$ and $Q_e$. From these, the corresponding pressures $P_a, P_1$, and $P_2$ are determined from (20.6) and (20.7), and plotted.

For this model, the filtration rate varies substantially as a function of arterial pressure. However, in reality (according to data shown in Fig. 20.5), the glomerular filtration rate remains relatively constant even when the arterial pressure varies between 75 to 160 mm Hg, indicating that there is some autoregulation of the flow rate.

## 20.1.1  The Juxtaglomerular Apparatus

The need for autoregulation of the glomerular filtration rate is apparent. If the flow rate of filtrate is too slow, then we expect reabsorption to be too high, and the kidney fails to eliminate necessary waste products. On the other hand, at too high a flow rate, the tubules are unable to reabsorb those substances that need to be preserved and not eliminated, so that valuable substances are lost into the urine.

The idea of how to regulate the flow of filtrate is simple to understand. If you had a leaky hose and wanted to control the leakage rate precisely, regardless of the total flow

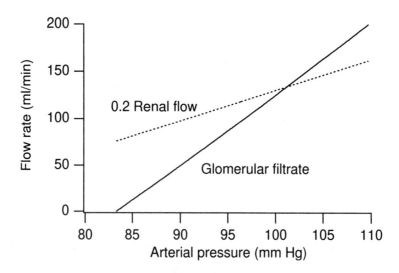

**Figure 20.4** Unregulated glomerular filtration and renal blood flow plotted as functions of arterial pressure, with $P_d = 18$, $P_e = 0$ mm Hg.

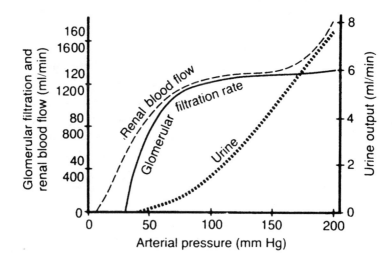

**Figure 20.5** Autoregulation of renal blood flow and glomerular filtration rate but lack of autoregulation of urine flow during changes in renal arterial pressure. (Guyton and Hall, 1996, Fig. 26-13, p. 327.)

rate, you could do so by regulating the outflow pressure at the end of the hose. The way the glomerulus controls the rate of filtration is similar. After its descent into the renal medulla, the long tubule returns to the proximity of the afferent and efferent arterioles at the glomerulus. The *juxtaglomerular complex* consists of *macula densa* cells in the

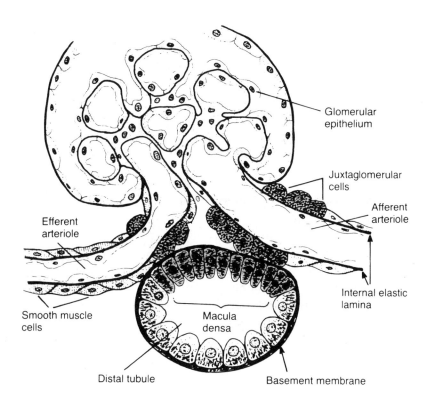

**Figure 20.6** Structure of the juxtaglomerular apparatus. (Guyton and Hall, 1996, Fig. 26-14, p. 328.)

distal tubule and *juxtaglomerular cells* in the walls of the afferent and efferent arterioles (as depicted in Fig. 20.6).

A low flow rate causes excessive reabsorption of $Na^+$ and chloride ions in the ascending limb of the loop of Henle, resulting in too large a decrease of these ionic concentrations at the end of the loop. The macula densa cells respond to decreases of $Na^+$ concentration (by a mechanism not completely understood), by releasing a vasodilator that decreases the resistance of the afferent arterioles. Simultaneously, the juxtaglomerular cells release renin, an enzyme that enables the formation of angiotensin II, which constricts the efferent arterioles. The simultaneous effect of these is to increase the flow of filtrate through the glomerulus.

A simple model to incorporate the effects of the vasodilator and vasoconstrictor (angiotensin) is to allow the arteriole resistances to depend on the rate of filtration, $Q_d = Q_i - Q_e$, via some functional dependence

$$R_a = f_a(Q_d - Q_t), \tag{20.9}$$

$$R_e = f_e(Q_d - Q_t), \tag{20.10}$$

where $Q_t$ is the target flow rate, about 125 ml/min. A more realistic model would take $R_a$ and $R_e$ to be functions of the $Na^+$ concentration at the distal end of the loop of Henle. However, since we do not yet have a model relating flow rate to $Na^+$ concentration, we leave this to interested readers to pursue on their own.

We take $f_a$ to be an increasing function of its argument, and we take $f_e$ to be a decreasing function of its argument. As a specific example, we take

$$R_a = r_a[1 + \tanh(\delta_a(Q_d - Q_t))], \qquad (20.11)$$

$$R_e = r_e[1 - \tanh(\delta_e(Q_d - Q_t))], \qquad (20.12)$$

where $\delta_a$ and $\delta_e$ are parameters that determine the sensitivity of the model to changes in flow rates, and $r_a$ and $r_e$ are "normal" values of the resistances. With $\delta_a$ and $\delta_e$ zero, the flow is unregulated. There is no direct evidence for these functional forms, so these results are qualitative at best. Plots of the functions $R_a/r_a$ and $R_e/r_e$ are shown in Fig. 20.7, with $\delta_a = 0.1$, $\delta_e = 0.01$, $Q_t = 125$ ml/min. With these parameters, control of afferent resistance is stronger than that of efferent resistance.

In Fig. 20.8 are shown the glomerular filtration and the renal blood flow as functions of the arterial pressure, in the case $\delta_a = 0.1$, $\delta_e = 0.01$. This simple model gives accept-

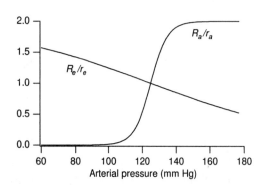

**Figure 20.7** Relative resistances $R_a/r_a$ and $R_e/r_e$ plotted as functions of glomerular flow rate $Q_d$ with $\delta_a = 0.1$, $\delta_e = 0.01$, $q_t = 125$ ml/min.

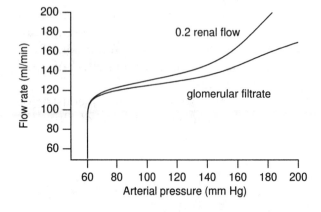

**Figure 20.8** Autoregulated glomerular filtration flow rate and renal blood flow rate, with $\delta_a = 0.1$, $\delta_e = 0.01$, $Q_t = 125$ ml/min, and with $P_d = 18$, $P_e = 0$ mm Hg.

able agreement with data, although there was no attempt to find a good quantitative fit.

## 20.2 Urinary Concentration: The Loop of Henle

The challenge of any model of urine formation is to see how concentrating and diluting mechanisms work together to determine the composition of the urine and to regulate the interstitial contents, and then to account quantitatively for the concentrating ability of particular species. The challenge is substantial. For example, for humans, the maximal urine concentrating ability is 1200 mOsm/liter, while some desert animals, such as the Australian hopping mouse, can concentrate urine to as high as 10,000 mOsm/liter. It is not understood how such high urine concentrations can be obtained. It is also necessary that the kidney be able to produce a dilute urine under conditions of high fluid intake.

A normal 70 kg human must excrete about 600 mOsm of solute (waste products of metabolism and ingested ions) every day. The minimal amount of urine to transport these solutes, called the *obligatory urine volume* is

$$\text{obligatory volume} = \frac{\text{total solute/day}}{\text{maximal urine concentration}} \tag{20.13}$$

$$= \frac{600 \text{ mOsm/day}}{1200 \text{ mOsm/L}} = 0.5\text{L/day}. \tag{20.14}$$

This explains why severe dehydration occurs from drinking seawater. The concentration of salt in the oceans averages 3% sodium chloride, with osmolarity between 2000 and 2400 mOsm/liter. Drinking 1 liter of water with a concentration of 2400 mOsm/liter provides 2400 mOsm of solute that must be excreted. If the maximal urine concentration is 1200 mOsm/liter, then 2 liters of urine are required to rid the body of this ingested solute, a deficit of 1 liter, which must be drawn from the interstitial fluid. This explains why shipwreck victims who drink seawater are rapidly dehydrated, while (as Guyton and Hall have kindly pointed out) the victim's pet Australian hopping mouse can drink all the seawater it wants with impunity.

Urinary concentration or dilution is accomplished primarily in the loop of Henle. After leaving Bowman's capsule, the glomerular filtrate flows into a tubule having five sections: the *proximal tubule*, the *descending limb of the loop of Henle*, the *ascending limb of the loop of Henle*, the *distal tubule*, and, finally, the *collecting duct*. These tubules are surrounded by capillaries, called the *peritubular capillaries*, that reabsorb the fluid that has been extracted from the tubules. In Fig. 20.9 are shown the relative concentrations of various substances at different locations along the tubular system.

The purpose of the proximal tubule is to extract much of the water and dissolved chemicals (electrolytes, glucose, various amino acids, etc.) to be reabsorbed into the bloodstream while concentrating the waste products of metabolism. It is this concentrate that eventually flows as urine into the bladder. The proximal tubular cells

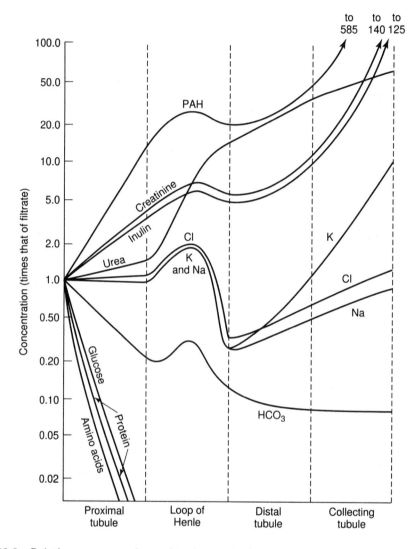

**Figure 20.9**  Relative concentrations of various substances as functions of distance along the renal tubule system. (Guyton and Hall, 1996, Fig. 27-11, p. 341.)

have large numbers of mitochondria to support rapid active transport processes. Indeed, about 65 percent of the glomerular filtrate is reabsorbed before reaching the descending limb of the loop of Henle. Furthermore, glucose, proteins, amino acids, acetoacetate ions, and the vitamins are almost completely reabsorbed by active cotransport processes through the epithelial cells that line the proximal tubule.

Any substance that is reabsorbed into the bloodstream must first pass through the tubular membrane into the interstitium and then into peritubular capillaries. There are three primary mechanisms by which this transport takes place, all of which we have

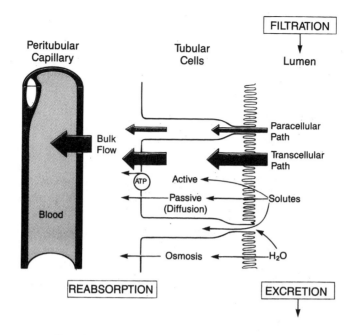

**Figure 20.10** Schematic diagram of the reabsorption of water and solutes in the proximal tubule. (Guyton and Hall, 1996, Fig. 27-1, p. 332.)

seen before (Fig. 20.10). First, there is active transport of $Na^+$ from the interior of the epithelial cells into the interstitium, mediated by a $Na^+$–$K^+$ ATPase pump. Although this pump actively pumps $K^+$ into the cell from the interstitium, both sides of the tubular epithelial cells are so permeable to $K^+$ that virtually all of the $K^+$ leaks back out of the cell almost immediately.

There are secondary transporters that use the gradient of $Na^+$ ions (established by the ATPase) to transport other substances from the tubular lumen into the interior of the epithelial cell. The most important of these are cotransporters of glucose and amino acid ions, but the epithelial cells of the proximal tubule also contain transporters of phosphate, calcium, and magnesium ions. There is also a transporter that exchanges hydrogen ions for $Na^+$ ions across the membrane of the epithelial cell membrane into the tubule. The third mechanism of transport is that of water across cell membranes, mediated by osmotic pressure (see Chapters 2 and 21).

The descending limb of the loop of Henle is lined with thin epithelial cells with few mitochondria, indicating minimal metabolic activity; it is highly permeable to water and moderately permeable to $Na^+$, urea, and most ions. The ascending limb of the loop of Henle begins with a thin wall but then about halfway up becomes grossly thickened. In contrast to the descending limb, the ascending limb is highly impermeable to water and urea. The cells of the thick ascending limb are similar to those of the proximal tubule, suited for strong active transport of $Na^+$ and $Cl^-$ ions from the tubular lumen into the interstitial fluid.

The thick segment travels back to the region of the glomerulus, where it passes between the afferent and efferent arterioles, forming the juxtaglomerular apparatus, where much of the feedback control of the flow rate takes places. Passing beyond this point, the tubule becomes the distal tubule, the function of which is similar to that of the ascending limb of the loop of Henle.

Finally, the flow enters the descending collecting duct, which gathers the flow from several nephrons and descends back through the cortex and into the outer and inner zones of the medulla. The flow from the collecting duct then flows out of the kidney through the ureter on the way to the bladder. The cells lining the collecting duct are sensitive to a number of hormones that act to regulate their function as well as the final chemical composition of the urine. Primary among these hormones are *aldosterone* and *antidiuretic hormone* (ADH). Aldosterone determines the rate at which $Na^+$ ions are transported out of the tubular lumen, and ADH determines the permeability of the collecting duct to water, and thereby determines the final concentration of the urine. When there is no ADH present, the collecting duct is impermeable to water, but with ADH present, the permeability of the collecting duct allows water to be reabsorbed out of the collecting duct, leaving behind a more highly concentrated urine.

Putting this all together, we arrive at a qualitative summary of how a nephron operates. Along the ascending limb of the loop of Henle portion of the tubule, $Na^+$ is absorbed into the interstitium, either passively (in the thin ascending limb) or actively (in the thick ascending limb). This creates a high $Na^+$ concentration in the interstitium, which then serves to draw water out of the descending limb and allows $Na^+$ to reenter the descending limb. Hence, fluid entering the descending limb is progressively concentrated until, at the turning point of the loop, the fluid osmolarity is about 1200 mOsm/liter (compared to the entering fluid, which is about 300 mOsm/liter). Clearly, because the fluid entering the ascending limb is so concentrated, $Na^+$ extraction from the ascending limb is enhanced, which further enhances water extraction from the descending limb, and so on. This positive feedback process is at the heart of the countercurrent mechanism, to be discussed in more detail below. As the fluid ascends the ascending limb, $Na^+$ is continually extracted until, at the level of the juxtaglomerular apparatus, the fluid in the tubule is considerably more dilute than the original filtrate. However (and this is the crucial part), the dilution process results in a steep gradient of $Na^+$ concentration in the interstitium, a gradient that can, when needed, concentrate the urine.

When there is no ADH present, the dilute urine formed by the loop of Henle proceeds through the collecting duct essentially unchanged, resulting in a large quantity of dilute urine. In the presence of large amounts of ADH, the collecting duct is highly permeable to water, so that by the time the filtrate reaches the level of the turning point of the loop of Henle, it is essentially at the same concentration as the interstitium, about 1200 mOsm/liter, thus giving a small quantity of concentrated urine.

It is important to emphasize that the principal functions of the loop of Henle are, first, the formation of dilute urine, which allows water to be excreted when necessary, and, second, the formation of the interstitial gradient in $Na^+$ concentration, which allows for the formation of a concentrated urine when necessary. The importance of the loop of Henle in creating the interstitial gradient of $Na^+$ concentration is underlined by the fact that although all vertebrates can produce dilute urine, only birds and mammals can produce hyperosmotic urine, and it is the kidneys of only these animals that contain loops of Henle.

## 20.2.1 The Countercurrent Mechanism

Solutes are exchanged between liquids by diffusion across their separating membranes. Since the rate of exchange is affected by the concentration difference across the membrane, the exchange rate is increased if large concentration differences can be maintained. One important way that large concentration differences can be maintained is by the *countercurrent mechanism*. As we will see, the countercurrent mechanism is important to renal function. Other examples of the countercurrent mechanism include the exchange of oxygen from water to blood through fish gills and the exchange of oxygen in the placenta between mother and fetus.

Suppose that two gases or liquids containing a solute flow along parallel tubes separated by a permeable membrane. We model this in the simplest possible way as a one-dimensional problem, and we assume that solute transport is a linear function of the concentration difference. Then the concentrations in the two one-dimensional tubes are given by

$$\frac{\partial C_1}{\partial t} + q_1 \frac{\partial C_1}{\partial x} = d(C_2 - C_1), \tag{20.15}$$

$$\frac{\partial C_2}{\partial t} + q_2 \frac{\partial C_2}{\partial x} = d(C_1 - C_2). \tag{20.16}$$

The mathematical problem is to find the outflow concentrations, given that the inflow concentrations, the length of the exchange chamber, and the flow velocities are known. It is a relatively easy matter to generalize this model to allow for an interstitium (see Exercise 4).

We assume that the flows are in steady state and that the input concentrations are $C_1^0$ and $C_2^0$. Then, if we add the two governing equations and integrate, we find that

$$q_1 C_1 + q_2 C_2 = k \text{ (a constant)}. \tag{20.17}$$

Pretending that $k$ is known, we eliminate $C_2$ from (20.16) and find the differential equation for $C_1$,

$$\frac{dC_1}{dx} = \frac{d}{q_1 q_2} \left( k - (q_1 + q_2) C_1 \right), \tag{20.18}$$

from which we learn that

$$C_1(x) = \kappa + (C_1(0) - \kappa)e^{-\lambda x}, \tag{20.19}$$

where $\kappa = \frac{k}{q_1 + q_2}$ and $\lambda = d\left(\frac{q_1 + q_2}{q_1 q_2}\right)$.

There are two cases to consider, namely when $q_1$ and $q_2$ are of the same sign and when they have different signs. If they have the same signs, say positive, then the input is on the left at $x = 0$, and it must be that $C_1(0) = C_1^0, C_2(0) = C_2^0$, from which, using (20.17), it follows that

$$\frac{C_1(L)}{C_1^0} = \frac{1 + \gamma\rho}{1 + \rho} + \rho\frac{1 - \gamma}{1 + \rho}e^{-\lambda L}, \qquad (20.20)$$

where $\gamma = C_2^0/C_1^0, \rho = q_2/q_1, \lambda = \frac{d}{q_1}(1 + \frac{1}{\rho})$.

Suppose that we are attempting to transfer material from vessel 1 to vessel 2, so that $\gamma < 1$. We learn from (20.20) that the output concentration from vessel 1 is an exponentially decreasing function of the residence length $dL/q_1$. Furthermore, the best that can be done (i.e., as $dL/q_1 \to \infty$) is $\frac{1+\gamma\rho}{1+\rho}$.

In the case that $q_1$ and $q_2$ are of opposite sign, say $q_1 > 0, q_2 < 0$, the inflow for vessel 1 is on the left at $x = 0$, but the inflow for vessel 2 is on the right at $x = L$. In this case we calculate that

$$\frac{C_1(L)}{C_1^0} = \frac{-\gamma\rho + (1 - \rho + \gamma\rho)e^{-\lambda L}}{e^{-\lambda L} - \rho}, \qquad (20.21)$$

where $\gamma = C_2(L)/C_1^0 = C_2^0/C_1^0, \rho = -q_2/q_1 > 0, \lambda = \frac{d}{q_1}(1 - \frac{1}{\rho})$, provided that $\rho \neq 1$. In the special case $\rho = 1$, we have

$$\frac{C_1(L)}{C_1^0} = \frac{q_1 + \gamma dL}{q_1 + dL}. \qquad (20.22)$$

Now we can see the substantial difference between a cocurrent ($q_1$ and $q_2$ of the same sign) and a countercurrent ($q_1$ and $q_2$ with the opposite sign). At fixed parameter values, if $\gamma < 1$, then the expression for $C_1(L)/C_1^0$ in (20.20) is always larger than that in (20.21), implying that the total transfer of solute is always more efficient with a countercurrent than with a cocurrent.

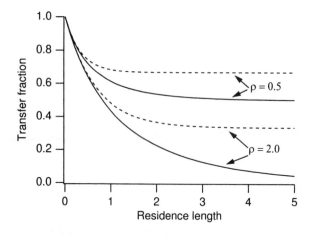

**Figure 20.11** Transfer fraction for a cocurrent (dashed) and a countercurrent (solid) when $\gamma = 0$ plotted as a function of residence length.

In Fig. 20.11 is shown a comparison between a countercurrent and a cocurrent. The dashed curves show the transfer fraction $C_1(L)/C_1^0$ for a cocurrent, plotted as a function of the residence length $dL/q_1$, with input in tube 2, $C_2(0) = 0$. The solid curves show the same quantity for a countercurrent, with input concentration $C_2(L) = 0$.

In the limit of a long residence time (large $dL/q_1$), the transfer fraction becomes $1 - \rho + \gamma\rho$ if $\rho < 1$, and $\gamma$ if $\rho > 1$. Indeed, this is always smaller than the result for a cocurrent, $\frac{1+\gamma\rho}{1+\rho}$.

## 20.2.2 The Countercurrent Mechanism in Nephrons

The countercurrent mechanism works slightly differently in nephrons because the two parallel tubes, the descending branch and the ascending branch of the loop of Henle, are connected at their bottom end. Thus the flow and concentration of solute out of the descending tube must match the flow and concentration of solute into the ascending tube.

Mathematical models of the urine-concentrating mechanism have been around for some time, but all make use of the same basic physical principles, namely, the establishment of chemical gradients via active transport processes, the movement of ions via diffusion, and the transport of water by osmosis. The unique feature of the nephron is its physical organization, which allows it to eliminate waste products while controlling other quantities. In what follows we present a model similar to that of Stephenson (1972, 1992) of urinary concentration that represents the gross organizational features of the loop of Henle. A number of other models are discussed in a special issue of the *Bulletin of Mathematical Biology* (volume 56, number 3, May 1994), while two useful reviews of mathematical work on the kidney are Knepper and Rector (1991) and Roy et al. (1992).

We view the loop of Henle as consisting of four compartments, including three tubules, the descending limb, the ascending limb, and the collecting duct, and a single compartment for the interstitium and peritubular capillaries (Fig. 20.12). The interstitium/capillary bed is treated as a one-dimensional tubule that accepts fluid from the other three tubules and loses it to the venules. It is an easy generalization to separate the peritubular capillaries and interstitium into separate compartments, but little is gained by doing so. In each of these compartments, one must keep track of the flow of water and the concentration of solutes. For the model presented here, we track only one solute, $Na^+$, because it is believed that the concentration of $Na^+$ in the interstitium determines over 90 percent of the osmotic pressure.

We assume that the flow in each of the tubes is a simple plug flow (positive in the positive $x$ direction) with flow rates $q_d, q_a, q_c, q_s$ for descending, ascending, collecting, and interstitial tubules, respectively. Similarly, the concentration of solute in each of these is denoted by $c_d, c_a, c_c, c_s$. The tubules are assumed to be one-dimensional, with glomerular filtrate entering the descending limb at $x = 0$, turning from the descending limb to the ascending limb at $x = L$, turning from the ascending limb to the collecting

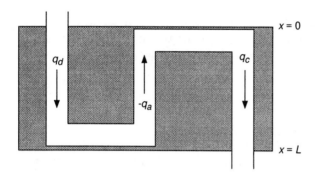

$x = 0$

$q_d$

$-q_a$

$q_c$

$x = L$

**Figure 20.12** Diagram of the simple four-compartment model of the loop of Henle.

duct at $x = 0$, and finally exiting the collecting duct at $x = L$. We assume that the interstitium/capillary compartment drains at $x = 0$ with no flow at $x = L$.

**Descending limb:** The flux of water from the descending limb to the interstitium is controlled by the pressure difference and the osmotic pressure difference; hence

$$\frac{1}{k_d}\frac{dq_d}{dx} = P_s - \pi_s - P_d + 2RT(c_d - c_s), \tag{20.23}$$

where $P_d$ and $P_s$ are the hydrostatic pressures in the descending tubule and interstitium, $\pi_s$ is the colloidal osmotic pressure of the interstitium, and $k_d$ is the filtration rate for the descending tubule. The factor two multiplying the osmotic pressure due to the solute is to take into account the fact that the fluid is electrically neutral, and the flow of $Na^+$ ions is followed closely by a flow of chloride ions, both of which contribute to the osmotic pressure. The transport of $Na^+$ ions from the descending limb is governed by simple diffusion, so that at steady state we have

$$\frac{d(q_d c_d)}{dx} = h_d(c_s - c_d), \tag{20.24}$$

where $h_d$ is the permeability of the descending limb to $Na^+$ ions.

**Ascending limb:** The ascending limb is assumed to be impermeable to water, so that

$$\frac{dq_a}{dx} = 0, \tag{20.25}$$

and the flow of $Na^+$ out of the ascending limb is by an active process, so that

$$\frac{d(q_a c_a)}{dx} = -p. \tag{20.26}$$

The pump rate $p$ certainly depends in nontrivial ways on the local concentrations of various ions. However, for this model we take $p$ to be a constant. This simplifying assumption causes problems with the behavior of the model at low $Na^+$ concentrations, because it allows the $Na^+$ concentration to become negative. Although the $Na^+$ ATPase is actually a $Na^+$–$K^+$ pump, the epithelial cells are highly permeable to $K^+$, and so we assume that $K^+$ can be safely ignored. For simplicity, we also ignore the fact that the $Na^+$ transport properties of the thin ascending limb are different from those of the thick ascending limb, and we assume active removal along the entire ascending limb.

**Collecting duct:** The flow of water from the collecting duct is also controlled by the hydrostatic and osmotic pressure differences, via

$$\frac{1}{k_c}\frac{dq_c}{dx} = P_s - \pi_s - P_c + 2RT(c_c - c_s), \tag{20.27}$$

and the transport of $Na^+$ from the collecting duct is governed by

$$\frac{d(q_c c_c)}{dx} = h_c(c_s - c_c). \tag{20.28}$$

Here, $k_c$ and $h_c$ are the permeability of the collecting duct to water and $Na^+$, and are controlled by ADH and aldosterone, respectively.

**Conservation equations:** Finally, because total fluid is conserved,

$$\frac{dq_s}{dx} = -\frac{d}{dx}(q_d + q_a + q_c), \tag{20.29}$$

and because total solute is conserved,

$$\frac{d(q_s c_s)}{dx} = -\frac{d}{dx}(q_d c_d + q_a c_a + q_c c_c). \tag{20.30}$$

To complete the description, we have the relationship between pressure and flow in a tube,

$$\frac{dP_j}{dx} = -R_j q_j, \tag{20.31}$$

for $j = d, a, c, s$. However, for renal modeling it is typical to take each pressure to be constant. Typical values for the pressures are $P_d = 14$–$18$ mm Hg, $P_a = 10$–$14$ mm Hg, $P_c = 0$–$10$ mm Hg, $P_s = 6$ mm Hg, and $\pi_s = 17$ mm Hg.

This description of the nephron consists of eight first-order differential equations in the eight unknowns $q_j$ and $c_j$, for $j = d, a, c, s$. To complete the description, we need boundary conditions. We assume that the inputs $q_d(0)$ and $c_d(0)$ are known and given. Then, because the flow from the descending limb enters the ascending limb, $q_d(L) = -q_a(L)$ and $c_d(L) = c_a(L)$. Furthermore, $q_s(L) = 0$. At $x = 0$, flow from the ascending limb enters the collecting duct, so that $q_a(0) = -q_c(0)$ and $c_a(0) = c_c(0)$. Finally, since total fluid must be conserved, what goes in must go out, so that $q_d(0) + q_s(0) = q_c(L)$.

It is useful to nondimensionalize the equations by normalizing the flows and solute concentrations. Thus, we let

$$x = Ly, Q_j = \frac{q_j}{q_d(0)}, C_j = \frac{c_j}{c_d(0)} \text{ for } j = d, a, c, s,$$

and the dimensionless parameters are

$$\rho_j = \frac{q_d(0)}{2LRTc_d(0)k_j}, \Delta P_j = \frac{P_j + \pi_s - P_s}{RT2c_d(0)}, H_j = \frac{Lh_j}{q_d(0)}, \text{ for } j = d, c.$$

In this scaling $Q_d(0) = C_d(0) = 1$.

Three of these equations are trivially solved. In fact, it follows easily from (20.25), (20.29), and (20.30) that

$$Q_a = Q_a(0) = Q_a(L), \tag{20.32}$$

$$Q_d + Q_a + Q_c + Q_s = Q_c(L), \tag{20.33}$$

$$Q_dC_d + Q_aC_a + Q_cC_c + Q_sC_s = Q_c(L)C_c(L). \tag{20.34}$$

Two more identities can be found. If we use (20.24) to eliminate $c_d - c_s$ from (20.23), we obtain

$$\rho_d \frac{dQ_d}{dy} + \Delta P_d = C_d - C_s = -\frac{1}{H_d} \frac{d(Q_dC_d)}{dy}, \tag{20.35}$$

from which it follows that

$$\rho_d(Q_d - 1) + \frac{1}{H_d}(Q_dC_d - 1) = -\Delta P_d y. \tag{20.36}$$

Similarly, we use (20.28) to eliminate $c_c - c_s$ from (20.27) to obtain

$$\rho_c \frac{dQ_c}{dy} + \frac{1}{H_c} \frac{d(Q_cC_c)}{dy} = -\Delta P_c, \tag{20.37}$$

which integrates to

$$\rho_c(Q_c - Q_c(0)) + \frac{1}{H_c}(Q_cC_c - Q_c(0)C_c(0)) = -\Delta P_c y. \tag{20.38}$$

As discussed above, we assume that the Na$^+$ concentration in the ascending limb is always sufficiently high so that the Na$^+$–K$^+$ pump is saturated and the pump rate is independent of concentration, in which case the solution of (20.26) (in nondimensional variables) is

$$Q_aC_a = Q_aC_a(0) - Py, \tag{20.39}$$

where $P = \frac{pL}{c_d(0)q_d(0)}$ is the dimensionless Na$^+$ pump rate.

Having solved six of the original eight differential equations, we are left with a system of two first-order equations in two unknowns. The two equations are

$$\rho_d \frac{dQ_d}{dy} = -\Delta P_d + C_d - C_s, \tag{20.40}$$

$$\rho_c \frac{dQ_c}{dy} = -\Delta P_c + C_c - C_s,$$ (20.41)

subject to boundary conditions $Q_d = 1, Q_c = -Q_a$ at $y = 0$, and $Q_d = -Q_a$ at $y = 1$, where $C_c, C_s$, and $C_d$ are functions of $Q_d$ and $Q_c$. Although there are three boundary conditions for two first-order equations, the number $Q_a$ is also unknown, so that this problem is well posed. Our goal in what follows is to understand the behavior of the solution of this system.

## Formation of urine without ADH

The primary control of renal dialysis is accomplished in the collecting duct, where the amount of ADH determines the permeability of the collecting duct to water and the amount of aldosterone determines the permeability of the collecting duct to $Na^+$. Impairment of normal kidney function is often related to ADH. For example, the inability of the pituitary to produce adequate amounts of ADH is called *"central" diabetes insipidus*, and results in the formation of large amounts of dilute urine. On the other hand, with *"nephrogenic" diabetes insipidus*, the abnormality resides in the kidney, either as a failure of the countercurrent mechanism to produce an adequately hyperosmotic interstitium, or as the inability of the collecting ducts to respond to ADH. In either case, large volumes of dilute urine are formed.

Various drugs and hormones can have similar effects. For example, alcohol, clonidine (an antihypertensive drug), and haloperidol (a dopamine blocker) are known to inhibit the release of ADH. Other drugs such as nicotine and morphine stimulate the release of ADH. Drugs such as lithium (used to treat manic-depressives) and the antibiotic tetracyclines impair the ability of the collecting duct to respond to ADH.

The second important controller of urine formation is the hormone aldosterone. Aldosterone, secreted by zona glomerulosa cells in the adrenal cortex, works by diffusing into the epithelial cells, where it interacts with several receptor proteins and diffuses into the cell nucleus. In the cell nucleus it induces the production of the messenger RNA associated with several important proteins that are ingredients of $Na^+$ channels. The net effect is that (after about an hour) the number of $Na^+$ channels in the cell membrane increases, with a consequent increase of $Na^+$ conductance. Aldosterone is also known to increase the $Na^+$–$K^+$ ATPase activity in the collecting duct, as well as in other places in the nephron (a feature not included in this model), thereby increasing $Na^+$ removal and also $K^+$ excretion into the urine. For persons with *Addison's disease* (severely impaired or total lack of aldosterone), there is tremendous loss of $Na^+$ by the kidneys and accumulation of $K^+$. Conversely, excess aldosterone secretion, as occurs in patients with adrenal tumors (Conn's syndrome), is associated with $Na^+$ retention and $K^+$ depletion.

To see the effect of these controls we examine the behavior of our model in two limiting cases. In the first case, we assume that there is no ADH present, so that $\rho_c = \infty$, and that there is no aldosterone present, so that $H_c = 0$. In this case it follows from (20.37) that $Q_c = Q_c(0) = -Q_a$ and that $C_c = C_c(0) = C_a(0)$. In other words, there is no loss of either water or $Na^+$ from the collecting duct: the collecting duct has effectively been removed from the model.

It remains to determine what happens in the descending and ascending tubules. The flow is governed by the single differential equation

$$\rho_d \frac{dQ_d}{dy} = C_d - C_s - \Delta P_d = f(Q_d, Q_a, y), \tag{20.42}$$

where, from (20.34), (20.36), and (20.39),

$$C_d = \frac{1}{Q_d}(1 + \rho_d H_d(1 - Q_d) - \Delta P_d H_d y), \tag{20.43}$$

$$C_s = \frac{(P + \Delta P_d H_d)(1 - y)}{Q_d + Q_a} - \rho_d H_d, \tag{20.44}$$

subject to the boundary conditions $Q_d(0) = 1, Q_d(1) = -Q_a$. As before, $Q_a$ is a constant, as the ascending limb is impermeable to water, and $C_a$ is a linearly decreasing function of $y$.

We view this problem as a nonlinear eigenvalue problem, since it is a single first-order differential equation with two boundary conditions. The unknown parameter $Q_a$ is the parameter that we adjust to make the solution satisfy the two boundary conditions. It is reasonable to take $\rho_d$ to be small, since the descending tubule is quite permeable to water. In this case, however, the differential equation (20.42) is singular, since a small parameter multiplies the derivative. We overcome this difficulty by seeking a solution in the form $y = y(Q_d, \rho_d)$ satisfying the differential equation

$$f(Q_d, Q_a, y)\frac{dy}{dQ_d} = \rho_d \tag{20.45}$$

subject to boundary conditions $y = 0$ at $Q_d = 1$ and $y = 1$ at $Q_d = -Q_a$.

With $\rho_d$ small we have a regular perturbation problem in which we seek $y$ as a function of $Q_d$ as a power series of $\rho_d$, which is solved as follows. We assume that $y$ has a power series representation of the form

$$y = y_0 + \rho_d y_1 + \rho_d^2 y_2 + O(\rho_d^2), \tag{20.46}$$

substitute into (20.45), expand in powers of $\rho_d$, collect like powers of $\rho_d$, and then solve these sequentially. We find that

$$y = 1 - \frac{Q_a + Q_d}{PQ_d - \Delta P_d H_d Q_a}\left[1 - \Delta P_d(Q_d + H_d)\right] + O(\rho_d). \tag{20.47}$$

Notice that $y = 1$ at $Q_d = -Q_a$. Now we determine $Q_a$ by setting $y = 0$, $Q_d = 1$ in (20.47), and solving for $Q_a$. To leading order in $\rho_d$ we find that

$$-Q_a = 1 - \frac{P + H_d \Delta P_d}{1 - \Delta P_d} + O(\rho_d). \tag{20.48}$$

It is now a straightforward matter to plot $y$ as a function of $Q_d$, and then rotate the axes so that we see $Q_d$ as a function of $y$. This is depicted with a dashed curve in Fig. 20.13 (using formulae that include higher-order correction terms for $\rho_d$). For comparison we also include the curves calculated for the case where ADH is present; the details of that calculation are given below. Note that in either the presence or absence of ADH, $Q_a$ is always independent of $y$, while in the absence of ADH, $Q_c$ is also independent of $y$. Once $Q_d$ is determined as a function of $y$, it is an easy matter to plot the concentrations $C_d$ and $C_a$ as functions of $y$, as shown in Fig. 20.14.

From these we can draw some conclusions about how the loop of Henle works in this mode. Sodium is extracted from the descending limb by simple diffusion and from the ascending loop by an active process. The $Na^+$ that is extracted from the ascending loop creates a large osmotic pressure in the interstitial region that serves to enhance the extraction of water from the descending loop. This emphasizes the importance of the countercurrent mechanism in the concentrating process. As the fluid proceeds down the descending loop, its $Na^+$ concentration is continually increasing, and during its passage along the ascending loop, its $Na^+$ concentration falls. At the lower end of the loop the relative concentration of the formed urine (i.e., of substances that are impermeable, such as creatinine) is $\frac{1}{Q_d(1)}$. This quantity represents the "concentrating ability" of the nephron in this mode. Since $C_a(0) < C_d(0)$, as can be seen from Fig. 20.14, by the time the fluid reaches the top of the ascending loop, it has been diluted. Furthermore, comparing the value of $Q_d(1)(= Q_c)$ in the absence of ADH (dashed curve in Fig. 20.13) to the value of $Q_c(1)$ in the presence of ADH (solid curve in Fig. 20.13) shows that the flux out of the collecting duct is higher in the absence of ADH. Hence, combining these two observations, we conclude that in the absence of ADH, the nephron produces a large quantity of dilute urine, while in the presence of ADH, it produces a smaller quantity of concentrated urine. This is consistent with the qualitative explanation of nephron function given earlier in the chapter.

In Fig. 20.15 are shown the solute concentration $C_a$ and the flow rate $Q$ at the upper end of the ascending tubule as functions of dimensionless pump rate $P$. The formed urine is dilute whenever this solute concentration is less than one. The fact that this concentration can become negative at larger pump rates is a failure of the model, since the pump rate in the model is not concentration dependent.

### Formation of urine with ADH

In the presence of ADH, the collecting tube is highly permeable to water, so that, since the concentration of $Na^+$ in the interstitium at the lower end of the tube is high, additional water can be extracted from the collecting duct, thereby concentrating the dilute urine formed by the loop of Henle.

To solve the governing equations in this case is much harder than in the case with no ADH. This is because the equations governing the flux (20.40) and (20.41) are both singular in the limit of zero $\rho_d$ and $\rho_c$. Furthermore, one can show that the quasi-steady solution (found by setting $\rho_d = \rho_c = 0$ in (20.40) and (20.41)) cannot be made to satisfy the boundary conditions at $y = 1$, suggesting that the solution has a boundary layer. To avoid the difficulties associated with boundary layers, it is preferable to formulate

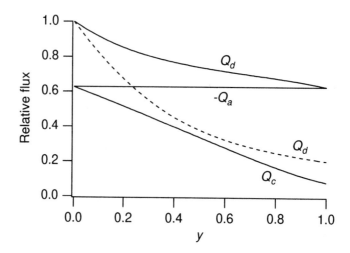

**Figure 20.13**   The flux of fluid in the loop of Henle, with ADH present (solid curve, $\rho_c = 2.0$) and without ADH present (dashed curve, $\rho_c = \infty$). Parameter values are $P = 0.9$, $\Delta P_d = 0.15$, $H_d = 0.1$, $\rho_d = 0.15$, $H_c = 0$.

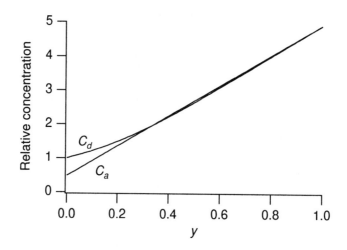

**Figure 20.14**   The solute concentration in the descending ($C_d$) and ascending ($C_a$) tubules with no ADH present ($\rho_c = \infty$), plotted as a function of distance $y$ for the parameter set as in Fig. 20.13.

the problem in terms of the solute flux $S_d = Q_d C_d$, because according to (20.35) this function is nearly linear and does not change rapidly when $\rho_d$ is small.

   In the case that ADH is present but there is no aldosterone ($H_c = 0$), the governing equations are

$$\frac{dS_d}{dy} = H_d \left( \frac{S_s}{Q_s} - \frac{S_d}{Q_d} \right) = H_d F_d(S_d, Q_c),  \tag{20.49}$$

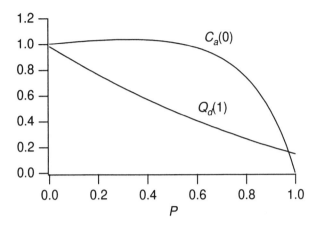

**Figure 20.15** The solute concentration and the flow rate at the upper end of the ascending tubule plotted as functions of pump rate $P$ when there is no ADH or aldosterone present ($\rho_c = \infty$, $H_c = 0$). Dilution occurs if the solute concentration is less than one. Parameter values are $\Delta P_d = 0.15$, $H_d = 0.1$, $\rho_d = 0.15$.

$$\rho_c \frac{dQ_c}{dy} = -\Delta P_c + \frac{S_c(0)}{Q_c} - \frac{S_s}{Q_s} = F_c(S_d, Q_c), \tag{20.50}$$

where

$$S_s = P(y - 1) + S_d(1) - S_d, \tag{20.51}$$

$$Q_s = -1 - Q_a - Q_c + Q_c(1) - \frac{1 - S_d - \Delta P_d H_d y}{\rho_d H_d}, \tag{20.52}$$

$$Q_d = 1 + \frac{1 - S_d - \Delta P_d H_d y}{\rho_d H_d}, \tag{20.53}$$

subject to boundary conditions $S_d(0) = 1$, $Q_c(0) = 1 + \frac{1-S_d-\Delta P_d H_d}{\rho_d H_d}$, and $Q_d(1) = -Q_a$.

These equations are difficult to solve because there are two unknown functions, $S_d$ and $Q_c$, and an unknown constant $Q_a$, subject to three boundary conditions. One way to solve them is to introduce the constants $Q_a$ and $Q_c(1)$ as unknown variables satisfying the obvious differential equations $\frac{dQ_a}{dy} = \frac{dQ_c(1)}{dy} = 0$, and to solve the expanded fourth-order system of equations in the four unknowns $S_d, Q_c, Q_a, Q_c(1)$ with four corresponding boundary conditions (adding the requirement that $Q_c = Q_c(1)$ at $y = 1$).

These equations were solved numerically using a centered difference scheme for the discretization and Newton's method to find a solution of the nonlinear equations (see Exercise 8). Typical results are shown in Fig. 20.16. Here we see what we expected (or hoped), namely that the collecting duct concentrates the dilute urine by extracting water. In fact, we see that the concentration increases on its path through the descending loop, decreases in the ascending loop, and then increases again in the collecting duct. This behavior is similar to the data for $Na^+$ concentration shown in Fig. 20.9.

The effect of the parameter $\rho_c$ is shown in Figs. 20.17 and 20.18. In these figures are shown the solute concentrations and the flow rates at the bottom and top of the loop of Henle and at the end of the collecting duct. Here we see that the effect of ADH is, as expected, to reconcentrate the solute and to further reduce the loss of water.

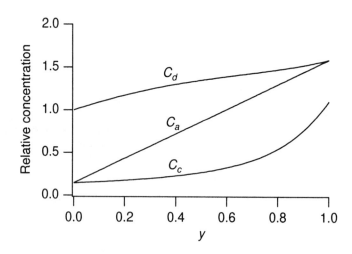

**Figure 20.16** Solute concentrations in the loop of Henle and the collecting duct, plotted as functions of $y$ for $P = 0.9, \Delta P_d = 0.15, \Delta P_c = 0.22, H_d = 0.1, \rho_d = 0.15$ and with $\rho_c = 2.0, H_c = 0$.

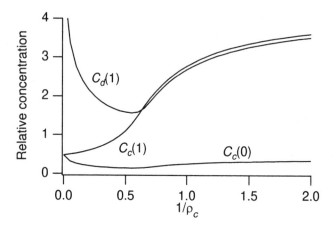

**Figure 20.17** Solute concentrations at the bottom and top of the loop of Henle and at the end of the collecting duct plotted as functions of inverse permeability $\frac{1}{\rho_c}$, with $P = 0.9, \Delta P_d = 0.15, \Delta P_c = 0.22, H_d = 0.1, \rho_d = 0.15$, and $H_c = 0$.

The asymptotic value of $C_c(1)$ as $\rho_c \to 0$ is the maximal solute concentration possible and determines, for example, whether or not the individual can safely drink seawater without dehydration. The asymptotic value of $1/Q_c(1)$ represents the highest possible relative concentration of impermeable substances such as creatinine.

### Further generalizations

This model shows the basic principles behind nephron function, but the model is qualitative at best, and there are many questions that remain unanswered and many generalizations that might be pursued. For example, the model could be improved by incorporating a better representation of the interstitial/capillary bed flow, taking into account that the peritubular capillaries issue directly from the efferent arteriole of the

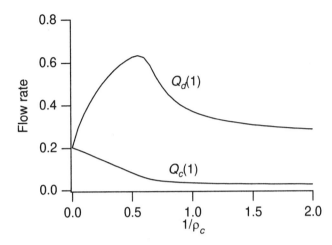

**Figure 20.18** Fluid flow rates at the bottom of the loop of Henle and at the end of the collecting duct plotted as functions of inverse permeability $\frac{1}{p_c}$, with $P = 0.9$, $\Delta P_d = 0.15$, $\Delta P_c = 0.22$, $H_d = 0.1$, $\rho_d = 0.15$, and $H_c = 0$.

glomerulus, thus determining the hydrostatic and osmotic pressures in the capillary bed. The model is also incorrect in that the active pumping of $Na^+$ out of the ascending limb is not concentration dependent, and as a result negative concentrations can occur for certain parameter values.

It is a fairly easy matter to add equations governing the flux of solutes other than $Na^+$, as the principles governing their flux are the same. One can also consider a time-dependent model in which the flow of water is not steady, by allowing the cross-sectional area of the tubules to vary. Nonsteady models are difficult to solve because they are stiff, and there is a substantial literature on the numerical analysis and simulation of time-dependent models (Layton et al., 1991).

Nephrons occur in a variety of lengths, and models describing kidney function have been devised that recognize that nephrons are distributed both in space and in length. These models are partial differential equations, and again, because of inherent stiffness, their simulation requires careful choice of numerical algorithms (Layton et al., 1995).

## 20.3 EXERCISES

1. The flow of glomerular filtrate and the total renal blood flow increase by 20 to 30 percent within 1 to 2 hours following a high-protein meal. How can you incorporate this feature into a model of renal function and regulation of glomerular function?
   Hint: Amino acids, which are released into the blood after a high protein meal, are cotransported with $Na^+$ ions from the filtrate in the proximal tubule. Thus, high levels of amino acids leads to high reabsorption of $Na^+$ in the proximal tubule, and therefore, lower than normal levels of $Na^+$ at the macula densa.

2. How much water must one drink to prevent any dehydration after eating a 1.5 oz bag of potato chips? (See Exercise 10 in Chapter 2.) Remark: A mole of NaCl is 58.5 grams and it dissociates in water into 2 osmoles.

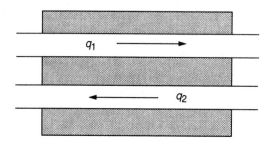

**Figure 20.19** Diagram of a countercurrent flow mediated by an interstitium, for Exercise 4.

3.  Why is alcohol a diuretic? What is the combined effect on urine formation of drinking beer (instead of water) while eating potato chips? What is the combined effect on urine formation of drinking beer while smoking cigarettes?
    Hint: Alcohol inhibits the release of ADH, while nicotine stimulates ADH release.

4.  Construct a simple model of the countercurrent mechanism that includes an interstitial compartment (Fig. 20.19). Show that inclusion of the interstitium has no effect on the overall rates of transport. Allow the solute to diffuse in the interstitium, but not escape the boundaries.
    Hint: View the interstitium as a tube with zero flow rate.

5.  Generalize the four-compartment model for the loop of Henle by separating the interstitium and peritubular capillaries into separate compartments, allowing no flow across $x = 0$ or $x = L$ for the interstitium.

6.  What changes in the exchange rates of the four compartment model for the loop of Henle might better represent the geometry of the loop of Henle, as depicted in Fig. 20.2?
    Remark: Some features you might want to consider include the location of the thickening of the ascending and descending limbs and the location of the junction of the peritubular capillaries with the arcuate vein.

7.  Formulate a time-dependent four-compartment model of urine concentration that tracks the concentration of both $Na^+$ ions and urea.

8.  Develop a numerical computer program to solve the equations of renal flow in the case that both ADH and aldosterone are present. It is preferable to formulate the problem in terms of the unknowns $S_d$ and $S_c$ and to expand the system of equations to a fourth-order system by allowing $S_d(1)$ and $S_c(1)$ to be unknowns that satisfy the simple differential equations $\frac{dS_d(1)}{dy} = 0$ and $\frac{dS_c(1)}{dy} = 0$. With the 4 unknowns, $S_d(y), S_c(y), S_d(1)$, and $S_c(1)$, the Jacobian matrix is a banded matrix, and numerical algorithms to solve banded problems are faster and more efficient than full matrix solvers.

9.  Generalize the renal model to include a concentration-dependent $Na^+$ pump in the ascending tubule. Does this change in the model guarantee that the flux and concentrations are nowhere negative?

# CHAPTER 21

# The Gastrointestinal System

Although the detailed structure of the gastrointestinal tract varies from region to region, there is a common basic structure, outlined in the cross-section shown in Fig. 21.1. It is surrounded by a number of heavily innervated muscle layers, arranged both circularly and longitudinally. Contraction of these muscle layers can mix the contents of the tract and move food in a controlled manner in the appropriate direction. Beneath the muscle layer is the *submucosa*, consisting mostly of connective tissue, and beneath that is a thin layer of smooth muscle called the *muscularis mucosae*. Finally, there is the *lamina propria*, a layer of connective tissue containing capillaries and many kinds of secreting glands, and then a layer of epithelial cells, whose nature varies in different regions of the tract.

## 21.1   Fluid Absorption

The primary function of the gastrointestinal tract is to absorb nutrients from the mix of food and liquid that moves through it. To accomplish this, the absorptive surface of the intestines consists of many folds and bends called *valvulae conniventes*, which increase the surface area of the absorptive mucosa about threefold. Located over the entire surface of the mucosa of the small intestine are millions of *villi*, which project about 1 mm from the surface of the mucosa and enhance the absorptive area another tenfold. The absorptive surface of the villi consists of epithelial cells that are characterized by a brush border, consisting of as many as 1000 *microvilli* 1 $\mu$m in length and 0.1 $\mu$m in diameter. The brush border increases the surface area exposed to the intestinal material by another twentyfold. The combination of all surface protrusions yields an absorptive surface area of about 250 square meters—about the surface area of a tennis court.

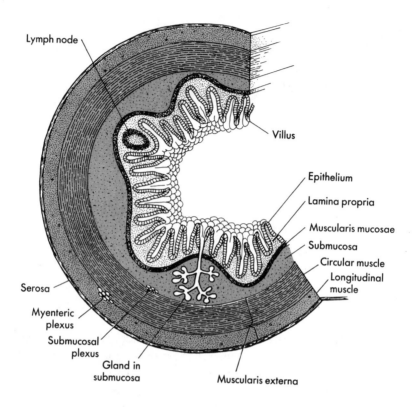

**Figure 21.1** Cross-section of the gastrointestinal tract. The outermost layers of the tract consist of smooth muscle, while the innermost layer consists of epithelial cells. The epithelial cell layer contains many gastric pits, glands that secrete hydrochloric acid, and thus the stomach lumen is highly acidic. (Berne and Levy, 1993, Fig. 38-1, p. 616.)

Epithelial cells are responsible for the absorption of nutrients and water from the intestine. The absorption of chemical nutrients, for example glucose and amino acids, is by the same process as in the kidney, via cotransporters with sodium. The absorption of water, however, is driven by osmosis.

The epithelial cells are not permeable to water on their lumenal side. However, there are 0.7–1.5 nm pores through the *tight junctions* between epithelial cells that permit water to diffuse readily between the lumen and the interstitium. The absorption of water through these pores is driven primarily by the sodium gradient between the lumen and the interstitium. Sodium is transported to the interior of the epithelial cell by passive transport and then is removed from the interior to the interstitium by a sodium–potassium ATPase. The sodium is transported from the interstitium by capillary blood flow.

To model the transport of water by the epithelial cell lining, we consider a small section of the epithelial gastrointestinal tract as two well-mixed compartments, the lumen and the interstitium, separated by a membrane (Fig. 21.2). We suppose that

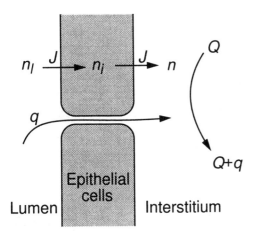

$n_l$ $\xrightarrow{J}$ $n_i$ $\xrightarrow{J}$ $n$

$Q$

$q$

$Q+q$

Epithelial cells

Lumen

Interstitium

**Figure 21.2** Diagram for osmotic transport of water across the epithelial cell wall. $J$ denotes the flow of Na$^+$, $q$ denotes the flow of water, and $Q$ denotes capillary blood flow.

the sodium concentration in the lumen is $n_l$ and in the cell interior is $n_i$, and that the concentration of all osmolites in the interstitium is $n$. The flow of sodium $J$ from the lumen to the interior of the cells is assumed to be passive (i.e., we ignore the effects of the membrane potential; see Exercise 1), and so

$$J = g(n_i - n_l), \tag{21.1}$$

for some constant $g$. Sodium flux from the cell interior to the interstitium is via an active sodium–potassium ATPase,

$$J = f(n_i), \tag{21.2}$$

for some saturating function $f$. The flow of water $q$ through the tight junctions is driven by the osmotic pressure difference between the lumen and the interstitium, so that

$$Rq = n - n_l, \tag{21.3}$$

where $R$ is the resistance (in appropriate units) of the tight junctions. Finally, we assume that there is a flow into and out of the interstitium provided by capillary flow. The influx of fluid is $Q$ with an incoming concentration of osmolites $n_0$, while the outflow of osmolites is $Q + q$ at concentration $n$. At steady state, the conservation of sodium implies that

$$g(n_l - n_i) = f(n_i) \tag{21.4}$$

and

$$(Q + q)n - Qn_0 = f(n_i). \tag{21.5}$$

The behavior of this system of three algebraic equations is relatively easy to sort out. Since $f$ is a positive, monotone increasing function of its argument, there is a one-to-one relationship between $n_l$ and $n_i$,

$$n_l = n_i + \frac{1}{g} f(n_i). \tag{21.6}$$

We can use (21.3) to eliminate $n$ from (21.5) and obtain

$$Rq^2 + (RQ + n_l)q + Q(n_l - n_0) - f(n_i) = 0. \tag{21.7}$$

Because the rate of sodium removal is dependent on the sodium concentration, we take $f(n) = \frac{Q_f n^3}{N^3 + n^3}$, for some constants $Q_f$ and $N$.

It is valuable to nondimensionalize this problem by scaling all concentrations by $N$, setting $u_j = n_j/N$ and $y = q/Q$. Then (21.7) becomes

$$\rho y^2 + (\rho + u_l)y + \kappa = 0, \tag{21.8}$$

where $\kappa = u_i - u_0 + (1 - \gamma)\beta F(u_i) = 0$, $\rho = RQ/N$, $\gamma = g/Q$, and $\beta = \frac{Q_f}{gN}$, and (21.6) becomes

$$u_l = u_i + \beta F(u_i), \tag{21.9}$$

where $F(u) = \frac{u^3}{1+u^3}$. There are four nondimensional parameters, namely $u_0$, the (relative) concentration of incoming interstitial osmolites; $\rho$, the resistance of the tight junctions to water; $\gamma$, the relative permeability of the lumenal cell wall to sodium; and $\beta$, the maximal velocity of active sodium transport (which depends primarily on the density of sodium pumps).

Observe that (21.8) is a quadratic polynomial in $y$ that has at most one positive root. In fact, the larger root of this polynomial is positive if and only if $\kappa < 0$. Furthermore, the positive root is a monotone decreasing function of $\kappa$.

There are several behaviors of the solution depending on the parameter values. However, the behavior that is of most interest here occurs when $\beta(\gamma - 1)$ is a large positive number. In this case, $\kappa$ is an "N"-shaped function of $u_i$, negative at $u_i = 0$, increasing for small values of $u_i$, then decreasing and finally increasing and eventually becoming positive for large $u_i$.

For much of parameter space this "N"-shaped behavior for $\kappa$ translates into "N"-shaped behavior for the positive root of (21.8). That is, with $u_l = 0$, there is a positive root. This root initially decreases to a minimal value and then increases to a maximal value, whereupon it decreases and eventually becomes negative, as a function of $u_l$. This behavior is depicted in Fig. 21.3, with parameter values $\rho = u_0 = \beta = 1, \gamma = 10$.

The implications of this are interesting. It implies that one can maximize the absorption of water by adjusting the sodium level. Thus, hydration occurs more quickly with fluids containing electrolytes than with pure water, as many high-performance athletes (such as road cyclists and long-distance runners) already know. However, too much sodium has the opposite effect of dehydrating the interstitium. This is a local effect only, as water is reabsorbed further along the tract.

When a person becomes dehydrated, large amounts of aldosterone are secreted by the adrenal glands. Aldosterone greatly enhances the transport of sodium by epithelial

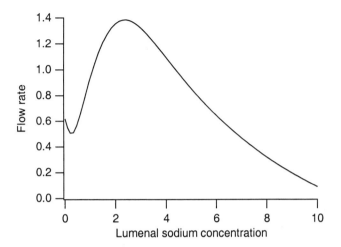

**Figure 21.3** Flux of water through the epithelial membrane plotted as a function of lumenal sodium concentration.

cells by activating the production of channel and pump proteins, which increases the passive and active transport of sodium. Indeed, a person can acclimatize to heavy exercise in hot weather, as over a period of weeks increased aldosterone secretion from the adrenal cortex will prevent excessive sodium loss in sweat, thus dispensing with the need for dietary sodium supplements. Loss of potassium can still, however, be a problem.

In this model the presence of aldosterone can be modeled by increasing $g$, the conductivity of sodium transport from the lumen, and/or by increasing $Q_f$, the maximal rate of active sodium pumping. It is easy to see that the total flux of sodium $J = f(n_i)$ and the flux of water $q$ both increase if either $g$ or $Q_f$ (or both) are increased. However, this increase is not without bound, since in the limit $g \to \infty$, we have $n_i \to n_l$, so that

$$\lim_{g \to \infty} J = f(n_l) \tag{21.10}$$

and

$$\lim_{g \to \infty} q = Q\left(\frac{n_0}{n_l} - 1\right) + \frac{f(n_i)}{n_l} \tag{21.11}$$

when $R = 0$. Thus, if a person is dehydrated, aldosterone production works to increase sodium absorption and decrease water loss.

Now we can construct a simple model of water content and sodium concentration as a function of distance along the intestinal length. We suppose that the *chyme* (the mixture of food, water, and digestive secretions entering from the stomach) moves as a plug flow with constant velocity. Water is removed from the chyme by osmosis and sodium is removed by the epithelial cells at local rates determined by the local sodium concentration. In steady state,

$$\frac{dQ_w}{dx} = -q(n_l), \qquad (21.12)$$

$$\frac{d(n_l Q_w)}{dx} = -J(n_l), \qquad (21.13)$$

where $Q_w$ is the flow of water in the intestine, $x$ is the distance along the intestine, and $q(n_l)$ and $J(n_l)$ are the removal rates of water and sodium, such as those suggested above. The analysis of this system of equations is straightforward and is left as an exercise (Exercise 2).

There are two common abnormalities that can occur in this process. *Constipation* occurs if the movement of feces through the large intestine is abnormally slow, allowing more time for the removal of water and therefore hardening and drying of the feces. Any pathology of the intestines that obstructs normal movement, including tumors, ulcers, or forced inhibition of normal defecation reflexes, can cause constipation.

The opposite condition, in which there is rapid movement of the feces through the large intestine, is known as *diarrhea*. There are several causes of diarrhea, the most common of which is infectious diarrhea, in which a viral or bacterial infection causes an inflammation of the mucosa. Wherever it is infected, the rate of secretion of the mucosa is greatly increased, with the net effect that large quantities of fluid are made available to aid in the elimination of the infectious agent.

For example, the toxins of cholera and other diarrheal bacteria stimulate immature epithelial cells (which are constantly being produced) to release large amounts of sodium and water, presumably to combat the disease by washing away the bacteria. However, if this excess secretion of sodium and water cannot be overcome by the absorption of mature, healthy cells, the result can be lethal because of serious dehydration. In most instances, the life of a cholera victim can be saved by intravenous administration of large amounts of sodium chloride solution to make up for the loss.

## 21.2 Gastric Protection

The inner surface of the gastrointestinal tract is a layer of columnar epithelial cells that actively secrete mucus and a fluid rich in bicarbonate. The mucus is highly viscous and coats the cells with a 0.5–1.0 mm thick layer that is insoluble by other gastric secretions and creates a lubricating boundary for the intestinal wall. In addition, this layer of cells is studded with a large number of gastric pits (Fig. 21.4). Each gastric pit contains *parietal cells* that secrete hydrochloric acid through an active transport process, leading to a pH of about 1 in the stomach lumen. Since the pH of the blood supplying the surface epithelium is about 7.4, there is a large $H^+$ concentration gradient (approximately a millionfold increase in concentration of hydrogen) across each epithelial cell. Clearly, the epithelial cells must be protected from the high lumenal acidity. It is believed that

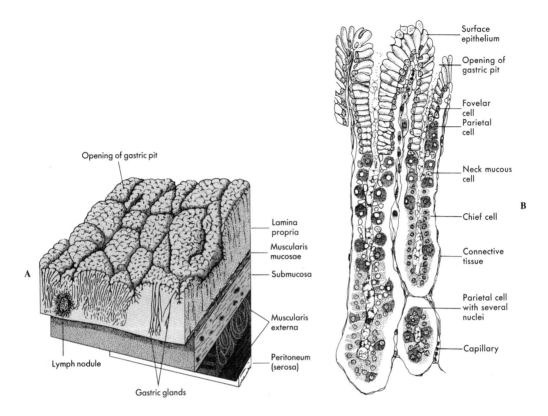

A

Opening of gastric pit

Lamina propria

Muscularis mucosae

Submucosa

Muscularis externa

Peritoneum (serosa)

Lymph nodule

Gastric glands

Surface epithelium

Opening of gastric pit

Fovelar cell
Parietal cell

Neck mucous cell

B

Chief cell

Connective tissue

Parietal cell with several nuclei

Capillary

**Figure 21.4** Closeup view of the gastric mucosa and two gastric pits. The epithelium of the gastric wall contains large numbers of gastric pits, each of which is lined by parietal cells that secrete HCl. (Berne and Levy, 1993, Fig. 39-9, p. 659.)

the secretion of mucus and bicarbonate by epithelial cells plays an important role in gastric protection.

## 21.2.1 A Steady-State Model

To model gastric protection (following Engel et al., 1984) we assume that the lumenal surface of the gastric mucosa is a plane located at $x = 0$, where $x$ is a coordinate measured perpendicular to the mucosal wall, while the mucus layer is of uniform thickness $l$. Thus, the mucus–lumen interface lies at $x = l$, as illustrated in Fig. 21.5. Inside the mucus layer $H^+$ and $HCO_3^-$ react according to

$$H^+ + HCO_3^- \underset{k_-}{\overset{k_+}{\rightleftarrows}} H_2O + CO_2. \tag{21.14}$$

This bicarbonate buffering system is one of the most important buffering systems in the body, and its role in the transport of carbon dioxide was discussed in Chapter 17.

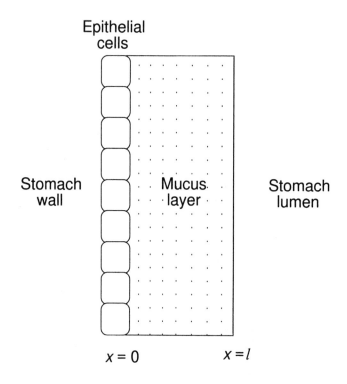

Epithelial
cells

Stomach
wall

Mucus
layer

Stomach
lumen

$x = 0$           $x = l$

**Figure 21.5**   Schematic diagram of the mucus layer in the model for gastric protection.

In the time-dependent problem, each species obeys a reaction–diffusion equation, such as

$$\frac{\partial [H^+]}{\partial t} = D_{H^+} \frac{\partial^2 [H^+]}{\partial x^2} - k_+ [H^+][HCO_3^-] + k_- [CO_2], \qquad (21.15)$$

where $D_{H^+}$ is the diffusion coefficient of $H^+$ in the mucus layer. However, at steady state the time derivatives are zero, and the partial derivatives with respect to $x$ become ordinary derivatives. Thus, at steady state,

$$D_{H^+} \frac{d^2 [H^+]}{dx^2} = D_{HCO_3^-} \frac{d^2 [HCO_3^-]}{dx^2} = k_+ [H^+][HCO_3^-] - k_- [CO_2], \qquad (21.16)$$

$$D_{CO_2} \frac{d^2 [CO_2]}{dx^2} = -k_+ [H^+][HCO_3^-] + k_- [CO_2]. \qquad (21.17)$$

To complete the formulation of the problem we add boundary conditions at the epithelial and lumenal boundaries of the mucus layer. On the lumenal side we assume that $[H^+] = [H^+]_l$ and $[CO_2] = [CO_2]_l$ are constant and known, determined by the concentration of the contents of the gastrointestinal tract, while $[HCO_3^-]$ is given by the equilibrium relation

$$[HCO_3^-]_l = \frac{k_- [CO_2]_l}{k_+ [H^+]_l}. \qquad (21.18)$$

On the epithelial side we assume that the fluxes of $HCO_3^-$ and $CO_2$ are known, as these chemicals are actively secreted by the epithelial cells, and thus, from Fick's law,

$$D_{HCO_3^-} \frac{d[HCO_3^-]}{dx} = -\bar{J}, \tag{21.19}$$

$$D_{CO_2} \frac{d[CO_2]}{dx} = -\bar{I}, \tag{21.20}$$

at $x = 0$, for some known constants $\bar{J}$ and $\bar{I}$. Finally, we assume that the flux of $H^+$ across the boundary at $x = 0$ is proportional to the concentration difference across the boundary; i.e.,

$$D_{H^+} \frac{d[H^+]}{dx} = P_{H^+} \left( [H^+] - [H^+]_{epi} \right), \tag{21.21}$$

where $P_{H^+}$ is the permeability and $[H^+]_{epi}$ is the concentration of $H^+$ in the epithelial cells. Since the concentration of $H^+$ in the epithelial cells is low compared to the concentration external to the cell, we set $[H^+]_{epi}$ to zero, and thus require

$$D_{H^+} \frac{d[H^+]}{dx} = P_{H^+} [H^+] \tag{21.22}$$

at $x = 0$.

To study this system of equations, we introduce nondimensional variables $y = x/l, u = [H^+]/[H^+]_l, v = [HCO_3^-]/[H^+]_l, w = [CO_2]/[H^+]_l$, in terms of which the model becomes

$$\epsilon \frac{d^2u}{dy^2} = uv - \zeta w, \tag{21.23}$$

$$\frac{d^2u}{dy^2} = \gamma \frac{d^2v}{dy^2} = -\beta \frac{d^2w}{dy^2}, \tag{21.24}$$

where $\beta = D_{CO_2}/D_{H^+}, \gamma = D_{HCO_3^-}/D_{H^+}, \epsilon = \frac{D_{H^+}}{k_+ l^2 [H^+]_l}, \zeta = \frac{k_-}{k_+ [H^+]_l}$. The boundary conditions at $y = 1$ $(x = l)$ are

$$u(1) = 1, \qquad v(1) = \zeta\alpha, \qquad w(1) = \alpha, \tag{21.25}$$

and the boundary conditions at $y = 0$ are

$$\frac{du}{dy}(0) = \lambda u(0), \qquad \gamma \frac{dv}{dy}(0) = -J, \qquad \beta \frac{dw}{dy}(0) = -I, \tag{21.26}$$

where $\alpha = \frac{[CO_2]_l}{[H^+]_l}, J = \frac{\bar{J}l}{D_{H^+}[H^+]_l}, I = \frac{\bar{I}l}{D_{H^+}[H^+]_l}, \lambda = \frac{P_{H^+}l}{D_{H^+}}$. Integrating (21.24) from 0 to $y$ and using the boundary conditions (21.26) we obtain

$$\frac{du}{dy} - \lambda u(0) = \gamma \frac{dv}{dy} + J = -\beta \frac{dw}{dy} - I. \tag{21.27}$$

Integrating (21.26) from $y$ to 1 and applying the boundary conditions (21.25) gives

$$u - 1 - \lambda u(0)(y - 1) = \gamma(v - \zeta\alpha) + J(y - 1) = -\beta(w - \alpha) - I(y - 1). \tag{21.28}$$

From this we obtain $v$ and $w$ as functions of $u$ and $y$:

$$v(y) = \zeta\alpha + \frac{1}{\gamma}[u(y) - 1 - (\lambda u(0) + J)(y - 1)] \tag{21.29}$$

and

$$w(y) = \alpha - \frac{1}{\beta}[u(y) - 1 + (I - \lambda u(0))(y - 1)]. \tag{21.30}$$

Thus, we can write the model as

$$\epsilon\frac{d^2u}{dy^2} = uv - \zeta w = f(u(y), y), \tag{21.31}$$

$$\frac{du}{dy}(0) = \lambda u(0), \qquad u(1) = 1. \tag{21.32}$$

From the molecular weights of the chemicals, we estimate $\beta \approx 0.14$ and $\gamma \approx 0.13$. The forward and reverse rates of the bicarbonate reaction are, respectively, $k_- = 11$ s$^{-1}$ and $k_+ = 2.6 \times 10^{10}$ cm$^3\cdot$mol$^{-1}\cdot$s$^{-1}$. Other experimentally determined quantities include $\bar{J} = 1.4 \times 10^{-10}$ mol$\cdot$cm$^{-2}\cdot$s$^{-1}$, $[H^+]_l = 140$ mM, $l = 0.05$ cm, $D_{H^+} = 1.75 \times 10^{-5}$cm$^2\cdot$s$^{-1}$, and $P_{H^+} = 1.3 \times 10^{-5}$cm $\cdot$ s$^{-1}$. From these parameter values we see that $\epsilon = O(10^{-7})$ and $\zeta = O(10^{-6})$ are small parameters, while $\lambda = 0.037$ and $J = 0.0003$.

We now use singular perturbation theory to solve this two-point boundary value problem. This approach is possible because $\epsilon$, which is the ratio of the rate of diffusion through the mucus to the rate of reaction, is small. Outside of a thin layer the bicarbonate reaction is in a pseudo-steady state at each point in space; in this region, diffusion of hydrogen ions or bicarbonate plays little role. It is only within the thin layer that the bicarbonate concentration is determined by the balance of reaction and diffusion. This allows the representation of the solution in two different spatial variables, one describing the solution outside this thin layer, and one describing the solution inside it. The solutions are then matched to obtain a uniformly valid solution. As we will see, although the bicarbonate reaction is in local chemical equilibrium outside the thin layer, the bulk of the reaction actually occurs within the thin layer. For the parameter values used here, the thin layer occurs at $y = 0$, but this need not necessarily be so. If the acidity of the lumen is low enough, the thin reaction layer occurs within the mucus layer (see Exercise 3).

### The outer solution

We look for a solution of the form

$$u = u_0 + \epsilon u_1 + \cdots, \tag{21.33}$$

substitute into the differential equation, and equate coefficients of powers of $\epsilon$. This gives a hierarchy of equations for the outer solution. To lowest order in $\epsilon$ we have

$$0 = f(u_0, y), \tag{21.34}$$

$$\frac{du_0}{dy}(0) = \lambda u_0(0), \qquad u_0(1) = 1. \tag{21.35}$$

Obviously, both boundary conditions cannot be satisfied, so we drop the boundary condition at $y = 0$ and keep the boundary condition at $y = 1$. There are good physical reasons for this choice. As we discussed above, the balance of reaction and diffusion is important only in a thin layer around $y = 0$. Thus, if we ignore diffusion (by setting $\epsilon = 0$) we do not expect to be able to satisfy the boundary condition at $y = 0$. (This is also the correct mathematical choice, because, as we will see, there is a "corner layer" at $y = 0$; the other choice, ignoring the boundary condition at $y = 1$, fails to produce a viable solution.)

The equation $f(u_0, y) = 0$ is the quadratic polynomial in $u_0$,

$$\beta u_0^2 + [\zeta\gamma(\alpha\beta + 1) - \beta + \beta(J + \lambda u_0(0))(1 - y)]u_0$$
$$+ \zeta\gamma[(\lambda u_0(0) - I)(1 - y) - \alpha\beta - 1] = 0, \tag{21.36}$$

so it can be solved exactly. The easiest way to represent this solution is to find $y$ as a function of $u_0$, since (21.36) is linear in $y$. However, because $\zeta$ is small, we find that

$$u_0(y) = 1 + (\lambda u_0(0) + J)(y - 1) + O(\zeta) \tag{21.37}$$

and

$$v = O(\zeta). \tag{21.38}$$

If we put $y = 0$ in (21.36), we can solve for $u_0(0)$ to get

$$u_0(0) = \frac{1 - J}{1 + \lambda} + O(\zeta), \tag{21.39}$$

from which it follows that

$$u_0(y) = 1 - \frac{\lambda + J}{\lambda + 1}(1 - y) + O(\zeta), \tag{21.40}$$

$$w(y) = \alpha + \frac{J + I}{\beta}(1 - y) + O(\zeta). \tag{21.41}$$

Hence, to leading order, there is no $HCO_3^-$ in the mucus layer, and $H^+$ and $CO_2$ vary linearly with distance through the mucus layer.

### The inner solution

The outer solution (21.40) does not satisfy the boundary condition at $y = 0$. A uniformly valid solution of this problem must include a "corner layer," that is, a solution with large second derivative, which therefore changes slope, but not value (at least to lowest order), in a small region close to $y = 0$. The corner layer here results from the fact that the boundary condition at $y = 0$ is expressed in terms of the derivative of $u$ at 0. Hence, to satisfy the boundary condition, the derivative of $u$ must change quickly. It is beyond the scope of this book to give a detailed description of the construction of this corner layer (see Engel et al., 1984, or, for a more general description, Keener, 1988, or Holmes, 1995). Suffice it to say that the corner layer is found by introducing a scaled variable $\tilde{y} = y/\sqrt{\epsilon}$ (which eliminates the $\epsilon$ from the second derivative term in (21.31))

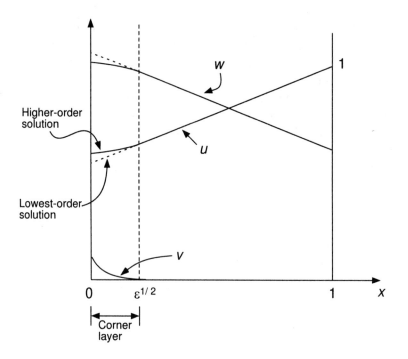

**Figure 21.6** Sketch (not to scale) of the solution to the model for gastric protection. (Adapted from Engel et al., 1984, Fig. 10.)

and then seeking a power series solution in powers of $\sqrt{\epsilon}$. The result is a modification of the outer solution by the addition of a term of the form

$$\sqrt{\epsilon}e^{-\mu y/\sqrt{\epsilon}}, \tag{21.42}$$

which is of small amplitude and satisfies the boundary condition at the origin. As a result, to leading order in $\epsilon$, the outer solution provides a uniformly valid representation of the solution on the entire interval $0 < y < 1$. A sketch of the solution is given in Fig. 21.6.

### Physical interpretation of the corner layer

There is an interesting interpretation of the corner layer in terms of the physiology of the problem. Recall that the boundary conditions at $y = 0$ for the original problem are

$$\frac{du}{dy}(0) = \lambda u(0), \qquad \gamma\frac{dv}{dy}(0) = -J, \qquad \beta\frac{dw}{dy}(0) = -I. \tag{21.43}$$

The outer solution does not satisfy these boundary conditions. Instead, if we evaluate the derivatives of the outer solution to leading order, we find that

$$\frac{du_0}{dy}(0) = \lambda u_0(0) + J, \qquad \frac{dv_0}{dy}(0) = 0, \qquad \beta\frac{dw_0}{dy}(0) = -I + J. \tag{21.44}$$

In other words, to lowest order, the bicarbonate flux $J$ can be replaced by a flux of hydrogen ions in the opposite direction with the same magnitude, and the $CO_2$ flux must be altered to compensate. Thus the original problem, which has a bicarbonate source at $y = 0$, is replaced by a simpler problem that has a $H^+$ sink and a $CO_2$ source at $y = 0$ and no $H^+$–$HCO_3^-$ reaction. This implies that each bicarbonate molecule that exits the epithelium reacts immediately with a hydrogen ion, with the consequent disappearance of the hydrogen ion. Hence, to lowest order, all the chemical reaction occurs within the corner layer.

In dimensional variables, the outer solution is

$$[H^+] = [H^+]_l - \frac{\bar{J} + P_{H^+}[H^+]_l}{D_{H^+} + P_{H^+}l}(l - x), \tag{21.45}$$

$$[CO_2] = [CO_2]_l + \frac{\bar{J} + \bar{I}}{D_{CO_2}}(l - x), \tag{21.46}$$

and thus, at the epithelial surface,

$$[H^+]_0 = \frac{D_{H^+}[H^+]_l - \bar{J}l}{D_{H^+} + P_{H^+}l}. \tag{21.47}$$

Using experimentally determined values for the parameters, we find that $[H^+]_0 = 135$ mM, a decrease of only 3.5%, which is too small to protect the epithelial cells from high lumenal acidity. Thus, this simple model of the mucus layer is insufficient to explain how the epithelial layer is protected.

## 21.2.2 Gastric Acid Secretion and Neutralization

The primary difficulty with the above model for gastric protection is that the flux of bicarbonate, $J$, is too small to cause a sufficient reduction of hydrogen ions at the surface of the epithelial cells. A model that addresses this shortcoming by examining the relationship between hydrochloric acid secretion and the release of bicarbonate was constructed by Lacker and his coworkers (de Beus et al., 1993).

Hydrochloric acid is secreted from the parietal cells of the oxyntic glands using a number of reactions. First, water in the cells is dissociated into hydrogen and hydroxyl ions in the cell cytoplasm. The hydrogen ions are actively secreted via a hydrogen–potassium ATPase. In addition, chloride ions are actively secreted and sodium ions are actively absorbed, via separate ATPases. The result is a high concentration of hydrochloric acid in the lumen. At the same time, carbon dioxide combines with hydroxyl ions (catalyzed by carbonic anhydrase) to form carbonic acid and thence bicarbonate. This bicarbonate diffuses out of the cell into the extracellular medium and is transported by the capillary blood flow. The direction of capillary blood flow is from the oxyntic cell in the gastric pit to the epithelial lining of the lumen. Since the epithelial cells are downstream of the oxyntic cells, they absorb bicarbonate from the blood and then secrete it into the mucus. Thus, as acid production increases, so does the rate at which bicarbonate is secreted into the lumen by the epithelial cells. According to de

Beus et al., the lack of this feature in the Engel model caused an underestimation of the rate of bicarbonate secretion from the epithelial layer.

De Beus et al. estimated the model parameters from the available experimental literature and showed that analytic solutions in certain simplified cases agreed well with the full solution. Of particular interest is their reproduction of the *alkaline tide*. As the rate of $H^+$ secretion into the lumen increases, the downstream $[H^+]$ (i.e., the gastric venous $[H^+]$) *decreases*. This reinforces the major idea behind this model, that secretion of $HCO_3^-$ by the epithelial cells is driven by $H^+$ secretion by the oxyntic cells, so that gastric protection is automatically increased as the lumenal $[H^+]$ increases.

# 21.3   Coupled Oscillators in the Small Intestine

One principal function of the gastrointestinal tract is to mix ingested food and move it through the tract in the appropriate direction. It does this by contraction of the layers of smooth muscle illustrated in Fig. 21.1, contractions that are controlled on a number of different levels. At the lowest level, each smooth muscle cell has intrinsic electrical activity, which can be oscillatory in nature. At higher levels, the properties of the local oscillations are modified by extrinsic and intrinsic neuronal stimulation, or chemical stimuli. Different parts of the tract have different kinds of contractile behavior. Here, we focus on the electrical activity of the smooth muscle of the small intestine. The small intestine is itself divided into three different sections: the first 25 cm or so after the pylorus (the passage from the stomach to the small intestine, controlled by the pyloric sphincter) is called the *duodenum*; the next section, comprising about 40% of the length of the small intestine, is called the *jejunum*; while the remainder is called the *ileum*. However, although this nomenclature is useful for understanding some of the experimental results we present here, we do not distinguish between the electrical activity of different sections of the small intestine.

## 21.3.1   Temporal Control of Contractions

Smooth muscle cells throughout the gastrointestinal tract exhibit oscillations in their membrane potential, with periods ranging from 2 to 40 cycles/min. A typical example of this *electrical control activity*, or ECA, is shown in Fig. 21.7. Although depolarization of the cell membrane potential can cause muscular contractions, this happens only if the membrane potential is depolarized past a threshold, in which case the potential begins to oscillate, or burst, at a much higher frequency. Whether or not bursting occurs depends on the level of neuronal or chemical stimulation. In this way, contractile activity depends on the local oscillatory properties of the smooth muscle cells, as well as on the higher-level control processes. Electrical bursts are termed *electrical response activity*, or ERA. Muscular contractions cannot occur with a frequency greater than that of the ECA, and thus the properties of the local ECA constrain the possible types of muscular contraction.

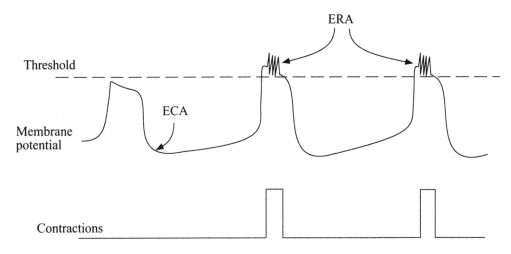

**Figure 21.7** Schematic diagram of electrical control activity (ECA), electrical response activity (ERA), and muscular contraction. (Adapted from Sarna, 1989, Fig. 2.)

Often, when faced with cellular oscillators, modelers seek to understand the cellular mechanisms that cause such behavior; this book contains many examples of this approach. Here, by contrast, we study what happens when a large number of oscillators are coupled to one another, without being concerned with the exact mechanisms underlying each oscillation. Although this approach cannot determine a direct relationship between cellular properties and global behavior, it gives much greater insight into how coupled oscillators can give rise to organized wave activity of the type that is frequently seen in the stomach and small intestine.

## 21.3.2 Waves of Electrical Activity

If each local oscillator were uncoupled from its neighbors, we expect there would be no organized waves of contraction moving along the intestine. However, the main point of this section is that weak coupling between the oscillators causes the propagation of waves of ECA and ERA along the intestine.

The importance of coupling between the local oscillators is demonstrated in the top panel of Fig. 21.8, where we show the experimentally measured frequency of segments of the small intestine in the intact intestine, and in segments that have been dissociated from one another by circumferential cuts across the intestine. In the intact intestine, the frequency of the ECA is constant over the entire region close to the pylorus, even though the intrinsic frequency is steadily decreasing over this region. At approximately 60 cm from the pylorus the ECA frequency begins to decrease. In the frequency plateau (the region of constant frequency) region, each oscillator is phase-locked to its neighbor, resulting in organized waves that move along the intestine away from the pylorus. This is illustrated in the top panel of Fig. 21.9, where the oscillation peaks in neighboring parts of the intestine are connected by solid lines. The slope of the solid line gives the speed of the wave along the intestine, and the fact that subsequent lines are regularly spaced, parallel and straight, shows that the waves are repetitive and highly organized. Following the frequency plateau is a region where the ECA frequency decreases along the intestine, and the corresponding waves are not phase-locked and therefore much less regular (lower panel of Fig. 21.9).

Note that phase locking (i.e., oscillation with the same frequency) does not necessarily imply that there is a regular wave. A *phase wave* occurs when there is a constant advance (or delay) of phase from one point to the next along the length of the intestine.

When the segments are uncoupled, each shows oscillatory ECA, but with an intrinsic frequency that decreases with distance from the pylorus; the frequency plateau disappears in the isolated segments. It appears that in the intact intestine, the highest-frequency segment closest to the pylorus entrains the nearby oscillators, which have similar but lower frequencies. However, when the difference in intrinsic frequency is too large, entrainment is not possible: the frequency plateau breaks down, and the waves lose regularity. This is illustrated further in the lower panel of Fig. 21.8, which shows the effect of a single cut in the intestine part of the way along the frequency plateau. To the left of the cut, the ECA frequency is entrained to the same high frequency as that of the frequency plateau in the intact intestine. To the right of the cut, a new frequency plateau emerges as the highest-frequency oscillator again entrains its neighbors. In this case the frequency of the second plateau is lower than that of the first, as it is entrained to an oscillator with a lower frequency, but it extends further to the right, into the region where the intact intestine has a variable ECA frequency.

There is some evidence to suggest that ECA frequency decreases along the intestine in a stepwise fashion, and this is illustrated in Fig. 21.10. The frequency plateaus are separated by regions where the amplitude of the oscillation is variable. Often, however, the wave activity in subsequent plateaus is less organized than in the first, as the oscillations are not so closely phase-locked.

### 21.3.3  Models of Coupled Oscillators

The two primary means by which the waves of electrical activity in the intestine have been studied are with numerical simulations of large coupled systems of oscillators and with rigorous mathematical analysis of approximating "phase equations."

### Numerical investigations

A number of investigators have used numerical simulations to study the behavior of chains of coupled oscillators in the small intestine (Nelsen and Becker, 1968; Diamant et al., 1970; Sarna et al., 1971; Robertson-Dunn and Linkens, 1974; Brown et al., 1975; Patton and Linkens, 1978). As a typical example, Diamant et al. (1970) coupled from 5 to 25 van der Pol oscillators with frequencies that decreased along the chain. Each oscillator was coupled to its nearest neighbor with the lower frequency in a procedure called forward coupling, and the coupling was assumed to be resistive. Numerical

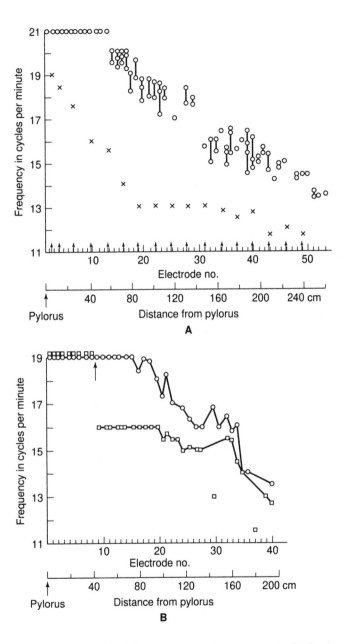

**Figure 21.8**  A: Intact (circles) and intrinsic frequency (crosses) of ECA in dog small intestine. The intrinsic frequencies were obtained by cutting across the small intestine at the places indicated by the arrows so as to disrupt oscillator coupling. B: The effect of a single cut (at the arrow) across the small intestine. To the right of the cut a frequency plateau still occurs, but now at a lower frequency than in the intact intestine. To the left of the cut the frequencies are unchanged. (Sarna, 1989, Fig. 10.)

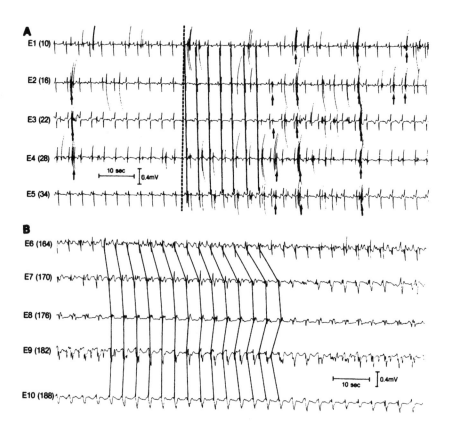

**Figure 21.9** Experimental recordings from dog small intestine. A: Recordings taken from the frequency plateau region close to the pylorus. Solid lines connect the peaks of the ERA; these lines are straight and parallel, thus indicating the propagation of regular wave trains in this region. B: Recordings taken from the variable frequency region. The peaks of the ERA are no longer well organized, indicating that regular wave propagation has broken down. (Sarna, 1989, Fig. 11 A and B.)

simulations showed that the oscillators are organized into frequency plateaus, whose lengths increased as the coupling strength increased. Because the coupling was in the forward direction only, the frequency plateaus lay above the intrinsic frequencies of the individual oscillators. The frequency plateaus were separated by regions in which the local frequency waxed and waned.

### The phase equations

The mathematical study of waves of electrical activity on the small intestine begins with the assumption that there are $n + 1$ coupled oscillators, described by the system of equations

$$\frac{du_i}{dt} = F_i(u_i) + \epsilon \sum_{j=1}^{n+1} a_{ij} H(u_j), \tag{21.48}$$

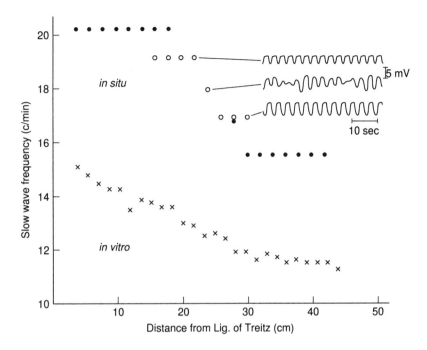

**Figure 21.10** Frequency of the ECA in the cat. As the distance from the ligament of Treitz increases, the *in vitro* measurements (crosses) show a steady decline in frequency, while the *in vivo* measurements (open and filled circles) show clear frequency plateaus. (Diamant and Bortoff, 1969, Fig. 2.) (The ligament of Treitz marks the beginning of the jejunum.)

where $u_i$ is the vector of independent variables describing the $i$th oscillator, and where $a_{ij}$ are the coupling coefficients. The oscillators are assumed to be nearly identical, so that the behavior of each oscillator is described approximately by some periodic function, denoted by $u_i = U(\omega(\epsilon)t + \delta\theta_i(t))$, where $\delta\theta_i$ is the *phase shift* of the oscillator and is presumed to be slowly varying. Then, the equations (21.48) can be reduced to equations describing the phase shifts of the individual oscillators, of the form

$$\frac{d}{d\tau}\delta\theta_i = \xi_i - \Omega_1 + \sum_{j=1}^{n+1} a_{ij}h(\delta\theta_j - \delta\theta_i) + O(\epsilon^2), \qquad i = 1,\ldots,n+1, \tag{21.49}$$

for some periodic function $h$, where $\tau = \epsilon t$ is a slow time. The phase equations are asymptotically valid in the limit that the coupling is weak and the oscillators are similar. A derivation of the phase equations is given in Section 14.5, where the function $h$ and the constants $\xi_i$ and $\Omega_1$ are determined. As a reminder, recall that $2\pi(1 + \epsilon\xi_i)$ is the natural (uncoupled) frequency of the $i$th oscillator, and that $\omega(\epsilon) = 1 + \epsilon\Omega_1 + O(\epsilon^2)$.

When each oscillator is coupled only to its nearest neighbors in a linear chain, the equations are

$$\frac{du_i}{dt} = F_i(u_i) + \epsilon(u_{i+1} - u_i) + \epsilon(u_{i-1} - u_i). \tag{21.50}$$

Here, the term $\epsilon(u_{i+1} - u_i)$ is deleted if $i = n + 1$, and the term $\epsilon(u_{i-1} - u_i)$ is deleted if $i = 1$. Then, the phase equations are of the form (21.49), where $a_{ij} = 1$ if $j = i + 1$ or if $j = i - 1$, $a_{ii} = -a_{i,i+1} - a_{i,i-1}$, and all other elements of $a_{ij}$ are zero. We find equations for the consecutive phase differences $\phi_i = \delta\theta_{i+1} - \delta\theta_i$ to be

$$\frac{d\phi_i}{d\tau} = [\Delta_i + h(\phi_{i+1}) + h(-\phi_i) - h(\phi_i) - h(-\phi_{i-1})] + O(\epsilon^2), \tag{21.51}$$

where $\Delta_i = \xi_{i+1} - \xi_i$ is a measure of the amount of detuning of the oscillators, i.e., how much the natural frequencies vary along the chain. The term $h(-\phi_{i-1})$ is omitted if $i = 1$, and the term $h(\phi_{i+1})$ is omitted if $i = n$. Finally, we take $h$ to be odd, in which case the phase difference equation becomes

$$\dot\phi = \beta\Delta + K\mathbf{H}(\phi), \tag{21.52}$$

where $\phi = (\phi_1, \ldots, \phi_n)$, $\beta\Delta = (\Delta_1, \ldots, \Delta_n)$, and $\mathbf{H} = (h(\phi_1), \ldots, h(\phi_n))$, and $K$ is a tridiagonal matrix with $-2$ on the diagonal and $1$ above and below the diagonal. Here, the dot denotes differentiation with respect to the slow time $\tau = \epsilon t$. The parameter $\beta$ has been introduced so that the gradient in the uncoupled oscillator frequency (i.e., the strength of the detuning) can be readily modified.

## Some simple solutions

Before discussing how frequency plateaus arise in the phase equation, it is useful to consider the solution in some simpler cases.

## Two coupled oscillators

For two coupled oscillators there is only a single phase equation,

$$\frac{d\phi}{d\tau} = \beta\Delta - 2h(\phi). \tag{21.53}$$

A phase-locked solution is one for which the phase difference between neighboring oscillators does not change, i.e., $\phi$ is constant. Thus, phase-locked solutions are found by setting $d\phi/d\tau = 0$ and solving for $\Delta$. This gives

$$\beta\Delta = 2h(\phi). \tag{21.54}$$

Since $h$ is $2\pi$-periodic and odd, we can solve (21.54) only if $\beta\Delta$ is not too large, as otherwise it would be greater than the maximum value of $2h$. In a common example, $h(\phi)$ is taken to be $\sin(\phi)$, in which case $|\beta\Delta|$ must be less than two for a phase-locked solution to exist. Since $\beta\Delta$ measures the amount of detuning, a phase-locked solution exists if and only if the two oscillators have natural frequencies that are not too different. If $\beta\Delta$ is small enough, the phase difference established between the oscillators (at least to lowest order in $\epsilon$) is given by the solution of (21.54).

### Three coupled oscillators

When three oscillators are coupled, the two phase difference equations are

$$\frac{d\phi_1}{d\tau} = \beta\Delta_1 - 2h(\phi_1) + h(\phi_2), \tag{21.55}$$

$$\frac{d\phi_2}{d\tau} = \beta\Delta_2 - 2h(\phi_2) + h(\phi_1), \tag{21.56}$$

and so a phase-locked solution occurs if

$$\frac{2\beta\Delta_1 + \beta\Delta_2}{3} = h(\phi_1), \tag{21.57}$$

$$\frac{2\beta\Delta_2 + \beta\Delta_1}{3} = h(\phi_2). \tag{21.58}$$

Clearly, this can be solved only if $|\beta(2\Delta_1 + \Delta_2)|$ and $|\beta(2\Delta_2 + \Delta_1)|$ are small enough.

It is important to note that if solutions for $\phi_1$ and $\phi_2$ exist, $\phi_1$ does not necessarily equal $\phi_2$. Thus, although the oscillators are phase-locked, the phase difference between the first and second oscillators is not necessarily the same as the phase difference between the second and third. If the phase differences are unequal, there is not a regular (constant speed) phase wave moving along the chain. Hence, phase locking does not necessarily imply a wavelike appearance.

To have a regular phase wave, the phase differences must be both constant and *equal* along the chain of oscillators. In this case, the peak of the wave moves at a constant speed down the oscillator chain. For the case of three coupled oscillators, phase wave solutions exist if $\Delta_1 = \Delta_2$. In other words, if the frequency difference between the first and second oscillators is the same as the difference between the second and third, and if this difference is not too large, then a phase wave solution exists.

This highlights the fact that in general, we can specify the frequency gradient along the oscillator chain and then solve for the phase differences, or we can specify the phase differences and then solve for the required frequency gradient, but we cannot specify the phase difference and the frequency gradient, expecting a phase wave.

### A chain of coupled oscillators

The equations for a phase-locked steady-state solution for a chain of $n + 1$ coupled oscillators are given by

$$\beta\Delta + K\mathbf{H}(\phi) = 0. \tag{21.59}$$

This is a system of $n$ equations in $n$ unknowns. In general, we can view the frequencies as given and the phase differences as unknown, or we can specify the phase differences and view the frequency differences as unknown. For example, if we seek a solution that is both phase-locked and has a phase wave, we need $\phi_1 = \phi_2 = \cdots = \phi_n$. Letting $h(\phi_i) = \eta$, we obtain $\Delta_1 = \Delta_n = \eta/\beta$ and $\Delta_i = 0$ for $i = 2,\ldots,n-1$. Thus, a phase wave solution exists only if all the middle oscillators have the same frequency, $\omega$ say, while the first oscillator is tuned to $\omega - \eta/\beta$ and the last oscillator to $\omega + \eta/\beta$. Note that $\eta$ can be either positive or negative, as different signs correspond to waves moving in opposite directions.

This observation poses a dilemma for the application of the phase equation to the electrical waves in the small intestine. Recall that ECA in the small intestine has a frequency plateau in the region close to the pylorus, and in this plateau, waves appear to be traveling away from the pylorus. These are phase-locked, with constant phase difference along the intestine. However, each segment of the small intestine has a natural oscillation frequency that decreases with distance from the pylorus. These two observations are inconsistent with the phase equations, for which a constant phase difference implies a constant natural frequency on the interior of the chain.

## Frequency plateaus

One partial solution to this dilemma was given by Ermentrout and Kopell (1984). They showed that on each plateau, the phase differences are not exactly constant, but make small oscillations, being locked only in an average sense, a phenomenon sometimes called *phase trapping*.

For simplicity, we assume that $h(\phi)$ is odd and $2\pi$-periodic, with a maximum $M$ at $\phi_M$ and a minimum $m$ at $\phi_m$, qualitatively like $\sin\phi$ (Fig. 21.11). Critical points of the differential equation (21.52) are solutions of

$$\mathbf{H}(\phi) = K^{-1}(-\beta\Delta),\tag{21.60}$$

which has a solution if and only if every component of $K^{-1}(-\beta\Delta)$ lies between $m$ and $M$. Let

$$\beta_0 = \max\{\beta : m \le (K^{-1}(-\beta\Delta))_i \le M, \text{ for every } i\}.\tag{21.61}$$

When $\beta < \beta_0$, for every $i$ there are two solutions to the scalar equation $h(\phi_i) = (K^{-1}(-\beta\Delta))_i$. These solutions, which we denote by $\phi_i^+$ and $\phi_i^-$, are shown in Fig. 21.11.

Since each component of $\phi$ can have one of two values, there are $2^n$ possible steady states when $\beta < \beta_0$. Because of the definition of $\beta_0$, there is some value $j$ such that the roots $\phi_j^+$ and $\phi_j^-$ coalesce and disappear as $\beta$ crosses $\beta_0$. Thus, as $\beta$ crosses $\beta_0$, all the critical points coalesce in pairs and disappear. This follows because every critical point can be matched with another that agrees with it in every component $i \ne j$, one having $j$th component $\phi_j^+$, the other having $j$th component $\phi_j^-$. When $\beta < \beta_0$ the members of the pair differ only in the $j$th component; when $\beta = \beta_0$ the members of the pair are identical, and when $\beta > \beta_0$ there is no solution for the $j$th component, and so both solutions fail to exist.

There is one particular pair of critical points that is of interest for reasons that are explained below. Let $\xi_j(\beta)$ denote the critical point whose $k$th component is $\phi_k^-$ for all $k \ne j$, and whose $j$th component is $\phi_j^+$. Also, let $\xi_0(\beta)$ denote the critical point with the $k$th component equal to $\phi_k^-$ for all $k$. Clearly, as $\beta \to \beta_0$, $\xi_j$ coalesces with $\xi_0$ and the two critical points disappear.

Finally, to complete the preliminaries, we restrict $\Delta$ to be of a particular form. Since experimental data suggest that the natural frequencies of the oscillators decrease ap-

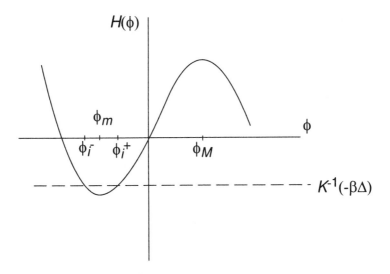

**Figure 21.11** The roots of the $i$th component of the phase difference equation.

proximately linearly along the small intestine, it is reasonable to take $\Delta = (-1,\ldots,-1)$, corresponding to a linear decrease in frequency along the oscillator chain. We also assume that there is an even number of oscillators in the chain (an odd number of phase differences), i.e., that $n$ is odd, with $n = 2j-1$, and thus the central phase difference is at position $j$. For this choice of $\Delta$ and $n$, the solution of (21.60) fails first at the $j$th component; that is, if $j$ is the position of the middle phase difference, then $\phi_j^+ \to \phi_j^-$ as $\beta \to \beta_0$. This is easily seen by noting that $K^{-1}(\beta\Delta)$ has $k$th component $-\beta k(n + 1 - k)/2 < 0$ (Exercise 4). Hence $\phi_k^\pm(\beta) < 0$ for all $k$ as long as $\beta < \beta_0$. Further, when $n = 2j - 1$, $k(n+1-k) = k(2j-k)$, which is greatest when $k = j$. Since the $j$th component of the solution to (21.60) is the one with the greatest modulus, it follows that the $j$th component will be the first to "hit" the minimum and disappear.

We now return to the particular pair of steady states, $\xi_0$ and $\xi_j$, defined above. Linear stability analysis of the system (21.52) shows that $\xi_0$ is a sink (a stable node), while $\xi_j$ is a saddle point with one positive and $n - 1$ negative eigenvalues. Further, both branches of the unstable manifold at $\xi_j$ tend to $\xi_0$ as $\tau \to \infty$, and thus the closure of the unstable manifold forms a closed loop.

We can get some insight into the meaning of this last statement and why it is true by considering the special case $n = 2$. It is convenient to introduce the change of variables $\psi = K^{-1}\phi$, in which case the system of differential equations (21.52) becomes

$$\dot{\psi} = K^{-1}\beta\Delta + \mathbf{H}(K\psi), \tag{21.62}$$

or, in the specific case that $n = 2$,

$$\dot{\psi}_1 = \beta + h(\psi_2 - 2\psi_1), \tag{21.63}$$

$$\dot{\psi}_2 = \beta + h(\psi_1 - 2\psi_2). \tag{21.64}$$

This is a two-dimensional system whose phase portrait is easily studied. First, note that since (21.52) is a flow on a torus, so also is this system. The torus for (21.52) is the domain $0 \leq \phi_i \leq 2\pi, i = 1, \ldots, n$, with the boundary at $\phi_i = 0$ "identified" with, or equivalent to, the boundary at $\phi_i = 2\pi$. Here, however, the boundaries of the torus are modified, being the four straight lines

$$\psi_1 - 2\psi_2 = 0, -2\pi, \tag{21.65}$$

$$\psi_2 - 2\psi_1 = 0, -2\pi. \tag{21.66}$$

These bounding lines are shown dashed in Fig. 21.12. Now, the flow on this torus can be understood by first examining the nullclines $\dot{\psi}_1 = 0$ and $\dot{\psi}_2 = 0$. There are four such curves,

$$\psi_2 - 2\psi_1 = -\phi_\pm, \tag{21.67}$$

$$\psi_1 - 2\psi_2 = -\phi_\pm, \tag{21.68}$$

where the numbers $-\phi_\pm$ satisfy $h(-\phi_\pm) = -\beta$, as depicted in Fig. 21.11. The nullclines are shown in Fig. 21.12 by solid lines.

Clearly, there are four critical points. By sketching in a few elements of the vector field, we can see that the critical point at the leftmost and lowest position is the only stable critical point (denoted by a filled circle), two of the critical points are saddle points (denoted by open circles), and the fourth is an unstable node. We also see that the stable critical point is a global attractor. That is, every trajectory, excluding the other critical points, tends to the unique stable critical point as time goes to infinity. It follows that the unstable manifold of each saddle point forms a closed loop; both closed loops are therefore invariant manifolds.

In general, for arbitrary $n$, when $\beta < \beta_0$ there is a closed invariant manifold containing the two steady states, $\xi_j$ and $\xi_0$. Furthermore, this loop is a smooth invariant attracting cycle on which $\phi_i$ completes a full rotation from 0 to $2\pi$ but on which the

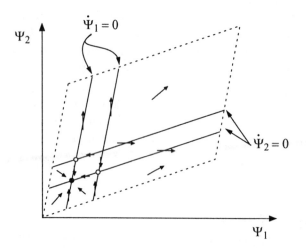

**Figure 21.12** Phase portrait for the two-dimensional system of equations (21.65)–(21.66). The stable critical point is denoted by a filled circle, and the two saddle points by open circles.

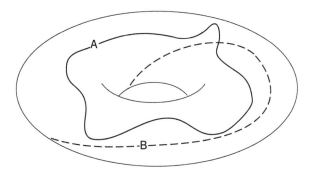

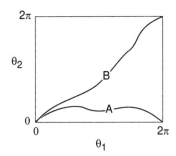

**Figure 21.13** Schematic diagram of two orbits on a torus, drawn two different ways. The upper panel shows the orbits on a torus, while the lower panel has unfolded the torus into a square, periodic in both directions. Orbit A is homotopic to the circle $\theta_2 = 0$, while orbit B is not.

other $\phi$s do not. In other words, as $\phi_j$ moves through the $2\pi$-cycle, all the other $\phi_k, k \neq j$, vary without making a full cycle. This is illustrated in Fig. 21.13 by orbit A. Orbit B, however, experiences a full $2\pi$ cycle in angle $\theta_2$ for every $2\pi$ cycle of $\theta_1$. It follows that the invariant attracting manifold formed by the two branches of the unstable manifold of $\xi_j$ is homotopic to (i.e., is continuously deformable into) the circle $\phi_k = 0, k \neq j$, $0 \leq \phi_j \leq 2\pi$.

The crucial result proved by Ermentrout and Kopell is that this invariant attracting manifold exists even when $\beta > \beta_0$. Thus, although steady states of the phase difference equations disappear when $\beta > \beta_0$, a smooth, invariant, attracting manifold that is homotopic to the circle $\phi_k = 0, k \neq j, 0 \leq \phi_j \leq 2\pi$, persists. Since it contains no critical points, this manifold is an attracting limit cycle. This stable limit cycle corresponds to a pair of frequency plateaus in the chain of coupled oscillators. To see this, define the average frequency of the $k$th oscillator to be

$$\omega(\epsilon) + \lim_{T \to \infty} \frac{1}{T} \int_0^T \delta\theta_k'(t)\, dt, \tag{21.69}$$

provided that the limit exists. Here, a prime denotes differentiation with respect to $t$. Note that if $\delta\theta_k'$ is constant, the frequency is exactly $\omega(\epsilon) + \delta\theta_k'$, as expected. Subtracting $\delta\theta_k'$ from $\delta\theta_{k+1}'$ gives the average phase difference as

$$\lim_{T \to \infty} \frac{\epsilon}{T} \int_0^T \phi_k'(\tau)\, d\tau. \tag{21.70}$$

Around the attracting limit cycle, this simplifies to

$$\frac{\epsilon}{T_0} \int_0^{T_0} \phi'_k(\tau)\,d\tau, \tag{21.71}$$

where $T_0(\beta)$ is the period of the limit cycle on the torus. However, we readily calculate that

$$\int_0^{T_0} \phi'_k(\tau)\,d\tau = \begin{cases} 0, & k \neq j, \\ 2\pi, & k = j. \end{cases} \tag{21.72}$$

It follows that the first $j$ oscillators all have the same average frequency, as do the oscillators from $j+1$ to $n$. Between the $j$th and the $(j+1)$st oscillators there is a frequency jump of $2\pi\epsilon/T_0(\beta)$.

It is important to note that the phase differences $\phi_k, k \neq j$, make small oscillations about the constant $\phi_k^-$, but are not identically constant. Thus, on each frequency plateau, the phases are not locked, but only "trapped" on average over each cycle. In contrast to some experimental data, one therefore does not expect to see exactly regular propagating waves appearing on each plateau, since such waves require phase locking at each instant and a constant phase difference along the plateau. The precise reasons for this discrepancy are, as yet, unknown.

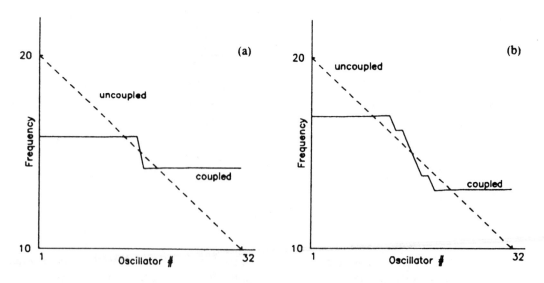

**Figure 21.14** Frequency plateaus in a chain of 32 coupled oscillators. For these numerical simulations the phase equation was chosen to be $\dot{\phi}_k = -10/31 + \delta(\sin\phi_{k+1} - 2\sin\phi_k + \sin\phi_{k-1})$. (a): $\delta = 32$, (b): $\delta = 18$. Note that decreasing $\delta$ while leaving the intrinsic oscillator frequencies unchanged is equivalent to increasing $\beta$. (Ermentrout and Kopell, 1984, Fig. 4.1.)

## Numerical solutions

Numerical solutions of the phase equation for two different values of $\beta$ are shown in Fig. 21.14. Once $\beta$ is greater than the critical value $\beta_0$ a pair of plateaus emerges, and as $\beta$ is increased still further, multiple plateaus appear. Comparison of Fig. 21.8 with Fig. 21.14 shows that the above model differs from experimental results in an important way. In the model, the frequency of the plateau lies between the maximum and minimum natural frequencies of the oscillators in the plateau; a plateau cannot have a higher frequency than all of its constituent oscillators, as is seen in the experimental data. However, the above simple model can be extended to obtain better qualitative agreement with experiment. For instance, the coupling between cells can be made stronger in one direction than the other (nonisotropic coupling) or it can be made nonuniform along the oscillator chain, in which case the phase difference equation can reproduce the asymmetrical behavior exhibited by the experimental system.

However, despite this qualitative agreement, it must be admitted that the simple model presented here does not give a quantitative explanation of the properties of frequency plateaus in the small intestine. It is an excellent example of how, to obtain an analytical understanding of a particular phenomenon, it is often necessary to study a model that has been reduced to caricature by successive approximations. Although hope of quantitative agreement is thereby lost, such simple models often permit a substantial understanding of the underlying structure.

## 21.4 EXERCISES

1. Modify (21.1) to account for the effects of the membrane potential on sodium flux. How does membrane potential affect the transport of water?

2. Use the model of local sodium removal and osmotic transport of water to analyze the removal of water and sodium along the length of the intestinal tract.

    (a) Give a phase-plane analysis for the system. What is the trajectory of sodium and water if sodium is initially quite high? What is the trajectory of sodium and water if sodium is initially quite low? How does this compare with the trajectory when there is no sodium?

    (b) If the flow of water from the intestine is assumed to depend solely on the sodium concentration, then the flow can become negative, which is clearly unphysiological. How might you modify this assumption and how might you justify it on physical grounds?
    Hint: As the chyme dries, one expects the continued extraction of water to become more difficult.

3. (a) Find two terms of the power series representation of the solution of (21.36) in powers of $\zeta$. Find $u(0)$ to the same order in $\zeta$.

    (b) It appears from the leading-order solution that $u(0)$ is negative if $J > 1$. Show that this is not correct, but that $u(0) > 0$ for all positive values of $J$. Show that when $J > 1$, the bulk of the reaction occurs in a thin layer contained within the mucus layer, and that the epithelial surface is completely protected. What is the physical interpretation of the condition $J > 1$?

4.  Show that the $k$th component of $K^{-1}(\beta\Delta)$ in (21.60) is $\alpha_k = -\beta k(n+1-k)/2$.
    Hint: Verify that $(K\alpha)_i = \beta$.

5.  What steady phase-locked solutions are possible for two coupled identical oscillators? What stable, steady, phase-locked solutions are possible?

6.  Describe the behavior expected from two coupled oscillators when $\beta\Delta = -2 - \epsilon$ and $h(\phi) = \sin\phi$ for $\epsilon \ll 1$. Check your prediction numerically. This behavior is called *rhythm splitting*, and is discussed in more detail in Murray (1989).

7.  In Exercise 17 of Chapter 14 the coupling function $h(\theta)$ was calculated for a collection of coupled FitzHugh–Nagumo equations. Using this coupling function, pick any initial conditions you wish, and solve the phase equation numerically to determine $\phi(t)$. How quickly do identical FitzHugh–Nagumo oscillators synchronize?

8.  Extend the previous question (and refer to Chapters 5 and 7) to study the synchronization of intracellular calcium oscillations. Suppose two cells, each with a well-mixed interior, are coupled by a membrane through which $Ca^{2+}$ can diffuse through gap junctions. Suppose further that each cell exhibits intracellular calcium oscillations of slightly different periods (i.e., each cell has a slightly different background $IP_3$ concentration). Using your favorite model for calcium oscillations, determine the coupling function numerically, and thus determine how fast such synchronization occurs, as a function of the intercellular permeability of $Ca^{2+}$. (As of writing, these calculations have not yet been performed, even though they have direct relevance to the measurement of the intercellular permeability of $Ca^{2+}$ in glial and epithelial cell cultures. If you do this soon enough, compare your results to experimental data and publish them!)

# CHAPTER 22

# The Retina and Vision

The visual system is arguably the most important system through which our brain gathers information about our surroundings, and forms one of our most complex physiological systems. In vertebrates, light entering the eye through the lens is detected by photosensitive pigment in the photoreceptors, converted to an electrical signal, and passed back through the layers of the retina to the optic nerve, and from there, through the visual nuclei, to the visual cortex of the brain. At each stage, the signal passes through an elaborate system of biochemical and neural feedbacks, the vast majority of which are poorly, if at all, understood.

There are a number of phenomena that we would like to understand:

- One striking feature of the visual system is its ability to adapt to widely varying levels of background light. Thus, it has the ability to operate in both dim and bright situations, from a starlit night to a bright sunny day.
- The eye is more sensitive to flashing light than to steady light. When a space-independent pulse of light is shone on the entire eye, the retina responds with a large-amplitude signal at the beginning followed by a decrease to a lower plateau. Similarly, at the end of the impulse, the off-transient is nearly the negative image of the on-transient, with a large negative transient followed by a return to a plateau. Response to transients with adaptation in steady conditions is characteristic of inhibitory feedback, or *self-inhibition*, which occurs at a number of levels in the retina.
- When a time-independent strip of light is applied to the retina, the response is greatest at the edges of the pattern. These response variations are known as *Mach bands* and are due to *lateral inhibition*, which plays a similar role in space as self-inhibition plays in time. For example, in the interior of a uniformly bright part

**Figure 22.1** Test pattern with which to observe the effects of self-inhibition and lateral inhibition.

of the visual field, neurons are inhibited from all sides, while regions near the edge receive little inhibition from their dimly illuminated neighbors and therefore appear brighter. The result is contour enhancement. The effect of lateral inhibition can be seen in the white intersections of Fig. 22.1. In particular, if one looks intently at one of the white intersections, the remaining intersections will appear to have a gray or darkened interior, and the center of the white strips will appear slightly darkened compared to their edges, because of lateral inhibition.

The visual system has been studied on very many levels, ranging from the biochemistry of the photopigments, to the cellular electrophysiology of the individual retinal cells, to the neural pathways responsible for image processing, to the large-scale structure of the visual cortex. Obviously, there is insufficient space here for a detailed study of all these aspects of the visual system. For a more comprehensive view of the visual system, the reader is referred to the excellent book by Nicholls, Martin, and Wallace (1992); other discussions of visual processing can be found in Blakemore (1990), Landy and Movshon (1991), and Spilmann and Werner (1990).

## 22.1 Retinal Light Adaptation

The first stage of visual processing occurs in the retina, a structure consisting of at least five major neuronal cell types (Fig. 22.2). After entering the eye, light passes through all the cell layers of the retina before being absorbed by photosensitive pigments in the photoreceptors in the final layer of cells. (A functional reason for this arrangement is not known.) Photoreceptors come in two varieties: rods, which operate in conditions

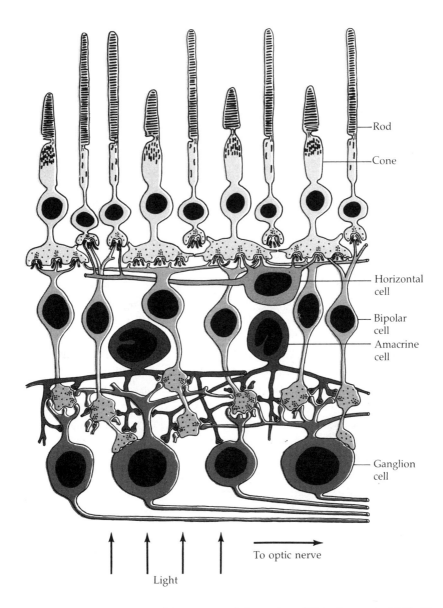

**Figure 22.2** Schematic diagram of the layers of the retina. (Nicholls et al., 1992, Chapter 16, Fig. 14, p. 583.)

of low light, and cones, which operate in bright light conditions and detect color. In the dark, photoreceptors have a resting membrane potential of around −40 mV, and they hyperpolarize in response to light. The light response is graded, with larger light stimuli resulting in larger hyperpolarizations. Note that this is different behavior from typical neurons, in which the action potential is a depolarization and is all-or-nothing, as described in Chapter 4. Photoreceptors make connections to both horizontal cells and bipolar cells. Each horizontal cell makes connections to many photoreceptors (and, often, to bipolar cells), and is coupled to other horizontal cells by gap junctions. The bipolar cells form a more direct pathway, coupling photoreceptor responses to ganglion cells, but this is also a simplification. Amacrine cells connect only to bipolar cells and ganglion cells, and their precise function is unknown. Ganglion cells (which fire action potentials, unlike photoreceptors and horizontal cells) are the output stage of the retina and form the optic nerve. The interconnections among the retinal cells are complex and not well understood; there has been a great deal of work done on how the retina detects features such as moving edges and orientation, while ignoring much of the information presented to it. Here, we study only the simplest models.

## 22.1.1  Weber's Law and Contrast Detection

One of the basic features of the retina is *light adaptation*, the ability to adapt to varying levels of background light. Over a wide range of light levels, the sensitivity of the retina is observed to be approximately inversely proportional to the background light level. This fact is known as *Weber's law*, or the *Weber–Fechner law*. There are three common definitions of sensitivity. It can mean *psychophysical sensitivity*, defined as 1/threshold, where the threshold is the minimal stimulus necessary to elicit an observable response when superimposed on a given background. Weber's law describes the fact that in psychophysical experiments (i.e., with human subjects who report what they detect) the threshold increases as the background light level increases.

A second definition of sensitivity is the one used most in this chapter. In response to a light step increase, the membrane potential of a photoreceptor (or horizontal cell) first decreases and then increases back to a steady level. If $V(I, I_0)$ is the maximum deviation of the membrane potential in response to changing the amplitude of the background light from $I_0$ to $I$, then the *peak sensitivity* is

$$S(I_0) = \left. \frac{\partial V}{\partial I} \right|_{I=I_0} . \tag{22.1}$$

The third definition of sensitivity is the *steady-state sensitivity*. If $V_0(I_0)$ is the steady response as a function of the background light level, then the steady-state sensitivity is defined to be $dV_0/dI_0$.

Light adaptation serves two fundamentally important purposes. First, it helps the retina handle the wide range of light levels in which the eye must operate. The eye functions in a range of light levels that spans about 10 log units, from a starlit night to bright sunlight. (Light intensities are typically plotted on a dimensionless logarithmic scale. For example, if $I_0$ is a standard unit of light intensity, and the intensity of the light stimulus is $I$, then $\log(\frac{I}{I_0}) = \log I - \log I_0$, so that on a logarithmic scale, the unit scale $I_0$ only shifts $\log \frac{I}{I_0}$). Further, the retina is so sensitive that it can reliably detect as few as 20 photons, and can even, although less reliably, detect single photons.

The two requirements of operation over a wide range of light levels and high sensitivity in the dark are in conflict. Without control mechanisms, a retina that can detect single photons would be saturated, and hence blinded, by bright light. In bright light, there is a *saturation catastrophe*, in which every photoreceptor is saturated, each sending the same signal to the brain, so that no contrast in the scene can be detected. However, for the human retina this saturation catastrophe is about 10 log units above the level of no response. This range of light sensitivity is achieved partly by the use of two different types of photoreceptors, rods and cones, having different sensitivities, rods operating in dim light and cones in bright light. However, by itself, two types of photoreceptors are inadequate to account for the observed range of light sensitivity.

The second effect of light adaptation is to send a signal to the brain that is dependent only on the contrast in the scene, not on the background light level. When a scene is observed with different background light levels, the amount of light reflected from an object in that scene varies considerably. Hence, the eye should measure something other than the total amount of light from an object. In fact, the eye measures the contrast in the scene, which, since it is dependent only on the reflectances of the objects, is independent of the background light level.

Contrast detection is a consequence of Weber's law, as can be seen from the following argument of Shapley and Enroth-Cugell (1984). Suppose we observe an object superimposed on a background, where the background reflectance is $R_b$, the object reflectance is $R_o$, and the background light level is $I$. As the receptive field of a retinal neuron moves across the boundary of the object, the stimulus it receives changes from $IR_b$ to $IR_o$, a difference of $IR_o - IR_b$. Since according to Weber's law, the sensitivity of the cell is inversely proportional to $IR_b$ (the amount of light reaching the cell from the background), the cell's response will be approximately proportional to $(IR_o - IR_b)/(IR_b) = (R_o - R_b)/R_b$, which is dependent only on the contrast in the scene.

## 22.1.2  Intensity–Response Curves and the Naka–Rushton Equation

Light adaptation in photoreceptors is elegantly summarized by intensity–response curves (Fig. 22.3), a set of curves that repays careful consideration. For a fixed background light level $I_0$, we consider the response to a family of superimposed light steps. For each light step, the photoreceptor membrane potential shows a large transient response, followed by a slower change to the steady-state level, as the photoreceptor adapts to the maintained stimulus. For instance, in response to a step increase in light, the photoreceptor membrane potential shows a large transient decrease, followed by a slower increase to the steady level (note that the vertical axis in Fig. 22.3 is reversed). We denote the final light amplitude after the step by $I$ (and thus the amplitude of the step is $I - I_0$), and the turning point of the voltage response (which may be either a maximum or a minimum) by $V(I, I_0)$. For each $I_0$, we then plot $V(I, I_0)$ against $I$ to get the set of curves shown in Fig. 22.3.

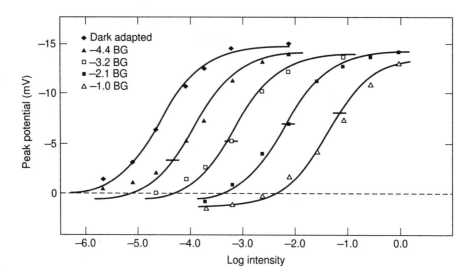

**Figure 22.3** Intensity response curves measured in a red-sensitive cone of the turtle. The peak of the response to a step of light (either increasing or decreasing) is plotted against the log of the intensity of the light at the end of the step. Each data set corresponds to a different background light level, and the smooth curves are drawn by using the Naka–Rushton equation, (22.2), as a template, and shifting it across and down for the higher background light levels. The short horizontal lines denote the resting hyperpolarization for each background light level. The membrane potential has been scaled to be zero in the dark. (Normann and Perlman, 1979, Fig. 7.)

It is observed experimentally that around each background light level, $V(I, I_0)$ can be approximately described by using the Naka–Rushton equation

$$\frac{V(I, I_0)}{V_{\max}} = \frac{I}{I + I_s(I_0)} \tag{22.2}$$

as a template and moving it across and down slightly to describe the higher background light levels. Here, $I_s$ is the light level at which the photoreceptor's response is half maximal. From Fig. 22.3 it can be seen that $I_s$ is an increasing function of the background light level. Thus, as the background light level increases, the response curve maintains its shape, but shifts to higher light levels and moves down slightly (although this shift is not accounted for in (22.2)). Note that since $I$ and $I_s$ are positive, the Naka–Rushton equation is always well defined.

In contrast to the peak response, the steady response, depicted by the small horizontal lines in Fig. 22.3, is a much shallower function of the background light level. In this way retinal neurons can detect contrast over a wide range of light levels without saturating. The steep Naka–Rushton intensity–response curves around each background light level give a high sensitivity to changes superimposed on that background, but the shallower dependence of the steady response on the background light level postpones saturation.

The fact that the Naka–Rushton equation provides a good template for the intensity–response curves has one particularly interesting consequence. Let $V_0(I_0)$ denote the steady response to a steady background light level $I_0$. Then, since this point must lie on the intensity–response curve corresponding to $I_0$, it follows that

$$V_0(I_0) = \frac{I_0}{I_0 + I_s(I_0)}. \tag{22.3}$$

Note that since $I_s$ is a function of $I_0$, this is *not* the Naka–Rushton equation, although it looks similar. Next, notice that for given a background light level $I_0$, the peak sensitivity is (from (22.2))

$$S(I_0) = \left. \frac{\partial V}{\partial I} \right|_{I=I_0} = \frac{I_s}{(I_0 + I_s)^2}, \tag{22.4}$$

where $I_s$ also depends on $I_0$. For simplicity, we have set $V_{max} = 1$. Clearly, the peak sensitivity $S(I_0)$ is inversely proportional to the background light level $I_0$ if and only if $I_s$ is a constant multiple of $I_0$, i.e., $I_s = kI_0$ for some constant $k$, in which case $V_0 = 1/(1+k)$. It thus follows that Weber's law is followed exactly when the steady-state response is independent of the background light level, a feature called *exact adaptation*, and one that was discussed previously in the context of adaptation of hormone receptors (Chapter 19). Of course, in reality peak sensitivity cannot always be proportional to $1/I_0$, as then it would be infinite in the dark, and the steady response of a photoreceptor is not independent of the stimulus. Thus, Weber's law is only an approximation to reality, but one that is extremely useful.

## 22.1.3   A Linear Input–Output Model

Turtle horizontal cells respond linearly to modulations about a mean light level, provided that the modulations are not too large. Thus, at each background light level $I_0$, their behavior can be described by a first-order transfer function (see Section 22.6 for a brief summary of linear systems theory and transfer functions). The transfer function depends on $I_0$, and so the behavior of the horizontal cell over all light levels is described by a family of transfer functions, parametrized by $I_0$.

Typical experimental data are shown in Fig. 22.4. The light input to the system is modulated around a mean level, and the output is the membrane potential of the horizontal cell. Gain is measured in units of mV photon$^{-1}$ and plotted relative to gain in the dark. Since the experiments were performed under conditions that minimized feedback from horizontal cells to photoreceptors, it is reasonable to suppose that the observed transfer functions are determined by the photoreceptor responses, rather than intrinsic to the horizontal cell.

As can be seen from Fig. 22.4, when the background light level increases by one log unit in intensity (indicated by moving from filled squares to open circles, or from open circles to filled triangles, etc.), the relative gain at low frequency decreases by approximately one log unit. Hence, over a range of light levels the low-frequency gain

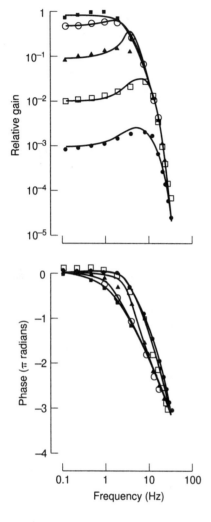

**Figure 22.4** A family of temporal frequency responses from turtle horizontal cells, measured at different mean light levels. (From top to bottom, the filled square, open circle, filled triangle, open square, and filled circle denote, respectively, −4, −3, −2, −1, and 0 log units.) Symbols are experimental data, and the smooth curves are from the model described in the text. The data are presented in a typical Bode plot format, with the amplitude plotted in the upper panel, the phase difference in the lower. (Tranchina et al., 1984, Fig. 1.)

is inversely proportional to $I_0$, and thus the steady-state sensitivity obeys Weber's law. At high frequencies and low background light levels, Weber's law breaks down, and the gain becomes nearly independent of the background light level. It is important to note that the steady-state sensitivity is not the same as the peak sensitivity. In photoreceptors, however, for reasons that are not clear, both the steady state and peak sensitivity follow Weber's law approximately.

A model to describe these data was constructed by Tranchina et al. (1984). Their model, shown in Fig. 22.5, consists of two linear filters, with transfer functions $P(\omega)$ and $Q(\omega)$, and a multiplicative feedback proportional to $I_0$, where the parameters of the linear filters are determined by fitting to the experimental data shown in Fig. 22.4. The transfer function of this linear system can be calculated as follows. For a light stimulus of the form $e^{i\omega t}$, the output from the first filter is $P(\omega)e^{i\omega t}$. If the output from the second filter is $G$, then the input into the second filter is $P(\omega)e^{i\omega t} - I_0 G$, so that

$$G = Q(\omega)(P(\omega)e^{i\omega t} - I_0 G). \tag{22.5}$$

Solving for $G$ we find that

$$G = \frac{P(\omega)Q(\omega)e^{i\omega t}}{1 + I_0 Q(\omega)}, \tag{22.6}$$

so that the transfer function for the feedback system is $T(\omega) = -P(\omega)Q(\omega)/(1+I_0 Q(\omega))$. The $I_0$ in the denominator gives Weber's law behavior over the range of light levels and frequencies for which $I_0 Q(\omega) \gg 1$, since then the transfer function is approximately $P(\omega)/I_0$.

The transfer functions $P$ and $Q$, were determined by fitting to the experimentally measured transfer function shown in Fig. 22.4. This gives $P(\omega)Q(\omega) = P_0(1+i\omega\tau_1)^{-3}(1+i\omega\tau_2)^{-7}$, $Q(\omega) = Q_0(1 + i\omega\tau_3)(1 + i\omega\tau_4)^3(1 + i\omega\tau_5)^{-1}(1 + i\omega\tau_6)^{-3}$, where $P_0 \approx 4 \times 10^{-10}$ mV s cm$^2$ per photon, and the time constants $\tau_1, \ldots, \tau_6$ are, respectively, 46, 5.4, 12.7, 6.8, 0.4, and 76.3 ms. $Q_0 I_0 \approx 900$ at the 0 log unit background light level. Results from the model are plotted in Fig. 22.4 as smooth curves.

### 22.1.4  A Nonlinear Feedback Model

The close agreement between the above model and the experimental data suggests that Weber's law behavior and light adaptation are the result of a feedback mechanism that is proportional to the background light level. However, we are now faced with the question of how to construct a model that incorporates the desired behavior at each light level but does not require feedback that depends explicitly on $I_0$. To answer

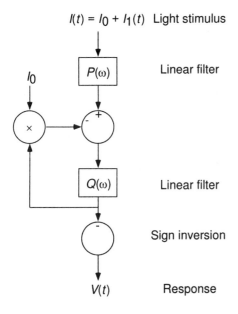

**Figure 22.5**  A family of linear models, which describes the linear responses of turtle horizontal cells around a mean light level. (Adapted from Tranchina and Peskin, 1988.)

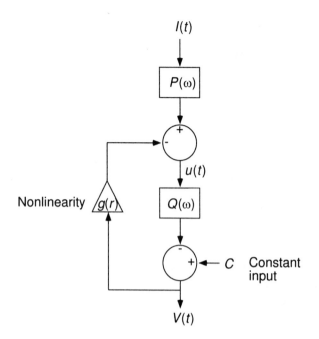

**Figure 22.6** A nonlinear model that embeds the family of linear models shown in the previous figure. (Adapted from Tranchina and Peskin, 1988.)

this question, Tranchina and Peskin (1988) constructed a nonlinear model that has the same linear behavior as the above model but does not depend explicitly on $I_0$. Their model, shown in Fig. 22.6, is similar to that in Fig. 22.5, but the feedback amplification is through a static nonlinearity rather than by the factor $I_0$.

Suppose that the input to this system is $I(t) = I_0 + \epsilon e^{i\omega t}$, and the output $V$ is given by

$$V(t) = V_0(I_0) + \epsilon V_1(\omega; I_0)e^{i\omega t}, \tag{22.7}$$

to leading order in $\epsilon$. $V_1(\omega; I_0)$ is the first-order transfer function of the system, and remains to be calculated.

The output from the first filter is $P(0)I_0 + \epsilon P(\omega)e^{i\omega t}$, and thus the input to the second filter is

$$u(t) = P(0)I_0 + \epsilon P(\omega)e^{i\omega t} - g(V_0 + \epsilon V_1 e^{i\omega t}) \tag{22.8}$$

$$= P(0)I_0 - g(V_0) + \epsilon[P(\omega) - g'(V_0)V_1]e^{i\omega t} + O(\epsilon^2). \tag{22.9}$$

Putting $u$ through the filter $Q$, changing its sign, and adding $C$ gives

$$V(t) = C - Q(0)[P(0)I_0 - g(V_0)] - \epsilon Q(\omega)[P(\omega) - g'(V_0)V_1]e^{i\omega t} + O(\epsilon^2). \tag{22.10}$$

Equating this with (22.7) gives

$$V_0 = C - Q(0)[P(0)I_0 - g(V_0)] \tag{22.11}$$

and

$$V_1(\omega; I_0) = \frac{-P(\omega)Q(\omega)}{1 - g'(V_0)Q(\omega)}. \tag{22.12}$$

Thus, for (22.12) to have the same frequency response as the linear model, we must have

$$g'(V_0) = -I_0. \tag{22.13}$$

Eliminating $I_0$ between (22.13) and (22.11), we find the differential equation for $g(V_0)$,

$$Q(0)P(0)g'(V_0) + Q(0)g(V_0) - V_0 + C = 0. \tag{22.14}$$

This linear differential equation can be solved for $g$, with the result that

$$g(V) = \frac{P(0)}{Q(0)}\{e^y - (1+y)\}, \tag{22.15}$$

where $y = (C - V)/P(0)$, and the constant of integration is chosen such that $g(C) = 0$ when $I_0 = 0$. The feedback function $g$ is an exponentially increasing function of its argument $y = (C - V)/P(0)$. Typically, $C - V$ is an increasing function of light intensity, because photoreceptors hyperpolarize in response to light, and $V$ is interpreted as the membrane potential.

To find the steady-state output $V_0$, we differentiate (22.11) with respect to $I_0$ and use that $g'(V_0) = -I_0$ to get

$$\frac{dV_0}{dI_0} = \frac{-P(0)Q(0)}{1 + I_0Q(0)}, \tag{22.16}$$

which can be integrated to give

$$V_0 = C - P(0)\ln(1 + I_0Q(0)). \tag{22.17}$$

Thus, the steady-state output is a logarithmic function of $I_0$, and is therefore much shallower than the Naka–Rushton equation (22.2) describing the peak response around each light level. This expression agrees well with experimental data when $I_0$ is not too large.

## 22.2 Photoreceptor Physiology

The previous models show that Weber's law can be duplicated by a nonlinear feedback control system. However, we would like some indication of how this control is established by biochemical processes, rather than the "black box" proposal of the previous models. This requires a more detailed, mechanistic, model of the molecular events underlying the light response. As a preliminary to the construction of such a model, we present a brief discussion of the physiology of the vertebrate photoreceptor. More detailed discussions are given in Fain and Matthews (1990), McNaughton (1990), and Pugh and Lamb (1990). A selection of detailed articles is given in Hargrave et al. (1992).

Vertebrate photoreceptors are composed of two principal segments: an outer segment that contains the photosensitive pigment, and an inner segment that contains the necessary cellular machinery. A connecting process, called an axon, connects the inner segment to a *synaptic pedicle*, which communicates with neurons (such as horizontal and bipolar cells) in the inner layers of the retina. In rods, the photosensitive pigment is located on a stack of membrane-enclosed disks that take up the majority of the space in the outer segment, while in cones, the pigment is located on invaginations of the outer segment membrane. The connecting process does not transmit action potentials, and hence the name "axon" is somewhat misleading. Photoreceptors respond in a graded manner to light, and give an analogue, rather than a digital, output.

In the dark, the resting membrane potential is about $-40$ mV. Current, carried by $Na^+$ and $Ca^{2+}$ ions, flows into the cell through light-sensitive channels in the outer segment, and is balanced by current, carried mostly by $K^+$ ions, flowing out through $K^+$ channels in the inner segment. Thus, in the dark there is a circulating current of about 35–60 pA. In the dark, the light-sensitive channels are held open by the binding of 3 molecules of cGMP. Ionic balance is maintained by a $Na^+$–$K^+$ pump in the inner segment that removes 3 $Na^+$ ions for the entry of 2 $K^+$ ions, and a $Na^+$–$Ca^{2+}$, $K^+$ exchanger in the outer segment that removes 1 $Ca^{2+}$ and 1 $K^+$ for the entry of 4 $Na^+$ ions. The $Na^+$–$Ca^{2+}$, $K^+$ exchanger is the principal method for $Ca^{2+}$ extrusion from the cytoplasm.

The light response begins when a photon of light strikes the photosensitive pigment, initiating a series of reactions (described in more detail below) that results in the activation of rhodopsin, and its subsequent binding to a G-protein, transducin. The bound transducin exchanges a molecule of GDP for GTP and then binds to cGMP-phosphodiesterase (PDE), thereby activating PDE to PDE*. Since the rate of hydrolysis of cGMP by PDE* is greater than by PDE, this leads to a decline in [cGMP] and subsequent closure of some of the light-sensitive channels. As the light-sensitive conductance decreases, the membrane potential moves closer to the reversal potential of the inner segment $K^+$ conductance (about $-65$ mV), hyperpolarizing the membrane.

Light adaptation is mediated by the cytoplasmic free $Ca^{2+}$ concentration. When the light-sensitive channels close, the entry of $Ca^{2+}$ is restricted, as about 10–15% of the light-sensitive current is carried by $Ca^{2+}$. However, since the $Na^+$–$Ca^{2+}$, $K^+$ exchanger continues to operate, the intracellular [$Ca^{2+}$] falls. This decrease in [$Ca^{2+}$] increases the activity of an enzyme called guanylate cyclase that makes cGMP from GTP. Thus, a decrease in [$Ca^{2+}$] results in an increase in the rate of production of cGMP, reopening the light-sensitive channels, completing the feedback loop. A schematic diagram of the reactions involved in adaptation is given in Fig. 22.8. Although it is likely that there are other important reactions involved in phototransduction (for instance, $Ca^{2+}$ may affect the activity of PDE), the above scheme incorporates many essential features of the light response.

The mechanisms of phototransduction are similar in rods and cones, with one important difference being the light-sensitive pigment contained in the cell. In rods, rhodopsin consists of retinal and a protein called *scotopsin*, while in cones, the

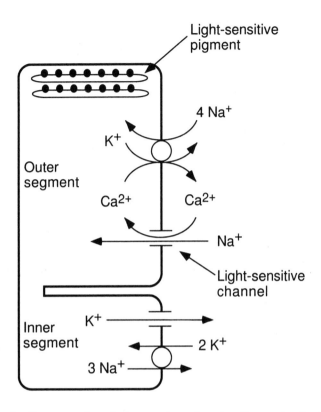

**Figure 22.7** Schematic diagram of a photoreceptor, showing the major ionic currents and pumps that regulate phototransduction. (Adapted from McNaughton, 1990.)

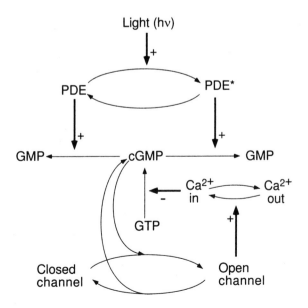

**Figure 22.8** Schematic diagram of the major reactions involved in light adaptation. The initial cascade, leading to the activation of PDE, is condensed into a single step here, but is considered in more detail in Fig. 22.9. (Adapted from Sneyd and Tranchina, 1989.)

rhodopsin consists of retinal and different proteins, called *photopsins*. The primary effect of this compositional difference is that in rods, rhodopsin absorbs light in a range of wavelengths centered at 505 nm, while the rhodopsin in cones absorbs light in a range of wavelengths centered at 445 (blue cones), 535 (green cones), and 570 (red cones) nm. Night blindness is caused by insensitivity of rods because of inadequate amounts of rhodopsin, often associated with vitamin A deficiency. Colorblindness, on the other hand, occurs when green or red cones are missing or when blue cones are underrepresented. Colorblindness is a genetically inherited disorder. Another important difference between rods and cones is that the light response of a cone is much faster than that of a rod, due principally to its smaller size.

There are a number of models of phototransduction, some of which (Baylor et al., 1974a,b; Carpenter and Grossberg, 1981) were constructed before the molecular events underlying adaptation were well known. More detailed recent models include those of Tranchina and his colleagues for turtle cones (Sneyd and Tranchina, 1989; Tranchina et al., 1991), Forti et al. (1989) for newt rods, and Tamura et al. (1991) for primate rods. These models have confirmed that feedback of $Ca^{2+}$ onto the activity of guanylate cyclase is indeed sufficient to explain many features of the light response in both rods and cones. Detailed models of the initial cascade and the activation process have been constructed by Cobbs and Pugh (1987) and Lamb and Pugh (1992).

## 22.2.1 The Initial Cascade

Although the main consequence of the absorption of light by a photoreceptor is the conversion of PDE to a more active form, and a resultant decline in the concentration of cGMP, there are many biochemical steps between these events (Fig. 22.9). Absorption of a photon causes the isomerization of 11-cis retinal to the all-trans form, and this in turn causes a series of isomerizations of rhodopsin, ending with metarhodopsin II. Metarhodopsin II is converted to metarhodopsin III, which is, in turn, hydrolyzed to opsin and all-trans retinal. The activated form of rhodopsin is metarhodopsin II, called $R^*$ here. In the dark, the G-protein transducin is in its deactivated form, T-GDP. After absorption of a photon, $R^*$ binds to T-GDP and catalyzes the exchange of GDP for GTP. This exchange reduces the affinity of $R^*$ for transducin, and also causes transducin to split into an $\alpha$ subunit, $T_\alpha$-GTP, and a $\beta\gamma$ subunit, $T_{\beta\gamma}$. It is $T_\alpha$-GTP that then binds to PDE, forming the complex PDE*-$T_\alpha$-GTP, which is the activated form of the PDE. The cycle is completed when the GTP is dephosphorylated to GDP, the PDE leaves the complex, and the $T_\alpha$ subunit recombines with the $T_{\beta\gamma}$ subunit forming the deactivated form of transducin again. All of these reactions occur in the membrane that contains rhodopsin, and are thus influenced by the speed at which the various proteins can diffuse within the membrane, an aspect that we do not consider (but see Lamb and Pugh, 1992).

Although it is possible to model this sequence of reactions in detail (Exercise 1), we do not do so here, as it appears that nonlinearities in the initial stages of the light response have little effect on adaptation at low light levels. This allows the considerable

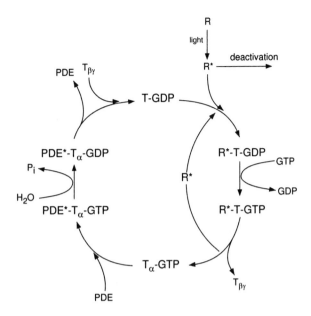

R

light

deactivation

R*

PDE $T_{\beta\gamma}$

T-GDP

PDE*-$T_\alpha$-GDP

$P_i$

R*-T-GDP

GTP

R*

GDP

$H_2O$

PDE*-$T_\alpha$-GTP

R*-T-GTP

$T_\alpha$-GTP

$T_{\beta\gamma}$

PDE

**Figure 22.9** Schematic diagram of the biochemical reactions involved in the initial steps of phototransduction. (Adapted from Stryer, 1986, Fig. 6.)

simplification of modeling the initial stages of the light response as a simple multistage linear filter.

We model the initial cascade as a sequence of linear reactions,

$$\frac{dr}{dt} = l_1 I(t) - l_2 r, \tag{22.18}$$

$$\frac{dg_1}{dt} = l_3 r - l_4 g_1, \tag{22.19}$$

$$\frac{dg_2}{dt} = l_5 g_1 - l_6 g_2, \tag{22.20}$$

$$\frac{dg}{dt} = l_7 g_2 - l_8 g, \tag{22.21}$$

where $l_1, \ldots, l_8$ are rate constants, $r$ is the concentration of R*, $g$ is the concentration of transducin, and $g_1$ and $g_2$ are hypothetical intermediate states between the formation of activated rhodopsin and the activation of transducin. These intermediate states need not occur in this specific location, but could be included anywhere preceding the activation of PDE. Two intermediate states are used because there is evidence that to get acceptable agreement with data, at least four stages are needed before the activation of PDE (Cobbs and Pugh, 1987). The transfer function of this linear system is

$$H(\omega) = \frac{\eta}{(1 + i\omega\tau_1)^4}, \tag{22.22}$$

where $\eta = l_1 l_3 l_5 l_7 \tau_1^4$, and where we have assumed that $l_2 = l_4 = l_6 = l_8 = 1/\tau_1$. The impulse response $K(t)$ of the system is given by

$$K(t) = \frac{\eta}{\tau_1 3!} \left( \frac{t}{\tau_1} \right)^3 e^{-t/\tau_1}. \tag{22.23}$$

Finally, we let $p$ denote the concentration of PDE*, and let $P_0$ denote the total concentration of PDE, to get

$$\frac{dp}{dt} = s(t)P_0 - k_1 p, \tag{22.24}$$

where

$$s(t) = \int_{-\infty}^{t} I(\tau)K(t-\tau)\,d\tau. \tag{22.25}$$

Note that we have assumed that the amount of PDE* is small enough so that $P_0 - p \approx P_0$, and that the deactivation of PDE* is linear (Hodgkin and Nunn, 1988).

## 22.2.2 Light Adaptation in Cones

We can now incorporate the model of the initial cascade into a complete model for excitation and adaptation by including equations for the concentrations of cGMP, $Ca^{2+}$, and $Na^+$, as well as the membrane potential. First, we scale $p$ by $P_0$, and then let $x = [\text{cGMP}]/[\text{cGMP}]_{\text{dark}}$, $y = [Ca^{2+}]/[Ca^{2+}]_{\text{dark}}$, and $z = [Na^+]/[Na^+]_{\text{dark}}$, so that $x = y = z = 1$ in the dark. We also shift the membrane potential $V$ so that $V = 0$ in the dark.

cGMP is produced at some rate dependent upon calcium concentration, given by $g(y)$, an unknown function to be determined. cGMP is hydrolyzed by both the active (at rate proportional to $xp$) and the inactive (at rate proportional to $x(1-p)$) forms of PDE, although the rate of hydrolysis by PDE* is faster than by PDE. Thus,

$$\frac{dx}{dt} = g(y) - \gamma xp - \delta x(1-p) = g(y) - (\gamma - \delta)xp - \delta x. \tag{22.26}$$

Note that the units of $g(y)$, $\gamma$, and $\delta$ are $s^{-1}$. (The assumptions behind (22.26) are not as simple as they may seem; see Exercise 2.)

The light-sensitive channel is held open by three cGMP molecules, and, in the physiological regime, has a current–voltage relation proportional to $e^{-V/V^*}$, for some constant $V^*$. In general, one expects the number of open light-sensitive channels to be a sigmoidal function of $x$, as in (1.45), with a Hill coefficient of 3. However, since $x$ is very small in the dark (and becomes even smaller in the presence of light), few light-sensitive channels ever open, and thus the light-sensitive current $J_{\text{ls}}$ is well represented by

$$J_{\text{ls}} = Jx^3 e^{-V/V^*}, \tag{22.27}$$

for some constant $J$, with units of current.

Calcium enters the cell via the light-sensitive current, of which approximately 15% is carried by $Ca^{2+}$, and is pumped out by the $Na^+-Ca^{2+}, K^+$ exchanger. Assuming that the $Na^+-Ca^{2+}, K^+$ exchanger removes $Ca^{2+}$ with first-order kinetics, the balance equation for calcium is

$$\beta \frac{dy}{dt} = \frac{\kappa}{2Fvy_d} Jx^3 e^{-V/V^*} - k_2 y, \tag{22.28}$$

where $v$ is the cell volume, $\kappa$ is the fraction of the light-sensitive current carried by $Ca^{2+}$, $F$ is Faraday's constant, $k_2$ is the rate of the exchanger, and $y_d$ denotes $[Ca^{2+}]_{dark}$. To incorporate $Ca^{2+}$ buffering, we assume that the ratio of bound to free $Ca^{2+}$ is $\beta$, and that the buffering is fast and linear. This means that the rate of change of $[Ca^{2+}]$ must be scaled by $\beta$ (Section 12.3). Typically, $\beta$ is approximately 99.

Similarly, the balance equation for $Na^+$ is derived by assuming that the exchanger brings 4 $Na^+$ ions in for each $Ca^{2+}$ ion it pumps out, and that the rate of the $Na^+ - K^+$ pump is a linear function of $Na^+$. Further, most of the light-sensitive current not carried by $Ca^{2+}$ is carried by $Na^+$. Thus,

$$\frac{dz}{dt} = \frac{1-\kappa}{Fvz_d} Jx^3 e^{-V/V^*} + \frac{4k_2 y_d}{z_d} y - k_3 z, \tag{22.29}$$

where $z_d$ denotes $[Na^+]_{dark}$.

Some parameter relationships can be determined and the equations for $y$ and $z$ simplified by using the fact that $x = y = 1$, $V = 0$ must be a steady state. From this it follows that

$$k_2 = \frac{J\kappa}{2Fvy_d} \tag{22.30}$$

and

$$k_3 = \frac{(1-\kappa)J}{Fvz_d} + \frac{4k_2 y_d}{z_d} = \frac{J(1+\kappa)}{Fvz_d}, \tag{22.31}$$

so that

$$\tau_y \frac{dy}{dt} = x^3 e^{-V/V^*} - y, \tag{22.32}$$

$$\tau_z \frac{dz}{dt} = \left(\frac{1-\kappa}{1+\kappa}\right) x^3 e^{-V/V^*} + \left(\frac{2\kappa}{1+\kappa}\right) y - z, \tag{22.33}$$

where $\tau_z = \frac{Fvz_d}{J(1+\kappa)}$ and $1/\tau_y = \beta k_2$.

Finally, we derive an equation for the membrane potential. Since the exchangers and pumps transfer net charge across the cell membrane, there are four sources of transmembrane current: the light-sensitive current, the $Na^+-K^+$ pump current, the $Na^+-Ca^{2+}, K^+$ exchange current, and the light-insensitive $K^+$ current, which is modeled as an ohmic conductance. Also note that for every 1 $Ca^{2+}$ ion pumped out of the cell, one positive charge enters, and for every 3 $Na^+$ ions pumped out, one positive charge leaves. Thus

$$C_m \frac{dV}{dt} = Jx^3 e^{-V/V^*} - \frac{Fk_3 z_d v}{3} z - G(V - E) + (Fk_2 y_d v)y, \tag{22.34}$$

where $G$ and $E$ are, respectively the conductance and reversal potential of the light-insensitive $K^+$ channel, and $C_m$ is the capacitance of the cell membrane. Recall that the potential $V$ is measured with respect to the potential in the dark. Using (22.30) and (22.31), the voltage equation becomes

$$C_m \frac{dV}{dt} = Jx^3 e^{-V/V^*} - \frac{J(1 + \kappa)}{3} z - G(V - E) + \frac{J\kappa}{2} y, \tag{22.35}$$

and then using that $V = 0$, $y = z = 1$ must be a steady state, we get

$$J = \frac{-6GE}{4 + \kappa}. \tag{22.36}$$

Substituting this expression back into the voltage equation gives

$$\tau_m \frac{dV}{dt} = -\left(\frac{6E}{4 + \kappa}\right) x^3 e^{-V/V^*} + 2\left(\frac{1 + \kappa}{4 + \kappa}\right) Ez - (V - E) - \left(\frac{3E\kappa}{4 + \kappa}\right) y, \tag{22.37}$$

where $\tau_m = C_m/G$ is the membrane time constant.

### Determination of the unknowns

The unknown function $g(y)$ is determined by forcing the steady-state membrane potential to be the logarithmic function

$$V_0 = -s_1 \log(1 + s_2 I_0), \tag{22.38}$$

for some constants $s_1, s_2$. The form of this steady-state relation is suggested by (22.17), and (22.38) gives very good agreement with experimental data (although it does not give exact Weber's law behavior). This results in a long and complicated expression that we do not give here, as its analytic form has no physiological significance (see Exercise 4). Its shape, however, is of interest, and that can be determined only after the parameters are determined by fitting to experimental data.

Some of the parameters are known from experiment. For instance, $\kappa$, the proportion of the light-sensitive current carried by $Ca^{2+}$, is known to be 0.1–0.15, while $\tau_z$, the time constant for $Na^+$ extrusion, is known to be around 0.04 s. Similarly, from measurements of the current/voltage relation of the light-sensitive channel, $V^*$ is known to be 35.7 mV. The remaining unknown parameters ($s_1$, $s_2$, $E$, $\tau_y$, $k_4$, $\gamma$, $\delta$, $\eta$, $\tau_1$, and $\tau_m$) are determined by fitting the first-order transfer function of the model to experimental data (typical experimental data are shown in Fig. 22.4). The results of this parameter estimation are given in Table 22.1.

**683**

**Table 22.1** Parameter values for the model of light adaptation in turtle cones. (Tranchina et al., 1991.)

| | | | | | |
|---|---|---|---|---|---|
| $s_1$ | = | 1.59 mV | $s_2$ | = | 1130 |
| $E$ | = | -13 mV | $V^*$ | = | 35.7 mV |
| $\tau_y$ | = | 0.07 s | $k_1$ | = | $35.4s^{-1}$ |
| $\gamma$ | = | $303\ s^{-1}$ | $\delta$ | = | $5s^{-1}$ |
| $\kappa$ | = | 0.1 | $\eta$ | = | $52.5s^{-1}$ |
| $\tau_1$ | = | 0.012 s | $\tau_m$ | = | 0.016 s |
| $\tau_z$ | = | 0.04 s | | | |

## Model predictions and behavior

The most interesting prediction of the model is the shape of the feedback function $g(y)$ that mediates light adaptation. A plot of $g$ is given in Fig. 22.10. In the physiological regime, $g(y)$ is well approximated by the function $A(y)$, where

$$A(y) = 4 + \frac{91}{1 + (y/0.34)^4}. \tag{22.39}$$

In other words, as $[Ca^{2+}]$ falls, the rate of cGMP production by guanylate cyclase rises along a sigmoidal curve, with a Hill coefficient of 4. This prediction of the model has been confirmed experimentally (Koch and Stryer, 1988), thus lending quantitative support to the hypothesis that the modulation of guanylate cyclase activity by $[Ca^{2+}]$ is sufficient to account for light adaptation in turtle cones.

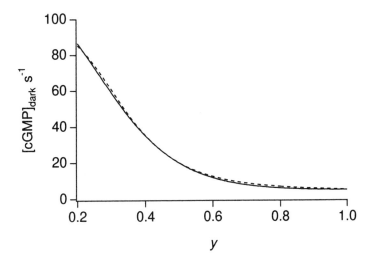

**Figure 22.10** Predicted and measured activities of guanylate cyclase as functions of the $Ca^{2+}$ concentration. The solid line denotes $g(y)$ (theoretical prediction), and the dotted line denotes $A(y)$ (experimental measurement). For convenience, the activity of guanylate cyclase is expressed in units of $[cGMP]_{dark}$ per second.

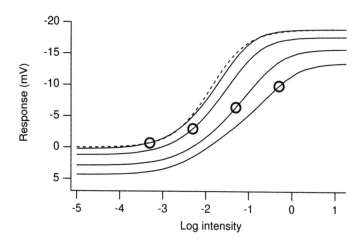

**Figure 22.11** Intensity–response curves from the model for adaptation in turtle cones. The dotted line is the Naka–Rushton equation, and the solid lines are the model results. The open symbols denote the steady states for 4 different light levels.

The model exhibits constant contrast sensitivity over a range of lower light levels. As $I_0$ increases, the contrast sensitivity first increases and then decreases slightly, in agreement with the results of Daly and Normann (1985); the impulse response becomes biphasic; and the time-to-peak decreases as the response speeds up. Further, the intensity–response curves agree well with the Naka–Rushton equation (22.2) and shift to the right and slightly down as $I_0$ increases (Fig. 22.11), again in good agreement with experimental data.

An unexpected prediction of the model is that [cGMP] does not fall much as the background light level is increased. For example, if the background light level is changed so that the sensitivity decreases by a factor of 1000, [cGMP] decreases by less than a factor of two (see Sneyd and Tranchina, 1989, Fig. 4). This gives a possible explanation for the puzzling observation that even though a decrease in [cGMP] is believed to underlie light adaptation, such decreases are sometimes not experimentally observed (DeVries et al., 1979; Dawis et al., 1988). In other words, the model predicts that even though a decrease in [cGMP] may indeed mediate light adaptation, the actual decrease may be too small to measure reliably.

The model agrees quantitatively with experiment in a number of other ways (discussed in detail by Tranchina et al., 1991), lending further support to the hypothesis that it provides an excellent description of many features of light adaptation. Similar conclusions have been reached by Forti et al. (1989), who modeled phototransduction in newt rods, and Tamura et al. (1991), who modeled adaptation in primate rods. It thus appears that although $Ca^{2+}$ feedback onto the activity of guanylate cyclase cannot explain all features of light adaptation in rods and cones, it is likely to be one of the principal mechanisms.

## 22.3 Photoreceptor and Horizontal Cell Interactions

Thus far we have considered only the responses of individual photoreceptors. However, spatial interactions in the retina also play an important role in regulation of the light response.

Photoreceptors and horizontal cells form layers of cells through which their potential can spread laterally. The output from the photoreceptors is directed toward the horizontal cells, but the response of the horizontal cells also influences the photoreceptors, forming a feedback loop with spatial interactions. Here we examine two models of this structure. The simpler, due to Peskin (1976), was originally constructed as a model for the horseshoe crab eye, but has wider applicability.

### 22.3.1 Lateral Inhibition: A Qualitative Model

We suppose that $E$ is the excitation of a receptor by light and that $I$ is the inhibition of the receptor from the layer of horizontal cells. The photoreceptor response is $R = E - I$. A light stimulus $L$ causes an excitation $E$ in the receptor and $E$ decays with time constant $\tau$. The response of the receptor $R$ provides an input into a layer of inhibitory cells, which are laterally connected, and so the inhibition spreads laterally by diffusion and decays with time constant 1. The model equations are

$$\tau \frac{\partial E}{\partial t} = L - E, \tag{22.40}$$

$$\frac{\partial I}{\partial t} = \nabla^2 I - I + \lambda R, \tag{22.41}$$

$$R = E - I. \tag{22.42}$$

**Space-independent behavior**

If we assume that the light stimulus is spatially uniform, then spatial dependence can be ignored, and the model equations reduce to the ordinary differential equations

$$\tau \frac{dE}{dt} = L - E, \tag{22.43}$$

$$\frac{dI}{dt} + (\lambda + 1)I = \lambda E. \tag{22.44}$$

If $L$ is a unit step applied at time $t = 0$, then the response at subsequent times is

$$R = E - I \tag{22.45}$$

$$= \frac{1}{\lambda + 1} - \frac{k - 1}{k - \lambda - 1} e^{-kt} + \frac{\lambda k}{(k - \lambda - 1)(\lambda + 1)} e^{-(\lambda + 1)t}, \tag{22.46}$$

where $k = 1/\tau$. $R$ is graphed in Fig. 22.12A, from where it can be seen that the response is an initial peak followed by a decay to a plateau.

### Time-independent behavior

If the input is steady, so that time derivatives vanish, then $E = L$ and

$$\nabla^2 I = (\lambda + 1)I - \lambda L. \tag{22.47}$$

Suppose there is an edge in the pattern of light, represented by

$$L(x,y) = \begin{cases} 1, & x > 0, \\ 0, & x < 0. \end{cases} \tag{22.48}$$

Then the solution for $I$ is

$$I = \begin{cases} \dfrac{\lambda}{\lambda + 1}\left(1 - \dfrac{1}{2}e^{-x\sqrt{\lambda+1}}\right), & x > 0, \\[2ex] \dfrac{\lambda}{\lambda + 1}\dfrac{1}{2}e^{x\sqrt{\lambda+1}}, & x < 0, \end{cases} \tag{22.49}$$

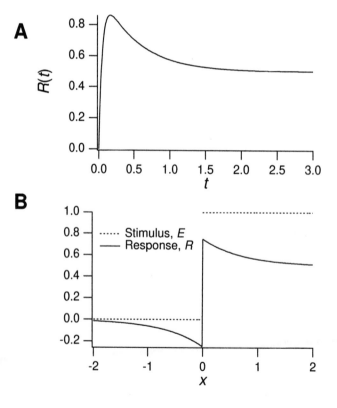

**Figure 22.12** Solutions of the qualitative model for lateral inhibition in the retina, calculated with the parameters $\lambda = 1$, $k = 20$. A: The space-independent response. B: The steady response to a band of light extending from $x = 0$ to $x = \infty$.

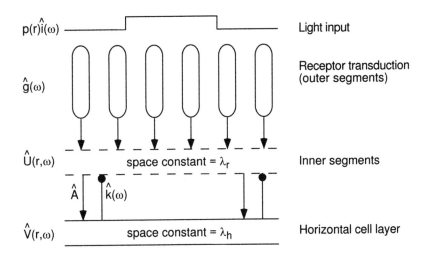

**Figure 22.13** Schematic diagram of the lateral inhibition model of Krausz and Naka. (Adapted from Krausz and Naka, 1980.)

where $I$ and $dI/dx$ are required to be continuous at $x = 0$. Graphs of $E$ and $R = E - I$ are shown in Fig. 22.12B. $R$ exhibits Mach bands at the edge of the light stimulus.

## 22.3.2 Lateral Inhibition: A Quantitative Model

A more detailed model of receptor/horizontal cell interactions was constructed by Krausz and Naka (1980), and the model parameters were determined by fitting to experimental data from the catfish retina. The model is depicted in Fig. 22.13.

In this model, receptor and horizontal cells are assumed to form continuous sheets, within which voltage spreads continuously. The coupling coefficient for voltage spread in the sheet of receptors differs from that in the horizontal cell sheet. The receptors feed forward to the horizontal cells, with transfer function $\hat{A}$, and the horizontal cells feed back to the receptors with transfer function $\hat{k}$. The receptor response is the excitation due to light minus that due to horizontal cell feedback.

We first consider the model where the voltage spreads laterally in the horizontal cell layer, but not in the photoreceptor layer. To specify this model we must first determine how voltage spreads within a cell layer. The primary assumption is that the horizontal cell layer is effectively a continuous two-dimensional sheet of cytoplasm, and spread of current within this layer can be modeled by the passive cable equation (Chapter 8), with a source term describing the current input from the photoreceptor layer. If the variations of light around the mean are small it is reasonable to assume that the ionic currents are passive and that the governing equation is linear. Thus, from (8.12) we have

$$\tau_h \frac{\partial V}{\partial t} + V = \lambda_h^2 \nabla^2 V + R_h I_{\text{ph}}, \tag{22.50}$$

where $\tau_h$ is the membrane time constant, $\lambda_h$ is the membrane space constant, $R_h$ is the membrane resistivity, and $I_{ph}$ is the current input from the photoreceptor layer.

To simplify the model, we assume that the light input, and all subsequent responses, are radially symmetric, functions only of the distance from the center of the stimulus. We suppose that the light input is $Ii(t)p(r)$, i.e., modulated temporally by $i(t)$ and spatially by $p(r)$. We let $\hat{U}(r, \omega)$ denote the Fourier transform of receptor potential at position $r$, and let $\hat{g}(\omega)$ denote the transfer function of the linear stages of receptor phototransduction. Then, in the frequency domain we have

$$\hat{U} = Ip(r)\hat{g}(\omega)\hat{i}(\omega) - \hat{k}(\omega)\hat{V}. \tag{22.51}$$

$U$ is influenced by two terms, the first due to excitation by light, and the second due to inhibitory feedback from horizontal cells, with transfer function $\hat{k}(\omega)$. Finally, taking the feedforward transfer function to be $\hat{A}(\omega)$, and taking $V = \hat{V}e^{i\omega t}$, we obtain

$$\lambda_h^2 \nabla^2 \hat{V} - (1 + i\omega\tau_h)\hat{V} = -\hat{A}(\omega)\hat{U}. \tag{22.52}$$

Although we expect the qualitative behavior of the Krausz–Naka model to be similar to that of Peskin's model, the goal of this model is to obtain quantitative agreement with experiment by fitting it directly to data.

We simplify (22.51) and (22.52) by a change of variables. We set

$$\Phi = \frac{\hat{U}}{I\hat{g}(\omega)\hat{i}(\omega)}, \tag{22.53}$$

$$\Psi = \frac{\hat{V}}{I\hat{g}(\omega)\hat{i}(\omega)\hat{A}(\omega)}, \tag{22.54}$$

and from (22.52) find that

$$\nabla^2\Psi - \frac{1}{\alpha^2(\omega)}\Psi = -\frac{p(r)}{\lambda_h^2}, \tag{22.55}$$

where

$$\alpha^2(\omega) = \frac{\lambda_h^2}{1 + i\omega\tau_h + \hat{A}(\omega)\hat{k}(\omega)}. \tag{22.56}$$

It is left as an exercise (Exercise 7) to show that the solution of (22.55), assuming an infinite domain, is

$$\Psi = \frac{1}{\lambda_h^2}\int_0^\infty p(s)G(r,s)\,s\,ds, \tag{22.57}$$

where $G$, the fundamental solution, is given by

$$G(r,s) = \begin{cases} I_0(r/\alpha)K_0(s/\alpha), & r < s, \\ K_0(r/\alpha)I_0(s/\alpha), & r > s. \end{cases} \tag{22.58}$$

Here $I_0$ and $K_0$ are modified Bessel functions of the first and second kind of order zero. (Unfortunately, $I_0$ is standard notation for the modified Bessel function of the first kind, but it should not be confused with the background light level.) Of particular interest is the case of a circular spot of light of radius R,

$$p(s) = \begin{cases} 1, & s < R, \\ 0 & s > R. \end{cases} \tag{22.59}$$

In this case, we can use the identities $\frac{d}{dz}(zK_1(z)) = zK_0(z)$ and $\frac{d}{dz}(zI_1(z)) = zI_0(z)$ to evaluate the integral (22.57) with the result that

$$\Psi(r, \omega) = \frac{1}{\lambda_h^2} F(r, R, \omega), \tag{22.60}$$

where

$$F(r, R, \omega) = \begin{cases} \alpha^2[1 - (R/\alpha)I_0(r/\alpha)K_1(R/\alpha)], & r < R, \\ \alpha R I_1(R/\alpha)K_0(r/\alpha), & r > R. \end{cases} \tag{22.61}$$

**Fitting to data**

Krausz and Naka determined the model parameters by fitting the ratio of the uniform field response to the spot response to experimental data (Fig. 22.14). The field response (i.e., taking $R \to \infty$) can be calculated from (22.55) by setting $p(r) \equiv 1$, in which case the constant solution for $\Psi$ is easily seen to be $\Psi(r, \omega)_{\text{field}} = \alpha^2/\lambda_h^2$. Thus,

$$\frac{\hat{V}_{\text{spot}}}{\hat{V}_{\text{field}}} = \frac{\Psi(r, \omega)_{\text{spot}}}{\Psi(r, \omega)_{\text{field}}} = \frac{1}{\alpha^2} F(r, R, \omega). \tag{22.62}$$

For fixed values of $r$ and $R$, $\alpha(\omega)$ can be determined to give good agreement between model and experiment. Note that by taking response ratios, any dependence on the feedforward steps within each photoreceptor is eliminated. Thus, attention is focused on the interactions between the horizontal cells and the photoreceptors. However, the model cannot distinguish between $\hat{A}$ and $\hat{k}$, since only the product of these terms appear. So, for simplicity, it is assumed that $\hat{A}$ is a constant gain, with no frequency dependence, and that $\hat{k}$ has unity gain. Values for $\hat{k}(\omega)$ are obtained at each frequency, and then $k(t)$ determined by an inverse Fourier transform. The result is well approximated by

$$k(t) = \frac{3}{\tau} e^{-(t-t_0)/\tau} [1 - e^{-(t-t_0)/\tau}]^2, \tag{22.63}$$

an S-shaped rising curve followed by exponential decay. The parameters $t_0$ and $\tau$ are, respectively, the feedback delay and the feedback time constant. The function $k(t)$ has an important physiological interpretation, as it describes the feedback from horizontal cells to photoreceptors; in response to a delta function input from the horizontal cells, the photoreceptor response is given by $-k(t)$. Parameters resulting from the fit are given in Table 22.2. The membrane time constant of the horizontal cells was found to be small in all cases, and so was set to zero. Hence, the potential of the horizontal cell layer responds essentially instantaneously to a stimulus. However, the time constant for the response of the photoreceptor layer to horizontal cell feedback is significant.

**Table 22.2**   Parameters of the Krausz–Naka model for catfish retinal neurons. These parameters correspond to the model in which there is no coupling between photoreceptors.

| | | |
|---|---|---|
| $\lambda_h$ | $=$ | 0.267 mm |
| $\tau_h$ | $=$ | 0 ms |
| $\hat{A}$ | $=$ | 3.77 |
| $\tau$ | $=$ | 24.8 ms |
| $t_0$ | $=$ | 0.022 ms |

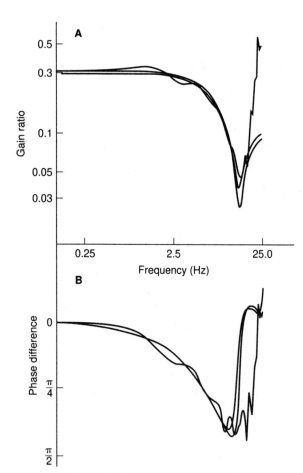

**Figure 22.14**   Spot-to-field transfer function measured in the catfish retina. The wavy traces are the experimental results, and the two smooth curves are results from the model, using two slightly different parameter sets, one of which is given in Table 22.2. (Krausz and Naka, 1980, Fig. 4.)

## Predicting the response to a moving grating

As a test of the model, Krausz and Naka calculated the response to a 1-dimensional moving grating, and compared the result to experimental data. A moving grating provides a light stimulus of the form $l = l(t - x/c)$, where $c$ is the speed of the grating. We look for solutions of the form $V = V(\xi)$, where $\xi = t - x/c$, in which case the differential equation (22.50) becomes

$$\lambda_h^2 V'' - \tau_h V' - V = -R_h I_{ph},\qquad(22.64)$$

where a prime denotes differentiation with respect to $\xi$. Taking Fourier transforms with respect to $\xi$, and recalling that in the frequency domain the input to the horizontal cell layer is given by $\hat{A}\hat{U}$, we find that

$$(1 + i\omega\tau_h)\hat{V} = -\frac{\omega^2\lambda_h^2}{c^2}\hat{V} + \hat{A}\hat{U}.\qquad(22.65)$$

As before, $\hat{U}$ satisfies (22.51). Assuming the light input is a periodic function of $\xi$, we can replace $p(x)i(t)$ by $e^{i\omega\xi}$, and thus replace $p(x)\hat{i}(\omega)$ by 1, since we are now taking Fourier transforms with respect to $\xi$. Hence,

$$\begin{aligned}
\Psi(\omega) &= \frac{\hat{V}}{\bar{I}\hat{g}(\omega)\hat{A}}\\[2mm]
&= \frac{1}{1 + i\omega\tau_h + \hat{A}\hat{k} + \omega^2\lambda_h^2/c^2}\\[2mm]
&= \left(\frac{\alpha^2}{\lambda_h^2}\right)\frac{c^2}{c^2 + \omega^2\alpha^2}.
\end{aligned}\qquad(22.66)$$

Krausz and Naka showed that the model predictions for the rectilinear stimulus predict experimental results accurately, confirming that the model provides a general quantitative description of the horizontal cell response that is not limited to the data upon which it was based.

## Receptor coupling

Receptors are electrically coupled by gap junctions, and the potential spreads through the receptor layer in a continuous fashion, as it does in the horizontal cell layer, but with a different space constant, $\lambda_r$ (Lamb and Simon, 1977; Detwiler and Hodgkin, 1979). To incorporate receptor coupling, we need only add spatial coupling for the receptor layer. In the frequency domain, we have

$$\lambda_r^2\nabla^2\Phi - (1 + i\omega\tau_r)\Phi - \hat{k}(\omega)\hat{A}(\omega)\Psi = -p(r),\qquad(22.67)$$
$$\lambda_h^2\nabla^2\Psi - (1 + i\omega\tau_h)\Psi = -\Phi,\qquad(22.68)$$

where $\tau_r$ is the membrane time constant for the receptor cell layer. Writing $q_r = (1 + i\omega\tau_r)/\lambda_r^2$ and $q_h = (1 + i\omega\tau_h)/\lambda_h^2$ and substituting (22.68) into (22.67) gives

$$\left(\nabla^2 - \frac{1}{\gamma^2}\right)\left(\nabla^2 - \frac{1}{\delta^2}\right)\Psi = \frac{p(r)}{\lambda_h^2\lambda_r^2},\qquad(22.69)$$

where $1/\gamma^2$ and $1/\delta^2$ are defined by

$$\frac{1}{\gamma^2} + \frac{1}{\delta^2} = q_r + q_h,\qquad(22.70)$$

$$\frac{1}{\gamma^2 \delta^2} = q_r q_h + \frac{\hat{A}(\omega)\hat{k}(\omega)}{\lambda_h^2 \lambda_r^2}. \tag{22.71}$$

Note that $\gamma$ and $\delta$ are analogous to $\alpha$ in (22.55).

Now we define $\chi(r, \gamma)$ to satisfy

$$\left(\nabla^2 - \frac{1}{\gamma^2}\right)\chi(r, \gamma) = -p(r). \tag{22.72}$$

Since the operator $\left(\nabla^2 - \frac{1}{\gamma^2}\right)$ has a unique inverse, we use $\chi(r, \gamma)$ to eliminate $p$ in (22.69) and find that

$$\left(\nabla^2 - \frac{1}{\delta^2}\right)\Psi = \frac{-\chi(r, \gamma)}{\lambda_h^2 \lambda_r^2}. \tag{22.73}$$

Similarly, by symmetry

$$\left(\nabla^2 - \frac{1}{\gamma^2}\right)\Psi = \frac{-\chi(r, \delta)}{\lambda_h^2 \lambda_r^2}. \tag{22.74}$$

Subtracting these two equations, we obtain

$$\Psi(r, \omega) = \frac{1}{\lambda_h^2 \lambda_r^2}\left(\frac{\gamma^2 \delta^2}{\gamma^2 - \delta^2}\right)[\chi(r, \gamma) - \chi(r, \delta)]. \tag{22.75}$$

Solving (22.73) for $\nabla^2 \Psi$, substituting into (22.68), and using the expression for $\Psi$ then gives

$$\Phi(r, \omega) = \frac{1}{\lambda_r^2}\left(\frac{\gamma^2 \delta^2}{\gamma^2 - \delta^2}\right)\left[\left(q_h - \frac{1}{\gamma^2}\right)\chi(r, \gamma) - \left(q_h - \frac{1}{\delta^2}\right)\chi(r, \delta)\right]. \tag{22.76}$$

Since (22.72) for $\chi$ is of the same form as (22.55) for $\Psi$, its solution takes the same form. Thus, (22.75) and (22.76) give an explicit solution to the general problem of electrical flow in two coupled cell layers connected by reciprocal pathways.

## 22.4  Receptive Fields

The output stage of the retina is the layer of ganglion cells, which extend through the optic nerve to the lateral geniculate nucleus, from there transmitting signals to the visual cortex. Each ganglion cell responds by a series of action potentials, with the information encoded in the frequency and duration of the wave train. Thus, ganglion cells transmit a digital signal typical of neurons. The input stage of ganglion cells is highly organized (Kuffler, 1953, 1973; Rodieck, 1965). Each ganglion cell responds only to light in a well-defined part of the retina, called the *receptive field* of the cell, and these receptive fields are organized into two concentric, mutually antagonistic regions, the center and the surround (Fig. 22.15).

The center can be either excitatory (*on-center*) or inhibitory (*off-center*). A white figure moved across the receptive field of an on-center cell gives the same response as

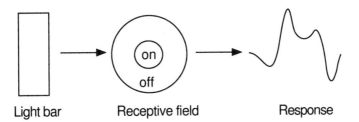

Light bar          Receptive field          Response

**Figure 22.15** Diagram of the center-surround arrangement of the receptive field of an on-center retinal ganglion cell, and its response to a wide bar moving across the receptive field. (Adapted from Rodieck, 1965, Fig. 1.)

a black figure moved across the receptive field of an off-center cell. A typical response curve for a bar moving across the receptive field is shown in Fig. 22.15. Note that the ganglion cell has a large response to the edges of the bar, but responds much less to the maintained stimulus in the middle of the bar. This is reminiscent of the Mach bands seen in the Krausz–Naka model. There are different types of on/off responses for ganglion cells. Some respond to both "on" and "off," while others respond only to one or the other. Some ganglion cells are directionally dependent, responding to a stimulus only if it enters the receptive field from a particular direction. Other ganglion cells are color dependent.

One of the earliest models of ganglion cell behavior was constructed by Rodieck (1965). In this model it is assumed that the response of a ganglion cell is a weighted sum of the responses from each part of the receptive field, with negative weights for the inhibitory part of the field and positive weights for the excitatory part. Here we consider only on-center cells, as the model is the same for off-center cells, with reversed signs. We also consider the model in one spatial dimension only, as the extension to two dimensions introduces greater algebraic complexity, but no new concepts.

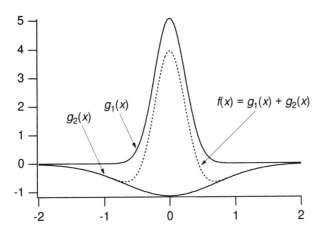

**Figure 22.16** The addition of two Gaussian distributions, one with positive sign denoting the excitatory center, and one with negative sign denoting the inhibitory surround, gives the response function $f(x)$, which weights the light stimulus according to its position in space. Computed using $\sigma_1 = 3, \sigma_2 = 1, g_1 = 3, g_2 = 1$.

Suppose that the steady response to a step change in illumination of a small area $dx$ centered at the point $x$ is given by $f(x)dx$. From consideration of experimental data, Rodieck showed that $f(x)$ can be described as the sum of two Gaussians, one contributing a positive component from the center and the other contributing a negative component from the surround. Thus,

$$f(x) = \frac{g_1 \sigma_1}{\sqrt{\pi}} e^{-\sigma_1^2 x^2} - \frac{g_2 \sigma_2}{\sqrt{\pi}} e^{-\sigma_2^2 x^2}, \tag{22.77}$$

which is plotted in Fig. 22.16. The constants $g_1$ and $g_2$ are the gains of the excitatory center and inhibitory surround, respectively, and the parameters $\sigma_1$ and $\sigma_2$ control their radial size. Now we suppose that the response of the ganglion cell is infinitely fast, and we stimulate the cell with a semi-infinite bar, extending from $x = -\infty$ to $x = ct$, so that the edge of the bar is moving from left to right with speed $c$. Then, the response of the ganglion cell, $R(t)$, is

$$R(t) = \int_{-\infty}^{ct} f(x)\,dx \tag{22.78}$$

$$= g_1 \left[ \frac{1}{2} + \frac{1}{\sqrt{\pi}} \mathrm{erf}(ct/\sigma_1) \right] + g_2 \left[ \frac{1}{2} + \frac{1}{\sqrt{\pi}} \mathrm{erf}(ct/\sigma_2) \right], \tag{22.79}$$

where $\mathrm{erf}(x)$, the error function, is defined by

$$\mathrm{erf}(x) = \frac{2}{\sqrt{\pi}} \int_0^{\infty} e^{-x^2}\,dx. \tag{22.80}$$

From the plot of $R$ given in Fig. 22.17 (solid line) it can be seen that the ganglion cell responds preferentially to the edge of the bar, as is seen in experimental data.

In reality, the response of the ganglion cell is not infinitely fast, but the response to a step of light has an initial peak followed by a decrease to a lower plateau. Thus, in Rodieck's model, the time-dependent response to a step input is taken to be

$$h(t) = \left[1 + te^{-t}\right] H(t), \tag{22.81}$$

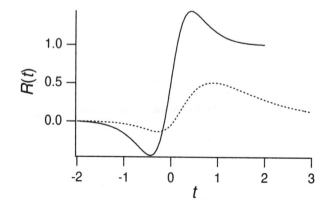

**Figure 22.17** Response of a ganglion cell to a moving bar of semi-infinite width. Solid line assuming that the response at each point $x$ is infinitely fast; dotted line assuming that each point $x$ responds to the light stimulus according to (22.81). Both curves were calculated with $\sigma_1 = 3$, $\sigma_2 = 1$, $g_1 = 3$, $g_2 = 1$, $c = 1$.

where $H(t)$ is the Heaviside function. This is a simple way to incorporate the dynamic behavior of the earlier retinal stages. More general forms for the response produce little difference in the overall response.

To illustrate the temporal behavior of the overall response, we calculate the response to the moving bar, extending from $x = -\infty$ to $x = ct$. Since the edge of the moving bar reaches an element at position $x$ at time $t = x/c$, the response of an element at position $x$ is $f(x)h(t - x/c)dx$. Integrating over the entire domain then gives

$$R(t) = \int_{-\infty}^{\infty} f(x)h(t - x/c)\,dx, \tag{22.82}$$

$$= \underbrace{\int_{-\infty}^{ct} f(x)\,dx}_{\text{steady term}} + \underbrace{\int_{-\infty}^{ct} f(x)(t - x/c)e^{x/c-t}\,dx}_{\text{transient term}}, \tag{22.83}$$

which is graphed in Fig. 22.17 (dotted line). If $f(x)$ decays sufficiently rapidly at $\pm\infty$, the transient term goes to zero as $t \to \infty$, leaving only the steady response. Of course, since real retinas are not infinite in extent, $f$ is zero outside a bounded domain, and so such decay is guaranteed. Further, in the limit as $c \to 0$, keeping $ct$ fixed, the transient term again approaches zero. That is, if the bar moves slowly, the effect of $h(t)$ is small, again as expected. On the other hand, as $c \to \infty$, $R(t)$ approaches $h(t)\int_{-\infty}^{\infty} f(x)\,dx$, which is exactly the response to a space-independent flash. Proofs of these statements are left for the exercises (Exercise 9).

Cells higher up in the visual pathway, in the lateral geniculate nucleus and the visual cortex, have progressively more complex receptive fields, designed to make particular cells respond maximally to stimuli of particular orientation or direction of movement. The above model serves as a brief introduction to the type of modeling involved in the analysis of receptive fields. A more detailed discussion of receptive fields is given by Kuffler et al. (1984) and the references therein.

## 22.5 The Pupil Light Reflex

The control of pupil size is yet another way in which the eye can adjust to varying levels of light intensity. While the adjustment of pupil size accounts for much less of visual adaptation than those mechanisms described earlier, it is nonetheless an important control mechanism.

The size of the pupil of the eye is determined by a balance between constricting and dilating mechanisms. Pupil constriction is caused by contraction of the circularly arranged pupillary constrictor muscle, which is innervated by parasympathetic fibers. The motor nucleus for this muscle is the Edinger–Westphal nucleus located in the oculomotor complex of the midbrain. Dilation is controlled by contraction of the radially arranged pupillary dilator muscle innervated by sympathetic fibers and by inhibition of the Edinger–Westphal nucleus.

The effect of the pupil light reflex is to control the retinal light flux

$$\phi = IA, \tag{22.84}$$

where $I$ is the illuminance (lumen mm$^{-1}$) and $A$ is the pupil area (mm$^2$). It performs this function by acting like the aperture of a camera. When light is shined on the retina, the pupil constricts, thereby decreasing $\phi$. However, there is a latency of $\approx 180$–$400$ ms following a change in light input before changes in pupil size are detected.

This combination of negative feedback with delay may lead to oscillations of pupil size. These oscillations were first observed by a British army officer, Major Stern, who noticed that pupil cycling could be induced by carefully focusing a narrow beam of light at the pupillary margin. Initially, the retina is exposed to light, causing the pupil to constrict, but this causes the iris to block the light from reaching the retina, so that the pupil subsequently dilates, reexposing the retina to light, and so on indefinitely.

Longtin and Milton (1989) developed a model for the dynamics of pupil contraction and dilation. In their model it is assumed that the light flux $\phi$ is transformed after a time delay $\tau_r$ into neural action potentials that travel along the optic nerve. The frequency of these action potentials is related to $\phi$ by

$$N(t) = \eta F \left( \ln \left[ \frac{\phi(t - \tau_r)}{\bar{\phi}} \right] \right), \tag{22.85}$$

where $F(x) = x$ for $x \geq 0$ and $F(x) = 0$ for $x < 0$, $\bar{\phi}$ is a threshold retinal light level (the light level below which there is no response), and $\eta$ is a rate constant. The notation $\phi(t - \tau_r)$ is used to indicate dependence on the flux at time $\tau_r$ in the past.

This afferent neural action potential rate is used by the midbrain nuclei, after an additional time delay $\tau_t$, to produce an efferent neural signal. This signal exits the midbrain along preganglionic parasympathetic nerve fibers, which terminate in the ciliary ganglion where the pupillary sphincter is innervated. Neural action potentials at the neuromuscular junction result in the release of neurotransmitter (ACh), which diffuses across the synaptic cleft, thus generating muscle action potentials and initiating muscle contraction. These events are assumed to require an additional time $\tau_m$.

The relationship between iris muscle activity $x$ and the rate of arriving action potentials $E(t)$ is not known. We take a simple differential relationship

$$\tau_x \frac{dx}{dt} + x = E(t), \tag{22.86}$$

where

$$E(t) = \gamma F \left( \ln \left[ \frac{\phi(t - \tau)}{\bar{\phi}} \right] \right), \tag{22.87}$$

and $\tau = \tau_r + \tau_t + \tau_m$ is the total time delay in the system.

Finally, we close the model by assuming some relationship between iris muscle activity $x$ and pupil area $A$ as $A = f(x)$. For example, one reasonable possibility is the

Hill equation

$$A = f(x) = \Lambda_0 + \frac{\Lambda \theta^n}{x^n + \theta^n}, \tag{22.88}$$

for which area is a decreasing function of activity, with maximal area $\Lambda + \Lambda_0$ and minimal area $\Lambda_0$. It follows that the differential equation governing iris muscle activity is

$$\tau_x \frac{dx}{dt} + x = \gamma F \left( \ln \left[ \frac{I(t - \tau) f(x(t - \tau))}{\bar{\phi}} \right] \right) \tag{22.89}$$

$$= g(x(t - \tau), I(t - \tau)). \tag{22.90}$$

## 22.5.1 Linear Stability Analysis

Because $f(x)$ is a decreasing function of $x$, a steady solution of (22.90) is assured when the input $I(t)$ is constant. We identify this value of $x$ as $x^*$, satisfying $x^* = g(x^*, I)$. Linearized about $x^*$, the delay differential equation becomes

$$\tau_x \frac{dX}{dt} + X = -GX(t - \tau), \tag{22.91}$$

where $G = -g_x(x^*, I) = -\gamma \frac{f'(x^*)}{f(x^*)}$ is called the *gain* of this negative feedback system. If we set $X = X_0 e^{\mu t}$, we find the characteristic equation for $\mu$ to be

$$\tau_x \mu + 1 = -G e^{-\mu \tau}. \tag{22.92}$$

If $|G| < 1$, there are no roots of this equation with positive real part; the solution is linearly stable. Since $G > 0$, there are no positive real roots of this characteristic equation. The only possible way for the solution to become unstable is through a Hopf bifurcation, whereby a root of (22.92) with nonzero imaginary part changes the sign of its real part. If we set $\mu = i\omega$, we can separate (22.92) into real and imaginary parts, obtaining

$$G \cos \omega \tau = -1, \tag{22.93}$$

$$G \sin \omega \tau = \tau_x \omega. \tag{22.94}$$

From these two expressions, we readily find a parametric representation of the critical stability curve to be

$$G = \frac{-1}{\cos \eta}, \qquad \frac{\tau}{\tau_x} = \frac{-\eta}{\tan \eta}. \tag{22.95}$$

The first instability curve is plotted in Fig. 22.18, with the gain $G$ plotted as a function of the dimensionless delay $\tau/\tau_x$. It is easily seen that on the critical curve $G$ is a decreasing function of $\tau/\tau_x$. If the delay is larger than the critical delay, the steady solution is unstable and there is a stable periodic solution of the full differential delay equation (22.90), corresponding to periodic cycling of pupil size with constant light stimulus.

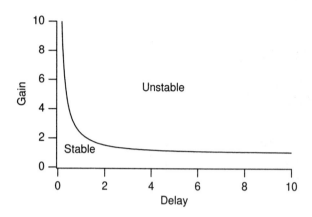

**Figure 22.18**  Critical stability curve for the pupil light reflex.

# 22.6   Appendix: Linear Systems Theory

One of the most widely used tools in the study of the visual system is linear systems analysis. Here we have assumed that the basic tools of linear function theory, such as Fourier transforms, delta functions, and the convolution theorem, are familiar to readers. There are numerous books that provide the necessary background, for example, Papoulis (1962).

The essential idea of linear systems theory is that for any linear differential equation $L[u] = f(t)$, where $L[\cdot]$ is a time-autonomous differential operator, the Fourier transform of the solution can be written as

$$\hat{u}(\omega) = T(\omega)\hat{f}(\omega), \tag{22.96}$$

where $\hat{f}(\omega)$ is the Fourier transform of the input function $f(t)$, and $T(\omega)$ is called the *transfer function* for this linear system. Note that if $f(t)$ is the delta function, then $\hat{u} = T(\omega)$, and thus the transfer function is the Fourier transform of the impulse response. The transfer function can also be found by assuming an input of the form $f = e^{i\omega t}$ and looking for an output of the form $u = T(\omega)e^{i\omega t}$. Thus, the amplitude and phase of the sinusoidal input are modulated by the amplitude and phase of the transfer function. Such sinusoidally varying inputs are commonly used in experimental studies of the visual system; by varying the frequency of the stimulus and measuring (at each fixed frequency) the amplitude and phase of the output, the transfer function can be experimentally determined. Typical experimental data collected in this way are shown in Fig. 22.4.

One significant problem is that most realistic systems are nonlinear, so a transfer function cannot be defined (since solutions are not the linear superposition of fundamental solutions). However, if the amplitude of the sinusoidal input is small, so that $I(t) = \epsilon e^{i\omega t}$, then the response should also be small, of the form $\epsilon T(\omega)e^{i\omega t} + O(\epsilon^2)$. If $\epsilon$ is small enough, and higher-order terms can be neglected, the response of the system is well described by the *first-order transfer function* $T(\omega)$.

Of course, the response of retinal cells to stimuli is not linear, and therefore one can determine only their first-order frequency response. An example of this is plotted in Fig. 22.4. However, measurement of the amplitudes of higher harmonics ($e^{2i\omega t}, e^{3i\omega t}, e^{4i\omega t}$, etc.) in response to an input of the form $e^{i\omega t}$ indicates that nonlinearities have little effect for the light stimuli used in the experiments. Thus, the behavior of retinal cells can be described well by their first-order frequency responses.

Suppose $V_0(x)$ denotes the response to a steady input $x$. If the input is of the form $I(t) = I_0 + \epsilon e^{i\omega t}$, then the output will be of the form $V(t) = V_0(I_0) + \epsilon V_1(w; I_0)e^{i\omega t} + O(\epsilon^2)$. The function $V_1(w; I_0)$ is the first-order transfer function, or first-order frequency response. When $\omega = 0$, this becomes $V_0(I_0 + \epsilon) = V_0(I_0) + \epsilon V_1(0; I_0) + O(\epsilon^2)$. However, expanding $V_0(I_0+\epsilon)$ in a Taylor series around $I_0$ gives $V_0(I_0+\epsilon) = V_0(I_0)+\epsilon V_0'(I_0)+O(\epsilon^2)$, from which it follows that the steady-state sensitivity is

$$\frac{dV_0}{dI_0} = V_1(0; I_0), \tag{22.97}$$

an identity that is of considerable use.

## 22.7 EXERCISES

1. Using the reaction scheme shown in Fig. 22.9, construct a more detailed model of the initial stages of the light response. Under what conditions does this model reduce to the linear model (22.18)–(22.21)? Compute the impulse response of the model at different background light levels, and compare to (22.23).

2. cGMP is hydrolyzed to GMP by PDE or PDE* in an enzymatic reaction, and as we saw in Chapter 1, such reactions do not necessarily follow the law of mass action. So, assume that cGMP reacts with PDE according to

$$\text{cGMP} + \text{PDE} \underset{k_{-1}}{\overset{k_1}{\rightleftharpoons}} \text{complex} \overset{k_2}{\rightarrow} \text{GMP} + \text{PDE}. \tag{22.98}$$

Derive the conditions under which (22.26) may be expected to apply, keeping in mind that [PDE] is much larger than [cGMP], and thus the usual approximation of enzyme kinetics does not apply. (See also Exercise 6 of Chapter 1, and Sneyd and Tranchina, 1989.)

3. By writing down the balance equations for $\text{Na}^+$, $\text{Ca}^{2+}$, and $\text{K}^+$, show that the outward current generated by the electrogenic $\text{Na}^+ - \text{K}^+$ pump must be approximately a third of the total inward light-sensitive current. Thus, show that the current generated by the electrogenic pumps cannot be ignored in the modeling. Must the electrogenic pumps be included in the model if the model is compared to photocurrent measurements and not to voltage measurements?

4. Calculate the explicit expression for $g(y)$ in the model for adaptation in turtle cones. Also calculate $g'(y)$ by using the identity (22.97). Show that $g'(y)$ can be calculated even if no explicit expression can be found for $g(y)$.

5. Calculate the response of the Peskin model (Section 22.3.1) when the light stimulus is a strip of width $2a$, modulated sinusoidally with frequency $\omega$. Show that as $a$ increases (i.e., as the stimulus goes from a spot to a uniform field), the gain at $x = 0$ becomes more band-pass in nature. Calculate the spot-to-field response ratio and compare it to that of the Krausz–Naka

model. Hint: Show that the solution for $\hat{I}$ is of the form

$$\hat{I} = \begin{cases} \dfrac{\lambda \hat{L}}{(i\omega\tau + 1)(1 + \lambda + i\omega)}[1 + A\cosh[x\sqrt{1 + \lambda + i\omega}], & |x| < a, \\ B\exp(-|x|\sqrt{1 + \lambda + i\omega}), & |x| > a, \end{cases} \qquad (22.99)$$

and then require that $\hat{I}$ and its derivative be continuous at $|x| = a$.

6. Calculate the response of the Peskin model (Section 22.3.1) to a moving step of light. Show that as the speed tends to zero, the response approaches the response to a steady step, while as the speed goes to infinity, the response behaves like the response to a step of light presented simultaneously to the entire retina.

7. Derive (22.58) and verify (22.60). Hints: Show first that everywhere except $r = s$, $G$ satisfies the modified Bessel equation of order zero, $z^2 \frac{d^2 u}{dz^2} + z \frac{du}{dz} - z^2 u = 0$. Two independent solutions of the modified Bessel equation are the modified Bessel functions of the first and second kind. Then show that the jump condition at $r = s$ is

$$\left.\frac{dG}{dr}\right|_{r=s^+} - \left.\frac{dG}{dr}\right|_{r=s^-} = -\frac{1}{s}, \qquad (22.100)$$

and use the fact that the Wronskian $W$ is

$$W(K_\nu, I_\nu) = K_\nu(z)\frac{d}{dz}I_\nu(z) - I_\nu(z)\frac{d}{dz}K_\nu(z) = \frac{1}{z}. \qquad (22.101)$$

8. Show that the Krausz–Naka model (Section 22.3.2) is essentially the same as the Peskin model (Section 22.3.1), with a few more details. Thus, investigate the response of the Krausz–Naka model to a space-independent light step and a time-independent bar of light. Demonstrate Mach bands and adaptation, as in the Peskin model.

9. Prove the statements made about (22.83).
   Hint: Suppose that $f(x)$ is bounded and absolutely integrable so that for any $\epsilon > 0$ there is a $C_0$ so that

$$\int_{C_0}^\infty |f(x)|\, dx < \epsilon. \qquad (22.102)$$

# CHAPTER 23

# The Inner Ear

In humans (and other mammals) the ear has three major components: the outer, middle, and inner ears (Fig. 23.1A). The outer ear consists of a cartilaginous flange, the *pinna*, incorporating a resonant cavity that connects to the *ear canal* and finally to the *tympanic membrane*. It performs an initial filtering of the sound waves, increasing the sound pressure gain at the tympanic membrane in the 2 to 7 kHz region. It also aids sound localization. Bats, for instance, have highly developed pinnae, with a high degree of directional selectivity. Although less efficient in humans, the outer ear accounts for our ability to distinguish whether sounds come from above or below, in front or behind.

The function of the middle ear is to transmit the sound vibrations from the tympanic membrane to the cochlea. Because of the much higher impedance of the cochlear fluid, the middle ear must also function as an impedance-matching device, focusing the energy of the tympanic membrane on the oval window of the cochlea. If not for impedance matching, much of the energy of the sound waves in air would be reflected by the cochlear fluid. This impedance matching is carried out by the *ossicles*, three small bones, the *malleus*, *incus*, and *stapes*, that connect the tympanic membrane to the oval window. The tympanic membrane has a much higher surface area than the oval window, and the ossicles act as levers that increase the force at the expense of velocity, resulting in the required concentration of energy at the oval window.

Most of the events central to hearing occur in the inner ear, in particular the *cochlea*. The vestibular apparatus (the semicircular canals and the otolith organs) are also in the inner ear, but their principal function is the detection of movement and acceleration, not sound. The cochlea is a tube, about 35 mm long, divided longitudinally into three compartments and twisted into a spiral (Fig. 23.1B). The three compartments are the *scala vestibuli*, the *scala tympani*, and the *scala media*, and they wind around the spiral

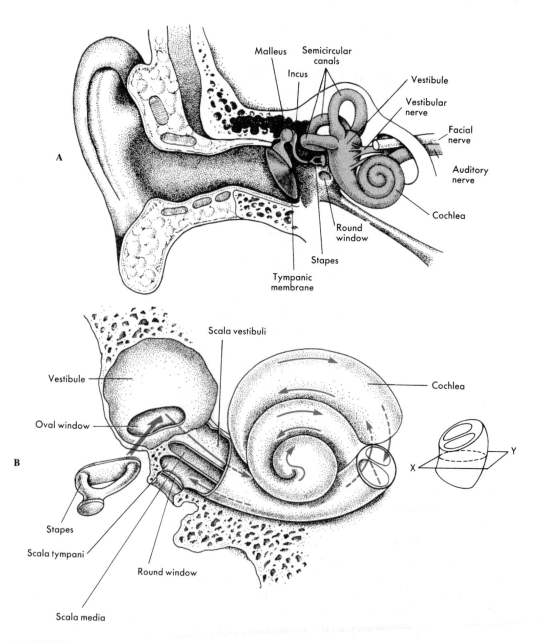

**Figure 23.1** Location and structure of the cochlea. A: Location of the cochlea in relation to the middle ear, the tympanic membrane, and the outer ear. B: Diagram of the cochlea at increased magnification, showing its spiral structure and the relative positions of the two larger internal compartments, the scala vestibuli and the scala tympani. (Berne and Levy, 1993, Fig. 10-6.)

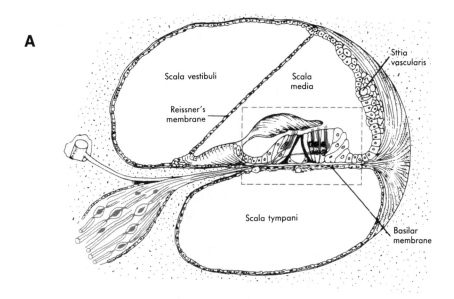

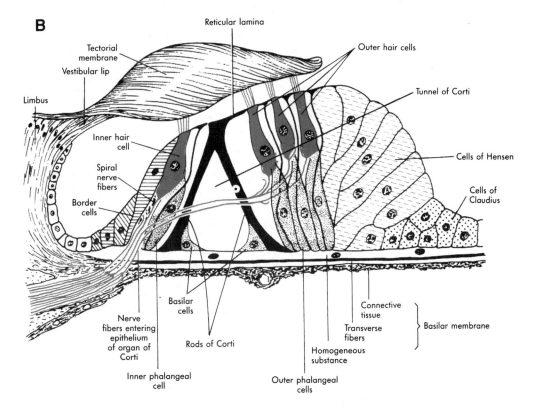

**Figure 23.2** Location and structure of the cochlea, continued. A: Cross-section of the cochlea in the plane indicated by the inset in Fig. 23.1B. B: Enlarged view of the organ of Corti, including the basilar membrane, the tectorial membrane, and the hair cells. (Berne and Levy, 1993, Fig. 10-6.)

together, preserving their spatial orientation. Reissner's membrane separates the scala vestibuli from the scala media, which in turn is separated from the scala tympani by the *spiral lamina* and the *basilar membrane* (Fig. 23.2A). The scala vestibuli and the scala tympani are filled with *perilymph*, a fluid similar to extracellular fluid, while the scala media is filled with *endolymph*, a fluid with a high $K^+$ concentration and a low $Na^+$ concentration. Sound waves transmitted through the middle ear are focused by the stapes onto the oval window, an opening into the scala vestibuli. The resultant waves in the perilymph travel along the length of the scala vestibuli, creating complementary waves in the basilar membrane and the scala tympani. At the end of the cochlea, an opening between the scala vestibuli and the scala tympani, the *helicotrema*, equalizes the local pressure in the two compartments. Because the perilymph is essentially in-compressible, it is necessary for the scala tympani also to have an opening analogous to the oval window; otherwise, conservation of mass would preclude movement of the stapes. The opening in the scala tympani is called the *round window*. Inward motion of the stapes at the oval window is compensated for by the corresponding outward motion of fluid at the round window.

Transduction of sound into electrical impulses is carried out by the *organ of Corti* (Fig. 23.2B), which sits on top of the basilar membrane. *Hair cells* in the organ of Corti have hairs projecting out the top, and these hairs are attached to a flap called the *tectorial membrane* that sits over the organ of Corti. Waves in the basilar membrane create a shear force on these hairs, which in turn causes a change in the membrane potential of the hair cell. This is transmitted to nerve cells, and from there to the brain.

## 23.1   Frequency Tuning

The task of the cochlea is to identify the constituent frequencies of a sound wave, and thus identify the sound. The different ways in which this is accomplished in different animals fall into three principal groupings: mechanical tuning of the hair cells, mechan-ical tuning of the basilar membrane, and electrical tuning of the hair cells (Hudspeth, 1985).

In many lizards, the length of the hair bundles on the hair cells increases system-atically from the base to the apex. In much the same way that a longer string produces notes of lower pitch, the longer hair cell bundles respond preferentially to inputs of lower frequency, while the short bundles are tuned to higher frequencies. Thus, the input frequency can be determined by the position of maximal stimulation. In mam-mals, the basilar membrane itself acts as a frequency analyzer, and this is discussed in the next section. The third tuning mechanism results from the properties of ionic channels in the hair cell membrane. Each hair cell is an electrical resonator, with a band-pass frequency response. The input frequency that gives the greatest response is a function of the biophysical properties of the hair cell, and the systematic variation of these properties along the length of the cochlea allows the cochlea to distinguish between frequencies based on the position of maximal response.

We do not discuss the first tuning mechanism further, but concentrate on the remaining two. We begin by looking at models of the basilar membrane that demonstrate mechanical tuning, and then briefly discuss models for resonance of the hair cell membrane potential.

## 23.1.1 Cochlear Mechanics and the Place Theory of Hearing

In mammals, vibrations of the stapes set up a wave with a particular shape on the basilar membrane. The amplitude envelope of the wave is first increasing, then decreasing, and the position of the peak of the envelope is dependent on the frequency of the stimulus (von Békésy, 1960), as illustrated in Fig. 23.3. The wave speed decreases as it moves along the membrane, resulting in a continual decrease in phase, and an apparent increase in frequency. Low-frequency stimuli have a wave envelope that peaks closer to the apex of the cochlea (i.e., near the helicotrema), and as the frequency of the stimulus increases, the envelope peak moves toward the base of the cochlea, as illustrated in Fig. 23.4.

The amplitude of the envelope is a two-dimensional function of distance from the stapes and frequency of stimulation; the curves shown in Fig. 23.4 are cross-sections of the function for fixed frequency. Another way to present the data is to give cross-sections for a fixed distance. This gives the envelope amplitude as a function of frequency, for a fixed distance from the stapes, i.e., the frequency response of the basilar membrane for that fixed distance. Frequency responses measured by von Békésy are shown in Fig. 23.5, from which it can be seen that each part of the basilar membrane responds maximally to a certain frequency, and as the frequency increases, the site of maximum

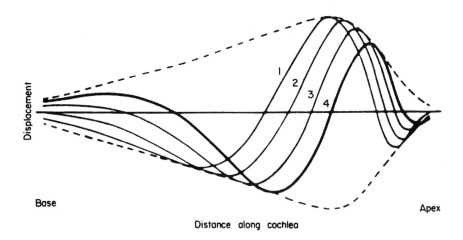

**Figure 23.3** Membrane waves and their envelope in the cochlea. The solid lines show the deflection of the basilar membrane at successive times, denoted (in order of increasing time) by 1, 2, 3, 4. The dashed line is the envelope of the membrane wave, and remains constant over time. (von Békésy, 1960, Fig. 12-17.)

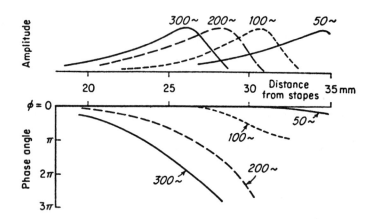

**Figure 23.4** Amplitude and phase of the cochlear membrane wave for four different frequencies. As the frequency of the wave increases, the peak of the wave envelope moves toward the base of the cochlea (i.e., toward the oval and round windows). (von Békésy, 1960, Fig. 11-58. p. 462.)

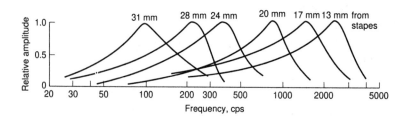

**Figure 23.5** Frequency responses of the basilar membrane measured at different distances from the stapes. Only the amplitude is shown. Close to the stapes, the basilar membrane responds preferentially to tones of high frequency, while farther away from the stapes, the membrane responds preferentially to tones of lower frequencies. (von Békésy, 1960, Fig. 11-49.)

response moves toward the stapes. In this way the cochlea determines the frequency of the incoming signal from the place on the basilar membrane of maximal amplitude, the so-called *place theory* of hearing.

Although questions have been raised concerning the accuracy of von Békésy's results (he performed his experiments, somewhat gruesomely, on cadavers, but it is believed that the properties of the basilar membrane in a living person are different), in all theoretical studies of cochlear mechanics, the experimental results of von Békésy, and the associated place theory, have been the gold standard by which model performance has been judged.

The name "basilar membrane" is misleading, as it is not a true membrane. This is shown by the fact that when it is cut, the edges do not retract. Thus it is not under tension; resistance to movement comes from the bending elasticity. The stiffness of the basilar membrane decreases exponentially from the base to the apex, with a length constant of about 7 mm. Although the width of the cochlea decreases from the base to the apex, the width of the basilar membrane increases in this direction.

Models for waves on the basilar membrane can be distinguished by the types of equations used for the membrane and the fluid. Early models by Ranke (1950) and Zwislocki (1965) assumed the perilymph to be incompressible and inviscid, and modeled the basilar membrane as a damped, forced harmonic oscillator, with no elastic coupling along the length of the membrane. Ranke used deep-water wave theory, while Zwislocki used shallow-water wave theory, leading to considerable controversy over which is the best approach. These models were developed by many authors, the best known being due to Peterson and Bogert (1950), Fletcher (1951), Lesser and Berkley (1972), and Siebert (1974). Subsequent models by Steele (1974), Inselberg and Chadwick (1976), Chadwick et al. (1976), Chadwick (1980), and Holmes (1980a,b, 1982) used more sophisticated representations of the basilar membrane as an elastic plate and incorporated fluid viscosity and the geometry of the plate. A survey of recent experimental and theoretical results can be found in Dallos et al. (1990). Here we give an overview of some of the earlier and simpler models, as they provide elegant demonstrations of how the basilar membrane and the perilymph can interact to give the types of waves observed by von Békésy.

## 23.2 Models of the Cochlea

### 23.2.1 Equations of Motion for an Incompressible Fluid

The fluid in the cochlea surrounding the basilar membrane is incompressible, and assumed to be inviscid. The equations of motion of this fluid are well known, and are derived in many places (e.g., Batchelor, 1967).

We let $\mathbf{u} = (u_1, u_2, u_3)$ be the fluid velocity, $p$ the pressure, and $\rho$ the constant density of the fluid. The mass of fluid in a fixed volume $V$ can change only in response to fluid flux across the boundary of the volume. Thus,

$$\frac{d}{dt} \int_V \rho\, dV = - \int_S \rho(\mathbf{u} \cdot \mathbf{n})\, dS = 0, \tag{23.1}$$

where $S$ is the surface of $V$, and $\mathbf{n} = (n_1, n_2, n_3)$ is the outward unit normal to $V$. Similarly, the momentum of the fluid in a fixed domain $V$ can change only in response to applied forces or to the flux of momentum across the boundary of the domain. Thus (for an inviscid fluid) conservation of momentum implies that

$$\frac{d}{dt} \int_V \rho u_i\, dV = - \int_S [(\mathbf{u} \cdot \mathbf{n})\rho u_i + p n_i]\, dS. \tag{23.2}$$

Using the divergence theorem to convert surface integrals to volume integrals, we obtain

$$\int_V \left( \rho \frac{\partial u_i}{\partial t} + \rho \nabla \cdot (u_i \mathbf{u}) + \frac{\partial p}{\partial x_i} \right) dV = 0, \tag{23.3}$$

$$\int_V \nabla \cdot \mathbf{u}\, dV = 0. \tag{23.4}$$

Finally, since $V$ is arbitrary, it follows that

$$\rho\frac{\partial \mathbf{u}}{\partial t} + \rho(\nabla \cdot \mathbf{u})\mathbf{u} + \nabla p = 0, \tag{23.5}$$

$$\nabla \cdot \mathbf{u} = 0. \tag{23.6}$$

When the fluid motions are of small amplitude, as we expect to be true in the cochlea, the nonlinear terms may be ignored, yielding

$$\rho\frac{\partial \mathbf{u}}{\partial t} + \nabla p = 0, \tag{23.7}$$

$$\nabla \cdot \mathbf{u} = 0. \tag{23.8}$$

An important special case is when $\mathbf{u} = \nabla\phi$ for some potential $\phi$ (an irrotational flow), in which case (23.7) and (23.8) become

$$\rho\frac{\partial \phi}{\partial t} + p = 0, \tag{23.9}$$

$$\nabla^2 \phi = 0, \tag{23.10}$$

where we let $p$ denote the deviation from the steady-state pressure.

### 23.2.2   The Basilar Membrane as a Harmonic Oscillator

One of the simplest models of the cochlea combines (23.9) and (23.10) with the equation of a damped, forced harmonic oscillator. One of the clearest presentations of

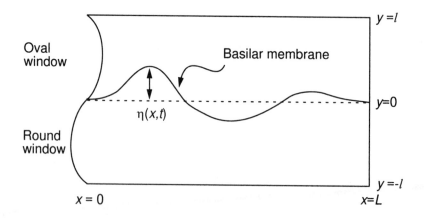

**Figure 23.6**   Schematic diagram of the cochlea, adapted from the model of Lesser and Berkley (1972). The cochlea is modeled as having two rectangular compartments filled with fluid, separated by the basilar membrane. The upper compartment corresponds to the scala vestibuli, and the lower compartment to the scala tympani. For simplicity, the scala media, shown in Fig. 23.1, is omitted from the model.

this model is due to Lesser and Berkley (1972). In their model, the cochlea is assumed to have a configuration as shown in Fig. 23.6. Thus, letting subscripts 1 and 2 denote quantities in the upper and lower compartments, respectively, we have two copies of (23.9) and (23.10),

$$\rho \frac{\partial \phi_1}{\partial t} + p_1 = \rho \frac{\partial \phi_2}{\partial t} + p_2 = 0, \tag{23.11}$$

$$\nabla^2 \phi_1 = \nabla^2 \phi_2 = 0, \tag{23.12}$$

where the pressure is determined only up to an arbitrary constant.

Each point of the basilar membrane is modeled as a simple damped harmonic oscillator with mass, damping, and stiffness that vary along the length of the membrane. The movement of any part of the membrane is assumed to be independent of the movement of neighboring parts of the membrane, as there is no direct lateral coupling. The deflection of the basilar membrane, $\eta(x,t)$, is specified by

$$m(x)\frac{\partial^2 \eta}{\partial t^2} + r(x)\frac{\partial \eta}{\partial t} + k(x)\eta = p_2(x, \eta(x,t), t) - p_1(x, \eta(x,t), t), \tag{23.13}$$

where $m(x)$ is the mass per unit area of the basilar membrane, $r(x)$ is its damping coefficient, and $k(x)$ is its stiffness (Hooke's constant) per unit area.

Since the vertical displacement of the membrane is small, the driving force is taken to be the pressure difference at $y = 0$, rather than at $y = \eta$. Thus, we have

$$m(x)\frac{\partial^2 \eta}{\partial t^2} + r(x)\frac{\partial \eta}{\partial t} + k(x)\eta = p_2(x, 0, t) - p_1(x, 0, t), \tag{23.14}$$

which is a considerable simplification.

Boundary conditions are specified as follows. Since $\partial \phi / \partial y$ is the $y$ component of the fluid velocity, the boundary conditions on the basilar membrane are

$$\frac{\partial \eta}{\partial t} = \frac{\partial \phi_1}{\partial y} = \frac{\partial \phi_2}{\partial y}, \qquad y = 0, \ 0 < x < L. \tag{23.15}$$

We further assume that there is no vertical motion at the top, so that

$$\frac{\partial \phi_1}{\partial y} = 0, \qquad y = l, \ 0 < x < L. \tag{23.16}$$

There are a number of ways to specify how the system is externally forced. One way, due to Lesser and Berkley, is to assume that the motion of the stapes in contact with the oval window determines the position of the oval window. Since $\partial \phi / \partial x$ is the $x$ component of the fluid velocity, the boundary condition at $x = 0$ is

$$\frac{\partial \phi_1}{\partial x} = \frac{\partial F(y, t)}{\partial t}, \qquad 0 < y < l, \tag{23.17}$$

where $F(y, t)$ is the specified horizontal displacement of the oval window. Further, we assume that there is no horizontal motion at the far end, so that at $x = L$

$$\frac{\partial \phi_1}{\partial x} = 0, \qquad 0 < y < l. \tag{23.18}$$

## 23.2.3   A Numerical Solution

Because of the inherent symmetry of the problem, we seek solutions that are odd in
$y$ (Exercise 1). Thus, we consider only the problem in the upper region and drop the
subscript 1.

When the input has a single frequency, $F(y, t) = \hat{F}(y)e^{i\omega t}$, then $\phi(x, y, t)$ is of the form
$\hat{\phi}(x, y; \omega)e^{i\omega t}$ and similarly for the other variables. Looking for solutions of this form for
all the variables, we obtain the equations

$$\nabla^2 \hat{\phi} = 0, \qquad \hat{p} + i\omega\rho\hat{\phi} = 0, \tag{23.19}$$

$$\frac{\partial \hat{\phi}}{\partial y} = i\omega\hat{\eta}, \qquad i\omega\hat{\eta}Z = -\hat{p} \qquad \text{on } y = 0,$$

$$\frac{\partial \hat{\phi}}{\partial x} = U_0 \qquad\qquad \text{on } x = 0,$$

$$\frac{\partial \hat{\phi}}{\partial x} = 0 \qquad\qquad \text{on } x = L, \tag{23.22}$$

$$\frac{\partial \hat{\phi}}{\partial y} = 0 \qquad\qquad \text{on } y = l, \tag{23.23}$$

where $Z = i\omega m + r + k/(i\omega)$ and $U_0 = i\omega\hat{F}$. By looking for solutions in the frequency
domain, we have transformed the differential equations on the basilar membrane into
algebraic equations. The term $i\omega Z$ is the frequency response of the damped harmonic
oscillator, and $Z$, the *impedance*, is a function of $x$. Also note that in (23.20) we have
assumed that the pressure is an odd function of $y$.

Finally, we nondimensionalize the model equations by scaling $x$ and $y$ by $L$, $Z$ by
$i\omega\rho L$, and $\hat{\phi}$ by $U_0L$; rearranging; and dropping the hats we get

$$\nabla^2 \phi = 0, \tag{23.24}$$

$$\frac{\partial \phi}{\partial y} = \frac{2\phi}{Z}, \qquad \text{on } y = 0, \tag{23.25}$$

$$\frac{\partial \phi}{\partial x} = 1 \qquad \text{on } x = 0, \tag{23.26}$$

$$\frac{\partial \phi}{\partial x} = 0 \qquad \text{on } x = 1, \tag{23.27}$$

$$\frac{\partial \phi}{\partial y} = 0 \qquad \text{on } y = \sigma, \tag{23.28}$$

where $\sigma = l/L$.

An analytical solution of this problem can be found using standard Fourier series
(Lesser and Berkley, 1972). We look for solutions of the form

$$\phi = x\left(1 - \frac{x}{2}\right) - \sigma y\left(1 - \frac{y}{2\sigma}\right) + \sum_{n=0}^{\infty} A_n \cosh[n\pi(\sigma - y)]\cos(n\pi x), \tag{23.29}$$

for some unknown constants $A_n$. Since $\phi$ satisfies all the boundary conditions except (23.25), we use (23.25) to determine the unknown coefficients $A_n$. This gives

$$\sigma + \sum_{n=0}^{\infty} n\pi A_n \sinh(n\pi\sigma)\cos(n\pi x)$$

$$-\frac{2}{Z}\left[x(1-x/2) + \sum_{n=0}^{\infty} A_n \cosh(n\pi\sigma)\cos(n\pi x)\right] = 0. \tag{23.30}$$

Truncating the series at $N$ terms, multiplying by $\cos(m\pi x)$, and integrating from 0 to 1, we obtain the system of linear equations

$$\sum_{n=0}^{N} A_n \alpha_{nm} = f_m, \tag{23.31}$$

where

$$\alpha_{mn} = 2\cosh(n\pi\sigma)\int_0^1 \frac{\cos(n\pi x)\cos(m\pi x)}{Z}\,dx - \frac{1}{2}n\pi \sinh(n\pi\sigma)\delta_{nm} \tag{23.32}$$

and

$$f_m = \sigma\delta_{m0} - \int_0^1 \frac{x(2-x)\cos(m\pi x)}{Z}\,dx. \tag{23.33}$$

Here, $\delta_{ij} = 1$ if $i = j$, and 0 otherwise. Since $f_m$ and $\alpha_{mn}$ can be evaluated explicitly, we get a set of $N$ linear equations for $A_n$, $1 \le n \le N$, which as $N \to \infty$ should give the solution of the model equations when substituted into (23.29). Typical results are shown in Fig. 23.7. The wave envelope has the same qualitative shape as von Békésy's results (Fig. 23.3), and the peak of the wave envelope moves toward the base of the cochlea as the frequency is increased.

## 23.2.4 Long-Wave and Short-Wave Models

Although the approximate Fourier solution of the Lesser and Berkley model shows that it is qualitatively correct, it would be nice to get a better analytic understanding of the behavior of the basilar membrane. There are two classic approximations of the model equations that allow further analytic investigation. The long-wave approximation, studied by Zwislocki and others, assumes that the wavelength is long compared to the depth of the cochlea, and the short-wave approximation of Ranke assumes the opposite, that the cochlea is effectively infinitely deep. Experiments suggest that the depth of the cochlea has little effect on the cochlear wave, supporting the short-wave theory. Indeed, even if one side of the cochlea is completely removed, there is little effect on the wave. However, neither model gives a complete description of cochlear behavior (Zwislocki, 1953).

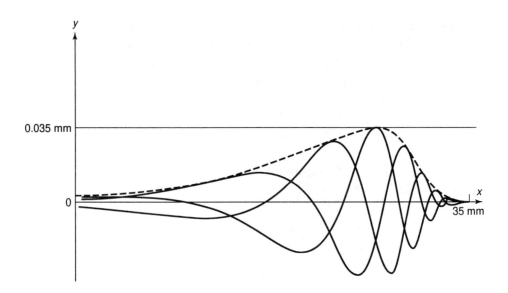

**Figure 23.7** Results from the Lesser and Berkley model, showing a typical wave on the basilar membrane and the wave envelope. Parameters are $m = 0.05$ g/cm$^2$, $k = 10^7 e^{-1.5x}$ dynes/cm$^3$, $r = 3000 e^{-1.5x}$ dynes sec/cm$^3$, $\omega = 1000$/sec. The perilymph was assumed to have the same density as water, 1 g/cm$^3$. (Lesser and Berkley, 1972, Fig. 6.)

Both short-wave and long-wave models can be derived as approximate cases of the model described in Section 23.2.2. To show this, we use a generalized form of the previous model (Siebert, 1974), as illustrated in Fig. 23.8. The only change is to assume that there is a direct mechanical forcing at the two ends of the basilar membrane. Modifying the equation of membrane motion (23.14) to include this direct forcing gives

$$m(x)\frac{\partial^2 \eta}{\partial t^2} + r(x)\frac{\partial \eta}{\partial t} + k(x)\eta = p_2(x,0,t) - p_1(x,0,t) + F_0(t)\delta(x) - F_L(t)\delta(x-L). \quad (23.35)$$

As before, we assume that the forcing is at a single frequency with $F_0(t) = F_0 e^{i\omega t}$ and $F_L(t) = F_L e^{i\omega t}$. It follows from (23.15), (23.35), and (23.19) that

$$\nabla^2 p(x,y) = 0, \quad (23.36)$$

$$-i\omega\eta = \frac{1}{i\omega\rho}\frac{\partial p(x,0)}{\partial y}, \quad (23.37)$$

$$Y(x)p(x,0) = -i\omega\eta(x) + \eta_0\delta(x) - \eta_L\delta(x-L), \quad (23.38)$$

where $Y = 2/Z$, $\eta_0 = \frac{F_0}{Z(0)}$, $\eta_L = \frac{F_L}{Z(L)}$, and where we have dropped the hats associated with the Fourier transform.

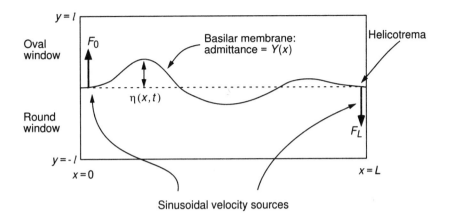

Figure 23.8 Schematic diagram of the cochlea model of Siebert (1974). It differs from the model of Lesser and Berkley in the boundary conditions at $x = 0$ and $x = L$, where it is assumed that there is direct mechanical forcing at both ends of the membrane.

Because Laplace's equation is separable on a rectangular domain, we use Fourier series to write the solution of (23.35) as

$$p(x,y) = \sum_{n=-\infty}^{\infty} \alpha_n \frac{\cosh[2\pi n(y - l)/L]}{\cosh(2\pi nl/L)} e^{2\pi inx/L}, \qquad (23.38)$$

where we have used the boundary condition $\partial p/\partial y = 0$ on $y = l$. It follows that

$$-i\omega\eta = \frac{1}{i\omega\rho} \frac{\partial p(x,0)}{\partial y} = -\frac{1}{i\omega\rho} \sum_{n=-\infty}^{\infty} \alpha_n \frac{2\pi n}{L} \tanh(2\pi nl/L) e^{2\pi inx/L}. \qquad (23.39)$$

Since our only interest is in the behavior of the basilar membrane, from now on we restrict our attention to $p(x, 0)$, which we denote by $p(x)$.

Equation (23.39) has been approximated in two principal ways. The first assumes that the depth of the cochlea is small compared to the wavelengths of the waves on the basilar membrane, the so-called *long-wave*, or *shallow-water*, approximation. The second approach, the *short-wave*, or *deep-water*, approximation, assumes that the wavelengths of the membrane waves are short compared to the cochlear depth.

## The shallow-water approximation

In the shallow-water approximation, we assume that the wavelengths of the waves on the basilar membrane are greater than the depth of the cochlea. As a consequence, we assume that $\alpha_n = 0$ for all $n > N$, for some integer $N$ such that $Nl/L \ll 1$. Since the sum over $n$ includes only those terms with $nl/L \ll 1$, it follows that for each term in the sum, $\tanh(2\pi nl/L)$ can be approximated by the lowest-order term in its Taylor expansion. Thus, $\tanh(2\pi nl/L) \approx 2\pi nl/L$, and so the sum becomes

$$-i\omega\eta \approx -\frac{l}{i\omega\rho} \sum_{n=-\infty}^{\infty} \alpha_n \left(\frac{2\pi n}{L}\right)^2 e^{2\pi inx/L}. \tag{23.40}$$

However, it follows from (23.38) that

$$p(x) = \sum_{n=-\infty}^{\infty} \alpha_n e^{2\pi inx/L}, \tag{23.41}$$

and thus, combining this with (23.40), we have

$$-i\omega\eta \approx \frac{l}{i\omega\rho}\frac{d^2p}{dx^2}. \tag{23.42}$$

Combining this with (23.36) and (23.37), we get a single equation for $p(x)$,

$$Y(x)p(x) = \frac{l}{i\omega\rho}\frac{d^2p(x)}{dx^2} + \eta_0\delta_0(x) - \eta_L\delta_L(x). \tag{23.43}$$

To convert the delta functions in this equation into boundary conditions, we integrate the differential equation (23.43) from $x = -\epsilon$ to $x = +\epsilon$ and let $\epsilon \to 0$ and find that

$$\frac{dp}{dx} = -i\omega\rho\eta_0 \quad \text{at } x = 0, \tag{23.44}$$

where we have assumed that $dp/dx = 0$ at $x = 0^-$, which is outside the boundaries of the cochlea. Similarly, integrating from $x = L - \epsilon$ to $x = L + \epsilon$ and letting $\epsilon \to 0$ gives

$$\frac{dp}{dx} = i\omega\rho\eta_L \quad \text{at } x = L. \tag{23.45}$$

Note that when $\eta_0 = 1$ and $\eta_L = 0$ (and since $p + i\omega\rho\phi = 0$), these boundary conditions are the same as (23.26) and (23.27) used in the Lesser–Berkley model.

The analysis of this equation exploits the fact that $Y(x)$ is a slowly varying function. To see what this means mathematically, note that $i\omega\rho Y(x)/l$ has dimensional units of length$^{-2}$, which determines the length scale (wavelength) of the spatial oscillations of $p(x)$. On the other hand, $Y(x)$ varies exponentially with a length constant of $\lambda^{-1} \approx 0.7$ cm. If the ratio of these two length constants is small, then we assert that $Y(x)$ is a slowly varying function. Furthermore, there is a rescaling of space, $x = z/q$, of (23.43), putting it into the dimensionless form

$$\frac{d^2p}{dz^2} + g^2(\epsilon z)p(z) = 0, \tag{23.46}$$

where $\epsilon$ is a small positive number and $g^2(\epsilon z) = \frac{-i\omega\rho Y(z/q)}{lq^2}$ is of order one in amplitude and slowly varying in $z$. Note that $q$ is an arbitrary length scale, chosen so that $g^2$ is of order one in amplitude; by assumption, $\lambda/q \ll 1$.

As a specific example, suppose that $m = 0, k(x) = k_0 e^{-\lambda x}, r(x) = r_0 e^{-\lambda x}$, in which case

$$\frac{i\omega \rho Y(x)}{l} = \frac{-2\omega^2 \rho}{lk_0} \frac{e^{\lambda x}}{1 + i\omega r_0/k_0}. \tag{23.47}$$

We set $q^2 = \frac{2\omega^2 \rho}{lk_0}$, and then define $\epsilon$ by

$$\epsilon = \frac{\lambda}{2q}. \tag{23.48}$$

If the parameters are such that $\epsilon \ll 1$, we have a slowly varying oscillation.

Problems of this type are well known in the theory of oscillations and can be solved approximately using multiscale analysis (Kevorkian and Cole, 1996; Keener, 1988). We wish to find approximate solutions of (23.46). If $g$ were a constant ($\epsilon = 0$), the solution of (23.46) would be simply

$$p(z) = Ae^{igz} + Be^{-igz}. \tag{23.49}$$

However, since $g$ is assumed to be slowly varying, we expect this basic solution to be a reasonable local (but not global) approximation. To find a solution that has a longer range of validity, we introduce two scales, a slow scale variable $\sigma = \epsilon z$ and a fast variable $\tau$ for which $\frac{d\tau}{dz} = f(\epsilon z)$, where $f$ is a function to be determined. It follows that the derivative $\frac{d}{dz}$ must be replaced by partial derivatives

$$\frac{d}{dz} = f(\sigma)\frac{\partial}{\partial\tau} + \epsilon\frac{\partial}{\partial\sigma}. \tag{23.50}$$

In terms of these two variables the original ordinary differential equation (23.46) becomes the partial differential equation

$$f^2(\sigma)\frac{\partial^2 p}{\partial\tau^2} + \epsilon f(\sigma)\frac{\partial^2 p}{\partial\sigma\partial\tau} + \epsilon\frac{\partial}{\partial\sigma}\left(f(\sigma)\frac{\partial p}{\partial\tau}\right) + \epsilon^2\frac{\partial}{\partial\sigma}\left(f(\sigma)\frac{\partial p}{\partial\sigma}\right) + g^2(\sigma)p = 0. \tag{23.51}$$

The obvious choice for $f$ is $f = g$, because then the solution to leading order (with $\epsilon = 0$) is simple, being

$$P_0 = Ae^{i\tau} + Be^{-i\tau}. \tag{23.52}$$

However, because the equation is a partial differential equation, the parameters $A$ and $B$ are allowed to be functions of the slow variable $\sigma$.

To determine the variation of $A$ and $B$, we set $p = P_0 + \epsilon P_1 + O(\epsilon^2)$, collect like powers of $\epsilon$, and determine that the equation for $P_1$ is

$$g^2(\sigma)\left(\frac{\partial^2 P_1}{\partial\tau^2} + P_1\right) = -g(\sigma)\frac{\partial^2 P_0}{\partial\sigma\partial\tau} - \frac{\partial}{\partial\sigma}\left(g(\sigma)\frac{\partial P_0}{\partial\tau}\right). \tag{23.53}$$

Now we require that $P_1$ be "nonsecular," that is, that the right-hand side of (23.53) contain no terms proportional to $e^{i\tau}$ or $e^{-i\tau}$. It follows that

$$\frac{\partial}{\partial\sigma}(gA^2) = 0, \qquad \frac{\partial}{\partial\sigma}(gB^2) = 0, \tag{23.54}$$

or that

$$A(\sigma) = \frac{A_0}{\sqrt{g(\sigma)}}, \qquad B(\sigma) = \frac{B_0}{\sqrt{g(\sigma)}}, \tag{23.55}$$

from which we obtain, to lowest order in $\epsilon$,

$$p = \frac{1}{\sqrt{g(\sigma)}}(A_0 e^{iG(z)} + B_0 e^{-iG(z)}), \tag{23.56}$$

where $G(z) = \int_0^z g(\epsilon z)dz$.

In terms of the original dimensioned variables, this is

$$p(x) = \phi^{-1/2}\left(A_1 \exp\left[i\int_0^x \phi(s)\,ds\right] + B_1 \exp\left[-i\int_0^x \phi(s)\,ds\right]\right), \tag{23.57}$$

where

$$\phi(x) = \sqrt{\frac{-i\omega\rho Y(x)}{l}}. \tag{23.58}$$

The constants $A_1$ and $B_1$ are determined from boundary conditions (23.44) and (23.45), and then the membrane displacement is found from the identity $i\omega\eta(x) = -Y(x)p(x)$.

The key feature of this solution is that it is oscillatory with an envelope, whose maximal amplitude and position are determined by the frequency $\omega$. We get some idea of this behavior in the special case $m = 0$, $k(x) = k_0 e^{-\lambda x}$, and $r(x) = r_0 e^{-\lambda x}$ in which case $\phi(x) = \alpha e^{\lambda x/2}$, where $\alpha^2 = \frac{2\omega^2\rho}{l(k_0+i\omega r_0)}$. If we let $\alpha = \alpha_r + i\alpha_i$ and suppose that $\frac{\alpha_i}{\lambda} \gg 1$ (not valid at low frequencies), then with $\eta_L = 0$, we find that

$$\eta(x) = -\frac{1}{i\omega}Y(x)p(x) \approx \hat{A}\exp\left(\frac{3\lambda x}{4} - \frac{2\alpha_i}{\lambda}e^{\lambda x/2} + \frac{2i\alpha_r}{\lambda}e^{\lambda x/2}\right). \tag{23.59}$$

This represents an oscillation with exponentially increasing phase and amplitude

$$|\eta| \approx |\hat{A}|\exp\left(\frac{3\lambda x}{4} - \frac{2\alpha_i}{\lambda}e^{\lambda x/2}\right). \tag{23.60}$$

The maximal value of this envelope occurs at

$$x_p = -\frac{2}{\lambda}\ln\left(\frac{4\alpha_i}{3\lambda}\right). \tag{23.61}$$

The location of this maximum is dependent on frequency, as

$$\alpha_i \approx \sqrt{\frac{\rho\omega}{lr_0}}, \tag{23.62}$$

provided that $\omega$ is sufficiently large. Thus, for large $\omega$ we have

$$x_p = -\frac{1}{\lambda}\ln\left(\frac{16\rho\omega}{9l\lambda^2 r_0}\right). \tag{23.63}$$

### The deep-water approximation

The second approach, the *short-wave*, or *deep-water*, approximation, assumes that the wavelength of the membrane waves is short compared to the cochlear depth. In this case the Fourier expansion of $p(x)$ includes only high frequencies, and so $\alpha_n = 0$ whenever $|n| < N$ for some large integer $N$. However, when $|n| > N$ and $l \gg L$, then $\tanh(2\pi n l/L) \approx \text{sign}(n)$. Thus, (23.39) becomes

$$-i\omega\eta \approx -\frac{1}{i\omega\rho}\sum_{n=-\infty}^{\infty}\alpha_n\frac{2\pi}{L}|n|e^{2\pi inx/L}. \qquad (23.64)$$

Now we separate the sum into two pieces by defining two functions,

$$p_+(x) = \sum_{n=0}^{\infty}\alpha_n e^{2\pi inx/L} \qquad (23.65)$$

and

$$p_-(x) = \sum_{n=-\infty}^{-1}\alpha_n e^{2\pi inx/L}, \qquad (23.66)$$

and then we observe that (23.37) becomes

$$Yp = Y(p_+ + p_-) \approx \frac{1}{\omega\rho}\left[\frac{dp_+}{dx} - \frac{dp_-}{dx}\right] + \eta_0\delta_0(x) - \eta_L\delta_L(x), \qquad (23.67)$$

which we take to be the governing equation for $p$.

We can remove the delta function influence from this equation by integrating across the boundaries at $x = 0$ and $x = L$, and assuming that outside the cochlea, $p = 0$. This gives

$$\frac{1}{\omega\rho}[p_+(0) - p_-(0)] = \eta_0, \qquad (23.68)$$

$$\frac{1}{\omega\rho}[p_+(L) - p_-(L)] = \eta_L. \qquad (23.69)$$

Although $p_+$ is a linear combination of only positive (spatial) frequencies, the same is not true of $Yp_+$. However, if we assume that $Y$ is a slowly varying function of $x$, then the Fourier series of $Y$ with Fourier coefficients $b_k$ has $b_k \approx 0$ whenever $|k| > k_0$, for some number $k_0$ that is small compared to the dominant frequency of $p_+$. It follows that

$$Yp_+ = \sum_{k=-\infty}^{\infty}c_k e^{2\pi ikx/L}, \qquad (23.70)$$

where $c_k = \sum_{j=0}^{\infty}\alpha_j b_{k-j}$. If the dominant frequencies of $p_+$ and $Y$ are separated, as stated above, then $c_k$ is small for $k \leq 0$. Thus we can approximate $Yp_+$ by its Fourier series with positive frequencies. A similar argument applies for $Yp_-$. With these approximations, (23.67) separates into a pair of differential equations for the positive and negative frequencies separately,

$$Y(x)p_\pm(x) \approx \frac{\pm 1}{\omega\rho}\frac{dp_\pm}{dx}, \qquad 0 < x < L. \tag{23.71}$$

These first-order linear equations can be integrated directly to get

$$p_\pm(x) = A_\pm \exp\left[\pm\omega\rho\int_0^x Y(\zeta)\,d\zeta\right] \tag{23.72}$$

for some constants $A_\pm$, so that

$$p = A_+ \exp\left[\omega\rho\int_0^x Y(\zeta)\,d\zeta\right] + A_- \exp\left[-\omega\rho\int_0^x Y(\zeta)\,d\zeta\right]. \tag{23.73}$$

We use the boundary conditions at $x = 0$ and $x = L$ to determine the constants $A_\pm$. From (23.68) and (23.69) it follows that these constants must satisfy the equations

$$A_+ - A_- = \omega\rho\eta_0, \qquad \gamma A_+ - \frac{1}{\gamma}A_- = \omega\rho\eta_L, \tag{23.74}$$

where $\gamma = \exp[\omega\rho\int_0^L Y(\zeta)\,d\zeta]$, from which it follows that

$$A_+ = \frac{\omega\rho}{\gamma^2 - 1}[\gamma\eta_L - \eta_0], \tag{23.75}$$

$$A_- = \frac{\omega\rho}{\gamma^2 - 1}[\gamma\eta_L - \gamma^2\eta_0]. \tag{23.76}$$

For physiological values of $Y$, $|\gamma| \gg 1$ for all except the lowest frequencies; for instance, for the parameter values in Fig. 23.7, $|\gamma| = 40$ when $\omega = 800$, and $|\gamma| = 10^9$ when $\omega = 1500$. Since on physical grounds $\eta_0$ and $\eta_L$ do not get large, it follows that $|A_-| \approx -\omega\rho\eta_0 \gg |A_+|$. Finally, from (23.37) and (23.73), we find that the membrane displacement is given by

$$\eta = -i\rho\eta_0 Y(x)\exp\left[-\omega\rho\int_0^x Y(\zeta)\,d\zeta\right]. \tag{23.77}$$

The amplitude of $\eta$ can be plotted as a function of $x$ to give the envelope of the wave on the basilar membrane. The frequency response is similarly obtained, by fixing $x$ and plotting $|\eta|$ as a function of $\omega$. Typical results are shown in Fig. 23.9. The qualitative agreement with data is good, with the peak of the wave envelope moving toward the stapes as the frequency increases.

In the special case $m = 0, k(x) = k_0 e^{-\lambda x}, r(x) = r_0 e^{-\lambda x}$, we can calculate the waveform (23.77) to be

$$\eta = 2\eta_0\xi\exp[\lambda x + \beta(1 - e^{\lambda x})], \tag{23.78}$$

where

$$\beta = \frac{2\omega^3\rho r_0}{\lambda(k_0^2 + \omega^2 r_0^2)} + i\frac{2\omega^2\rho k_0}{\lambda(k_0^2 + \omega^2 r_0^2)} = \beta_r + i\beta_i \tag{23.79}$$

and

$$\xi = \frac{\omega\rho}{i\omega\rho + k_0}. \tag{23.80}$$

Here we again see an oscillatory waveform with an envelope of amplitude

$$|\eta| = 2\eta_0|\xi| \exp[\lambda x + \beta_r(1 - e^{\lambda x})], \tag{23.81}$$

the maximum of which occurs at

$$x_p = \frac{-1}{\lambda} \ln\left(\frac{2\omega^3\rho r_0}{\lambda(k_0^2 + \omega^2 r_0^2)}\right). \tag{23.82}$$

According to this expression, the peak of the envelope moves to the left (toward the base of the cochlea) as $\omega$ increases. The principal fault of the short-wave model is that the phase of the model waves increases much more than is observed experimentally.

A similar solution was found by Peskin (1976, 1981), who calculated an exact solution to a special case of the cochlear model. In his model the cochlear membrane was taken to be infinitely long, with $r(x)$ and $k(x)$ chosen to be decaying exponential functions with decay rate $\lambda$ and a fluid container of height $\lambda l = \pi/2$. With these assumptions and simplifications, Peskin found the exact solution using conformal mapping and contour integration techniques.

## 23.2.5  More Complex Models

In this chapter we have concentrated on the simpler models of the basilar membrane that assume that the cochlea is two-dimensional and that the basilar membrane can be described by a point impedance function, i.e., that each point of the basilar membrane acts as a damped harmonic oscillator, with no coupling along the length of the membrane except for that imposed indirectly via the fluid motion.

Although the wave motion on the basilar membrane is an important component of the hearing process, many other factors are involved (Pickles, 1982; Rhode, 1984; Hudspeth, 1985). Nonlinearities in the cochlear response and acoustic emissions suggest the presence of active feedback processes that modulate the waveform. This feedback may occur in the outer hair cells and the organ of Corti. Simple hydrodynamic models do not reproduce the degree of tuning observed in the mammalian cochlea, and the precise tuning mechanism is still controversial. Many other, more complex, models have been constructed (see, for instance, Steele, 1974; Steele and Tabor, 1979a,b; Inselberg and Chadwick, 1976; Chadwick et al., 1976; Chadwick, 1980; Holmes, 1980a,b, 1982). In general, these models use similar equations for the fluid flow, but model the basilar membrane in greater detail, including spatial coupling in the membrane. The resultant membrane equations are of fourth order in space, and heavy use is made of asymptotic expansions in the solution of the model equations. Perhaps the most detailed study was performed by Steele (1974), who constructed a series of models ranging from a plate in an infinite body of fluid right up to a tapered elastic basilar membrane, a cochlea with rigid walls, and flexible arches of Corti.

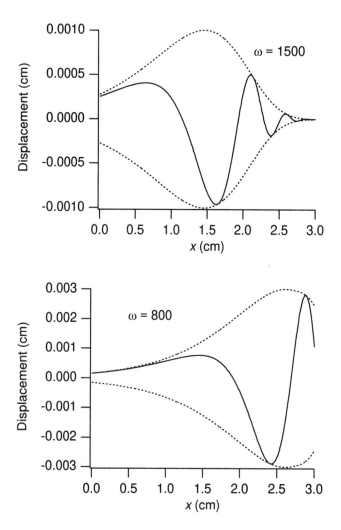

**Figure 23.9** Plots of the amplitude of the wave on the basilar membrane for two different frequencies. The envelope of the wave is shown as a dotted line. Calculated from the deep-water approximation (23.73) using the same parameter values as in the Lesser and Berkley model (given in the caption to Fig. 23.7).

## 23.3   Electrical Resonance in Hair Cells

In many lower vertebrates frequency decomposition is performed, not by a wave on the basilar membrane, but by the hair cells themselves. Hair cells in the turtle cochlea and the bullfrog sacculus (to name but two examples) respond preferentially to stimuli of a certain frequency, and this band-pass response is mediated by the ionic channels in the hair cell membrane. At the top of each hair cell is the hair bundle, a group of stereocilia connected to each other at the tips by a thin *fiber*, called a tip link. Each stereocilium is rigid and, in response to a force applied at the tip, pivots around its base rather than bending. It is postulated that the tip links act like elastic springs connected directly to ionic channels such that when the hair bundle is deflected in one direction, the tip links pull channels open, while when the hair bundle is deflected in the opposite direction, the tip links relax and allow channels to close. The mechanically sensitive ion channels are nonselective, and the modulation of current flow through these channels results in hyperpolarization or depolarization of the hair cell membrane. The membrane potential is then modulated by other ionic channels, including $K^+$ channels, $Ca^{2+}$-sensitive $K^+$ channels, and voltage-sensitive $Ca^{2+}$ channels. The structure, tuning, sensitivity, and function of hair cells are reviewed by Hudspeth (1985, 1989; Hudspeth and Gillespie, 1994), and these papers give a readable summary of recent work.

In response to a step current input, the membrane potential of hair cells exhibits damped oscillations, with a period and amplitude dependent on the size of the step. Thus, each cell has a natural frequency of oscillation and responds best to a stimulus at a similar frequency. Crawford and Fettiplace (1981) and Ashmore and Attwell (1985) have developed simple models for electrical resonance that while not based on the details of known mechanisms, provide a good description of the experimental results. Later work by Lewis and Hudspeth (1988a,b), using a more detailed model, showed that the measured properties of the ionic conductances are sufficient to explain resonance in the hair cells of the bullfrog sacculus.

## 23.3.1 An Electrical Circuit Analogue

The models of Crawford and Fettiplace (1981) and Ashmore and Attwell (1985) are based on the electrical circuit shown in Fig. 23.10A. In response to a current input $I$,

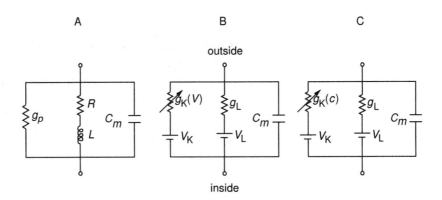

**Figure 23.10** Electrical circuits for electrical tuning, adapted from Ashmore and Attwell (1985). A: The basic model, with an inductance in place of ionic currents. B: Voltage-gated $K^+$ current, with a conductance that is increased by membrane depolarization. C: $Ca^{2+}$-gated current, with the $K^+$ conductance controlled by the intracellular concentration of $Ca^{2+}$, denoted by $c$.

the voltage $V$ is given by

$$\frac{d^2V}{dt^2} + \gamma\frac{dV}{dt} + \omega_0^2 V = f(t), \tag{23.83}$$

where

$$\gamma = \frac{g_p}{C_m} + \frac{R}{L}, \tag{23.84}$$

$$\omega_0^2 = \frac{g_p R + 1}{LC_m}, \tag{23.85}$$

$$f(t) = \frac{1}{C_m}\frac{dI}{dt} + \frac{IR}{LC_m}. \tag{23.86}$$

It is simplest to demonstrate resonance when $f(t) = e^{i\omega t}$, in which case $V = V_1(\omega)e^{i\omega t}$, where

$$V_1(\omega) = \frac{1}{\omega_0^2 - \omega^2 + i\gamma\omega}. \tag{23.87}$$

Thus, $|V_1|$ has a band-pass frequency response, with a maximum at $\hat{\omega}$, where $\hat{\omega}^2 = \omega_0^2 - \gamma^2/2$. Solutions of (23.83) are of the form $\exp(-\gamma t/2)\exp\left(\pm i\sqrt{\omega_0^2 - \gamma^2/4}\right)$, and thus $\hat{\omega}$ is slightly smaller than the natural frequency of oscillation of the system. However, if damping is small (i.e., if $\gamma$ is small), the maximum amplitude of the frequency response occurs at approximately the natural frequency of oscillation. The sharpness of the peak of $|V_1|$ is a measure of the degree of tuning of the electrical circuit, with sharper peaks giving greater frequency selectivity. Since

$$\left.\frac{d^2}{d\omega^2}[(\omega_0^2 - \omega^2)^2 + \gamma^2\omega^2]\right|_{\omega=\hat{\omega}} = 4\gamma^2(2Q^2 - 1), \tag{23.88}$$

where

$$Q = \frac{\omega_0}{\gamma}, \tag{23.89}$$

it follows that $Q$, often called the *quality factor*, is a useful measure of the degree of tuning. As $Q$ increases, so does the frequency selectivity of the circuit.

We now consider the response of the circuit when the input is a sinusoidally varying current. When $I = e^{i\omega t}$,

$$V_1(\omega) = \frac{\frac{R}{LC_m} + \frac{i\omega}{C_m}}{\omega_0^2 - \omega^2 + i\gamma\omega}, \tag{23.90}$$

which again corresponds to a band-pass filter, with the maximum response occurring at $\hat{\omega}$, where

$$(\hat{\omega}^2)^2 + 2\left(\frac{R}{L}\right)^2\hat{\omega}^2 + \left(\frac{R}{L}\right)^2(\gamma^2 - 2\omega_0^2) = 0. \tag{23.91}$$

Crawford and Fettiplace (1981) used a model of this type (without the leak conductance $g_p$) to determine the electrical tuning characteristics of hair cells from the turtle cochlea. By comparison of these tuning curves with tuning curves obtained by acoustic stimulation of the hair cells they were able to determine that electrical resonance can account for most of the frequency selectivity of the hair cell.

Although the above circuit exhibits the required resonance, it would be much more satisfactory to explain electrical resonance in terms of components that have a more direct connection to the hair cell. This can be done in at least two ways. In Fig. 23.10B and C we show two circuits, one involving a voltage-sensitive $K^+$ conductance, the other a $Ca^{2+}$-sensitive $K^+$ conductance, that, formally at least, are equivalent to the circuit in Fig. 23.10A.

We consider the circuit in Fig. 23.10B first. If the leak has a constant conductance, but the $K^+$ conductance is a function of time and voltage, then

$$I = C_m \frac{dV}{dt} + g_L(V - V_L) + fg_K(V - V_K), \tag{23.92}$$

$$\tau \frac{df}{dt} = f_\infty - f, \tag{23.93}$$

where we take a linear approximation for $f_\infty$,

$$f_\infty = f_r + \mu(V - V_r). \tag{23.94}$$

Here, $V_r$ is assumed to be the resting membrane potential, $f_r$ is the value of $f$ when $V = V_r$, and $\mu$ is the slope of the activation curve at the steady state. Note that at steady state,

$$0 = g_L(V_r - V_L) - g_K f_r(V_K - V_r), \tag{23.95}$$

and thus $V_L$ can be eliminated.

It follows that

$$\frac{dI}{dt} + \frac{I}{\tau} = C_m \frac{d^2\tilde{V}}{dt^2} + \left(g_L + g_K f_r + \frac{C_m}{\tau}\right)\frac{d\tilde{V}}{dt} + \left(\frac{g_L + g_K f_r + g_K(V_r - V_K)\mu}{\tau}\right)\tilde{V}, \tag{23.96}$$

where $\tilde{V} = V - V_r$ and where we have linearized the equation around $V_r$ by assuming that $V \approx V_r$. Equation (23.96) is equivalent to (23.83)–(23.86) if

$$L = \frac{\tau}{g_K(V_r - V_K)\mu}, \tag{23.97}$$

$$R = \frac{1}{g_K(V_r - V_K)\mu}, \tag{23.98}$$

$$g_p = g_L + g_K f_r. \tag{23.99}$$

A similar procedure can be followed for the circuit in Fig. 23.10C, in which the $K^+$ conductance is $Ca^{2+}$-dependent rather than voltage-dependent, but extra assumptions about the $Ca^{2+}$ kinetics must be made. As a first approximation, it is assumed that $Ca^{2+}$ enters the cell through channels at a rate that is a linear function of voltage, with slope $\theta$, and is removed with first-order kinetics, i.e., at a rate proportional to its concentration. Finally, it is assumed that the proportion of open $K^+$ channels is linearly related to the

Ca$^{2+}$ concentration. Thus,

$$I = C_m \frac{dV}{dt} + g_L(V - V_L) + g_K kc(V - V_K), \qquad (23.100)$$

$$W \frac{dc}{dt} = \frac{I_r + \theta(V - V_r)}{F} - pc, \qquad (23.101)$$

where $c$ denotes Ca$^{2+}$ concentration, $F$ is Faraday's constant, $W$ is the cell volume, $p$ is the rate of Ca$^{2+}$ pumping, and $I_r$ is the steady Ca$^{2+}$ current when $V = V_r$. The constant $k$ is the rate at which Ca$^{2+}$ activates the K$^+$ current. Again, linearizing this system about the steady state gives a system that is equivalent to (23.83)–(23.86), provided that

$$L = \frac{WF}{g_K(V_r - V_K)k\theta}, \qquad (23.102)$$

$$R = \frac{pF}{g_K(V_r - V_K)k\theta}, \qquad (23.103)$$

$$g_p = g_L + g_K kc_r, \qquad (23.104)$$

where $c_r$ is the steady Ca$^{2+}$ concentration at the resting potential.

Ashmore and Attwell showed that although the model with the voltage-sensitive K$^+$ conductance can generate a wide range of optimal frequencies, physiological values for the parameters result in values for the quality factor $Q$ that are an order of magnitude too low. Thus, for reasonable parameters, the model can distinguish between frequencies, but not sharply enough. Experimental values for $Q$ are often 5 or more, while $Q$ values in the model are not above 0.7. This, they argue, is the result of the low value of $\mu$: a physiological value for $\mu$ is about 0.33 mV$^{-1}$, but $Q$ is large enough in the model with $\mu$ about 3 mV$^{-1}$. Thus, it appears that the activation of the K$^+$ current by voltage is not steep enough to account for the observed resonance in hair cells.

In the third model, however, the activation of the K$^+$ current by Ca$^{2+}$ can be made much steeper. Here, the effective activation slope of the K$^+$ channel is $k\theta/(pF)$, which can be made large by decreasing the pump rate $p$ or by increasing the sensitivity of the K$^+$ channel to Ca$^{2+}$. Ashmore and Attwell conclude that frequency tuning in hair cells is more likely the result of a Ca$^{2+}$-sensitive K$^+$ conductance than of a voltage-sensitive conductance.

## 23.3.2  A Mechanistic Model of Frequency Tuning

This conclusion has been upheld by the more recent, and more detailed, work of Hudspeth and Lewis (1988a,b). Based on a series of experiments in which they measured the kinetic properties of the ionic conductances in saccular hair cells of the bullfrog, Hudspeth and Lewis constructed a detailed model for electrical resonance in these cells. They concluded that the observed properties of the Ca$^{2+}$-sensitive K$^+$ conductance, in concert with a voltage-sensitive Ca$^{2+}$ conductance and a leak, are a sufficient quantitative explanation of frequency tuning in these cells.

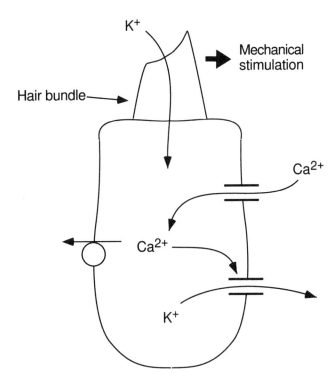

K$^+$

Mechanical stimulation

Hair bundle

Ca$^{2+}$

Ca$^{2+}$

K$^+$

**Figure 23.11** Schematic diagram of a model for electrical tuning in hair cells, adapted from Hudspeth (1985).

A schematic diagram of their model is given in Fig. 23.11. Mechanical deflection of the hair bundle opens transduction channels in the hair bundle allowing the entry of positive ions, mostly K$^+$. The consequent depolarization of the cell activates voltage-gated Ca$^{2+}$ channels, and the intracellular Ca$^{2+}$ concentration rises. This, in turn, opens Ca$^{2+}$-sensitive K$^+$ channels. K$^+$ ions flow out of the cell, and the cell repolarizes. Ca$^{2+}$ balance is maintained by pumps that remove Ca$^{2+}$ from the hair cell. One crucial, and rather unusual, feature of the model is that K$^+$ can both enter and leave the cell passively. Since the hair bundle projects into the scala media, the fluid surrounding the hair bundle (the endolymph in the case of hair cells in the cochlea) is of different composition from that surrounding the base of the hair cell, having a high K$^+$ and a low Na$^+$ concentration.

We do not present all the many details of the model here. Suffice it to say that it is assumed that there are three significant ionic currents contributing to resonance in the hair cell: a voltage-gated Ca$^{2+}$ current, a Ca$^{2+}$-activated K$^+$ current, and a leak current. Thus, for an applied current $I$,

$$I = C_m \frac{dV}{dt} + I_c + I_{kc} + I_L. \tag{23.105}$$

The voltage-gated Ca$^{2+}$ current $I_c$ and the leak currents are described by similar equations as in the Hodgkin–Huxley model (Chapter 4). The model for the Ca$^{2+}$-activated K$^+$ channel, $I_{kc}$, is considerably more complicated. It is assumed that the channel has

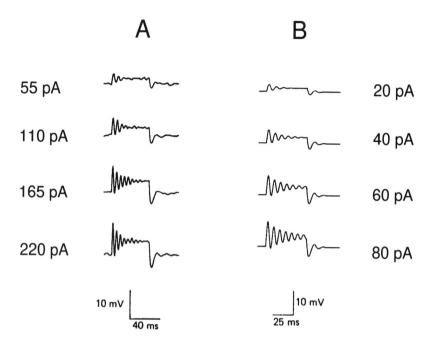

**Figure 23.12** A: Responses of bullfrog saccular hair cells to depolarizing current steps. As the current step increases in size, the hair cells show more pronounced oscillatory behavior. (Adapted from Hudspeth and Lewis, 1988b, Fig. 3.) B: Responses of the model to current steps. (Adapted from Hudspeth and Lewis, 1988b, Fig. 6.)

three closed states and two open states: binding of two $Ca^{2+}$ ions converts the channel into a state in which it can spontaneously open, while the binding of an additional $Ca^{2+}$ ion can prolong opening. The transition rate constants are dependent on $Ca^{2+}$ and voltage. Finally, $Ca^{2+}$ handling is treated simply by assuming that $Ca^{2+}$ comes in through the $Ca^{2+}$ channel and is removed by a first-order process.

The parameters (of which there are about 30) were determined by constraining the model to agree with voltage-clamp data from a single cell, and then the response of the model to current pulses was investigated. It was found that depolarizing current steps induced damped membrane potential oscillations in the model, with a frequency and amplitude dependent on the magnitude of the current step, in close agreement with experimental data (Fig. 23.12).

To simulate a transduction current, a term $I_T = g_T(V - V_T)$ is added to the right-hand side of (23.105). The transduction conductance is assumed to be a function of hair cell displacement, which, in turn, is assumed to vary sinusoidally. The resultant model frequency response is band-pass in nature, with the maximal response at frequency 112 Hz and a quality factor of 3. The frequency at which the response is maximal (the resonant frequency) is a function of the model parameters, and realistic changes in the model parameters can account for the range of experimentally observed resonant frequencies in the bullfrog sacculus. In particular, because the resonant frequency is sensitive to $g_{kc}$, the model predicts that controlling the number of $Ca^{2+}$-sensitive $K^+$ channels is one simple way in which cells could tune their frequency response.

## 23.4  Exercises

1. Show that if $F(y, t)$ in the Lesser and Berkley model is an odd function of $y$, it is sufficient to consider only the solution in the region $0 < y < l$. Hint: show that $\phi_2(y, t) = -\phi_1(-y, t)$ and $p_2(y, t) = -p_1(-y, t)$ satisfy the differential equations for $\phi_2$ and $p_2$. Hence the potential and pressure are odd functions of $y$.

2. Formulate a model of the cochlea in which the basilar membrane is spatially coupled in the $x$ direction as if it were a damped string. How does the Fourier transform of this problem differ from the Lesser–Berkley model (23.24)–(23.28)?

3. Show that (23.73) describes two waves moving in opposite directions. Hint: $Y$ has an imaginary component, and thus the terms in (23.73) have phases with opposite signs. What are the envelopes of the waves?

4. If the functions $m(x)$, $r(x)$, and $k(x)$ are proportional to the same exponential, then the solution of (23.43) can be found exactly.

   (a)  Suppose that $m = m_0 e^{-\lambda x}$, $k(x) = k_0 e^{-\lambda x}$, $r(x) = r_0 e^{-\lambda x}$, and show that (23.43) becomes

   $$\frac{d^2 p}{dx^2} + \alpha^2 e^{\lambda x} p = 0, \tag{23.106}$$

   where $\alpha^2 = \frac{2\omega^2 \rho}{l(i\omega r_0 + k_0 - \omega^2 m_0)}$.

   (b)  Show that the transformation $s = \frac{2\alpha}{\lambda} e^{\lambda x/2}$ transforms (23.106) into

   $$s\frac{d}{ds}\left(s\frac{dp}{ds}\right) + s^2 p = 0, \tag{23.107}$$

   which is Bessel's equation of order zero. Thus, the general solution of (23.106) is

   $$p(x) = AJ_0\left(\frac{2\alpha}{\lambda} e^{\lambda x/2}\right) + BY_0\left(\frac{2\alpha}{\lambda} e^{\lambda x/2}\right), \tag{23.108}$$

   where $J_0$ and $Y_0$ are the zeroth-order Bessel functions of the first and second kind, or equivalently,

   $$p(x) = \tilde{A}H_0^{(1)}(s) + \tilde{B}H_0^{(2)}(s), \tag{23.109}$$

   where $H_0^{(1)}(s)$ and $H_0^{(2)}(s)$ are the zeroth-order Hankel functions of first and second kind. Use the boundary conditions (23.44) and (23.45) to determine the coefficients $\tilde{A}$ and $\tilde{B}$.

   (c)  Use the asymptotic behavior of the Hankel functions (Keener, 1988)

   $$H_0^{(1)}(s) \sim \left(\frac{2}{\pi s}\right)^{1/2} e^{i(s-\pi/4)}, \qquad H_0^{(2)}(s) \sim \left(\frac{2}{\pi s}\right)^{1/2} e^{-i(s-\pi/4)} \tag{23.110}$$

   to approximate $p(x)$ in the case that $\tilde{B} = 0$. Show that $\eta(x)$ is approximated by (23.59). Under what conditions is this approximation valid?

5. For the long-wave model it was claimed that the phase of $\eta$, where $\eta$ is given by (23.78), does not agree with experimental data. Confirm this by showing that for a fixed $\omega$, the phase increases exponentially with $x$, and that the phase does not go to zero as $\omega$ goes to zero.

6. Compare the location of the envelope maximum $x_p$ for the shallow-water approximation (23.63) in the case $l\lambda = \pi/2$ with that for the deep-water approximation (23.82), for large $\omega$.

7. Let $\eta_0 = 1$, $\eta_L = 0$, and $m = 0.05e^{-1.5x}$, and let the other parameters be the same as in the Lesser and Berkley model. Use the boundary conditions to solve for $A_1$ and $B_1$ in (23.57), and thus calculate the displacement of the basilar membrane for the shallow-water model. Compare the shallow-water and deep-water models by plotting the displacements for high and low frequencies. How does the behavior of the long-wave model change as $l$ is decreased?

# Appendix: Units and Physical Constants

| Quantity | Name | Symbol | Units |
|---|---|---|---|
| Amount | mole | mol | |
| Electric charge | coulomb | C | |
| Mass | kilogram | kg | |
| Temperature | kelvin | K | |
| Time | second | s | |
| Length | meter | m | |
| Force | newton | N | $kg \cdot m \cdot s^{-2}$ |
| Energy | joule | J | $N \cdot m$ |
| Pressure | pascal | Pa | $N \cdot m^{-2}$ |
| Capacitance | farad | F | $A \cdot s \cdot V^{-1}$ |
| Resistance | ohm | $\Omega$ | $V \cdot A^{-1}$ |
| Electric current | ampere | A | $C \cdot s^{-1}$ |
| Conductance | siemen | S | $A \cdot V^{-1} = \Omega^{-1}$ |
| Potential difference | volt | V | $N \cdot m \cdot C^{-1}$ |
| Concentration | Molar | M | $mol \cdot L^{-1}$ |

| Physical Constant | Symbol | Value |
|---|---|---|
| Boltzmann's constant | $k$ | $1.381 \times 10^{-23}$ J $\cdot$ K$^{-1}$ |
| Planck's constant | $h$ | $6.626 \times 10^{-34}$ J $\cdot$ s |
| Avogadro's number | $N_A$ | $6.02257 \times 10^{23}$ mol$^{-1}$ |
| unit charge | $q$ | $1.6 \times 10^{-19}$ C |
| gravitational constant | $g$ | $9.78049$ m/s$^2$ |
| Faraday's constant | $F$ | $9.649 \times 10^4$ C $\cdot$ mol$^{-1}$ |
| permittivity of free space | $\epsilon_0$ | $8.854 \times 10^{-12}$ F/m |
| universal gas constant | $R$ | $8.315$ J mol$^{-1} \cdot$ K$^{-1}$ |
| atmosphere | atm | $1.01325 \times 10^5$ N $\cdot$ m$^{-2}$ |

**Lumen:** 1 lm = quantity of light emitted by $\frac{1}{60}$ cm$^2$ surface area of pure platinum at its melting temperature (1770° C), within a solid angle of 1 steradian.

**Angstrom:** 1 Å $= 10^{-10}$ m.

**Liter:** 1 L $= 10^{-3}$ m$^3$.


| Other Identities |
|---|
| 1 atm $= 760$ mmHg |
| $R = kN_A$ |
| $F = qN_A$ |
| pH $= -\log_{10}[\text{H}^+]$ with $[\text{H}^+]$ in moles per liter |
| 273.15 K $= 0°$C (ice point) |

# References

Aharon, S., H. Parnas, and I. Parnas (1994) The magnitude and significance of $Ca^{2+}$ domains for release of neurotransmitter, *Bulletin of Mathematical Biology*. 56: 1095–1119.

Akin, E. and H. M. Lacker (1984) Ovulation control: the right number or nothing, *Journal of Mathematical Biology*. 20: 113–132.

Alberts, B., D. Bray, J. Lewis, M. Raff, K. Roberts, and J. D. Watson, (1994) *Molecular Biology of the Cell*: Garland Publishing, Inc., New York, London.

Aldrich, R. W., D. P. Corey, and C. F. Stevens (1983) A reinterpretation of mammalian sodium channel gating based on single channel recording, *Nature*. 306: 436–441.

Allbritton, N. L., T. Meyer, and L. Stryer (1992) Range of messenger action of calcium ion and inositol 1,4,5-trisphosphate, *Science*. 258: 1812–1815.

Alt, W. and D. A. Lauffenberger (1987) Transient behavior of a chemotaxis system modelling certain types of tissue inflammation, *Journal of Mathematical Biology*. 24: 691–722.

Anliker, M., R. L. Rockwell, and E. Ogden (1971a) Nonlinear analysis of flow pulses and shock waves in arteries. Part I: derivation and properties of mathematical model, *Z. Ang. Math. Phys*. 22: 217–246.

Anliker, M., R. L. Rockwell, and E. Ogden (1971b) Nonlinear analysis of flow pulses and shock waves in arteries. Part II: parametric study related to clinical problems., *Z. Ang. Math. Phys*. 22: 563–581.

Armstrong, C. M. and F. Bezanilla (1973) Currents related to movement of the gating particles of the sodium channels, *Nature*. 242: 459–461.

Armstrong, C. M. and F. Bezanilla (1974) Charge movement associated with the opening and closing of the activation gates of the Na channels, *Journal of General Physiology*. 63: 533–552.

Armstrong, C. M. and F. Bezanilla (1977) Inactivation of the sodium channel II Gating current experiments, *Journal of General Physiology*. 70: 567–590.

Armstrong, C. M. (1981) Sodium channels and gating currents, *Physiological Reviews*. 61: 644–683.

Arnold, V. I. (1983) *Geometric Methods in the Theory of Ordinary Differential Equations*: Springer-Verlag, New York.

Aronson, D. G. and H. F. Weinberger (1975) Nonlinear diffusion in population genetics, combustion and nerve pulse propagation, *Lecture Notes in Mathematics*. 446: 5–49.

Asdell, S. A. (1946) *Patterns of Mammalian Reproduction*: Comstock Publishing Company, New York.

Ashmore, J. F. and D. Attwell (1985) Models for electrical tuning in hair cells, *Proc. Roy. Soc. Lond. B*. 226: 325–344.

Atri, A., J. Amundson, D. Clapham, and J. Sneyd (1993) A single-pool model for intracellular calcium oscillations and waves in the *Xenopus laevis* oocyte, *Biophysical Journal*. 65: 1727–1739.

Atwater, I., C. M. Dawson, A. Scott, G. Eddlestone, and E. Rojas (1980) The nature of the oscillatory behavior in electrical activity from pancreatic $\beta$-cell, *J. of Horm. Metabol. Res.* 10 (suppl.): 100–107.

Atwater, I., L. Rosario, and E. Rojas (1983) Properties of calcium-activated potassium channels in the pancreatic $\beta$-cell, *Cell Calcium*. 4: 451–461.

Atwater, I. and J. Rinzel (1986), *The $\beta$-cell bursting pattern and intracellular calcium*. In: Ionic Channels in Cells and Model Systems, Ed: R. Latorre, Plenum Press, New York, London.

Av-Ron, E., H. Parnas, and L. Segel (1993) A basic biophysical model for bursting neurons, *Biol. Cybern*. 69: 87–95.

Baan, J., A. Noordergraaf and J. Raines, Eds. (1978) *Cardiovascular System Dynamics*. The MIT Press, Cambridge, MA.

Backx, P. H., P. P. d. Tombe, J. H. K. V. Deen, B. J. M. Mulder, and H. E. D. J. t. Keurs (1989) A model of propagating calcium-induced calcium release mediated by calcium diffusion, *Journal of General Physiology*. 93: 963–977.

Bar, M. and M. Eiswirth (1993) Turbulence due to spiral breakup in a continuous excitable medium, *Phys. Rev. E*. 48: 1635–1637.

Barcilon, V. (1992) Ion flow through narrow membrane channels: Part I, *SIAM Journal on Applied Mathematics*. 52: 1391–1404.

Barcilon, V., D. P. Chen, and R. S. Eisenberg (1992) Ion flow through narrow membrane channels: Part II, *SIAM Journal on Applied Mathematics*. 52: 1405–1425.

Barkley, D. (1994) Euclidean symmetry and the dynamics of rotating spiral waves, *Phys. Rev. Lett*. 72: 164–167.

Batchelor, G. K. (1967) *An Introduction to Fluid Dynamics*: Cambridge University Press, Cambridge.

Baylor, D. A., A. L. Hodgkin, and T. D. Lamb (1974a) The electrical response of turtle cones to flashes and steps of light, *Journal of Physiology*. 242: 685–727.

Baylor, D. A., A. L. Hodgkin, and T. D. Lamb (1974b) Reconstruction of the electrical responses of turtle cones to flashes and steps of light, *Journal of Physiology*. 242: 759–791.

Beck, J. S., R. Laprade, and J.-Y. Lapointe (1994) Coupling between transepithelial Na transport and basolateral K conductance in renal proximal tubule, *American Journal*

of Physiology (Renal Fluid and Electrolyte Physiology). 266: F517–F527.

Beeler, G. W. and H. J. Reuter (1977) Reconstruction of the action potential of ventricular myocardial fibers, *Journal of Physiology*. 268: 177–210.

Begenisich, T. B. and M. D. Cahalan (1980) Sodium channel permeation in squid axons I: reversal potential experiments, *Journal of Physiology*. 307: 217–242.

Belair, J., M. C. Mackey, and J. M. Mahaffy (1995) Age-structured and two-delay models for erythropoiesis, *Mathematical Biosciences*. 128: 317–346.

Beltrami, E. and J. Jesty (1995) Mathematical analysis of activation thresholds in enzyme-catalyzed positive feedbacks: Application to the feedbacks of blood coagulation, *Proc. Natl. Acad. Sci. USA*. 92: 8744–8748.

Bergstrom, R. W., W. Y. Fujimoto, D. C. Teller, and C. D. Haën (1989) Oscillatory insulin secretion in perifused isolated rat islets, *American Journal of Physiology (Endocrin. Metab. 20)*. 257: E479–E485.

Berman, N., H.-F. Chou, A. Berman, and E. Ipp (1993) A mathematical model of oscillatory insulin secretion, *American Journal of Physiology (Reg. Int. Comp. Physiol.)*. 264: R839–R851.

Berne, R. M. and M. N. Levy, Eds. (1993) *Physiology*. St. Louis, Mosby Year Book.

Bernstein, J. (1902) Untersuchungen zur Thermodynamik der bioelektrischen Ströme. Erster Theil., *Pflügers Arch*. 82: 521–562.

Berridge, M. J. and A. Galione (1988) Cytosolic calcium oscillators, *FASEB Journal*. 2: 3074–3082.

Bertram, R., M. J. Butte, T. Kiemel, and A. Sherman (1995) Topological and phenomenological classification of bursting oscillations, *Bulletin of Mathematical Biology*. 57: 413–439.

Bertram, R., A. Sherman, and E. F. Stanley (1996) Single-domain/bound calcium hypothesis of transmitter release and facilitation, *Journal of Neurophysiology*. 75: 1919–1931.

Bezprozvanny, I., J. Watras, and B. E. Ehrlich (1991) Bell-shaped calcium-response curves of Ins(1,4,5)P$_3$- and calcium-gated channels from endoplasmic reticulum of cerebellum, *Nature*. 351: 751–754.

Blakemore, C., Ed. (1990) *Vision: Coding and Efficiency*. Cambridge University Press, Cambridge, UK.

Bluman, G. W. and H. C. Tuckwell (1987) Methods for obtaining analytical solutions for Rall's model neuron, *J. Neurosci. Methods*. 20: 151–166.

Bogumil, R. J., M. Ferin, J. Rootenberg, L. Speroff, and R. L. Van de Wiele (1972) Mathematical studies of the human menstrual cycle. I. formulation of a mathematical model, *J. Clin. Endocrinol. Metab*. 35: 126–156.

Boitano, S., E. R. Dirksen, and M. J. Sanderson (1992) Intercellular propagation of calcium waves mediated by inositol trisphosphate, *Science*. 258: 292–294.

Borghans, J. A. M., R. J. de Boer, and L. A. Segel (1996) Extending the quasi-steady state approximation by changing variables, *Bulletin of Mathematical Biology*. 58: 43–63.

Bowen, J. R., A. Acrivos, and A. K. Oppenheim (1963) Singular perturbation refinement to quasi-steady state approximation in chemical kinetics, *Chem. Eng. Sci*. 18: 177–188.

Boyce, W. E. and R. C. DiPrima (1997) *Elementary Differential Equations and Boundary Value Problems (sixth edition)*: John Wiley and Sons, New York.

Boyd, I. A. and A. R. Martin (1956) The end-plate potential in mammalian muscle, *Journal of Physiology*. 132: 74–91.

Brabant, G., K. Prank, and C. Schöfl (1992) Pulsatile patterns in hormone secretion, *Trends in Endocrinology and Metabolism*. 3: 183–190.

Braun, M., (1993), *Differential Equations and their Applications. Fourth edition:* Springer-Verlag, New York.

Briggs, G. E. and J. B. S. Haldane (1925) A note on the kinematics of enzyme action, *Biochemical Journal*. 19: 338–339.

Brink, P. R. and S. V. Ramanan (1985) A model for the diffusion of fluorescent probes in the septate giant axon of earthworm: axoplasmic diffusion and junctional membrane permeability, *Biophysical Journal*. 48: 299–309.

Britton, N. F (1986) *Reaction-Diffusion Equations and their Applications to Biology*: Academic Press, London.

Brown, B. H., H. L. Duthie, A. R. Horn, and R. H. Smallwood (1975) A linked oscillator model of electrical activity of human small intestine, *American Journal of Physiology*. 229: 384–388.

Burger and Milaan (1948) Heart vector and leads. Part III, Geometrical representation, *British Heart Journal*. 10: 229–233.

Burton, W. K., N. Cabrera, and F. C. Frank (1951) The growth of crystals and the equilibrium structure of their surfaces, *Phil. Trans. Roy. Soc. Lond. A*. 243: 299–358.

Campbell, D. T. and B. Hille (1976) Kinetic and pharmacological properties of the sodium channel of frog skeletal muscle, *Journal of General Physiology*. 67: 309–323.

Carpenter, G. (1977) A geometric approach to singular perturbation problems with applications to nerve impulse equations, *Journal of Differential Equations*. 23: 335–367.

Carpenter, G. A. and S. Grossberg (1981) Adaptation and transmitter gating in vertebrate photoreceptors, *Journal of Theoretical Neurobiology*. 1: 1–42.

Cartwright, M. and M. Husain (1986) A model for the control of testosterone secretion, *Journal of Theoretical Biology*. 123: 239–250.

Casten, R. G., H. Cohen, and P. A. Lagerstrom (1975) Perturbation analysis of an approximation to the Hodgkin–Huxley theory, *Quarterly of Applied Mathematics*. 32: 365–402.

Chadwick, R. S., A. Inselberg, and K. Johnson (1976) Mathematical model of the cochlea. II: results and conclusions, *SIAM Journal on Applied Mathematics*. 30: 164–179.

Chadwick, R. (1980), *Studies in cochlear mechanics*. In: Mathematical Modeling of the Hearing Process. Lecture Notes in Biomathematics, 43., Ed: M. H. Holmes and L. A. Rubenfeld, Springer-Verlag, Berlin, Heidelberg, New York.

Changeux, J. P. (1965) The control of biochemical reactions, *Scientific American*. 212: 36–45.

Chapman, R. A. and C. H. Fry (1978) An analysis of the cable properties of frog ventricular myocardium, *Journal of Physiology*. 283: 263–282.

Charles, A. C., J. E. Merrill, E. R. Dirksen, and M. J. Sanderson (1991) Intercellular signaling in glial cells: calcium waves and oscillations in response to mechanical stimulation and glutamate, *Neuron*. 6: 983–992.

Charles, A. C., C. C. G. Naus, D. Zhu, G. M. Kidder, E. R. Dirksen, and M. J. Sanderson (1992) Intercellular calcium signaling via gap junctions in glioma cells, *Journal of Cell Biology*. 118: 195–201.

Chay, T. R. and J. Keizer (1983) Minimal model for membrane oscillations in the pancreatic $\beta$-cell, *Biophysical Journal*. 42: 181–190.

Chay, T. R. (1986) On the effect of the intracellular calcium-sensitive $K^+$ channel in the bursting pancreatic $\beta$-cell, *Biophysical Journal*. 50: 765–777.

Chay, T. R. (1987) The effect of inactivation of calcium channels by intracellular $Ca^{2+}$ ions in the bursting pancreatic β-cell, *Cell Biophysics*. 11: 77–90.

Chay, T. R. and H. S. Kang (1987), *Multiple oscillatory states and chaos in the endogeneous activity of excitable cells: pancreatic β-cell as an example*. In: Chaos in Biological Systems, Ed: H. Degn, A. V. Holden, and L. F. Olsen, Plenum Press, New York.

Chay, T. R. and D. L. Cook (1988) Endogenous bursting patterns in excitable cells, *Mathematical Biosciences*. 90: 139–153.

Cheer, A., R. Nuccitelli, G. F. Oster, and J.-P. Vincent (1987) Cortical waves in vertebrate eggs I: the activation waves, *Journal of Theoretical Biology*. 124: 377–404.

Chen, P. S., P. D. Wolf, E. G. Dixon, N. D. Danieley, D. W. Frazier, W. M. Smith, and R. E. Ideker (1988) Mechanism of ventricular vulnerability to single premature stimuli in open-chest dogs, *Circ. Res*. 62: 1191–1209.

Chen, D. P., V. Barcilon, and R. S. Eisenberg (1992) Constant fields and constant gradients in open ionic channels, *Biophysical Journal*. 61: 1372–1393.

Chen, D. and R. Eisenberg (1993) Charges, currents, and potentials in ionic channels of one conformation, *Biophysical Journal*. 64: 1405–1421.

Chen, L. and M. Q. Meng (1995) Compact and scattered gap junctions in diffusion mediated cell–cell communication, *Journal of Theoretical Biology*. 176: 39–45.

Christ, G. J., P. R. Brink, and S. V. Ramanan (1994) Dynamic gap junctional communication: a delimiting model for tissue responses, *Biophysical Journal*. 67: 1335–1344.

Ciani, S. and B. Ribalet (1988) Ion permeation and rectification in ATP-sensitive channels from insulin-secreting cells (RINm5F): effects of $K^+$, $Na^+$, and $Mg^{2+}$, *Journal of Membrane Biology*. 103: 171–180.

Civan, M. M. and R. J. Podolsky (1966) Contraction kinetics of striated muscle fibres following quick changes in load, *Journal of Physiology*. 184: 511–534.

Civan, M. M. and R. J. Bookman (1982) Transepithelial $Na^+$ transport and the intracellular fluids: a computer study, *Journal of Membrane Biology*. 65: 63–80.

Clapham, D. (1995) Calcium signaling, *Cell*. 80: 259–268.

Cobbs, W. H. and E. N. Pugh (1987) Kinetics and components of the flash photocurrent of isolated retinal rods of the larval salamander *Ambystoma Tigrinum*, *Journal of Physiology*. 394: 529–572.

Coddington, E. A. and N. Levinson (1984) *Ordinary Differential Equations*: Robert E. Krieger Publishing Company, Malabar, Florida.

Cole, K. S. and H. J. Curtis (1940) Electric impedance of the squid giant axon during activity, *Journal of General Physiology*. 22: 649–670.

Cole, K. S., H. A. Antosiewicz, and P. Rabinowitz (1955) Automatic computation of nerve excitation, *SIAM Journal on Applied Mathematics*. 3: 153–172.

Colli-Franzone, P., L. Guerri, and S. Rovida (1990) Wavefront propagation in an activation model of the anisotropic cardiac tissue: asymptotic analysis and numerical simulations, *Journal of Mathematical Biology*. 28: 121–176.

Colli-Franzone, P., L. Guerri, and B. Taccardi (1993) Spread of excitation in a myocardial volume: simulation studies in a model of anisotropic ventricular muscle activated by a point stimulation, *J. Cardivasc. Phys*. 4: 144–160.

Colli-Franzone, P. and L. Guerri (1993) Spreading of excitation in 3-D models of the anisotropic cardiac tissue I. Validation of the eikonal model, *Mathematical Biosciences*. 113: 145–209.

Cornish-Bowden, A. and R. Eisenthal (1974) Statistical considerations in the estimation of enzyme kinetic parameters by the direct linear plot and other methods, *Biochemical Journal*. 139: 721–730.

Courant, R. and D. Hilbert (1953) *Methods of Mathematical Physics*: Wiley-Interscience, New York.

Courtemanche, M. and A. T. Winfree (1991) Re-entrant rotating waves in a Beeler–Reuter-based model of 2-dimensional cardiac electrical activity, *Int. J. Bif. and Chaos*. 1: 431–444.

Courtemanche, M., J. P. Keener, and L. Glass (1993) Instabilities of a propagating pulse in a ring of excitable media, *Phys. Rev. Letts*. 70: 2182–2185.

Courtemanche, M., J. P. Keener, and L. Glass (1996) A delay equation representation of pulse circulation on a ring in excitable media, *SIAM Journal on Applied Mathematics*. 56: 119–142.

Crawford, A. C. and R. Fettiplace (1981) An electrical tuning mechanism in turtle cochlear hair cells, *Journal of Physiology*. 312: 377–412.

Cronin, J. (1981) *Mathematics of Cell Electrophysiology*: M. Dekker, New York.

Dallos, P., C. D. Geisler, J. W. Matthews, M. A. Ruggero, and C. R. Steele, Eds. (1990) *The Mechanics and Biophysics of Hearing. Lecture Notes in Biomathematics, 87.* Springer-Verlag, Berlin, Heidelberg, New York.

Daly, S. J. and R. I. Normann (1985) Temporal information processing in cones: effects of light adaptation on temporal summation and modulation, *Vision Res.* 25: 1197–1206.

Dani, J. A. and D. G. Levitt (1990) Diffusion and kinetic approaches to describe permeation in ionic channels, *Journal of Theoretical Biology.* 146: 289–301.

Davis, B. O., N. Holtz, and J. C. Davis (1985) *Conceptual Human Physiology*: C.E. Merrill Pub. Co., Columbus.

Dawis, S. M., R. M. Graeff, R. A. Heyman, T. F. Walseth, and N. D. Goldberg (1988) Regulation of cyclic GMP metabolism in toad photoreceptors, *Journal of Biological Chemistry.* 263: 8771–8785.

Dawson, D. C. and N. W. Richards (1990) Basolateral K conductance: role in regulation of NaCl absorption and secretion, *American Journal of Physiology (Cell Physiology).* 259: C181–C195.

Dawson, D. C. (1992), *Water transport: principles and perspectives.* In: The Kidney: Physiology and Pathophysiology, Ed: D. W. Seldin and G.Giebisch, Raven Press, New York.

de Beus, A. M., T. L. Fabry, and H. M. Lacker (1993) A gastric acid secretion model, *Biophysical Journal.* 65: 362–378.

de Vries, G. (1995) *Analysis of models of bursting electrical activity in pancreatic beta cells. Ph.D. thesis, Department of Mathematics*: University of British Columbia, Vancouver.

De Young, G. W. and J. Keizer (1992) A single pool IP$_3$-receptor based model for agonist stimulated Ca$^{2+}$ oscillations, *Proc. Natl. Acad. Sci. USA.* 89: 9895–9899.

del Castillo, J. and B. Katz (1954) Quantal components of the end-plate potential, *Journal of Physiology.* 124: 560–573.

Demer, L. L., C. M. Wortham, E. R. Dirksen, and M. J. Sanderson (1993) Mechanical stimulation induces intercellular calcium signalling in bovine aortic endothelial cells, *American Journal of Physiology (Heart & Circ. Physiol.).* 33: M2094–M2102.

Denbigh, K. (1981) *The Principles of Chemical Equilibrium*: Cambridge University Press, Cambridge, London, New York.

Detwiler, P. B. and A. L. Hodgkin (1979) Electrical coupling between cones in the turtle retina, *Journal of Physiology.* 291: 75–100.

DeVries, G. W., A. I. Cohen, O. Lowry, and J. A. Ferendelli (1979) Cyclic nucleotide in the cone-dominant ground-squirrel retina, *Expl. Eye Res.* 29: 315–321.

Diamant, N. E. and A. Bortoff (1969) Nature of the intestinal slow-wave frequency gradient, *American Journal of Physiology.* 216: 301–307.

Diamant, N. E., P. K. Rose, and E. J. Davison (1970) Computer simulation of intestinal slow-wave frequency gradient, *American Journal of Physiology.* 219: 1684–1690.

DiFrancesco, D. and D. Noble (1985) A model of cardiac electrical activity incorporating ionic pumps and concentration changes, *Phil. Trans. R. Soc. B.* 307: 353–398.

Dixon, M. and E. C. Webb (1979) *Enzymes*: Academic Press, New York.

Dockery, J. D. and J. P. Keener (1989) Diffusive effects on dispersion in excitable media, *SIAM Journal on Applied Mathematics.* 49: 539–566.

Doedel, E. (1986) *Software for continuation and bifurcation problems in ordinary differential equations*: California Institute of Technology.

Duffy, M. R., N. F. Britton, and J. D. Murray (1980) Spiral wave solutions of practical reaction-diffusion systems, *SIAM Journal on Applied Mathematics.* 39: 8–13.

Dufour, J.-F., I. M. Arias, and T. J. Turner (1997) Inositol 1,4,5-trisphosphate and calcium regulate the calcium channel function of the hepatic inositol 1,4,5-trisphosphate receptor, *Journal of Biological Chemistry.* 272: 2675–2681.

Dupont, G. and A. Goldbeter (1993) One-pool model for Ca$^{2+}$ oscillations involving Ca$^{2+}$ and inositol 1,4,5-trisphosphate as co-agonists for Ca$^{2+}$ release, *Cell Calcium.* 14: 311–322.

Dupont, G. and A. Goldbeter (1994) Properties of intracellular Ca$^{2+}$ waves generated by a model based on Ca$^{2+}$-induced Ca$^{2+}$ release, *Biophysical Journal.* 67: 2191–2204.

Durand, D. (1984) The somatic shunt cable model for neurons, *Biophysical Journal.* 46: 645–653.

Ebihara, L. and E. A. Johnson (1980) Fast sodium current in cardiac muscle, a quantitative description, *Biophysical Journal.* 32: 779–790.

Edelstein-Keshet, L. (1988) *Mathematical Models in Biology*: McGraw-Hill, New York.

Einstein, A. (1906) Eine neue Bestimmung der Moleküldimensionen, *Ann. Phys.* 19: 289.

Eisenthal, R. and A. Cornish-Bowden (1974) A new graphical method for estimating enzyme kinetic parameters, *Biochemical Journal.* 139: 715–720.

Endo, M., M. Tanaka, and Y. Ogawa (1970) Calcium-induced release of calcium from the sarcoplasmic reticulum of skinned skeletal muscle fibres, *Nature*. 228: 34–36.

Engel, E., A. Peskoff, G. L. Kauffman, and M. I. Grossman (1984) Analysis of hydrogen ion concentration in the gastric gel mucus layer, *American Journal of Physiology*. 247 (Gastrointest. Liver Physiol. 10): G321–G338.

Ermentrout, G. B. and N. Kopell (1984) Frequency plateaus in a chain of weakly coupled oscillators, *SIAM J. Math. Anal*. 15: 215–237.

Eyring, H., R. Lumry, and J. W. Woodbury (1949) Some applications of modern rate theory to physiological systems, *Record Chem. Prog*. 10: 100–114.

Fabiato, A. (1983) Calcium-induced release of calcium from the cardiac sarcoplasmic reticulum, *American Journal of Physiology*. 245: C1–C14.

Fabiato, A. (1985) Time and calcium dependence of activation and inactivation of calcium-induced release of calcium from the sarcoplasmic reticulum of a skinned cardiac Purkinje cell, *Journal of General Physiology*. 85: 247–289.

Fabiato, A. (1992), *Two kinds of calcium-induced release of calcium from the sarcoplasmic reticulum of skinned cardiac cells*. In: Excitation-contraction coupling in skeletal, cardiac, and smooth muscle, Ed: G. B. G. Frank, P. Bianchi, and H. Keurs, Plenum Press, New York.

Faddy, M. J. and R. G. Gosden (1995) A mathematical model of follicle dynamics in the human ovary, *Human Reproduction*. 10: 770–775.

Fain, G. and H. R. Matthews (1990) Calcium and the mechanism of light adaptation in vertebrate photoreceptors, *Trends Neurosci*. 13: 378–384.

Fatt, P. and B. Katz (1952) Spontaneous subthreshold activity at motor nerve endings, *Journal of Physiology*. 117: 109–128.

Fife, P. C. and J. B. McLeod (1977) The approach of solutions of nonlinear diffusion equations to travelling front solutions, *Arch. Rat. Mech. Anal*. 65: 335–361.

Fife, P. (1979) *Mathematical Aspects of Reacting and Diffusing Systems*: Springer-Verlag, Berlin.

Finch, E. A., T. J. Turner, and S. M. Goldin (1991) Calcium as a coagonist of inositol 1,4,5-trisphosphate-induced calcium release, *Science*. 252: 443–446.

Finkelstein, A. and C. S. Peskin (1984) Some unexpected consequences of a simple physical

mechanism for voltage-dependent gating in biological membranes, *Biophysical Journal*. 46: 549–558.

FitzHugh, R. (1960) Thresholds and plateaus in the Hodgkin–Huxley nerve equations, *Journal of General Physiology*. 43: 867–896.

FitzHugh, R. (1961) Impulses and physiological states in theoretical models of nerve membrane, *Biophysical Journal*. 1: 445–466.

FitzHugh, R. (1969), *Mathematical models of excitation and propagation in nerve*. In: Biological Engineering, Ed: H. P. Schwan, McGraw-Hill, New York.

Fletcher, H. (1951) On the dynamics of the cochlea, *J. Acoust. Soc. Am*. 23: 637–645.

Foerster, P., S. Muller, and B. Hess (1989) Critical size and curvature of wave formation in an excitable chemical medium, *Proc. Natl. Acad. Sci. USA*. 86: 6831–6834.

Fogelson, A. L. and R. S. Zucker (1985) Presynaptic calcium diffusion from various arrays of single channels, *Biophysical Journal*. 48: 1003–1017.

Fogelson, A. L. (1992) Continuum models of platelet aggregation: Formulation and mechanical properties, *SIAM Journal on Applied Mathematics*. 52: 1089–1110.

Fogelson, A. and A. Kuharsky (1998) Membrane binding-site density can modulate activation thresholds in enzyme systems, *Journal of Theoretical Biology*. : in press.

Forti, S., A. Menini, G. Rispoli, and V. Torre (1989) Kinetics of phototransduction in retinal rods of the newt *Triturus Cristatus*, *Journal of Physiology*. 419: 265–295.

Frank, O. (1899) Die Grundform des Arteriellen Pulses (see translation by Sagawa et al., 1990), *Zeitschrift für Biologie*. 37: 483-526.

Frankel, M. L. and G. I. Sivashinsky (1987) On the nonlinear diffusive theory of curved flames, *J. Physique (Paris)*. 48: 25–28.

Frankel, M. L. and G. I. Sivashinsky (1988) On the equation of a curved flame front, *Physica D*. 30: 28–42.

Frankenhaeuser, B. (1960a) Quantitative description of sodium currents in myelinated nerve fibres of *Xenopus laevis*, *Journal of Physiology*. 151: 491–501.

Frankenhaeuser, B. (1960b) Sodium permeability in toad nerve and in squid nerve, *Journal of Physiology*. 152: 159–166.

Frankenhaeuser, B. (1963) A quantitative description of potassium currents in myelinated nerve fibres of *Xenopus laevis*, *Journal of Physiology*. 169: 424–430.

Frenzen, C. L. and P. K. Maini (1988) Enzyme kinetics for a two-step enzymic reaction with comparable initial enzyme-substrate ratios, *Journal of Mathematical Biology*. 26: 689–703.

Friel, D. (1995) [$Ca^{2+}$]$_i$ oscillations in sympathetic neurons: an experimental test of a theoretical model, *Biophysical Journal*. 68: 1752–1766.

Gatti, R. A., W. A. Robinson, A. S. Denaire, M. Nesbit, J. J. McCullogh, M. Ballow, and R. A. Good (1973) Cyclic leukocytosis in chronic myelogenous leukemia, *Blood*. 41: 771–782.

Gerhardt, M., H. Schuster, and J. J. Tyson (1990) A cellular automaton model of excitable media including curvature and dispersion, *Science*. 247: 1563–1566.

Girard, S., A. Lückhoff, J. Lechleiter, J. Sneyd, and D. Clapham (1992) Two-dimensional model of calcium waves reproduces the patterns observed in *Xenopus* oocytes, *Biophysical Journal*. 61: 509–517.

Glass, L. and M. C. Mackey (1979) A simple model for phase locking of biological oscillators, *Journal of Mathematical Biology*. 7: 339–352.

Glass, L. and M. C. Mackey (1988) *From Clocks to Chaos*: Princeton University Press, Princeton.

Glass, L. and D. Kaplan (1995) *Understanding Nonlinear Dynamics*: Springer-Verlag, New York.

Goldberger, A. L. and E. Goldberger (1994) *Clinical Electrocardiography: A Simplified Approach*: Mosby, St. Louis.

Goldbeter, A. and R. Lefever (1972) Dissipative structures for an allosteric model; application to glycolytic oscillations, *Biophysical Journal*. 12: 1302–1315.

Goldbeter, A., G. Dupont, and M. J. Berridge (1990) Minimal model for signal-induced $Ca^{2+}$ oscillations and for their frequency encoding through protein phosphorylation, *Proc. Natl. Acad. Sci. USA*. 87: 1461–1465.

Goldbeter, A. (1996) *Biochemical Oscillations and Cellular Rhythms: The Molecular Bases of Periodic and Chaotic Behaviour*: Cambridge University Press, Cambridge.

Gomatam, J. and P. Grindrod (1987) Three dimensional waves in excitable reaction-diffusion systems, *Journal of Mathematical Biology*. 25: 611–622.

Goodner, C. J., B. C. Walike, D. J. Koerker, J. W. Ensinck, A. C. Brown, E. W. Chideckel, J. Palmer, and L. Kalnasy (1977) Insulin, glucagon, and glucose exhibit synchronous, sustained oscillations in fasting monkeys, *Science*. 195: 177–179.

Gordon, A. M., A. F. Huxley, and F. J. Julian (1966) The variation in isometric tension with sarcomere length in vertebrate muscle fibres, *Journal of Physiology*. 184: 170–192.

Griffith, J. S. (1971) *Mathematical Neurobiology*: Academic Press, London.

Grindrod, P. (1991) *Patterns and Waves: the Theory and Application of Reaction–Diffusion Equations*: Clarendon Press, Oxford.

Guckenheimer, J. and P. Holmes (1983) *Nonlinear Oscillations, Dynamical Systems, and Bifurcations of Vector Fields*: Springer-Verlag, New York, Heidelberg, Berlin.

Guevara, M. R. and L. Glass (1982) Phase locking, period doubling bifurcations and chaos in a mathematical model of a periodically driven oscillator: a theory for the entrainment of biological oscillators and the generation of cardiac dysrhythmias, *Journal of Mathematical Biology*. 14: 1–23.

Guneroth, W. G. (1965) *Pediatric Electro-cardiography*: W.B. Saunders Co., Philadelphia.

Gurney, C. W., E. L. Simmons, and E. O. Guston (1981) Cyclic erythropoiesis in W/W$^V$ mice following a single small dose of $^{89}$SR, *Exp. Haematol*. 9: 118–122.

Guyton, A. C. (1963) *Circulatory Physiology: Cardiac Output and its Regulation*: W.B. Saunders, Philadelphia.

Guyton, A. C. and J. E. Hall (1996) *Textbook of Medical Physiology*: W.B. Saunders, Philadelphia.

Györke, S. and M. Fill (1993) Ryanodine receptor adaptation: control mechanism of $Ca^{2+}$-induced $Ca^{2+}$ release in heart, *Science*. 260: 807–809.

Hale, J. K. and H. Koçak (1991) *Dynamics and Bifurcations*: Springer-Verlag, New York.

Hamill, O. P., A. Marty, E. Neher, B. Sakmann, and F. J. Sigworth (1981) Improved patch-clamp techniques for high-resolution current recording from cells and cell-free membrane patches, *Pflügers Arch*. 391: 85–100.

Hargrave, P. A., K. P. Hoffman, and U. B. Kaupp, Eds. (1992) *Signal Transduction in Photoreceptor Cells*. Springer-Verlag, Berlin.

Hastings, S. P. (1975) The existence of progressive wave solutions to the Hodgkin-Huxley equations, *Arch. Rat. Mech. Anal*. 60: 229–257.

Heineken, F. G., H. M. Tsuchiya, and R. Aris (1967) On the mathematical status of the pseudo-steady state hypothesis of biochemical kinetics, *Mathematical Biosciences*. 1: 95–113.

Henriquez, C. S. (1993) Simulating the electrical behavior of cardiac tissue using the bidomain model, *CRC Crit. Revs. Biomed. Eng.* 21: 1–77.

Hess, B. and A. Boiteux (1973) *Substrate control of glycolytic oscillations*. In: B. Chance, E. K. Pye, A. K. Ghosh, and B. Hess, Academic Press, New York.

Hill, A. V. (1938) The heat of shortening and the dynamic constants of muscle, *Proc. Roy. Soc. Lond.* B126: 136–195.

Hill, T. L. (1974) Theoretical formalism for the sliding filament model of contraction of striated muscle. Part I, *Progress in Biophysics and Molecular Biology.* 28: 267–340.

Hill, T. L. (1975) Theoretical formalism for the sliding filament model of contraction of striated muscle. Part II, *Progress in Biophysics and Molecular Biology.* 29: 105–159.

Hille, B. (1975) Ionic selectivity, saturation, and block in sodium channels, *Journal of General Physiology.* 66: 535–560.

Hille, B. and W. Schwartz (1978) Potassium channels as multi-ion single-file pores, *Journal of General Physiology.* 72: 409–442.

Hille, B. (1992) *Ionic Channels of Excitable Membranes*: Sinauer, Sunderland, MA.

Himmel, D. M. and T. R. Chay (1987) Theoretical studies on the electrical activity of pancreatic $\beta$-cells as a function of glucose, *Biophysical Journal.* 51: 89–107.

Hindmarsh, J. L. and R. M. Rose (1982) A model of the nerve impulse using two first order differential equations, *Nature.* 296: 162–164.

Hindmarsh, J. L. and R. M. Rose (1984) A model of neuronal bursting using three coupled first order differential equations, *Proc. R. Soc. Lond. B.* 221: 87–102.

Hirsch, M. W. and S. Smale (1974), *Differential Equations, Dynamical Systems and Linear Algebra*: Academic Press, New York.

Hirsch, M. W., C. C. Pugh, and M. Shub (1977) *Invariant Manifolds*: Springer-Verlag, New York.

Hodgkin, A. L. and W. A. H. Rushton (1946) The electrical constants of a crustacean nerve fibre, *Proc. Roy. Soc. London B.* 133: 444–479.

Hodgkin, A. L. and B. Katz (1949) The effect of sodium ions on the electrical activity of the giant axon of the squid, *Journal of Physiology.* 108: 37–77.

Hodgkin, A. L., A. F. Huxley, and B. Katz (1952) Measurement of current–voltage relations in the membrane of the giant axon of *Loligo*, *Journal of Physiology.* 116: 424–448.

Hodgkin, A. L. and A. F. Huxley (1952a) Currents carried by sodium and potassium ions through the membrane of the giant axon of *Loligo*, *Journal of Physiology.* 116: 449–472.

Hodgkin, A. L. and A. F. Huxley (1952b) The components of membrane conductance in the giant axon of *Loligo*, *Journal of Physiology.* 116: 473–496.

Hodgkin, A. L. and A. F. Huxley (1952c) The dual effect of membrane potential on sodium conductance in the giant axon of *Loligo*, *Journal of Physiology.* 116: 497–506.

Hodgkin, A. L. and A. F. Huxley (1952d) A quantitative description of membrane current and its application to conduction and excitation in nerve, *Journal of Physiology, London.* 117: 500–544.

Hodgkin, A. L. and R. D. Keynes (1955) The potassium permeability of a giant nerve fibre, *Journal of Physiology.* 128: 61–88.

Hodgkin, A. L. (1976) Chance and design in electrophysiology: an informal account of certain experiments on nerve carried out between 1934 and 1952, *Journal of Physiology.* 263: 1–21.

Hodgkin, A. L. and B. J. Nunn (1988) Control of light-sensitive current in salamander rods, *Journal of Physiology.* 403: 439–471.

Holmes, M. H. (1980a) An analysis of a low-frequency model of the cochlea, *J. Acoust. Soc. Am.* 68: 482–488.

Holmes, M. H. (1980b) Low frequency asymptotics for a hydroelastic model of the cochlea, *SIAM Journal on Applied Mathematics.* 38: 445–456.

Holmes, M. H. (1982) A mathematical model of the dynamics of the inner ear, *J. Fluid Mech.* 116: 59–75.

Holmes, M. H. (1995) *Introduction to Perturbation Methods*: Springer-Verlag, New York.

Hoppensteadt, F. C. and J. P. Keener (1982) Phase locking of biological clocks, *Journal of Mathematical Biology.* 15: 339–349.

Hoppensteadt, F. C. and C. S. Peskin (1992) *Mathematics in Medicine and the Life Sciences*: Springer-Verlag, New York.

Hudspeth, A. J. (1985) The cellular basis of hearing: the biophysics of hair cells, *Science.* 230: 745–752.

Hudspeth, A. J. and R. S. Lewis (1988a) Kinetic analysis of voltage- and ion-dependent conductances in saccular hair cells of the bull-frog, *Rana Catesbeiana*, *Journal of Physiology.* 400: 237–274.

Hudspeth, A. J. and R. S. Lewis (1988b) A model for electrical resonance and frequency tuning in saccular hair cells of the bull-frog, *Rana Catesbeiana*, *Journal of Physiology*. 400: 275–297.

Hudspeth, A. J. (1989) How the ear's works work, *Nature*. 341: 397–404.

Hudspeth, A. J. and P. G. Gillespie (1994) Pulling springs to tune transduction: adaptation by hair cells, *Neuron*. 12: 1–9.

Huntsman, L. L., E. O. Attinger, and A. Noordergraaf (1978), *Metabolic autoregulation of blood flow in skeletal muscle*. In: Cardiovascular System Dynamics, Ed: J. Baan, A. Noordergraaf, and J. Raines, MIT Press, Cambridge, MA.

Huxley, A. F. (1957) Muscle structure and theories of contraction, *Progress in Biophysics*. 7: 255–318.

Huxley, A. F. and R. M. Simmons (1971) Proposed mechanism of force generation in striated muscle, *Nature*. 233: 533–538.

Huxley, A. F. (1980) *Reflections on muscle*: Princeton University Press, Princeton, NJ.

Inselberg, A. and R. S. Chadwick (1976) Mathematical model of the cochlea. I: formulation and solution, *SIAM Journal on Applied Mathematics*. 30: 149–163.

Irving, M., J. Maylie, N. L. Sizto, and W. K. Chandler (1990) Intracellular diffusion in the presence of mobile buffers: application to proton movement in muscle, *Biophysical Journal*. 57: 717–721.

Jack, J. J. B., D. Noble, and R. W. Tsien (1975) *Electric Current Flow in Excitable Cells*: Oxford University Press, Oxford.

Jaffe, L. F. (1991) The path of calcium in cytosolic calcium oscillations: a unifying hypothesis, *Proc. Natl. Acad. Sci. USA*. 88: 9883–9887.

Jafri, M. S. and J. Keizer (1995) On the roles of $Ca^{2+}$ diffusion, $Ca^{2+}$ buffers and the endoplasmic reticulum in $IP_3$-induced $Ca^{2+}$ waves, *Biophysical Journal*. 69: 2139–2153.

Jafri, M. S. (1995) A theoretical study of cytosolic calcium waves in *Xenopus* oocytes, *Journal of Theoretical Biology*. 172: 209–216.

Jafri, M. S. and J. Keizer (1997) Agonist-induced calcium waves in oscillatory cells: a biological example of Burger's equation, *Bulletin of Mathematical Biology*. 59: 1125–1144.

Jahnke, W., C. Henze, and A. T. Winfree (1988) Chemical vortex dynamics in three-dimensional excitable media, *Nature*. 336: 662–665.

Jahnke, W. and A. T. Winfree (1991) A survey of spiral-wave behaviors in the Oregonator model, *Int. J. Bif. Chaos*. 1: 445–466.

Jakobsson, E. (1980) Interactions of cell volume, membrane potential, and membrane transport parameters, *American Journal of Physiology (Cell Physiology)*. 238: C196–C206.

Jesty, J., E. Beltrami, and G. Willems (1993) Mathematical analysis of a proteolytic positive-feedback loop: dependence of lag time and enzyme yields on the initial conditions and kinetic parameters, *Biochemistry*. 32: 6266–6274.

Jewell, B. R. and D. R. Wilkie (1958) An analysis of the mechanical components in frog's striated muscle, *Journal of Physiology*. 143: 515–540.

Johnston, D. and S. M.-S. Wu (1995) *Foundations of Cellular Neurophysiology*: The MIT Press, Cambridge, MA.

Jones, C. K. R. T. (1984) Stability of the traveling wave solutions of the FitzHugh–Nagumo system, *Trans. Amer. Math. Soc*. 286: 431–469.

Jones, K. C. and K. G. Mann (1994) A model for the tissue factor pathway to thrombin. II. A mathematical simulation, *Journal of Biological Chemistry*. 269: 23367–23373.

Julian, F. J. (1969) Activation in a skeletal muscle contraction model with a modification for insect fibrillar muscle, *Biophysical Journal*. 9: 547–570.

Kaplan, W. (1981) *Advanced Engineering Mathematics*: Addison-Wesley, Reading, MA.

Kargacin, G. J. (1994) Calcium signaling in restricted diffusion spaces, *Biophysical Journal*. 67: 262–272.

Karma, A. (1993) Spiral breakup in model equations of action potential propagation in cardiac tissue, *Phys. Rev. Lett*. 71: 1103–1106.

Karma, A. (1994) Electrical alternans and spiral wave breakup in cardiac tissue, *Chaos*. 4: 461–472.

Katz, B. and R. Miledi (1968) The role of calcium in neuromuscular facilitation, *Journal of Physiology*. 195: 481–492.

Keener, J. P. (1980a) Waves in excitable media, *SIAM Journal on Applied Mathematics*. 39: 528–548.

Keener, J. P. (1980b) Chaotic behavior in piecewise continuous difference equations, *Trans. AMS*. 261: 589–604.

Keener, J. P., F. C. Hoppensteadt, and J. Rinzel (1981) Integrate and fire models of nerve membrane response to oscillatory input, *SIAM Journal on Applied Mathematics*. 41: 503–517.

Keener, J. P. (1981) On cardiac arrhythmias: AV conduction block, *Journal of Mathematical Biology*. 12: 215–225.

Keener, J. P. (1983) Analog circuitry for the van der Pol and FitzHugh–Nagumo equation, *IEEE Trans. Sys. Man. Cybernetics*. SMC-13: 1010–1014.

Keener, J. P. and L. Glass (1984) Global bifurcations of a periodically forced oscillator, *Journal of Mathematical Biology*. 21: 175–190.

Keener, J. P. (1986) A geometrical theory for spiral waves in excitable media, *SIAM Journal on Applied Mathematics*. 46: 1039–1056.

Keener, J. P. and J. J. Tyson (1986) Spiral waves in the Belousov–Zhabotinsky reaction, *Physica D*. 21: 307–324.

Keener, J. P. (1987) Propagation and its failure in coupled systems of discrete excitable cells, *SIAM Journal on Applied Mathematics*. 47: 556–572.

Keener, J. P. (1988) *Principles of Applied Mathematics*: Addison-Wesley, Reading, Massachusetts.

Keener, J. P. (1988b) The dynamics of three dimensional scroll waves in excitable media, *Physica D*. 31: 269–276.

Keener, J. P. (1991a) An eikonal-curvature equation for action potential propagation in myocardium, *Journal of Mathematical Biology*. 29: 629–651.

Keener, J. P. (1991b) The effects of discrete gap junctional coupling on propagation in myocardium, *Journal of Theoretical Biology*. 148: 49–82.

Keener, J. P. (1992) The core of the spiral, *SIAM Journal on Applied Mathematics*. 52: 1372–1390.

Keener, J. P. and J. J. Tyson (1992) The dynamics of scroll waves in excitable media, *SIAM Review*. 34: 1–39.

Keener, J. P. (1994) Symmetric spirals in media with relaxation kinetics and two diffusing species, *Physica D*. 70: 61–73.

Keener, J. P. and A. V. Panfilov (1995) *Three-dimensional propagation in the heart: the effects of geometry and fiber orientation on propagation in myocardium*. In: Cardiac Electrophysiology. From Cell to Bedside, Ed: D. P. Zipes and J. Jalife, Saunders, Philadelphia, PA.

Keener, J. P. and A. V. Panfilov (1996) A biophysical model for defibrillation of cardiac tissue, *Biophysical Journal*. 71: 1335–1345.

Keener, J. P. and A. V. Panfilov (1997) *The effects of geometry and fibre orientation on propagation and extracellular potentials in myocardium*. In: Computational Biology of the Heart, Ed: A. V. Panfilov and A. V. Holden, John Wiley and Sons, New York.

Keizer, J. and G. Magnus (1989) ATP-sensitive potassium channel and bursting in the pancreatic beta cell, *Biophysical Journal*. 56: 229–242.

Keizer, J. and P. Smolen (1991) Bursting electrical activity in pancreatic $\beta$-cells caused by $Ca^{2+}$ and voltage-inactivated $Ca^{2+}$ channels, *Proc. Natl. Acad. Sci. USA*. 88: 3897–3901.

Keizer, J. and G. DeYoung (1994) Simplification of a realistic model of $IP_3$-induced $Ca^{2+}$ oscillations, *Journal of Theoretical Biology*. 166: 431–442.

Keizer, J., G. D. Smith, S. Ponce-Dawson and J. E. Pearson (1998) Saltatory propagation of $Ca^{2+}$ waves by $Ca^{2+}$ sparks, *Biophysical Journal*. : in press.

Keizer, J. and L. Levine (1996) Ryanodine receptor adaptation and $Ca^{2+}$-induced $Ca^{2+}$ release-dependent $Ca^{2+}$ oscillations, *Biophysical Journal*. 71: 3477–3487.

Keller, E. F. and L. A. Segel (1971) Models for chemotaxis, *Journal of Theoretical Biology*. 30: 225–234.

Kessler, D. A. and R. Kupferman (1996) Spirals in excitable media: the free-boundary limit with diffusion, *Physica D*. 97: 509–516.

Kevorkian, J. and J. D. Cole (1996) *Multiple Scale and Singular Perturbation Methods*: Springer-Verlag, New York, Berlin, Heidelberg.

Knepper, M. A. and F. C. Rector, Jr. (1991), *Urinary concentration and dilution*. In: The Kidney (fourth edition) Volume 1, Ed: B. M. Brenner and F. C. Rector, Jr., Saunders, Philadelphia, PA.

Knight, B. W. (1972) Dynamics of encoding a population of neurons, *Journal of General Physiology*. 59: 734–766.

Knobil, E. (1981) Patterns of hormonal signals and hormone action, *New England Journal of Medicine*. 305: 1582–1583.

Knorre, W. A. (1968) Oscillations of the rate of synthesis of $\beta$-galactosidase in *Escherichia coli* ML30 and ML308, *Biochem. Biophys. Res. Com*. 31: 812–817.

Knox, B. E., P. N. Devreotes, A. Goldbeter, and L. A. Segel (1986) A molecular mechanism for sensory adaptation based on ligand-induced receptor modification, *Proc. Natl. Acad. Sci. USA*. 83: 2345–2349.

Koch, K.-W. and L. Stryer (1988) Highly cooperative feedback control of retinal rod guanylate cyclase by calcium ions, *Nature*. 334: 64–66.

Koch, C. and I. Segev, Eds. (1989) *Methods in Neuronal Modeling*. MIT Press, Cambridge, MA.

Koefoed-Johnsen, V. and H. H. Ussing (1958) The nature of the frog skin potential, *Acta Physiologica Scandinavica*. 42: 298–308.

Kohler, H.-H. and K. Heckman (1979) Unidirectional fluxes in saturated single-file pores of biological and artificial membranes I: pores containing no more than one vacancy, *Journal of Theoretical Biology*. 79: 381–401.

Kopell, N. and L. N. Howard (1973) Plane wave solutions to reaction-diffusion equations, *Studies in Applied Mathematics*. 52: 291–328.

Kopell, N. and G. B. Ermentrout (1986) Subcellular oscillations and bursting, *Mathematical Biosciences*. 78: 265–291.

Kramers, H. A. (1940) Brownian motion in a field of force and the diffusion model of chemical reactions, *Physica*. 7: 284–304.

Krausz, H. I. and K.-I. Naka (1980) Spatiotemporal testing and modeling of catfish retinal neurons, *Biophysical Journal*. 29: 13–36.

Kreyszig, E. (1994) *Advanced Engineering Mathematics*. *(Seventh edition)*: John Wiley and Sons, New York.

Kuba, K. and S. Takeshita (1981) Simulation of intracellular $Ca^{2+}$ oscillation in a sympathetic neurone, *Journal of Theoretical Biology*. 93: 1009–1031.

Kuffler, S. W. (1953) Discharge patterns and functional organization of the mammalian retina, *J. Neurophysiol*. 16: 37–68.

Kuffler, S. W. (1973) The single-cell approach in the visual system and the study of receptive fields, *Invest. Ophthalmol*. 12: 794–813.

Kuffler, S. W., J. G. Nicholls, and R. Martin (1984) *From Neuron to Brain*: Sinaeur Associates, Sunderland, MA.

Kuramoto, Y. and T. Tsuzuki (1976) Persistent propagation of concentration waves in dissipative media far from thermal equilibrium, *Prog. Theor. Phys.*. 55: 356.

Kuramoto, Y. and T. Yamada (1976) Pattern formation in oscillatory chemical reactions, *Prog. Theor. Phys*. 56: 724.

Lacker, H. M. (1981) Regulation of ovulation number in mammals: a follicle interaction law that controls maturation, *Biophysical Journal*. 35: 433–454.

Lacker, H. M. and C. S. Peskin (1981), *Control of ovulation number in a model of ovarian follicular maturation*. In: Lectures on Mathematics in the Life Sciences, Ed: S.

Childress, American Mathematical Society, Providence.

Lacker, H. M. and C. S. Peskin (1986) A mathematical method for unique determination of cross-bridge properties from steady-state mechanical and energetic experiments on macroscopic muscle, *Lectures on Mathematics in the Life Sciences*. 16: 121–153.

Lacy, A. H. (1967) The unit of insulin, *Diabetes*. 16: 198–200.

Laidler, K. J. (1969) *Theories of Chemical Reaction Rates*: McGraw-Hill, New York.

Lamb, T. D. and E. J. Simon (1977) Analysis of electrical noise in turtle cones, *Journal of Physiology*. 272: 435–468.

Lamb, T. D. and E. N. Pugh (1992) A quantitative account of the activation steps involved in phototransduction in amphibian photoreceptors, *Journal of Physiology*. 449: 719–758.

Landy, M. S. and J. A. Movshon, Eds. (1991) *Computational Models of Visual Processing*. The MIT Press, Cambridge, MA.

Lane, D. C., J. D. Murray and V. S. Manoranjan (1987) Analysis of wave phenomena in a morphogenetic mechanochemical model and an application to post-fertilisation waves on eggs, *IMA J. Math. Applied in Medic. and Biol.*. 4: 309–331.

Langer, G. A. and A. Peskoff (1996) Calcium concentration and movement in the diadic left space of the cardiac ventricular cell, *Biophysical Journal*. 70: 1169–1182.

Layton, H. E., E. B. Pitman, and L. C. Moore (1991) Bifurcation analysis of TGF-mediated oscillations in SNGFR, *American Journal of Physiology*. 261: F904-F919.

Layton, H. E., E. B. Pitman, and M. A. Knepper (1995) A dynamic numerical method for models of the urine concentrating mechanism, *SIAM Journal on Applied Mathematics*. 55: 1390–1418.

Lechleiter, J., S. Girard, D. Clapham, and E. Peralta (1991a) Subcellular patterns of calcium release determined by G protein-specific residues of muscarinic receptors, *Nature*. 350: 505–508.

Lechleiter, J., S. Girard, E. Peralta, and D. Clapham (1991b) Spiral calcium wave propagation and annihilation in *Xenopus laevis* oocytes, *Science*. 252: 123–126.

Lechleiter, J. and D. Clapham (1992) Molecular mechanisms of intracellular calcium excitability in *X. laevis* oocytes, *Cell.* 69: 283–294.

Lesser, M. B. and D. A. Berkley (1972) Fluid mechanics of the cochlea. Part I., *J. Fluid Mech.* 51: 497–512.

Levine, I. N. (1978) *Physical Chemistry*: McGraw-Hill Kogokusha Ltd., Tokyo.

Lew, V. L., H. G. Ferreira, and T. Moura (1979) The behaviour of transporting epithelial cells. I. Computer analysis of a basic model, *Proc. Roy. Soc.Lond. B.* 206: 53–83.

Li, Y.-X. and A. Goldbeter (1989) Frequency specificity in intercellular communication: influence of patterns of periodic signaling on target cell responsiveness, *Biophysical Journal.* 55: 125–145.

Li, Y.-X. and J. Rinzel (1994) Equations for InsP$_3$ receptor-mediated [Ca$^{2+}$] oscillations derived from a detailed kinetic model: a Hodgkin-Huxley-like formalism, *Journal of Theoretical Biology.* 166: 461–473.

Lighthill, J. (1975) *Mathematical Biofluiddynamics*: SIAM, Philadelphia, PA.

Lin, C. C. and L. A. Segel (1988) *Mathematics Applied to Deterministic Problems in the Natural Sciences*: SIAM, Philadelphia, PA.

Liu, B.-Z. and G.-M. Deng (1991) An improved mathematical model of hormone secretion in the hypothalamo-pituitary-gonadal axis in man, *Journal of Theoretical Biology.* 150: 51–58.

Llinás, R., I. Z. Steinberg, and K. Walton (1976) Presynaptic calcium currents and their relation to synaptic transmission: voltage clamp study in squid giant synapse and theoretical model for the calcium gate, *Proc. Natl. Acad. Sci. USA.* 73: 2918–2922.

Loeb, J. N. and S. Strickland (1987) Hormone binding and coupled response relationships in systems dependent on the generation of secondary mediators, *Molecular Endocrinology.* 1: 75–82.

Longtin, A. and J. G. Milton (1989) Modelling autonomous oscillations in the human pupil light reflex using non-linear delay-differential equations, *Bulletin of Mathematical Biology.* 51: 605–624.

Lugosi, E. and A. T. Winfree (1988) Simulation of wave propagation in three dimensions using Fortran on the Cyber 205, *J. Comput. Chem.* 9: 689–701.

Luo, C. H. and Y. Rudy (1991) A model of the ventricular cardiac action potential; depolarization, repolarization and their interaction, *Circ. Res.* 68: 1501–1526.

Luo, C. H. and Y. Rudy (1994a) A dynamic model of the cardiac ventricular action potential; I: Simulations of ionic currents and concentration changes, *Circ. Res.* 74: 1071–1096.

Luo, C. H. and Y. Rudy (1994b) A dynamic model of the cardiac ventricular action potential; II: Afterdepolarizations, triggered activity and potentiation, *Circ. Res.* 74: 1097–1113.

Lytton, J., M. Westlin, S. E. Burk, G. E. Shull, and D. H. MacLennan (1992) Functional comparisons between isoforms of the sarcoplasmic or endoplasmic reticulum family of calcium pumps, *Journal of Biological Chemistry.* 267: 14483–14489.

Läuger, P. (1973) Ion transport through pores: a rate-theory analysis, *Biochimica et Biophysica Acta.* 311: 423–441.

Macknight, A. D. C. (1988) Principles of cell volume regulation, *Renal Physiol. Biochem..* 3–5: 114–141.

Maginu, K. (1985) Geometrical characteristics associated with stability and bifurcations of periodic travelling waves in reaction-diffusion equations, *SIAM Journal on Applied Mathematics.* 45: 750–774.

Magleby, K. L. and C. F. Stevens (1972) A quantitative description of end-plate currents, *Journal of Physiology.* 223: 173–197.

Maki, L. W. and J. Keizer (1995a) Analysis of possible mechanisms for *in vitro* oscillations of insulin secretion, *American Journal of Physiology (Cell Physiol.).* 268: C780–C791.

Maki, L. W. and J. Keizer (1995b) Mathematical analysis of a proposed mechanism for oscillatory insulin secretion in perifused HIT-15 cells, *Bulletin of Mathematical Biology.* 57: 569–591.

Marland, E. (1998) Dynamics of the sarcomere, *PhD thesis, Univ. of Utah.* Salt Lake City, UT. : .

McAllister, R. E., D. Noble, and R. W. Tsien (1975) Reconstruction of the electrical activity of cardiac Purkinje fibres, *Journal of Physiology.* 251: 1–59.

McDonald, D. A. (1974) *Blood Flow in Arteries*: Arnold, London.

McKean, H. P. (1970) Nagumo's equation, *Advances in Mathematics.* 4: 209–223.

McKenzie, A. and J. Sneyd (1998) On the formation and breakup of spiral waves of calcium, *Int. J. Bif. and Chaos.* : in press.

McNaughton, P. A. (1990) Light response of vertebrate photoreceptors, *Physiological Reviews.* 70: 847–883.

McQuarrie, D. A. (1967) *Stochastic Approach to Chemical Kinetics*: Methuen and Co., London.

Michaelis, L. and M. I. Menten (1913) Die Kinetik der Invertinwirkung, *Biochem. Z.* 49: 333–369.

Miller, R. N. and J. Rinzel (1981) The dependence of impulse propagation speed on firing frequency, dispersion, for the Hodgkin–Huxley model, *Biophysical Journal.* 34: 227–259.

Milton, J. G. and M. C. Mackey (1989) Periodic haemotological diseases: mystical entities or dynamical disorders?, *J. Roy. Coll. Physicians (London).* 23: 236–241.

Mines, G. R. (1914) On circulating excitations in heart muscle and their possible relation to tachycardia and fibrillation, *Trans. Roy. Soc. Can.* 4: 43–53.

Minorsky, N. (1962) *Nonlinear Oscillations*: Van Nostrand, New York

Miura, R. M. (1981), *Nonlinear waves in neuronal cortical structures.* In: Nonlinear Phenomena in Physics and Biology, Ed: R. H. Enns, B. L. Jones, R. M. Miura, and S. S. Rangnekar, Plenum Press, New York.

Moe, G. K., W. C. Rheinbolt, and J. A. Abildskov (1964) A computer model of atrial fibrillation, *Am. Heart J.* 67: 200–220.

Monod, J., J. Wyman, and J. P. Changeux (1965) On the nature of allosteric transition: A plausible model, *Journal of Molecular Biology.* 12: 88–118.

Morris, C. and H. Lecar (1981) Voltage oscillations in the barnacle giant muscle fiber, *Biophysical Journal.* 35: 193–213.

Mountcastle, V. B. Ed. (1974) Medical Physiology, C. V. Mosby, St. Louis.

Murray, J. D. (1984) *Asymptotic Analysis*: Springer-Verlag, New York.

Murray, J. D. (1989) *Mathematical Biology*: Springer-Verlag, New York.

Nagumo, J., S. Arimoto, and S. Yoshizawa (1964) An active pulse transmission line simulating nerve axon, *Proc. IRE.* 50: 2061–2070.

Nelsen, T. S. and J. C. Becker (1968) Simulation of the electrical and mechanical gradient of the small intestine, *American Journal of Physiology.* 214: 749–757.

Nesheim, M. E., R. P. Tracy, and K. G. Mann (1984) "Clotspeed," a mathematical simulation of the functional properties of prothrombinase, *Journal of Biological Chemistry.* 259: 1447–1453.

Nesheim, M. E., R. P. Tracy, P. B. Tracy, D. S. Boskovic, and K. G. Mann (1992) Mathematical simulation of prothrombinase, *Methods Enzymol.* 215: 316–328.

Neu, J. C. (1979) Chemical waves and the diffusive coupling of limit cycle oscillators, *SIAM Journal on Applied Mathematics.* 36: 509–515.

Neu, J. C. and W. Krassowska (1993) Homogenization of syncytial tissues, *Crit. Rev. Biomed. Eng.* 21: 137–199.

Nicholls, J. G., A. R. Martin, and B. G. Wallace (1992) *From Neuron to Brain*: Sinauer Associates, Inc., Sunderland, MA.

Nielsen, P. M. F., I. J. LeGrice, and B. H. Smaill (1991) A mathematical model of geometry and fibrous structure of the heart, *American Journal of Physiology.* 260: H1365–H1378.

Nielsen, K., P. G. Sørensen, and F. Hynne (1997) Chaos in glycolysis, *Journal of Theoretical Biology.* 186: 303–306.

Noble, D. (1962) A modification of the Hodgkin–Huxley equations applicable to Purkinje fiber action and pacemaker potential, *Journal of Physiology.* 160: 317–352.

Noble, D. and S. J. Noble (1984) A model of S.A. node electrical activity using a modification of the DiFrancesco-Noble (1984) equations, *Proc. Roy. Soc. Lond.* 222: 295–304.

Norman, R. A. and I. Perlman (1979) The effects of background illumination on the photoresponses of red and green cones, *Journal of Physiology.* 286: 491–507.

Novak, B. and J. J. Tyson (1993) Numerical analysis of a comprehensive model of M-phase control in *Xenopus* oocyte extracts and intact embryos, *J. Cell Science.* 106: 1153–1168.

Nowycky, M. C. and M. J. Pinter (1993) Time courses of calcium and calcium-bound buffers following calcium influx in a model cell, *Biophysical Journal.* 64: 77–91.

O'Neill, P. V. (1983) *Advanced Engineering Mathematics*: Wadsworth, Belmont, CA.

Ohta, T., M. Mimura, and R. Kobayashi (1989) Higher dimensional localized patterns in excitable media, *Physica D.* 34: 115–144.

Orr, J. S., J. Kirk, K. G. Gray, and J. R. Anderson (1968) A study of the interdependence of red blood cell and bone marrow stem cell populations, *Brit. J. Haematol.* 15: 23–34.

Ortoleva, P. and J. Ross (1973) Phase waves in oscillatory chemical reactions, *J. Chem. Phys.* 58: 5673–5680.

Ortoleva, P. and J. Ross (1974) On a variety of wave phenomena in chemical reactions, *J. Chem. Phys.* 60: 5090–5107.

Osher, S. and J. A. Sethian (1988) Fronts propagating with curvature-dependent speed: algorithms based on Hamilton–Jacobi formulations, *J. Comp. Phys.* 79: 12–49.

Pace, N., E. Strajman, and E. Walker (1950) Acceleration of carbon monoxide elimination in man by high pressure oxygen, *Science*. 111: 652.

Panfilov, A. V. and A. V. Holden (1990) Self-generation of turbulent vortices in a two-dimensional model of cardiac tissue, *Phys. Lett. A*. 147: 463–466.

Panfilov, A. V. and P. Hogeweg (1995) Spiral break-up in a modified FitzHugh–Nagumo model, *Phys. Lett. A*. 176: 295–299.

Panfilov, A. V. and J. P. Keener (1995) Re-entry in an anatomical model of the heart, *Chaos, Solitons, and Fractals*. 5: 681–689.

Papoulis, A. (1962) *The Fourier Integral and its Applications*: McGraw-Hill, New York.

Parker, I. and I. Ivorra (1990) Inhibition by $Ca^{2+}$ of inositol trisphosphate-mediated $Ca^{2+}$ liberation: a possible mechanism for oscillatory release of $Ca^{2+}$, *Proc. Natl. Acad. Sci. USA*. 87: 260–264.

Parker, I., Y. Yao, and V. Ilyin (1996) Fast kinetics of calcium liberation induced in *Xenopus* oocytes by photoreleased inositol trisphosphate, *Biophysical Journal*. 70: 222–237.

Parnas, H. and L. A. Segel (1980) A theoretical explanation for some effects of calcium on the facilitation of neurotransmitter release, *Journal of Theoretical Biology*. 84: 3–29.

Parnas, H., G. Hovav, and I. Parnas (1989) Effect of $Ca^{2+}$ diffusion on the time course of neurotransmitter release, *Biophysical Journal*. 55: 859–874.

Parys, J. B., S. W. Sernett, S. DeLisle, P. M. Snyder, M. J. Welsh, and K. P. Campbell (1992) Isolation, characterization, and localization of the inositol 1,4,5-trisphosphate receptor protein in *Xenopus laevis* oocytes, *Journal of Biological Chemistry*. 267: 18776–18782.

Pate, E. (1997) *Mathematical modeling of muscle crossbridge mechanics*. In: Case Studies in Mathematical Biology, Ed: H. Othmer, F. Adler, M. Lewis, and J. Dallon, Prentice Hall, Upper Saddle River, NJ.

Patton, R. J. and D. A. Linkens (1978) Hodgkin–Huxley type electronic modelling of gastrointestinal electrical activity, *Med. Biol. Engrg. Computing*. 16: 195–202.

Pauwelussen, J. P. (1981) Nerve impulse propagation in a branching nerve system: a simple model, *Physica D*. 4: 67–88.

Pedley, T. J. (1980) *The Fluid Mechanics of Large Blood Vessels*: Cambridge University Press, Cambridge.

Pelce, P. and J. Sun (1991) Wave front interaction in steadily rotating spirals, *Physica D*. 48: 353–366.

Pernarowski, M., R. M. Miura, and J. Kevorkian (1991) *The Sherman-Rinzel-Keizer model for bursting electrical activity in the pancreatic β-cell*. In: Differential Equations Models in Biology, Epidemiology and Ecology, Ed: S. Busenberg and M. Martelli, Springer-Verlag, New York.

Pernarowski, M., R. M. Miura, and J. Kevorkian (1992) Perturbation techniques for models of bursting electrical activity in pancreatic β-cells, *SIAM Journal on Applied Mathematics*. 52: 1627–1650.

Pernarowski, M. (1994) Fast subsystem bifurcations in a slowly varying Liénard system exhibiting bursting, *SIAM Journal on Applied Mathematics*. 54: 814–832.

Peskin, C. S. (1975) *Mathematical Aspects of Heart Physiology*: Courant Institute of Mathematical Sciences Lecture Notes, New York.

Peskin, C. S. (1976) *Partial Differential Equations in Biology*: Courant Institute of Mathematical Sciences Lecture Notes, New York.

Peskin, C. S. (1981) Lectures on mathematical aspects of physiology, *AMS Lectures in Applied Mathematics*. 19: 38–69.

Peskin, C. S. (1991) *Mathematical Aspects of Neurophysiology*: Courant Institute of Mathematical Sciences Lecture Notes, New York.

Peskoff, A., J. A. Post, and G. A. Langer (1992) Sarcolemmal calcium binding sites in heart: II. Mathematical model for diffusion of calcium released from the sarcoplasmic reticulum into the diadic region, *Journal of Membrane Biology*. 129: 59–69.

Peskoff A. and G. A. Langer (1998) Calcium concentration and movement in the ventricular cardiac cell during an excitation–contraction cycle, *Biophysical Journal*. 74: 153–174.

Peterson, L. C. and B. P. Bogert (1950) A dynamical theory of the cochlea, *J. Acoust. Soc. Am*. 22: 369–381.

Pickles, J. O. (1982) *An Introduction to the Physiology of Hearing*: Academic Press, London.

Plant, R. E. (1981) Bifurcation and resonance in a model for bursting nerve cells, *Journal of Mathematical Biology*. 11: 15–32.

Podolsky, R. J., A. C. Nolan, and S. A. Zaveler (1969) Cross-bridge properties derived from muscle isotonic velocity transients, *Proc. Natl. Acad. Sci. USA*. 64: 504–511.

Podolsky, R. J. and A. C. Nolan (1972) *Cross-bridge properties derived from physiological studies of frog muscle fibres*. In: Contractility of Muscle Cells and Related Processes, Ed: R. J. Podolsky, Prentice Hall, Englewood Cliffs, NJ.

Podolsky, R. J. and A. C. Nolan (1973) *Muscle contraction transients, cross-bridge kinetics and the Fenn effect*: 37th Cold Spring Harbor Symposium of Quantitative Biology, Cold Spring Harbor, NY.

Pugh, E. N. and T. D. Lamb (1990) Cyclic GMP and calcium: messengers of excitation and adaptation in vertebrate photoreceptors, *Vision Research*. 30: 1923–1948.

Rall, W. (1957) Membrane time constant of motoneurons, *Science*. 126: 454.

Rall, W. (1959) Branching dendritic trees and motoneuron membrane resistivity, *Expt. Neurology*. 2: 491–527.

Rall, W. (1960) Membrane potential transients and membrane time constant of motoneurons, *Expt. Neurology*. 2: 503–532.

Rall, W. (1969) Time constants and electrotonic length of membrane cylinders and neurons, *Biophysical Journal*. 9: 1483–1508.

Rall, W. (1977) *Core conductor theory and cable properties of neurons*. In: Handbook of Physiology. The Nervous System I, Ed: J. M. Brookhart and V. B. Mountcastle, American Physiological Society, Bethesda, MD.

Ramanan, S. V. and P. R. Brink (1990) Exact solution of a model of diffusion in an infinite chain or monolayer of cells coupled by gap junctions, *Biophysical Journal*. 58: 631–639.

Rand, R. H. and P. J. Holmes (1980) Bifurcation of periodic motions in two weakly coupled van der Pol oscillators, *J. Non-linear Mechanics*. 15: 387–399.

Ranke, O. F. (1950) Theory of operation of the cochlea: A contribution to the hydrodynamics of the cochlea, *J. Acoust. Soc. Am*. 22: 772–777.

Rauch, J. and J. Smoller (1978) Qualitative theory of the FitzHugh–Nagumo equations, *Advances in Mathematics*. 27: 12–44.

Reeve, E. B. and A. C. Guyton, Eds. (1967) *Physical Bases of Circulatory Transport: Regulation and Exchange*. W.B. Saunders, Philadelphia, PA.

Reuss, L. (1988) Cell volume regulation in nonrenal epithelia, *Renal Physiol. Biochem*. 3–5: 187–201.

Rhode, W. S. (1984) Cochlear mechanics, *Ann. Rev. Physiol*. 46: 231–246.

Rinzel, J. and J. B. Keller (1973) Traveling wave solutions of a nerve conduction equation, *Biophysical Journal*. 13: 1313–1337.

Rinzel, J. (1978) On repetitive activity in nerve, *Federation Proc*. 37: 2793–2802.

Rinzel, J. and J. P. Keener (1983) Hopf bifurcation to repetitive activity in nerve, *SIAM Journal on Applied Mathematics*. 43: 907–922.

Rinzel, J. and K. Maginu (1984) *Kinematic analysis of wave pattern formation in excitable media*. In: Non-equilibrium Dynamics in Chemical Systems, Ed: A. Pacault and C. Vidal, Springer-Verlag, Berlin.

Rinzel, J. (1985) *Bursting oscillations in an excitable membrane model*. In: Ordinary and Partial Differential Equations, Ed: B. D. Sleeman and R. J. Jarvis, Springer-Verlag, New York.

Rinzel, J. and Y. S. Lee (1986) *On different mechanisms for membrane potential bursting*. In: Nonlinear Oscillations in Biology and Chemistry. Lecture Notes in Biomathematics, Vol. 66, Ed: H. G. Othmer, Springer-Verlag, New York.

Rinzel, J. (1987) *A formal classification of bursting mechanisms in excitable systems*. In: Mathematical Topics in Population Biology, Morphogenesis, and Neurosciences. Lecture Notes in Biomathematics, Vol. 71, Ed: E. Teramoto and M. Yamaguti, Springer-Verlag, Berlin.

Rinzel, J. and Y. S. Lee (1987) Dissection of a model for neuronal parabolic bursting, *Journal of Mathematical Biology*. 25: 653–675.

Rinzel, J. (1990) Electrical excitability of cells, theory and experiment: review of the Hodgkin–Huxley foundation and an update, *Bulletin of Mathematical Biology*. 52: 5–23.

Robb-Gaspers, L. D. and A. P. Thomas (1995) Coordination of $Ca^{2+}$ signaling by intercellular propagation of $Ca^{2+}$ waves in the intact liver, *Journal of Biological Chemistry*. 270: 8102–8107.

Robello, M., M. Fresia, L. Maga, A. Grasso, and S. Ciani (1987) Permeation of divalent cations through $\alpha$-Latrotoxin channels in lipid bilayers: steady-state current-voltage relationships, *Journal of Membrane Biology*. 95: 55–62.

Roberts, D. and A. M. Scher (1982) Effect of tissue anisotropy on extracellular potential fields in canine myocardium *in situ*, *Circ. Res*. 50: 342.

Robertson-Dunn, B. and D. A. Linkens (1974) A mathematical model of the slow-wave electrical activity of the human small intestine, *Med. Biol. Engrg*. 12: 750–757.

Rodieck, R. W. (1965) Quantitative analysis of cat retinal ganglion cell response to visual stimuli, *Vision Research*. 5: 583–601.

Rooney, T. A. and A. P. Thomas (1993) Intracellular calcium waves generated by Ins(1,4,5)P$_3$-dependent mechanisms, *Cell Calcium.* 14: 674–690.

Rorsman, P. and G. Trube (1986) Calcium and delayed potassium currents in mouse pancreatic $\beta$-cells under voltage clamp conditions, *Journal of Physiology.* 375: 531–550.

Roughton, F. J. W., E. C. DeLand, J. C. Kernohan, and J. W. Severinghaus (1972), *Some recent studies of the oxyhaemoglobin dissociation curve of human blood under physiological conditions and the fitting of the Adair equation to the standard curve.* In: Oxygen Affinity of Hemoglobin and Red Cell Acid Base States, Ed: M. Rorth and P. Astrup, Academic Press, New York.

Roy, D. R., H. E. Layton, and R. L. Jamison (1992), *Countercurrent mechanism and its regulation.* In: The Kidney: Physiology and Pathophysiology, Ed: D. W. Seldin and G. Giebisch, Raven Press, New York.

Rubinow, S. I. (1973) *Mathematical Problems in the Biological Sciences*: SIAM, Philadelphia, PA

Rubinow, S. I. (1975) *Introduction to Mathematical Biology*: John Wiley and Sons, New York.

Rushmer, R. F. (1976) *Structure and Function of the Cardiovascular System (second edition)*: W.B. Saunders Co., Philadelphia, PA.

Sabah, N. H. and R. A. Spangler (1970) Repetitive response of the Hodgkin–Huxley model for the squid giant axon, *Journal of Theoretical Biology.* 29: 155–171.

Sagawa, K., H. Suga, and K. Nakayama (1978) *Instantaneous pressure–volume ratio of the left ventricle versus instantaneous force–length relation of papillary muscle.* In: Cardiovascular System Dynamics, Ed: J. Baan, A. Noordergraaf, and J. Raines, M.I.T. Press, Cambridge, MA.

Sagawa, K., R. K. Lie, and J. Schaefer (1990) Translation of Otto Frank's Paper "Die Grundform des Arteriellen Pulses," *Zeitschrift für Biologie*, 37:483–526 (1899), *J. Mol. Cell. Cardiol.* 22: 253–277.

Sakmann, N. and E. Neher, Eds. (1983) *Single Channel Recording.* Plenum Press, New York.

Sala, F. and A. Hernández-Cruz (1990) Calcium diffusion modeling in a spherical neuron: relevance of buffering properties, *Biophysical Journal.* 57: 313–324.

Sanderson, M. J., A. C. Charles, and E. R. Dirksen (1990) Mechanical stimulation and intercellular communication increases intracellular Ca$^{2+}$ in epithelial cells, *Cell Regulation.* 1: 585–596.

Sanderson, M. J., A. C. Charles, S. Boitano, and E. R. Dirksen (1994) Mechanisms and function of intercellular calcium signaling, *Molecular and Cellular Endocrinology.* 98: 173–187.

Sarna, S. K., E. E. Daniel, and Y. J. Kingma (1971) Simulation of the slow wave electrical activity of small intestine, *American Journal of Physiology.* 221: 166–175.

Sarna, S. K. (1989) *In vivo myoelectric activity: methods, analysis and interpretation.* In: Handbook of Physiology. Section 6: The Gastrointestinal System, Ed: S. G. Schultz, J. D. Wood, and B. B. Rauner, American Physiological Society, Bethesda, Maryland.

Schultz, S. G. (1981) Homocellular regulatory mechanisms in sodium-transporting epithelia: avoidance of extinction by "flush-through", *American Journal of Physiology (Renal Fluid and Electrolyte Physiology).* 241: F579–F590.

Schumaker, M. F. and R. MacKinnon (1990) A simple model for multi-ion permeation, *Biophysical Journal.* 58: 975–984.

Schwartz, N. B. (1969) A model for the regulation of ovulation in the rat, *Recent Progress in Hormone Research.* 25: 1–53.

Segel, L. A. (1977) *Mathematics Applied to Continuum Mechanics*: Macmillan, New York.

Segel, L. A., A. Goldbeter, P. N. Devreotes, and B. E. Knox (1986) A mechanism for exact sensory adaptation based on receptor modification, *Journal of Theoretical Biology.* 120: 151–179.

Segel, L. A. (1988) On the validity of the steady state assumption of enzyme kinetics, *Bulletin of Mathematical Biology.* 50: 579–593.

Segel, L. A. and M. Slemrod (1989) The quasi-steady state assumption: a case study in perturbation, *SIAM Review.* 31: 446–447.

Segel, L. A. and A. S. Perelson (1992) Plasmid copy number control: a case study of the quasi-steady state assumption, *Journal of Theoretical Biology.* 158: 481–494.

Segel, L. and A. Goldbeter (1994) Scaling in biochemical kinetics: dissection of a relaxation oscillator, *Journal of Mathematical Biology.* 32: 147–160.

Segev, I., J. Rinzel, and G. M. Shepherd (1995) *The Theoretical Foundation of Dendritic Function*: MIT Press, Cambridge, MA.

Sel'kov, E. E. (1968) Self-oscillations in glycolysis, *European J. Biochem.* 4: 79–86.

Shapley, R. M. and C. Enroth-Cugell (1984) *Visual adaptation and retinal gain controls.* In: Progress in Retinal Research, Ed: N. Osborne and G. Chader, Pergamon Press, London.

Sherman, A., J. Rinzel, and J. Keizer (1988) Emergence of organized bursting in clusters of pancreatic $\beta$-cells by channel sharing, *Biophysical Journal*. 54: 411–425.

Sherman, A. and J. Rinzel (1991) Model for synchronization of pancreatic $\beta$-cells by gap junction coupling, *Biophysical Journal*. 59: 547–559.

Sherman, A. and J. Rinzel (1992) Rhythmogenic effects of weak electrotonic coupling in neuronal models, *Proc. Natl. Acad. Sci. USA*. 89: 2471–2474.

Sherman, A. (1994) Anti-phase, asymmetric and aperiodic oscillations in excitable cells – I. coupled bursters, *Bulletin of Mathematical Biology*. 56: 811–835.

Shotkin, L. M. (1974a) A model for LH levels in the recently-castrated adult rat and its comparison with experiment, *Journal of Theoretical Biology*. 43: 1–14.

Shotkin, L. M. (1974b) A model for the effect of daily injections of gonadal hormones on LH levels in recently-castrated adult rats and its comparison with experiment, *Journal of Theoretical Biology*. 43: 15–28.

Siebert, W. M. (1974) Ranke revisited—a simple short-wave cochlear model, *J. Acoust. Soc. Am.* 56: 594–600.

Smith, W. R. (1980) Hypothalamic regulation of pituitary secretion of luteinizing hormone - II. Feedback control of gonadotropin secretion, *Bulletin of Mathematical Biology*. 42: 57–78.

Smith, W. R. (1983) Qualitative mathematical models of endocrine systems, *American Journal of Physiology. (Regulatory Integrative Comp. Physiol.)*. 245: R473–R477.

Smith, J. M. and R. J. Cohen (1984) Simple finite element model accounts for wide range of cardiac dysrhythmias, *Proc. Natl. Acad. Sci. USA*. 81: 233–237.

Smolen, P. and J. Keizer (1992) Slow voltage inactivation of $Ca^{2+}$ currents and bursting mechanisms for the mouse pancreatic $\beta$-cell, *Journal of Membrane Biology*. 127: 9–19.

Smolen, P. (1995) A model for glycolytic oscillations based on skeletal muscle phosphofructokinase kinetics, *Journal of Theoretical Biology*. 174: 137–148.

Sneyd, J. and D. Tranchina (1989) Phototransduction in cones: an inverse problem in enzyme kinetics, *Bulletin of Mathematical Biology*. 51: 749–784.

Sneyd, J., S. Girard, and D. Clapham (1993) Calcium wave propagation by calcium-induced calcium release: an unusual excitable system, *Bulletin of Mathematical Biology*. 55: 315–344.

Sneyd, J. and A. Atri (1993) Curvature dependence of a model for calcium wave propagation, *Physica D*. 65: 365–372.

Sneyd, J., A. C. Charles, and M. J. Sanderson (1994) A model for the propagation of intercellular calcium waves, *American Journal of Physiology (Cell Physiology)*. 266: C293–C302.

Sneyd, J., B. Wetton, A. C. Charles, and M. J. Sanderson (1995a) Intercellular calcium waves mediated by diffusion of inositol trisphosphate: a two-dimensional model, *American Journal of Physiology (Cell Physiology)*. 268: C1537–C1545.

Sneyd, J., J. Keizer, and M. J. Sanderson (1995b) Mechanisms of calcium oscillations and waves: a quantitative analysis, *FASEB Journal*. 9: 1463–1472.

Sneyd, J., P. Dale, and A. Duffy (1998) Traveling waves in buffered systems: applications to calcium waves, *SIAM Journal on Applied Mathematics*. In press.

Spilmann, L. and J. S. Werner, Eds. (1990) *Visual Perception: The Neurophysiological Foundations*. Academic Press. London.

Steele, C. R. (1974) Behavior of the basilar membrane with pure-tone excitation, *J. Acoust. Soc. Am.* 55: 148–162.

Steele, C. R. and L. Tabor (1979a) Comparison of WKB and finite difference calculations for a two-dimensional cochlear model, *J. Acoust. Soc. Am.* 65: 1001–1006.

Steele, C. R. and L. Tabor (1979b) Comparison of WKB calculations and experimental results for three-dimensional cochlear models, *J. Acoust. Soc. Am.* 65: 1007–1018.

Stephenson, J. L. (1972) Concentration of the urine in a central core model of the counterflow system, *Kidney Int*. 2: 85–94.

Stephenson, J. L. (1992), *Urinary concentration and dilution: models*. In: Handbook of Physiology. Section 8: Renal Physiology, Ed: E. E. Windhager, American Physiological Society, Bethesda, Maryland.

Stern, M. D. (1992) Theory of excitation–contraction coupling in cardiac muscle, *Biophysical Journal*. 63: 497–517.

Stern, M. D., G. Pizarro, and E. Ríos (1997) Local control model of excitation–contraction coupling in skeletal muscle, *Journal of General Physiology*. 110: 415–440.

Stoker, J. J. (1950) *Nonlinear Vibrations*: Interscience, New York.

Strang, G. (1986) *Introduction to Applied Mathematics*: Wellesley-Cambridge Press, Wellesley, MA.

Strieter, J., J. L. Stephenson, L. G. Palmer, and A. M. Weinstein (1990) Volume-activated chloride permeability can mediate cell volume regulation in a mathematical model of a tight epithelium, *Journal of General Physiology*. 96: 319–344.

Strogatz, S. H. (1994) *Nonlinear Dynamics and Chaos*: Addison-Wesley, Reading, Massachusetts.

Stryer, L. (1986) Cyclic GMP cascade of vision, *Annu. Rev. Neurosci*. 9: 87–119.

Stryer, L. (1988) *Biochemistry*: W.H. Freeman, New York.

Sturis, J., K. S. Polonsky, E. Mosekilde, and E. V. Cauter (1991) Computer model for mechanisms underlying ultradian oscillations of insulin and glucose, *American Journal of Physiology (Endocrinol. Metab.)*. 260: E801–E809.

Tamura, T., K. Nakatani, and K.-W. Yau (1991) Calcium feedback and sensitivity regulation in primate rods, *Journal of General Physiology*. 98: 95–130.

Tang, Y. and H. G. Othmer (1994) A model of calcium dynamics in cardiac myocytes based on the kinetics of ryanodine-sensitive calcium channels, *Biophysical Journal*. 67: 2223–2235.

Tang, Y., J. L. Stephenson, and H. G. Othmer (1996) Simplification and analysis of models of calcium dynamics based on $IP_3$-sensitive calcium channel kinetics, *Biophysical Journal*. 70: 246–263.

Tang, Y. and J. L. Stephenson (1996) Calcium dynamics and homeostasis in a mathematical model of the principal cell of the cortical collecting tubule, *Journal of General Physiology*. 107: 207–230.

Tosteson, D. C. and J. F. Hoffman (1960) Regulation of cell volume by active cation transport in high and low potassium sheep red cells, *Journal of General Physiology*. 44: 169–194.

Tranchina, D., J. Gordon, and R. Shapley (1984) Retinal light adaptation—evidence for a feedback mechanism, *Nature*. 310: 314–316.

Tranchina, D. and C. S. Peskin (1988) Light adaptation in the turtle retina: embedding a parametric family of linear models in a single non-linear model, *Visual Neuroscience*. 1: 339–348.

Tranchina, D., J. Sneyd, and I. D. Cadenas (1991) Light adaptation in turtle cones: testing and analysis of a model for phototransduction, *Biophysical Journal*. 60: 217–237.

Tranquillo, R. and D. Lauffenberger (1987) Stochastic models of leukocyte chemosensory movement, *Journal of Mathematical Biology*. 25: 229–262.

Troy, W. C. (1976) Bifurcation phenomena in FitzHugh's nerve conduction equations, *Journal of Mathematical Analysis and Applications*. 54: 678–690.

Troy, W. C. (1978) The bifurcation of periodic solutions in the Hodgkin-Huxley equations, *Quaterly of Applied Mathematics*. 36: 73–83.

Tuckwell, H. C. and R. M. Miura (1978) A mathematical model for spreading cortical depression, *Biophysical Journal*. 23: 257–276.

Tuckwell, H. C. (1988) *Introduction to Theoretical Neurobiology*: Cambridge University Press, Cambridge, New York.

Tung, L. (1978) *A bi-domain model for describing ischemic myocardial D-C potentials. Ph.D. thesis*: MIT, Cambridge, MA.

Tyson, J. J. and P. C. Fife (1980) Target patterns in a realistic model of the Belousov–Zhabotinskii reaction, *J. Chem. Phys*. 73: 2224–2237.

Tyson, J. J. and J. P. Keener (1988) Singular perturbation theory of traveling waves in excitable media, *Physica D*. 32: 327–361.

Tyson, J. J., B. Novak, K. Chen, and J. Val (1995) Checkpoints in the cell cycle from a modeler's perspective, *Prog. Cell Cycle Res*. 1: 1–8.

Urban, B. W. and S. B. Hladky (1979) Ion transport in the simplest single file pore, *Biochimica et Biophysica Acta*. 554: 410–429.

Ussing, H. H. (1949) The distinction by means of tracers between active transport and diffusion, *Acta Physiologica Scandinavica*. 19: 43–56.

Ussing, H. H. (1982) Volume regulation of frog skin epithelium, *Acta Physiologica Scandinavica*. 114: 363–369.

van der Pol, B. and J. van der Mark (1928) The heartbeat considered as a relaxation oscillation, and an electrical model of the heart, *Phil. Mag*. 6: 763–775.

von Békésy, V. (1960) *Experiments in Hearing*: McGraw-Hill, New York.

von Euler, C. (1980) Central pattern generation during breathing, *Trends in Neuroscience*. 3: 275–277.

Wagner, J. and J. Keizer (1994) Effects of rapid buffers on $Ca^{2+}$ diffusion and $Ca^{2+}$ oscillations, *Biophysical Journal*. 67: 447–456.

Wang, S.-Y., A. Peskoff and G. A. Langer (1996) Inner sarcolemmal leaflet $Ca^{2+}$ binding; its role in cardiac Na/Ca exchange, *Biophysical Journal*. 70: 2266–2274.

Wang, X.-J. and J. Rinzel (1995) *Oscillatory and bursting properties of neurons*. In: The handbook of Brain Theory and Neural Networks, Ed: M. Arbib, MIT Press, Cambridge, Mass.

Weinstein, A. (1992) Analysis of volume regulation in an epithelial cell model, *Bulletin of Mathematical Biology*. 54: 537–561.

Weinstein, A. M. (1994) Mathematical models of tubular transport, *Annual Review of Physiology*. 56: 691–709.

Weinstein, A. (1996) Coupling of entry to exit by peritubular $K^+$ permeability in a mathematical model of rat proximal tubule, *American Journal of Physiology (Renal Fluid Electrolyte Physiol.)*. 271: F158–F168.

Wenckebach, K. F. (1904) *Arrhythmia of the Heart: a Physiological and Clinical Study*: Green, Edinburgh.

West, J. B., Ed. (1980) *Pulmonary Gas Exchange*, Academic Press, New York.

White, D. C. S. and J. Thorson (1975) *The kinetics of muscle contraction*: Pergamon Press. Originally published in Progress in Biophysics and Molecular Biology, volume 27, 1973, Oxford, New York.

Whitham, G. B. (1974) *Linear and Nonlinear Waves*: Wiley-Interscience, New York.

Wildt, L., A. Häusler, G. Marshall, J. S. Hutchison, T. M. Plant, P. E. Belchetz, and E. Knobil (1981) Frequency and amplitude of gonadotropin-releasing hormone stimulation and gonadotropin secretion in the Rhesus monkey, *Endocrinology*. 109: 376–385.

Willems, G. M., T. Lindhout, W. T. Hermens, and H. C. Hemker (1991) Simulation model for thrombin generation in plasma, *Haemostasis*. 21: 197–207.

Williams, M. M. (1990) *Hematology*: McGraw-Hill, New York.

Winfree, A. T. (1967) Biological rhythms and the behavior of populations of coupled oscillators, *Journal of Theoretical Biology*. 16: 15–42.

Winfree, A. T. (1972) Spiral waves of chemical activity, *Science*. 175: 634–636.

Winfree, A. T. (1973) Scroll-shaped waves of chemical activity in three dimensions, *Science*. 181: 937–939.

Winfree, A. T. (1974) Rotating chemical reactions, *Scientific American*. 230: 82–95.

Winfree, A. T. (1980) *The Geometry of Biological Time*: Springer-Verlag, Berlin, Heidelberg, New York.

Winfree, A. T. and S. H. Strogatz (1983a) Singular filaments organize chemical waves in three dimensions: 1. Geometrically simple waves, *Physica D*. 8: 35–49.

Winfree, A. T. and S. H. Strogatz (1983b) Singular filaments organize chemical waves in three dimensions: 2. twisted waves, *Physica D*. 9: 65–80.

Winfree, A. T. and S. H. Strogatz (1983c) Singular filaments organize chemical waves in three dimensions: 3. knotted waves, *Physica D*. 9: 333–345.

Winfree, A. T. and S. H. Strogatz (1984) Singular filaments organize chemical waves in three dimensions: 4. wave taxonomy, *Physica D*. 13: 221–233.

Winfree, A. T. (1987) *When Time Breaks Down*: Princeton University Press, Princeton, NJ.

Winfree, A. T. (1991) Varieties of spiral wave behavior: an experimentalist's approach to the theory of excitable media, *Chaos*. 1: 303–334.

Wittenberg, J. B. (1966) The molecular mechanism of haemoglobin-facilitated oxygen diffusion, *Journal of Biological Chemistry*. 241: 104–114.

Wong, A. Y. K., A. Fabiato, and J. B. Bassingthwaigthe (1992) Model of calcium-induced calcium release mechanism in cardiac cells, *Bulletin of Mathematical Biology*. 54: 95–116.

Woodbury, J. W. (1971), *Eyring rate theory model of the current–voltage relationship of ion channels in excitable membranes*. In: Chemical Dynamics: Papers in Honor of Henry Eyring, Ed: J. Hirschfelder, John Wiley and Sons, Inc., New York.

Wyman, J. (1966) Facilitated diffusion and the possible role of myoglobin as a transport mechanism, *Journal of Biological Chemistry*. 241: 115–121.

Wyman, R. J. (1977) Neural generation of breathing rhythm, *Ann. Rev. Physiol*. 39: 417–448.

Yamada, W. M. and R. S. Zucker (1992) Time course of transmitter release calculated from simulations of a calcium diffusion model, *Biophysical Journal*. 61: 671–682.

Yanagida, E. (1985) Stability of fast travelling pulse solutions of the FitzHugh–Nagumo equation, *Journal of Mathematical Biology*. 22: 81–104.

Yanagihara, K., A. Noma, and H. Irisawa (1980) Reconstruction of sino-atrial node pacemaker potential based on voltage clamp experiments, *Jap. J. Physiology*. 30: 841–857.

Yue, D. T. (1997) Quenching the spark in the heart, *Science*. 276: 755.

Zigmond, S. H. (1977) Ability of polymorphonuclear leukocytes to orient in gradients of chemotactic factors, *Journal of Cell Biology*. 75: 606–616.

Zigmond, S. H., H. I. Levitsky, and B. J. Kreel (1981) Cell polarity: an examination of its behavioral expression and its consequences for polymorphonuclear leukocyte chemotaxis, *Journal of Cell Biology*. 89: 585–592.

Zinner, B. (1992) Existence of traveling wavefront solutions for the discrete Nagumo equation, *Journal of Differential Equations*. 96: 1–27.

Zipes, D. P. and J. Jalife (1995) *Cardiac Electrophysiology; From Cell to Bedside*: W.B. Saunders Co, Philadelphia, PA.

Zucker, R. S. and A. L. Fogelson (1986) Relationship between transmitter release and presynaptic calcium influx when calcium enters through discrete channels, *Proc. Natl. Acad. Sci. USA*. 83: 3032–3036.

Zucker, R. S. and L. Landò (1986) Mechanism of transmitter release: voltage hypothesis and calcium hypothesis, *Science*. 231: 574–579.

Zwislocki, J. (1953) Review of recent mathematical theories of cochlear dynamics, *J. Acoust. Soc. Am*. 25: 743–751.

Zwislocki, J. J. (1965) *Analysis of some auditory characteristics*. In: Handbook of Mathematical Psychology, Ed: R. D. Luce, R. R. Buck, and E. Galanter, Wiley, New York.

# Index

# Interdisciplinary Applied Mathematics